AF293833

Universitext

D. B. Fuks   V. A. Rokhlin

Beginner's Course in Topology

Geometric Chapters

Translated from the Russian by A. Iacob
With 17 Figures

Springer-Verlag
Berlin Heidelberg New York Tokyo 1984

Dmitrij Borisovich Fuks
Laboratory of Bioorganic Chemistry
Moscow State University, Moscow, USSR

Vladimir Abramovich Rokhlin
Department of Mathematics and Mechanics
Leningrad State University, Leningrad, USSR

Andrei Iacob
Department of Theoretical Mathematics. The Weizmann Institute of
Science. Rehovot 76100, Israel

Title of the Russian original edition: *Nachal'nyj Kurs Topologii: Geome
tricheskie Glavy.* Publisher Nauka, Moscow 1977

This volume is part of the *Springer Series in Soviet Mathematics*
Advisers: L. D. Faddeev (Leningrad), R. V. Gamkrelidze (Moscow)

AMS Subject Classification (1980): 54-01, 54A05, 54Bxx, 54Cxx,
54Dxx, 54Exx, 55Pxx, 55Qxx, 55Rxx, 57-01, 57Mxx, 57Nxx,
57Rxx, 57Sxx, 58Axx, 58C25, 58D10, 58D15, 58E05

ISBN 3-540-13577-4 Springer-Verlag Berlin Heidelberg New York Tokyo
ISBN 0-387-13577-4 Springer-Verlag New York Heidelberg Berlin Tokyo

Second Printing 2004

Library of Congress Cataloging in Publication Data. Rokhlin, V. A. Beginner's course
in topology. (Springer series in Soviet mathematics) (Universitext)
Translation of: Nachal'nyĭ kurs topologii / V. A. Rokhlin, D. B. Fuks. Bibliography: p.
Includes index. 1. Topology. I. Fuks, D. B. II. Title. III. Series.
QA611.R6513 1984 514 84-10657 ISBN 0-387-13577-4 (U.S.)

Springer-Verlag is a part of Springer Science+Business Media

springeronline.com

© Springer-Verlag Berlin Heidelberg 1984
Printed in Germany

Printing and bookbinding: AZ Druck und Datentechnik GmbH, Kempten
2141/3140-543210

Preface

This book is the result of reworking part of a rather lengthy course of lectures of which we delivered several versions at the Leningrad and Moscow Universities. In these lectures we presented an introduction to the fundamental topics of topology: homology theory, homotopy theory, theory of bundles, and topology of manifolds. The structure of the course was well determined by the guiding term elementary topology, whose main significance resides in the fact that it made us use a rather simple apparatus. In this book we have retained those sections of the course where algebra plays a subordinate role. We plan to publish the more algebraic part of the lectures as a separate book.

Reprocessing the lectures to produce the book resulted in the profits and losses inherent in such a situation: the rigour has increased to the detriment of the intuitiveness, the geometric descriptions have been replaced by formulas needing interpretations, etc. Nevertheless, it seems to us that the book retains the main qualities of our lectures: their elementary, systematic, and pedagogical features. The preparation of the reader is assumed to be limited to the usual knowledge of set theory, algebra, and calculus which mathematics students should master after the first year and a half of studies. The exposition is accompanied by examples and exercises. We hope that the book can be used as a topology textbook.

The most essential difference between the book and the corresponding part of our lectures is the arrangement of the material: here we have followed a much more orderly succesion of topics. However, from our experience, a lecture course in elementary topology which exaggerates in the last respect is rather tedious and less efficient than one which mixes geometry with algebra and applications. This remark may serve as a warning to the teacher who would like to use our book as a guide. In fact, it is by no means necessary to read the book in its order; a reader who is interested in getting to the homotopy groups or to any other topic sooner, can easily do so.

Concerning the terminology and notation, we have tried to stick to standard usage, and have permitted ourselves only a few reforms. For example, we do not use the terms "simplicial complexes" or "CW-complexes", but simplicial spaces and cellular spaces; not "cofibrations", but Borsuk pairs; not "fiber bundles" (or "fibered products"), but Steenrod bundles. There is even one term which we do not use in the generally accepted way: for us, a connected space refers to what usually is called a linearly connected (or path-connected) space (we do not have a special name for the spaces which are usually called connected). Furthermore, we have avoided using non-standardized notations for standard objects. In fact, in the majority of cases, our notation is just an abbreviation of the corresponding term and can be understood by itself: for example, pr stands for projection, in - for inclusion, dim - for dimension, ske - for skeleton, bs - for base, etc.

Topology requires a very precise set-theoretic language, and this compelled us to devote a special attention to this language; this is illustrated by pp. 1 - 4. We emphasize that on these pages we only list the terms and notations, assuming that the objects themselves are known.

In this book we rarely refer to the history of topology. We have even departed from the tradition that some theorems bear the names of their real or imaginary authors. In return, we willingly have used names of topologists in the terminology and notations.

The organization of the text and the system of references may be briefly described as follows. Each Chapter is divided into Sections, each Section - into Subsections, each Subsection - into Numbers. The chapters, sections and subsections have numbers and titles, while the numbers are denoted only by their numbers. Each fact announced without proof is called Information, and is distinguished from the rest of the text by this title. To refer to a section, subsection, or number within the same chapter, we do not indicate the number of the chapter, and references within a section or subsection are similarly abbreviated. Examples: the entries § 1.2 (Section 2 of Chapter 1), Subsection 1.2.3 (Subsection 3 of Section 2 of Chapter 1), and 1.2.3.4 (No. 4 of Subsection 3 of Section 2 of Chapter 1) are abbreviated, within Chapter 1, as § 2, Subsection 2.3, and 2.3.4, respectively; the second of these entries is abbreviated within § 1.2 as Subsection 3; the third entry is abbreviated within § 1.2 and Subsection 1.2.3 as 3.4 and 4, respectively.

The Authors

Contents

Set-Theoretical Terms and Notations Used in this Book, but not Generally Adopted

Mathematicians manage with a surprisingly modest collection of set-theoretic terms and notations, which can be roughly divided into three groups. The first contains terms and notations which have attained general recognition. The terms and notations in the second group are equally well-known, but can be understood differently or have varying connotations. The third group consists of terms and notations used less frequently.

There is no need to define terms from the first group. For example, the notations $X \cup Y$, $X \cap Y$, and $X_1 \times \ldots \times X_n$ for the union, intersection, and product of sets, or the notations $f: X \to Y$, $\operatorname{Im} f$ and $f|_A : A \to Y$ for a map, its image, and its restriction are understood in the same way by all people. The same is true for the notation $x \in X$ and the terms <u>one-to-one</u> (<u>injective</u>) <u>map</u> and <u>map onto</u> (<u>surjective map</u>).

For the sake of precision, we must say a few words about our usage of terms and notations from the second group. We denote the empty set by $\emptyset$. We understand the notation $X \subset Y$ in the broadest sense, i.e., the equality $X = Y$ is not excluded. The same is true for the term <u>countable set:</u> we use it both for infinite countable and finite sets. The identity map of the set X is denoted by $\operatorname{id} X$, or, when there is no ambiguity about X, simply by id. We shall say that a map is <u>invertible</u> if it has an inverse, i.e., it is simultaneously injective and surjective. We let $\{x \in X \mid \ldots\}$ denote the set of points x of the set X which satisfy the condition appearing instead of the three dots. A <u>family</u> $\{X_\mu\}_{\mu \in M}$ is a map of a set M onto a set of objects X_μ with $\mu \in M$, defined by the formula $\mu \mapsto X_\mu$.

Next our main task is to list the terms and notations appearing in this book and belonging to the third group.

Maps

If A is a subset of a set X, then the inclusion of A in
X may be considered as the map defined by the formula $x \mapsto x$. We
denote it by in: $A \to X$. If there is no ambiguity about A and X,
we simply write in.

If A is a subset of X and B is a subset of Y, then
each map $f: X \to Y$ such that $f(A) \subset B$ induces a map ab $f : A \to B$,
$x \mapsto f(x)$, and called here the <u>abridgment</u> (or <u>compression</u>) <u>of the map</u>
f <u>to</u> A,B. When there is no ambiguity about A and B, one can write
ab f   instead of ab $f : A \to B$. If B = Y, then ab f   is just the
usual restriction of f to A.

By a map of a sequence $(X, A_1, \ldots, A_n)$ into a sequence
$(Y, B_1, \ldots, B_n)$, where $A_1, \ldots, A_n$ $(B_1, \ldots, B_n)$ are subsets of X
(respectively, Y), we mean a sequence of maps

$$(\varphi: X \to Y, \ \varphi_1: A_1 \to B_1, \ldots, \ \varphi_n: A_n \to B_n)$$

such that $\varphi_i = \mathrm{ab}\,\varphi$. We denote such a map by

$$(\varphi, \varphi_1, \ldots, \varphi_n): (X, A_1, \ldots, A_n) \to (Y, B_1, \ldots, B_n).$$

If the subsets $A_1, \ldots, A_n$ and $B_1, \ldots, B_n$ are fixed, then the map
$f = (\varphi, \varphi_1, \ldots, \varphi_n)$ and its first component φ are uniquely determined
by each other and usually we do not distinguish between them. For
example, the notation $f: (X, A_1, \ldots, A_n) \to (Y, B_1, \ldots, B_n)$ may be also
used to say that f is a map of X into Y such that $f(A_1) \subset B_1$,
$\ldots, f(A_n) \subset B_n$. When we wish to emphasize explicitly this relationship
between f and φ, we shall write:

$$f = \mathrm{rel}\,\varphi, \quad \varphi = \mathrm{abs}\,f.$$

Sometimes we simply write rel instead of rel id.

Quotients

We denote the quotient (or factor) set of a set X by a
partition p of X by X/p. The map $X \to X/p$ which takes each point
into the element of p containing it is called <u>projection</u> and is
denoted by pr. A subset of X which is a union of elements of the
partition is said to be <u>saturated.</u> The smallest saturated set
containing a given subset A of X (i.e., $\mathrm{pr}^{-1}(\mathrm{pr}(A))$) is called

the <u>saturation</u> of A.

If p and q are partitions of X and Y, respectively, then each map $f: X \to Y$ which transforms elements of p into elements of q induces a map $X/p \to Y/q$, which takes each element A of p into the element of q which contains $f(A)$. This map is denoted by fact f. In particular, fact f is defined when q is the partition of Y into its points, and f is constant on the elements of p. Thus, for each map $f: X \to Y$ constant on the elements of the partition p of X we have the corresponding map fact $f: X/p \to Y$.

Given a map $f: X \to Y$, the partition of X into the nonempty preimages of the points of Y is denoted by zer(f). The corresponding map fact $f: X/\mathrm{zer}(f) \to Y$ is injective and is called the <u>injective factor</u> (or <u>injective quotient</u>) <u>of the map</u> f.

Sums

The <u>sum of the family of sets</u> $\{X_\mu\}_{\mu \in M}$ is the union of disjoint copies of the sets X_μ, i.e., the set of pairs (x,μ) such that $x_\mu \in X_\mu$. Notation: $\bigsqcup_{\mu \in M} X_\mu$. The map of X_ν ($\nu \in M$) into $\bigsqcup_{\mu \in M} X_\mu$, defined by the formula $x \mapsto (x,\nu)$, is denoted by in_ν.

We note that the maps in_ν are injective and their images $\mathrm{in}_\nu(X_\nu)$ are pairwise disjoint and cover $\bigsqcup_{\mu \in M} X_\mu$. Therefore, for any family $\{Y_\mu\}_{\mu \in M}$ indexed by the same set M, and each family of maps $\{f_\mu: X_\mu \to Y_\mu\}_{\mu \in M}$, there is a unique map $f: \bigsqcup_{\mu \in M} X_\mu \to \bigsqcup_{\mu \in M} Y_\mu$ which satisfies the relations $f \circ \mathrm{in}_\nu = \mathrm{in}_\nu \circ f$. f is called the <u>sum of the maps</u> f and is denoted by $\bigsqcup_{\mu \in M} f_\mu$.

If M consists of the numbers $1,\ldots,n$, we use, along with $\bigsqcup X_\mu$ and $\bigsqcup f_\mu$, the notations

$$X_1 \sqcup \ldots \sqcup X_n \quad \text{and} \quad f_1 \sqcup \ldots \sqcup f_n.$$

Products

The product $X_1 \times \ldots \times X_n$ is mapped naturally onto its factor X_i following the rule

$$(x_1,\ldots,x_n) \mapsto x_i.$$

This map is called the <u>i-th projection</u> and is denoted by pr_i.

Given maps $f_1: X_1 \to Y_1,\ldots,f_n: X_n \to Y_n$, the rule

$$(x_1, \ldots, x_n) \mapsto (f_1(x_1), \ldots, f_n(x_n))$$

defines a map of the product $X_1 \times \ldots \times X_n$ into the product $Y_1 \times \ldots \times Y_n$, called the _product of the maps_ $f_1, \ldots, f_n$, and denoted $f_1 \times \ldots \times f_n$.

If p is a partition of the set X and q is a partition of the set Y, we let $p \times q$ denote the partition of the product $X \times Y$ into the sets $A \times B$, where A is an element of p and B is an element of q.

There is a natural map of X into its square $X \times X$, given by $x \mapsto (x,x)$. This is the _diagonal_ map and is denoted by diag. Its image $\mathrm{diag}(X \times X) \subset X \times X$ is called the _diagonal_ of $X \times X$.

Chapter 1. Topological Spaces

1.   Topologies

1. We say that a <u>topological structure</u> or, simply, a <u>topology</u>, is defined on a set X if there is given a class of subsets of X which contains the union of any collection in the class and the intersection of any finite collection in the class. A set endowed with a topological structure is called a <u>topological space</u>, its elements - <u>points</u>, and the sets of the given class - <u>open sets.</u>

A collection of sets for which we take the union or the intersection may be empty. The union of the empty collection is $\emptyset$, while the intersection of the empty collection of subsets of X is the entire set X. Hence $\emptyset$ and X are open sets (in any topology).

Two examples of topological structures are: the <u>trivial</u> topology, whose only open sets are $\emptyset$ and X, and the <u>discrete</u> topology, in which all the subsets of X are open. If X has more than one element, it is possible to define other topologies on X. For example, if X consists of two elements, a and b, then it will admit two topologies aside from the trivial and discrete ones. In one, the open sets are $\emptyset$, a, and X, while in the second the open sets are $\emptyset$, b, and X. More serious examples will appear in the sequel.

2. A subset of a topological space is <u>closed</u> if its complement is open. The class of closed sets contains the intersection of any collection of sets from the class, and it contains the union of any finite collection of sets from the class. Moreover, given any class of subsets of a set X with these properties, there exists a unique topology on X such that the given class is the class of all closed

sets.

3.   A <u>neighborhood</u> of a point of a topological space is any open set containing the given point. A neighborhood of a subset of a topological space is any open set containing the given subset.

Derived Concepts

4.   If we consider the open sets contained in a given subset A of a topological space X, there is one which is the largest, namely the union of all such sets. It is called the <u>interior part</u> or, simply, the <u>interior</u> of the set A. We denote it by $\text{Int}\,A$ or $\text{Int}_X A$. Similarly, among the closed sets containing A, there is one which is the smallest, namely the intersection of all such sets. It is called the <u>closure</u> of the set A and is denoted by $\text{Cl}\,A$ or $\text{Cl}_X A$. The difference $\text{Cl}\,A \smallsetminus \text{Int}\,A$ can be represented as the intersection of the closed sets $\text{Cl}\,A$ and $X \smallsetminus \text{Int}\,A$, and is therefore closed. This set is called the <u>boundary</u> or <u>frontier</u> of the set A and is denoted by $\text{Fr}\,A$ or $\text{Fr}_X A$. We remark that $X \smallsetminus \text{Int}\,A = \text{Cl}(X \smallsetminus A)$ and that A and $X \smallsetminus A$ have the same boundary.

5.   Relative to the set A, the points of the sets $\text{Int}\,A$, $\text{Cl}\,A$, $\text{Fr}\,A$, and $X \smallsetminus \text{Cl}\,A = \text{Int}(X \smallsetminus A)$ are called <u>interior</u>, <u>adherent</u> <u>boundary</u> (or <u>frontier</u>) and <u>exterior</u> points, respectively. They can be characterized more explicitly in terms of neighborhoods. A point is: an interior point if it has a neighborhood entirely contained in A; an adherent point if each of its neighborhoods intersects A; a boundary point if each of its neighborhoods intersects both A and $X \smallsetminus A$; and an exterior point if it has a neighborhood which does not intersect A.

Clearly, a set is open (closed) if and only if it coincides with its interior (respectively, closure), i.e., if it consists only of interior points (respectively, if it contains all its boundary points).

6. A subset A of a topological space X is said to be <u>dense in</u> X (or <u>everywhere dense</u>) if $\text{Cl}\,A = X$, i.e., if A intersects any nonempty open set in X. A set A is <u>nowhere dense</u> if $X \smallsetminus \text{Cl}\,A$ is everywhere dense.

Bases and Prebases

7.   A <u>base of a topological space</u> is a collection of open

sets such that any open set can be represented as a union of sets from this collection. Equivalently, a collection Γ of open sets is a base if for any open set U and any point $x \in U$ there is $V \in \Gamma$ such that $x \in V \subset U$.

A base completely determines the topology: the open sets are exactly those which can be expressed as union of elements of the base.

8. The following proposition provides us with a standard method of introducing a topology on a set. Let Γ be a collection of subsets of a set X. Then <u>there exists a topology on</u> X <u>with base</u> Γ <u>if and only if the intersection of every finite subcollection of sets from</u> Γ <u>can be expressed as a union of sets from</u> Γ.

The intersection of a finite collection of sets belonging to a base is open; hence the necessity of the condition. Sufficiency follows from the fact that the class of subsets of X which are representable as unions of sets from Γ satisfies the conditions of Definition 1.

9. The previous proposition can be reformulated in the following useful way. <u>There is a topology on</u> X <u>with base</u> Γ <u>if and only if the sets of</u> Γ <u>cover</u> X, <u>and for any</u> $U,V \in \Gamma$ <u>and any point</u> $x \in U \cap V$, <u>there exists</u> $W \in \Gamma$ <u>such that</u> $x \in W \subset U \cap V$.

10. A collection of subsets of a topological space is said to be a <u>prebase</u> of the space if the intersections of finite subcollections of sets from the given collection form a base.

The proposition in 8 shows that <u>any collection</u> Γ <u>of subsets of a set</u> X <u>is the prebase of a unique topology on</u> X.

11. A <u>base at the point</u> x of a topological space X is a collection of neighborhoods of x such that any neighborhood of x contains a neighborhood from this collection. A <u>prebase at the point</u> x is a collection of sets such that the intersections of finite subcollections form a base at x.

Covers

12. As a rule, the covers we shall encounter will be either covers of a topological space by some of its subsets, or covers of a subset of a topological space by other subsets of this space. If we need to emphasize that a certain cover of a subset A of a topological space X consists of subsets of X which are not necessarily included in A, we shall refer to it as a <u>cover of the set</u> A <u>in</u> X.

A cover Γ is a _refinement_ of a cover Δ if any element of Γ is contained in an element of Δ.

A cover is _locally finite_ if any point of the space has a neighborhood which intersects only a finite number of elements of the cover.

A cover is _open_ (_closed_) if all its elements are open (respectively, closed) sets.

13. _Every open cover of a topological space_ X _has a refinement whose sets belong to a given base of_ X.

For example, the sets of the base contained in the sets of the given cover yield such a refinement.

2. Metrics

1. A nonnegative real function ρ defined on the square $X \times X$ of the set X is a _metric_ on X if it satisfies three conditions: $\rho(x,y) = 0$ if and only if $x = y$; $\rho(x,y) = \rho(y,x)$ for all $x,y \in X$; $\rho(x,z) \leqslant \rho(x,y) + \rho(y,z)$ for any $x,y,z \in X$. A _metric space_ is a set equipped with a metric. We use the symbol dist as the standard notation for metrics.

The values taken by a metric are called _distances,_ and the _inequality_ they satisfy according to the definition is the _triangle inequality._

2. The standard n-dimensional Euclidean space $\mathbb{R}^n$ ($n \geqslant 0$) is the fundamental example of metric space. $\mathbb{R}^n$ is the set of all sequences $\{x_i\}_1^n$ of real numbers, where the distance between two sequences $\{x_i\}_1^n$ and $\{y_i\}_1^n$ is defined as $[\sum_1^n (x_i - y_i)^2]^{1/2}$. The line $\mathbb{R}^1$ is usually identified with the field of real numbers, denoted by $\mathbb{R}$.

We obtain the definition of the standard Hilbert space ℓ_2 by replacing the n-term sequences $\{x_i\}_1^n$ with infinite sequences $\{x_i\}_1^\infty$ satisfying the condition $\sum_1^\infty x_i^2 < \infty$, and writing $\sum_1^\infty$ instead of $\sum_1^n$ in the distance formula.

3. The _ball with center_ $x_0 \in X$ _and radius_ $r > 0$ in the metric space X is the set of points $x \in X$ such that $\text{dist}(x_0,x) \leqslant r$. If we write $<$ (respectively $=$) instead of $\leqslant$, we obtain the definition of the _open ball_ (respectively, of the _sphere_). The unit

ball and the unit sphere of $\mathbb{R}^n$, i.e., the ball and sphere with center $(0,\ldots,0)$ and radius 1, are simply called the <u>n-dimensional ball</u> D^n and the <u>(n-1)-dimensional sphere</u> S^{n-1}. In particular, D^0 is just a point, S^0 - a pair of points, and $S^{-1} = \emptyset$. Moreover, we set $D^n = \emptyset$ for $n \leqslant -1$ and $S^n = \emptyset$ for $n \leqslant -2$.

4. By definition, the <u>distance between two sets</u>, A and B, is the number $\inf_{x \in A, y \in B} \mathrm{dist}(x,y)$, and we denote it by $\mathrm{Dist}(A,B)$. In particular, if a is a point, $\mathrm{Dist}(a,B) = \inf_{y \in B} \mathrm{dist}(a,y)$.

The <u>diameter</u> of the set A is the number $\sup_{x,y \in A} \mathrm{dist}(x,y)$, denoted by $\mathrm{diam}\, A$. A set is <u>bounded</u> if its diameter is finite.

The Metric Topology

5. As a consequence of the triangle inequality, if the open ball with center at x_0 and radius r contains a point x_1, then it also contains the open ball with center at x_1 and radius $r - \mathrm{dist}(x_0,x_1)$. Therefore, in any metric space, the intersection of two open balls contains, together with each point, some open ball centered at that point. Moreover, since the open balls cover the space, they constitute the base of a certain topology (see 1.9). In this way, every metric space becomes a topological one.

The resulting topology is called the <u>metric topology.</u> From now on we shall tacitly regard the metric spaces as topological spaces, having in mind the metric topology. In particular, this refers to $\mathbb{R}^n$ and ℓ_2.

A topological space whose topology is the metric topology relative to some metric is said to be <u>metrizable.</u>

6. Clearly, the open balls centered at a given point of a metric space constitute a base at that point. The part of this base consisting of the balls of radii $1/n$ $(n = 1,2,\ldots)$ is also a base.

7. If A is a subset of a metric space X, its <u>metric neighborhood of radius</u> $r > 0$ is, by definition, the set of all points $x \in X$ such that $\mathrm{Dist}(A,x) < r$. Since this set is the union of all open balls of radius r centered at the points of A, it is open, i.e., a genuine neighborhood of A.

3. Subspaces

1. We shall now discuss the <u>relative topology</u>, which transforms any subset A of a topological space X into an independent topological space. This topology is defined by taking its open sets to be those of the form A ∩ B, where B is an open subset of X. It is evident that all the conditions of Definition 1.1 are satisfied. Moreover, the closed subsets of A in this topology are exactly the intersections A ∩ B where B is a closed subset of X. The subsets of the space X, equipped with the relative topology, are called <u>subspaces</u> of X.

If X is a topological space and A is a subspace of X, the pair (X,A) is called a <u>topological pair.</u> A <u>topological triple</u> is a triple (X,A,B) consisting of a topological space X and two subspaces A, B of X, such that $B \subset A$.

2. Let A be a subspace of X. It is clear that any subset of A which is open or closed in X has the same property in A. If A is open, then every set open in A is also open in X. If A is closed, then any set closed in A is also closed in X. In any case, if $B \subset A \subset X$, then $Cl_A B = (Cl_X B) \cap A$.

Obviously, if Γ is a base (prebase) of the space X, then the sets A ∩ B with $B \in \Gamma$ yield a base (respectively, prebase) of the space A.

As a direct consequence of its definition, the relative topology is transitive: if B is a subset of the subspace A of X, the topologies induced on B by the inclusions $B \subset A$ and $B \subset X$ coincide.

3. If X is a metric space and A is a subset of X, then the restriction of the function dist to $A \times A$ is clearly a metric on A. Consequently, any subset of a metric space is itself a metric space. In addition, it is obvious that the metric topology of the latter coincides with the relative topology induced on A by the metric topology of the ambient space X.

4. The previous constructions greatly increase the supply of nontrivial examples of topological spaces: we can now include all the subsets of $\mathbb{R}^n$ and ℓ_2. In particular, the balls and spheres of $\mathbb{R}^n$ are topological spaces. Here we shall add only the <u>cubes</u> of $\mathbb{R}^n$, defined by inequalities of the form $\alpha_i \leqslant x_i \leqslant \alpha_i + a$ $(i = 1,\ldots,n)$, with real numbers $\alpha_1,\ldots,\alpha_n, a$, and $a > 0$. If $\alpha_1 = \ldots = \alpha_n = 0$

and $a = 1$, the cube is called the <u>unit cube</u> I^n. The unit segment I^1 is also denoted by I.

Fundamental Covers

5. A cover Γ of a topological space X is <u>fundamental</u> if each subset A of X such that $A \cap B$ is open in B for all $B \in \Gamma$ is itself open. Equivalently, Γ is fundamental if each subset A of X such that $A \cap B$ is closed in B for all $B \in \Gamma$ is itself closed.

Obviously, a cover which admits a fundamental refinement is itself fundamental.

6. <u>All open covers, and all finite or locally finite closed covers are fundamental.</u>

Clearly, the claim is true for both open and finite closed covers. Now suppose that Γ is a locally finite closed cover of a space X. Consider a cover Δ of X consisting of open sets which intersect only a finite number of elements of Γ. Since Δ is fundamental, it is enough to check that, given any set $U \in \Delta$, the cover of U with elements $U \cap B$, $B \in \Gamma$, is fundamental. But this results from the fact that the latter cover is finite and closed.

7. A triple (X,A,B), where X is a topological space and A and B are subsets of X which constitute a fundamental cover of X, is termed a <u>triad.</u> If $\operatorname{Int} A \cup \operatorname{Int} B = X$, or if $A \cup B = X$ with A and B closed, then (X,A,B) is a triad.

4. <u>Continuous Maps</u>

1. A map f of a topological space X into a topological space Y is <u>continuous</u> if the preimage of each open subset of Y is open in X. Equivalently, f is continuous if the preimage of each closed set is closed.

A map $f: (X,A_1,\ldots,A_n) \to (Y,B_1,\ldots,B_n)$, where $A_1,\ldots,A_n$ and $B_1,\ldots,B_n$ are subsets of the spaces X and Y, respectively, is said to be continuous, if the map $\operatorname{abs} f : X \to Y$ is continuous.

A useful comment: in order for a map $X \to Y$ to be continuous, it is enough that the preimages of the sets comprising some prebase of Y be open.

2. If $f: X \to Y$ and $g: Y \to Z$ are continuous, then the composition $g \circ f: X \to Z$ is obviously continuous. Trivially, the identity map $\mathrm{id}\,X: X \to X$ is continuous for any topological space X.

According to the definition of the relative topology, if $f: X \to Y$ is continuous and $A \subset X$, $B \subset Y$ are subsets with $f(A) \subset B$, then the map $\mathrm{ab}\,f: A \to B$ is also continuous. In particular, the restriction $f\big|_A: A \to Y$ of a continuous map $f: X \to Y$ to an arbitrary subset A of X is continuous. For example, the inclusion of a subspace into its ambient space is always continuous.

If a compression $\mathrm{ab}\,f$ of the map $f: X \to Y$ is defined on the entire space X, then its continuity is equivalent to the continuity of f. In particular, $f: X \to Y$ is continuous if and only if the map $f: X \to f(X)$ is continuous.

3. It is clear that if Γ is a fundamental cover of X, then a map $f: X \to Y$ is continuous whenever all the restrictions $f\big|_A$, $A \in \Gamma$, are continuous. An equivalent formulation: <u>let Γ be a fundamental cover of the topological space X, and assume that for each $A \in \Gamma$ there is a continuous map $f_A : A \to Y$, such that</u> $f_A(x) = f_B(x)$ <u>for all</u> $x \in A \cap B$ $(A, B \in \Gamma)$; <u>then the map</u> $f: X \to Y$ <u>defined by</u>

$$f(x) = f_A(x) \quad \underline{for} \quad x \in A \quad (A \in \Gamma)$$

<u>is continuous.</u>

4. A continuous map is <u>open</u> if the images of the open sets are open, and <u>closed</u> if the images of the closed sets are closed.

Obviously, a composition of open maps is open, and a composition of closed maps is closed.

Here we note one useful sufficient condition for a map to be open: $f: X \to Y$ is certainly open if for each $x \in X$ there is a neighborhood U_x of $f(x)$ and a continuous map $g_x : U_x \to X$ such that the composition $f \circ g_x$ coincides with $\mathrm{in}: U_x \to Y$. Proof: if this is the case, then $f(A) = U_{x \in X}\, g_x^{-1}(A)$ for any subset A of X.

Continuity at a Point

5. A map $f: X \to Y$ is <u>continuous at the point</u> $x \in X$ if for any neighborhood V of the point $f(x)$ there is a neighborhood U of x such that $f(U) \subset V$.

One can reformulate this definition using fewer open sets.

In fact, let us assume that, along with the map $f: X \to Y$, we are given an arbitrary prebase at the point $x \in X$, Δ, and an arbitrary prebase at the point $f(x) \in Y$, E. Then one readily sees that f is continuous at x if and only if each neighborhood $V \in E$ contains the image of some neighborhood $U \in \Delta$.

When X and Y are metric spaces, and Δ and E consists of open balls centered at the points x and $f(x)$, respectively, the last statement reduces to the usual numerical formulation given in calculus: the map $f: X \to Y$ is continuous at the point $x \in X$ if for each $\varepsilon > 0$ there is $\delta > 0$ such that $\mathrm{dist}_X(x,x') < \delta$ implies $\mathrm{dist}_Y(f(x),f(x')) < \varepsilon$.

6. <u>A map</u> $f: X \to Y$ <u>is continuous if and only if it is continuous at each point of</u> X.

PROOF. If f is continuous and V is a neighborhood of the point $f(x)$, then $f^{-1}(V)$ is a neighborhood of the point x, and $f(f^{-1}(V)) \subset V$.

If f is continuous at each point and V is open in Y, then each point of the set $f^{-1}(V)$ is an interior point, since it has a neighborhood whose image lies in V.

Homeomorphisms and Embeddings

7. If $f: X \to Y$ is an invertible continuous map, the inverse map $f^{-1}: Y \to X$ is not necessarily continuous. For example, consider the identity map of a set with the discrete topology onto the same set, but equipped with a different topology; its inverse is not continuous.

An invertible map f such that both f and f^{-1} are continuous is a <u>homeomorphism.</u> If there is a homeomorphism $X \to Y$, the space Y is said to be <u>homeomorphic</u> to the space X.

The following maps are obviously homeomorphisms: the identity transformation of a space, the map inverse to a homeomorphism, and the composition of two homeomorphisms. Thus, the homeomorphism of spaces is an equivalence relation.

8 (EXAMPLES). The open ball $\mathrm{Int}\, D^n$ is homeomorphic to $\mathbb{R}^n$. The standard homeomorphism $\mathbb{R}^n \to \mathrm{Int}\, D^n$ is given by the formula

$$x \mapsto \begin{cases} 2x\, \arctan(\mathrm{dist}(0,x))/\pi\, \mathrm{dist}(0,x)\, , & \text{if } x \neq 0, \\ \\ 0, & \text{if } x = 0. \end{cases}$$

The cube I^n is homeomorphic to D^n; its interior $\operatorname{Int} I^n$ is homeomorphic to $\operatorname{Int} D^n$, and its boundary $\operatorname{Fr} I^n$ is homeomorphic to $\operatorname{Fr} D^n$, i.e., to S^{n-1}. The standard homeomorphisms $D^n \to I^n$, $\operatorname{Int} D^n \to \operatorname{Int} I^n$, and $\operatorname{Fr} D^n \to \operatorname{Fr} I^n$ are realized by translation with the vector $(\operatorname{ort}_1 + \ldots + \operatorname{ort}_n)/2$, followed by central projection (here $\operatorname{ort}_1, \ldots, \operatorname{ort}_n$ denote the vectors $(1,0,\ldots,0), \ldots, (0,\ldots,0,1)$).

The punctured sphere S^n (i.e., S^n with one point removed) is homeomorphic to $\mathbb{R}^n$. A homeomorphism $\mathbb{R}^n \to S^n \smallsetminus \operatorname{ort}_1$ is given by the composition of the homeomorphism $\{x_1,\ldots,x_n\} \mapsto \{0,x_1,\ldots,x_n\}$ of $\mathbb{R}^n$ onto a subspace of $\mathbb{R}^{n+1}$ with the stereographic projection, i.e., the central projection of this subspace onto $S^n \smallsetminus \operatorname{ort}_1$ from the point ort_1.

9. A map $f: X \to Y$ is an _embedding_ or, more specifically, a _topological embedding_, if $\operatorname{ab} f : X \to f(X)$ is a homeomorphism. For example, the inclusion of a subspace in its ambient space is an embedding.

Retractions

10. A _retraction_ is a continuous map of a space X onto a subspace A which is the identity map on A. A subset onto which a space can be retracted is a _retract_ of the space.

Each point of a topological space is a retract of this space. However, a pair of points is already not necessarily a retract. For example, a segment cannot be retracted onto its boundary since any such retraction would be a real continuous function taking two values but no intermediate ones.

11. _A subspace_ A _of a topological space_ X _is a retract of_ X _if and only if every continuous map_ $A \to Y$ _can be extended to a continuous map_ $X \to Y,$ _for any topological space_ Y.

PROOF. If $\rho: X \to A$ is a retraction and $f: A \to Y$ is continuous, then the composition $f \circ \rho$ extends f to X.

If every continuous map $A \to Y$ extends to a continuous map $X \to Y$, then extending the identity map $A \to A$ to a continuous map $X \to A$ yields a retraction.

Numerical Functions

12. The well-known theorem of calculus asserting that the arithmetic operations performed upon continuous functions again produce continuous functions is obviously true for the numerical functions defined on an arbitrary topological space. Similarly, the theorem asserting the continuity of the limit of a uniformly convergent sequence of continuous functions holds for numerical function on a topological space.

13. <u>If</u> X <u>is a metric space and</u> A <u>is a subset of</u> X, <u>then the function</u> X $\to$ $\mathbb{R}$, x $\mapsto$ Dist(x,A), <u>is continuous.</u>

PROOF. Let $x,y \in X$ and $z \in A$. Then Dist$(x,A) \leqslant$ dist(x,z) $\leqslant$ dist(x,y) + dist(y,z). Hence Dist$(x,A) \leqslant$ dist(x,y) + Dist(y,A) for any $x,y \in X$, and since x and y appear symmetrically, we obtain $|$Dist(x,A) - Dist$(y,A)| \leqslant$ dist(x,y).

14. A subset A of a topological space X is said to be <u>distinguishable</u> if there is a continuous function $f: X \to I$ such that $f(x) = 0$ for $x \in A$ and $f(x) > 0$ for $x \in X \smallsetminus A$. Any function with this property is said to <u>distinguish</u> the set A.

A distinguishable set is obviously closed. It is also clear that any closed subset of a metric space is distinguishable: for example, the function $x \mapsto \min(1,$Dist$(x,A))$ distinguishes the closed subset A.

5. <u>Separation Axioms</u>

1. In this subsection and the two that follow, we formulate additional restriction which are often imposed on a topological structure in order to bring the properties of the corresponding topological space closer to those that characterize the subsets of the spaces $\mathbb{R}^n$.

2. More than ten "separation axioms" are known. We need the following four.

T_1. <u>Given two arbitrary points</u> a <u>and</u> b, a $\neq$ b, <u>there is a</u> <u>neighborhood of</u> a <u>which does not contain</u> b. Equivalent formulations: <u>each point is a closed set; finite sets are closed.</u>

T_2. <u>Two arbitrary distinct points have disjoint neighborhoods.</u>

16

T_3. <u>Any point and any closed set not containing this point</u>
<u>have disjoint neighborhoods.</u> An equivalent formulation:
<u>every neighborhood of an arbitrary point contains the closure of a</u>
<u>neighborhood of this point.</u>

T_4. <u>Any two disjoint closed sets have disjoint neighborhoods.</u>
An equivalent formulation: <u>every neighborhood of an arbitrary closed</u>
<u>set contains the closure of a neighborhood of this set.</u> Another
equivalent formulation: <u>given an arbitrary finite collection of pairwise</u>
<u>disjoint closed sets, there are neighborhoods of these sets with</u>
<u>pairwise disjoint closures.</u>

3. Axiom T_1 is a consequence of T_2, but simple examples
show that it is not a consequence of T_3 or T_4.

Spaces which satisfy axiom T_2 are called <u>Hausdorff,</u> those
which satisfy the axioms T_1 and T_3 - <u>regular,</u> and those which satisfy
the axioms T_1 and T_4 - <u>normal.</u>

Every normal space is regular, and every regular space is
Hausdorff.

4. Obviously, every subspace of a Hausdorff space is Hausdorff,
every subspace of a regular space is regular, and every closed subspace
of a normal space is normal.

INFORMATION. A nonclosed subspace of a normal space is not
necessarily normal; see [11].

5. <u>Every retract of a Hausdorff space is closed.</u>

PROOF. Let A be a retract of a topological space X and
let $\rho: X \to A$ be a retraction. If $b \in X \smallsetminus A$, then since X is
Hausdorff and ρ is continuous, the points b and $\rho(b)$ have disjoint
neighborhoods, U and V, such that $\rho(U) \subset V$. This implies that
$\rho(x) \neq x$ for $x \in U$, i.e., $U \cap A = \emptyset$. Thus any point which is not
contained in A is an exterior point for A.

6. <u>Every metric space is normal.</u>

PROOF. Clearly, every metric space satisfies axiom T_1. Let
us verify T_4. Suppose A and B are disjoint closed subsets of a
metric space, and set $U = \{x \mid Dist(x,A) < Dist(x,B)\}$ and
$V = \{x \mid Dist(x,B) < Dist(x,A)\}$. Since $Dist(x,A)$ and $Dist(x,B)$
depend continuously on x (see 4.13), U and V are open. Trivially,
$U \cap V = \emptyset$, $A \subset U$, and $B \subset V$.

Urysohn Functions

7 (LEMMA). <u>Let</u> A <u>and</u> B <u>be two closed subsets of a topological space</u> X. <u>Let</u> Γ <u>be the collection of all neighborhoods of</u> A <u>which do not intersect</u> B, <u>and let</u> Δ <u>be the set of dyadic rational numbers of the interval</u> I <u>(i.e., the numbers</u> $m/2^q$ <u>with arbitrary nonnegative integers</u> m,q <u>satisfying</u> $m \leqslant 2^q$). <u>If</u> X <u>is normal, then there exists a mapping</u> $\varphi: \Delta \to \Gamma$ <u>such that</u>

$$\mathrm{Cl}\,\varphi(r_1) \subset \varphi(r_2) \quad \underline{\text{for}} \quad r_1 < r_2. \tag{1}$$

PROOF. Set $\varphi(1) = X \smallsetminus B$ and let $\varphi(0)$ be any neighborhood of A which is contained, together with its closure, in $X \smallsetminus B$ (see the second formulation of axiom T_4). If $\varphi(r)$ is already defined in such a way that (1) holds for the numbers $r = m/2^q \in \Delta$ with $q = n$, we can extend the definition to $r = m/2^{n+1} \in \Delta$, keeping (1) valid: if $r = m/2^{n+1} \in \Delta$ with odd $m = 2k+1$, we take $\varphi(r)$ to be any open set containing $\mathrm{Cl}\,\varphi(k/2^n)$ and contained along with its closure in $\varphi((k+1)/2^n)$. This induction yields a mapping $\varphi: \Delta \to \Gamma$ with property (1).

8. <u>Given two arbitrary disjoint closed subsets</u> A <u>and</u> B <u>of a normal space</u> X, <u>there is a continuous function</u> $X \to I$, <u>equal to</u> 0 <u>on</u> A <u>and equal to</u> 1 <u>on</u> B.

PROOF. Using the previous lemma and its notations, define a function $f: X \to I$ by the formula

$$f(x) = \begin{cases} \inf\{r \mid \varphi(r) \ni x\}, & \text{if } x \in \varphi(1), \\ \\ 1, & \text{if } x \in X \smallsetminus \varphi(1). \end{cases}$$

It is evident that f is equal to 0 on A and to 1 on B. To show that f is continuous, note that the intervals $[0,r)$ and $(r,1]$ with $r \in \Delta$ constitute a prebase of the segment I, and that $f^{-1}([0,r)) = \bigcup_{r' < r} \varphi(r')$, while $f^{-1}([0,r]) = \bigcap_{r' > r} \varphi(r')$. Using property (1), we see that the last intersection is just $\bigcap_{r' > r} \mathrm{Cl}\,\varphi(r')$; hence $f^{-1}((r,1]) = X \smallsetminus f^{-1}([0,r]) = X \smallsetminus \bigcap_{r' > r} \mathrm{Cl}\,\varphi(r')$. Therefore, the intervals $[0,r)$ and $(r,1]$ have open preimages, and f is continuous.

9. A continuous function $f: X \to I$ such that $f(x) = 0$ for $x \in A \subset X$ and $f(x) = 1$ for $x \in B \subset X$ is referred to as a <u>Urysohn function for the pair</u> A, B.

A Urysohn function for a pair A, B may also take the value
0 outside of A. However, if A is distinguishable, f is any Urysohn
function for the pair A, B, and g distinguishes A, then
$x \mapsto \min(f(x)+g(x),1)$ provides a Urysohn function for A, B which is
positive outside A.

One may note that the proof of Theorem 8 does not use axiom
T_4 and conclude that this theorem is true for any T_4-space. The
converse is also true: if any pair of disjoint closed subsets of X
admits a Urysohn function, then X satisfies axiom T_4.

Finally, we remark that one can prove a slightly stronger
version of Theorem 8, by composing the Urysohn function with a linear
transformation $t \mapsto a + (b-a)t$, where a and b are arbitrary real
numbers. The composition is a continuous function $X \to [a,b]$, equal
to a on A and to b on B.

Extension Theorems

10 (LEMMA). <u>Let F be a closed subset of the topological
space X, and let $\phi: F \to \mathbb{R}$ be a continuous function bounded in
absolute value by a number L > 0. If X is normal, then there exists
a continuous function $\psi: X \to \mathbb{R}$ such that</u>

$$|\psi(x)| \leqslant L/3 \quad \underline{\text{for}} \ \ x \in X,$$

<u>and</u>

$$|\psi(x) - \phi(x)| \leqslant 2L/3 \quad \underline{\text{for}} \ \ x \in F. \qquad (2)$$

PROOF. The subsets of F determined by the inequalities
$\phi(x) \leqslant -L/3$ and $\phi(x) \geqslant L/3$ are closed in F, and hence in X, and
disjoint. Therefore, there is a continuous function $\psi: X \to [-L/3,L/3]$,
equal to -L/3 on the first set, and equal to L/3 on the second
(see 9). It is clear that ψ satisfies the requirements (2).

11. <u>If A is a closed subset of the normal space X, then
every continuous function $A \to \mathbb{R}$ extends to a continuous function
$X \to \mathbb{R}$. This claim remains true if one takes an interval instead of the
real line $\mathbb{R}$.</u>

First, let us show that every continuous map f of A into
an interval can be extended to a continuous map g of X into the same
interval. Without loss of generality, we may take the interval [-1,1].
Define g as the sum of a series of continuous functions $g_k: X \to \mathbb{R}$
which satisfy the conditions

$$|g_k(x)| \leqslant 2^{k-1}/3^k, \quad \text{if} \quad x \in X, \tag{3}$$

and

$$|f(x) - \sum_0^k g_i(x)| < (2/3)^k, \quad \text{if} \quad x \in A. \tag{4}$$

The functions g_k are constructed inductively: take $g_0 = 0$ and, assuming that $g_0, \ldots, g_n$ are already constructed and satisfy (3) and (4) for $k \leqslant n$, define g_{n+1} to be the function obtained when one applies the previous lemma to $\phi = f - \sum_0^n (g_i|_A)$, $F = A$, and $L = (2/3)^n$. Inequality (3) shows that the series $\sum_0^\infty g_k$ converges uniformly on X, and hence its sum g is a continuous function (see 4.12). Inequality (4) implies that $g|_A = f$.

To prove the first part of the theorem, notice that $\mathbb{R}$ is homeomorphic to the open interval $(-1,1)$. We have just showed that given any continuous function $g: A \to (-1,1)$, there exists a continuous $g: X \to [-1,1]$ such that $g(x) = f(x)$ for all $x \in A$. Let $B = g^{-1}(-1) \cup g^{-1}(1)$. The sets A and B are closed and disjoint; hence the pair A, B has a Urysohn function. If we multiply the latter by g, we get the desired extension $X \to (-1,1)$ of the function f.

12. _If_ A _is a closed subset of the normal space_ X, _then every continuous map_ $A \to \mathbb{R}^n$ _extends to a continuous map_ $X \to \mathbb{R}^n$. _This claim remains true if one takes a cube instead of_ $\mathbb{R}^n$.

To see this, it suffices to apply Theorem 12 to the coordinate functions of the given map $A \to \mathbb{R}^n$.

6. Countability Axioms

1. A topological space is said to satisfy the _second axiom of countability_ (or to be a _second countable space_) if it has a countable base. A topological space is said to satisfy the _first axiom of countability_ (or to be a _first countable space_) if it has a countable base at each point. A topological space is _separable_ if it has a countable dense subset.

2. _The second axiom of countability implies the first axiom of countability and the separability. A metric space is always first countable, and is second countable if and only if it is separable._

It is immediate that a second countable space is first countable, and 2.6 shows that every metric space is first countable.

To produce a contable dense set in a space with countable base, just
pick a point in each set of a given countable base. Given a separable
metric space, the open balls centered at the points of a countable dense
set and with radii $1/n$ $(n = 1,2,...)$ constitute a countable base.

  3. $\mathbb{R}^n$ <u>and</u> ℓ_2 <u>are separable, and hence have a countable</u>
<u>base.</u>

  The collection of all sequences $\{x_i\}_1^n$ with rational x_i's
is a countable dense set in $\mathbb{R}^n$. The set of all finitely supported
(i.e., having only a finite number of nonzero terms) sequences $\{x_i\}_1^\infty$
with rational x_i's is countable and dense in ℓ_2.

  4. <u>Every subspace of a second countable space is second</u>
<u>countable. In particular, all subspaces of $\mathbb{R}^n$ and ℓ_2 have countable</u>
<u>base.</u>

  Indeed, a countable base of the space induces a countable base
of each of its subspaces; see 3.2.

  5. <u>In a separable space, every collection of pairwise</u>
<u>disjoint open subsets is countable.</u>

  In fact, let S be a countable dense subset of the given
space. For any set in the given collection, pick a point of S
contained in this set. This yields an injective mapping of the given
collection into S.

  6. Obviously, a continuous surjective map of topological
spaces carries every dense set into a dense set. It is also clear that
an open map between two topological spaces transforms each base into a
base, and each base at a point into a base at the image of this point.
Therefore, the image of a separable space under a continuous map is
separable, and the image of a first or a second countable space under
an open map is first, respectively second countable.

  7. <u>Every regular second countable space is normal.</u>

  PROOF. Let A and B be closed disjoint subset of a regular
second countable space. According to the second formulation of axiom
T_3, each point of any of the sets A and B has a neighborhood whose
closure does not intersect the other set. Picking such neighborhoods,
we get open covers of A and B, and we may assume that these covers
are countable; if not, we may refine them by covers made of sets
belonging to a countable base (see 1.13). Let us index these two
covers, writing them as $U_1, U_2, \ldots$ and $V_1, V_2, \ldots$, and then set
$U'_n = U_n \smallsetminus \cup_1^n \mathrm{Cl}\, V_i$ and $V'_n = V_n \smallsetminus \cup_1^n \mathrm{Cl}\, U_i$. The sets $U = \cup_1^\infty U'_n$ and

$V = \bigcup_1^\infty V_n'$ are open and clearly disjoint. Since $\operatorname{Cl} U_i \cap B = \emptyset$ and $\operatorname{Cl} V_i \cap A = \emptyset$, we have $U \supset A$ and $V \supset B$.

Embedding and Metrization Theorems

8. <u>Every regular second countable space can be embedded in ℓ_2.</u>

PROOF. Let X be a regular space with countable base Γ. We index the pairs (U,V), $U,V \in \Gamma$, satisfying $\operatorname{Cl} U \subset V$, writing them as a sequence (U_1,V_1), (U_2,V_2),.... Now define $f: X \to \ell_2$ by the rule $f(x) = \{k^{-1}\phi_k(x)\}_1^\infty$, where ϕ_k is an arbitrary Urysohn function for the pair $\operatorname{Cl} U_k$, $X \smallsetminus V_k$ (see 7). If $x \neq y$, then there exists an index k such that $x \in U_k$, $y \in X \smallsetminus V_k$ (indeed, X is regular), and so f is injective. We show next that f is an embedding.

Since the ϕ_k's are continuous, given $x_0 \in X$, $\varepsilon > 0$, and n, there is a neighborhood U of x_0 such that $\sum_1^n (|\phi_k(x) - \phi_k(x_0)|/k)^2 < \varepsilon^2/2$ for all $x \in U$. Choosing n such that $\sum_{n+1}^\infty k^{-2} < \varepsilon^2/2$, we see that $\operatorname{dist}(f(x_0),f(x)) < \varepsilon$, and so f is continuous.

Let g denote the inverse of $\operatorname{ab} f: X \to f(X)$. Given a point $y_0 \in f(X)$ and a neighborhood U of $g(y_0)$, find n such that $g(y_0) \in U_n$ and $V_n \subset U$. If $y \in f(X)$ and $\operatorname{dist}(y_0,y) < 1/n$, then clearly $|\phi_n(g(y)) - \phi_n(g(y_0))| < 1$, which in turn implies that $g(y) \in V_n$. In conclusion, for $y \in f(X)$ and $\operatorname{dist}(y_0,y) < 1/n$, we have $g(y) \in U$, proving the continuity of g.

9. <u>A second countable topological space is metrizable if and only if it is regular.</u>

The necessity of this condition is contained in 5.6, and its sufficiency - in 8.

7. <u>Compactness</u>

1. A topological space is <u>compact</u> if any of its open covers contains a finite cover. For example, a finite set endowed with an arbitrary topology is compact, whereas an infinite set endowed with the discrete topology is not compact.

It is clear that a subspace A of a topological space X is

compact if only if from each open cover of A in X one can extract
a finite cover.

    2.   <u>Every closed subset of a compact space is compact.</u>

    PROOF. Let A be a closed subset of the compact space X,
and let Δ be a cover of A in X. We add the set $X \smallsetminus A$ to Δ,
extract a finite cover from the resulting open cover of X, and then
delete the set $X \smallsetminus A$ from the latter, if it still remains. This
obviously yields a finite cover of A in X which is contained in Δ.

    3.   <u>In a Hausdorff space, any two compact disjoint sets have
disjoint neighborhoods.</u>

    PROOF. Let A and B denote the given sets. If B is a
point, then for each point $x \in A$ consider disjoint neighborhoods U_x
and V_x of x and B, and extract a finite cover $U_{x_1}, \ldots, U_{x_s}$ from
the open cover of A given by the neighborhoods U_x; then
$\cup_1^s U_{x_i}$ and $\cap_1^s V_{x_i}$ are disjoint neighborhoods of the set A and the
point B. In the general case, pick for each $x \in B$ disjoint
neighborhoods U_x and V_x of A and of x, and then extract a finite
cover $V_{x_1}, \ldots, V_{x_s}$ from the resulting cover of B by neighborhoods V_x;
then $\cap_1^s U_{x_i}$ and $\cup_1^s V_{x_i}$ are disjoint neighborhoods of A and B.

    4.   <u>Every compact subset of a Hausdorff space is closed.</u>

    Indeed, from 3 we see that a point which is not contained in
a given compact subset of a Hausdorff space has a neighborhood which
does not intersect this subset.

    5.   <u>Every compact Hausdorff space is normal.</u>

    This is a consequence of 2 and 3.

Compactness and Fundamental Covers

    6.   <u>Suppose</u> A <u>is a compact subset of a</u> T_1<u>-space</u> X. <u>Then
from every countable fundamental cover of</u> X <u>one can extract a finite
cover of</u> A.

    PROOF. Let $U_1, U_2, \ldots$ be the given cover. If none of the
sets $\cup_1^m U_i$ covers A, pick a point from each set $A \smallsetminus \cup_1^m U_i$, and
denote the set of all these points by Y. It is obvious that Y is
infinite and that each intersection $Y \cap U_i$ is finite. The latter
shows that Y and all its subsets are closed; hence Y is at the same

time compact (see 2) and discrete. But this contradicts the fact that
Y is infinite.

7. The countability assumption in Theorem 6 is essential.
For example, the cover of a segment by all its countable subsets is
fundamental, but one cannot extract from it a countable cover.

Compactness and Maps

8. <u>The image of a compact space under a continuous map is
compact.</u>

PROOF. Let f be a continuous map of the compact space X
onto a topological space Y, and let Δ be an open cover of Y. The
sets $f^{-1}(V)$, $V \in \Delta$, form an open cover of X, and clearly a subcover
$U_1,\ldots,U_s$ of this cover yields a subcover $f(U_1),\ldots,f(U_s)$ of Δ.

9. <u>Every continuous map of a compact space into a Hausdorff
space is closed.</u>

This is a corollary of 2, 8, and 4.

10. <u>Every invertible continuous map of a compact space onto
a Hausdorff space is a homeomorphism. Every injective continuous map
of a compact space into a Hausdorff space is an embedding.</u>

These are consequences of proposition 9 and of the obvious
fact that a closed invertible map is a homeomorphism.

Compactness and Metrics

11. <u>Every compact subset of a metric space can be covered by
a finite number of open balls having radius</u> ε, <u>for any positive</u> ε.

Such a cover can be extracted from any cover consisting of
balls of radius ε.

12. <u>Every compact metric space has a countable base.</u>

To obtain such a base, it suffices to construct, for each
positive integer n, a finite cover of open balls of radius $1/n$, and
then take the union of these covers.

13. <u>Every compact metric space is bounded.</u>

This is a consequence of 11.

14. <u>Let</u> X <u>be a compact topological space. Then every</u> <u>continuous function</u> $X \to \mathbb{R}$ <u>attains its absolute maximum and absolute</u> <u>minimum.</u>

PROOF. Propositions 8 and 13 show that the image of X in $\mathbb{R}$ is bounded. Proposition 9 shows that this image is closed, which in turn implies that it contains its adherent points, including its greatest lower bound and its least upper bound.

15. <u>Let</u> A <u>and</u> B <u>be disjoint subsets of a metric space.</u> <u>If</u> A <u>is compact and</u> B <u>is closed, then</u> $\mathrm{Dist}(A,B) > 0$.

PROOF. Since A is compact and $\mathrm{Dist}(x,B)$ depends continuously on $x \in A$ (see 4.13), there exists $a \in A$ such that $\mathrm{Dist}(a,B) = \inf_{x \in A} \mathrm{Dist}(x,B) = \mathrm{Dist}(A,B)$ (see 14). Since B is closed and $a \notin B$, $\mathrm{Dist}(a,B) > 0$, and thus $\mathrm{Dist}(A,B) > 0$.

16. <u>Suppose that</u> f <u>is a continuous map of a metric space</u> X <u>into a topological space</u> Y <u>and</u> Δ <u>is an open cover of</u> Y. <u>If</u> X <u>is compact, then there is</u> $\varepsilon > 0$ <u>such that for any set</u> $A \subset X$ <u>with</u> <u>diameter</u> $\mathrm{diam}\,A < \varepsilon$, $f(A)$ <u>is contained in some element of</u> Δ.

It is enough to show that there is an $\varepsilon > 0$ such that any two points $x,y \in X$ with $\mathrm{dist}(x,y) < \varepsilon$ are both contained in one of the sets of the cover $\Gamma = f^{-1}(\Delta)$. For each $x \in X$, pick a ball centered at x and contained in one of the sets of Γ, and let U_x be the concentric ball with half the radius. Now extract a finite cover $U_{x_1}, \ldots, U_{x_s}$ from the cover of X by the balls U_x. Let ε_i denote the radius of U_{x_i}, and let $\varepsilon = \min\{\varepsilon_1, \ldots, \varepsilon_s\}$. If $x,y \in X$ and $\mathrm{dist}(x,y) < \varepsilon$, then $\mathrm{dist}(x_i,x) < \varepsilon_i$ and $\mathrm{dist}(x_i,y) < \mathrm{dist}(x_i,x) + {} + \mathrm{dist}(x,y) < 2\varepsilon_i$ for some i. Therefore, x and y belong to one and the same set of the cover Γ.

Compactness in Euclidean Space

17. <u>The cubes of</u> $\mathbb{R}^n$ <u>are compact.</u>

PROOF. Obviously, any cube in $\mathbb{R}^n$ can be divided into 2^n cubes of half the edge, and if some cover Γ of the original cube by open subsets of $\mathbb{R}^n$ does not contain a finite subcover, then it retains the same property as a cover of one of the smaller cubes. An iteration of this argument yields a decreasing sequence of cubes $Q_1, Q_2, \ldots,$ each of them being half the size of the preceding one, and such that none of them is covered by a finite collection of sets from Γ. But the point common to all these cubes is certainly covered by some set from Γ, which must also cover all the cubes Q_k with k large enough.

18. <u>A subset of $\mathbb{R}^n$ is compact if and only if it is bounded and closed.</u>

The necessity of these conditions is implied in 13 and 4. The sufficiency is a consequence of 17 and 2, since any bounded subset of $\mathbb{R}^n$ is contained in some cube.

Local Compactness

19. A topological space is <u>locally compact</u> if each of its points has a neighborhood with compact closure.

Compact spaces are obviously locally compact. The most important examples of noncompact, locally compact spaces are $\mathbb{R}^n$ with $n > 0$.

20. <u>Every closed subset of a locally compact space is locally compact.</u>

Indeed, if a is a point of a closed subset A of the locally compact space X, and U is a neighborhood of a in X with compact closure $Cl_X U$, then $U \cap A$ is a neighborhood of a in A with compact closure $Cl_A(U \cap A)$. $[Cl_A(U \cap A)$, being closed in X, is closed in the compact subset $Cl_X U$ of X, and hence is compact; see 2].

21. <u>Every open subset of a locally compact Hausdorff space is locally compact.</u>

PROOF. Let a be a point of the open subset A of the locally compact space X, and let U be a neighborhood of a in X with compact closure $Cl_X U$. Since the space $Cl_X U$ is regular (see 5), a has a neighborhood V in $Cl_X U$ such that $Cl_{Cl_X U} V \subset U \cap A$. We show that V is a neighborhood of a in A with compact closure $Cl_A V$.

The set V is open in $Cl_X U$, hence in $U \cap A$, which in turn implies that V is open in A. To verify that $Cl_A V$ is compact, note that the closure $Cl_{Cl_X U} V$ is compact (see 2) and contained in $U \cap A$. This implies that it equals $Cl_{U \cap A} V$ and that the latter is closed in A (see 4), which finally shows that $Cl_{U \cap A} V = Cl_A V$.

22. <u>Let</u> U <u>be a neighborhood of the point</u> a <u>of the</u>
<u>locally compact space</u> X. <u>If</u> X <u>is Hausdorff, then</u> a <u>has a</u>
<u>neighborhood whose closure is compact and contained in</u> U.

PROOF. Since U is a locally compact space (see 21), a has
a neighborhood V in U with compact closure. Since U is open, V
is open in U. Finally, since $Cl_U V$ is compact and X is Hausdorff,
$Cl_U V$ is closed in X (see 4) and thus it coincides with $Cl_X V$. We
see that V is the desired neighborhood of a.

23. <u>Locally compact Hausdorff spaces are regular.</u>

This is a consequence of 22.

Information: Paracompactness

24. A Hausdorff space is <u>paracompact</u> if each of its open
covers has a locally finite refinement. The compact Hausdorff spaces
are (obviously) paracompact, and so are all the metric spaces. All
paracompact spaces are normal. For details, see [11].

One can show that a paracompact space which can be covered by
open metrizable sets is metrizable; see [13].

§2. CONSTRUCTIONS

1. <u>Sums</u>

1. The sum $\bigsqcup_{\mu \in M} X_\mu$ of a family $\{X_\mu\}_{\mu \in M}$ of topological
spaces becomes a topological space if we declare a subset to be open
if its preimages under all maps $in_\nu: X_\nu \to \bigsqcup X_\mu$ are open.
Equivalently, a subset of $\bigsqcup X_\mu$ is closed if its preimages under all
maps in_ν are closed. It is evident that each map $in_\nu: X_\nu \to \bigsqcup X_\mu$
is an embedding and that the images $in_\nu(X_\nu)$ are both open and closed
in $\bigsqcup X_\mu$.

Let $\{Y_\mu\}$ be another family of topological spaces, indexed
by the same set M, and let $f_\mu: X_\mu \to Y_\mu$ be continuous maps. Then
the map $\bigsqcup f_\mu: \bigsqcup X_\mu \to \bigsqcup Y_\mu$ is obviously continuous.

2. If all spaces X_μ satisfy one of the axioms $T_1, T_2, T_3,$

or T_4, then their sum satisfies the same axiom. The same hold for the first axiom of countability, and also for the properties of local compactness and metrizability [if X_μ are metric spaces, one can define a metric on $\bigsqcup X_\mu$ by the formulas: $\mathrm{dist}(\mathrm{in}_\nu(x),\mathrm{in}_{\nu'}(x')) = 1$ when $\nu \neq \nu'$ or when $\nu = \nu'$ and $\mathrm{dist}(x,x') \geqslant 1$; $\mathrm{dist}(\mathrm{in}_\nu(x),\mathrm{in}_\nu(x')) = \mathrm{dist}(x,x')$ when $\mathrm{dist}(x,x') < 1$]. If each X_μ has a countable base and M is countable, then $\bigsqcup X_\mu$ has a countable base too. Similarly, when M is countable and each X_μ is separable, $\bigsqcup X_\mu$ is separable too. Finally, $\bigsqcup X_\mu$ is compact whenever all X_μ are compact and M is finite.

2. Products

1. Let $X_1,\dots,X_n$ be topological spaces. We define a topology on $X_1 \times \dots \times X_n$ by taking as a base the collection of all sets $U_1 \times \dots \times U_n \subset X_1 \times \dots \times X_n$, where U_i is open in X_i, $i = 1,\dots,n$. The conditions of 1.1.8 are satisfied by virtue of the relation $(U_1' \times \dots \times U_n') \cap (U_1'' \times \dots \times U_n'') = (U_1' \cap U_1'') \times \dots \times (U_n' \cap U_n'')$. The resulting topological space is called the <u>product of the spaces</u> $X_1,\dots,X_n$.

If $A_1,\dots,A_n$ are subspaces of $X_1,\dots,X_n$, then the topology of the product $A_1 \times \dots \times A_n$ is obviously identical with the topology induced by the inclusion $A_1 \times \dots \times A_n \subset X_1 \times \dots \times X_n$.

Actually, we have met with some products already. Indeed, $\mathbb{R}^n$ is the product of n copies of the real line $\mathbb{R}$, while I^n is the product of n copies of the unit segment I.

The product $X \times I$, where X is a topological space, is known as the <u>cylinder over</u> X.

2. Forming the product of spaces is a commutative and associative operation: there are obvious canonical homeomorphisms $X_1 \times X_2 \to X_2 \times X_1$ and $(X_1 \times X_2) \times X_3 \to X_1 \times (X_2 \times X_3)$. Similarly, $(X_1 \times \dots \times X_{n-1}) \times X_n = X_1 \times \dots \times X_n$.

Moreover, sums and products of spaces satisfy the distributive law: there is a canonical homeomorphism $X \times (\bigsqcup_{\mu \in M} Y_\mu) \to \bigsqcup_{\mu \in M} (X \times Y_\mu)$.

3. <u>If the sets $A_1,\dots,A_n$ are open in $X_1,\dots,X_n$, then</u> $A = A_1 \times \dots \times A_n$ <u>is open in $X_1 \times \dots \times X_n$. If $A_1,\dots,A_n$ are closed, then A is closed. In all cases,</u> $\mathrm{Cl}\,A = \mathrm{Cl}\,A_1 \times \dots \times \mathrm{Cl}\,A_n$.

The first statement is a direct result of the definition of

the topology of $X_1 \times \ldots \times X_n$, while the second is a consequence of the third; so let us verify the third statement. A point $(x_1, \ldots, x_n) \in$ $\in X_1 \times \ldots \times X_n$ is an adherent point of A if and only if each of its base neighborhoods $U_1 \times \ldots \times U_n$ has a nonempty intersection with A, i.e., if and only if for any neighborhoods $U_1, \ldots, U_n$ of the points $x_1, \ldots, x_n$, $U_i \cap A_i \neq \emptyset$, $i = 1, \ldots, n$. That is to say, $(x_1, \ldots, x_n)$ is an adherent point of A if and only if x_i is an adherent point of A_i, $i = 1, \ldots, n$. Therefore, $\mathrm{Cl}\, A = \mathrm{Cl}\, A_1 \times \ldots \times \mathrm{Cl}\, A_n$.

4. We note that the projections $\mathrm{pr}_i \colon X_1 \times \ldots \times X_n \to X_i$ are continuous and open for any topological spaces $X_1, \ldots, X_n$.

The sets of the form $x_1^0 \times \ldots \times x_{i-1}^0 \times X_i \times x_{i+1}^0 \times \ldots \times x_n^0$, $x_j^0 \in X_j$ $(j \neq i)$ are called the <u>fibers</u> of the product $X_1 \times \ldots \times X_n$. Clearly, the restriction of $\mathrm{pr}_i \colon X_1 \times \ldots \times X_n \to X_i$ to any fiber $x_1^0 \times \ldots \times x_{i-1}^0 \times X_i \times x_{i+1}^0 \times \ldots \times x_n^0$ is a homeomorphism. Hence the fibers are canonically homeomorphic to the factors of the product.

For any map $f \colon Y \to X_1 \times \ldots \times X_n$, where $Y, X_1, \ldots, X_n$ are arbitrary sets, we have the corresponding maps $\mathrm{pr}_i \circ f \colon Y \to X_i$; conversely, given arbitrary $f_i \colon Y \to X_i$, there is a unique map $f \colon Y \to X_1 \times \ldots \times X_n$ such that $\mathrm{pr}_i \circ f = f_i$. Clearly, if $Y, X_1, \ldots, X_n$ are topological spaces, then f is continuous if and only if all the maps $\mathrm{pr}_i \circ f$ are continuous.

In particular, it follows that the map $\mathrm{diag} \colon X \to X \times X$ is continuous for every topological space X. We make the (obvious) remark that the diagonal $\mathrm{diag}(X)$ is closed if and only if X is Hausdorff.

5. Obviously, every product $f_1 \times \ldots \times f_n \colon X_1 \times \ldots \times X_n \to Y_1 \times \ldots \times Y_n$ of continuous maps $f_1 \colon X_1 \to Y_1, \ldots, f_n \colon X_n \to Y_n$ is continuous. Moreover, $f_1 \times \ldots \times f_n$ is open whenever $f_1, \ldots, f_n$ are open.

6. If X is a metric space, then clearly $|\mathrm{dist}(x_1', x_2') - \mathrm{dist}(x_1, x_2)| \leqslant \mathrm{dist}(x_1', x_1) + \mathrm{dist}(x_2', x_2)$ for any points $x_1, x_2, x_1', x_2' \in X$. This inequality shows that the function $\mathrm{dist} \colon X \times X \to \mathbb{R}$ is continuous.

Properties of Products

7. <u>Every product of</u> T_1<u>-spaces is a</u> T_1<u>-space. Every product of Hausdorff spaces is Hausdorff. Every product of regular spaces is regular.</u>

The first and second assertions are immediate. We show that a product of T_3-spaces $X_1,\ldots,X_n$ is a T_3-space. Let U be a neighborhood of $(x_1,\ldots,x_n)$ in $X_1 \times \ldots \times X_n$. Pick neighborhoods $U_1,\ldots,U_n$ of the points $x_1,\ldots,x_n$, such that $U_1 \times \ldots \times U_n \subset U$, and fix neighborhoods $V_1,\ldots,V_n$ of the same points with $\mathrm{Cl}\, V_1 \subset U_1$, $\ldots,\mathrm{Cl}\, V_n \subset U_n$. Since $\mathrm{Cl}(V_1 \times \ldots \times V_n) = \mathrm{Cl}\, V_1 \times \ldots \times \mathrm{Cl}\, V_n$ (see 3), one has $\mathrm{Cl}(V_1 \times \ldots \times V_n) \subset U$.

INFORMATION. There are product of normal spaces which are not normal; see [14].

8. If $S_1,\ldots,S_n$ are dense sets in the spaces $X_1,\ldots,X_n$, then $S_1 \times \ldots \times S_n$ is obviously dense in $X_1 \times \ldots \times X_n$. Consequently, a product of separable spaces is separable.

If $\Gamma_1,\ldots,\Gamma_n$ are bases of $X_1,\ldots,X_n$, then the sets $U_1 \times \ldots \times U_n$ with $U_1 \in \Gamma_1,\ldots,U_n \in \Gamma_n$ form a base of the space $X_1 \times \ldots \times X_n$. Consequently, a product of second countable spaces is second countable.

If now $\Gamma_1,\ldots,\Gamma_n$ are bases of $X_1,\ldots,X_n$ at the points $x_1,\ldots,x_n$, then the sets $U_1 \times \ldots \times U_n$ with $U_1 \in \Gamma_1,\ldots,U_n \in \Gamma_n$, form a base of $X_1 \times \ldots \times X_n$ at the point $(x_1,\ldots,x_n)$. Consequently, a product of first countable spaces is first countable.

9. Every product of metrizable spaces is metrizable.

In fact, we can say more: if $X_1,\ldots,X_n$ are metric spaces, then the formula $\mathrm{dist}((x_1,\ldots,x_n),(x_1',\ldots,x_n')) = [\sum_{i=1}^{n}(\mathrm{dist}(x_i,x_i'))^2]^{1/2}$ defines a canonical metric on the product $X_1 \times \ldots \times X_n$.

10. Every product of compact spaces is compact.

It suffices to consider a product of two spaces. So let X and Y be compact topological spaces, and let Γ be an open cover of $X \times Y$. Consider an arbitrary refinement Δ of Γ, consisting of open sets of the form $U \times V$ (see 1.1.13). Since the fibers $x \times Y$ are homeomorphic to Y and Y is compact, one can find, for each point $x \in X$, a finite collection $\Delta_x = \{U_i(x) \times V_i(x)\}_{i=1}^{n(x)}$ of elements of Δ which covers $x \times Y$ (see 1.7.2); one may assume that $x \in U_i(x)$ for all $i = 1,\ldots,n(x)$. Since the sets $U_x = \cap_{i=1}^{n(x)} U_i(x)$ are open and cover the compact space X, there exists a finite collection $U_{x_1},\ldots,U_{x_m}$ covering X. It is clear that $\Delta' = \cup_{j=1}^{m} \Delta_{x_j}$ is a cover of $X \times Y$. Finally, replacing each set $W \in \Delta'$ by a set of Γ containing W, we produce a finite subcover of Γ.

11. <u>Every product of locally compact spaces is locally compact.</u>

PROOF. Let $U_1,\ldots,U_n$ be neighborhoods of the points $x_1 \in X_1$, $\ldots,x_n \in X_n$. Then $U_1 \times \ldots \times U_n$ is a neighborhood of $(x_1,\ldots,x_n)$ in $X_1 \times \ldots \times X_n$. Furthermore, its closure $\mathrm{Cl}(U_1 \times \ldots \times U_n)$ is just $\mathrm{Cl}\, U_1 \times \ldots \times \mathrm{Cl}\, U_n$ (see 3), and so is compact whenever $\mathrm{Cl}\, U_1,\ldots,\mathrm{Cl}\, U_n$ are compact (see 10).

An Application:

A Method for Constructing Continuous Maps

12. Proposition 14 below allows us to establish continuity of a map in some situations similar to those treated by Theorem 1.4.3, but where the latter is not applicable.

13 (LEMMA). <u>Suppose that the map</u> $f: X \times Q \to Z$ <u>is continuous and transforms the fiber</u> $x_0 \times Q$ <u>into a point. If the space</u> Q <u>is compact, then given any neighborhood</u> W <u>of the point</u> $f(x_0 \times Q)$, <u>there is a neighborhood</u> U <u>of</u> x_0 <u>such that</u> $f(U \times Q) \subset W$.

PROOF. Given any point $q \in Q$, fix a neighborhood U_q of x_0 and a neighborhood V_q of q with $f(U_q \times V_q) \subset W$. Since Q is compact, one can cover it with a finite collection $V_{q_1},\ldots,V_{q_s}$. Now set $U = \cap_{i=1}^{s} U_{q_i}$.

14. <u>Suppose that</u> X,Y,Z <u>and</u> Q <u>are topological spaces,</u> A <u>is a subset of</u> X, B <u>is a closed subset of</u> Y, <u>and</u> $f: X \times Q \to Z$ <u>and</u> $g: Y \to X$ <u>are continuous maps such that</u> $f(x \times Q)$ <u>reduces to a point for each</u> $x \in A$, <u>and</u> $g(B) \subset A$. <u>If</u> Q <u>is compact, then for each continuous map</u> $\phi: Y \smallsetminus B \to Q$, <u>the map</u> $h: Y \to Z$ <u>given by</u>

$$h(y) = \begin{cases} f(g(y)),\phi(y)), & \text{for } y \in Y \smallsetminus B, \\[2ex] f(g(y) \times Q), & \text{for } y \in B, \end{cases}$$

<u>is continuous.</u>

PROOF. The map h is clearly continuous at the points of $Y \smallsetminus B$; let us verify its continuity at the points $y \in B$. By virtue of Lemma 13, given any neighborhood W of the point $h(y)$, there is a neighborhood U of the point $g(y)$ such that $f(U \times Q) \subset W$. The last inclusion shows that $h(g^{-1}(U)) \subset W$, and finally note that $g^{-1}(U)$ is a neighborhood of y.

Information

15. The notion of a product of an infinite number of topological spaces can be defined in a natural way; in this case too a product of compact spaces remains compact; see [3] for details.

3. Quotients

1. The quotient set X/p of a topological space X by any of its partitions p is equipped with a natural topology: a subset of X/p is open if its preimage under the map $\mathrm{pr}\colon X \to X/p$ is open. Equivalently, a subset of X/p is closed if its preimage is closed. This topology is called the quotient topology, and the set X/p with the quotient topology is the quotient space of the space X by its partition p.

It is clear that $\mathrm{pr}\colon X \to X/p$ is continuous.

In the special case of a partition p whose elements are a single set A and the points of $X \smallsetminus A$, X/p is called the quotient of the space X by A and is denoted by X/A.

2. Given two topological spaces X and Y with respective partitions p and q, and a continuous map $f\colon X \to Y$ which takes the elements of p into elements of q, the map $\mathrm{fact}\, f\colon X/p \to Y/q$ is continuous. This is a straightforward consequence of the definition of the quotient topology. Indeed, if $U \to Y/q$ is open, then the set $f^{-1}(\mathrm{pr}^{-1}(U))$ is open in X, and so the identity $f^{-1}(\mathrm{pr}^{-1}(U)) = \mathrm{pr}^{-1}((\mathrm{fact}\, f)^{-1}(U))$ implies that $(\mathrm{fact}\, f)^{-1}(U)$ is open in X/p.

If q is the partition of Y into single points, then $Y/q = Y$ and $\mathrm{pr}\colon Y \to Y/q$ is the identity map. In this case, $f \mapsto \mathrm{fact}\, f$ defines a one-to-one correspondence between continuous maps $X \to Y$ which are constant on the elements of the partition p, and continuous maps $X/p \to Y$.

3. In particular, the discussion above shows that given a continuous map $f\colon X \to Y$, its injective factor $\mathrm{fact}\, f\colon X/\mathrm{zer}(f) \to Y$ is continuous too. The converse is also true: every map $f\colon X \to Y$ can be represented as the composition $X \xrightarrow{\ \mathrm{pr}\ } X/\mathrm{zer}(f) \xrightarrow{\ \mathrm{fact}\, f\ } Y$, and so f is continuous whenever $\mathrm{fact}\, f$ is continuous.

4. A continuous map whose injective factor is a homeomorphism will be referred to as a factorial map (or a quotient map).

An equivalent definition: a map f of a topological space X
into a topological space Y is factorial if $f(X) = Y$, and the preimage
$f^{-1}(B)$ of a set $B \subset Y$ is open if and only if B is open. If we
substitute closed sets for open ones, we obtain another equivalent
definition.

Obviously, the composition of two factorial maps is factorial,
and any injective factorial map is a homeomorphism. Moreover, it is
plain that if f is factorial and the composition $g \circ f$ is continuous,
then the map g is continuous. Also, f continuous and $g \circ f$
factorial imply g factorial.

The projections onto quotient spaces form the main class of
factorial maps. A crude necessary condition for a map $f: X \to Y$ with
$f(X) = Y$ to be factorial is that f be an open or a closed map.

5.   Taking quotients is a transitive operation: if p is a
partition of X and p' is a partition of X/p, then the quotient
space $(X/p)/p'$ is canonically homeomorphic to X/q, where q
partitions X into the preimages of the elements of p' under the
projection $X \to X/p$. This canonical homeomorphism is defined as the
injective factor of the composite map $X \to X/p \to (X/p)p'$, and is truly
a homeomorphism, because this composition is factorial (see 4).

6.   <u>If the sets A and B constitute a fundamental cover of</u>
<u>the space X, then</u>

fact[in:A → X]: A/A∩B → X/B

<u>is a homeomorphism.</u>

Given an open subset U of the quotient A/A∩B, it is enough
to show that $V = [pr: X \to X/B]^{-1}(fact\,in(U))$ is open in X. But this
is a consequence of the equalities

$$V \cap A = [pr: A \to A/A{\cap}B]^{-1}(U)$$
and
$$V \cap B = \begin{cases} B, & \text{if} \quad pr(A \cap B) \in U, \\ \varnothing, & \text{if} \quad pr(A \cap B) \notin U. \end{cases}$$

Properties of Quotient Spaces

7.   Obviously, a qutient space X/p satisfies axiom T_1 if
and only if the elements of the partition p are closed. Also, X/p

is Hausdorff if and only if any two distinct elements of p have disjoint saturated neighborhoods. Similarly, X/p is a T_3-space (T_4-space) if and only if for any element A of p and any saturated closed subset B of X (respectively, for any saturated, closed subsets A and B of X) such that $A \cap B = \emptyset$, A and B have disjoint saturated neighborhoods in X.

Moreover, it is readily seen that X/p is second countable if and only if there is a countable collection of open saturated sets in X such that any saturated set can be expressed as the union of one of its subcollections.

It is immediate from 1.6.6 that a quotient of a separable space is separable.

Similarly, 1.7.8 implies that a quotient of a compact space is compact.

Closed Partitions

8. A partition p of the space X is <u>closed</u> if $\mathrm{pr}: X \to X/p$ is a closed map. An equivalent condition: saturations of closed sets are closed.

Obviously, a partition which has only one element that is not reduced to a point is closed if and only if this element is closed.

9. <u>The quotient of a T_1-space by a closed partition is a T_1-space. The quotient of a normal space by a closed partition is normal.</u>

Since the first assertion is straightforward, all we have to show is that the quotient X/p of a T_4-space X by a closed partition p is a T_4-space. Let F_1 and F_2 be disjoint, saturated, closed subsets of X. Since X is normal, F_1 and F_2 have disjoint neighborhoods. Furthermore, since p is a closed partition, the saturations of the complements of these neighborhoods are closed, and now it is clear that the complements of these saturations are disjoint saturated neighborhoods of F_1 and F_2.

Open Partitions

10. A partition p of the space X is <u>open</u> if $\mathrm{pr}: X \to X/p$ is an open map. An equivalent condition: saturations of open sets are open.

If p is an open partition and A is a saturated set, then the saturation of Int A is open, and hence equals Int A ; passing to complements, we see that the saturation of Cl A is just Cl A . Therefore, in the case of an open partition, the interior and the closure of a saturated set are saturated.

As it follows from 1.6.6, the quotient of a first countable (second countable) space by an open partition is first countable (respectively, second countable).

11. <u>Let</u> p <u>and</u> q <u>be open partitions of the respective</u> <u>spaces</u> X <u>and</u> Y. <u>The product</u> $(X/p) \times (Y/q)$ <u>is canonically homeo-</u> <u>morphic to the quotient</u> $(X \times Y)/(p \times q)$.

The injective factor of the map $\mathrm{pr} \times \mathrm{pr}\colon X \times Y \to (X/p) \times (Y/q)$ defines this canonical map, which is a homeomorphism because $\mathrm{pr} \times \mathrm{pr}$ is open (see 2.5).

4. Glueing

1. <u>Glueing</u> (or <u>pasting</u>) topological spaces is a composite operation which consists of taking a sum and subsequently passing to a quotient. More precisely, suppose that $\{X_\mu\}_{\mu \in M}$ is a family of topological spaces and p is a partition of the space $X = \bigsqcup X_\mu$. Then we say that <u>the quotient space</u> X/p <u>is obtained by glueing the spaces</u> X_μ <u>by</u> (or <u>according to</u>) p . The composite map

$$X_\nu \xrightarrow{\;\mathrm{in}_\nu\;} X \xrightarrow{\;\mathrm{pr}\;} X/p$$

is termed the <u>ν-th immersion</u> and is denoted by imm_ν . Clearly, the sets $\mathrm{imm}_\nu(X_\nu)$ yield a fundamental cover of X/p , and a map $f\colon X/p \to Y$, where Y is an arbitrary topological space, is continuous if and only if all the compositions $f \circ \mathrm{imm}_\nu$ are continuous.

Unions

2. Let $\{X_\mu\}_{\mu \in M}$ be a family of topological spaces. Suppose that for each pair $(\mu, \mu') \in M \times M$ there is given a subset $A_{\mu\mu'} \subset X_\mu$. In addition, suppose that for each pair $(\mu, \mu') \in M \times M$ there is given an invertible map $\phi_{\mu\mu'}\colon A_{\mu\mu'} \to A_{\mu'\mu}$, such that:

(i) $A_{\mu\mu} = X_\mu$ and $\phi_{\mu\mu} = \mathrm{id}\, X_\mu$, for any $\mu \in M$;

(ii) $\phi_{\mu\mu'}(A_{\mu\mu'} \cap A_{\mu\mu''}) = A_{\mu'\mu} \cap A_{\mu'\mu''}$ and the diagram

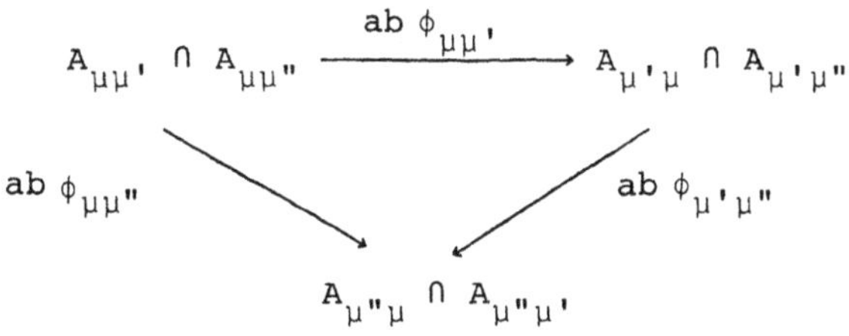

is commutative for every $\mu, \mu', \mu'' \in M$. For $x \in \bigsqcup X_\mu$, denote by B_x the subset of $\bigsqcup X_\mu$ consisting of all the points $in_\mu(\phi_{\nu\mu}(x))$, where $x \in A_{\nu\mu}$. The sets B_x are pairwise disjoint and define a partition of $\bigsqcup X_\mu$. The corresponding quotient space is called the <u>union of the spaces</u> X_μ <u>by</u> (or <u>along</u>) <u>the maps</u> $\phi_{\mu\mu'}$.

This construction is a special case of glueing, when all the immersions imm_μ are injective. Moreover, assuming that all the maps $\phi_{\mu\mu'}$ are homeomorphisms and that the sets $A_{\mu\mu'}$ are all open or all closed, we see at once that all the maps imm_μ are embeddings.

In the general case, a union of T_1-spaces is clearly a T_1-space.

3. Often the union construction is employed when all the spaces X_μ are subsets of a set X and cover X, while $A_{\mu\mu'}$ and $\phi_{\mu\mu'}$ are given by $A_{\mu\mu'} = X_\mu \cap X_{\mu'}$ and $\phi_{\mu\mu'} = id$. In this situation, conditions 2(i) and 2(ii) are automatically fulfilled, and one may describe the union of the X_μ's simply as the set X equipped with the following topology: a set $C \subset X$ is open (closed) if and only if the intersection $C \cap X_\mu$ is open (respectively, closed) in X_μ for any $\mu \in M$.

We devote some special attention to the case where the topology of each set X_μ is induced by some topology already given on X. Then our construction produces a new topology on X. It is clear that the sets open (closed) in the old topology remain open (respectively, closed) in the new topology. Moreover, if all the intersections $X_\mu \cap X_{\mu'}$ are open in their sets X_μ (endowed with the initial topology), then the new topology on X induces the initial topology back on each set X_μ; the same holds whenever all the intersections $X_\mu \cap X_{\mu'}$ are closed in their sets X_μ.

Limits and Filtrations

4. Let $X_0, X_1, \ldots$ be topological spaces, and let

$\phi_0: X_0 \to X_1,\ \phi_1: X_1 \to X_2,\ldots,$ be embeddings. Set

$$A_{kk'} = \begin{cases} \phi_{k-1} \circ \cdots \circ \phi_{k'}(X_{k'}), & \text{if } k' < k, \\[2ex] X_{k'}, & \text{if } k' \geqslant k, \end{cases}$$

and

$$\phi_{kk'} = \begin{cases} ab(\phi_{k-1} \circ \cdots \circ \phi_{k'}), & \text{if } k' < k, \\[1.5ex] id\, X_k, & \text{if } k' = k, \\[1.5ex] [ab(\phi_{k'-1} \circ \cdots \circ \phi_k)]^{-1}, & \text{if } k' > k. \end{cases}$$

The union of the spaces X_s is well-defined because conditions 2(i) and 2(ii) are obviously satisfied. This union is called the <u>limit of the sequence</u> $\{X_k\}$, and is denoted by $\lim(X_k, \phi_k)$ or $\lim X_k$.

A specific property of the limit construction is that the maps $imm_k: X_k \to \lim X_k$ are embeddings: indeed, every closed subset A of X_k is the preimage under imm_k of some closed subset of $\lim X_k$, for example, of

$$\bigcup_{k'=k+1}^{\infty} imm_{k'}(Cl_{X_{k'}}(\phi_{k'-1} \circ \cdots \circ \phi_k(A))).$$

Obviously, if $\phi_k(X_k)$ is open (closed) in X_{k+1} for all k, then all the sets $imm_k(X_k)$ are open (respectively, closed) in $\lim(X_k, \phi_k)$.

Suppose that $\{X_k', \phi_k': X_k' \to X_{k+1}'\}$ is another sequence of topological spaces and embeddings, and that for each k there is given a continuous map $f_k: X_k \to X_k'$, so that all the diagrams

$$\begin{array}{ccc} X_k & \xrightarrow{\;\;f_k\;\;} & X_k' \\ {\scriptstyle\phi_k}\downarrow & & \downarrow{\scriptstyle\phi_k'} \\ X_{k+1} & \xrightarrow{\;f_{k+1}\;} & X_{k+1}' \end{array}$$

are commutative. Then the rule $f(imm_k(x)) = imm_k(f_k(x))$ defines a continuous map $f: \lim(X_k, \phi_k) \to \lim(X_k', \phi_k')$ (see 3.2); f is called the <u>limit of the sequence</u> $f_0, f_1, \ldots,$ and is denoted by $\lim f_k$.

5. <u>If</u> $X_0, X_1, \ldots$ <u>are</u> T_1<u>-spaces, then every compact subset of</u> $\lim X_k$ <u>is contained in one of the sets</u> $imm_k(X_k)$.

$$37$$

This is a consequence of 1.7.6.

6. __If__ $X_0, X_1, \ldots$ __are normal spaces and__ $\phi_k(X_k)$ __is closed in__ X_{k+1} __for every__ k, __then__ $\lim(X_k, \phi_k)$ __is normal.__

PROOF. Since we already know that $\lim(X_k, \phi_k)$ is a T_1-space, it suffices to show that there exists a Urysohn function for any pair A, B of closed disjoint subsets of this space (see 1.5.9). To see this, we merely have to produce a sequence $f_k : \mathrm{imm}_k(X_k) \to I$, such that each f_k is a Urysohn function for the pair $A \cap \mathrm{imm}_k(X_k)$, $B \cap \mathrm{imm}_k(X_k)$, and $f_{k+1} \big| \mathrm{imm}_k(X_k) = f_k$ for each k.

As f_0 take any Urysohn function for the pair $A \cap \mathrm{imm}_0(X_0)$, $B \cap \mathrm{imm}_0(X_0)$. Given f_k, we define $f_{k+1} : \mathrm{imm}_{k+1}(X_{k+1}) \to I$ as the (continuous) extension of the function

$$g_k : [\mathrm{imm}_k(X_k)] \cup [A \cap \mathrm{imm}_{k+1}(X_{k+1})] \cup [B \cap \mathrm{imm}_{k+1}(X_{k+1})] \to I,$$

defined by the formula

$$g_k(x) = \begin{cases} f_k(x), & \text{if } x \in \mathrm{imm}_k(X_k), \\ 0, & \text{if } x \in A \cap \mathrm{imm}_{k+1}(X_{k+1}), \\ 1, & \text{if } x \in B \cap \mathrm{imm}_{k+1}(X_{k+1}) \end{cases}$$

(see 1.5.11). The functions g_k are continuous because the sets $\mathrm{imm}_k(X_k)$, $A \cap \mathrm{imm}_{k+1}(X_{k+1})$, and $B \cap \mathrm{imm}_{k+1}(X_{k+1})$ are closed (see 1.4.3), which in turn is a consequence of the fact that $\phi_k(X_k)$, $\phi_{k+1}(X_{k+1}), \ldots$ are closed.

7. A sequence $X_0, X_1, \ldots$ of subsets of a topological space X is a __filtration__ of X if, firstly, $X_0 \subset X_1 \subset \ldots$, and, secondly, the sets X_k form a fundamental cover of X.

The first condition shows that the inclusions $\mathrm{in} : X_k \to X_{k+1}$ and the limit $\lim(X_k, \mathrm{in})$ are meaningful, while the second condition is equivalent to the following: the map $X \to \lim(X_k, \mathrm{in})$, which equals $\mathrm{imm}_k : X_k \to \lim(X_k, \mathrm{in})$ on each X_k, is a homeomorphism. Using this canonical homeomorphism, we may identify $\lim(X_k, \mathrm{in})$ with X.

Attaching

8. Let X_1, X_2, C, and $\varphi : C \to X_2$ be two topological spaces, a subset of X_1, and a continuous map, respectively. Denote by p the partition of $X_1 \bigsqcup X_2$ into the points of $\mathrm{in}_1(X_1 \smallsetminus C)$ and

$in_2(X_2 \smallsetminus \varphi(C))$, and the sets $in_1(\varphi^{-1}(x)) \cup in_2(x)$ with $x \in \varphi(C)$. The quotient space $(X_1 \sqcup X_2)/p$ is written $X_2 \cup_\varphi X_1$. We say that $X_2 \cup_\varphi X_1$ is obtained by <u>attaching the space</u> X_1 <u>to the space</u> X_2 <u>by</u> (or <u>along</u>) φ.

This construction is clearly a special kind of glueing, and it is plain that $imm_2\colon X_2 \to X_2 \cup_\varphi X_1$ is an embedding.

If X_2 reduces to a point, then $X_2 \cup_\varphi X_1$ is canonically homeomorphic to the quotient space X_1/C; the canonical homeomorphism is just $fact[imm_1\colon X_1 \to X_2 \cup_\varphi X_1]$.

9. <u>If X_1 and X_2 are normal and C is closed, then $X_2 \cup_\varphi X_1$ is normal.</u>

PROOF. Since we already know that $X_2 \cup_\varphi X_1$ is a T_1-space (see 2), it suffices to show that there exists a Urysohn function for any pair of closed disjoint subsets A, B of $X_2 \cup_\varphi X_1$. Let $f_2\colon X_2 \to I$ be a Urysohn function for the pair $imm_2^{-1}(A)$, $imm_2^{-1}(B)$. Define $g\colon C \cup [imm_1^{-1}(A)] \cup [imm_1^{-1}(B)] \to I$ by

$$
g(x) = \begin{cases}
f_2(\varphi(x)), & \text{if } x \in C, \\
0, & \text{if } x \in imm_1^{-1}(A), \\
1, & \text{if } x \in imm_1^{-1}(B),
\end{cases}
$$

and extend it to a continuous function $f_1\colon X_1 \to I$ (see 1.5.11). The function $X_2 \cup_\varphi X_1 \to I$, defined as

$$
y \to \begin{cases}
f_1(x), & \text{if } y \in imm_1(x) \quad [x \in X_1], \\
f_2(x), & \text{if } y \in imm_2(x) \quad [x \in X_2],
\end{cases}
$$

is obviously a Urysohn function for the pair A, B.

5. Projective Spaces

1. In this subsection we shall describe the real, complex, quaternionic, and Cayley projective spaces. These may be considered as examples illustrating the previous definitions, but are also important spaces in their own right.

We denote the field of complex numbers by $\mathbb{C}$, the field of quaternions by $\mathbb{H}$, and the algebra of Cayley numbers by $\mathbb{C}a$. The corresponding n-dimensional spaces, i.e., the products of n copies

$\mathbb{C} \times \ldots \times \mathbb{C}$, $\mathbb{H} \times \ldots \times \mathbb{H}$, and $\mathbb{C}a \times \ldots \times \mathbb{C}a$, are denoted by $\mathbb{C}^n$, $\mathbb{H}^n$, and $\mathbb{C}a^n$. Since every complex number is a pair of real numbers, every quternion a quadruplet of real numbers, and every Caylely number an octuplet of real numbers, one can naturally identify $\mathbb{C}^n$, $\mathbb{H}^n$, and $\mathbb{C}a^n$ with $\mathbb{R}^{2n}$, $\mathbb{R}^{4n}$, and $\mathbb{R}^{8n}$, respectively. In particular, the former are endowed with natural topologies and metrics. The vector operations in $\mathbb{C}^n$, $\mathbb{H}^n$, and $\mathbb{C}a^n$ (addition of vectors and left or right multiplication by scalars) are continuous in these topologies.

2. The <u>n-dimensional real projective space</u> $\mathbb{R}P^n$ is defined as the quotient space of S^n by its partition into pairs of diametrically opposed points. One may equivalently describe $\mathbb{R}P^n$ as the quotient space of D^n by its partition into the points of $\mathrm{Int}\, D^n$ and the pairs of diametrically opossed points of $\mathrm{Fr}\, D^n = S^{n-1}$. The canonical homeomorphism permitting us to identify these two quotient spaces is $\mathrm{fact}\, f$, where $f: D^n \to S^n$ is defined by

$$(x_1,\ldots,x_n) \mapsto (x_1,\ldots,x_n,(1-x_1^2-\ldots-x_n^2)^{1/2}).$$

One may also identify the points of $\mathbb{R}P^n$ with the lines of $\mathbb{R}^{n+1}$ which pass through the point $0 = (0,\ldots,0)$ [the line passing through the points x and $-x$ corresponds to the pair of points $x,-x \in S^n$]. The set of all these lines, equipped with the angular metric (i.e., the distance between two lines is defined as the angle between them which is less than $\pi/2$), is a metric space, and the above natural map of $\mathbb{R}P^n$ onto this space is clearly a homeomorphism. This provides a third description of the real projective space. The fourth, a coordinate description, can be obtained if one remarks that every line passing through 0 is uniquely determined by any of its nonzero points, and that the coordinates of any two nonzero points of such a line are proportional. This enables us to interpret the points of $\mathbb{R}P^n$ as classes of proportional nonzero real sequences $(x_1,\ldots,x_{n+1})$; the point determined by the sequence $(x_1,\ldots,x_{n+1})$ is denoted by $(x_1:\ldots:x_{n+1})$, and the numbers $x_1,\ldots,x_{n+1}$ are called its <u>homogeneous coordinates.</u> The description of the topology of $\mathbb{R}P^n$ in terms of the homogeneous coordinates is plain.

3. All the above discussion of the space $\mathbb{R}P^n$ can be repeated naturally for the complex case, leading to four equivalent description of the <u>n-dimensional complex projective space</u> $\mathbb{C}P^n$. First description: $\mathbb{C}P^n$ is the quotient space of the unit sphere S^{2n+1} of $\mathbb{C}^{n+1}$ by its partition into the circles obtained by intersecting S^{2n+1} with the (complex) lines of $\mathbb{C}^n$ which pass through the point 0. The second description: $\mathbb{C}P^n$ is the quotient space of the unit ball

D^{2n} of $\mathbb{C}^n$ by its partition into the points of $\operatorname{Int} D^{2n}$ and the circles on $\operatorname{Fr} D^{2n} = S^{2n-1}$ obtained by intersecting S^{2n-1} with the lines of $\mathbb{C}^n$ passing through 0. The canonical homeomorphism between these two quotient spaces is $\operatorname{fact} f$, where $f: D^{2n} \to S^{2n+1}$ is given by $f(x_1,\ldots,x_{2n}) = (x_1,\ldots,x_{2n}, (1-x_1^2-\ldots-x_{2n}^2)^{1/2},0)$. The third description: $\mathbb{C}P^n$ is the set of lines of $\mathbb{C}^{n+1}$ passing through the point 0, equipped with the topology induced by the angular metric. The fourth description: $\mathbb{C}P^n$ is the space of classes of (complex) proportional nonzero complex sequences $(x_1,\ldots,x_{n+1})$.

The notation $(x_1:\ldots:x_{n+1})$ introduced in 2 also extends to the complex case, and the numbers $x_1,\ldots,x_{n+1}$ are known, as in the real case, as the homogeneous coordinates of the point $(x_1,\ldots,x_{n+1})$.

4. Since the field of quaternions is not commutative, one has to distinguish between the left and the right lines in $\mathbb{H}^n$. But as soon as we have chosen one type of lines, we can automatically repeat the discussion in 3 for the quaternionic case, and so obtain four equivalent descriptions of the corresponding (left or right) n-dimensional quaternionic projective space $\mathbb{H}P^n$. Since the anti-automorphism $x \mapsto x^{-1}$ of the multiplicative group of the field $\mathbb{H}$ takes left lines into right ones, the right space $\mathbb{H}P^n$ is homeomorphic to the left one.

Henceforth, we shall regard $\mathbb{H}^n$ as a left vector space, and accordingly $\mathbb{H}P^n$ will be regarded as the left projective space.

5. Since the Cayley algebra is not associative, we cannot successfully define lines in $\mathbb{C}a^n$ for $n > 2$. In the Cayley plane $\mathbb{C}a^2$ one can define a line passing through the point $(0,0)$ as a set $\{(x_1,x_2) \mid x_2 = cx_1\}$, where $c \in \mathbb{C}a$; in addition, there is the vertical coordinate line $\{(x_1,x_2) \mid x_1 = 0\}$. If one identifies $\mathbb{C}a^2$ with $\mathbb{R}^{16}$, it becomes clear that these lines are 8-dimensional subspaces. Moreover, every point different from $(0,0)$ of $\mathbb{C}a$ sits on exactly one of these lines, and each line intersects S^{15} along a 7-dimensional sphere. One can define the projective Cayley line $\mathbb{C}aP^1$ as the quotient space of S^{15} by its partition into these 7-dimensional spheres. Of course, there are three more description of this projective line, which are appropriately modified versions of those given in 2, 3, and 4. In addition, we can define the projective Cayley plane $\mathbb{C}aP^2$ as the quotient of D^{16} by its partition into the points of $\operatorname{Int} D^{16}$ and the 7-dimensional spheres just described. However, an attempt to describe the projective plane $\mathbb{C}aP^2$ in the spirit of the alternatives of 2, 3, and 4 fails. Projective Cayley spaces of higher dimensions

are not defined.

6. The spaces $\mathbb{R}P^1$, $\mathbb{C}P^1$, $\mathbb{H}P^1$, and $\mathbb{C}aP^1$ are canonically homeomorphic to S^1, S^2, S^4, and S^8. The homeomorphism $\mathbb{R}P^1 \to S^1$ transforms the line $x_1 = 0$ into ort_1, and each line $x_2 = cx_1$ into the point of the punctured sphere $S^1 \smallsetminus \mathrm{ort}_1$ which corresponds to c via the homeomorphism $\mathbb{R}^1 \to S^1 \smallsetminus \mathrm{ort}_1$ described in 1.4.8. The homeomorphisms $\mathbb{C}P^1 \to S^2$, $\mathbb{H}P^1 \to S^4$, and $\mathbb{C}aP^1 \to S^8$ are similarly defined, when we substitute the homeomorphisms $\mathbb{C}^1 = \mathbb{R}^2 \to S^2 \smallsetminus \mathrm{ort}_1$, $\mathbb{H}^1 = \mathbb{R}^4 \to S^4 \smallsetminus \mathrm{ort}_1$, and $\mathbb{C}a^1 = \mathbb{R}^8 \to S^8 \smallsetminus \mathrm{ort}_1$ for the homeomeorphism $\mathbb{R}^1 \to S^1 \smallsetminus \mathrm{ort}_1$.

7. The canonical embedding $(x_1,\ldots,x_k) \mapsto (x_1,\ldots,x_k,0)$ of $\mathbb{R}^k$ in $\mathbb{R}^{k+1}$ permits identification of $\mathbb{R}^k$ with the subspace $x_{k+1} = 0$ of $\mathbb{R}^{k+1}$ and may be regarded as an inclusion. This map induces inclusions $D^k \to D^{k+1}$, $S^{k-1} \to S^k$, and $\mathbb{R}P^{k-1} \to \mathbb{R}P^k$. Similarly, the inclusions $\mathbb{C}^k \to \mathbb{C}^{k+1}$ and $\mathbb{H}^k \to \mathbb{H}^{k+1}$ induce inclusions $\mathbb{C}P^{k-1} \to \mathbb{C}P^k$ and $\mathbb{H}P^{k-1} \to \mathbb{H}P^k$.

Set

$$\mathbb{R}^\infty = \lim \mathbb{R}^k, \quad \mathbb{C}^\infty = \lim \mathbb{C}^k, \quad \mathbb{H}^\infty = \lim \mathbb{H}^k,$$

$$D^\infty = \lim D^k, \quad S^\infty = \lim S^k,$$

and

$$\mathbb{R}P^\infty = \lim \mathbb{R}P^k, \quad \mathbb{C}P^\infty = \lim \mathbb{C}P^k, \quad \mathbb{H}P^\infty = \lim \mathbb{H}P^k.$$

The points of the spaces $\mathbb{R}^\infty$, $\mathbb{C}^\infty$, and $\mathbb{H}^\infty$ can be naturally identified with the real, complex, and quaternionic finitely-supported (i.e., having only a finite number of nonzero terms) sequences $\{x_k\}_1^\infty$. The sphere S^∞ is included in the ball D^∞, which in turn is included in the space $\mathbb{R}^\infty$. The projective spaces $\mathbb{R}P^\infty$, $\mathbb{C}P^\infty$, and $\mathbb{H}P^\infty$ are constructed from S^∞ and D^∞ by taking quotients that are limits of the quotients described in 2, 3, and 4.

8. The previous description of $\mathbb{C}P^n$ and $\mathbb{H}P^n$, and also of $\mathbb{C}aP^1$, as quotient spaces of spheres, define projections $S^{2n+1} \to \mathbb{C}P^n$, $S^{4n+3} \to \mathbb{H}P^n$, and $S^{15} \to \mathbb{C}aP^1$, which play a distinguished role. They are called _Hopf maps._ The most important Hopf maps are $S^3 \to \mathbb{C}P^1 = S^2$, $S^7 \to \mathbb{H}P^1 = S^4$, and $S^{15} \to \mathbb{C}aP^1 = S^8$.

6. <u>More Special Constructions</u>

1. Let X be a topological space. The quotient space
$(X \times I)/(X \times 0)$ is called the <u>cone over</u> X and is denoted by con X .
The point $\mathrm{pr}(X \times 0)$ is the <u>vertex</u> of the cone, the set $\mathrm{pr}(X \times 1)$ is
the <u>base</u> of the cone, and each set $\mathrm{pr}(x \times I)$, $x \in X$, is a <u>generatrix</u>
of the cone. The base of con X is canonically homeomorphic to X and
is usually identified with X . The generatrices are obviously canonically homeomeorphic to I .

For each map f from X into another topological space Y,
we have the map $\mathrm{fact}(f \times \mathrm{id}\,I) : \mathrm{con}\,X \to \mathrm{con}\,Y$, which is continuous
whenever f is so. This map is denoted by con f .

2. The quotient of the product $X \times I$ by its partition whose
elements are the sets $X \times 0$ and $X \times 1$, and the points of the set
$(X \times I) \smallsetminus [(X \times 0) \cup (X \times 1)]$, is called the <u>suspension of</u> X and is
denoted by su X . The points $\mathrm{pr}(X \times 0)$ and $\mathrm{pr}(X \times 1)$ are the
<u>vertices</u> of the suspension, the set $\mathrm{pr}(X \times \tfrac{1}{2})$ is its <u>base</u>, and the
sets $\mathrm{pr}(x \times I)$, $x \in X$, are its <u>generatrices.</u> The base of su X is
canonically homeomorphic to X, while each of its generatrices is
canonically homeomorphic to I .

For each map $f: X \to Y$ there is the corresponding map
$\mathrm{fact}(f \times \mathrm{id}\,I) : \mathrm{su}\,X \to \mathrm{su}\,Y$, which we denote by su f ; su f is
continuous whenever f is so.

Notice that the suspension su X can be alternatively
described as con X/X .

3. Let X_1 and X_2 be topological spaces. The quotient
space of $X_1 \times X_2 \times I$ by its partition into the sets $x_1 \times X_2 \times 0$
(with $x_1 \in X_1$), $X_1 \times x_2 \times 1$ (with $x_2 \in X_2$), and the points of
$(X_1 \times X_2 \times I) \smallsetminus [(X_1 \times X_2 \times 0) \cup (X_1 \times X_2 \times 1)]$, is called the <u>join</u> of
X_1 and X_2 and is denoted by $X_1 * X_2$. The sets $\mathrm{pr}(X_1 \times X_2 \times 0)$
and $\mathrm{pr}(X_1 \times X_2 \times 1)$ are the <u>bases</u> of the join, and the sets
$\mathrm{pr}(x_1 \times x_2 \times I)$ with $x_1 \in X_1$ and $x_2 \in X_2$ are its <u>generatrices.</u>
The bases are obviously canonically homeomorphic to X_1 and X_2, and
are usually identified with X_1 and X_2. The generatrices are
canonically homeomorphic to I .

For each pair of maps $f_1: X_1 \to Y_1$ and $f_2: X_2 \to Y_2$ we have
the map $\mathrm{fact}(f_1 \times f_2 \times \mathrm{id}\,I) : X_1 * X_2 \to Y_1 * Y_2$, denoted by $f_1 * f_2$;
$f_1 * f_2$ is continuous whenever both f_1 and f_2 are so.

The $*$ - operation is commutative, i.e., there exists a

canonical homeomorphism $X_2 * X_1 \to X_1 * X_2$.

We remark that the join $X_1 * X_2$ may be alternatively defined as $(X_1 \sqcup X_2) \cup_\varphi (X_1 \times X_2 \times I)$, where $\varphi: X_1 \times X_2 \times (0 \cup 1) \to X_1 \sqcup X_2$ is given by $\varphi(x_1, x_2, 0) = in_1(x_1)$, $\varphi(x_1, x_2, 1) = in_2(x_2)$. It is also clear that the quotient space of $X_1 * X_2$ by its partition whose elements are the bases X_1 and X_2, and the points of the set $(X_1 * X_2) \smallsetminus (X_1 \cup X_2)$ is the suspension $su(X_1 \times X_2)$.

4. The iterated join $(\dots((X_1 * X_2) * X_3)\dots) * X_n$ may be canonically embedded in the product $con X_1 \times \dots \times con X_n$. This embedding is denoted by jc or, more precisely, by $jc_{X_1,\dots,X_n}$, and is defined inductively: for $n = 1$ it takes $x_1 \in X_1$ into $pr(x_1, 1) \in con X_1$, while for $n \geqslant 2$ it is given by

$$jc_{X_1,\dots,X_n}(pr(x, x_n, t)) = ((1-t)jc_{X_1,\dots,X_{n-1}}(x), pr(x_n, t)),$$

where $x \in (\dots(X_1 * X_2) * X_3)\dots) * X_{n-1}$, $x_n \in X_n$, and $t \in I$; the multiplication of a point of $con X_1 \times \dots \times con X_{n-1}$ by $1-t$ is defined by the rule

$$(1-t)(pr(x_1, t_1), \dots, pr(x_{n-1}, t_{n-1})) =$$

$$= (pr(x_1, (1-t)t_1), \dots, pr(x_{n-1}, (1-t)t_{n-1})).$$

Clearly, the image of the embedding $jc_{X_1,\dots,X_n}$ is precisely

$$\{(pr(x_1, t_1), \dots, pr(x_n, t_n)) \in con X_1 \times \dots \times con X_n \mid t_1 + \dots + t_n = 1\},$$

which allows us to identify the iterated join $(\dots((X_1 * X_2) * X_3)\dots) * X_n$ with this set.

5. The $*$-operation is associative, meaning, as usual, that the two joins $(X_1 * X_2) * X_3$ and $X_1 * (X_2 * X_3)$ are canonical homeomorphic. The canonical homeomorphism $(X_1 * X_2) * X_3 \to X_1 * (X_2 * X_3)$ is the composition of the canonical homeomorphism $X_1 * (X_2 * X_3) \to (X_2 * X_3) * X_1$ with the suitable compression of the canonical homeomorphism $con X_2 \times con X_3 \times con X_1 \to$ $\to con X_2 \times con X_3 \times con X_1$.

A consequence of the associativity of the $*$-operation is that the multiple join $X_1 * \dots * X_n$ is meaningful for any topological spaces $X_1, \dots, X_n$.

6. <u>The product</u> $con X_1 \times \dots \times con X_n$ <u>is canonically</u>

<u>homeomorphic to</u> $\text{con}(X_1 * \ldots * X_n)$.

The canonical homeomorphism

$$\text{con}(X_1 * \ldots * X_n) \to \text{con}\,X_1 \times \ldots \times \text{con}\,X_n$$

is defined as

$$\text{pr}(jc_{X_1,\ldots,X_n}^{-1}(\text{pr}(x_1,t_1),\ldots,\text{pr}(x_n,t_n)),t) \mapsto$$

$$\mapsto (\text{pr}(x_1,tt_1/\max(t_1,\ldots,t_n)),\ldots,\text{pr}(x_n,tt_n/\max(t_1,\ldots,t_n))).$$

7. <u>$\text{con}\,S^m$ and $\text{su}\,S^m$ are canonically homeomorphic to</u> D^{m+1} <u>and</u> S^{m+1}.

The canonical homeomorphisms $\text{con}\,S^m \to D^{m+1}$ and $\text{su}\,S^m \to S^{m+1}$ are defined by the formulas

$$\text{pr}((x_1,\ldots,x_{m+1}),t) \mapsto (tx_1,\ldots,tx_{m+1})$$

and

$$\text{pr}((x_1,\ldots,x_{m+1}),t) \mapsto (x_1\sin\pi t,\ldots,x_{m+1}\sin\pi t,\cos\pi t).$$

8. <u>The join</u> $X * D^0$ <u>is canonically homeomorphic to</u> $\text{con}\,X$. <u>The join</u> $X * S^0$ <u>is canonically homeomorphic to</u> $\text{su}\,X$. <u>The join</u> $X * S^k$ <u>is canonically homeomorphic to the iterated suspension</u> $\text{su}^{k+1}X$; <u>in particular,</u> $S^{m_1} * S^{m_2}$ <u>is canonically homeomorphic to</u> $S^{m_1+m_2+1}$.

The canonical homeomorphism $X * D^0 \to \text{con}\,X$ is given by $\text{pr}(x,0,t) \mapsto \text{pr}(x,t)$. The canonical homeomorphism $X * S^0 \to \text{su}\,X$ is given by the formulas $\text{pr}(x,1,t) \mapsto \text{pr}(x,(1+t)/2)$, $\text{pr}(x,-1,t) \mapsto$ $\mapsto \text{pr}(x,(1-t)/2)$. Finally, the canonical homeomorphism $X * S^k \to \text{su}^{k+1}X$ is the composite map

$$X * S^k \to \text{su}\,X * S^{k-1} \to \ldots \to \text{su}^k X * S^0 \to \text{su}^{k+1}X,$$

where the last arrow denotes the canonical homeomorphism, and the r-th arrow, with $r \leqslant k$, denotes the composite canonical homeomorphism

$$\text{su}^{r-1}X * S^{k-r+1} \to \text{su}^{r-1}X * \text{su}\,S^{k-r} \to \text{su}^{r-1}X * S^{k-r} * S^0 \to$$

$$\to \text{su}^{r-1}X * S^0 * S^{k-r} \to \text{su}^r X * S^{k-r}.$$

9. Combining the canonical homeomorphisms constructed in 6-8, we obtain the composite homeomorphisms

$$S^{m_1} * \ldots * S^{m_n} \to S^{m_1} * \ldots * S^{m_{n-2}} * S^{m_{n-1}+m_n+1} \to \ldots \to$$

$$\to S^{m_1} * S^{m_2+\ldots+m_n+n-2} \to S^{m_1+\ldots+m_n+n-1},$$

$$D^{m_1} \times \ldots \times D^{m_n} \to \mathrm{con}\, S^{m_1-1} \times \ldots \times \mathrm{con}\, S^{m_n-1} \to$$

$$\to \mathrm{con}(S^{m_1-1} * \ldots * S^{m_n-1}) \to \mathrm{con}\, S^{m_1+\ldots+m_n-1} \to D^{m_1+\ldots+m_n}$$

and

$$D^{m_1} * \ldots * D^{m_n} \to \mathrm{con}\, S^{m_n-1} * \ldots * \mathrm{con}\, S^{m_n-1} \to$$

$$\to S^{m_1-1} * D^0 * \ldots * S^{m_n-1} * D^0 \to$$

$$\to \mathrm{con}(S^{m_1-1} * D^0 * \ldots * S^{m_{n-1}-1} * D^0 * S^{m_n-1}) \to$$

$$\to \mathrm{con}\, S^{m_1-1} \times \mathrm{con}\, D^0 \times \ldots \times \mathrm{con}\, S^{m_{n-1}-1} \times \mathrm{con}\, D^0 \times \mathrm{con}\, S^{m_n-1} \to$$

$$\to D^{m_1} \times I \times \ldots \times D^{m_{n-1}} \times I \times D^{m_n} \to$$

$$\to D^{m_1} \times D^1 \times \ldots \times D^{m_{n-1}} \times D^1 \times D^{m_n} \to D^{m_1+\ldots+m_n+n-1}.$$

Therefore, the join $S^{m_1} * \ldots * S^{m_n}$, the product $D^{m_1} \times \ldots \times D^{m_n}$, and the join $D^{m_1} * \ldots * D^{m_n}$ are canonically homeomorphic to the sphere $S^{m_1+\ldots+m_n+n-1}$, the ball $D^{m_1+\ldots+m_n}$, and the ball $D^{m_1+\ldots+m_n+n-1}$, respectively.

The Mapping Cylinder and the Mapping Cone

10. Let $f: X_1 \to X_2$ be a continuous map. The result of attaching the product $X_1 \times I$ to X_2 by the map $X_1 \times 1 \to X_2$, $(x,1) \mapsto f(x)$, is called the <u>mapping cylinder of</u> f, and is denoted by Cyl f. The sets $\mathrm{imm}_1(X_1 \times 0)$ and $\mathrm{imm}_2(X_2)$ are the <u>lower</u> and <u>upper bases</u> of Cyl f, and the sets $\mathrm{imm}_1(x \times I)$ with $x \in X_1$ are its <u>generatrices.</u> Clearly, the bases are canonically homeomorphic to X_1 and X_2, and they are usually identified with these two spaces; the generatrices are canonically homeomorphic to I. Moreover, there is a canonical retraction rt $f :$ Cyl $f \to X_2$, defined on $\mathrm{imm}_1(X \times I)$ as rt $f\,(\mathrm{imm}_1(x,t)) = \mathrm{imm}_1(x,1)\ [\,= f(x)]$. It is evident that the composite map

$$X_1 \xrightarrow{\;\text{in}\;} \text{Cyl } f \xrightarrow{\;\text{rt } f\;} X_2$$

equals f.

If $X_2 = X_1$ and $f = \mathrm{id}\, X_1$, then Cyl f is canonically homeomorphic to the cylinder over X_1, $X_1 \times I$.

11. The <u>mapping cone</u> of the continuous map $f: X_1 \to X_2$ is the space $X_2 \cup_f \mathrm{con}\, X$, denoted by Con f (do not confuse it with con f, defined in 1). Equivalent definition: Con f $= $ Cyl f/X_1.

7. Spaces of Continuous Maps

1. Let $C(X,Y)$ be the set of all continuous maps of a topological space X into a topological space Y. The set of all maps $\phi \in C(X,Y)$ such that $\phi(A_1) \subset B_1, \ldots, \phi(A_n) \subset B_n$, where $A_1, \ldots, A_n$ and $B_1, \ldots, B_n$ are given subsets of X and Y, respectively, is denoted by $C(X, A_1, \ldots, A_n; Y, B_1, \ldots, B_n)$. It may be interpreted as the set of all continuous maps $(X, A_1, \ldots, A_n) \to (Y, B_1, \ldots, B_n)$.

We equip $C(X,Y)$ with the <u>compact-open topology</u>: by definition, this is the topology with the prebase consisting of all sets $C(X,A;Y,B)$ with A compact and B open. Together with $C(X,Y)$, all the sets $C(X, A_1, \ldots, A_n; Y, B_1, \ldots, B_n)$ become topological spaces.

If Y is a point, then $C(X,Y)$ reduces to a point. If X is discrete and consists of the points $x_1, \ldots, x_n$, then $C(X,Y)$ is canonically homeomorphic to the product $Y \times \ldots \times Y$ of n copies of the space Y; this homeomorphism is given by $\phi \to (\phi(x_1), \ldots, \phi(x_n))$.

To each pair of continuous maps $f: X' \to X$ and $g: Y \to Y'$ there corresponds a mapping $C(X,Y) \to C(X',Y')$, given by the rule $\phi \mapsto g \circ \phi \circ f$. This mapping is continuous, and we shall denote it by $C(f,g)$.

2. <u>If Y is a Hausdorff space, then so is $C(X,Y)$.</u>

Indeed, if $\phi, \psi \in C(X,Y)$ and $\phi \neq \psi$, then there is $x \in X$ such that $\phi(x) \neq \psi(x)$. Let U and V be disjoint neighborhoods of the points $\phi(x)$ and $\psi(x)$. Then $C(X,x;Y,U)$ and $C(X,x;Y,V)$ are disjoint neighborhoods of the points ϕ and ψ.

3. <u>If X is compact and Y is metrizable, then $C(X,Y)$ is metrizable. Moreover, if Y is equipped with a metric, then</u> $\mathrm{dist}(\phi,\psi) = \sup_{x \in X} \mathrm{dist}(\phi(x), \psi(x))$ <u>defines a metric on $C(X,Y)$, compatible with its topology.</u>

PROOF. Given $\phi \in C(X,Y)$, the set $\phi(X)$ can be covered by a finite number of balls $U_1, \ldots, U_s$ of an arbitrarily small radius ε (see 1.7.11). It is clear that $W = \cap_{i=1}^{s} C(X, \phi^{-1}(U_i); Y, U_i)$ is a neighborhood of the point ϕ, contained in the ball of radius 2ε

centered at ϕ. Therefore, every ball in $C(X,Y)$ contains a neighborhood of its center.

On the other hand, if $A \subset X$ is compact and $B \subset Y$ is open, with $\phi(A) \subset B$, then $C(X,A;Y,B)$ contains the ball with radius $\text{Dist}(\phi(A), Y \smallsetminus B)$ centered at ϕ (see 1.7.15). Therefore, every neighborhood of ϕ belonging to the prebase considered in 1 contains a ball centered at ϕ.

4. <u>For any topological spaces</u> X <u>and</u> $Y_1, \ldots, Y_n$, <u>the space</u> $C(X, Y_1 \times \ldots \times Y_n)$ <u>is canonically homeomorphic to the product</u> $C(X, Y_1) \times \ldots \times C(X, Y_n)$.

This canonical homeomorphism takes each $\phi \in C(X, Y_1 \times \ldots \times Y_n)$ into $(\text{pr}_1 \circ \phi, \ldots, \text{pr}_n \circ \phi) \in C(X, Y_1) \times \ldots \times C(X, Y_n)$ [cf. 2.4].

5. <u>Let</u> p <u>be a closed partition of the compact Hausdorff space</u> X, <u>and let</u> Y <u>be an arbitrary topological space. Then</u> $C(\text{pr}, \text{id}\,Y) : C(X/p, Y) \to C(X, Y)$ <u>is an embedding.</u>

It suffices to show that given a compact subset A of X/p and an open subset B of Y, the set $C(\text{pr}, \text{id}\,Y)[C(X/p, A; Y, B)]$ is open in $C(\text{pr}, \text{id}\,Y)[C(X/p, Y)]$. Since X/p is Hausdorff (see 3.9), A is closed. It follows that $\text{pr}^{-1}(A)$ is closed, and hence compact. Consequently, $C(X, \text{pr}^{-1}(A); Y, B)$ is open in $C(X, Y)$, and it remains to note that

$$C(\text{pr}, \text{id}\,Y)[C(X/p, A; Y, B)] = C(X, \text{pr}^{-1}(A); Y, B) \cap C(\text{pr}, \text{id}\,Y)[C(X/p; Y)].$$

The Mappings $X \times Y \to Z$ and $X \to C(Y, Z)$

6. <u>Suppose that</u> X, Y <u>and</u> Z <u>are topological spaces, and</u> $\phi: X \times Y \to Z$ <u>is continuous. Then the formula</u> $[\phi^{\vee}(x)](y) = \phi(x, y)$ <u>defines a continuous mapping</u> $\phi^{\vee}: X \to C(Y, Z)$.

<u>Let</u> $\psi: X \to C(Y, Z)$ <u>be a continuous mapping, and suppose that</u> Y <u>is Hausdorff and locally compact. Then the formula</u> $\psi^{\wedge}(x, y) = [\psi(x)](y)$ <u>defines a continuous mapping</u> $\psi^{\wedge}: X \times Y \to Z$.

To prove the first assertion, pick a point $x_0 \in X$, a compact set $B \subset Y$, and an open set $C \subset Z$. Then it is enough to exhibit a neighborhood U of x_0 such that $\phi^{\vee}(U) \subset C(Y, B; Z, C)$. For each point $y \in B$ fix neighborhoods U_y and V_y of x_0 and y such that $\phi(U_y \times V_y) \subset C$, and then extract a finite cover $V_{y_1}, \ldots, V_{y_s}$ of B from the collection $\{V_y\}_{y \in B}$. It is clear that $U = \cap_{i=1}^{s} U_{y_i}$ is a neighborhood of x_0 and that $\phi(U \times B) \subset \bigcup_{i=1}^{s} \phi(U_{y_i} \times V_{y_i}) \subset C$. It remains to remark that the inclusion $\phi(U \times B) \subset C$ is equivalent to $\phi^{\vee}(U) \subset C(Y,B;Z,C)$.

To prove the second assertion, pick a point $(x_0,y_0) \in X \times Y$ and a neighborhood W of the point $\psi^{\wedge}(x_0,y_0)$. Now let us find a neighborhood V of y_0 with compact closure $\operatorname{Cl} V$ satisfying $\operatorname{Cl} V \subset [\psi(x_0)]^{-1}(W)$ (see 1.7.22), and then a neighborhood U of x_0 satisfying $\psi(U) \subset C(Y,\operatorname{Cl} V;Z,W)$. Obviously, $U \times V$ is a neighborhood of the point (x_0,y_0) and $\psi^{\wedge}(U \times V) \subset W$.

7. <u>The mapping</u> $C(X \times Y,Z) \to C(X,C(Y,Z))$ <u>defined by the rule</u> $\phi \mapsto \phi^{\vee}$ <u>(see 6) is continuous for any topological spaces</u> X, Y <u>and</u> Z. <u>If</u> X <u>is Hausdorff and</u> Y <u>is Hausdorff and locally compact, then this mapping is a homeomorphism, and its inverse is given by the rule</u> $\psi \mapsto \psi^{\wedge}$.

The continuity of the mapping $\phi \mapsto \phi^{\vee}$ results from the fact that the preimage of $C(X,A;C(Y,Z),C(Y,B;Z,C))$ under this mapping is just $C(X \times Y,A \times B;Z,C)$. Assume that X is Hausdorff and Y is Hausdorff and locally compact. Consider a point $\psi_0 \in C(X,C(Y;Z))$, a compact subset Q of $X \times Y$, a neighborhood W of the set $\psi_0^{\wedge}(Q)$, and a point $q \in Q$. Now find a neighborhood $U_q \times V_q$ of q such that $\psi_0^{\wedge}(U_q \times \operatorname{Cl} V_q) \subset W$. Since Q is compact, its images $\operatorname{pr}_1(Q)$ and $\operatorname{pr}_2(Q)$ in X and Y are also compact (see 1.7.8). Moreover, they are Hausdorff spaces together with X and Y, and hence normal (see 1.7.5). Consequently, there exist open subsets U_q' of $\operatorname{pr}_1(Q)$ and V_q' of $\operatorname{pr}_2(Q)$ such that

$$\operatorname{pr}_1(q) \in U_q', \quad \operatorname{Cl}_{\operatorname{pr}_1(Q)} U_q' \subset U_q,$$

and

$$\operatorname{pr}_2(q) \in V_q', \quad \operatorname{Cl}_{\operatorname{pr}_2(Q)} V_q' \subset V_q,$$

and it is plain that the intersection $(U_q' \times V_q') \cap Q$ is open in Q. Being compact, Q can be covered by a finite number of such intersections, say $U_{q_1}' \times V_{q_1}',\ldots,U_{q_s}' \times V_{q_s}'$. Now set

$$T = \bigcap_{i=1}^{s} C(X,\operatorname{Cl}_{\operatorname{pr}_1(Q)} U_{q_i}';C(Y,Z),C(Y,\operatorname{Cl}_{\operatorname{pr}_2(Q)} V_{q_i}';Z,W)).$$

It is clear that T is a neighborhood of ψ_0 and that the image of T under the mapping $\psi \to \psi^{\wedge}$ is contained in $C(X \times Y,Q;Z,W)$. We conclude that $\psi \mapsto \psi^{\wedge}$ is continuous. It is readily see that the mappings $\phi \mapsto \phi^{\vee}$ and $\psi \mapsto \psi^{\wedge}$ are inverses of one another.

A Surprising Application

8. <u>Let</u> $f: X \to X'$ <u>be a factorial map. If the space</u> Y <u>is Hausdorff and locally compact, then the map</u> $f \times \mathrm{id}\, Y: X \times Y \to X' \times Y$ <u>is factorial.</u>

One can assume that $X' = X/\mathrm{zer}(f)$ and that f is the projection $X \to X/\mathrm{zer}(f)$. Consider the projection
$\mathrm{pr}: X \times Y \to (X \times Y)/(\mathrm{zer}(f) \times \mathrm{zer}(\mathrm{id}\, Y))$. The mapping
$\mathrm{pr}^{\vee}: X \to C(Y,(X \times Y)/(\mathrm{zer}(f) \times \mathrm{zer}(\mathrm{id}\, Y))$ is constant on the elements of the partition $\mathrm{zer}(f)$, and hence it induces continuous mappings

$$\mathrm{fact}\, \mathrm{pr}^{\vee} : X' \to C(Y,(X \times Y)/(\mathrm{zer}(f) \times \mathrm{zer}(\mathrm{id}\, Y))$$
and
$$(\mathrm{fact}\, \mathrm{pr}^{\vee})^{\wedge} : X' \times Y \to (X \times Y)/(\mathrm{zer}(f) \times \mathrm{zer}(\mathrm{id}\, Y)).$$

It it clear that the second of these mappings is the inverse of the injective factor of $f \times \mathrm{id}\, Y: X \times Y \to X' \times Y$. Thus the injective factor of $f \times \mathrm{id}\, Y: X \times Y \to X' \times Y$ is a homeomorphism.

9. <u>Let</u> $f: X \to X'$ <u>and</u> $g: Y \to Y'$ <u>be factorial maps. If</u> X' <u>and</u> Y <u>are Hausdorff and locally compact, then the map</u>
$f \times g: X \times Y \to X' \times Y'$ <u>is factorial.</u>

In fact, one can express $f \times g$ as the composition
$$X \times Y \xrightarrow{\ f \times \mathrm{id}\ } X' \times Y \xrightarrow{\ \mathrm{id} \times g\ } X' \times Y'$$

and recall that a composition of factorial maps is again factorial.

8. The Case of Pointed Spaces

1. In the sequel, the class of topological spaces equipped with a simple additional structure - a distinguished point (i.e., topological pairs (X,x_0) , where x_0 is a point) will play an important role; we call these spaces <u>pointed spaces</u>, and call the distinguished point a <u>base point.</u> The constructions described in the previous subsections must be naturally modified when applied to such spaces. For some of these construction, the modification entails merely the addition of a base point to the resulting space: for example, the quotient space of pointed space (X,x_0) has the natural base point

$\mathrm{pr}(x_0)$, the product of the pointed spaces $(X_1,x_1),\ldots,(X_n,x_n)$ has the natural base point $(x_1,\ldots,x_n)$, and the space of continuous maps from X into a pointed topological space (Y,y_0) contains the constant map const: $X \to Y$, $x \mapsto y_0$, and hence has the natural base point const. Other constructions such as the sum, suspension, and join need more serious modifications.

We shall describe these modified constructions below, and also introduce a new one - the <u>tensor product of pointed spaces.</u> In every case, pointed spaces produce pointed spaces, and base point-preserving maps again produce base point-preserving maps. We remark that the maps fact f , $C(f,g)$, and $f_1 \times \ldots \times f_n$ preserve base points whenever the initial maps have this property.

We use the symbol bp as a general notation for the base points.

Bouquets and Tensor Products

2. The construction below replaces the sum construction for pointed spaces.

Let $\{X_\mu\}_{\mu \in M}$ be a family of topological spaces with base points x_μ. The quotient space of the sum $\bigsqcup_{\mu \in M} X_\mu$ by the subset consisting of all points $\mathrm{in}_\mu(x_\mu)$ is called the <u>bouquet</u> (or the <u>wedge</u>) <u>of the spaces</u> X_μ, and is denoted by $V_{\mu \in M}(X_\mu,x_\mu)$. If M consists of the numbers $1,\ldots,n$, we also write $(X_1,x_1) \vee \ldots \vee (X_n,x_n)$. The point $\mathrm{pr} \circ \mathrm{in}_\nu(x_\nu) \in V(X_\mu,x_\mu)$ does not depend on ν; it is called the <u>center of the bouquet</u> $V(X_\mu,x_\mu)$, and is taken as its base point.

The bouquet $V(X_\mu,x_\mu)$ is obviously a union of the spaces X_μ (see 4.2), and so there exist the embeddings $\mathrm{imm}_\nu: X_\nu \to V(X_\mu,x_\mu)$. The maps $\mathrm{pr}_\nu: V(X_\mu,x_\mu) \to X_\nu$, defined by

$$\mathrm{pr}_\nu(\mathrm{imm}_{\nu'}(x)) = \begin{cases} x_\nu, & \text{if } \nu' \neq \nu, \\ x, & \text{if } \nu' = \nu, \end{cases}$$

are specific to the bouquet construction. Clearly, $\mathrm{pr}_\nu \circ \mathrm{imm}_\nu = \mathrm{id}\, X_\nu$ and $\mathrm{pr}_\nu \circ \mathrm{imm}_{\nu'} = \text{const}$ if $\nu' \neq \nu$.

If M also indexes another family of pointed spaces (Y_μ,y_μ) and a family of continuous maps $f_\mu: X_\mu \to Y_\mu$ such that $f_\mu(x_\mu) = y_\mu$, then the map $\mathrm{fact}(\bigsqcup f_\mu): V(X_\mu,x_\mu) \to V(Y_\mu,y_\mu)$ is well-defined and continuous; we denote it by $V f_\mu$.

3. Let $(X_1,x_1),\ldots,(X_n,x_n)$ be pointed spaces. The rules $x \mapsto (x,x_2,\ldots,x_n)$ $[x \in X_1]$, $\ldots$, $x \mapsto (x_1,\ldots,x_{n-1},x)$ $[x \in X_n]$, define canonical embeddings $X_1 \to X_1 \times \ldots \times X_n$, $\ldots$, $X_n \to X_1 \times \ldots \times X_n$, denoted by $in_1,\ldots,in_n$. Moreover, the rule $x \mapsto (pr_1(x),\ldots,pr_n(x))$ defines a canonical embedding $(X_1,x_1) \vee \ldots \vee (X_n,x_n) \to X_1 \times \ldots \times X_n$, which allows us to regard the bouquet $(X_1,x_1) \vee \ldots \vee (X_n,x_n)$ as a subspace of $X_1 \times \ldots \times X_n$. Clearly, $in_i: X_i \to X_1 \times \ldots \times X_n$ is the composition of the embedding $imm_i: X_i \to (X_1,x_1) \vee \ldots \vee (X_n,x_n)$ with the inclusion $(X_1,x_1) \vee \ldots \vee (X_n,x_n) \to X_1 \times \ldots \times X_n$, while the projection $pr_i: (X_1,x_1) \vee \ldots \vee (X_n,x_n) \to X_i$ is the restriction of $pr_i: X_1 \times \ldots \times X_n \to X_i$.

The quotient space $(X_1 \times \ldots \times X_n)/[(X_1,x_1) \vee \ldots \vee (X_n,x_n)]$ is called the <u>tensor product of the spaces</u> $X_1,\ldots,X_n$, and is denoted by $(X_1,x_1) \otimes \ldots \otimes (X_n,x_n)$. The point $pr[(X_1,x_1) \vee \ldots \vee (X_n,x_n)] \in$ $\in (X_1,x_1) \otimes \ldots \otimes (X_n,x_n)$ is called the <u>center of the tensor product</u> $(X_1,x_1) \otimes \ldots \otimes (X_n,x_n)$, and is taken as its base point.

The tensor product is a commutative and associative operation: there are obvious canonical homeomorphisms $(X_1,x_1) \otimes (X_2,x_2) \to$ $\to (X_2,x_2) \otimes (X_1,x_1)$ and $(X_1,x_1) \otimes [(X_2,x_2) \otimes (X_3,x_3),bp] \to$ $[(X_1,x_1) \otimes (X_2,x_2),bp] \otimes (X_3,x_3)$; this is also the way we understand the more general equality $[(X_1,x_1) \otimes \ldots \otimes (X_{n-1},x_{n-1}),bp] \otimes (X_n,x_n) =$ $= (X_1,x_1) \otimes \ldots \otimes (X_n,x_n)$.

If $(Y_1,y_1),\ldots,(Y_n,y_n)$ are other pointed spaces and $f_1: X_1 \to Y_1,\ldots,f_n: X_n \to Y_n$ are continuous, base point-preserving maps, then the map $fact(f_1 \times \ldots \times f_n): (X_1,x_1) \otimes \ldots \otimes (X_n,x_n) \to$ $\to (Y_1,y_1) \otimes \ldots \otimes (Y_n,y_n)$ is well defined and continuous; we denote it by $f_1 \otimes \ldots \otimes f_n$.

Cones, Suspensions, and Joins

4. The <u>cone over the pointed space</u> (X,x_0) is defined as the quotient of the usual cone $con\,X$ by its generatrix $pr(x_0 \times I)$, and is denoted by $con(X,x_0)$. The image of $pr(x_0 \times I)$ under the projection $con\,X \to con(X,x_0)$ is the <u>vertex</u> of $con(X,x_0)$, and is taken as its base point. The image of the base of $con\,X$ under the projection $con\,X \to con(X,x_0)$ is the <u>base</u> of $con(X,x_0)$; this projection carries the first base onto the second one, and thus allows us to identify the base of $con(X,x_0)$ with X.

If (Y,y_0) is another pointed space and $f: X \to Y$ is

continuous, with $f(x_0) = y_0$, then the map fact con f: $con(X,x_0) \to con(Y,y_0)$ is well defined and continuous, and we denote it simply by con f .

Equivalently, one may describe $con(X,x_0)$ as the quotient space of the cylinder $X \times I$ by $(X \times 0) \cup (x_0 \times I)$.

5. The <u>suspension of the pointed space</u> (X,x_0) is defined as the quotient of the usual suspension su X by its generatrix $pr(x_0 \times I)$, and is denoted by $su(X,x_0)$. The image of this generatrix under the projection $su X \to su(X,x_0)$ is the <u>vertex</u> of $su(X,x_0)$ and is taken as its base point.

If (Y,y_0) is another pointed space and f: $X \to Y$ is continuous, with $f(x_0) = y_0$, then the map fact su f: $su(X,x_0) \to su(Y,y_0)$ is well defined and continuous, and we denote it simply by su f .

Equivalently, we may decribe $su(X,x_0)$ as the quotient space of the cylinder $X \times I$ by $(X \times (0 \cup 1)) \cup (x_0 \times I)$, i.e., as $(X,x_0) \otimes (I/(0 \cup 1),bp) = (X,x_0) \otimes (S^1,ort_1)$. Another equivalent description: $su(X,x_0) = con(X,x_0)/X$.

6. The <u>join of the pointed spaces</u> (X_1,x_1) <u>and</u> (X_2,x_2) is defined as the quotient space of the usual join $X_1 * X_2$ by its generatrix $pr(x_1 \times x_2 \times I)$, and is denoted by $(X_1,x_1) * (X_2,x_2)$. The image of $pr(x_1 \times x_2 \times I)$ under the projection $X_1 * X_2 \to$ $\to (X_1,x_1) * (X_2,x_2)$ is the <u>center</u> of $(X_1,x_1) * (X_2,x_2)$, and is taken as its base point.

If (Y_1,y_1) and (Y_2,y_2) are another pointed spaces, and $f_1: X_1 \to Y_1$ and $f_2: X_2 \to Y_2$ are continuous maps such that $f_1(x_1) =$ $= y_1$ and $f_2(x_2) = y_2$, then the map fact$(f_1 * f_2)$: $(X_1,x_1) * (X_2,x_2) \to$ $\to (Y_1,y_1) * (Y_2,y_2)$ is well defined and continuous, and we denote it simply by $f_1 * f_2$.

7. <u>For any two pointed spaces</u> (X_1,x_1) <u>and</u> (X_2,x_2), <u>the</u> <u>bouquet</u> $(su(X_1,x_1),bp) \vee (su(X_2,x_2),bp)$ <u>is canonically homeomorphic</u> <u>to</u> $su((X_1,x_1) \vee (X_2,x_2),bp)$.

The canonical homeomorphism $su((X_1,x_1) \vee (X_2,x_2),bp) \to$ $\to (su(X_1,x_1),bp) \vee (su(X_2,x_2),bp)$ is given by

$$pr(imm_i(x),t) \mapsto imm_i(pr(x,t)) \qquad [x \in X_i, \quad i = 1,2].$$

8. $con(S^m,ort_1)$, $su(S^m,ort_1)$, <u>and</u> $(S^m,ort_1) * (S^n,ort_1)$ <u>are canonically homeomorphic to</u> D^{m+1}, S^{m+1}, <u>and</u> S^{m+n+1}, <u>respectively.</u>

The canonical homeomorphism $su(S^m, \mathrm{ort}_1) \to D^{m+1}$ is defined as $pr((x_1, \ldots, x_{m+1}), t) \mapsto (tx_1 + (1-t), tx_2, \ldots, tx_{m+1})$.

The canonical homeomorphism $su(S^m, \mathrm{ort}_1) \to S^{m+1}$ transforms the generatrix passing through the point $x \in S^m$ onto the circle on S^{m+1} with center $(\mathrm{ort}_1 + x)/2$ (which degenerates to the point ort_1 if $x = \mathrm{ort}_1$); as t varies from 0 to 1, the image of the point $pr(x, t)$ moves uniformly on this circle, starting from ort_1, and continuing into the half space $x_{m+1} \leqslant 0$ (see Fig. 1).

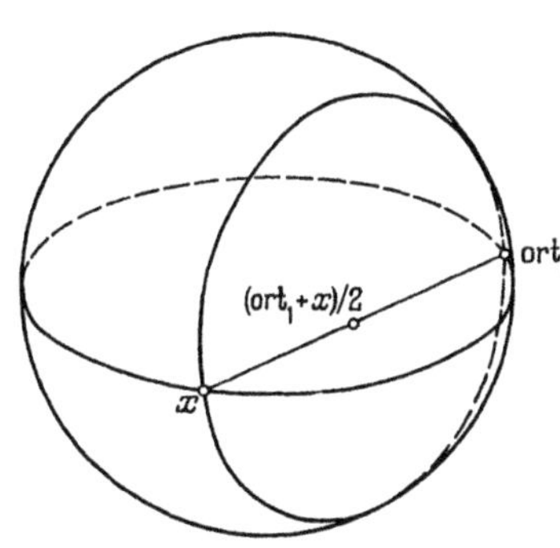

Fig. 1 (m = 2)

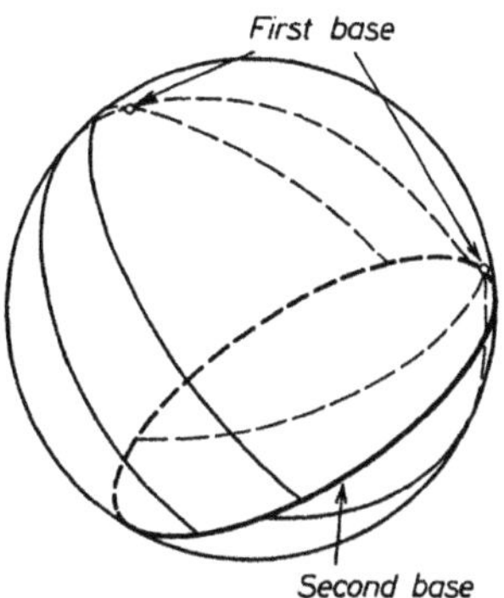

Fig. 2 (m = 0, n = 1)

Finally, the canonical homeomorphism $(S^m, \mathrm{ort}_1) * (S^n, \mathrm{ort}_1) \to S^{m+n+1}$ is defined on the bases S^m and S^n by the formulas

$$(x_1, \ldots, x_{m+1}) \mapsto \left(\frac{3x_1+1}{4}, \frac{x_2\sqrt{3}}{2}, \ldots, \frac{x_{m+1}\sqrt{3}}{2}, 0, \ldots, 0, \frac{\sqrt{3}(1-x_1)}{4}\right)$$

and

$$(x_1, \ldots, x_{n+1}) \mapsto \left(\frac{3x_1+1}{4}, 0, \ldots, 0, \frac{x_2\sqrt{3}}{2}, \ldots, \frac{x_{n+1}\sqrt{3}}{2}, \frac{\sqrt{3}(x_1-1)}{4}\right),$$

and maps the generatrix joining the points $x \in S^m$ and $x' \in S^n$ onto the arc of the great circle on S^{m+n+1} which joins the images of x and x', in such a way that the lengths are linearly transformed (see Fig. 2).

9. Since $con(S^m, \mathrm{ort}_1) = (S^m, \mathrm{ort}_1) \otimes (I, 0)$ (see 4), and $su(S^m, \mathrm{ort}_1) = (S^m, \mathrm{ort}_1) \otimes (S^1, \mathrm{ort}_1)$ (see 5), the homeomorphisms $con(S^m, \mathrm{ort}_1) \to D^{m+1}$ and $su(S^m, \mathrm{ort}_1) \to S^{m+1}$, defined in 8, lead for $n \geqslant 1$ to the canonical homeomorphisms

$$S^n = \underbrace{(S^1, \mathrm{ort}_1) \otimes \ldots \otimes (S^1, \mathrm{ort}_1)}_{n}$$

and

$$D^n = \underbrace{(S^1,ort_1) \otimes \ldots \otimes (S^1,ort_1)}_{n-1} \otimes (I,0).$$

Now one can define the maps

$$id \otimes \ldots \otimes id \otimes pr: D^n = (S^1,ort_1) \otimes \ldots \otimes (S^1,ort_1) \otimes (I,0) \to$$
$$\to (S^1,ort_1) \otimes \ldots (S^1,ort_1) \otimes (I/(0 \cup 1),bp) = S^n,$$

and

$$(pr \otimes \ldots \otimes pr \otimes id\, I) \circ pr: I^n = I \times \ldots \times I \to$$
$$\to (I/(0 \cup 1),bp) \otimes \ldots \otimes (I/(0 \cup 1),bp) \otimes (I,0) = D^n.$$

We denote them by DS and ID. It is clear that DS takes $Int\,D^n$ homeomorphically onto $S^n \smallsetminus ort_1$ and takes $Fr\,D^n$ into ort_1, while ID takes homeomorphically $Int\,I^n$ onto $Int\,D^n$, and $Int\,I^{n-1}$ onto $S^{n-1} \smallsetminus ort_1$, and carries $Fr\,I^n \smallsetminus Int\,I^{n-1}$ into ort_1.

Since the map DS is closed (see 1.7.9), its injective factor fact $DS : D^n/S^{n-1} \to S^n$ is a homeomorphism. Consequently, <u>for</u> $n \geq 1$ <u>the quotient space</u> D^n/S^{n-1} <u>is canonically homeomorphic to</u> S^n.

The Mappings $(X,x_0) \otimes (Y,y_0) \to Z$ and $X \to C(Y,y_0;Z,z_0)$

10. <u>Let</u> (X,x_0), (Y,y_0) <u>and</u> (Z,z_0) <u>be pointed topological spaces. If</u> $\phi: (X,x_0) \otimes (Y,y_0) \to Z$ <u>is continuous and preserves base points, then the formula</u> $[\phi^U(x)](y) = \phi(pr(x,y))$ <u>defines a continuous, base point-preserving mapping</u> $\phi^U: X \to C(Y,y_0;Z,z_0)$.

<u>Let</u> $\psi: X \to C(Y,y_0;Z,z_0)$ <u>be a continuous, base point-preserving mapping, and suppose that</u> Y <u>is Hausdorff and locally compact. Then the formula</u> $\psi^\cap(pr(x,y)) = [\psi(x)](y)$ <u>defines a continuous, base point-preserving mapping</u> $\psi^\cap: (X,x_0) \otimes (Y,y_0) \to Z$.

Indeed,

$$\phi^U = ab[(\phi \circ pr: X \times Y \to Z)^\vee],$$

while

$$\psi^\cap \circ (pr: X \times Y \to (X,x_0) \otimes (Y,y_0)) =$$
$$= [\psi \circ (in: C(Y,y_0;Z,z_0) \to C(Y,Z))]^\wedge.$$

11. <u>Given arbitrary pointed topological spaces</u> (X,x_0), (Y,y_0), <u>and</u> (Z,z_0), <u>the mapping</u>

$$C((X,x_0) \otimes (Y,y_0),bp;Z,z_0) \to C(X,x_0;C(Y,y_0;Z,z_0),\text{const})$$

<u>given by the formula</u> $\phi \mapsto \phi^U$ (<u>see</u> 10) <u>is continuous. If</u> X <u>and</u> Y <u>are Hausdorff and compact, then this mapping is a homeomorphism and its inverse is given by the formula</u> $\psi \mapsto \psi^\cap$.

The preimage of $C(X,A,x_0;C(Y,y_0;Z,z_0),C(Y,B,y_0;Z,C,z_0),\text{const})$ under the mapping $\phi \mapsto \phi^U$ is just $C((X,x_0) \otimes (Y,y_0),\text{pr}(A \times B),bp;Z,C,z_0)$ which shows that $\phi \mapsto \phi^U$ is continuous. Assume now that X and Y are Hausdorff and compact, and consider the mapping

$$C(\text{pr},\text{id } Z) : C((X,x_0) \otimes (Y,y_0),Z) \to C(X \times Y,Z).$$

By 7.5, this mapping is an embedding. Consider the diagram

$$
\begin{array}{ccc}
C(X,x_0;C(Y,y_0;Z,z_0),\text{const}) & \longrightarrow & C((X,x_0) \otimes (Y,y_0),bp;Z,z_0) \\
\downarrow & & \downarrow \\
C(X,C(Y,Z)) & \longrightarrow & C(X \times Y,Z)
\end{array}
$$

where the horizontal arrows denote the mappings $\psi \mapsto \psi^\cap$ and $\psi \mapsto \psi^\wedge$, and the vertical arrows the composite mappings

$$C(X,x_0;C(Y,y_0;Z,z_0),\text{const}) \xrightarrow{\text{in}} C(X,C(Y,y_0;Z,z_0)) \xrightarrow{C(\text{id } X,\text{in})}$$

$$\longrightarrow C(X,C(Y,Z))$$

and

$$C((X,x_0) \otimes (Y,y_0),bp;Z,z_0) \xrightarrow{\text{in}} C((X,x_0) \otimes (Y,y_0),Z) \longrightarrow$$

$$\xrightarrow{C(\text{pr},\text{id } Z)} C(X \times Y,Z).$$

Since this diagram is commutative, the fact that $C(\text{pr},\text{id } Z)$ is an embedding implies the continuity of $\psi \mapsto \psi^\cap$. That the mappings $\phi \mapsto \phi^U$ and $\psi \mapsto \psi^\cap$ are inverses of one another is plain.

9. <u>Exercises</u>

1. Show that for each topological space X and each compact topological space Y the map $\text{pr}_1: X \times Y \to X$ is closed.

2. Show that the subset of S^n defined by the inequality

$x_1^2 + \ldots + x_k^2 \leqslant x_{k+1}^2 + \ldots + x_n^2$ in the standard coordinates of $\mathbb{R}^n$, is homeomorphic to $D^k \times S^{n-k}$.

3. Let M, X_μ, and $\phi_{\mu\mu'}$ be as in 4.2, and let X denote the union of the spaces X_μ defined by the homeomorphisms $\phi_{\mu\mu'}$. Show that the maps $\mathrm{imm}_\mu \colon X_\mu \to X$ are topological embeddings whenever M has only two elements, but when M has three elements this is not necessarily so.

4. Show that for $n \geqslant 1$ the spaces $C(I,0,1;S^n,\mathrm{ort}_1,\mathrm{ort}_2)$ and $C(I,0,1;S^n,\mathrm{ort}_1,\mathrm{ort}_1)$ are homeomorphic.

5. Let T be the set of all real sequences $\{x_i\}_1^\infty$, with the topology defined by the prebase consisting of all sets of the form $\{\{x_i\}_1^\infty \mid a < x_s < b\}$. Further, let S be the quotient space of $T \smallsetminus 0$ (where $0 = \{x_i = 0\}_1^\infty$) by its partition into rays, i.e., into the sets $\{\{tx_i^0\}_1^\infty \mid 0 < t < \infty\}$ with $\{x_i^0\}_1^\infty \in T \smallsetminus 0$. Show that T is metrizable, while S is regular, but has the peculiar property that every continuous map $S \to \mathbb{R}$ is constant.

§3. HOMOTOPIES

1. General Definitions

1. A continuous map $f' \colon X \to Y$ is _homotopic_ to the continuous map $f \colon X \to Y$ if there is a continuous map $F \colon X \times I \to Y$ such that $F(x,0) = f(x)$ and $F(x,1) = f'(x)$, for all $x \in X$. Every such map F is called a _homotopy from_ f _to_ f' (or _connecting_ f _to_ f'). One says also that F is a _homotopy of_ f. A map homotopic to a constant map is also said to be _null homotopic._

Often a homotopy $F \colon X \times I \to Y$ is interpreted as a family of continuous maps $f_t \colon X \to Y$, related to F via $f_t(x) = F(x,t)$ $(0 \leqslant t \leqslant 1)$. Acoording to 2.7.6, the continuity of F implies that this family is continuous as a map of the segment I into $C(X,Y)$. Moreover, if X is Hausdorff, then the continuity of the family f_t is equivalent to that of the map F.

Obviously, the _constant_ homotopy F of a continuous map $f \colon X \to Y$, given by $F(x,t) = f(x)$, connects f to f; if the homotopy F connects f to f', then the <u>inverse</u> homotopy, F', defined by F'(x,t) = F(x,1-t), connects f' to f; if the homotopy F connects f to f' and the homotopy F' connects f' to f", then their <u>product</u> F", defined as

$$F''(x,t) = \begin{cases} F(x,2t), & \text{for } t \leqslant 1/2, \\ \\ F'(x,2t-1), & \text{for } t \geqslant 1/2, \end{cases}$$

is a homotopy connecting f to f". Thus <u>homotopy is an equivalence relation</u>, which yields a partition of $C(X,Y)$ into equivalence classes, called <u>homotopy classes</u>. We denote the set of these classes by $\pi(X,Y)$.

2. An example is the <u>rectilinear</u> homotopy. Namely, let f and f' be continuous maps of a space X into a subspace Y of $\mathbb{R}^n$. If for each $x \in X$ the segment joining $f(x)$ to $f'(x)$ is entirely contained in Y, then $F(x,t) = (1-t)f(x) + tf'(x)$ defines a homotopy from f to f', referred to as rectilinear.

Obviously, any two maps of an arbitrary space into $\mathbb{R}^n$ or D^n are rectilinearly homotopic.

3. <u>Let the maps</u> $f,f': X \to Y$ <u>be homotopic. Then given any continuous maps</u> $g: Y \to Y'$ <u>and</u> $h: X' \to X,$ <u>the maps</u> $g \circ f \circ h$ <u>and</u> $g \circ f' \circ h$ <u>are homotopic.</u>

In fact, let $F: X \times I \to Y$ be a homotopy from f to f'. Then $g \circ F \circ (h \times \text{id}\, I)$ is a homotopy from $g \circ f \circ h$ to $g \circ f' \circ h$.

4. As 3 shows, the mapping $C(h,g) : C(X,Y) \to C(X',Y')$ induced by two continuous maps $h: X' \to X$ and $g: Y \to Y'$ transforms homotopy classes into homotopy classes. The resulting mapping fact $C(h,g) : \pi(X,Y) \to \pi(X',Y')$ is denoted by $\pi(h,g),$ and 3 implies that it depends only on the homotopy classes of h and g.

Stationary Homotopies

5. Let A be a subset of the space X. A homotopy $F: X \times I \to Y$ is said to be <u>stationary on</u> A or, simply, to be an <u>A-homotopy</u> if $F(x,t) = F(x,0)$ for all $x \in A$ and $t \in I$. Two maps which can be connected by an A-homotopy are <u>A-homotopic.</u>

As with usual homotopy, A-homotopy defines an equivalence relation, dividing the set of continuous maps $X \to Y$ which coincide on A with a given map $f: A \to Y,$ into equivalence classes. The latter are called A-homotopy classes or, in full, <u>homotopy classes of continuous</u>

extensions of the map f to X. We denote the set of these classes by
$\pi(X,A;f)$.

Notice that a rectilinear homotopy from f to f' (see 2)
is stationary on the set of points where f and g agree.

If one wants to specify that a certain homotopy is ordinary,
i.e., not stationary, then one says that it is free.

Homotopy Equivalence of Spaces

6. A continuous map $g: Y \to X$ is a homotopy inverse of the
continuous map $f: X \to Y$ if the composition $g \circ f$ is homotopic to
id X and the composition $f \circ g$ is homotopic to id Y. A continuous
map which has a homotopy inverse is called a homotopy equivalence. If
there is a homotopy equivalence $X \to Y$, then one says that the space
Y is homotopy equivalent to the space X.

The following are obviously homotopy equivalences: the
identity map of any space, a map which is a homotopy inverse of a
homotopy equivalence, and the composition of two homotopy equivalences.
Thus, homotopy equivalence among topological spaces is an equivalence
relation. It divides the topological spaces into classes called
homotopy types (instead of saying that Y is homotopy equivalent to X,
one says also that X and Y have the same homotopy type).

Every homeomorphism is clearly a homotopy equivalence.

7. If one of the continuous maps $f: X \to Y$ and $g: Y \to Z$,
and their composition $g \circ f: X \to Z$ are homotopy equivalences, then
the other map is also a homotopy equivalence.

Indeed, let h be a homotopy inverse of $g \circ f$, and suppose
that f is a homotopy equivalence. Then $f \circ h$ is a homotopy inverse
of g. Similarly, if g is a homotopy equivalence, then $h \circ g$ is a
homotopy inverse of f.

8. $\pi(X,Y)$ is a homotopy invariant. That is to say, if
$g: Y \to Y'$ and $f: X \to X'$ are homotopy equivalences, then
$\pi(f,g): \pi(X,Y) \to \pi(X',Y')$ is invertible.

Evidently, if f' (respectively, g') is a homotopy
inverse of f (respectively, of g), then the map $\pi(f',g')$ is the
inverse of $\pi(f,g)$.

Contractible Spaces

9. A space X is <u>contractible</u> if the map $\mathrm{id}\,X$ is homotopic to a constant map.

$\mathbb{R}^n$ and D^n are examples of contractible spaces (see 2).

10. <u>A space is contractible if and only if it is homotopy equivalent to a point.</u>

If $\mathrm{id}\,X$ is homotopic to a constant map ϕ, then the map $f: D^0 \to X$ taking the value $\phi(X)$, and the map $g: X \to D^0$ are homotopy inverses of one another: indeed, $f \circ g = \phi$ and $g \circ f = \mathrm{id}\,D^0$.

If now $f: D^0 \to X$ and $g: X \to D^0$ are homotopy inverses of one another, then $\mathrm{id}\,X$ is homotopic to the constant map $f \circ g$.

11. <u>If</u> X <u>is contractible, then any two continuous maps of an arbitrary topological space into</u> X <u>are homotopic. In particular, $\mathrm{id}\,X$ is homotopic to any constant map</u> $X \to X$.

This is a straightforward consequence of 10 and 8.

Deformation Retractions

12. A retraction ρ of a topological space X onto one of its subspaces A (see 1.4.10) is called a <u>deformation</u> (<u>strong deformation</u>) <u>retraction</u> if the composition $X \xrightarrow{\rho} A \xrightarrow{\text{in}} X$ is homotopic (respectively, A-homotopic) to $\mathrm{id}\,X$. If the space X admits a deformation retraction (a strong deformation retraction) onto A, then A is called a <u>deformation retract</u> (respectively, a <u>strong deformation retract</u>) of X.

Obviously, if $\rho: X \to A$ is a deformation retraction, then ρ and the inclusion $A \to X$ are homotopy equivalences, each being a homotopy inverse of the other. It is clear also that any space which admits a deformation retraction onto one of its points is contractible, and that every point of a contractible space is a deformation retract of the ambient space.

Relative Homotopies

13. Let X (respectively Y) be a space with a distinguished sequence of subsets $A_1, \ldots, A_n$ (respectively, $B_1, \ldots, B_n$). A map

$F: (X \times I, A_1 \times I, \ldots, A_n \times I) \to (Y, B_1, \ldots, B_n)$ is called a homotopy connecting the continuous maps $f, f' : (X, A_1, \ldots, A_n) \to (Y, B_1, \ldots, B_n)$ if $abs\, F$ is a homotopy connecting the maps $abs\, f$ and $abs\, f'$. In this case, it is evident that $ab\, abs\, F: A_i \times I \to B_i$ is a homotopy connecting the maps $ab\, abs\, f$, $ab\, abs\, f': A_i \to B_i$. Moreover, it is readily seen that the homotopies $(X \times I, A_1 \times I, \ldots, A_n \times I) \to$ $\to (Y, B_1, \ldots, B_n)$ yield an equivalence relation. This relation divides $C(X, A_1, \ldots, A_n; Y, B_1, \ldots, B_n)$ into homotopy classes forming a set denoted by $\pi(X, A_1, \ldots, A_n; Y, B_1, \ldots, B_n)$. We may give an analogous definition of the map $\pi(h, g)$ from 4.

A continuous map $g: (Y, B_1, \ldots, B_n) \to (X, A_1, \ldots, A_n)$ is said to be a homotopy inverse of the continuous map $f: (X, A_1, \ldots, A_n) \to$ $\to (Y, B_1, \ldots, B_n)$ if $g \circ f$ is homotopic to $rel\, id\, X$ and $f \circ g$ is homotopic to $rel\, id\, Y$. A continuous map possessing a homotopy inverse is called a homotopy equivalence. Two sequences $(X, A_1, \ldots, A_n)$ and $(Y, B_1, \ldots, B_n)$ are said to be homotopy equivalent, or to have the same homotopy type, if they are related by a homotopy equivalence. Propositions 7 and 8, as they stand, apply to the case of relative homotopy.

14. The situation discussed in 13 encompasses the case when X and Y are pointed spaces (in this case A_1 and B_1 are points, $n = 1$, and the homotopies defined in 13 are just the homotopies stationary at A_1). Moreover, the definition of a contractible space given in 9 extends to pointed spaces (however, the homotopy from $id\, X$ to a constant map must be stationary at the base point). The same is true for theorems 10 an 11, as well as for the definitions of a deformation retraction and deformation retract, given in 12 (X and A must have the same base point, and the homotopy from the composition $X \xrightarrow{\rho} A \xrightarrow{in} X$ to $id\, X$ must be stationary at this point). Also, the remarks in 12 remain valid, while the definition of strong deformation retraction is entirely unaffected by the presence of a base point.

2. Paths

1. A _path_ in a topological space X is any continuous map of the closed unit interval I into X. The points $s(0)$ and $s(1)$ are called the _origin_ and the _end_ of the path s. Closed path are also termed _loops._

Given a path s, the formula $t \mapsto s(1-t)$ defines a new path, called the <u>inverse</u> of s and denoted by s^{-1}. Given two paths s_1 and s_2 with $s_1(1) = s_2(0)$, the formula

$$t \mapsto \begin{cases} s_1(2t), & \text{for } t \leqslant 1/2, \\ \\ s_2(2t-1), & \text{for } t \geqslant 1/2, \end{cases}$$

defines a path, called the <u>product</u> of the paths s_1 and s_2, and denoted by $s_1 s_2$. Obviously, $(s^{-1})^{-1} = s$ and $(s_1 s_2)^{-1} = s_2^{-1} s_1^{-1}$.

Since $I = D^0 \times I$, any path can be considered as a homotopy of a map $D^0 \to X$. If one adopts such an interpretation, then the inverse path becomes the inverse homotopy, while the product of paths becomes the product of homotopies.

On the other hand, every homotopy between two continuous maps $f, f' : X \to Y$ defines a path in $C(X,Y)$, joining f and f' (see 2.7.6), and again the inverse path corresponds to the inverse homotopy, and the product of paths to the product of homotopies. If X is Hausdorff and locally compact, then a homotopy connecting two maps $f, f': X \to Y$ may be even defined as a path in $C(X,Y)$ joining f and f'.

3. Since any path is a continuous map, it can be also subjected to homotopies. Unfortunately, the generally accepted terminology for such homotopies is not in complete agreement with our definitions in subsection 1 (which are also generally accepted). More precisely, when we consider paths, the homotopies and the homotopy relation are understood always as $(0 \cup 1)$-homotopies (i.e., homotopies stationary at the extremities of the interval I) and $(0 \cup 1)$-homotopy relation, respectively. Moreover, a free homotopy of a loop is understood always as a usual free homotopy whereby the path remains a loop all the time (i.e., as a continuous map $F: I \times I \to X$ such that $F(0,t) = F(1,t)$ for all $t \in I$).

3.   Connectedness and k-Connectedness

1. The properties of topological spaces we study in this subsection represent weaker versions of the contractibility in the absolute case, and of deformation retractability in the relative case.

Connectedness

2. A topological space is <u>connected</u> (see the Preface) if each pair of its points can be joined by a path. Equivalently, X is connected if the set $\pi(D^0,X)$ contains just an element; see 2.2.

Since $\pi(D^0,X)$ is a homotopy invariant, connectedness is a homotopically invariant property. In particular, all contractible spaces are connected. For example, $\mathbb{R}^n$ and D^n are connected for every n.

For $n > 0$, S^n is also connected: any two points of S^n can be joined by a path, which in fact is contained in $S^n \smallsetminus p$, where p is a third point (recall that the punctured sphere $S^n \smallsetminus p$ is homeomorphic to $\mathbb{R}^n$). S^0 is not connected: a path joining -1 and 1 would be a continuous function on $[0,1]$, taking two distinct values but no intermediate ones.

The only connected subsets of the real line $\mathbb{R}$ are the empty set, the finite or infinite intervals, the finite or infinite semi-intervals, and the closed intervals. Indeed, if α and β are the exact lower and upper bounds of a connected subset A of $\mathbb{R}$, then A contains the interval (α,β).

3. Given an arbitrary topological space X, the property of being joined by a path defines a relation between its points, which obviously satisfies all the requirements for an equivalence relation . This relation defines a partition of X into subsets which are the maximal connected subsets of X, and are called the <u>components</u> of X. Clearly, the set of components may be identified with $\pi(D^0,X)$. We denote it by comp X .

Every continuous map $f: X \to Y$ induces the map fact f $= \pi(\mathrm{id}\, D^0,f)$: comp X $\to$ comp Y . This map does not change when we replace f by an arbitrary homotopic map, and fact f is invertible whenever f is a homotopy equivalence (see 1.4 and 1.8). It is also plain that if $f(X) = Y$, then fact f (comp X) $=$ comp Y . In particular, the image of a connected space under a continuous map is connected.

4. <u>If X can be written as the union of two connected subsets A_1 and A_2 with $A_1 \cap A_2 \neq \emptyset$, then X is connected.</u>

Indeed, a component of X which contains a point $x_0 \in A_1 \cap A_2$ contains also A_1 and A_2, i.e., contains X.

5. <u>Consider a partition of X into open sets. Then every connected subset of X is contained in one of the elements of this</u>

partition. In particular, every subset of a connected space which is both open and closed is either empty or the whole space X.

PROOF. Let A be a connected subset of X, and let U be an element of the partition, such that $U \cap A \neq \emptyset$. Consider the map $f\colon X \to S^0$ which takes U into 1 and $X \smallsetminus U$ into -1. Since f is continuous, f(A) is connected, whence f(A) = 1 and $A \subset U$.

k-Connectedness

6. <u>The following properties of a continuous map</u> $f\colon S^r \to X$ <u>with</u> $r \geqslant 0$ <u>are equivalent:</u>

(i) f <u>is homotopic to a constant map;</u>

(ii) f <u>extends to a continuous map</u> $D^{r+1} \to X$;

(iii) <u>the compositions</u> $f \circ DS_+$, $f \circ DS_-\colon D^r \to X$ <u>are</u> S^{r-1}-<u>homotopic, where</u> DS_+ <u>and</u> DS_- <u>are the embeddings of</u> D^r <u>in</u> S^r, <u>defined by</u>

$$DS_+(x_1,\ldots,x_r) = (x_1,\ldots,x_r,(1-x_1^2-\ldots-x_r^2)^{1/2})$$

<u>and</u>

$$DS_-(x_1,\ldots,x_r) = (x_1,\ldots,x_r,-(1-x_1^2-\ldots-x_r^2)^{1/2});$$

(iv) f <u>is ort$_1$-homotopic to a constant map.</u>

The proof follows the following scheme:

$$(i) \Longleftrightarrow (iv) \quad\Longrightarrow\quad (ii) \rightleftarrows (iii)$$

(i) $\Rightarrow$ (ii). A homotopy $F\colon S^r \times I \to X$ from f to a constant map takes the upper base of the cylinder $S^r \times I$ into one point. Consequently, F may be expressed as the composition of the map $S^r \times I \to D^{r+1}$, defined by $((x_1,\ldots,x_{r+1}),t) \mapsto (x_1(1-t),\ldots,x_{r+1}(1-t))$, and a continuous map $g\colon D^{r+1} \to X$ (see 2.3.4 and 1.7.9), and it is clear that $g\big|_{S^r} = f$.

(ii) $\Rightarrow$ (iii) and (ii) $\Rightarrow$ (iv). Suppose $g\colon D^{r+1} \to X$ is a continuous extension of f. Then the formulas

$$((x_1,\ldots,x_r),t) \mapsto g(x_1,\ldots,x_r,(1-2t)(1-x_1^2-\ldots-x_r^2)^{1/2})$$

and

$$((x_1,\ldots,x_{r+1}),t) \mapsto g(t+(1-t)x_1,(1-t)x_2,\ldots,(1-t)x_{r+1})$$

define an S^{r-1}-homotopy $D^r \times I \to X$ from $f \circ DS_+$ to $f \circ DS_-$, and an ort_1-homotopy $S^r \times I \to X$ from f to a constant map.

(iii) $\Rightarrow$ (ii). An S^{r-1}-homotopy $f: D^r \times I \to X$ from $f \circ DS_+$ to $f \circ DS_-$ takes every generatrix of the cylinder $S^r \times I$ into one point. Consequently, F can be expressed as the composition of the map $D^r \times I \to D^{r+1}$, defined by

$$((x_1,\ldots,x_r),t) \mapsto (x_1,\ldots,x_r,(2t-1)(1-x_1^2-\ldots-x_r^2)^{1/2}),$$

and some continuous map $g: D^{r+1} \to X$ (see 2.3.4 and 1.7.9), and it is clear that $g\big|_{S^r} = f$.

The implication (iv) $\Rightarrow$ (i) is trivial.

7. A nonempty space X is said to be <u>k-connected</u> ($0 \leqslant k \leqslant \infty$), if any continuous map $S^r \to X$ with $r \leqslant k$ is homotopic to a constant map, i.e., satisfies condition 6(i). Theorem 6 shows that this definition has three more equivalent formulations, based on conditions 6(ii), 6(iii) and 6(iv). Moreover, since for any continuous maps $f_1,f_2: D^r \to X$ which agree on S^{r-1} there is a continuous map $f: S^r \to X$ such that $f \circ DS_+ = f_1$ and $f \circ DS_- = f_2$, we conclude that a nonempty space X is k-connected if and only if any continuous maps $f_1,f_2: D^r \to X$, $r \leqslant k$, which agree on S^{r-1} are S^{r-1}-homotopic.

Obviously, for nonempty spaces 0-connectedness is nothing else but connectedness. The 1-connected spaces are usually called <u>simply connected.</u> Note that a 0-connected space is simply connected if and only if any two paths with common extremities are homotopic.

The homotopy invariance of the sets $\pi(S^r,X)$ implies that a space which is homotopy equivalent to a k-connected space is itself k-connected. In particular, every contractible space is ∞-connected.

The Relative Case

8. <u>The following properties of a continuous map</u> $f: (D^r,S^{r-1}) \to (X,A)$ <u>with</u> $r > 0$ <u>are equivalent:</u>

(i) f <u>is homotopic to a constant map;</u>

(ii) abs f <u>is S^{r-1}-homotopic to a map which carries</u> D^r <u>into a subset of</u> $A;$

PROOF. (i) ⇒ (ii). If $F: (D^r \times I, S^r \times I) \to (X,A)$ is a homotopy from f to a constant map, then the formula

$$(x,t) \mapsto \begin{cases} F(x/\mathrm{dist}(0,x), 2(1-\mathrm{dist}(0,x))), & \text{if } \mathrm{dist}(0,x) \geq (2-t)/2, \\[2ex] F(2x/(2-t), t), & \text{if } \mathrm{dist}(0,x) \leq (2-t)/2, \end{cases}$$

defines an S^{r-1}-homotopy $D^r \times I \to X$ from abs f to a map which carries D^r into a subset of A.

(ii) ⇒ (i). If $G: D^r \times I \to X$ is a homotopy stationary on S^{r-1} from abs f to a map which carries D^r into a subset of A, consider the map $F: D^r \times I \to X$ given by

$$F((x_1, \ldots, x_r), t) = \begin{cases} G((x_1, \ldots, x_r), 2t), & \text{if } t \leq 1/2, \\[2ex] G((2x_1(1-t), \ldots, 2x_r(1-t), 1), & \text{if } t \geq 1/2 \end{cases}$$

Then rel $F: (D^r \times I, S^{r-1} \times I) \to (X,A)$ is a homotopy from f to a constant map.

9. A pair (X,A) is <u>k-connected</u> $(0 \leq k \leq \infty)$ if for any map $f: (D^r, S^{r-1}) \to (X,A)$ with $r \leq k$, abs f is S^{r-1}-homotopic to a map whose image is contained in A.

It is clear that the pair (X,A) is 0-connected if and only if each component of the space X intersects A. If $k > 0$, then (X,A) is k-connected if and only if every continuous map $f: (D^r, S^{r-1}) \to (X,A)$ with $r \leq k$ is homotopic to a constant map; see 8.

A pair which is homotopy equivalent to a k-connected pair is k-connected. As a consequence, we see that when A is a strong deformation retract of X, the pair (X,A) is ∞-connected; indeed, (X,A) is homotopy equivalent to (X,X). It will be clear later that the pair (X,A) is already ∞-connected if A is a deformation retract of X, or even when the inclusion $A \to X$ is a homotopy equivalence; see 5.1.6.5.

4. Local Properties

1. A topological space X is <u>locally contractible at the point</u> $x_0 \in X$ if each neighborhood U of x_0 contains another neighborhood V of x_0 such that the inclusion $V \to U$ is homotopic to the constant map $V \to x_0$. A topological space is <u>locally contractible</u> if it is locally contractible at any of its points.

If we replace in these definitions the homotopies by x_0-homotopies, the we get the definitions of a space X which is <u>strongly locally contractible at the point</u> x_0, and of a <u>strongly locally contractible</u> space X.

$\mathbb{R}^n$, D^n and S^n are examples of strongly locally contractible spaces.

2. A topological space X is <u>locally connected at the point</u> $x_0 \in X$ if each neighborhood U of x_0 contains another neighborhood V of x_0, such that any two points in V can be joined by a path in U. A topological space is <u>locally connected</u> if it is locally connected at any of its points.

It is clear that a locally contractible space is locally connected. As an example of a connected space which is not locally connected we may take the subset of $\mathbb{R}^2$ consisting of the lines $m_1 x_1 + m_2 x_2$, with m_1 and m_2 integers.

3. <u>A space is locally connected if and only if the components of its open sets are open. In particular, in a locally connected space every neighborhood of an arbitrary point contains a connected neighborhood of this point.</u>

PROOF. Suppose that X is locally connected, U is an open subset of X, A a component of U, and $x_0 \in A$ an arbitrary point. Then U contains a neighborhood V of x_0 such that any two points in V can be joined by a path in U. Hence $V \subset A$ and $x_0 \in \text{Int}\,A$. This proves that in a locally connected space the components of the open sets are open.

5. Borsuk Pairs

1. A topological pair (X,A) is a <u>Borsuk pair</u> if given any topological space Y, any continuous map $f\colon X \to Y$, and any homotopy $F\colon A \times I \to Y$ of the map $f\big|_A$, there is a homotopy $X \times I \to Y$ of f which extends F.

If (X,A,B) is a topological triple such that (X,A) and (A,B) are Borsuk pairs, then (X,B) is obviously a Borsuk pair.

2. <u>Let</u> (X,A) <u>be a topological pair. Then in order for</u> (X,A) <u>to be a Borsuk pair it is necessary that</u> $(X \times 0) \cup (A \times I)$ <u>be a retract of the cylinder</u> $X \times I$. <u>When</u> A <u>is closed, this condition is also sufficient.</u>

PROOF OF THE NECESSITY. Any homotopy of the map in: $X = X \times 0 \to (X \times 0) \cup (A \times I)$ which extends the homotopy in: $A \times I \to (X \times 0) \cup (A \times I)$ is a retraction of the cylinder $X \times I$ onto $(X \times 0) \cup (A \times I)$.

PROOF OF THE SUFFICIENCY. Let $\rho: X \times I \to (X \times 0) \cup (A \times I)$ be a retraction. Then given any topological space Y, any continuous map $f: X \to Y$, and any homotopy $F: A \times I \to Y$ of the map $f|_A$, the composition

$$X \times I \xrightarrow{\rho} (X \times 0) \cup (A \times I) \xrightarrow{G} Y,$$

where G is defined by

$$G(x,t) = \begin{cases} f(x), & \text{if } t = 0, \\[2mm] F(x,t), & \text{if } x \in A, \end{cases}$$

is a homotopy of f which extends F.

3. The following statement completes Theorem 2 in an essential way. If X is Hausdorff, then the assumption that $(X \times 0) \cup (A \times I)$ is a retract of the cylinder $X \times I$ implies automatically that A is closed. Indeed, it suffices to note that the above hypothesis implies that $(X \times 0) \cup (A \times I)$ is closed in $X \times I$ (see 1.5.5), and that A is the preimage of this set under the map $X \to X \times I$, $x \mapsto (x,1)$.

4. If the sets A and B form a closed cover of the space X and $(A, A \cap B)$ is a Borsuk pair, then (X,B) is also a Borsuk pair.

This is a consequence of 2; in fact, any retraction

$$\rho: A \times I \to [A \times 0] \cup [(A \cap B) \times I]$$

defines a retraction $X \times I \to (X \times 0) \cup (B \times I)$ by

$$(x,t) \mapsto \begin{cases} \rho(x,t), & \text{if } x \in A, \\[2mm] (x,t), & \text{if } x \in B. \end{cases}$$

5. If (X,A) is a Borsuk pair and A is closed, then $(Z \times X, Z \times A)$ is a Borsuk pair for every topological space Z.

PROOF. If ρ is a retraction of the cylinder $X \times I$ onto $(X \times 0) \cup (A \times I)$, then $\mathrm{id}\, Z \times \rho$ is a retraction of the cylinder $(Z \times X) \times I = Z \times (X \times I)$ onto $[(Z \times X) \times 0] \cup [(Z \times A) \times I] = Z \times [(X \times 0) \cup (A \times I)]$.

Borsuk Pairs and Deformation Retractions

      6. <u>If</u> (X,A) <u>is a Borsuk pair and the inclusion</u> $A \to X$ <u>is a homotopy equivalence, then</u> A <u>is a deformation retract of</u> X.

      PROOF. Let $\pi : X \to A$ be a homotopy inverse of the inclusion $A \to X$. Extend the homotopy from $\pi |_A = \pi \circ$ in$: A \to A$ to id A to a homotopy of the map π; this yields a homotopy from π to a retraction of X onto A, which we denote by ρ. Since the composition $X \xrightarrow{\pi} A \xrightarrow{\text{in}} X$ is homotopic to id X, the composition $X \xrightarrow{\rho} A \xrightarrow{\text{in}} X$ is also homotopic to id X, and thus ρ is a deformation retraction.

      7. <u>If</u> A <u>is a deformation retract of</u> X <u>and</u> $(X \times I, (X \times 0) \cup (A \times I) \cup (X \times 1))$ <u>is a Borsuk pair, then</u> A <u>is a strong deformation retract of</u> X.

      PROOF. Let $\rho : X \to A$ be a deformation retraction, and let $f : X \times I \to X$ be a homotopy from id X to the composite map $X \xrightarrow{\rho} A \xrightarrow{\text{in}} X$. Define a homotopy

$$g : [(X \times 0) \cup (A \times I) \cup (X \times 1)] \times I \to X$$

by

$$g((x,t_1),t_2) = \begin{cases} x, & \text{if } t_1 = 0, \\ f(x,(1-t_2)t_1), & \text{if } x \in A, \\ f(\rho(x),1-t_2), & \text{if } t_1 = 1, \end{cases}$$

and extend it to some homotopy $G : (X \times I) \times I \to X$ of the map $f : X \times I \to X$. It is clear that $(x,t) \mapsto G((x,t),1)$ yields an A-homotopy $X \times I \to X$ from id X to in $\circ \rho$.

      8. <u>If</u> (X,A) <u>is a Borsuk pair and</u> B <u>is a strong deformation retract of the space</u> A, <u>then the map</u> rel$: (X,B) \to (X,A)$ <u>is a homotopy equivalence.</u>

      PROOF. Consider a B-homotopy from id A to the composition of a strong deformation retraction $A \to B$ and the inclusion $B \to A$. Now extend it to a homotopy G of id X. It is clear that the map $(X,A) \to (X,B)$, $x \mapsto G(x,1)$, is a homotopy inverse of rel.

Local Characteristics of Borsuk Pairs

9. **Suppose that (X,A) is a Borsuk pair with X normal, Y is any topological space, and $f: X \to Y$ is any continuous map. Then given any homotopy F of the map $f|_A$, and any neighborhood U of A, there is an $(X \smallsetminus U)$-homotopy of f extending F.**

PROOF. Let G be a homotopy of f extending F, and let ϕ be any Urysohn function for the pair $X \smallsetminus U$, A. Then the formula $(x,t) \mapsto G(x,t\phi(x))$ defines an $(X \smallsetminus U)$-homotopy of f extending F.

10. **If (X,A) is a Borsuk pair, then there exists a neighborhood U of A such that the inclusion $U \to X$ is A-homotopic to a map which takes U into a subset of A. If X is normal and A is distinguishable (in particular, if X is metrizable and A is closed), then this condition is also sufficient, i.e., the converse of the above statement is valid.**

PROOF OF THE NECESSITY. If $\sigma: X \times I \to (X \times 0) \cup (A \times I)$ is a retraction, then the set U of all points $x \in X$ with $\sigma(x,1) \in A \times (0,1]$ is open, and the composition

$$U \times I \xrightarrow{\text{in}} X \times I \xrightarrow{\sigma} (X \times 0) \cup (A \times I) \xrightarrow{\text{in}} X \times I \xrightarrow{\text{pr}_1} X$$

is an A-homotopy from the inclusion $U \to X$ to a map which takes U into a subset of A.

PROOF OF THE SUFFICIENCY. Let $F : U \times I \to X$ be an A-homotopy such that $F(x,0) = x$ and $f(x,1) \in A$ for all $x \in U$, and let $\phi: X \to I$ be a Urysohn function for the pair A, $X \smallsetminus U$, which distinguishes A (see 1.5.9). The formula

$$G(x,t) = \begin{cases} F(x,\min(t/\phi(x),1)), & \text{if } x \in U \smallsetminus A, \\[2mm] x, & \text{if } x \in A, \end{cases}$$

defines a map $G: U \times I \to X$, and Theorem 2.2.14 shows that G is continuous. This in turn implies the continuity of the map $H: X \times I \to X \times I$ defined by

$$H(x,t) = \begin{cases} (G(x,\max(0,t-\phi(x))),\max(0,t-2\phi(x))), & \text{if } x \in U, \\[2mm] (x,0), & \text{if } x \in X \smallsetminus U. \end{cases}$$

It is readily seen that $H(X \times I) = (X \times 0) \cup (A \times I)$ and that ab H : X × I → (X × 0) ∪ (A × I)   is a retraction.

11. <u>Let</u> (X,A) <u>be a topological pair such that</u> A <u>is a</u>
<u>strong deformation retract of one of its neighborhoods. If</u> X <u>is</u>
<u>normal and</u> A <u>is distinguishable (in particular, if</u> X <u>is metrizable</u>
<u>and</u> A <u>is closed), then</u> (X,A) <u>is a Borsuk pair.</u>

This is a corollary of 10.

12. <u>If</u> (X,A) <u>is a Borsuk pair, then given any neighborhood</u>
V <u>of</u> A, <u>there is another neighborhood</u> W <u>of</u> A, <u>such that</u> W ⊂ V
<u>and the inclusion</u> W → V <u>is A-homotopic to a map which takes</u> W <u>into</u>
<u>a subset of</u> A.

PROOF. By 10, there exists a neighborhood U of A and an
A-homotopy F: U × I → X such that F(x,0) = x and F(x,1) ∈ A for
all x ∈ U. Now 2.2.13 shows that every point x ∈ U has a neighborhood
W_x in U with $F(W_x × I) ⊂ V$. Set $W = U_{x∈A} W_x$. It is clear that
W ⊂ V, F(W × I) ⊂ V, and that ab F : W × I → V is an A-homotopy from
the inclusion W → I to a map which takes W into A.

13. <u>If</u> X <u>is a topological space and</u> x ∈ X <u>is such that</u>
(X,x) <u>is a Borsuk pair, then</u> X <u>is strogly locally contractible at</u> x.
<u>If</u> X <u>is normal and locally contractible at a distinguishable point</u> x,
<u>then</u> (X,x) <u>is a Borsuk pair.</u>

This is a consequence of 12 and 10.

6. <u>CNRS-spaces</u>

1. A subset A of a topological space X is said to be a
<u>neighborhood retract of</u> X if A is a retract of one of its neighbor-
hoods in X.

The retracts and the open sets are trivial examples of
neighborhood retracts.

2. <u>If</u> A <u>is a neighborhood retract of</u> X <u>and</u> B <u>is a</u>
<u>neighborhood retract of</u> A, <u>then</u> B <u>is a neighborhood retract of</u> X.

Indeed, let ρ: U → A be a neighborhood retraction for A
in X, and let σ: V → B be a neighborhood retraction for B in A.
Then $σ ∘ (ρ|_W): W → B$, where $W = ρ^{-1}(V)$, is a neighborhood
retraction for B in X.

3. A topological space is a <u>CNRS-space</u> or, simply, a <u>CNRS,</u>
if it is compact and can be embedded in a Euclidean space (of a certain

dimension) as a neighborhood retract; CNRS is the abbreviation of compact neighborhood retract of a sphere.

D^n and S^n are obvious examples of CNRS's.

4. A compact neighborhood retract of a CNRS is a CNRS.

This is a result of 2.

5. The image of any embedding of a CNRS in a normal space is a neighborhood retract.

PROOF. Let $f: X \to Y$ be an embedding of the CNRS X in the normal space Y, and let $g: X \to \mathbb{R}^n$ be an embedding of X such that $g(X)$ is a neighborhood retract of $\mathbb{R}^n$. Further, consider $f_1 = [\mathrm{ab}\, f: X \to f(X)]$ and $g_1 = [\mathrm{ab}\, g: X \to g(X)]$, and let $\rho: V \to g(X)$ be a neighborhood retraction. Since $f(X)$ is closed (see 1.7.9), $g \circ f_1^{-1}: f(X) \to \mathbb{R}^n$ extends to a continuous map $h: Y \to \mathbb{R}^n$ (see 1.5.12). It is clear that $U = h^{-1}(V)$ is a neighborhood of $f(X)$, and that $f_1 \circ g_1^{-1} \circ \rho \circ [\mathrm{ab}\, h: U \to V]$ is a retraction of U onto $f(X)$.

6. Given any compact neighborhood retract X of $\mathbb{R}^n$, there is a number $\varepsilon > 0$ such that any two maps, f and g, of an arbitrary space Y into X which satisfy

$$\sup_{y \in Y} \mathrm{dist}(f(y), g(y)) < \varepsilon$$

are homotopic. Moreover, one may choose a homotopy stationary on the set where f and g agree.

PROOF. Let $\sigma: U \to X$ be a neighborhood retraction. We show that one may take ε to be the distance between X and $\mathbb{R}^n \smallsetminus U$ (which is positive by 1.7.15).

Let $f, g: Y \to X$ be continuous and satisfy $\sup_{y \in Y} \mathrm{dist}(f(y), g(y)) < \varepsilon$. For any point $y \in Y$, the segment with the extremities $f(y)$ and $g(y)$ lies in U. Consequently, the composite maps $Y \xrightarrow{f} X \xrightarrow{\mathrm{in}} U$ and $Y \xrightarrow{g} X \xrightarrow{\mathrm{in}} U$ can be connected by a rectilinear homotopy $F: Y \times I \to U$, and it is plain that $\sigma \circ F: Y \times I \to X$ is a homotopy from f to g. Furthermore, $\sigma \circ F$ is is stationary on the set where f and g agree.

7. If A is a neighborhood retract of a CNRS X, then (X, A) is a Borsuk pair.

PROOF. Let $\sigma: U \to A$ be a neighborhood retraction. Consider X as a neighborhood retract of $\mathbb{R}^n$ and pick ε as in 6. Denote by V the neighborhood of A in X consisting of all the point $x \in U$ for which $\mathrm{dist}(x, \sigma(x)) < \varepsilon$, and let ϕ be the composition

$$V \xrightarrow{\sigma|V} A \xrightarrow{\text{in}} X.$$

Then $\text{dist}(\phi(x),x) < \varepsilon$ for $x \in V$, and $\phi(x) = x$ for $x \in A$. Hence the inclusion $V \to X$ is A-homotopic to ϕ, and since $\phi(V) = A$, we can apply Theorem 5.10.

8. <u>Every CNRS is strongly locally contractible.</u>

This is a consequence of 5, 7, and 5.13.

Information

9. The converse of Theorem 8 is also true: every locally contractible compact subspace of a Euclidean space is a neighborhood retract of this space. For a proof, see [14].

7. Homotopy Properties of Topological Constructions

1. In this subsection we establish the homotopy invariance of some of the constructions described in § 2 and study the homotopy properties of the resulting spaces.

Products

2. Obviously, two continuous maps $f,g: Y \to X_1 \times \ldots \times X_n$ are homotopic if and only if $\text{pr}_i \circ f, \text{pr}_i \circ g: Y \to X_i$ are homotopic for all $i = 1,\ldots,n$ (see 2.2.4). In particular, $X_1 \times \ldots \times X_n$ is k-connected if and only if all the X_i's are k-connected ($0 \leqslant k \leqslant \infty$).

It is also clear that if the maps $g_1: X_1 \to Y_1, \ldots, g_n: X_n \to Y_n$ are homotopic to the maps $f_1: X_1 \to Y_1, \ldots, f_n: X_n \to Y_n$, then $g_1 \times \ldots \times g_n: X_1 \times \ldots \times X_n \to Y_1 \times \ldots \times Y_n$ and $f_1 \times \ldots \times f_n: X_1 \times \ldots \times X_n \to Y_1 \times \ldots \times Y_n$ are homotopic; moreover, if $f_1,\ldots,f_n$ are homotopy equivalences, then so is $f_1 \times \ldots \times f_n$.

3. If A is a deformation retract (strong deformation retract) of X, then $A \times Y$ is a deformation retract (respectively, a strong deformation retract) of the product $X \times Y$, for any space Y. In particular, if X is contractible, then the fibers $x \times Y$ of the product $X \times Y$ are deformation retracts of $X \times Y$.

Quotients

4. Since the projection $X \to X/p$ is continuous, the quotient space of a connected space is connected (see 3.3). Moreover, in order that the quotient X/A of X by its subspace A be connected, it is even enough that the pair (X,A) be 0-connected; if the components of X are open, then the connectedness of X/A implies the 0-connectedness of (X,A) .

We shall see later that neither the k-connectedness with $k > 0$, nor the contractibility are, generally speaking, preserved when one takes quotients.

5. <u>Let p and q be partitions of the spaces X and Y . If the maps $f_t: X \to Y$ form a homotopy and take the elements of p into elements of q , then the maps $\operatorname{fact} f_t : X/p \to Y/q$ also form a homotopy.</u>

We have to verify that the map $G: (X/p) \times I \to Y/q$, given by $G(x,t) = (\operatorname{fact} f_t)(x)$, is continuous. To do this, it suffices to note (see 2.3.4) that the composition $X \times I \xrightarrow{\operatorname{pr} \times \operatorname{id} I} (X/p) \times I \xrightarrow{G} Y/q$ is continuous, and that the map $\operatorname{pr} \times \operatorname{id} I$ is factorial. The first is a consequence of the commutativity of the diagram

$$
\begin{array}{ccc}
X \times I & \xrightarrow{\operatorname{pr} \times \operatorname{id} I} & (X/p) \times I \\
{\scriptstyle F}\downarrow & & \downarrow{\scriptstyle G} \\
Y & \xrightarrow{\qquad \operatorname{pr} \qquad} & Y/q
\end{array}
$$

where F is $F(x,t) = f_t(x)$, while the second follows from Theorem 2.7.8.

6. <u>If $f,f': (X,A) \to (Y,B)$ are homotopic, then the maps $\operatorname{rel} \operatorname{fact} f, \operatorname{rel} \operatorname{fact} f': (X/A,\operatorname{pr}(A)) \to (Y/B,\operatorname{pr}(B))$ are also homotopic. If f is a homotopy equivalence, then so is $\operatorname{rel} \operatorname{fact} f$.</u>

The first assertion is a corollary of 5, while the second is a consequence of the first: if $g: (Y,B) \to (X,A)$ is a homotopy inverse of f , then $\operatorname{rel} \operatorname{fact} g: (Y/B,\operatorname{pr}(B)) \to (X/A,\operatorname{pr}(A))$ is a homotopy inverse of $\operatorname{rel} \operatorname{fact} f$.

7. <u>If (X,A) is a Borsuk pair and A is contractible, then $\operatorname{rel} \operatorname{pr} : (X,A) \to (X/A,\operatorname{pr}(A))$ is a homotopy equivalence.</u>

PROOF. Let $F: A \times I \to A$ be a homotopy from $\operatorname{id} A$ to a constant map, and let $G: X \times I \to X$ be a homotopy of $\mathrm{id}\,X$ extending F. Denote by $g: X \to X$ the map $x \mapsto G(x,1)$, which is homotopic to $\mathrm{id}\,X$. Since g is constant on A, the map $\mathrm{fact}\,g: X/A \to X$ is meaningful; moreover, since $\mathrm{fact}\,g\,(\mathrm{pr}(A)) \subset A$, the map $\mathrm{rel}\,\mathrm{fact}\,g: (X/A; \mathrm{pr}(A)) \to (X,A)$ is also meaningful. Let us check that $\mathrm{rel}\,\mathrm{fact}\,g$ is a homotopy inverse of $\mathrm{rel}\,\mathrm{pr}$. Consider the homotopies $\mathrm{rel}\,G: (X \times I, A \times I) \to (X,A)$ and $\mathrm{rel}\,\mathrm{fact}\,G: ((X/A) \times I, \mathrm{pr}(A) \times I) \to (X/A, \mathrm{pr}(A))$. The first connects the maps

$$\mathrm{rel}\,\mathrm{id}\,X,\ \mathrm{rel}\,g: (X,A) \to (X,A)$$

while the second connects the maps

$$\mathrm{rel}\,\mathrm{id}(X/A),\ \mathrm{rel}\,\mathrm{fact}\,g: (X/A, \mathrm{pr}(A)) \to (X/A, \mathrm{pr}(A)).$$

It is clear that $[\mathrm{rel}\,g: (X,A) \to (X,A)] =$

$$= [\mathrm{rel}\,\mathrm{fact}\,g: (X/A, \mathrm{pr}(A)) \to (X,A)] \circ [\mathrm{rel}\,\mathrm{pr}: (X,A) \to (X/A, \mathrm{pr}(A))] \quad \text{and}$$

$$[\mathrm{rel}\,\mathrm{fact}\,g: (X/A, \mathrm{pr}(A)) \to (X/A, \mathrm{pr}(A))] =$$

$$= [\mathrm{rel}\,\mathrm{pr}: (X,A) \to (X/A, \mathrm{pr}(A))] \circ [\mathrm{rel}\,\mathrm{fact}\,g: (X/A, \mathrm{pr}(A)) \to (X,A)].$$

Attachings

8. **If** (X_1, C) **is a Borsuk pair and** $\varphi, \varphi': C \to X_2$ **are homotopic, then the spaces** $X_2 \cup_\varphi X_1$ **and** $X_2 \cup_{\varphi'} X_1$ **are homotopy equivalent. Moreover, there is a homotopy equivalence** $f: X_2 \cup_\varphi X_1 \to X_2 \cup_{\varphi'} X_1$ **such that the following diagram is commutative:**

$$
\begin{array}{ccc}
 & X_2 & \\
{\scriptstyle \mathrm{imm}_2}\nearrow & & \searrow{\scriptstyle \mathrm{imm}_2} \\
X_2 \cup_\varphi X_1 & \xrightarrow{\ f\ } & X_2 \cup_{\varphi'} X_1
\end{array}
\qquad (1)
$$

PROOF. Let $\Phi: C \times I \to X_2$ be a homotopy from φ to φ', and let $\sigma: X_1 \times I \to (X_1 \times 0) \cup (C \times I)$ be a retraction. Define the maps $h: (X_1 \times 0) \cup (C \times I) \to X_2 \cup_\varphi X_1$ and $h': (X_1 \times 0) \cup (C \times I) \to X_2 \cup_{\varphi'} X_1$ by

$$
h(x,t) = \begin{cases} \mathrm{imm}_1(x), & \text{if } t = 0, \\[2ex] \mathrm{imm}_2 \circ \Phi(x,t), & \text{if } x \in C, \end{cases}
$$

and

$$h'(x,t) = \begin{cases} imm_1(x), & \text{if } t = 0, \\[2ex] imm_2 \circ \Phi(x,1-t), & \text{if } x \in C. \end{cases}$$

Now define two more maps, $f: X_2 \cup_\varphi X_1 \to X_2 \cup_{\varphi'} X_1$ and $g: X_2 \cup_{\varphi'} X_1 \to X_2 \cup_\varphi X_1$ via

$$f \circ imm_1(x) = h' \circ \sigma(x,1), \quad f \circ imm_2(x) = imm_2(x),$$

and

$$g \circ imm_1(x) = h \circ \sigma(x,1), \quad g \circ imm_2(x) = imm_2(x).$$

It is clear that all these maps are continuous and that the diagram (1) is commutative. Moreover, it is readily seen that the map $(X_2 \cup_\varphi X_1) \times I \to X_2 \cup_{\varphi'} X_1$ given by

$$(imm_1(x),t) \mapsto h \circ \sigma \circ \psi(\sigma(x,t),t), \quad (imm_2(x),t) \mapsto imm_2(x),$$

where $\psi: (X_1 \times I) \times I \to X_1 \times I$, $\psi((x,u),t) = (x,\max(0,t-u))$, is a homotopy from $id(X_2 \cup_\varphi X_1)$ to $g \circ f$. Thus, $g \circ f$ is homotopic to $id(X_2 \cup_\varphi X_1)$ and, similarly, $f \circ g$ is homotopic to $id(X_2 \cup_{\varphi'} X_1)$.

9 (LEMMA). <u>If the maps</u> $f: Y \to Y'$ <u>and</u> $f': Y' \to Y$ <u>are homotopy inverses of one another, then given any homotopy</u> $F': Y \times I \to Y$ <u>from</u> $id\,Y$ <u>to</u> $f' \circ f$, <u>there is a homotopy</u> $F': Y' \times I \to Y'$ <u>from</u> $f \circ f'$, <u>such that the maps</u> $f \circ F$, $F' \circ (f \times id\,I): Y \times I \to Y'$ <u>are</u> $[(Y\times 0) \cup (Y\times 1)]$<u>-homotopic.</u>

PROOF. Let $G: Y' \times I \to Y'$ be an arbitrary homotopy from $id\,Y'$ to $f \circ f'$, and let F' be the product of the following three homotopies: G, $f \circ F \circ (f' \times id\,I)$, and the inverse of the homotopy $G \circ (f \times id\,I) \circ (f' \times id\,I)$. Divide the square I^2 into eight pieces, as shown in Fig. 3 (the points A_1,A_2,A_3,A_4 have the abscissae $0,1/2,3/4,1$, and ordinate 0, the points $B_1,\ldots,B_6$ the abscissae $0,1/8,1/4,1/2,3/4,1$, and ordinate $1/2$, and the points C_1,C_2 the abscissae $0,1$, and ordinate 1).

Now define affine maps $\alpha_1: I^2 \to P_1$ and $\alpha_2: I^2 \to P_2$ with the following properties: $\alpha_1(0,0) = B_3$, $\alpha_1(1,0) = B_4$, $\alpha_1(0,1) = A_2$, $\alpha_2(0,0) = B_5$, $\alpha_2(1,0) = B_4$, and $\alpha_2(0,1) = A_4$. Further, for $y \in Y$ define a map $\phi_y: I^2 \to Y'$ through the conditions:

$$\phi_y(\alpha_1(t_1,t_2)) = f(F(F(y,t_2),t_1)$$

and

$$\phi_y(\alpha_2(t_1,t_2)) = G(f \circ F(y,t_2),t_1);$$

if $0 \leqslant t \leqslant 1/2$, then $\phi_y(t,0) = G(f(y),2t)$; the restriction $\phi_y|_{Q_1}$
is constant on all segments parallel to the line A_2B_2; $\phi_y|_{Q_2}$ is
constant on all segments passing through the point D_1; $\phi_y|_{Q_3}$ is
constant on the vertical segments; $\phi_y|_{Q_4}$ is constant on all segments
parallel to the line C_1B_5; $\phi_y|_{Q_5}$ is constant on all segments passing
through the point D_2; and, at last, $\phi_y|_{Q_6}$ is constant on the
vertical segments (those segments on which ϕ_y is constant are depicted
in Fig. 4). It is clear that the formula $((y,t_1),t_2) \mapsto \phi_y(t_1,t_2)$
defines an $[(Y{\times}0){\cup}(Y{\times}1)]$-homotopy $(Y \times I) \times I \to Y'$ from $f \circ F$ to
$F' \circ (f \times \mathrm{id}\,I)$.

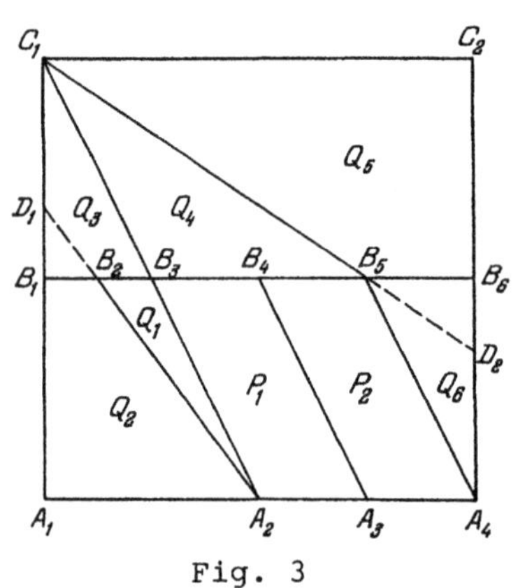

Fig. 3

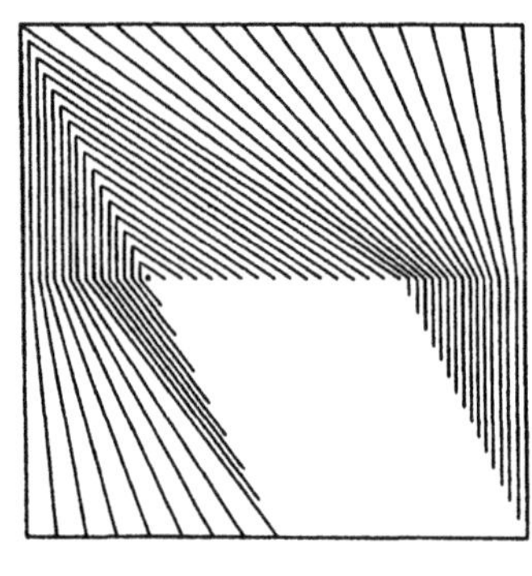

Fig. 4

10. <u>Let (X,C) be a Borsuk pair with C closed, and let</u>
$\varphi: C \to Y$ <u>be a continuous map. If $f: Y \to Y'$ is a homotopy equivalence,</u>
<u>then</u>

$$\mathrm{fact}(\mathrm{id}\,X \sqcup f) : Y \cup_\varphi X \to Y' \cup_{f\circ\varphi} X$$

<u>is also a homotopy equivalence.</u>

PROOF. Let us fix: (i) a homotopy inverse of f,
$f': Y' \to Y$; (ii) a homotopy $F: Y \times I \to Y$ from $\mathrm{id}\,Y$ to $f' \circ f$;
(iii) a homotopy $F': Y' \times I \to Y'$ from $\mathrm{id}\,Y'$ to $f \circ f'$, together
with an $[(Y{\times}0){\cup}(Y{\times}1)]$-homotopy $G: (Y \times I) \times I \to Y'$ from $f \circ F$ to
$F' \circ (f \times \mathrm{id}\,I)$ (see 9); (iv) a retraction $\rho: X \times I \to (X \times 0) \cup (C \times I)$
Further, define $g: (X \times 0) \cup (C \times I) \to Y \cup_\varphi X$ and
$g': [(X \times 0) \cup (C \times I)] \times I \to Y' \cup_{f\circ\varphi} X$ by

$$g(x,t) = \begin{cases} \mathrm{imm}_1(x), & \text{if } t = 0, \\[2ex] \mathrm{imm}_2 \circ F(\varphi(x),t), & \text{if } x \in C \end{cases}$$

and

$$g'((x,t_1),t_2) = \begin{cases} \mathrm{imm}_1(x), & \text{if } t_1 = 0, \\[2ex] \mathrm{imm}_2 \circ G((\varphi(x),t_1),t_2), & \text{if } x \in C. \end{cases}$$

Together, these maps yield a map $h: Y' \cup_{f\circ\varphi} X \to Y \cup_\varphi X$ which satisfies

$$h(\mathrm{imm}_1(x)) = g \circ \rho(x,1), \quad h(\mathrm{imm}_2(y')) = \mathrm{imm}_2(f'(y')).$$

Now it is clear that the formulas

$$H(\mathrm{imm}_1(x),t) = g \circ \rho(x,t), \quad H(\mathrm{imm}_2(y),t) = \mathrm{imm}_2(F(y,t))$$

define a homotopy $H: (Y \cup_\varphi X) \times I \to Y \cup_\varphi X$ from $\mathrm{id}(Y \cup_\varphi X)$ to $h \circ \mathrm{fact}(\mathrm{id}\,X \sqcup f)$. Also, we see that the formulas

$$H_1'(\mathrm{imm}_1(x),t) = g'(\rho(x,t),1), \quad H_1'(\mathrm{imm}_2(y'),t) = \mathrm{imm}_2(F'(y',t)),$$

and

$$H_2'(\mathrm{imm}_1(x),t) = g'(\rho(x,1),t), \quad H_2'(\mathrm{imm}_2(y'),t) = \mathrm{imm}_2(f\circ f'(y')),$$

define two homotopies $H_1',H_2': (Y' \cup_{f\circ\varphi} X) \times I \to Y' \cup_{f\circ\varphi} X$, whose product is a homotopy from $\mathrm{id}(Y' \cup_{f\circ\varphi} X)$ to $\mathrm{fact}(\mathrm{id}\,X \sqcup f) \circ h$. Therefore, h is a homotopy inverse of $\mathrm{fact}(\mathrm{id}\,X \sqcup f)$.

More Special Constructions

11. $\mathrm{con}\,X$ __is contractible for any__ X. $\mathrm{su}\,X$ __is connected for any__ X. $X_1 * X_2$ __is connected for any__ X_1 __and__ X_2.

The contractibility of the cone is obvious. The connectedness of the suspension is a consequence of the fact that it is a quotient space of the cone. To prove that the join $X_1 * X_2$ is connected, we may check that any two points $\mathrm{pr}(x_1,x_2,t)$, $\mathrm{pr}(x_1',x_2',t') \in X_1 * X_2$ $[(x_1,x_2,t), (x_1',x_2',t') \in X_1 \times X_2 \times I]$ can be joined by the path

$$\tau \mapsto \begin{cases} \mathrm{pr}(x_1,x_2,3\tau(1-t)+t), & \text{if } \tau \leqslant 1/3, \\[1ex] \mathrm{pr}(x_1',x_2,2-3\tau), & \text{if } 1/3 \leqslant \tau \leqslant 2/3, \\[1ex] \mathrm{pr}(x_1',x_2',t'(3\tau-2)), & \text{if } \tau \geqslant 2/3. \end{cases}$$

12. Given two homotopic maps $f, g: X \to Y$, a straightforward application of Theorem 5 shows that the maps $su\, f, su\, g: su\, X \to su\, Y$ are also homotopic. As usual, we may deduce that $su\, f$ is a homotopy equivalence whenever f is one.

Moreover, the same Theorem 5 shows that given homotopic maps $f_1, g_1: X_1 \to Y_1$ and homotopic maps $f_2, g_2: X_2 \to Y_2$, the maps $f_1 * f_2, g_1 * g_2: X_1 * X_2 \to Y_1 * Y_2$ are also homotopic. Similarly, we conclude that $f_1 * f_2$ is a homotopy equivalence whenever f_1 and f_2 are homotopy equivalences.

13. Given a continuous map $f: X_1 \to X_2$, $rt\, f: Cyl\, f \to X_2$ (see 2.6.10) is obviously a strong deformation retraction of the mapping cylinder $Cyl\, f$ onto X_2. Hence X_2 is a strong deformation retract of $Cyl\, f$.

The inclusion $X_1 \to Cyl\, f$ is a homotopy equivalence if and only if f is a homotopy equivalence. In fact, the composition
$$X_1 \xrightarrow{\ \ in\ \ } Cyl\, f \xrightarrow{\ \ rt\, f\ \ } X_2$$
coincides with f, and $rt\, f$ is a homotopy equivalence.

The Case of Pointed Spaces

14. The results obtained in 2, 3, 6, 7, 10, 11, and 12 have obvious analogs for pointed spaces. Let us add that those theorems which are the analogs of 12 are valid for both bouquets and tensor products: if $f_\mu, g_\mu: (X_\mu, x_\mu) \to (Y_\mu, y_\mu)$ are homotopic for each μ, then $Vf_\mu, Vg_\mu: (V(X_\mu, x_\mu), bp) \to (V(Y_\mu, y_\mu), bp)$ are homotopic, and if all the f_μ's are homotopy equivalences, then so is Vf_μ. Similarly, if $f_1, g_1: (X_1, x_1) \to (Y_1, y_1)$ are homotopic, and $f_2, g_2: (X_2, x_2) \to (Y_2, x_2)$ are homotopic, then $f_1 \otimes f_2, g_1 \otimes g_2: ((X_1, x_1) \otimes (X_2, x_2), bp) \to ((Y_1, y_1) \otimes (Y_2, y_2), bp)$ are homotopic; if f_1 and f_2 are homotopy equivalences, then so is $f_1 \otimes f_2$.

8. <u>Exercises</u>

1. Show that the sphere S^∞ is contractible.

2. Show that if (X, A) is a Borsuk pair and X is contractible, then the quotient X/A is homotopy equivalent to $su\, A$.

3. Suppose that the product of two topological spaces is homeomorphic to the suspension of some other topological space. Prove

that either both factors of the product are contractible, or one of them reduces to a point.

4.   Let $f_1\colon X_1 \to X_2$ be a homotopy equivalence. Show that X_1 is a strong deformation retract of Cyl f .

5.   Show that if X is metrizable and $x \in X$ is such that (X,x) is a Borsuk pair, then the projection su X $\to$ su(X,x) is a homotopy equivalence.

6.   Given an arbitrary connected topological space X and two arbitrary points $x,y \in X$, show that the subset of $C(I,0;X,x)$ consisting of all paths passing through the point y is contractible.

Chapter 2. Cellular Spaces

1.   Fundamental Concepts

1. A decomposition* p of a topological space X is called
cellular if there is a function d taking the set of elements of p
into the nonnegative integers, such that for every element e of p
there exists a continuous map $D^{d(e)} \to X$ with the following two
properties: (i) it maps $\operatorname{Ind} D^{d(e)}$ homeomorphically onto e;
(ii) it maps $S^{d(e)-1}$ onto a union of elements of p on which d
takes values smaller than $d(e)$.

The elements of a cellular decomposition and their closures
are called cells and closed cells, respectively. The number $d(e)$ is
the dimension of the cell e and is usually denoted by $\dim e$; the
n-dimensional cells are also termed n-cells. Any continuous map
$D^{d(e)} \to X$ with the properties (i) and (ii) is said to be characteristic
for e; we use the symbol cha_e as a standard notation for such a map.

Obviously, $\operatorname{cha}_e(D^{\dim e}) \subset \operatorname{Cl} e$, and if X is Hausdorff,
then $\operatorname{cha}_e(D^{\dim e}) = \operatorname{Cl} e$. In particular, every closed cell in a
cellular decomposition of a Hausdorff space is compact. Moreover, in
the case of a Hausdorff space, given any cell e, $\operatorname{Cl} e \smallsetminus e$ is covered
by cells of dimension lower than $\dim e$.

2. A cellular decomposition is said to be rigged (or equipped)
if for each of its cells there is fixed a characteristic map. The
resulting family $\{\operatorname{cha}_e: D^{\dim e} \to X\}$ is called a rigging (or an
equipment) of the decomposition, and the map

---

* Translator's note: in this chapter we use instead of "partition" the
equivalent term "decomposition", which is generally accepted in the cellular case.

$$\text{cha:} \bigsqcup_{e \in X/p} D^{\dim e} \to X$$

defined by the relations $\text{cha} \circ \text{in}_e = \text{cha}_e$ is called the <u>total characteristic map</u>.

3. According to the general definitions of Chapter 1 (see 1.2.4.3), the cover of the space X by the closed cells of a cellular decomposition p defines a new topology on X. A subset of X is closed in this topology if and only if its intersections with the closed cells of the decomposition p are closed in the initial topology of X. The new topology on X is called the <u>weak</u> or <u>cellular topology</u>, and the process by which we pass from the initial to the cellular topology of X is called a cellular weakening of the initial topology. The weakening of a topology can only enlarge the supply of open and closed sets; in particular, a Hausdorff space remains Hausdorff. In any case, it does not affect the topology of the closed cells, and so the decomposition p remains cellular and retains its characteristic maps.

If X is Hausdorff and the decomposition p is endowed with a rigging $\{\text{cha}_e\}$, then the cellular topology can be described effectively in terms of the corresponding total characteristic map: a subset A of X is open (closed) if and only if its preimage $\text{cha}^{-1}(A)$ is open (respectively, closed). In other words, the cellular topology is exactly that topology on X which transforms the injective factor of the map cha into a homeomorphism. Recall that the injective factor of cha is a map of the quotient space of the sum $\bigsqcup_{e \in X/p} D^{\dim e}$ by its partition $\text{zer}(\text{cha})$ onto X. The equivalence of these two definitions of the cellular topology results from the fact that the maps cha_e are closed (see 1.1.7.9).

4. A <u>cellular space</u> is a Hausdorff topological space endowed with a cellular decomposition which satisfies the following two conditions: (C) <u>every closed cell intersects only a finite number of cells</u>; (W) <u>the closed cells constitute a fundamental cover of the space.</u> Obviously, condition (W) implies that the cellular topology coincides with the initial one. The notations (C) and (W) are generally accepted, and originate from the terms <u>closure finiteness</u> and <u>weak topology.</u>

Property (C) is clearly preserved under the cellular weakening of the topology. Therefore, a Hausdorff topological space having a cellular decomposition satisfying (C) becomes a cellular space via cellular weakening.

Usually the terminology specific to cellular decompositions

is applied to cellular spaces too. In particular, a cellular space may be finite, countable, and rigged. Thus, a finite cellular space is one that has a finite number of cells, but not necessarily a finite number of points.

The _dimension_ of a cellular space is the supremum of the dimensions of its cells; the dimension of the empty space (which is not excluded from the family of cellular spaces) is taken equal to -1. The dimension - finite or infinite - of the cellular space X is denoted by $\dim X$.

We let $\operatorname{cell}_r X$ denote the set of all r-cells of X.

5. The simplest cellular spaces are the discrete spaces decomposed into 0-cells (isolated points). It is clear that all 0-dimensional cellular spaces are of this type: a decomposition of a nondiscrete Hausdorff space into 0-cells does not satisfy condition (W).

The decomposition of the ball D^n (n > 1) into the n-cell $\operatorname{Int} D^n$ and the 0-cells which cover $\operatorname{Fr} D^n = S^{n-1}$ is an example of cellular decomposition which satisfies condition (W), but not condition (C).

The Locally Finite Case

6. In agreement with the general definitions given in 1.1.1.12, a cellular decomposition is _locally finite_ if every point of the given space has a neighborhood which intersects only a finite number of cells. An equivalent condition: every point has a neighborhood which intersects only a finite number of closed cells.

Clearly, in a space possessing a locally finite cellular decomposition, every compact subset has a neighborhood which intersects only a finite number of cells. As a consequence, every locally finite cellular decomposition of a Hausdorff space satisfies condition (C). Theorem 1.1.3.6 shows that condition (W) is satisfied by _any_ locally finite cellular decomposition. We conclude that _every Hausdorff space endowed with a finite or locally finite cellular decomposition is a cellular space._

7. _A cellular space is locally finite if and only if every cell intersects only a finite number of closed cells._

PROOF. In a locally finite cellular space the closure of an arbitrarily given cell has a neighborhood which intersects only a finite number of cells (see 6). It is clear that this neighborhood, and hence the given cell, do not intersect the closure of other cells.

Conversely, if every cell of a cellular space intersects only a finite number of closed cells, then the axioms (C) and (W) imply that the union of any collection of closed cells is closed. Consequently, the complement of the union of all closed cells which do not intersect an arbitrarily given cell e is a neighborhood of e. Since this complement cannot intersect the closed cells which do not intersect e, it intersects only a finite number of closed cells.

Subspaces

8 (LEMMA). <u>Let</u> A <u>be a subset of the cellular space</u> X <u>with the following property: if</u> $x \in A$, <u>then</u> A <u>contains the closure of the cell in which</u> x <u>lies. Then any part of</u> A <u>whose intersections with all the closed cells contained in</u> A <u>are closed is itself closed.</u>

Indeed, if $B \subset A$ is such a part, and e is an arbitrary cell, then one can write the intersection $B \cap Cl\,e$ as $\bigcup_{i=1}^{s} [(B \cap Cl\,e_i) \cap Cl\,e]$, where $e_1, \ldots, e_s$ are all the cells of A which intersect $Cl\,e$, and this shows that $B \cap Cl\,e$ is closed.

9. A subset of a cellular space which contains together with each point the closure of its cell is called a <u>subspace</u> of the given cellular space. Every subspace is a cellular space, with the cellular decomposition induced by the cellular decomposition of the ambient space. By Lemma 8, this decomposition satisfies condition (W), and it obviously satisfies condition (W).

As another consequence of Lemma 8, we see that every subspace of a cellular space is closed. Notice also that the union and the intersection of any collection of subspaces are again subspaces, and that every cover of a cellular space by subspaces is fundamental.

A pair consisting of a cellular space and one of its subspaces is called a <u>cellular pair.</u> <u>Cellular triples</u> and <u>cellular triads</u> are defined similarly.

Warning: a closed cell is not necessarily a subspace. For an example, consider the bouquet $(D^1,0) \vee (S^2,ort_1)$, with the decomposition into four cells: the 0-cells $imm_1(-1)$ and $imm_1(1)$, the 1-cell $imm_1(Int\,D^1)$, and the 2-cell $imm_2(S^2 \smallsetminus ort_1)$. This is obviously a cellular space; however, the closure of the 2-cell touches the 1-cell, but does not contain it.

10. The most important subspaces of a cellular space X are

its <u>skeletons</u> $\mathrm{ske}_0 X$, $\mathrm{ske}_1 X, \ldots,$ $\mathrm{ske}_r X$, $\ldots,$ defined as $\mathrm{ske}_r X = \cup_{\dim e \, \leqslant \, r} e$. If $X \neq \emptyset$, then all the skeletons are nonempty (since the presence of cells of a certain positive dimension implies the presence of cells of lower dimension). For formal reasons, we add the empty skeleton $\mathrm{ske}_{-1} X$ and the skeleton $\mathrm{ske}_\infty X = X$. The sequence $\{\mathrm{ske}_r X\}_{0 \leqslant r < \infty}$ is clearly a filtration of X.

We remark that any map $D^{\dim e} \to X$ which is characteristic for the cell e takes $S^{\dim e - 1}$ into $\mathrm{ske}_{\dim e - 1} X$ (in fact, we saw this already in 1). If X carries a rigging $\{\mathrm{cha}_e\}$, then the map ab $\mathrm{cha}_e : S^{\dim e - 1} \to \mathrm{ske}_{\dim e - 1} X$ is called an <u>attaching map</u> for e, and is denoted by att_e.

11. <u>Every cell of a cellular space is included in a finite subspace.</u>

PROOF. Use induction on the dimension of the cell. A 0-cell is itself a subspace. If e is a cell of positive dimension, then $\mathrm{Cl}\, e \smallsetminus e$ may be covered by a finite number of lower dimensional cells, and the union of e with a collection of finite subspaces which contain these cells is a finite subspace containing e.

Compact Subsets

12. <u>A compact subset of a cellular space intersects only a finite number of cells.</u>

PROOF. Every subset A of a cellular space contains a part B which intersects at only one point each cell intersecting A. Since B intersects any closed cell at a finite number of points, B and all its subsets are closed. Therefore, B must be discrete. When A is compact, B as a discrete, closed, and compact set is finite.

13. <u>Every compact subset of a cellular space is contained in a finite subspace.</u>

For each cell intersected by the given subset, pick a finite subspace containing this cell. The union of these subspaces is the desired finite subspace.

14. <u>Every compact subset of a locally finite cellular space is contained in the interior of a finite subspace.</u>

Indeed, such a subset has a neighborhood which intersects only a finite number of cells (cf. 6). For each such cell, pick a finite subspace which contains it; the union of these subspace is again a finite subspace and contains the above neighborhood.

Cellular Maps

15. A map of a cellular space X into a cellular space Y is said to be <u>cellular</u> if it is continuous and maps the skeleton $ske_r X$ into $ske_r Y$, for each r.

A cellular map obviously transforms 0-cells into 0-cells. However, a cell of positive dimension is not necessarily transformed into a single cell: for example, consider the identity map of the segment D^1, decomposed into the 0-cells $-1,1$ and the 1-cell $(-1,1)$, onto the same segment, but decomposed now into the 0-cells $-1,0,1$ and the 1-cells $(-1,0)$, $(0,1)$; this cellular map takes the 1-cell $(-1,1)$ into the union of a 0-cell and two 1-cells.

16. A cellular map is a <u>cellular equivalence</u> if it is invertible and its inverse is also cellular. An equivalent formulation: a cellular equivalence is a homeomorphism which transforms the cellular decomposition of the domain space exactly into the cellular decomposition of the image space. If there is a cellular equivalence between two cellular spaces, then they are said to be <u>cellular equivalent.</u> Two rigged cellular spaces related by a cellular equivalence which transforms one rigging into the other are said to be <u>rigged-equivalent.</u>

If X and Y are cellular spaces, a map $f: X \to Y$ is a <u>cellular embedding</u> if $f(X)$ is a subspace of Y (defined as in 9) and ab $f: X \to f(X)$ is a cellular equivalence.

Warning: there are cellular homeomorphisms which are not cellular equivalences. An example is the homeomorphism described in 15.

2. <u>Glueing Cellular Spaces From Balls</u>

1. <u>If</u> $r \geqslant 0$, <u>then the skeleton</u> $ske_r X$ <u>of the rigged cellular space</u> X <u>is canonically homeomorphic to the space</u>

$$(ske_{r-1}X) \cup_\varphi (\textstyle\coprod_{e \in M_r} (D_e = D^r)),$$

<u>where</u> $M_r = cell_r X$ <u>and the map</u> $\varphi: \coprod_{e \in M_r} (S_e = S^{r-1}) \to ske_{r-1}X$ <u>is defined by</u> $\varphi \circ in_e = att_e$ $(e \in M_r)$.

This canonical homeomorphism between the above spaces is

the injective factor of the map $(\mathrm{ske}_{r-1}X) \bigsqcup (\bigsqcup_{e \in M_r} D_e) \to \mathrm{ske}_r X$ defined by the inclusion $\mathrm{ske}_{r-1}X \to \mathrm{ske}_r X$ and the maps ab cha$_e : D_e \to \mathrm{ske}_r X$.

2. The decription in 1.3 of the weak topology in terms of the total characteristic map shows that one can glue any cellular space from balls, and even do it in a nice way. Theorem 1 above reduces this glueing to a sequence of attaching processes: the r-th process transforms $\mathrm{ske}_{r-1}X$ into $\mathrm{ske}_r X$ $(r = 0,1,\ldots)$, and X is defined, starting from the sequence $\{\mathrm{ske}_r X\}$, as $X = \lim \mathrm{ske}_r X$ (see 1.10).

The following formal procedure transforms this description of cellular spaces into a useful inductive method of constructing such spaces. First of all, note that if we are given a topological space A with a rigged cellular decomposition into cells of dimensions $< q$, and to A we attach a sum $\bigsqcup_{\mu \in M}(D_\mu = D^q)$ of q-dimensional balls by some continuous map $\varphi: \bigsqcup_{\mu \in M}(S_\mu = S^{q-1}) \to A$, then we obtain a space endowed with an obvious rigged cellular decomposition into cells of dimensions $< q+1$. This space satisfies condition (W) whenever A satisfies it, and Theorem 1.2.4.9 shows that it is normal if A is normal. Moreover, it follows from Theorem 1.12 that this space satisfies condition (C) provided A is cellular. Finally, we conclude that if A is a normal rigged cellular space, then $A \cup_\varphi (\bigsqcup_{\mu \in M} D_\mu)$ is a normal rigged cellular space too.

These observations form the basis of our inductive construction. We start with $q = 0$, i.e., take $A = \emptyset$, and at the r-th step we attach the space $\bigsqcup_{\mu \in M_r} (D_\mu = D^r)$ to the previously constructed normal rigged cellular space X_{r-1}, $\dim X_{r-1} \leqslant r-1$, by a continuous map $\varphi_r: \bigsqcup_{\mu \in M_r} (S_\mu = S^{r-1}) \to X_{r-1}$. The r-th step yields a normal rigged cellular space $X_r = X_{r-1} \cup_{\varphi_r} (\bigsqcup_{\mu \in M_r} D_\mu)$, with $\dim X_r \leqslant r$. The result of the whole process is a sequence $\emptyset = X_{-1}, X_0, X_1, \ldots,$ with natural cellular embeddings $X_r \to X_{r+1}$, and limit space $X = \lim X_r$. According to Theorem 1.2.4.6, X is normal and is endowed with an obvious cellular decomposition satisfying properties (C) and (W). Therefore, X is a normal rigged cellular space, and clearly $\mathrm{ske}_r X = X_r$.

We say that X is an <u>inductively glued cellular space.</u> The discussion above demonstrates that every rigged cellular space is rigged-equivalent to an inductively glued cellular space.

3 (COROLLARY). <u>Every cellular space is normal.</u>

3. The Canonical Cellular Decompositions of Spheres, Balls, and Projective Spaces

1. The spheres, balls, and projective spaces admit canonical cellular decompositions making them into cellular spaces. These are all rigged cellular decompositions and will be described in the present subsection. It will be evident in each case that properties (C) and (W) are satisfied.

2. The canonical cellular decomposition of the sphere S^n with $0 \leqslant n < \infty$, consists of the 0-cell ort_1 and the n-cell $S^n \smallsetminus \mathrm{ort}_1$. As the characteristic map of the cell $S^n \smallsetminus \mathrm{ort}_1$ we take $DS: D^n \to S^n$.

3. The canonical cellular decomposition of the ball D^n with $1 \leqslant n < \infty$, is given by the 0-cell ort_1, the $(n-1)$-cell $S^{n-1} \smallsetminus \mathrm{ort}_1$, and the n-cell $\mathrm{Int}\, D^n$. For the characteristic maps of the cells $S^{n-1} \smallsetminus \mathrm{ort}_1$ and $\mathrm{Int}\, D^n$ we take the composite map

$$D^{n-1} \xrightarrow{\ DS\ } S^{n-1} \xrightarrow{\ in\ } D^n$$

and $\mathrm{id}\, D^n$, respectively.

4. The canonical cellular decomposition of the real projective space $\mathbb{R}P^n$ $(0 \leqslant n \leqslant \infty)$ consists of the r-cells $e_r = \mathbb{R}P^r \smallsetminus \mathbb{R}P^{r-1}$, where $0 \leqslant r \leqslant n$ for $n < \infty$, and $0 \leqslant r < \infty$ for $n = \infty$. The composition

$$D^r \xrightarrow{\ pr\ } \mathbb{R}P^r \xrightarrow{\ in\ } \mathbb{R}P^n$$

(where pr is the projection arising from the characterization of $\mathbb{R}P^r$ as a quotient space of D^r, given in 1.2.5.2) is taken as the characteristic map of the cell e_r.

It is clear that att_{e_r} is simply $pr: S^{r-1} \to \mathbb{R}P^{r-1}$, and that $\mathrm{ske}_r \mathbb{R}P^n = \mathbb{R}P^r$, $r \leqslant n$.

5. The canonical cellular decomposition of the complex projective space $\mathbb{C}P^n$ $(0 \leqslant n \leqslant \infty)$ consists of the 2r-cells $e_r = \mathbb{C}P^r \smallsetminus \mathbb{C}P^{r-1}$, where $0 \leqslant r \leqslant n$ for $n < \infty$, and $0 \leqslant r < \infty$ for $n = \infty$. The composition

$$D^{2r} \xrightarrow{\ pr\ } \mathbb{C}P^r \xrightarrow{\ in\ } \mathbb{C}P^n$$

is taken as the characteristic map of the cell e_{2r}. Obviously,

$$\mathrm{ske}_r \mathbb{C}P^n = \mathbb{C}P^{[r/2]}, \quad \text{for} \quad r \leqslant 2n.$$

The canonical rigged cellular decompositions of the projective spaces $\mathbb{H}P^n$ $(0 \leqslant n \leqslant \infty)$ and $\mathbb{C}aP^n$ $(0 \leqslant n \leqslant 2)$ are defined similarly. The decomposition of $\mathbb{H}P^n$ is given by the 4r-cells $e_r = \mathbb{H}P^r \smallsetminus \mathbb{H}P^{r-1}$, where $0 \leqslant r \leqslant n$ for $n < \infty$, and $0 \leqslant r < \infty$ for $n = \infty$. For $\mathbb{C}aP^n$ the cells are $e_r = \mathbb{C}aP^r \smallsetminus \mathbb{C}aP^{r-1}$ with $0 \leqslant r \leqslant n$ and $\dim e_r = 8r$.

4. More Topological Properties of Cellular Spaces

1. Our task in this subsection is to examine what connections exist between properties of cellular spaces such as compactness, local compactness, separability, second countability, and metrizability, and properties of their cellular decompositions such as finiteness, countability, and local finiteness. Incidentally, we prove that cellular spaces are CNRS's. Moreover, conditions for the connectedness of a cellular space, as well as its partition into components are studied.

Compactness and Local Compactness

2. **A cellular space is compact if and only if it is finite.**

The necessity of this condition is a result of 1.12. Since every finite cellular space can be covered by a finite number of closed cells, the condition is also sufficient.

3. **A cellular space is locally compact if and only if it is locally finite.**

By Theorem 1.12, every neighborhood with compact closure of an arbitrarily given point intersects only a finite number of cells, and hence the condition is necessary. It is also sufficient, because the closure of a neighborhood which intersects only a finite number of cells is contained in the union of the closures of these cells.

Embedding Theorems

4. **Every cellular space can be embedded in a Euclidean space of sufficiently high dimension.**

We shall proceed by induction. Given a finite cellular space

X of dimension $n \geqslant 0$, and an embedding $j: \mathrm{ske}_{n-1}X \to \mathbb{R}^q$, we shall construct an embedding $J: X \to \mathbb{R}^{q+n+1}$ (since a cellular space of dimension -1 is empty, the first step of the induction is trivial). Pick a rigging of X and arrange the n-cells of X in a sequence $e_1, \ldots, e_s$. Now define the maps $\phi_1, \ldots, \phi_s: D^n \to \mathbb{R}^{q+n+1}$ by

$$\phi_k(y_1, \ldots, y_k) =$$

$$= \begin{cases} (0, \ldots, 0, y_1, \ldots, y_{n-1}, y_n + 2k, 1), & \text{if } \rho = (y_1^2 + \ldots + y_n^2)^{1/2} \leqslant 1/2, \\[2ex] (2\rho - 1)\, j \circ \mathrm{cha}_{e_k}\!\left(\dfrac{y_1}{\rho}, \ldots, \dfrac{y_n}{\rho}\right) + (2 - 2\rho)\left(0, \ldots, 0, \dfrac{y_1}{2\rho}, \ldots, \dfrac{y_{n-1}}{2\rho}, \dfrac{y_n}{2\rho} + 2k, 1\right), \\[2ex] \hspace{6cm} \text{if } \rho \geqslant 1/2, \end{cases}$$

and then set

$$J(x) = \begin{cases} j(x), & \text{if } x \in \mathrm{ske}_{n-1}X, \\[2ex] \phi_k(y), & \text{if } x \in \mathrm{Cl}\, e_k \text{ and } x = \mathrm{cha}_{e_k}(y). \end{cases}$$

This yields a continuous map $J: X \to \mathbb{R}^{q+n+1}$ (see 1.1.4.3), and since the plane $y_1 = 0, \ldots, y_q = 0, y_{q+n+1} = 1$ contains no lines parallel to $\mathbb{R}^q$, J is injective. Hence, J is an embedding (see 1.1.7.10).

5. Every finite cellular space is a CNRS.

The proof is a continuation of the previous one and also requires an induction on n. Namely, we show that $\mathbb{R}^q \cup J(X)$ is a neighborhood retract of $\mathbb{R}^{q+n+1}$. This is enough: if we assume that $j(\mathrm{ske}_{n-1}X)$ is a neighborhood retract of $\mathbb{R}^q$, then $J(X)$ is obviously a neighborhood retract of $\mathbb{R}^q \cup J(X)$.

Set

$$A_k = \phi_k(\mathrm{Int}\, D^n) \quad \text{and} \quad B_k = A_k \cap \{(x_1, \ldots, x_{q+n+1}) \mid x_{q+n+1} \neq 1\}.$$

Since the sets $A_1, \ldots, A_s$ are pairwise disjoint and closed in $\mathbb{R}^{q+n+1} \smallsetminus \bigcup_{k=1}^s \phi_k(S^{n-1})$, they have pairwise disjoint neighborhoods $U_1, \ldots, U_s$ in $\mathbb{R}^{q+n+1} \smallsetminus \bigcup_{k=1}^s \phi_k(S^{n-1})$, and clearly $U_1, \ldots, U_s$ are open in $\mathbb{R}^{q+n+1}$. Moreover, since $A_1, \ldots, A_s$ are all homeomorphic to $\mathbb{R}^n$, the identity maps $A_1 \to A_1, \ldots, A_s \to A_s$ extend to continuous maps $\psi_1: U_1 \to A_1, \ldots, \psi_s: U_s \to A_s$ (see 1.1.5.12). Now let

$$V_k = \{x \in U_k \mid \mathrm{dist}(x, \psi_k(x)) < \mathrm{Dist}(x, \mathbb{R}^q)\},$$

$$V = \{x \in \mathbb{R}^{q+n+1} \mid \text{Dist}(x, \mathbb{R}^q) < 1/2\},$$

and denote by ψ the orthogonal projection $V \to \mathbb{R}^q$. It is clear that $W = V \cup V_1 \cup \ldots \cup V_s$ is a neighborhood of $\mathbb{R}^q \cup J(X)$ in $\mathbb{R}^{q+n+1}$ and that the sets

$$\left.\begin{array}{c}\{[(\cup_{k=1}^{s} \text{Cl } V_k) \cap \text{Fr } V] \cup (\cup_{k=1}^{s} A_k)\} \cap W, \\[2ex] [(\cup_{k=1}^{s} \text{Fr } V_k) \cap \text{Cl } V \cap W] \smallsetminus \cup_{k=1}^{s} \phi_k(S^{n-1}), \end{array}\right\} \qquad (1)$$

are disjoint and closed in $Y \smallsetminus \cup_{k=1}^{s} \phi_k(S^{n-1})$, where

$$Y = (\cup_{k=1}^{s} \text{Cl } V_k) \cap \text{Cl } V \cap W.$$

Let $f: Y \smallsetminus \cup_{k=1}^{s} \phi_k(S^{n-1}) \to I$ be a Urysohn function for the pair (1). Now given $y \in \text{Cl } B_k$ and $z \in \mathbb{R}^q$, let u_{yz} be the path in $\mathbb{R}^q \cup J(X)$ defined as the product of the rectilinear path between y and $\phi_k(\phi_k^{-1}(y)/\text{dist}(0,\phi_k^{-1}(y)))$ with the rectilinear path between $\phi_k(\phi_k^{-1}(y)/\text{dist}(0,\phi_k^{-1}(y)))$ and z. Define

$$\sigma: (\cup_{k=1}^{s} \text{Cl } B_k) \times \mathbb{R}^q \times I \to \mathbb{R}^q \cup J(X)$$

by $\sigma(y,z,t) = u_{yz}(t)$. According to 1.2.2.14, the map $\tau: Y \to \mathbb{R}^q \cup J(X)$,

$$\tau(x) = \begin{cases} \sigma(\psi_k(x),\psi(x),f(x)), & \text{if } x \in (\text{Cl } V_k \cap \text{Cl } V \cap W) \smallsetminus \phi_k(S^{n-1}), \\[2ex] x, & \text{if } x \in \cup_{k=1}^{s} \phi_k(S^{n-1}), \end{cases}$$

is continuous. By 1.1.4.3, the map $W \to \mathbb{R}^q \cup J(X)$ defined by the formula

$$x \mapsto \begin{cases} \psi_k(x), & \text{if } x \in \text{Cl } V_k \cap (W \smallsetminus V), \\[1ex] \psi(x), & \text{if } x \in W \smallsetminus \cup_{k=1}^{s} V_k, \\[1ex] \tau(x), & \text{if } x \in Y, \end{cases}$$

is also continuous, and it is clearly a retraction.

Connectedness. Components

6. <u>The components of a cellular space are open subspaces.</u>

PROOF. Every closed cell is connected, as the image of a ball under a characteristic map. Therefore, a component of a cellular space contains along with each point the closure of the cell in which the point lies, i.e., it is a subspace. The complement of a component is the union of the remaining components and so it is a subspace too. Consequently, this complement is closed, and hence the component is open.

7. **If** $r \geqslant 1$, **then the r-th skeleton** $\mathrm{ske}_r A$ **of a component** A **of the cellular space** X **is a component of** $\mathrm{ske}_r X$. **In particular, a cellular space** X **is connected if and only if** $\mathrm{ske}_1 X$ **is connected.**

PROOF. Obviously, $\mathrm{ske}_r A = A \cap \mathrm{ske}_r X$ for all r, and if B is another component of X, then $\mathrm{ske}_r A$ and $\mathrm{ske}_r B$ sit in different components of $\mathrm{ske}_r X$. Therefore, all we have to show is that the skeletons $\mathrm{ske}_r A$ with $r \geqslant 1$ are connected or, equivalently, that given a connected cellular space X, all its skeletons $\mathrm{ske}_r X$ $(r \geqslant 1)$ are connected. But this is plain if one notes that the contruction of an inductively glued cellular space described in 2.2 cannot result in a connected space if one of the spaces X_r with $r \geqslant 1$ is not connected.

8. **Every connected locally finite cellular space is countable.**

PROOF. Given a connected, locally finite cellular space X, fix an arbitrary point $x_0 \in X$ and consider the set A_m of all points of X that can be joined to x_0 by a path intersecting at most m cells. Since any path intersects only a finite number of cells (see 1.12), $X = \cup_{m=1}^{\infty} A_m$, and it is clear that each A_m consists of whole cells. Therefore, all we have to verify is that each set A_m with $m \geqslant 1$ contains only a finite number of cells, and we do it by induction on m.

It is readily seen that the closure of any cell contained in A_{m+1} intersects the closure of some cell from A_m. On the other hand, since X is locally finite, the closure of any cell from A_m (being a compact subset of X) can intersect only a finite number of closed cells. Thus the number of cells in A_{m+1} is finite provided that the number of cells in A_m is so, and to complete the proof note that A_1 is just one cell.

Countability Axioms and Metrizability

9. **A cellular space is separable if and only if it is countable.**

PROOF. If the cellular space X is countable, pick a countable dense set in each cell and then take the union of all these sets to produce a countable dense set in X.

Now suppose that X is separable. Then every point of X
lies in a finite subspace (see 1.11). Pick such a subspace for each
point of a fixed countable dense set in X. The union of these subspaces is a counatble subspace and actually coincides with X.

10 (LEMMA). <u>If the cellular space X has a countable base
at a point $x_0 \in X$, then this base contains a neighborhood of x_0 which
intersects only a finite number of cells of X.</u>

Suppose that this is not true. Write the elements of the
given base in a sequence $U_1, U_2, \ldots$, and (using a trivial induction),
select a sequence of points $x_1, x_2, \ldots$ in $X \smallsetminus x_0$, such that:
(i) $x_i \in U_i$; (ii) if $i \neq j$, then x_i and x_j sit in distinct
cells. Since any closed cell contains only a finite number of the x_i's,
the set of all x_i's is closed, and its complement is a neighborhood
of x_0. We reached a contradiction, because this complement contains
none of the neighborhoods $U_1, U_2, \ldots$.

11. <u>A cellular space is first countable if and only if it is
locally finite.</u>

The necessity of this condition is a corollary of Lemma 10.
To prove its sufficiency, use Theorem 1.14 to deduce that every point
x_0 of a locally finite cellular space has a neighborhood U contained
in a finite subspace. By Theorem 4, such a subspace, and hence U, are
first countable spaces, and it is obvious that a countable base of U
at x_0 is also a countable base of the given space at x_0.

12. <u>A cellular space is second countable if and only if it is
countable and locally finite.</u>

The necessity of this condition is a corollary of 9 and 11.
To prove its sufficiency, for each closed cell we fix a neighborhood
contained in a finite subspace (see 1.14). As a result of Theorem 4,
these neighborhoods are second countable, and it is clear that the union
of their countable bases is a countable base of the given space.

13. <u>A cellular space is metrizable if and only if it is
locally finite.</u>

The necessity of this condition is a corollary of 11. The
metrizability of a connected locally finite cellular space follows from
theorems 8, 12, 2.3, and 1.1.6.9. Since every locally finite cellular
space is homeomorphic to the sum of its components, and these are
metrizable, the space is metrizable too (see 6 and 1.2.1.2).

5. <u>Cellular Constructions</u>

1. When applied to cellular spaces, the constructions described in § 1.2 need to be appropriately modified. For certain constructions the modification consists merely of observing that the resulting space is endowed with a cellular decomposition and becomes cellular; an obvious example is the sum. For other examples, such as the product, the modification also affects the topology of the resulting space.

Below we describe the main modifications of both types. We emphasize that all these constructions, when applied to rigged cellular spaces, produce again rigged cellular spaces.

Cellular Product

2. Let X_1 and X_2 be topological spaces with cellular decompositions p_1 and p_2. Then the product $X_1 \times X_2$ has a natural cellular decomposition, namely $p_1 \times p_2$, with $\dim(e_1 \times e_2) = \dim e_1 + \dim e_2$. As a characteristic map for the cell $e_1 \times e_2$ one may take the composition of the canonical homeomorphism

$$D^{\dim e_1 + \dim e_2} \longrightarrow D^{\dim e_1} \times D^{\dim e_2}$$

with the product $\mathrm{cha}_{e_1} \times \mathrm{cha}_{e_2} : D^{\dim e_1} \times D^{\dim e_2} \to X_1 \times X_2$ of arbitrary characteristic maps $\mathrm{cha}_{e_1} : D^{\dim e_1} \to X_1$ and $\mathrm{cha}_{e_2} : D^{\dim e_2} \to X_2$. When the decompositions p_1 and p_2 are rigged, $p_1 \times p_2$ takes on a canonical rigging.

If both p_1 and p_2 fulfil property (C), then (C) holds for $p_1 \times p_2$ too. However, there are situations where $p_1 \times p_2$ does not have property (W), even when X_1 and X_2 are cellular spaces; see Exercise 6.6. The cellular space arising from the product $X_1 \times X_2$ of the cellular spaces X_1 and X_2 through cellular weakening of its topology is called the <u>cellular product</u> of X_1 and X_2, and is denoted by $X_1 \times_c X_2$.

We note that the cellular weakening does not alter the topology of the compact parts of the space $X_1 \times X_2$. Indeed, every compact subset of $X_1 \times X_2$ has compact images under the projections $X_1 \times X_2 \to X_1$ and $X_1 \times X_2 \to X_2$, and hence it can be covered by a finite number of cells.

3. If X_1 is locally finite, then $X_1 \times_c X_2 = X_1 \times X_2$ for any cellular space X_2.

PROOF. Let cha^1 and cha^2 be the total characteristic maps corresponding to some riggings of the cellular decompositions p_1 and p_2 of X_1 and X_2. It is clear that the total characteristic map corresponding to $p_1 \times p_2$ can be expressed as the composition

$$\bigsqcup D^{\dim(e_1 \times e_2)} \longrightarrow \bigsqcup (D^{\dim e_1} \times D^{\dim e_2}) =$$

$$= (\bigsqcup D^{\dim e_1}) \times (\bigsqcup D^{\dim e_2}) \xrightarrow{\ cha^1 \times cha^2\ } X_1 \times X_2, \tag{2}$$

where the first map is the sum of the canonical homeomorphisms $D^{\dim(e_1 \times e_2)} \to D^{\dim e_1} \times D^{\dim e_2}$. Since p_1 and p_2 satisfy condition (W), the maps cha^1 and cha^2 are factorial (see 1.3). Furthermore, since $\bigsqcup D^{\dim e_2}$ and X_1 are locally compact (see 4.3), the map $cha^1 \times cha^2$ is factorial too (see 1.2.7.9), which in turn implies that the composite map (2) is factorial. Therefore, the decomposition $p_1 \times p_2$ has property (W).

4 (INFORMATION). If every point of each of the cellular spaces X_1 and X_2 has a neighborhood which intersects only a countable family of cells, then $X_1 \times_c X_2 = X_1 \times X_2$; see [6] for a proof.

Attaching

5. Consider two cellular spaces X_1 and X_2, a subspace C of X_1, and a cellular map $\varphi: C \to X_2$. According to 1.2.4.8, $X_2 \cup_\varphi X_1$ is a well-defined topological space, while 2.3 and 1.2.4.9 imply that $X_2 \cup_\varphi X_1$ is normal. Now decompose $X_2 \cup_\varphi X_1$ into the sets $imm_1 e_1$ and $imm_2 e_2$, where e_1 and e_2 run over the cells in $X_1 \smallsetminus C$ and X_2, respectively, and put $\dim(imm_1 e_1) = \dim e_1$ and $\dim(imm_2 e_2) = \dim e_2$. This is a cellular decomposition: as a characteristic map for $imm_i e_i$ one may take the composition of an arbitrary characteristic map cha_{e_i} with imm_i. Clearly, the only cells that the closure of the cell $imm_1 e_1$ intersects are either $imm_1 \varepsilon_1$, where ε_1 is a cell in X_1 intersecting $Cl\, e_1$, or $imm_2 \varepsilon_2$, where ε_2 is a cell in X_2

intersecting $\varphi(Cl\, e_1 \cap C)$. Moreover, we see that $Cl\, imm_2 e_2$ intersects only the cells $imm_2 \varepsilon_2$, where ε_2 is a cell in X_2 intersecting $Cl\, e_2$. Consequently, our decomposition has property (C). To see that it has property (W) too, let F be a subset of $X_2 \cup_\varphi X_1$ having closed intersections with all the cells in $X_2 \cup_\varphi X_1$. The equality

$$imm_2^{-1}(F) \cap Cl\, e_2 = imm_2^{-1}(F \cap Cl(imm_2 e_2))$$

shows that $imm_2^{-1}(F)$ is closed, and now the equality

$$imm_1^{-1}(F) \cap Cl\, e_1 = \begin{cases} \varphi^{-1}(imm_2^{-1}(F) \cap \varphi(Cl\, e_1)), & \text{if } e_1 \subset C, \\[2ex] imm_1^{-1}(F \cap Cl(imm_1 e_1)), & \text{if } e_1 \subset X_1 \smallsetminus C, \end{cases}$$

proves that $imm_1^{-1}(F)$ is closed too. We conclude that the space $X_2 \cup_\varphi X_1$ is cellular. It is immediate that $imm_2(X_2)$ is a subspace of $X_2 \cup_\varphi X_1$, that imm_2 is a cellular embedding, and that the map imm_1 is cellular.

If $X_2 = D^0$, then $\varphi: C \to X_2$ is a cellular map for any cellular pair (X_1, C), and $X_2 \cup_\varphi X_1 = X_1/C$. Thus, the previous definition implies that the quotient space of a cellular space by a subspace is cellular.

Limits

6. Suppose that $X_0, X_1, \ldots$ are cellular spaces and $\phi_0: X_0 \to X_1$, $\phi_1: X_1 \to X_2, \ldots$ are cellular embeddings. By 1.2.4.4, the limit $\lim(X_k, \phi_k)$ is a well-defined topological space, which is also normal (see 2.3 and 1.2.4.6). Now consider the decomposition of $\lim(X_k, \phi_k)$ into the sets $imm_k e_k$, where e_k is a cell in $X_k \smallsetminus \phi_{k-1}(X_{k-1})$, $k = 0, 1, \ldots$, and put $dim(imm_k e_k) = dim\, e_k$. If we take the composition of an arbitrary characteristic map cha_{e_k} with imm_k as a characteristic map for the cell $imm_k e_k$, we see that this decomposition is cellular. Since it obviously satisfies conditions (C) and (W), $\lim(X_k, \phi_k)$ becomes a cellular space, and imm_k become cellular embeddings.

Notice that this definition of the limit includes as a special case the inductive process of gluing a cellular space from balls that we discussed in subsection 2.

More Special Constructions

7. Since decomposing the segment I into the cells 0, 1, and $\mathrm{Int}\,I$ makes I into a finite cellular space, the cylinder $X \times I$ is cellular for any cellular space X; see 2 and 3. The bases of $X \times I$ are cellular subspaces (in the sense of 1.9); hence when we pass to the quotient space $\mathrm{con}\,X$ of $X \times I$, and then to the quotient space $\mathrm{su}\,X$ of $\mathrm{con}\,X$, we find ourselves in the situation covered by the construction 5. Therefore, the cone and the suspension over a cellular space are also cellular spaces.

If $f\colon X_1 \to X_2$ is a cellular map, then the attaching processes which transform $X_1 \times I$ into $\mathrm{Cyl}\,f$, and $\mathrm{con}\,X_1$ into $\mathrm{Con}\,f$ fall again into the category described in 5. Therefore, the mapping cylinder and the mapping cone of a cellular map are cellular spaces.

8. The <u>cellular join</u> $X_1 *_c X_2$ of two cellular spaces X_1 and X_2 is defined as

$$X_1 *_c X_2 = (X_1 \sqcup X_2) \cup_\varphi [(X_1 \times_c X_2) \times I],$$

where $\varphi\colon [(X_1 \times_c X_2) \times 0] \cup [(X_1 \times_c X_2) \times 1] \to X_1 \sqcup X_2$ is given by $\varphi(x_1,x_2,0) = \mathrm{in}_1(x_1)$, $\varphi(x_1,x_2,1) = \mathrm{in}_2(x_2)$; cf. 1.2.6.3. Since φ is cellular, the space $X_1 *_c X_2$ is cellular.

According to 3, when X_1 is locally finite, $X_1 *_c X_2$ is topologically the same as $X_1 * X_2$. In general, the cellular decomposition of $X_1 *_c X_2$ is cellular for $X_1 * X_2$ too, and so the cellular weakening of the topology of $X_1 * X_2$ yields $X_1 *_c X_2$. However, this process does not affect the topology of the compact subsets of $X_1 * X_2$; cf. 2.

The Case of Pointed Spaces

9. Suppose that X is a cellular space and x_0 is a 0-cell that we take as a base point. The cone $\mathrm{con}(X,x_0)$ and the suspension $\mathrm{su}(X,x_0)$ are quotients of $\mathrm{con}\,X$ and $\mathrm{su}\,X$ by subspaces, and as such they are cellular spaces. Similarly, the bouquet of a family of cellular spaces with 0-cells as base points is the quotient of the sum of this family by a subspace, and hence is a cellular space.

Finally, we define the <u>cellular tensor product</u> and the <u>cellular join</u> of the cellular spaces X_1 and X_2 with the 0-cells x_1 and x_2 taken as base points, as the quotient spaces

$$(X_1 \times_c X_2)/[(X_1 \times x_2) \cup (x_1 \times X_2)] \quad \text{and} \quad (X_1 *_c X_2)/(x_1 * x_2),$$

respectively. These are cellular spaces, denoted by $(X_1,x_1) \otimes_c (X_2,x_2)$ and $(X_1,x_1) *_c (X_2,x_2)$. If X_1 is locally finite, then they are identical with $(X_1,x_1) \otimes (X_2,x_2)$ and $(X_1,x_1) * (X_2,x_2)$ as topological spaces. In the general case, the cellular decompositions of $(X_1,x_1) \otimes_c (X_2,x_2)$ and $(X_1,x_1) *_c (X_2,x_2)$ are cellular for $(X_1,x_1) \otimes (X_2,x_2)$ and $(X_1,x_1) * (X_2,x_2)$ too. Thus, $(X_1,x_1) \otimes_c (X_2,x_2)$ and $(X_1,x_1) *_c (X_2,x_2)$ arise from the cellular weakening of the topologies of $(X_1,x_1) \otimes (X_2,x_2)$ and $(X_1,x_1) * (X_2,x_2)$, respectively, and it is clear that this process does not affect the topology of the compact subsets of $(X_1,x_1) \otimes (X_2,x_2)$ and $(X_1,x_1) * (X_2,x_2)$.

6. Exercises

1. Show that given an arbitrary cellular space X and an arbitrary point $x \in X$, there exists a cellular space Y together with a cellular homeomorphism $f: X \to Y$ such that $f(x) \in ske_0 Y$.

2. Show that the sphere S^∞ and the ball D^∞ are homeomorphic to cellular spaces.

3. Show that every connected, locally finite cellular space can be topologically embedded in $\mathbb{R}^\infty$.

4. Show that every connected, finite dimensional, locally finite cellular space can be embedded in $\mathbb{R}^q$, for large enough q.

5. Show that every finite cellular space admits a cellular embedding in a cellular space homeomorphic to D^q, for large enough q.

INFORMATION. Every finite celular space of dimension n can be embedded in a cellular space homeomorphic to D^{2n+1}.

6. Consider the bouquet $B = V_{t \in \mathbb{R}}(I_t = I, 0)$ as a cellular space (see 5.9) and show that the map $id: B \times_c B \to B \times B$ is not a homeomorphism.

§2. SIMPLICIAL SPACES

1. Euclidean Simplices

1. Let A be a subset of $\mathbb{R}^n$ consisting of $r+1$ points $(r \geqslant 0)$ which are not contained in any $(r-1)$-dimensional plane. The convex hull of A (i.e., the smallest convex set containing A) is called the Euclidean simplex spanned by A, and is denoted by $\mathrm{Esi}\,A$. The points of A are the vertices of the simplex $\mathrm{Esi}\,A$, and the number r is its dimension. $\mathrm{Esi}\,A$ is also called a Euclidean r-simplex.

Obviously, a point of $\mathrm{Esi}\,A$ is a vertex if and only if $\mathrm{Esi}\,A$ contains no nondegenerate segment whose midpoint falls on the given point. Therefore, the set A is uniquely determined by $\mathrm{Esi}\,A$.

Every simplex spanned by a subset of A is called a face of the simplex $\mathrm{Esi}\,A$. It is clear that $\mathrm{Esi}\,A_1 \cap \mathrm{Esi}\,A_2 = \mathrm{Esi}(A_1 \cap A_2)$, for any $A_1, A_2 \subset A$.

Two faces spanned by complementary subsets A_1 and A_2 of A are said to be opposite. In this case, the formula

$$\mathrm{pr}(x_1, x_2, t) \mapsto (1-t)x_1 + tx_2 \quad (x_1 \in \mathrm{Esi}\,A_1 ,\ \ x_2 \in \mathrm{Esi}\,A_2 ,\ \ t \in I)$$

defines a homeomorphism of the join $\mathrm{Esi}\,A_1 * \mathrm{Esi}\,A_2$ onto $\mathrm{Esi}\,A$. Thus, every Euclidean simplex is canonically homeomorphic to the join of any of the pairs of its opposite faces.

Since D^r is canonically homeomorphic to any join $D^p * D^q$ with $p+q = r-1$ (see 1.2.6.9), a trivial induction proves that both the spaces $\mathrm{Esi}\,A$ and D^r are homeomorphic to a join of $r+1$ points. We conclude that every Euclidean r-simplex is homeomorphic to D^r.

It is clear that the boundary of the simplex $\mathrm{Esi}\,A$ in the r-plane that it determines is precisely the union of its $(r-1)$-faces. Usually, this boundary and its complement in $\mathrm{Esi}\,A$ are simply referred to as the boundary and the interior of the simplex $\mathrm{Esi}\,A$.

2. We may equivalently describe the simplex $\mathrm{Esi}\,A$ as the set of all sums $\sum_{a \in A} t_a a$, where $t_a \geqslant 0$ and $\sum_{a \in A} t_a = 1$. Since there is no $(r-1)$-plane containing A, the numbers t_a are determined uniquely for any point $x = \sum_{a \in A} t_a a$; t_a is called the a-th bary -

centric coordinates of x and is denoted by $\mathrm{bar}_a(x)$.

Obviously, a face Esi B of the simplex Esi A is defined in the barycentric coordinates of Esi A by the equations $\mathrm{bar}_a(x) = 0$ for $a \in A \smallsetminus B$. Moreover, if $x \in$ Esi B , then the coordinates $\mathrm{bar}_a(x)$ computed in Esi A and Esi B coincide for all $a \in B$.

The point of Esi A having all barycentric coordinates equal, i.e., equal to $1/(r+1)$, is the <u>center</u> of the simplex Esi A .

3. A map Esi A $\rightarrow$ Esi B is called <u>simplicial</u> if it is affine and takes A into B . It is clear that such a map takes each face of Esi A simplicially into a face of Esi B , and takes the interior of Esi A onto the interior of the simplex which is its image.

Obviously, every map A $\rightarrow$ B extends uniquely to a simplicial map Esi A $\rightarrow$ Esi B . If the given map A $\rightarrow$ B is injective (invertible), then its simplicial extension Esi A $\rightarrow$ Esi B is an embedding (respectively, a homeomorphism).

4. Esi A is said to be an <u>ordered</u> simplex if the set A is ordered. Since the subsets of an ordered set inherit a natural order, all the faces of an ordered simplex are ordered simplices.

If Esi A and Esi B are ordered r-simplices, then the orders of A and B define an invertible map A $\rightarrow$ B , and hence a simplicial homeomorphism Esi A $\rightarrow$ Esi B . Consequently, all ordered Euclidean simplices of the same dimension are canonically simplicial homeomorphic.

5. The simplex spanned by the points $\mathrm{ort}_1, \ldots, \mathrm{ort}_r$ of $\mathbb{R}^{r+1}$ is called the <u>unit r-simplex</u> and is denoted by T^r . This simplex is notable due to the fact that its barycentric coordinates are the usual coordinates in $\mathbb{R}^{r+1}$. The given order of its vertices transforms T^r into an ordered simplex, and thus every ordered Euclidean r-simplex is canonically simplicial homeomorphic to T^r .

Note that given an ordered simplex Esi A , the homeomorphism Esi A $\rightarrow D^r$ discussed in 1 is now canonical. The canonical homeomorphism $T^r \rightarrow D^r$ and its inverse are denoted by TD and DT, respectively. That TD maps the boundary (the interior) of T^r onto S^{r-1} (respectively, onto Int D^r) is plain.

Topological Simplices

6. A topological space X is an <u>ordered topological simplex of dimension</u> r (or an <u>ordered topological r-simplex</u>) if there exists

a homeomorphism $T^r \to X$; this is called a <u>characteristic homeomorphism</u> <u>of the simplex</u> X, while X is sometimes referred to as the <u>support</u> of the simplex. For example, all ordered Euclidean r-simplices and the ball D^r are ordered topological r-simplices; see 5.

The standard way to detroy an order is to introduce simultaneously all possible orders. Accordingly, we say that the topological space X is a <u>topological simplex of dimension</u> r (or a <u>topological r-simplex</u>) if there are given (r+1)! homeomorphisms $T^r \to X^r$, which can be transformed into each other by simplicial homeomorphisms $T^r \to T^r$. The terms <u>characteristic homeomorphism</u> and <u>support</u> are employed in this situation too; however, now we have at our disposal (r+1)! equally rightful characteristic homeomorphisms.

If X is a topological r-simplex (an ordered topological r-simplex), and Y is a topological space, then every homeomorphism $X \to Y$ transforms Y into a topological r-simplex (respectively, into an ordered topological r-simplex). Consequently, every homeomorphic image of a Euclidean r-simplex (ordered Euclidean r-simplex) is a topological r-simplex (respectively, an ordered topological r-simplex).

The vertices, faces, boundary, interior, barycentric coordinates, center, and simplicial maps are defined in an obvious fashion for topological simplices. The faces of a topological simplex (ordered topological simplex) are topological simplices (respectively, ordered topological simplices). As with a Euclidean simplex, a topological simplex becomes an ordered one as soon as we fix an order of its vertices.

2. <u>Simplicial Spaces and Simplicial Maps</u>

1. A <u>triangulation</u> of a set X is a cover Δ of X by topological simplices such that:

(i) every face of an arbitrary simplex in Δ is again a simplex in Δ;

(ii) if a simplex in Δ is contained in another simplex of Δ, then the first is a face of the second;

(iii) the intersection of the supports of two overlaping simplices of Δ is again the support of a simplex in Δ.

A set X endowed with a triangulation is known as a <u>simplicial space</u>; the simplices of the triangulation are called <u>simplices of the space</u>, and the 0-simplices are its <u>vertices</u>. The smallest simplex in the triangulation which contains a given point x $\in$ X is denoted by si x .

According to 1.2.4.3, a triangulation transforms the given set into a topological space, and 1.2.4.1 shows that the supports of the simplices of the triangulation yield a fundamental cover of this space. Since the intersection of two simplices in the triangulation is closed in each of them, the simplices in the triangulation keep the same topology when considered as subspaces of this topological space (see 1.2.4.2).

Let a be a vertex of the simplicial space X. Then the a-th barycentric coordinate $\mathrm{bar}_a(x)$ is well defined for any point x belonging to any simplex which has a as one of its vertices (see 1.2 and 1.6), and we obtain a continuous function $\mathrm{bar}_a: X \to \mathbb{R}$ if we set $\mathrm{bar}_a(x) = 0$ for those points x $\in$ X contained in simplices which do not have a as a vertex. bar_a is called the <u>a-th barycentric function.</u> Given two arbitrary distinct points x,y $\in$ X, there obviously is a vertex a such that $\mathrm{bar}_a(x) \neq \mathrm{bar}_a(y)$. Consequently, <u>every simplicial space is Hausdorff.</u>

When a set X endowed with a triangulation already has a topology, it is useful to find conditions ensuring that the topology defined by the triangulation is identical with the initial one. We have an immediate necessary and sufficient condition: the topology of each simplex in the triangulation coincides with the topology induced by the initial topology of X, and the cover of X by the supports of these simplices is fundamental in the initial topology. If this condition is satisfied, then the given triangulation is said to be a triangulation of the initial topological space X. Example: the cover of a topological simplex by all its faces is a triangulation of this simplex.

A simplicial space is <u>ordered</u> if its simplices are ordered in such a way that the orders of the faces of any simplex agree with the order of the simplex itself. In particular, this holds whenever the order of the simplices is induced by some order on the set of all vertices of the given space, which incidentally shows that a simplicial space can be always ordered.

2. We shall presently describe a fundamental class of simplicial spaces. Given an arbitrary nonempty set A, we let Si A denote the set of all nonnegative, finitely supported functions $\phi: A \to \mathbb{R}$ such that $\sum_{a \in A} \phi(a) = 1$. If $B \subset A$, then we identify Si B with the subset of Si A consisting of all functions $\phi \in$ Si A such that $\phi(x) = 0$ for x $\in$ A $\smallsetminus$ B. If A is finite and has r+1 elements, then Si A is obviously a topological simplex: indeed, Si A

is a subset of the $(r+1)$-dimensional Euclidean space of all functions
$A \to \mathbb{R}$. Moreover, corresponding to the $(r+1)!$ orders on A there are
$(r+1)!$ homeomorphisms $T^r \to Si\,A$, each transforming the point
$(x_1, \ldots, x_{r+1})$ into a function taking the values $x_1, \ldots, x_{r+1}$; one may
transform one homeomorphism into another by composition with a simplicial
homeomorphism $T^r \to T^r$. In the general case, $Si\,A$ is covered by the
topological simplices $Si\,B$ corresponding to all finite subsets B of
A, and it is clear that this yields a triangulation of $Si\,A$. $Si\,A$ is
therefore a simplicial space, and we call it the <u>simplex spanned by</u> A.
Ordering $Si\,A$ is equivalent to ordering the set A.

3. The interiors of the simplices of a simplicial space X
constitute a decomposition of the set X. If we define the dimension of
the interior e of the simplex s by $\dim e = \dim s$, and take as a
characteristic map for e the composition

$$D^{\dim s} \xrightarrow{\ DT\ } T^{\dim s} \xrightarrow{\ \phi\ } s \xrightarrow{\ in\ } X\ , \tag{1}$$

where ϕ is any characteristic homeomorphism for the simplex s, then
the above decomposition becomes cellular. Since conditions (C) and (W)
are clearly satisfied in this situation, and we already know that every
simplicial space is Hausdorff, we see that this cellular decomposition
transforms X into a cellular space. Thus, <u>every simplicial space,
decomposed into the interiors of its simplices, is a cellular space.</u>

Since an r-simplex has $(r+1)!$ characteristic homeomorphisms,
formula (1) distinguishes $(r+1)!$ privileged maps in the family of all
maps that are characteristic for an r-cell in a simplicial space; we
call them <u>simplicial characteristic maps</u> and note that they are
topological embeddings. Fixing a simplicial characteristic map is
equivalent to fixing an order of the simplex s; hence every ordered
simplicial space is canonically rigged.

It is clear that the skeleton $ske_r X$ of a simplicial space
X is simply the union of all its simplices of dimension $\leqslant r$, and that
$\dim(ske_r X) = r$ for $r \leqslant \dim X$. In particular, $ske_0 X$ is the set of
vertices of X. Of course, X is finite if and only if $ske_0 X$ is
finite as a set. A simplicial space is locally finite if and only if
each of its vertices is contained in only a finite number of simplices
(see 1.1.7).

Subspaces

4. A <u>subspace</u> of a simplicial space is any subset which is a union of whole simplices. Every subspace has a natural triangulation, and hence is a simplicial space. The subspaces of an ordered simplicial space are also ordered simplicial spaces.

Obviously, a subset of a simplicial space X is a subspace of X if and only if it is a cellular subspace of the cellular space X.

A subspace of a simplicial space X is <u>complete</u> if its intersection with the support of any simplex of X is either the support of some simplex of X or empty. An equivalent formulation: a subspace is complete if it contains, along with the vertices of a simplex, the simplex itself.

Of course, the simplices of a simplicial space are complete subspaces. A subspace of $Si\,A$ is complete if and only if it is of the form $Si\,B$, where $B \subset A$.

Simplicial Maps

5. A map $X \to Y$, where X and Y are simplicial spaces, is called <u>simplicial</u> if it transforms every simplex of X simplicially into a simplex of Y. It is clear that such a map is also cellular and maps X onto a subspace of Y.

The following facts are also immediate.

An invertible simplicial map is a homeomorphism, and its inverse is also simplicial. Every injective simplicial map is a topological embedding. A simplicial map $f: X \to Y$ is uniquely defined by the map $ab\,f : ske_0 X \to ske_0 Y$ from the set of vertices of X into the set of vertices of Y. A map $ske_0 X \to ske_0 Y$ extends to a simplicial map $X \to Y$ if and only if it carries the vertices of each simplex of X into the vertices of a simplex of Y. A simplicial map $f: X \to Y$ is injective (invertible) if and only if $ab\,f : ske_0 X \to ske_0 Y$ is injective (respectively, invertible).

Two simplicial spaces which can be transformed one into another by a simplicial homeomorphism are said to be <u>simplicial homeomorphic.</u>

6. A simplicial map $f: X \to Y$, where X and Y are ordered simplicial spaces, is <u>monotone</u> if $f(a) \leqslant f(b)$ for any pair of vertices, a and b, of X which belong to the same simplex and satisfy $a \leqslant b$.

Every simplicial map between simplicial spaces can be made monotone by suitably ordering the spaces. Moreover, if X and Y are

simplicial spaces and Y is ordered, then one can transform a given
simplicial map $f: X \to Y$ into a monotone one my suitably ordering X;
indeed, it suffices to order arbitrarily the preimage of each vertex of
Y, and then order the simplices of X by the rule: $a < b$ whenever
$f(a) < f(b)$ or $f(a) = f(b)$ and $a < b$ in $f^{-1}(f(a))$.

3. Simplicial Schemes

1. A _simplicial scheme_ (or _schema_) is a pair (M,S), where
M is a set and S is a cover of M by finite subsets, such that S
contains, along with each set $A \in S$, all the parts of A.

A _map of the simplicial scheme_ (M,S) _into the simplicial
scheme_ (M',S') is a pair of maps, $\phi: M \to M'$ and $\Phi: S \to S'$, such
that $\Phi(A) = \phi(A)$ for all $A \in S$. The last condition shows that the
map (ϕ,Φ) of (M,S) into (M',S') is actually uniquely determined
by ϕ. Obviously, $\phi: M \to M'$ defines a map of the scheme (M,S) into
the scheme (M',S') if and only if $\phi(A) \in S'$ for all $A \in S$. If
ϕ and Φ are invertible, i.e., ϕ is invertible and $\Phi(S) = S'$,
then the map (ϕ,Φ) is called an _isomorphism._ Two simplicial schemes
which can be related by an isomorphism are _isomorphic._

A simplicial scheme (M,S) is a _subscheme_ of the simplicial
scheme (M',S') if $M \subset M'$ and $S \subset S'$. The subscheme (M,S) is
complete if $A \in S'$ and $A \subset M$ imply $A \in S$.

2. The simplicial scheme given by the skeleton $\mathrm{ske}_0 X$ of a
simplicial space X and the cover of $\mathrm{ske}_0 X$ by the 0-skeletons of the
simplices of X is termed the _scheme of the space_ X and is denoted by
sch X . For example, the scheme of Si A (see 2.2) consists of the set
A and of the cover of A by all its finite subsets.

The map $\mathrm{ab}\, f : \mathrm{ske}_0 X \to \mathrm{ske}_0 X'$ induced by a simplicial map
$f: X \to X'$ takes the 0-skeleton of each simplex of X into the
0-skeleton of a simplex of X'. Hence it defines a map of sch X into
sch X' , called the _scheme of the map_ f and denoted by sch f . The
discussion in 2.5 implies that a simplicial map is uniquely determined
by its scheme, that every map of sch X into sch X' is the scheme of
some simplicial map $X \to X'$, for any simplicial spaces X and X',
and that a simplicial map is invertible if and only if its scheme is an
isomorphism. In particular, _two simplicial spaces_ X _and_ X' _are
simplicial homeomorphic if and only if their schemes_ sch X _and_ sch X'
are isomorphic.

3. If X is a subspace of the simplicial space X', then

sch X is a subscheme of sch X' , and sch X is complete if and only
if X is complete. Moreover, it is clear that every subscheme of
sch X' is the scheme of a subspace of X'.

In particular, let (M,S) be an arbitrary simplicial scheme,
and consider the simplex Si M . Obviously, (M,S) is a subscheme of
sch Si M, and so (M,S) is the scheme of a subspace of Si M . Thus,
<u>every simplicial scheme is the scheme of a simplicial space.</u> Moreover,
given an arbitrary simplicial space X, we may take (M,S) to be the
scheme of X and conclude that every simplicial space X can be
simplicially embedded in Si ske_0X .

4. A simplicial scheme (M,S) is <u>ordered</u> if the sets of S
are ordered and the order of each set $A \in S$ is compatible with the
orders of the subsets of A. A map (ϕ,Φ) between ordered simplicial
schemes (M,S) and (M',S') is <u>monotone</u> if $\phi(a) < \phi(b)$ whenever
a < b. Therefore, ordering the scheme of a simplicial space is
equivalent to ordering the space itself, and the scheme of a simplicial
map between two ordered simplicial spaces is monotone if and only if the
map itself is monotone.

4. <u>Polyhedra</u>

1. A <u>polyhedron</u> is a subset of Euclidean space which admits
a finite triangulation by Euclidean simplices. Of course, the simplest
polyhedra are the Euclidean simplices.

A subspace of a polyhedron is obviously a polyhedron.

Now any simplicial space can be simplicially embedded in the
simplex spanned by its 0-skeleton (see 3.3), and when the initial space
is finite and has q vertices, this skeleton is simplicial homeomorphic
to T^{q-1}. Therefore, every finite simplicial space admits a simplicial
embedding in a Euclidean simplex with the same number of vertices.
Hence every finite simplicial space is simplicial homeomorphic to a
polyhedron.

2. <u>Every finite n-dimensional simplicial space is simplicial</u>
<u>homeomorphic to a polyhedron contained in</u> $\mathbb{R}^{2n+1}$.

PROOF. Since every finite n-dimensional simplicial space can
be simplicially embedded in $\text{ske}_n T^q$ for q large enough (see 1), it
suffices to construct, for arbitrarily given q and n, a linear
mapping $f: \mathbb{R}^{q+1} \to \mathbb{R}^{2n+1}$ which is injective on $\text{ske}_n T^q$.

If $q \leqslant 2n$, we may take f to be the inclusion $\mathbb{R}^{q+1} \to \mathbb{R}^{2n+1}$.

If $q > 2n$, define f by

$$f(\{x_j\}_{j=1}^{q+1}) = \{\textstyle\sum_{j=1}^{q+1} j^i x_j\}_{i=1}^{2n+1} \ .$$

All that remains is to verify that if $x = (x_1,\ldots,x_{q+1})$ and $x' = (x_1',\ldots,x_{q+1}')$ belong to $\text{ske}_n T^q$ and $f(x) = f(x')$, then $x = x'$. Since each of the points x and x' lies in an n-dimensional face of T^q, at most $n+1$ of the numbers $x_1,\ldots,x_{q+1}$, and $n+1$ of the numbers $x_1',\ldots,x_{q+1}'$, are different from zero. Consequently, no more that $2n+2$ numbers $x_1-x_1',\ldots,x_{q+1}-x_{q+1}'$ are different from zero, i.e., there are positive integers $j_1,\ldots,j_{2n+2}$ such that $j_1 < \ldots < j_{2n+2} \leqslant 2q+1$ and $x_j = x_j'$ for $j \neq j_1,\ldots,j_{2n+2}$. Since

$$\textstyle\sum_{j=1}^{q+1} x_j = \sum_{j=1}^{q+1} x_j' \ (= 1) \quad \text{and} \quad \sum_{j=1}^{q+1} j^i x_j = \sum_{j=1}^{q+1} j^i x_j'$$

for $i = 1,\ldots,2n+1$, we have

$$\textstyle\sum_{r=1}^{2n+2} j_r^i (x_{j_r} - x_{j_r}') = 0, \quad i = 0,\ldots,2n+1.$$

The determinant of the matrix $\{j_r^i\}_{i=0,r=1}^{i=2n+1,r=2n+2}$ does not vanish, and so $x_{j_r} = x_{j_r}'$ for $r = 1,\ldots,2n+2$, and finally $x = x'$.

3 (INFORMATION). For any n there are n-dimensional polyhedra which cannot be topologically embedded in $\mathbb{R}^{2n}$. An example is $\text{ske}_n T^{2n+2}$; see [10] for a proof.

5. Simplicial Constructions

1. Many of the topological and cellular constructions described in § 1.2 and Subsection 1.5 can be replaced by parallel constructions which produce simplicial spaces out of simplicial ones. The simplest examples are the $\bigsqcup$ and $\vee$ operations: a sum of simplicial spaces and a bouquet of pointed simplicial spaces with vertices as base points are obviously simplicial spaces. There are also more elaborate constructions, the more important ones being discussed below. The main one is the barycentric subdivision construction, which refines triangulations and has no analogs in § 1.2 and Subsection 1.5.

2 (LEMMA). <u>Let</u> Γ <u>be a fundamental cover of the topological space</u> X <u>by triangulated subspaces. Suppose that for any</u> $A,B \in \Gamma$ <u>the intersection</u> $A \cap B$ <u>is a complete subspace of both</u> A <u>and</u> B <u>(considered as simplicial spaces) and inherits from</u> A <u>and</u> B <u>the</u>

same triangulation. Then there exists a unique triangulation of X relative to which the elements of Γ become simplicial subspaces.

This triangulation of X is simply the union of the triangulations of the elements of Γ. One may check directly that this union satisfies conditions (i), (ii), and (iii) in 2.1 [the completeness of the intersections A $\cap$ B is necessary for (iii)]. Uniqueness is also evident.

Barycentric Subdivision

3. The construction below produces a new simplicial space, ba X , from any simplicial space X, such that ba X is identical to X as a topological space, but has a finer triangulation, called the barycentric subdivision of the initial triangulation.

Consider first a Euclidean simplex X. For an arbitrary numeration $a_0,\ldots,a_r$ of the vertices of X, form the set

$$\{x \in X \mid \operatorname{bar}_{a_0}(x) \leqslant \operatorname{bar}_{a_1}(x) \leqslant \ldots \leqslant \operatorname{bar}_{a_r}(x)\}. \tag{2}$$

It is readily seen that (2) is the Euclidean simplex whose vertices are the centers of the simplices $\operatorname{Esi} A_0,\ldots,\operatorname{Esi} A_r$, where $A_i = \{a_0,\ldots,a_i\}$. Furthermore, the simplices of the form (2) corresponding to all possible numerations of the vertices of X and their faces clearly yield a triangulation of X. This is precisely the barycentric subdivision of the standard triangulation of the simplex X, and it transforms X into ba X. An obvious property of this construction is that the inclusion ba X $\rightarrow$ ba X' is a simplicial embedding whenever X is a face of the simplex X'.

Now if X is a topological simplex, we define the barycentric subdivision of its standard triangulation as the image of the barycentric subdivision of the standard triangulation of the unit simplex T^r, r = dim X , under a simplicial homeomorphism $T^r \rightarrow X$. This is clearly a correct definition, i.e., the triangulation of X thus obtained does not depend on the choice of the simplicial homeomorphism $T^r \rightarrow X$ among the (r+1)! available ones.

Finally, let X be an arbitrary simplicial space, and consider the cover of X by its simplices, each subdivided as above. It is easy to verify that this cover satisfies the conditions of Lemma 2, and hence we obtain a new triangulation of X, which is precisely the barycentric subdivision of the initial triangulation of X.

We note that the barycentric subdivision transforms a finite (locally finite) simplicial space into a finite (respectively, locally finite) one. Moreover, if X is a polyhedron, then so is $\operatorname{ba} X$.

4. The set of vertices of the space $\operatorname{ba} X$ equals exactly the set of centers of the simplices of X. The centers of the simplices $s_1, \ldots, s_m$ of X are the vertices of a simplex of $\operatorname{ba} X$ if and only if $s_1, \ldots, s_m$ can be reindexed to form an increasing sequence. This observation enables us to give a concise description of the barycentric subdivision in the language of schemes: if $\operatorname{sch} X = (M, S)$, then $\operatorname{sch} \operatorname{ba} X = (S, \operatorname{ba} S)$, where $\operatorname{ba} S$ is precisely the collection of those finite parts of S that can be ordered by inclusion. At the same time, we obtain a canonical order of $\operatorname{ba} X$: if $a, a' \in \operatorname{ske}_0 \operatorname{ba} X$, then $a < a'$ whenever the simplex (of X) with center a is contained in the simplex with center a'.

In particular, the above description of $\operatorname{sch} \operatorname{ba} X$ shows that $\operatorname{ba} X$ is a complete subspace of $\operatorname{ba} X'$ whenever X is a subspace of X'. Indeed, $\operatorname{sch} \operatorname{ba} X$ is clearly a complete subscheme of $\operatorname{sch} \operatorname{ba} X'$.

In general, given a simplicial map $f: X \to X'$, the map $f: \operatorname{ba} X \to \operatorname{ba} X'$ is not simplicial (the simplest example: take $X = T^2$, $X' = T^1$, $f(\operatorname{ort}_1) = \operatorname{ort}_1$, $f(\operatorname{ort}_2) = f(\operatorname{ort}_3) = \operatorname{ort}_2$). However, the map $\operatorname{sch} f: \operatorname{sch} X \to \operatorname{sch} X'$ naturally induces a map $\operatorname{sch} \operatorname{ba} X \to \operatorname{sch} \operatorname{ba} X'$, and hence a simplicial map $\operatorname{ba} X \to \operatorname{ba} X'$. The latter is denoted by $\operatorname{ba} f$ and is clearly always monotone.

5. <u>If</u> X <u>is a polyhedron, then the maximal diameter of the simplices of the polyhedron</u> $\operatorname{ba} X$ <u>does not exceed the maximal diameter of the simplices of</u> X <u>times</u> $n/(n+1)$, <u>where</u> $n = \dim X$.

It is enough to show that if X is the Euclidean simplex with vertices $a_0, \ldots, a_r$, then the diameter of the simplex (2) is no bigger than $[r/(r+1)] \operatorname{diam} X$. Consider the part X' of X defined by the inequality $\operatorname{bar}_{a_r}(x) \geqslant r/(r+1)$. X' is the Euclidean simplex obtained by contracting X towards the vertex a_r by a factor of $r/(r+1)$. Consequently, $\operatorname{diam} X' \leqslant [r/(r+1)] \operatorname{diam} X$, and we finally note that X' contains the simplex (2).

6 (COROLLARY). <u>For any polyhedron</u> X <u>and any</u> $\varepsilon > 0$ <u>there is a positive integer</u> m <u>such that every simplex of the polyhedron</u> $\operatorname{ba}^m X$ <u>has diameter</u> $< \varepsilon$.

Simplicial Products

7. If X_1 and X_2 are simplicial spaces with $\dim X_1 > 0$ and $\dim X_2 > 0$, then it is readily seen that their cellular product $X_1 \times_c X_2$ does not admit a triangulation such that the interiors of its simplices are products of interiors of simplices of X_1 and X_2. However, we shall presently show that $X_1 \times_c X_2$ admits triangulations, and we shall construct a canonical triangulation when X_1 and X_2 are ordered. This construction produces a simplicial space out of $X_1 \times_c X_2$, called the <u>simplicial product</u> of X_1 and X_2, and denoted by $X_1 \times_s X_2$.

To begin with, let X_1 be the Euclidean simplex in $\mathbb{R}^m$ with vertices $a_0, \ldots, a_q$, and let X_2 be the Euclidean simplex in $\mathbb{R}^n$ with vertices $b_0, \ldots, b_r$. Set for $x_1 \in X_1$ and $x_2 \in X_2$:

$$\alpha_i(x_1) = \sum_{k=0}^{i} \mathrm{bar}_{a_k}(x_1), \quad \beta_j(x_2) = \sum_{l=0}^{j} \mathrm{bar}_{b_l}(x_2),$$

and arrange the numbers $\alpha_0(x_1), \ldots, \alpha_{q-1}(x_1), \beta_0(x_2), \ldots, \beta_{r-1}(x_2)$ in a nondecreasing sequence $\gamma_1(x_1, x_2), \ldots, \gamma_{q+r}(x_1, x_2)$. Further, let M_{qr} denote the collection of subsets with q elements of $\{1, \ldots, q+r\}$, and let $s(\mu)$, where $\mu \in M_{qr}$, denote the set of all points $(x_1, x_2) \in X_1 \times X_2$ such that each of the numbers $\gamma_p(x_1, x_2)$, $p \in \mu$, is equal to one of the numbers $\gamma_0(x_1), \ldots, \gamma_{q-1}(x_1)$. One may check directly that there is no $(q+r-1)$-dimensional plane containing the $q+r+1$ points

$$(a_0, b_0), \ldots, (a_0, b_{j_1-1});$$
$$(a_1, b_{j_1-1}), \ldots, (a_1, b_{j_2-2});$$
$$\cdots\cdots\cdots\cdots\cdots\cdots\cdots\cdots\cdots\cdots \quad (3)$$
$$(a_{q-1}, b_{j_q-(q-1)}), \ldots, (a_{q-1}, b_{j_q-q});$$
$$(a_q, b_{j_q-q}), \ldots, (a_q, b_r),$$

where $j_1, \ldots, j_q \in \mu$, $j_1 < \ldots < j_q$. Also, one may verify that

$$\sum_{k=0}^{q} \sum_{l=j_k-k}^{j_{k+1}-(k+1)} [\gamma_{k+l+1}(x_1, x_2) - \gamma_{k+l}(x_1, x_2)](a_k, b_l) = (x_1, x_2),$$

for $(x_1, x_2) \in s(\mu)$, $j_0 = 0$, $j_{q+1} = q+r+1$, $\gamma_0(x_1, x_2) =$ and $\gamma_{q+r+1}(x_1, x_2) = 1$. Moreover,

$$\sum_{k=0}^{q} \sum_{l=j_k-k}^{j_{k+1}-(k+1)} [\gamma_{k+1+1}(x_1,x_2) - \gamma_{k+1}(x_1,x_2)] =$$

$$= \sum_{p=0}^{q+r} [\gamma_{p+1}(x_1,x_2) - \gamma_p(x_1,x_2)] = 1$$

and $\gamma_{p+1}(x_1,x_2) - \gamma_p(x_1,x_2) \geqslant 0$. Consequently, the set $s(\mu)$ is contained in the Euclidean simplex spanned by the points (3), and since $s(\mu)$ is obviously convex and contains all the points (3), it equals this simplex. Now it is clear that the sets $s(\mu)$ cover $X_1 \times X_2$, and that $s(\mu_1) \cap s(\mu_2)$ is just the Euclidean simplex spanned by the vertices common to the simplices $s(\mu_1)$ and $s(\mu_2)$, for any $\mu_1, \mu_2 \in M_{qr}$. As a result, the simplices $s(\mu)$ and their faces constitute a triangulation of the product $X_1 \times X_2$, and this triangulation transforms $X_1 \times X_2$ into $X_1 \times_s X_2$. Moreover, it is readily seen that if X_1 and X_2 are faces of the ordered Euclidean simplices X_1' and X_2', then the inclusion $X_1 \times_s X_2 \to X_1' \times_s X_2'$ is a simplicial embedding.

Next let X_1 and X_2 be ordered topological simplices. To define the simplicial product $X_1 \times_s X_2$, use the previous prescription to triangulate the product of the unit simplices T^q and T^r, with $q = \dim X_1$ and $r = \dim X_2$, and then employ the product $T^q \times T^r \to X_1 \times X_2$ of canonical simplicial homeomorphisms $T^q \to X_1$ and $T^r \to X_2$ to carry this triangulation to $X_1 \times X_2$. With the resulting triangulation, $X_1 \times X_2$ becomes $X_1 \times_s X_2$.

Finally, let X_1 and X_2 be arbitrary ordered simplicial spaces. It is readily verified that the cover of $X_1 \times_c X_2$ by the product $s_1 \times s_2$ of simplices s_1 of X_1 and s_2 of X_2, where $s_1 \times s_2$ is triangulated as above, satisfies the conditions of Lemma 2. This lemma yields a triangulation which transforms $X_1 \times_c X_2$ into the simplicial product $X_1 \times_s X_2$.

We remark that each cell e of $X_1 \times_c X_2$ can be represented as the union of a finite number of cells of $X_1 \times_s X_2$, having dimensions $\leqslant \dim e$. In particular, the map $\mathrm{id}: X_1 \times_c X_2 \to X_1 \times_s X_2$ is cellular.

8. A straightforward corollary of the simplicial product construction is that the product $f_1 \times f_2: X_1 \times_s X_2 \to X_1' \times_s X_2'$ of two monotone simplicial maps, $f_1: X_1 \to X_1'$ and $f_2: X_2 \to X_2'$, is also simplicial. It is also plain that if X_1 and X_2 are subspaces of the ordered simplicial spaces X_1' and X_2', then $X_1 \times_s X_2$ is a subspace

of $X_1' \times_s X_2'$.

Let us conclude with a description of the simplicial product in terms of schemes. Suppose $\operatorname{sch} X_1 = (M_1, S_1)$ and $\operatorname{sch} X_2 = (M_2, S_2)$. Then $\operatorname{sch}(X_1 \times_s X_2) = (M_1 \times M_2, S)$, where S is the collection of sets $A \subset X_1 \times X_2$ such that: (i) $\operatorname{pr}_1(A) \in S_1$, $\operatorname{pr}_2(A) \in S_2$; (ii) if $(a_1, a_2) \in A$, $(a_1', a_2') \in A$, and $a_1 < a_1'$, then $a_2 < a_2'$.

Limits

9. Let $X_0, X_1, X_2, \ldots$ be simplicial spaces, together with simplicial embeddings $\phi_0: X_0 \to X_1$, $\phi_1: X_1 \to X_2, \ldots$. Consider the cover of $\lim(X_k, \phi_k)$ by the sets $\operatorname{imm}_k(s_k)$, where s_k is a simplex of X_k ($k = 0, 1, \ldots$). If we take for a characteristic homeomorphism of $\operatorname{imm}_k(s_k)$ the composition of a characteristic homeomorphism $T^{\dim s_k} \to s_k$ with the homeomorphism $\operatorname{ab} \operatorname{imm}_k : s_k \to \operatorname{imm}_k(s_k)$, then $\operatorname{imm}_k(s_k)$ becomes a topological simplex. It is clear that in this way the cover $\{\operatorname{imm}_k(s_k)\}$ becomes a triangulation of $\lim(X_k, \phi_k)$; hence $\lim(X_k, \phi_k)$ is a simplicial space. As a cellular space, this space coincides with the limit defined in 1.5.6. Moreover, imm_k are simplicial embeddings.

If X_k are ordered spaces and ϕ_k are monotone, then the space $\lim(X_k, \phi_k)$ is ordered, and the embeddings imm_k are monotone.

Joins, Cones, and Suspensions

10. Let X_1 and X_2 be topological simplices. Then the join $X_1 * X_2$ naturally becomes a topological simplex if we define the characteristic homeomorphisms as

$$T^{\dim X_1 + \dim X_2 + 1} = T_1 * T_2 \xrightarrow{\phi_1 * \phi_2} X_1 * X_2 ,$$

where T_1 and T_2 are opposite faces of the simplex $T^{\dim X_1 + \dim X_2 + 1}$, and ϕ_1 and ϕ_2 are simplicial homeomorphisms (the equality $T^{\dim X_1 + \dim X_2 + 1} = T_1 * T_2$ stands for the simplicial homeomorphism established in 1.1). Now one may canonically triangulate the cellular join $X_1 *_c X_2$ of two arbitrary simplicial spaces X_1 and X_2: its simplices are the images of the simplices of X_1 and X_2 under the inclusions $X_1 \to X_1 *_c X_2$ and $X_2 \to X_1 *_c X_2$, and also the images of the simplices $s_1 * s_2$ under the inclusions $in * in: s_1 * s_2 \to X_1 * X_2$, where s_1 is a simplex of X_1 and s_2 a simplex of X_2. The resulting simplicial space is called the __simplicial join__ of X_1 and X_2, and is denoted by $X_1 *_s X_2$. As a cellular space, $X_1 *_s X_2$ is identical to $X_1 *_c X_2$.

If $\operatorname{sch} X_1 = (M_1, S_1)$ and $\operatorname{sch} X_2 = (M_2, S_2)$, then clearly $\operatorname{sch}(X_1 *_s X_2) = (M_1 \sqcup M_2, S)$, where S is the collection of nonempty subsets A of $M_1 \sqcup M_2$ such that $in_1^{-1}(A) \in S_1$ or $in_1^{-1}(A) = \emptyset$, while $in_2^{-1}(A) \in S_2$ or $in_2^{-1}(A) = \emptyset$. Here $in_1 : M_1 \to M_1 \sqcup M_2$ and $in_2 : M_2 \to M_1 \sqcup M_2$ are the canonical mappings.

In particular, since

$$\operatorname{con} X = X * D^0 \quad \text{and} \quad \operatorname{su} X = X * S^0$$

for any topological space X (see 1.2.6.8), we see that the simplicial join construction transforms the cone and the suspension over an arbitrary simplicial space into simplicial spaces.

Simplicial Mapping Cylinders

11. To a given monotone simplicial map $f : X_1 \to X_2$, this construction associates a simplicial space $\operatorname{Scyl} f$, called the __simplicial mapping cylinder of__ f. Generally speaking, $\operatorname{Scyl} f$ is not homeomorphic to the usual mapping cylinder $\operatorname{Cyl} f$ of f (see 4.6.6.10), but has similar properties.

The most suitable language for describing the space $\operatorname{Scyl} f$ is that of schemes. Thus, let $\operatorname{sch} X_1 = (M_1, S_1)$, $\operatorname{sch} X_2 = (M_2, S_2)$, and $\operatorname{sch} f = (\phi, \Phi)$. We define $\operatorname{Scyl} f$ by the formula $\operatorname{sch} \operatorname{Scyl} f = (M_1 \sqcup M_2, S)$, where S is the collection of (finite) subsets $A \subset M_1 \sqcup M_2$ such that: (i) $in_1^{-1}(A) \in S_1$ or $in_1^{-1}(A) = \emptyset$; (ii) $\phi(in_1^{-1}(A)) \cup in_2^{-1}(A) \in S_2$; (iii) if $in_1^{-1}(A) \neq \emptyset$, then $in_2^{-1}(A) \in \Phi(S_1)$; (iv) if $a_1 \in in_1^{-1}(A)$ and $a_2 \in in_2^{-1}(A)$, then $a_2 < \phi(a_1)$. The maps in_1 and in_2 define two maps $\operatorname{sch} X_1 \to \operatorname{sch} \operatorname{Scyl} f$ and $\operatorname{sch} X_2 \to \operatorname{sch} \operatorname{Scyl} f$, and hence two simplicial embeddings, $X_1 \to \operatorname{Scyl} f$ and $X_2 \to \operatorname{Scyl} f$. The images of these embeddings are called the (lower and upper) __bases__ of the cylinder $\operatorname{Scyl} f$ and can be identified with X_1 and X_2. Moreover, the map $M_1 \times (0 \cup 1) \to M_1 \sqcup M_2$ defined by $(a, 0) \mapsto in_1(a)$, $(a, 1) \mapsto in_2 \circ \phi(a)$, induces a certain map $\operatorname{sch}(X_1 \times_s I) \to \operatorname{sch} \operatorname{Scyl} f$, and hence a simplicial map $X_1 \times_s I \to \operatorname{Scyl} f$. Clearly, together with the inclusion $X_2 \to \operatorname{Scyl} f$, this simplicial map yields a continuous map $(X_1 \times I) \bigsqcup X_2 \to \mathrm{Scyl}\, f$, which in turn induces a continuous map $\mathrm{csc}\, f : \mathrm{Cyl}\, f \to \mathrm{Scyl}\, f$. Moreover, we see that $\mathrm{csc}\, f\, (\mathrm{Cyl}\, f) = \mathrm{Scyl}\, f$, and that the canonical retraction $\mathrm{rt}\, f : \mathrm{Cyl}\, f \to X_2$ (see 1.2.6.10) is constant on the elements of the partition $\mathrm{zer}(\mathrm{csc}\, f)$. Also, the canonical X_2-homotopy from $\mathrm{id}(\mathrm{Cyl}\, f)$ to the composite map $\mathrm{Cyl}\, f \xrightarrow{\ \mathrm{rt}\, f\ } X_2 \xrightarrow{\ \mathrm{in}\ } \mathrm{Cyl}\, f$ is constant on the elements of the partition $\mathrm{zer}(\mathrm{csc}\, f) \times \mathrm{zer}(\mathrm{id}\, I)$. Consequently, $\mathrm{rt}\, f$ defines a strong deformation retraction $\mathrm{Scyl}\, f \to X_2$, and the composition of the inclusion $X_1 \to \mathrm{Scyl}\, f$ with this retraction obviously equals f. We conclude that the inclusion $X_2 \to \mathrm{Scyl}\, f$ is always a homotopy equivalence, whereas the inclusion $X_1 \to \mathrm{Scyl}\, f$ is a homotopy equivalence if and only if f is a homotopy equivalence.

6. <u>Stars. Links. Regular Neighborhoods</u>

1. The <u>star</u> of a simplex s in a simplicial space X is the union of all simplices of X which contain s. Notation: $\mathrm{St}\, s$ or $\mathrm{St}(s,X)$. Clearly, $\mathrm{St}\, s$ is a subspace of X.

The <u>open star</u> of the simplex s is the union of the interiors of all simplices containing s. Notation: $\mathrm{st}\, s$ or $\mathrm{st}(s,X)$. It is readily seen that $\mathrm{st}\, s$ is the open set defined by the inequalities

$$\mathrm{bar}_{a_0}(x) > 0,\ldots,\mathrm{bar}_{a_q}(x) > 0,$$

where $a_0,\ldots,a_q$ are the vertices of s. Moreover, $\mathrm{Cl}\,\mathrm{st}\, s = \mathrm{St}\, s$.

The <u>link</u> of the simplex s is the union of all simplices in $\mathrm{St}\, s$ which do not intersect s. Notation: $\mathrm{lk}\, s$ or $\mathrm{lk}(s,X)$. Clearly, $\mathrm{lk}\, s$ is a subspace of the spaces X and $\mathrm{St}\, s$.

The following are obvious facts.

If s' is a face of s, then

$$\mathrm{St}\, s' \subset \mathrm{St}\, s\,,\quad \mathrm{st}\, s' \subset \mathrm{st}\, s\,,\quad \mathrm{lk}\, s' \subset \mathrm{lk}\, s\,.$$

If $a_0,\ldots,a_q$ are vertices which do not sit in the same simplex, then the intersection $\cap_{i=0}^{q}\, \mathrm{st}\, a_i$ is empty. However, if $a_0,\ldots,a_q$ are vertices of a simplex s, then

$$\cap_{i=0}^{q}\, \mathrm{st}\, a_i = \mathrm{st}\, s\,.$$

If X is a subspace of X', then

$$\mathrm{st}(s,X) = \mathrm{st}(s,X') \cap X$$

for any simplex s of X. Moreover, if X is complete, then

$$St(s,X) = St(s,X') \cap X \quad \text{and} \quad lk(s,X) = lk(s,X') \cap X.$$

Finally, if s' is a simplex of lk(s,X), then

$$lk(s',lk(s,X)) = lk(s'',X),$$

where s" is the smallest simplex containing s and s'.

2. We can extend the definition of the star, open star, and link to points of a simplicial space X: for $x \in X$, the star St x = St(x,X), the open star st x = st(x,X), and the link lk x = lk(x,X), are defined as

$$St\,x = St\,si\,x, \quad st\,x = st\,si\,x, \quad \text{and} \quad lk\,x = St\,si\,x \smallsetminus st\,si\,x.$$

Obviously, st x is a neighborhood of x, and lk x = = Fr St x = Fr st x. In addition, the star St x is homeomorphic to the cone over lk x . In fact, the formula

$$pr\,(y,t) \mapsto \phi((1-t)\phi^{-1}(x) + t\phi^{-1}(y)),$$

where $y \in lk\,x$ and $t \in I$, defines a canonical homeomorphism con lk x $\rightarrow$ St x ; here ϕ is any characteristic homeomorphism of any simplex containing x and y.

Warning: the equality lk x = lk si x holds only when x is a vertex.

If s' is a simplex of lk(s,X), then by 1.1 and 5.10, the join s * s' is canonically simplicial homeomorphic to the smallest simplex of X containing both s and s'. It is clear that all these simplicial homeomorphisms together yield a simplicial homeomorphism s * lk(s,X) $\rightarrow$ St(s,X). If follows that the star St(x,X) of any point $x \in X$ is canonically simplicial homeomorphic to si x * lk(si x,X) , and we readily see that this simplicial homeomorphism maps the join of the boundary of si x with the link lk(si x,X) onto lk(x,X). Moreover, since Fr si x is homeomorphic to $S^{\dim si\,x-1}$ and the join $S^{\dim si\,x-1}$ * lk(si x,X) is homeomorphic to the iterated suspension $su^{\dim si\,x} lk(si\,x,X)$, we conclude that lk(x,X) is homeomorphic to $su^{\dim si\,x} lk(si\,x,X)$.

The link of a point is a homotopy invariant

3 (LEMMA). <u>Let A and B be retracts of a topological space Y. If the inclusions $i: A \to Y$ and $j: B \to Y$ are homotopic to some maps $f: A \to Y$ and $g: B \to Y$ such that $f(A) \subset B$ and $g(B) \subset A$, then A and B are homotopy equivalent.</u>

PROOF. Consider two arbitrary retractions, $\sigma: Y \to B$ and $\rho: Y \to A$. Then the restrictions $\sigma|_A$ and $\rho|_B$ are homotopy equivalences $A \to B$ and $B \to A$, and inverses of one another. Indeed, $\sigma|_A = \sigma \circ i$, $\rho|_B = \rho \circ j$, and the composition $\rho \circ j \circ \sigma \circ i$ is homotopic to $\rho \circ j \circ \sigma \circ f = \rho \circ f$, which in turn is homotopic to $\rho \circ i = \mathrm{id}\, A$. Therefore, $\rho|_B \circ \sigma|_A$ is homotopic to $\mathrm{id}\, A$, and a similar argument proves that $\sigma|_A \circ \rho|_B$ is homotopic to $\mathrm{id}\, B$.

4. <u>Let T_1 and T_2 be subspaces of the topological space X, and assume that both T_1 and T_2 are endowed with finite triangulations. If $x_0 \in X$ is an interior point for both T_1 and T_2, then the links $\mathrm{lk}(x_0, T_1)$ and $\mathrm{lk}(x_0, T_2)$ have the same homotopy type.</u>

PROOF. Let F_i denote a homotopy $\mathrm{St}(x_0, T_i) \times I \to X$ from the inclusion $\mathrm{St}(x_0, T_i) \to X$ to the constant map $\mathrm{St}(x_0, T_i) \to x_0 \xrightarrow{\text{in}} X$, such that F_i is rectilinear on each simplex of $\mathrm{St}(x_0, T_i)$ $(i = 1, 2)$. Set $C_i(t) = F_i(\mathrm{St}(x_0, T_i) \times t)$ $(i = 1, 2)$. Since $\mathrm{st}(x_0, T_i)$ is open and $\mathrm{St}(x_0, T_i)$ is compact $(i = 1, 2)$, there exists $\varepsilon > 0$ and $\delta > 0$ such that $C_1(\varepsilon) \subset \mathrm{St}(x_0, T_2)$, $C_2(\varepsilon) \subset \mathrm{St}(x_0, T_1)$, $C_1(\delta) \subset C_2(\varepsilon)$, and $C_2(\delta) \subset C_1(\varepsilon)$. Moreover, since $C_i(\varepsilon) \smallsetminus x_0$ is a retract of $\mathrm{St}(x_0, T_i) \smallsetminus x_0$ $(i = 1, 2)$, $C_1(\varepsilon) \smallsetminus x_0$ and $C_2(\varepsilon) \smallsetminus x_0$ are retracts of $Y = [\mathrm{St}(x_0, T_1) \cup \mathrm{St}(x_0, T_2)] \smallsetminus x_0$. Finally, the formulas

$$(y, t) \mapsto F_1(y, \delta t/\varepsilon) \quad \text{and} \quad (y, t) \mapsto F_2(y, \delta t/\varepsilon)$$

define homotopies $(C_1(\varepsilon) \smallsetminus x_0) \times I \to Y$ and $(C_2(\varepsilon) \smallsetminus x_0) \times I \to Y$ from the inclusions $C_1(\varepsilon) \smallsetminus x_0 \to Y$ and $C_2(\varepsilon) \smallsetminus x_0 \to Y$ to maps whose images lie in $C_2(\varepsilon) \smallsetminus x_0$ and $C_1(\varepsilon) \smallsetminus x_0$, respectively. Consequently, $C_2(\varepsilon) \smallsetminus x_0$ and $C_1(\varepsilon) \smallsetminus x_0$ have the same homotopy type (see 3), and it remains to note that $C_i(\varepsilon) \smallsetminus x_0$ has the same homotopy type as $\mathrm{lk}(x_0, T_i)$ $(i = 1, 2)$.

Regular Neighborhoods

5. The _regular neighborhood_ of the subspace A of a
simplicial space X is the union of the open stars $st(a,X)$ with
$a \in A$ or, equivalently, the union of the open stars $st(a,X)$ with
$a \in ske_0 A$.

If the subspace A _is complete, then_ A _is a deformation_
retract of its regular neighborhood U. In fact, there is even a
canonical A-homotopy $h: U \times I \to U$ from $id\, U$ to the composition of
a retraction $U \to A$ with the inclusion $A \to U$. This homotopy is given
by

$$
bar_a(h(x,t)) = \begin{cases} \dfrac{\left(1 - t \displaystyle\sum_{b \in ske_0 A} bar_b(x)\right) bar_a(x)}{1 - \displaystyle\sum_{b \in ske_0 A} bar_b(x)}, & \text{if } a \in ske_0 A, \\[2em] t\, bar_a(x), & \text{if } a \in ske_0 X \smallsetminus ske_0 A. \end{cases}
$$

In particular, this shows that _every subspace of a simplicial_
space X _is a deformation retract of its regular neighborhood in_ $ba\, X$
(see 5.4).

Barycentric Stars and Barycentric Links

6. The _barycentric star_ of the simplex s of a simplicial
space X is the union of all simplices of $ba\, X$ which have as their
first vertex the center of s. Notation: $bst\, s$ or $bst(s,X)$. An
equivalent description: $bst\, s$ is the set of all points $x \in X$ such
that

$$bar_a(x) = bar_b(x), \quad \text{if } a,b \in s \cap ske_0 X,$$

and

$$bar_a(x) \geqslant bar_b(x), \quad \text{if } a \in ske_0 X, \ b \in (X \smallsetminus s) \cap ske_0 X.$$

It is clear that the barycentric stars of the simplices of X
cover X and are subspaces of $ba\, X$. Moreover, $bst\, s \neq bst\, s'$
whenever $s \neq s'$, and $bst\, s \subset bst\, s'$ whenever $s \supset s'$.

7. The union of those simplices of the barycentric star
$bst\, s$ which do not contain the center of s is the _barycentric link_

of the simplex s, and is denoted by blk s . The star bst s   is
clearly simplicial homeomorphic to the cone over blk s . Moreover,
the rectilinear projection from the center of s induces a homeomorphism
of blk s   onto the link lk s   of the simplex s in X (and the
barycentric subdivision of lk s   transforms this homeomorphism into s
simplicial one). Therefore, the pairs (bst s,blk s) and
(con lk s,lk s)   are homeomorphic.

7. Simplicial Approximation of Continuous Maps

1. Let $f\colon X \to Y$ be a continuous map, where X and Y are
simplicial spaces. A simplicial map $g\colon X \to Y$ is a __simplicial
approximation of__ f if $g(x) \in \text{si}\, f(x)$ for any point $x \in X$.

__Every simplicial approximation__ g __of the map__ $f\colon X \to Y$ __is
canonically homotopic to__ f: the canonical homotopy $X \times I \to Y$ from
f to g is an affine mapping from each generatrix $x \times I$ of the
cylinder $X \times I$ onto the (possibly degenerate) rectilinear segment
joining $f(x)$ and $g(x)$. It is clear that this homotopy is stationary
on the set of the points $x \in X$ where $g(x) = f(x)$.

2. __A simplicial map__ $g\colon X \to Y$ __is a simplicial approximation
of the continuous map__ $f\colon X \to Y$ __if and only if__ $f(\text{st}\, a) \subset \text{st}\, g(a)$ __for
every vertex__ a __of__ X.

PROOF. Assume first that g is a simplicial approximation
of f, and let $x \in \text{st}\, a$. Recalling that $g(x) \in \text{si}\, f(x)$, that g is
simplicial, and that x lies in the interior of a simplex with vertex
a, we conclude that $g(x)$ lies in the interior of a simplex with
vertex $g(a)$ (see 1.3). Thus, $g(a)$ is a vertex of $\text{si}\, f(x)$, and
hence $f(x) \subset \text{st}\, g(a)$.

Now suppose that $f(\text{st}\, a) \subset \text{st}\, g(a)$ for every vertex a of
X. Pick $x \in X$; if $a_0,\dots,a_q$ are the vertices of $\text{si}\, f(x)$, then
$x \in \bigcap_{i=0}^{q} \text{st}\, a_i$, whence

$$f(x) \in f(\cap_{i=0}^{q} \text{st}\, a_i) \subset \cap_{i=0}^{q} f(\text{st}\, a_i) \subset \cap_{i=0}^{q} \text{st}\, g(a_i).$$

Therefore, the points $g(a_0),\dots,g(a_q)$ are among the vertices of the
simplex $\text{si}\, f(x)$, and since $g(x)$ lies in the simplex with vertices
$g(a_0),\dots,g(a_q)$, $g(x) \in \text{si}\, f(x)$.

3. __A continuous map__ $f\colon X \to Y$ __of simplicial spaces has a
simplicial approximation if and only if for each vertex__ a __of__ X __there__

<u>is a vertex</u> b <u>of</u> Y <u>such that</u> $f(\operatorname{st} a) \subset \operatorname{st} b$.

The necessity of this condition is an immediate consequence of 2. To prove its sufficiency, fix a map $\varphi\colon \operatorname{ske}_0 X \to \operatorname{ske}_0 Y$ such that $f(\operatorname{st} a) \subset \operatorname{st} \varphi(a)$ for every vertex $a \in \operatorname{ske}_0 X$. If $a_0,\dots,a_q$ are the vertices of a simplex of X, then $\cap_{i=0}^{q} \operatorname{st} a_i \neq \emptyset$, and the inclusions

$$\cap_{i=0}^{q} \operatorname{st} \varphi(a_i) \supset \cap_{i=0}^{q} f(\operatorname{st} a_i) \supset f(\cap_{i=0}^{q} \operatorname{st} a_i)$$

demonstrate that $\cap_{i=0}^{q} \operatorname{st} \varphi(a_i) \neq \emptyset$ too. This in turn implies that $\varphi(a_0),\dots,\varphi(a_q)$ are among the vertices of a simplex of Y (see 6.1), Therefore, φ extends to a simplicial map $X \to Y$ (see 2.5) and, applying 2, this extension is a simplicial approximation of f.

4. <u>For each continuous map</u> f <u>of a finite simplicial space</u> X <u>into a simplicial space</u> Y <u>there is a positive integer</u> m <u>such that</u> <u>the map</u> $f\colon \operatorname{ba}^m X \to Y$ <u>admits a simplicial approximation.</u>

Without loss of generality, we may assume that X is a polyhedron (see 4.1). Since the open stars of the vertices of Y constitute an open cover, there is $\varepsilon > 0$ such that, given any subset A of X with $\operatorname{diam} A < \varepsilon$, f(A) is contained in one of these open stars (see 1.1.7.16). Let m be large enough so that the simplices of $\operatorname{ba}^m X$ have diameters less than $\varepsilon/2$ (see 5.6). Then given any vertex of $\operatorname{ba}^m X$, the diameter of its star is less than ε, and 3 shows that $f\colon \operatorname{ba}^m X \to Y$ has a simplicial approximation.

8. Exercises

1. Let X be a simplicial space. Show that the formula

$$\operatorname{dist}(x,y) = \Big[\sum_{a \in \operatorname{ske}_0 X} (\operatorname{bar}_a(y) - \operatorname{bar}_a(x))^2 \Big]^{1/2}$$

defines a metric on X, and verify that the resulting metric topology coincides with the initial topology if and only if X is locally finite.

2. Show that for every polyhedron $X \subset \mathbb{R}^n$ there is a triangulation of $\mathbb{R}^n$ by Euclidean simplices, relative to which X becomes a simplicial subspace of $\mathbb{R}^n$.

3. Show that every connected, locally finite, n-dimensional simplicial space can be simplicially embedded in $\mathbb{R}^{2n+1}$ triangulated by Euclidean simplices.

4. Let $f\colon X \to Y$ be continuous, where X and Y are

simplicial spaces. Produce a new triangulation of X with the
following two properties: a) each of its simplices is contained in
one of the simplices of the original triangulation; b) f has a
simplicial approximation relative to the new triangulation.

§3. HOMOTOPY PROPERTIES OF CELLULAR SPACES

1. Cellular Pairs

1. Suppose that X is a rigged cellular space and A is a
subspace of X. Let h_0: A $\cup$ ske$_0$X $\to$ I denote the function equal to
zero on A and equal to 1 on (A $\cup$ ske$_0$X) $\smallsetminus$ A, and define inductively
a sequence of functions h_r: A $\cup$ ske$_r$X $\to$ I (r = 1,2,...), such that

$$h_r(x) = \begin{cases} h_{r-1}(x), & \text{if } x \in A \cup \text{ske}_{r-1}X, \\ 1 - \tau[1 - h_{r-1}(\text{att}_e(y))], & \text{if } x = \text{cha}_e(\tau y), \end{cases}$$

where e $\in$ cell$_r$X $\smallsetminus$ cell$_r$A, $\tau \in$ I, and $y \in$ S^{r-1}. Since the functions
h_r are continuous and each of them extends the preceding one, together
they yield a continuous function X $\to$ I. This function is called the
<u>characteristic function of the pair</u> (X,A), and the neighborhood of A
consisting of all points of X where the characteristic function is
less than 1 is called the <u>neat neighborhood</u> of the subspace A.
Obviously, the characteristic function of the pair (X,A)
vanishes on A, and only on A; hence, <u>every subspace of a cellular
space is distinguishable.</u>

If X is a simplicial space, then we may construct a
characteristic function starting with a simplicial rigging of X, and
it is readily seen that such a function does not depend upon the choice
of the rigging. In this case, the neat neighborhood of a subspace A
is simply the regular neighborhood of A in baX.

2. <u>Every subspace A of a rigged cellular space X is a
strong deformation retract of its neat neighborhood.</u>

PROOF. Let U denote the neat neighborhood of A in X.
Since the products (A $\cup$ ske$_r$X) $\times$ I are subspaces of the cylinder
X $\times$ I and cover it, they constitute a fundamental cover of X $\times$ I.

Therefore, their intersections with $U \times I$, i.e., the cylinders $U_r \times I$, where $U_r = U \cap (A \cup \mathrm{ske}_r X)$, constitute a fundamental cover of $U \times I$. Let G_0 be the constant homotopy of the inclusion $A \to X$, and define homotopies $F_r : U_r \times I \to U$, $r \geqslant 1$, by the formula

$$F_r(x,t) = \begin{cases} x, & \text{if } x \in U_{r-1}, \\ \mathrm{cha}_e(((1-t)\tau + t)y), & \text{if } x = \mathrm{cha}_e(\tau y), \end{cases}$$

where $e \in \mathrm{cell}_r X \smallsetminus \mathrm{cell}_r A$, $\tau \in (0,1]$, and $y \in S^{r-1}$. Now construct homotopies $G_r : U_r \times I \to U$, $r \geqslant 1$, by

$$G_r(x,t) = \begin{cases} x, & \text{if } 0 \leqslant t \leqslant 2^{-r}, \\ F_r(x, 2^r t - 1), & \text{if } 2^{-r} \leqslant t \leqslant 2^{-r+1}, \\ G_{r-1}(F_r(x,1), t), & \text{if } 2^{-r+1} \leqslant t \leqslant 1. \end{cases}$$

Each homotopy G_r extends the preceding one, and together they yield an A-homotopy $U \times I \to U$ from $\mathrm{id}\, U$ to a map which takes U into A. The compression of the last map to a map $U \to A$ is the desired strong deformation retraction.

3. <u>Every cellular pair is a Borsuk pair.</u>

This is a consequence of 2 combined with 1.3.5.11, because cellular spaces are normal, and their subspaces are distinguishable.

4. <u>If</u> (X,A) <u>is a cellular pair and the inclusion</u> $A \to X$ <u>is a homotopy equivalence, then</u> A <u>is a strong deformation retract of</u> X.

In order to prove this, first apply theorems 3 and 1.3.5.6 to the pair (X,A), then apply Theorem 3 to the pair $(X \times I, (X \times 0) \cup (A \times I) \cup (X \times 1))$ and, finally, apply Theorem 1.3.5.7 to the pair (X,A).

Cellular Pairs and k-Connectedness

5. <u>Let</u> k <u>be a nonnegative integer or</u> ∞. <u>Suppose that</u> (X,A) <u>is a cellular pair such that all the cells in</u> $X \smallsetminus A$ <u>have dimension at most</u> k, <u>and let</u> (Y,B) <u>be an arbitrary k-connected topological pair. Then every continuous map</u> $f : X \to Y$ <u>such that</u> $f(A) \subset B$ <u>is A-homotopic to a map which takes</u> X <u>into a subset of</u> B. <u>In particular, every continuous map of a k-dimensional cellular space</u>

into a k-connected topological space is homotopic to a constant map.

PROOF. We exhibit a sequence of A-homotopies $\{F_r: (A \cup \text{ske}_r X) \times I \to Y\}_{r=-1}^{\infty}$, each extending the preceding one, and satisfying the conditions:

(i) $F_r(x,0) = f(x)$ for all $x \in A \cup \text{ske}_r X$;

(ii) $F_r((A \cup \text{ske}_r X) \times (1-2^{-r-1})) \subset B$;

(iii) $F_r(x,t)$ does not depend upon t for $t \geqslant 1 - 2^{-r-1}$.

Then the map $F: X \times I \to Y$ which equals F_r on $(A \cup \text{ske}_r X) \times I$ will be a homotopy from f to a map which takes X into a subset of B.

We proceed by induction. Define F_{-1} as the constant homotopy of $f|_A$, and assume that homotopies $F_{-1}, \ldots, F_{q-1}$, each extending its predecessor and satisfying (i)-(iii), are already constructed. If $q > k$, then $A \cup \text{ske}_q X = X$ and we simply take $F_q = F_{q-1}$. So suppose now that $q \leqslant k$. Since the pair $((A \cup \text{ske}_q X) \times I, (A \cup \text{ske}_{q-1} X) \times I)$ is Borsuk (see 3), there is a homotopy G of the map $f|_{A \cup \text{ske}_q X}$, such that $G|_{(A \cup \text{ske}_{q-1} X) \times I} = F_{q-1}$. Using the fact that $F_{q-1}((A \cup \text{ske}_{q-1} X) \times (1-2^{-q})) \subset B$, the formula $h_e(y) = G(\text{cha}_e(y), 1-2^{-q})$ defines a map $h_e: D^q \to Y$ which takes S^{q-1} into B, for each cell $e \in \text{cell}_q X \smallsetminus \text{cell}_q A$. Now take advantage of the k-connectedness of the pair (Y,B) to deduce that, given any cell $e \in \text{cell}_q X \smallsetminus \text{cell}_q A$, there is an S^{q-1}-homotopy $H_e: D^q \times I \to Y$ from h_e to a map whose image is a subset of B. We put

$$F_q(x,t) = \begin{cases} F_{q-1}(x,t), & \text{if } x \in A \cup \text{ske}_{q-1} X, \\ G(x,t), & \text{if } 0 \leqslant t \leqslant 1-2^{-q}, \\ H_e(y, 2^{q+1}(t - (1-2^{-q}))), & \text{if } x = \text{cha}_e(y) \text{ and} \\ & \quad 1-2^{-q} \leqslant t \leqslant 1-2^{-q-1}, \\ H_e(y,1), & \text{if } x = \text{cha}_e(y) \text{ and} \\ & \quad 1-2^{-q-1} \leqslant t \leqslant 1. \end{cases}$$

Then it is immediate that the map F_q is continuous, extends F_{q-1}, and fulfils properties (i)-(iii) with $r = q$.

6. <u>Let k be a nonnegative integer or ∞. If the cellular pair (X,A) is k-connected and every cell in $X \smallsetminus A$ is of dimension at most k, then A is a strong deformation retract of X. In particular, every k-connected k-dimensional cellular space is</u>

contractible.

Indeed, $\operatorname{id} X$ is A-homotopic to a map $g: X \to X$ such that $g(X) \subset A$ (see 5), and hence $\operatorname{ab} g: X \to A$ is a strong deformation retraction.

2. Cellular Approximation of Continuous Maps

1 (LEMMA). <u>Let</u> $A = A \cup_\varphi [\bigsqcup_{\mu \in M}(D_\mu = D^{k+1})]$, <u>where</u> A <u>is a topological space and</u> φ <u>is a continuous map</u> $\bigsqcup_{\mu \in M}(S_\mu = S^k) \to A$. <u>Let</u> $f: D^{r+1} \to X$ <u>be a continuous map such that</u> $f(S^r) \subset A' = \operatorname{imm}_2(A)$. <u>Then:</u>

(I) <u>if</u> $r < k$, f <u>is</u> S^r-<u>homotopic to a map</u> g <u>such that</u> $g(D^{r+1}) \subset A'$;

(II) <u>if</u> $r = k$, f <u>is</u> S^r-<u>homotopic to a map</u> g <u>such that there are affine maps</u> $\alpha_1, \ldots, \alpha_s: D^{k+1} \to D^{k+1}$ <u>with four properties:</u> (i) <u>the images</u> $d_i = \alpha_i(D^{k+1})$ <u>are pairwise disjoint balls lying in</u> $\operatorname{Int} D^{k+1}$; (ii) <u>each of the compositions</u> $g \circ \alpha_i$ <u>coincides with one of the composite maps</u>

$$D_\mu^{k+1} \xrightarrow{\operatorname{in}_\mu} \bigsqcup_{\nu \in M} D_\nu \xrightarrow{\operatorname{imm}_2} X \; ; \tag{1}$$

(iii) $g(D^{k+1} \smallsetminus \cup_{i=1}^s \operatorname{Int} d_i) \subset A'$; (iv) <u>for</u> $k \geqslant 1$, <u>the point of the ball</u> d_i <u>having the largest value of the first coordinate is just</u> $\alpha_i(\operatorname{ort}_1)$, <u>and the segment joining this point with</u> ort_1 <u>is entirely contained in</u> $D^{k+1} \smallsetminus \cup_{j=1}^s \operatorname{Int} d_j$ $(i = 1, \ldots, s)$.

(Part (I) of this lemma, i.e., the case $r < k$, merely asserts that the pair (X, A') is k-connected, and this is the only information that we shall actually use in the present section; part (II) is needed in §5.3 .)

PROOF. Denote the composite map (1) by h_μ, fix an arbitrary Euclidean $(k+1)$-simplex σ in D^{k+1}, and then fix in $\operatorname{Int} \sigma$ an arbitrary $(k+1)$-simplex τ of the triangulation $\operatorname{ba}^2 \sigma$. The sets $h_\mu(\operatorname{Int} \sigma)$, $\mu \in M$, and $X \smallsetminus \cup_{\mu \in M} h_\mu(\tau)$ constitute an open cover of X, and so there is a triangulation of D^{r+1} which is fine enough to ensure that the image of any of its simplices under f lies in one of the sets of this cover (see 2.5.6 and 1.1.7.6). Let K_μ (respectively, L) be the union of those simplices whose images are contained in $h_\mu(\mathrm{Int}\,\sigma)$ (respectively, in $X \smallsetminus \cup_\mu h_\mu(\tau)$). Obviously, the sets K_μ are pairwise disjoint, only a finite number of them are nonempty, K_μ and L are simplicial subspaces of the simplicial space D^{r+1}, and $L \cup (\cup_\mu K_\mu) = {} = D^{r+1}$.

Now apply Theorem 2.7.4 to the composite maps

$$K_\mu \xrightarrow{\ \mathrm{ab}\,f\ } h_\mu(\sigma) \xrightarrow{\ (\mathrm{ab}\,h_\mu)^{-1}\ } \mathrm{ba}^2\sigma \ . \tag{2}$$

This theorem tells us that there is an m such that the maps (2) admit simplicial approximations when one replaces K_μ by $\mathrm{ba}^m K_\mu$. We let F_μ denote the canonical homotopy from (2) to the above simplicial approximation. Since τ is not a face of any other simplex, $F_\mu((L \cap K_\mu) \times I) \cap \mathrm{Int}\,\tau = \emptyset$, and so together the homotopies F_μ define a homotopy $F\colon (L \cap (\cup_\mu K_\mu)) \times I \to X \smallsetminus \cup_\mu h_\mu(\mathrm{Int}\,\tau)$. By 1.3 and 1.3.5.9, F extends to a homotopy $G\colon L \times I \to X \smallsetminus \cup_\mu h_\mu(\mathrm{Int}\,\tau)$ of the map $\mathrm{ab}\,f\colon L \to X \smallsetminus \cup_\mu h_\mu(\mathrm{Int}\,\tau)$, stationary on $B = D^{r+1} \smallsetminus \cup_\mu f^{-1}(h_\mu(\mathrm{Int}\,\tau))$. It is evident that the composite maps

$$K_\mu \times I \xrightarrow{\ F_\mu\ } \sigma \xrightarrow{\ h_\mu|\sigma\ } X$$

and

$$L \times I \xrightarrow{\ G\ } X \smallsetminus \cup_\mu h_\mu(\mathrm{Int}\,\tau) \xrightarrow{\ \mathrm{in}\ } X$$

yield together a B-homotopy of f. This homotopy connects f to a map $f_1\colon D^{r+1} \to X$ such that, for every μ, the composition

$$\mathrm{ba}^m K \xrightarrow{\ \mathrm{ab}\,f_1\ } h_\mu(\sigma) \xrightarrow{\ (\mathrm{ab}\,h_\mu)^{-1}\ } \mathrm{ba}^2\sigma$$

is simplicial and $h_\mu(\tau) \subset f_1(K_\mu) \subset h_\mu(\sigma)$. Since $B \supset S^r$, f_1 is S^r-homotopic to f, and to complete the proof of our lemma it suffices to examine (I) and (II) for f_1 rather than f.

Consider an arbitrary ball $\delta \subset \mathrm{Int}\,\tau$, and let ψ denote the homeomorphism $\delta \to D^{k+1}$, $\psi(x) = (x - a)/\rho$, where a and ρ are the center and the radius of δ. Moreover, let $\Psi\colon D^{k+1} \times I \to D^{k+1}$ be defined as

$$\Psi(x,t) = \frac{x - t_0 a}{1 - t_0(1-\rho)} \ ,$$

where t_0 is the largest of the numbers $\theta \in [0,t]$ such that $(x - \theta a)/(1 - \theta(1-\rho)) \in D^{k+1}$. Now the family of mappings $p_t\colon X \to X$, given by

$$p_t(x) = \begin{cases} x, & \text{if } x \in A', \\ h_\mu \circ \Psi(h_\mu^{-1}(x),t), & \text{if } x \in h_\mu(\operatorname{Int} D^{k+1}), \end{cases}$$

obviously yields an A'-homotopy of $\operatorname{id} X$ such that $p_1(X \smallsetminus \cup_\mu h_\mu(\delta)) \subset A'$ and $p_1 \circ h_\mu = h_\mu \circ \psi$. If $r < k$, then $f_1(D^{r+1}) \subset X \smallsetminus \cup_\mu h_\mu(\delta)$, and to complete the proof in this case we only need to note that $(y,t) \mapsto p_t(f_1(y))$ is an S^r-homotopy from f_1 to a map whose image is included in A'. If $r = k$, then $f_1^{-1}(\cup_\mu h_\mu(\delta))$ can be decomposed into pairwise disjoint ellipsoids $\delta_1,\ldots,\delta_s$, each being affinely mapped onto one of the sets $h_\mu(\delta)$ by f_1. Let $\{q_t\}$ be an S^r-homotopy of $\operatorname{id} D^{k+1}$, with the following properties: the preimages $d_i = q_1^{-1}(\delta_i)$ are balls; the maps $ab\, q_1 : d_i \to \delta_i$ are affine; and for $k \geqslant 1$ the point of the ball d_i having the largest value of the first coordinate is carried by $f_1 \circ q_1$ into one of the points $h_\mu(\psi^{-1}(\operatorname{ort}_1))$, and the segment joining this point with ort_1 is entirely contained in $D^{k+1} \smallsetminus \cup_{j=1}^s \operatorname{Int} d_j$. Obviously, the formula $(y,t) \mapsto (f_1 \circ q_t(y),t)$ defines an S^r-homotopy from f_1 to a map g satisfying conditions (i)-(iv).

2. **Every cellular pair** (X,A) **with** $A \supset \operatorname{ske}_k X$ **is** **k-connected** $(0 < k \leqslant \infty)$. **In particular, every cellular space whose k-skeleton reduces to a point is k-connected.**

PROOF. Since any continuous map of a ball into X takes the ball into a subset of one of the skeletons $\operatorname{ske}_r X$, it suffices to show that all the pairs $(A \cup \operatorname{ske}_r X, A)$ with $r > k$ are k-connected. But this is an immediate consequence of the k-connectedness of the pairs $(A \cup \operatorname{ske}_{q+1} X, A \cup \operatorname{ske}_q X)$ with $q \geqslant k$, which in turn follows at once from Lemma 1 (see 1.2.1).

3 (COROLLARIES). S^n **is $(n-1)$-connected** $(n \geqslant 1)$. $\mathbb{C}P^n$ **is simply connected** $(0 \leqslant n \leqslant \infty)$. $\mathbb{C}aP^2$ **is 7-connected.**

4. **Every continuous map** f **from a cellular space** X **into a cellular space** Y **is homotopic to a cellular map. If, in addition, f is cellular on a subspace** A **of** X, **then** f **is A-homotopic to a cellular map.**

PROOF. Given $f: X \to Y$, continuous on X and cellular on A, we shall construct a sequence of maps $\{f_r: X \to Y\}_{r=-1}^\infty$ and a sequence of homotopies $\{F_r: X \times I \to Y\}_{r=0}^\infty$, such that:

(i) $f_{-1} = f$;

(ii) f_r is cellular on $A \quad \mathrm{ske}_r X$;

(iii) F_r is a homotopy from f_{r-1} to f_r, stationary on $A \cup \mathrm{ske}_{r-1} X$.

Then the formula

$$
(x,t) \mapsto \begin{cases} F_r(x,2 - 2^{r+1}(1-t)), & \text{if} \quad 1-2^{-r} \leqslant t \leqslant 1-2^{-r-1}, \\[2ex] F_r(x,t), & \text{if} \quad x \in A \cup \mathrm{ske}_r X \quad \text{and} \quad t = 1, \end{cases}
$$

will define an A-homotopy from f to a cellular map.

We proceed by induction. If f_r and F_r are already constructed for $r < k$ and satisfy (i)-(iii), then by 2 there is a $(\mathrm{ske}_k A \cup \mathrm{ske}_{k-1} X)$-homotopy from $f_{k-1}|\mathrm{ske}_k X$ to a map whose image is contained in $\mathrm{ske}_k Y$. This homotopy together with the constant homotopy of $f_{k-1}|_{A \cup \mathrm{ske}_{k-1} X}$ yield some $(A \cup \mathrm{ske}_{k-1} X)$-homotopy of $f_{k-1}|_{A \cup \mathrm{ske}_k X}$. Applying 1.3, the last homotopy extends to some $(A \cup \mathrm{ske}_{k-1} X)$-homotopy of the map f_{k-1}, which we take as F_k. Finally, set $f_k(x) = F_k(x,1)$, $x \in X$.

5. <u>Two homotopic cellular maps</u> $f,g: X \to Y$ <u>are cellular homotopic. If, in addition,</u> f <u>and</u> g <u>are A-homotopic and</u> A <u>is a subspace of</u> X, <u>then</u> f <u>and</u> g <u>are cellular A-homotopic.</u>

Surely, every A-homotopy from f to g is a continuous map of the cylinder $X \times I$ into Y, and is cellular on $(X \times (0 \cup 1)) \cup (A \times I)$. By Theorem 4, this map is $[(X \times (0 \cup 1)) \cup (A \times I)]$-homotopic to a cellular map.

3. k-Connected Cellular Pairs

1. <u>Every k-connected cellular pair</u> (X,A) $(0 \leqslant k \leqslant \infty)$ <u>is homotopy equivalent to a cellular pair</u> (Y,B) <u>such that</u> $B \supset \mathrm{ske}_k Y$.

PROOF. If $k = \infty$, then, by 1.6, A is a strong deformation retract of X, and hence the pair (X,A) is homotopy equivalent to (X,X).

Turning now to the case $k < \infty$, we may assume that $A \supset \mathrm{ske}_{k-1} X$; indeed, one reduces to this case by induction on k, because for $k \geqslant 1$ every k-connected pair is also $(k-1)$-connected, while the condition $A \supset \mathrm{ske}_{-1} X$ is trivially fulfilled. According to 1.5 and 2.5, there is a cellular $\mathrm{ske}_k A$-homotopy $f: \mathrm{ske}_k X \times I \to X$ from the inclusion $\mathrm{ske}_k X \to X$ to a map which takes $\mathrm{ske}_k X$ into A. Define

$F: \mathrm{ske}_k X \times I \times I \to X$ by $F(x,t_1,t_2) = f(x,t_1)$, and set

$$C = (\mathrm{ske}_k X \times I \times 0) \cup (\mathrm{ske}_k X \times (0 \cup 1) \times I) \cup (\mathrm{ske}_k A \times I \times I)$$

and

$$D = \mathrm{ske}_k X \times I \times 1.$$

Obviously, C and D are subspaces of the cellular space $\mathrm{ske}_k X \times I \times I$, and the map F is cellular. Now define Y and B by

$$Y = X \cup_{F|_C} (\mathrm{ske}_k X \times I \times I), \quad B = \mathrm{imm}_2(A) \cup \mathrm{imm}_1(D).$$

By 1.5.5, Y is a cellular space, and it is clear that B is a subspace of Y containing $\mathrm{ske}_k Y$. To verify that (X,A) and (Y,B) have the same homotopy type, note that $\mathrm{imm}_2(A)$ is a strong deformation retract of B and $\mathrm{imm}_2(X)$ is a strong deformation retract of Y. In fact, the formula

$$\begin{cases} (\mathrm{imm}_1(x,t_1,1),t) \mapsto \mathrm{imm}_1(x,tt_1,1) & [x \in \mathrm{ske}_k X, \; t,t_1 \in I], \\[2ex] (\mathrm{imm}_2(x),t) \mapsto \mathrm{imm}_2(x) & [x \in A, \; t \in I], \end{cases}$$

defines a homotopy $B \times I \to B$, stationary on $\mathrm{imm}_2(A)$, from $\mathrm{id}\,B$ to a retraction $B \to \mathrm{imm}_2(A)$. Similarly, the formula

$$\begin{cases} (\mathrm{imm}_1(x,t_1,t_2),t) \mapsto \mathrm{imm}_1(x,t_1,tt_2) & [x \in \mathrm{ske}_k X, \; t,t_1,t_2 \in I], \\[2ex] (\mathrm{imm}_2(x),t) \mapsto \mathrm{imm}_2(x) & [x \in X, \; t \in I], \end{cases}$$

defines a homotopy $Y \times I \to Y$, stationary on $\mathrm{imm}_2(Y)$, from $\mathrm{id}\,Y$ to a retraction $Y \to \mathrm{imm}_2(X)$. Consequently, the pair $(Y,\mathrm{imm}_2(A))$ is homotopy equivalent to both the pairs (Y,B) (see 1.3.5.8) and $(\mathrm{imm}_2(X),\mathrm{imm}_2(A))$, and it remains to observe that the pairs $(\mathrm{imm}_2(X),\mathrm{imm}_2(A))$ and (X,A) are homeomorphic.

2. <u>Every k-connected cellular space</u> $(0 \leqslant k \leqslant \infty)$ <u>is homotopy equivalent to a cellular space whose k-skeleton reduces to a point.</u>

PROOF. Let X be a k-connected cellular space and choose a 0-cell x_0 in X. The pair (X,x_0) is homotopy equivalent to a cellular pair (X,A) such that $A \supset \mathrm{ske}_k X$ (see 1). Set $Y = X/A$. Since A is contractible, Y has the same homotopy type as X (see 1.3.7.7 and 1.3), and it is clear that $\mathrm{ske}_k Y$ is just a point.

3. Theorem 2 says nothing about the dimension of the space Y which replaces the given space X. However, its proof demonstrates that one can always choose Y to satisfy $\dim Y \leqslant \max(\dim X, k+2)$. Our

next task is to prove that for $k = 0$ the last equality may be sharpened to $\dim Y \leqslant \dim X$ (see 6).

4 (LEMMA). <u>Let</u> Y <u>be a topological space, and let</u> $\{Y_k\}_{k=0}^{\infty}$ <u>be a fundamental cover of</u> Y <u>such that</u> $Y_k \cap Y_l = \emptyset$ <u>whenever</u> $k-l > 1$. <u>If</u> $Y_{k-1} \cap Y_k$ <u>is a strong deformation retract of</u> Y_k <u>for all</u> $k \geqslant 1$, <u>then</u> Y_0 <u>is a strong deformation retract of</u> Y.

PROOF. If $F_k : Y_k \times I \to Y_k$ is a homotopy, stationary on $Y_{k-1} \cap Y_k$, from $\mathrm{id}\, Y_k$ to a map which takes Y_k into $Y_{k-1} \cap Y_k$, then the formula

$$(y,t) \mapsto \begin{cases} y, & \text{if } y \in Y_k \text{ and } 0 \leqslant t \leqslant 2^{-k}, \\[2ex] F_l(F_{l+1}(\ldots F_k(x,1)\ldots,1),2^l t-1), & \text{if } y \in Y_k \text{ and} \\[1ex] & \quad 2^{-l} \leqslant t \leqslant 2^{-l+1} \quad (l \leqslant k), \end{cases}$$

defines an Y_0-homotopy $Y \times I \to Y$ from $\mathrm{id}\, Y$ to a map which takes Y into Y_0.

5. <u>Given any connected cellular space</u> X, <u>there is a</u> <u>contractible one-dimensional subspace of</u> X <u>containing all the 0-cells.</u>

PROOF. Fix an arbitrary 0-cell x_0 in X and let A_k be t the set of all 0-cells that can be joined to x_0 by a path $I \to \mathrm{ske}_1 X$ which touches at most k 1-cells. Since $\mathrm{ske}_1 X$ is connected (see 1.4.7), and a path can touch only a finite number of cells, $\bigcup_{k=0}^{\infty} A_k = $ $ = \mathrm{ske}_0 X$. Now given any 0-cell $x \in A_k \smallsetminus A_{k-1}$ with $k \geqslant 1$, pick a closed 1-cell $c(x)$ joining x to some cell in $A_{k-1} \smallsetminus A_{k-2}$, and set

$$Y_k = \begin{cases} x_0, & \text{if } k = 0, \\[2ex] \bigcup_{y \in A_k \smallsetminus A_{k-1}} c(y), & \text{if } k > 0, \end{cases}$$

and $Y = \bigcup_{k=0}^{\infty} Y_k$. Obviously, Y is a one-dimensional subspace of X containing $\mathrm{ske}_0 X$, and the cover $\{Y_k\}$ of Y satisfies the conditions of Lemma 4. Therefore, Y_0 is a strong deformation retract of Y, i.e., Y is contractible.

6. <u>Every connected n-dimensional cellular space is homotopy</u> <u>equivalent to a cellular space of dimension at most</u> n, <u>and having</u> <u>only one 0-cell. In particular, every connected one-dimensional</u> <u>cellular space is homotopy equivalent to a bouquet of circles.</u>

This results from 5, 1.3, and 1.3.7.7.

Applications to Cellular Constructions

7. <u>If the cellular space</u> X <u>is k-connected, then</u> $su\,X$ <u>and</u> $su(X,x_0)$, <u>where</u> x_0 <u>is a 0-cell, are (k+1)-connected.</u>

The proof reduces to three remarks. First, since $su\,X$ and $su(X,x_0)$ have the same homotopy type (see 1.4.5, 1.3.6.8, and 1.3.7.7), the (k+1)-connectedness of one is equivalent to the (k+1)-connectedness of the other. Secondly, according to 2 and 1.3.7.12, it is enough to verify that $su(X,x_0)$ is (k+1)-connected when $ske_k X = x_0$. And thirdly, if $ske_k X = x_0$, then the (k+1)-connectedness of $su(X,x_0)$ is a corollary of Theorem 2, because under this assumption $ske_{k+1}su(X,x_0)$ also reduces to a point.

8. <u>Suppose</u> X_i <u>is a</u> k_i-<u>connected cellular space and</u> x_i <u>is a 0-cell of</u> X_i, $i = 1,2$. <u>Then the tensor products</u> $(X_1,x_1) \otimes (X_2,x_2)$ <u>and</u> $(X_1,x_1) \otimes_c (X_2,x_2)$ <u>are</u> (k_1+k_2+1)-<u>connected.</u>

Again, the proof reduces to three remarks. First, since $(X_1,x_1) \otimes (X_2,x_2)$ induces on its compact subsets topologies which are identical to those induced by the topology of $(X_1,x_1) \otimes_c (X_2,x_2)$, the (k_1+k_2+1)-connectedness of one of these spaces implies the (k_1+k_2+1)-connectedness of the other. Secondly, using 1 and 1.3.7.12, it is enough to verify that $(X_1,x_1) \otimes_c (X_2,x_2)$ is (k_1+k_2+1)-connected when $ske_{k_1} X_1 = x_1$ and $ske_{k_2} X_2 = x_2$. Thirdly, under these circumstances, the (k_1+k_2+1)-connectedness of $(X_1,x_1) \otimes_c (X_2,x_2)$ follows from Theorem 2, because $ske_{k_1+k_2+1}((X_1,x_1) \otimes_c (X_2,x_2))$ also reduces to a point.

9 (LEMMA). <u>For any cellular spaces</u> X_1 <u>and</u> X_2 <u>with 0-cells</u> x_1 <u>and</u> x_2 <u>taken as base points, the cellular join</u> $(X_1,x_1) *_c (X_2,x_2)$ <u>is homotopy equivalent to</u> $su((X_1,x_1) \otimes_c (X_2,x_2),bp)$.

PROOF. By definition, the spaces $(X_1,x_1) *_c (X_2,x_2)$ and $su((X_1,x_1) \otimes_c (X_2,x_2),bp)$ are obtained from $(X_1 \times_c X_2) \times I$ by taking quotients two times, and the projection $(X_1 \times_c X_2) \times I \to$ $\to su((x_1,x_1) \otimes_c (X_2,x_2),bp)$ is constant on the elements of the partition $zer(pr\colon (X_1 \times_c X_2) \times I \to (X_1,x_1) *_c (X_2,x_2))$. The resulting map

$$f = fact[pr\colon (X_1 \times_c X_2) \times I \to (X_1,x_1) *_c (X_2,x_2)]\colon$$

$$\colon su((X_1,x_1) \otimes_c (X_2,x_2),bp) \to (X_1,x_1) *_c (X_2,x_2)$$

is factorial (see 1.2.3.4). Since the only element of the partition zer(f) which does not reduce to a point is f^{-1}(bp), we see that su$((X_1,x_1) \otimes_c (X_2,x_2),bp) = [(X_1,x_1) *_c (X_2,x_2)]/f^{-1}$(bp). Finally, note that f^{-1}(bp) $= [(X_1,x_1) *_c (x_2,x_2)] \cup [(x_1,x_1) *_c (X_2,x_2)]$, and since this union is contractible, the quotient space $[(X_1,x_1) *_c (X_2,x_2)]/f^{-1}$(bp) is homotopy equivalent to $(X_1,x_1) *_c (X_2,x_2)$.

10. <u>Let the cellular spaces X_1 and X_2 be k_1- and respectively k_2-connected. Then the joins $X_1 * X_2$, $X_1 *_c X_2$, $(X_1,x_1) * (X_2,x_2)$, and $(X_1,x_1) *_c (X_2,x_2)$, where x_1 and x_2 are 0-cells, are (k_1+k_2+2)-connected.</u>

The proof reduces to four remarks. First, since $(X_1,x_1) *_c (X_2,x_2)$ is a quotient of $X_1 *_c X_2$ by a contractible space (the closed 1-cell $x_1 * x_2$), $(X_1,x_1) *_c (X_2,x_2)$ and $X_1 *_c X_2$ have the same homotopy type. Secondly, $X_1 *_c X_2$ induces on its compact subsets the same topologies as does $X_1 * X_2$, and hence the (k_1+k_2+2)-connectedness of one of these spaces implies the (k_1+k_2+2)-connectedness of the other. Thirdly, and from the same reason, $(X_1,x_1) * (X_2,x_2)$ is (k_1+k_2+2)-connected if and only if $(X_1,x_1) *_c (X_2,x_2)$ is (k_1+k_2+2)-connected. Fourthly, the (k_1+k_2+2)-connectedness of $(X_1,x_1) *_c (X_2,x_2)$ is an immediate consequence of 9, 8, and 7.

4. Simplicial Approximation of Cellular Spaces

1 (LEMMA). <u>Suppose that X and Y are cellular spaces and $\{X_r\}_{r=0}^{\infty}$ and $\{Y_r\}_{r=0}^{\infty}$ are filtrations of X and Y by subspaces. Let $f: X \to Y$ be a cellular map such that $f(X_r) \subset Y_r$ $(0 \leqslant r \leqslant \infty)$. If all the maps ab$f: X_r \to Y_r$ are homotopy equivalence, then so is f.</u>

PROOF. Since the $Z_r = $ Cyl(ab$f: X_r \to Y_r)$ are cellular subspaces of $Z = $ Cyl$f$ and satisfy the conditions $Z_r \subset Z_{r+1}$ and $\cup_{r=0}^{\infty} Z_r = Z$, they yield a filtration of Z (see 1.1.9). Thus, the image of any continuous map $D^k \to Z$ is contained in one of the sets Z_r (see 1.2.4.5), and so the pair (Z,X) is ∞-connected provided that all the pairs (Z_r,X_r) are ∞-connected. Now note that (Z_r,X_r) is ∞-connected if and only if ab$f: X_r \to Y_r$ is a homotopy equivalence; similarly, (Z,X) is ∞-connected if and only if f is a homotopy equivalence (see 1.3, 1.6, 1.3.3.9, and 1.3.7.13).

2. <u>Given any cellular space</u> X, <u>there is a simplicial space</u> <u>which has the same homotopy type and the same dimension as</u> X, <u>and is</u> <u>finite or countable together with</u> X.

The proof consists of producing three sequences: one of simplicial spaces $\{Y_r\}_{r=0}^{\infty}$, one of simplicial embeddings $\{i_r: Y_r \to Y_{r+1}\}_{r=0}^{\infty}$, and one of cellular homotopy equivalences $\{f_r: \mathrm{ske}_r X \to Y_r\}_{r=0}^{\infty}$, with the following four properties:

(i) $f_r\big|\mathrm{ske}_{r-1}X = i_{r-1} \circ f_{r-1}$;

(ii) $\dim Y_r = \dim \mathrm{ske}_r X$;

(iii) if $\mathrm{ske}_r X$ is finite (countable), then Y_r is finite (respectively, countable);

(iv) if $\mathrm{ske}_r X = \mathrm{ske}_{r-1}X$, then $Y_r = Y_{r-1}$ and $i_{r-1} = \mathrm{id}\, Y_{r-1}$.

This will enable us to define the simplicial space $\lim(Y_r, i_r)$ having dimension $\dim X$, and finite or countable together with X, as well as a cellular map $f: X \to \lim(Y_r, i_r)$ such that $f\big|_{\mathrm{ske}_r X} = \mathrm{imm}_r \circ f$. Finally, we shall use Lemma 1 to show that f is a homotopy equivalence.

Define Y_0 and f_0 as $\mathrm{ske}_0 X$ and $\mathrm{id}\,\mathrm{ske}_0 X$, and assume that simplicial spaces Y_r, cellular homotopy equivalences f_r, and simplicial embeddings i_{r-1}, satisfying (i)–(iv), are already constructed for $r < q$. By 1.2.1, we may represent $\mathrm{ske}_q X$ as $\mathrm{ske}_{q-1}X \cup_{\varphi} \Delta$, where $\Delta = \bigsqcup_{e \in \mathrm{cell}_q X} (D_e = D^q)$ and φ is a continuous map of $\Sigma = \bigsqcup_{e \in \mathrm{cell}_q X} (S_e = S^{q-1})$ into $\mathrm{ske}_{q-1}X$. Next triangulate Δ so that Σ becomes a complete subspace and the map $f_{q-1} \circ \varphi: \Sigma \to Y_{q-1}$ admits a simplicial approximation $g: \Sigma \to Y_{q-1}$. Further, order Y_{q-1} and Σ in such a manner that the map g becomes monotone. Applying successively Theorems 1.3.7.10, 1.3.7.8, and then again 1.3.7.10, we obtain three homotopy equivalences:

– a homotopy equivalence $\mathrm{ske}_q X \to Y_{q-1} \cup_{f_{q-1} \circ \varphi} \Delta$ which agrees with f_{q-1} on $\mathrm{ske}_{q-1}X$;

– a homotopy equivalence $Y_{q-1} \cup_{f_{q-1} \circ \varphi} \Delta \to Y_{q-1} \cup_g \Delta$ which is the identity on Y_{q-1};

– and a homotopy equivalence $Y_{q-1} \cup_g \Delta \to (\mathrm{Scyl}\, g) \cup_{\mathrm{in}} \Delta$, where $\mathrm{in} = [\mathrm{in}: \Sigma \to \mathrm{Scyl}\, g]$ (see 2.5.11), which agrees with the inclusion $Y_{q-1} \to \mathrm{Scyl}\, g$ on Y_{q-1}.

At last, we may define Y_q as $(\mathrm{Scyl}\, g) \cup_{\mathrm{in}} \Delta$, f_q as the composition of the three homotopy equivalences above, and i_{q-1} as the composite embedding $Y_{q-1} \to \text{Scyl } g \to Y_q$. The triangulations of Δ and of the cylinder Scyl g yield together a triangulation of Y_q (see 2.5.2). It is plain that i_{q-1} is a simplicial embedding and that Y_q , f_q , and i_{q-1} satisfy conditions (i)-(iv) for $r = q$.

3. <u>Let</u> X <u>and</u> Y <u>be cellular spaces with</u> X <u>finite and</u> Y <u>countable. Then the set</u> $\pi(X,Y)$ <u>is countable.</u>

PROOF. By theorems 2 and 1.3.1.8, we need only consider the case when X and Y are simplicial spaces. Under this assumption, Theorem 2.4.7 shows that the cardinal of $\pi(X,Y)$ does not exceed the cardinal of the set of all simplicial mappings $\text{ba}^m X \to Y$ (m = 0,1,...), and the latter is obviously countable.

5. <u>Exercises</u>

1. Suppose that the cellular spaces X_i and X_i' are homotopy equivalent, i = 1,2. Show that $X_1 \times_c X_2$ and $X_1' \times_c X_2'$ are homotopy equivalent, and that the same is true for the spaces $X_1 *_c X_2$ and $X_1' *_c X_2'$.

2. Show that every cellular space is homotopy equivalent to a locally finite cellular space.

3. Show that every cellular pair is homotopy equivalent to a simplicial pair, and that every finite cellular pair is homotopy equivalent to a finite simplicial pair.

4. Show that there is no cellular space having the same homotopy type as the subspace of the real line consisting of the points 0 and 1/n, n = 1,2,... .

Chapter 3. Smooth Manifolds

§1.   FUNDAMENTAL CONCEPTS

1. <u>Topological Manifolds</u>

1.   This chapter comprises an elementary introduction to differential topology. The basic objects of this theory are the smooth manifolds. They are defined in the next subsection and represent (as do cellular and simplicial spaces) topological spaces with an additional structure. The present subsection is devoted to topological manifolds, which occupy an intermediate position between smooth manifolds and topological spaces, and do not carry an additional structure.

Locally Euclidean Spaces

2.   A topological space is said to be a an <u>n-dimensional locally Euclidean space</u> if each of its points has a neighborhood homeomorphic to the space $\mathbb{R}^n$ or to the half space $\mathbb{R}^n_-$, where $\mathbb{R}^n_-$ is the set of all points $(x_1,\ldots,x_n) \in \mathbb{R}^n$ with $x_1 \leqslant 0$. The half space $\mathbb{R}^n_-$ is defined for $n \geqslant 1$; we do not define it for $n = 0$ and, accordingly, a 0-dimensional locally Euclidean space is simply a topological space such that each of its points has a neighborhood homeomorphic to $\mathbb{R}^0$, i.e., a discrete space.

In a locally Euclidean space X, the points having a neighborhood homeomorphic to $\mathbb{R}^n$ are called <u>interior points</u>, while the remaining ones are called <u>boundary points.</u> The interior (boundary) points form the <u>interior</u> (respectively, the <u>boundary</u>) of the locally Euclidean space X, denoted by $\mathrm{int}\,X$ (respectively ∂X). (The difference between the notations int, ∂ and $\mathrm{Int}, \mathrm{Fr}$ should prevent us, in each context, from confusing the interior and boundary points, and the interior part and boundary defined here with the interior and boundary points and the interior part and boundary of a set in a topological space.) Clearly, the interior of X is a dense open set, whereas the boundary of X is closed.

If each point of a topological space has a neighborhood homeomorphic to an open subset of $\mathbb{R}^n$ or $\mathbb{R}^n_-$, then obviously it already is an n-dimensional locally Euclidean space. Consequently, every open subset of an n-dimensional locally Euclidean space is also an n-dimensional locally Euclidean space. In particular, the interior of an n-dimensional Euclidean space is an n-dimensional locally Euclidean space without boundary. Moreover, the interior and boundary of an open subset U of a locally Euclidean space X are given by $\operatorname{int} U =$ $= U \cap \operatorname{int} X$ and $\partial U = U \cap \partial X$.

Since a locally Euclidean space is locally connected, its components are open (see 1.3.4.3), and hence also closed.

Obvious examples of n-dimensional locally Euclidean spaces are $\mathbb{R}^n$, $\mathbb{R}^n_-$, S^n, and D^n. It is clear that $\partial \mathbb{R}^n = \emptyset$ and $\partial S^n = \emptyset$. Furthermore, all the boundary points of the half space $\mathbb{R}^n_-$ lie in the limiting hyperplane $\mathbb{R}^{n-1}_1$, consisting of the points $(x_1,\ldots,x_n)$ such that $x_1 = 0$, and all the boundary points of the ball D^n lie in the limiting sphere S^{n-1}.

3. Since the product $\mathbb{R}^{n_1} \times \mathbb{R}^{n_2}$ is homeomorphic to $\mathbb{R}^{n_1+n_2}$, we see that the product $X_1 \times X_2$ of two locally Euclidean spaces X_1 and X_2 of dimensions n_1 and n_2, and without boundary, is an (n_1+n_2)-dimensional locally Euclidean space. This is true in general, i.e., a product $X_1 \times \ldots \times X_s$ of boundaryless locally Euclidean spaces $X_1,\ldots,X_s$ of dimensions $n_1,\ldots,n_s$, is an $(n_1+\ldots+n_s)$-dimensional boundaryless locally Euclidean space. Turning to locally Euclidean spaces with boundary, note that the formula $((x_1,\ldots,x_{n_1}),(y_1,\ldots,y_{n_2}))$ $\mapsto (x_1,\ldots,x_{n_1},y_1,\ldots,y_{n_2})$, which gives the canonical homeomorphism $\mathbb{R}^{n_1} \times \mathbb{R}^{n_2} \to \mathbb{R}^{n_1+n_2}$, also defines a homeomorphism $\mathbb{R}^{n_1}_- \times \mathbb{R}^{n_2} \to \mathbb{R}^{n_1+n_2}_-$ for $n_1 > 0$. Similarly, the formula $((x_1,\ldots,x_{n_1}),(y_1,\ldots,y_{n_2})) \mapsto$ $\mapsto (y_1,\ldots,y_{n_2},x_1,\ldots,x_{n_1})$ defines a homeomorphism $\mathbb{R}^{n_1} \times \mathbb{R}^{n_2}_- \to \mathbb{R}^{n_1+n_2}_-$ for $n_2 > 0$, and the formula $((x_1,\ldots,x_{n_1}),(y_1,\ldots,y_{n_2})) \mapsto$ $(-2x_1y_1,x_1^2 - y_1^2,x_2,\ldots,x_{n_1},y_2,\ldots,y_{n_2})$ defines a homeomorphism $\mathbb{R}^{n_1}_- \times \mathbb{R}^{n_2}_- \to \mathbb{R}^{n_1+n_2}_-$ for $n_1 > 0$, $n_2 > 0$. Thus, each of the products

$\mathbb{R}_-^{n_1} \times \mathbb{R}^{n_2}$, $\mathbb{R}^{n_1} \times \mathbb{R}_-^{n_2}$, and $\mathbb{R}_-^{n_1} \times \mathbb{R}_-^{n_2}$ is homeomorphic to $\mathbb{R}_-^{n_1+n_2}$. We conclude that given locally Euclidean spaces X_1 and X_2 of dimensions n_1 and n_2, the product $X_1 \times X_2$ is an (n_1+n_2)-dimensional locally Euclidean space. In general, the product $X_1 \times \ldots \times X_s$ of arbitrary locally Euclidean spaces of dimensions $n_1, \ldots, n_s$ is an $(n_1+\ldots+n_s)$-dimensional locally Euclidean space.

4. The discussion in 2 raises two nontrivial questions.

The first one is whether a nonempty topological space can be a locally Euclidean space of dimension n and, simultaneously, a locally Euclidean space of a different dimension n' : $n \neq n'$? In Chapter 4 this question is answered negatively (see 4.6.5.10). The answer is obvious when $n = 1$, $n' > 1$ or $n' = 1$, $n > 1$. In fact, any connected subset of a one-dimensional locally Euclidean space becomes disconnected after one removes two suitably choosen points (for example, two points belonging to an open subset which is homeomorphic to $\mathbb{R}^1$); in contrast, every nonempty locally Euclidean space of dimension $n' > 1$ contains a nonempty open subset which cannot be disconnected by removing two points (any open subset homeomorphic to $\mathbb{R}^{n'}$ has this property). The picture is cristal clear when $n = 0$ or $n' = 0$. However, when $n > 1$, $n' > 1$, the proof requires a technique which we will develop only later.

The second question is whether we can formulate more efficient definitions of the interior and boundary points, which would permit us to actually recognize them. For example, consider the half space $\mathbb{R}_-^n$. At this point we can show only the trivial inclusion $\partial \mathbb{R}_-^n \subset \mathbb{R}_1^{n-1}$ (see 2), and we are forced to settle for one of the extreme equalities $\partial \mathbb{R}_-^n = \mathbb{R}_1^{n-1}$ or $\partial \mathbb{R}_-^n = \emptyset$ (which obviously are the only possible ones). We shall prove in Chapter 4 that $\partial \mathbb{R}_-^n = \mathbb{R}_1^{n-1}$ (see 4.6.5.12). This equality is plain for $n = 1$ (assuming that the point 0 has a neighborhood in $\mathbb{R}_-^1$ which is homeomorphic to $\mathbb{R}^1$, then by removing 0 we would disconnect this neighborhood; this is absurd, because the latter cannot happen to a connected neighborhood of 0 in $\mathbb{R}_-^1$). But for $n > 1$, we again need techniques which are to be developed. The equality $\partial \mathbb{R}_-^n = \mathbb{R}_1^{n-1}$ settles satisfactorily the general problem of recognizing the interior and boundary points too. Indeed, it follows that <u>a point</u> x <u>of the n-dimensional locally Euclidean space</u> X <u>which has a neighborhood</u> U <u>with a homeomorphism</u> $U \to \mathbb{R}_-^n$, <u>is a boundary point of</u> U, <u>and hence, of</u> X, <u>if and only if this homeomorphism takes</u> x <u>into a point of the hyperplane</u> $\mathbb{R}_1^{n-1}$. For D^n this theorem asserts that $\partial D^n = S^{n-1}$. Finally, we note that the alternative equality $\partial \mathbb{R}_-^n = \emptyset$ would obviously imply that $\partial X = \emptyset$ for any n-dimensional locally Euclidean space.

5. In general, it would be more prudent not to use the theorems formulated in 4, i.e., the theorem on dimensions and the equality $\partial \mathbb{R}^n_- = \mathbb{R}^{n-1}_1$, as long as they have not been proven. This indeed is the way we shall deal with the theorem on dimensions - the only exception is a harmless remark in 2.3. However, we have already used the equality $\partial \mathbb{R}^n_- = \mathbb{R}^{n-1}_1$, and we will take advantage of it again, before its proof, in 7 and in 2.6, 2.7. But these are the only instances where these theorems and their corollaries will be used before their proofs.

6. <u>The boundary of an n-dimensional locally Euclidean space is an (n-1)-dimensional locally Euclidean space without boundary.</u>

Let x be a boundary point of the locally Euclidean space X and let U be a neighborhood of x, homeomorphic to $\mathbb{R}^n_-$. Then the boundary ∂U is a neighborhood of x in ∂X, since $\partial U = U \cap \partial X$, and is homeomorphic to $\mathbb{R}^{n-1}$, since $\partial \mathbb{R}^n_- = \mathbb{R}^{n-1}_1$ (here the reference to Chapter 4 for the equality $\partial \mathbb{R}^n_- = \mathbb{R}^{n-1}_1$ is unnecessary: the alternative $\partial \mathbb{R}^n_- = \emptyset$ is excluded, because $\partial X \neq \emptyset$).

7. <u>For any locally Euclidean spaces</u> $X_1, \ldots, X_s$

$$\mathrm{int}(X_1 \times \ldots \times X_s) = \mathrm{int}\, X_1 \times \ldots \times \mathrm{int}\, X_s$$

<u>and</u>

$$\partial(X_1 \times \ldots \times X_s) = (\partial X_1 \times X_2 \times \ldots \times X_s) \cup \ldots \cup (X_1 \times \ldots \times X_{s-1} \times \partial X_s).$$

It is enough to prove the statement for $s = 2$. Let $x_i \in X_i$ and let φ_i be a homeomorphism of a neighborhood U_i of x_i onto $\mathbb{R}^{n_i}$ or $\mathbb{R}^{n_i}_-$, $i = 1,2$. Then $\varphi_1 \times \varphi_2$ is a homeomorphism of the neighborhood $U_1 \times U_2$ of the point (x_1, x_2) onto one of the products $\mathbb{R}^{n_1} \times \mathbb{R}^{n_2}$, $\mathbb{R}^{n_1} \times \mathbb{R}^{n_2}_-$, $\mathbb{R}^{n_1}_- \times \mathbb{R}^{n_2}$ or $\mathbb{R}^{n_1}_- \times \mathbb{R}^{n_2}_-$, and composing it with one of the homeomorphisms exhibited in 3, we obtain a homeomorphism of $U_1 \times U_2$ onto $\mathbb{R}^{n_1+n_2}$ or $\mathbb{R}^{n_1+n_2}_-$. We denote this composition by φ and analyse the four possible cases. If $\varphi_1(U_1) = \mathbb{R}^{n_1}$ and $\varphi_2(U_2) = \mathbb{R}^{n_2}$, then $\varphi(U_1 \times U_2) = \mathbb{R}^{n_1+n_2}$ and x_1, x_2, and (x_1, x_2) are all interior points. If $\varphi_1(U_1) = \mathbb{R}^{n_1}$ and $\varphi_2(U_2) = \mathbb{R}^{n_2}_-$, then $\varphi(U_1 \times U_2) = \mathbb{R}^{n_1+n_2}_-$, and $\varphi(x_1, x_2) \in \mathbb{R}^{n_1+n_2-1}_1$ if and only if $\varphi_2(x_2) \in \mathbb{R}^{n_2-1}_1$. Thus (x_1, x_2) is an interior (boundary) point if and only if x_2 is an interior (respectively, boundary) point, while x_1 is an interior point. Similarly, if $\varphi_1(U_1) = \mathbb{R}_-^{n_1}$ and $\varphi_2(U_2) = \mathbb{R}^{n_2}$, then (x_1,x_2) is an interior or boundary point simultaneously with x_1, while x_2 is an interior point. Finally, if $\varphi_1(U_1) = \mathbb{R}_-^{n_1}$ and $\varphi_2(U_2) = \mathbb{R}_-^{n_2}$, then $\varphi(U_1 \times U_2) = \mathbb{R}_1^{n_1+n_2-1}$ if and only if $\varphi_1(x_1) \in \mathbb{R}_1^{n_1-1}$ or $\varphi_2(x_2) \in \mathbb{R}_1^{n_2-1}$. That is to say, (x_1,x_2) is a boundary point if and only if at least one of the points x_1,x_2 is boundary. Our conclusion is that in all cases (x_1,x_2) is an interior (boundary point) if x_1 and x_2 are interior points (respectively, if x_1 or x_2 are boundary points).

8. <u>A locally Euclidean space is connected if and only if its interior is connected.</u>

This condition is obviously sufficient. Now let us show that it is also necessary. Let X be a connected locally Euclidean space, let A be a component of $\mathrm{int}\,X$, and let B be the union of the remaining components. Since the closed sets $\mathrm{Cl}\,A$ and $\mathrm{Cl}\,B$ cover X and X is connected, $\mathrm{Cl}\,A \cap \mathrm{Cl}\,B \neq \emptyset$ whenever $B \neq \emptyset$, and obviously $\mathrm{Cl}\,A \cap \mathrm{Cl}\,B \subset \partial X$. Let $x \in \mathrm{Cl}\,A \cap \mathrm{Cl}\,B$, and let U be a neighborhood of x homeomorphic to $\mathbb{R}_-^n$. Since $\mathrm{int}\,\mathbb{R}_-^n$ is connected, its homeomorphic image $U \cap \mathrm{int}\,X = \mathrm{int}\,U$ is also connected, which is impossible if $B \neq \emptyset$. Consequently, $B = \emptyset$ and $\mathrm{int}\,X$ is connected.

9. The topological space $X \cup_{\mathrm{in}:\partial X \to X} X$ constructed from a locally Euclidean space X is called the <u>double</u> of X and is denoted by $\mathrm{dopp}\,X$. The double of an n-dimensional locally Euclidean space is an n-dimensional locally Euclidean space without boundary.

From now on, we shall identify $\mathrm{imm}_1(X)$ with X and we shall denote the map $\mathrm{ab}\,\mathrm{imm}_2 : X \to \mathrm{imm}_2(X)$ by cop, and $\mathrm{imm}_2(X)$ - by $\mathrm{cop}\,X$. Note that X and $\mathrm{cop}\,X$ are closed in $\mathrm{dopp}\,X$.

Manifolds

10. A locally Euclidean space is called a <u>topological manifold</u> or, briefly, a <u>manifold</u> if it is a Hausdorff topological space with countable base. A manifold is <u>closed</u> if it is compact and has no boundary, and <u>open</u> if it has no compact components.

Comparing what was said in 2, 3, 6, and 9 with the

corresponding properties of Hausdorff, second countable, and compact spaces we see that: every open subset of an n-dimensional manifold is an n-dimensional manifold; the interior of an n-dimensional manifold is an n-dimensional manifold without boundary; the boundary of an n-dimensional manifold is an $(n-1)$-dimensional manifold without boundary; the boundary of a compact manifold is a closed manifold; the product of s manifolds of dimensions $n_1, \ldots, n_s$ is an $(n_1 + \ldots + n_s)$-dimensional manifold; the double of an n-dimensional manifold is an n-dimensional manifold without boundary; and the double of a compact manifold is a closed manifold.

Since the components of a manifold constitute an open cover, their number is finite in the compact case and countable in general (see 1.1.6.5).

Clearly $\mathbb{R}^n$, $\mathbb{R}^n_-$, S^n, and D^n, which we gave above as examples of locally Euclidean spaces, are manifolds.

11. Manifolds are locally compact.

PROOF. Let x be a point of an n-dimensional manifold X. Fix a homeomorphism φ of a neighborhood U of x onto $\mathbb{R}^n$ or $\mathbb{R}^n_-$, and a neighborhood V of $\varphi(x)$ in $\varphi(U)$ with compact closure $\mathrm{Cl}\, V$. Let $U' = \varphi^{-1}(V)$. Then obviously U' is a neighborhood of x in X, $U' \subset \varphi^{-1}(\mathrm{Cl}\, V)$, and $\varphi^{-1}(\mathrm{Cl}\, V)$ is compact. Thus $\varphi^{-1}(\mathrm{Cl}\, V)$ is closed and contains both the neighborhood U' and its closure $\mathrm{Cl}\, U'$. We conclude that $\mathrm{Cl}\, U'$ is compact.

12. Manifolds are metrizable.

PROOF. Every locally compact Hausdorff space is regular (see 1.1.7.23), and every regular second countable space is metrizable (see 1.1.6.9).

13. The following example shows that for $n \geqslant 1$ there are n-dimensional locally Euclidean spaces with countable base which are not Hausdorff. Consider $X = \mathbb{R}^n \cup_i \mathbb{R}^n$, where $i = [\mathrm{in}: \mathbb{R}^n \smallsetminus \mathbb{R}^n_- \to \mathbb{R}^n]$. Then X is obviously an n-dimensional, second countable, locally Euclidean space, but for $n \geqslant 1$ and $x \in \mathbb{R}^{n-1}_1$, any two neighborhoods of the points $\mathrm{imm}_1(x)$ and $\mathrm{imm}_2(x)$ in X intersect.

14 (INFORMATION). For $n \geqslant 1$ there are connected, Hausdorff, n-dimensional locally Euclidean spaces that are not second countable. A two-dimensional example can be found in [5], and a one-dimensional one - in [11], p. 164 (the transfinite line, or "Alexandrov's line"). Higher-dimensional examples can be constructed from these by taking direct products with Euclidean spaces.

One-dimensional Manifolds

15. A zero-dimensional connected manifold obviously reduces to a point. Theorems 17 and 19 below provide the topological classification of connected one-dimensional manifolds. The two-dimensional case will be analyzed in § 5 (see 5.3). The topological classification of manifolds of higher dimensions is a very difficult problem.

16 (LEMMA). <u>If a connected Hausdorff space</u> X <u>can be represented as the union of two open subsets homeomorphic to</u> $\mathbb{R}^1$, <u>then</u> X <u>is homeomorphic to either</u> $\mathbb{R}^1$ <u>or</u> S^1.

PROOF. Let $X = U \cup V$ be the above representation and $\varphi: U \to \mathbb{R}^1$, $\psi: V \to \mathbb{R}^1$ - the corresponding homeomorphisms. We exclude the trivial cases $U \subset V$ and $V \subset U$, where X is homeomorphic to $\mathbb{R}^1$, and examine the sets $\varphi(U \cap V)$ and $\psi(U \cap V)$.

Since the intersection $U \cap V$ is open in both U and V, $\varphi(U \cap V)$ and $\psi(U \cap V)$ are open in $\mathbb{R}^1$ and their components are intervals. None of these intervals is bounded: indeed, suppose that $\varphi(U \cap V)$ contains a bounded interval (a,b). Then $\varphi^{-1}((a,b))$ is both closed in V (as the intersection of the compact, and hence closed set $\varphi^{-1}([a,b])$ with V) and open in V, which implies $V = \varphi^{-1}((a,b)) \subset U$; contradiction. Moreover, $\varphi(U \cap V) \neq \mathbb{R}^1$, because, if not, $U \subset V$. Similarly, $\psi(U \cap V) \neq \mathbb{R}^1$. Finally, we are left with only two possible cases: (i) each of the sets $\varphi(U \cap V)$ and $\psi(U \cap V)$ is an open half line; (ii) each of the sets $\varphi(U \cap V)$ and $\psi(U \cap V)$ is the union of two disjoint open half lines.

Since we may multiply both φ and ψ by -1, we may assume that, in case (i), $\varphi(U \cap V)$ has the form $(-\infty,a)$, and $\psi(U \cap V)$ - the form (b,∞). Consider the composition

$$(-\infty,a) = \varphi(U \cap V) \xrightarrow{\text{ab }\varphi^{-1}} U \cap V \xrightarrow{\text{ab }\psi} \psi(U \cap V) = (b,\infty).$$

This map is injective and continuous, and hence monotone and obviously increasing (if it were to decrease, the points $\varphi^{-1}(a)$ and $\psi^{-1}(b)$ would have no disjoint heighborhoods in X). Thus,

$$X = \psi^{-1}((-\infty,\psi(x_0)]) \cup \varphi^{-1}([\varphi(x_0),\infty)),$$

for some point $x_0 \in U \cap V$, and so in case (i) X is homeomorphic to

$\mathbb{R}^1$.

In case (ii), $\varphi(U \cap V) = (-\infty, a_1) \cup (a_2, \infty)$ and $\psi(U \cap V) = (-\infty, b_1) \cup (b_2, \infty)$, for some a_1, a_2, b_1, b_2 $(a_1 < a_2, \ b_1 < b_2)$, and we may assume that the composite homeomorphism

$$\varphi(U \cap V) \xrightarrow{\text{ab } \varphi^{-1}} U \cap V \xrightarrow{\text{ab } \psi} \psi(U \cap V)$$

maps $(-\infty, a_1)$ onto (b_2, ∞), and (a_2, ∞) onto $(-\infty, b_1)$. Both functions $(-\infty, a_1) \to (b_2, \infty)$ and $(a_2, \infty) \to (-\infty, b_1)$, which represent compressions of this composite homeomorphism, are increasing (if, for example, the first were to decrease, then the points $\varphi^{-1}(a_1)$ and $\psi^{-1}(b_2)$ would have no disjoint neighborhoods in X). We can thus write

$$X = \psi^{-1}([\psi(x_2), \psi(x_1)]) \cup \varphi^{-1}([\varphi(x_1), \varphi(x_2)]),$$

with some points $x_1 \in \varphi^{-1}((-\infty, a_1)) = \psi^{-1}((b_2, \infty))$ and $x_2 \in \varphi^{-1}((a_2, \infty)) = \psi^{-1}((-\infty, b_1))$. Therefore, in case (ii) X is homeomorphic to S^1.

17. <u>Every compact, connected, one-dimensional manifold is homeomorphic to either S^1 or D^1.</u>

PROOF. For a start, assume that the given manifold is closed. Then it can be covered by a finite number of open subsets homeomorphic to $\mathbb{R}^1$, and we may arrange these in a sequence $U_1, \ldots, U_s$ such that each union $V_k = U_1 \cup \ldots \cup U_k$ is connected. According to Lemma 16, the first of the sets $V_1, \ldots, V_s$ not homeomorphic to $\mathbb{R}^1$ is homeomorphic to S^1, and being both open and closed, it is the entire manifold, which is thus homeomorphic to S^1.

Assume now that the manifold has a boundary. Then its double is a closed, connected, one-dimensional manifold, and as such is homeomorphic to S^1. Therefore, the original manifold is homeomorphic to a subset of S^1. Since this subset is connected, closed, nonempty, different from S^1, and not reduced to a point, it is homeomorphic to D^1.

18 (LEMMA). <u>If a topological space X can be represented as the union of a nondecreasing sequence of open subsets, all homeomorphic to $\mathbb{R}^1$, then X is homeomorphic to $\mathbb{R}^1$.</u>

PROOF. Let $X = \cup V_i$ be the given representation. Clearly, any homeomorphism of V_i onto some interval (a,b) extends to a homeomorphism of V_{i+1} onto one of the intervals (a,b), $(a-1,b)$, $(a,b+1)$, or $(a-1,b+1)$. Hence one can construct inductively a sequence of intervals $\Delta_1, \Delta_2, \ldots$ and a sequence of homeomorphisms $\varphi_1: V_1 \to \Delta_1$, $\varphi_2: V_2 \to \Delta_2, \ldots$, such that $\varphi_i = \text{ab } \varphi_{i+1}$. The map of X onto the interval $\cup \Delta_i$, which agrees with φ_i on V_i, is obviously a homeomorphism.

19. <u>Every noncompact, connected, one-dimensional manifold is homeomorphic to either</u> $\mathbb{R}^1$ <u>or</u> $\mathbb{R}_-^1$.

PROOF. First, assume that the given manifold X has no boundary. Then X can be covered by a countable family of open subsets, all homeomorphic to $\mathbb{R}^1$, and we can arrange these in a sequence $U_1, U_2, \ldots,$ such that all unions $U_1 \cup \ldots \cup U_k$ are connected. Then all these unions are homeomorphic to $\mathbb{R}^1$. Indeed, if not, the first of them not homeomorphic to $\mathbb{R}^1$ is, according to Lemma 16, homeomorphic to S^1, and being open and closed must coincide with X; contradiction. Therefore, one can apply Lemma 18 to our manifold and deduce that it is homeomorphic to $\mathbb{R}^1$.

Now assume that X has a boundary. Then $\operatorname{dopp} X$ is a noncompact, connected, one-dimensional, boundaryless manifold, and must be homeomorphic to $\mathbb{R}^1$. It follows that X is homeomorphic to a connected, closed, noncompact subset of $\mathbb{R}^1$, different from $\mathbb{R}^1$; as such, it is homeomorphic to $\mathbb{R}_-^1$.

2. Differentiable Structures

1. Recall that a real function defined on an open subset of $\mathbb{R}^n$ is of <u>class</u> C^r (or a C^r-<u>function</u>) if it has continuous partial derivatives of all orders up to and including r. The definition implies that $0 \leqslant r \leqslant \infty$, that C^0 is the class of all continuous functions, and that C^∞ is the class of all functions which have continuous partial derivatives of all orders. In addition, we say that the real analytic functions are of <u>class</u> C^a (or C^a-<u>functions</u>). It is convenient to consider $a > \infty$ and thus encompass all the classes listed above by the inequality $0 \leqslant r \leqslant a$.

Obviously, these definitions can be extended to real functions defined on an open subset of the half space $\mathbb{R}_-^n$. To do this, we consider the derivatives with respect to the first coordinate at the points of the boundary hyperplane $\mathbb{R}^1$ to be left derivatives, and analyticity at such points is understood as the existence of an analytic continuation to an open set in $\mathbb{R}^n$. We further extend the definitions to maps of an open subset of $\mathbb{R}^n$ or of $\mathbb{R}_-^n$, into any subset of $\mathbb{R}^q$: such a map is of <u>class</u> C^r, or, simply, a C^r-<u>map</u>, if its coordinate functions are of class C^r.

2. A map f of an open subset of $\mathbb{R}^n$ or $\mathbb{R}_-^n$ into an open

subset of $\mathbb{R}^p$ or $\mathbb{R}^p_-$ is a <u>diffeomorphism</u> if it is invertible and both f and f^{-1} are of class C^1. Two sets which can be transformed into each other by a diffeomorphism are said to be <u>diffeomorphic.</u>

The following facts are contained in well-known theorems of calculus:

(i) If an open subset of $\mathbb{R}^p$ or $\mathbb{R}^p_-$ is diffeomorphic to an open subset of $\mathbb{R}^n$ or $\mathbb{R}^n_-$, then $p = n$.

(ii) An open subset of $\mathbb{R}^n_-$ which is diffeomorphic to an open subset of $\mathbb{R}^n$ is open in $\mathbb{R}^n$.

(iii) A diffeomorphism which is the inverse of a diffeomorphism of class C^r is itself of class C^r.

C^r-structures and C^r-spaces

3. The definitions below refer to a given set X.

A <u>chart of dimension</u> n on X is an invertible map of a subset of X onto an open subset of $\mathbb{R}^n$ or $\mathbb{R}^n_-$. The domain of a chart φ is called the <u>support</u> of φ and is denoted by supp φ.

Two charts, φ and ψ, are C^r-<u>compatible</u> (or have a C^r-<u>overlap</u>) $(0 \leqslant r \leqslant \dot{a})$ if the set $\varphi(\text{supp}\,\varphi \cap \text{supp}\,\psi)$ is open in $\text{Im}\,\varphi$, the set $\psi(\text{supp}\,\varphi \cap \text{supp}\,\psi)$ is open in $\text{Im}\,\psi$, and the maps

$$\varphi(\text{supp}\,\varphi \cap \text{supp}\,\psi) \xrightarrow{\text{ab}\,\varphi^{-1}} \text{supp}\,\varphi \cap \text{supp}\,\psi \xrightarrow{\text{ab}\,\psi} \psi(\text{supp}\,\varphi \cap \text{supp}\,\psi)$$

and

$$\psi(\text{supp}\,\varphi \cap \text{supp}\,\psi) \xrightarrow{\text{ab}\,\psi^{-1}} \text{supp}\,\varphi \cap \text{supp}\,\psi \xrightarrow{\text{ab}\,\varphi} \varphi(\text{supp}\,\varphi \cap \text{supp}\,\psi),$$

which are inverses of one another, are of class C^r (i.e., C^r-diffeomorphisms for $r > 1$ and homeomorphisms for $r = 0$). This condition is trivially satisfied whenever $\text{supp}\,\varphi \cap \text{supp}\,\psi = \emptyset$. If $\text{supp}\,\varphi \cap \text{supp}\,\psi \neq \emptyset$, then the C^1-compatibility of the charts φ and ψ implies the equality of their dimensions. In fact, this equality results also from the C^0-compatibility of φ and ψ, as shown by 1.4.

A collection of charts is an <u>n-dimensional C^r-atlas of the set</u> X if these charts cover X, are n-dimensional, and each two of them are C^r-compatible. Two C^r-atlases of X are C^r-<u>equivalent</u> if their union is again a C^r-atlas. This is clearly an equivalence relation, and the equivalence classes of n-dimensional C^r-atlases of the set X are called <u>n-dimensional</u> C^r-<u>structures</u>. The C^r-structures with $r > 0$ are called <u>differentiable structures.</u>

Clearly, if $0 \leqslant q \leqslant r$, then each n-dimensional C^r-atlas is also an n-dimensional C^q-atlas, and two equivalent C^r-atlases are also C^q-equivalent. Thus, when $0 \leqslant q \leqslant r$ every n-dimensional C^r-structure uniquely extends to a C^q-structure.

Every C^r-structure contains a maximal atlas, namely the union of all its atlases. The latter is called the <u>complete atlas of the structure</u>, and its charts are called the <u>charts of the structure.</u> When we pass from a C^r-structure to its C^q-extension, the complete atlas extends too.

4. A set endowed with an n-dimensional C^r-structure is called an <u>n-dimensional C^r-space.</u> The charts and atlases of the structure are refered to as the charts and atlases of the space. The complete atlas of a C^r-space X is denoted by $\operatorname{Atl} X$.

The coordinate functions of a chart φ of the C^r-space X are called <u>coordinates on</u> $\operatorname{supp} \varphi$ or, alternatively, <u>local coordinates in</u> X.

We shall denote by $C^q X$ the C^q-space obtained from the C^r-space X by extending its C^r-structure to a C^q-structure, $0 \leqslant q \leqslant r$. The C^r-spaces with $r \geqslant q$ are also termed <u>$C^{\geqslant q}$-spaces.</u>

For examples of C^r-spaces we may look at all the open subsets X of $\mathbb{R}^n$ or $\mathbb{R}^n_-$, with the C^r-structure defined by the atlas reduced to the single chart $\operatorname{id} \colon X \to X$. In particular, for any r, the charts $\operatorname{id} \mathbb{R}^n$ and $\operatorname{id} \mathbb{R}^n_-$ transform $\mathbb{R}^n$ and $\mathbb{R}^n_-$ into n-dimensional C^r-spaces.

5. Every n-dimensional locally Euclidean space has an obvious C^0-structure: its complete atlas consists of all possible homeomorphisms $U \to U'$, where U is an open subset of the space and U' is an open subset of $\mathbb{R}^n$ or $\mathbb{R}^n_-$. On the other hand, applying the "union of topological spaces" construction (see 1.2.4.3) to the complete atlas of a given n-dimensional C^0-space, we obtain an n-dimensional locally Euclidean space, and this transition is the inverse of the previous one. Therefore, C^0-spaces are just locally Euclidean spaces.

Since any differentiable structure extends uniquely to a C^0-structure, every C^r-space with $r > 0$ is also a locally Euclidean space. Its topology may be described in a more direct fashion as the topology of the union constructed from any atlas of the structure.

6. Obviously, every point of an n-dimensional C^r-space X can be covered by a chart φ of X such that $\operatorname{Im} \varphi = \mathbb{R}^n$ or $\mathbb{R}^n_-$. The points with $\operatorname{Im} \varphi = \mathbb{R}^n$ are called <u>interior points</u> and form an open dense set, called the <u>interior</u> of the space X, denoted by $\operatorname{int} X$. The remaining points are called <u>boundary points</u> and they form a closed set, called the boundary of X, denoted by ∂X. These notations are

in agreement with those introduced at 1.2. In fact, when $r = 0$, the previous and present definitions of the interior and boundary points coincide.

When $r > 0$, we use Proposition 2 (ii) in order to recognize the interior and boundary points. According to this proposition, when $r > 0$ a point x of an n-dimensional C^r-space is a boundary point if and only if $\operatorname{Im}\varphi \subset \mathbb{R}^n_-$ and $\varphi(x) \in \mathbb{R}^{n-1}_1$, where φ is a chart on this space with $x \in \operatorname{supp}\varphi$. In particular, if we regard $\mathbb{R}^n_-$ as a C^r-space with $r > 0$, then $\partial\mathbb{R}^n_- = \mathbb{R}^{n-1}_1$. Recall that the corresponding statement for $r = 0$ appeared in 1.4 and its proof was postponed until Chapter 4.

The above characterization of the boundary points shows that the interior and the boundary of a C^r-space do not change when we extend its C^r-structure to a C^q-structure, for any $q \leqslant r$. In other words, for $0 \leqslant q \leqslant r$, $\operatorname{int}(C^q X) = \operatorname{int} X$ and $\partial(C^q X) = \partial X$. We emphasize that the equalities $\operatorname{int}(C^0 X) = \operatorname{int} X$ and $\partial(C^0 X) = \partial X$ were proved by a reference to 1.4, i.e., they depended upon results from Chapter 4, whereas the equalities $\operatorname{int}(C^q X) = \operatorname{int} X$ and $\partial(C^q X) = \partial X$ for $q > 0$ need no such reference.

Using the relation $\partial X = \partial(C^0 X)$, we see that Theorem 1.8 is valid for C^r-spaces with $r > 0$ too. That is to say, a C^r-space is connected if and only if its interior is connected. However, this C^r-variant of Theorem 1.8 can be proved by merely repeating the proof of the original theorem, and therefore we can eliminate the reference to Chapter 4.

7. Suppose A is an open subset of an n-dimensional C^r-space X. Then the charts of X whose supports are included in A yield a C^r-atlas of the set A, and define an n-dimensional C^r-structure on A. In this way, any open subset of an n-dimensional C^r-space is an n-dimensional C^r-space. In particular, the interior of any n-dimensional C^r-space is an n-dimensional C^r-space without boundary. Moreover, the interior and the boundary of an open subset U of the C^r-space X are obviously given by $\operatorname{int} U = U \cap \operatorname{int} X$ and $\partial U = U \cap \partial X$.

Suppose φ is a chart on an n-dimensional C^r-space X. Then $\operatorname{ab}\varphi : \partial X \cap \operatorname{supp}\varphi \to \varphi(\partial X \cap \operatorname{supp}\varphi)$ is an $(n-1)$-dimensional chart on ∂X. In this way we may construct a C^r-atlas of the set ∂X, and so define a C^r-structure on ∂X. Thus, the boundary of an n-dimensional C^r-space is an $(n-1)$-dimensional C^r-space without boundary.

If φ_1 (φ_2) is a chart on the C^{r_1}-space X_1 $(C^{r_2}$-space $X_2)$ such that $\operatorname{Im}\varphi_1 = \mathbb{R}^{n_1}$ or $\mathbb{R}^{n_1}_-$ (respectively, $\operatorname{Im}\varphi_2 = \mathbb{R}^{n_2}$ or $\mathbb{R}^{n_2}_-$), then the composition of $\varphi_1 \times \varphi_2$ with one of the homeomorphisms

$\mathbb{R}^{n_1} \times \mathbb{R}^{n_2} \to \mathbb{R}^{n_1+n_2}$ and $\mathbb{R}_-^{n_1} \times \mathbb{R}^{n_2} \to \mathbb{R}_-^{n_1+n_2}$, defined in 1.3, provides an (n_1+n_2)-dimensional chart on $X_1 \times X_2$. If $\partial X_2 = \emptyset$, then the charts constructed as above form a C^r-atlas of the set $X_1 \times X_2$, with $r = \min(r_1,r_2)$, and hence define a C^r-structure on $X_1 \times X_2$. Thus the product of the n_1-dimensional C^{r_1}-space X_1 and the n_2-dimensional C^{r_2}-space X_2 with $\partial X_2 = \emptyset$ is an (n_1+n_2)-dimensional C^r-space, where r is the smallest of the numbers r_1 and r_2. In general, the product of the n_i-dimensional C^{r_i}-spaces X_i, $i = 1,\ldots,s$, such that at most one of them has a boundary, is an $(n_1+\ldots+n_s)$-dimensional C^r-space, where $r = \min(r_1,\ldots,r_s)$. Moreover, $\mathrm{int}(X_1 \times \ldots \times X_s) = \mathrm{int}\,X \times \ldots \times \mathrm{int}\,X_s$, and if X_i is the only space having a boundary, then $\partial(X_1 \times \ldots \times X_s) = X_1 \times \ldots \times X_{i-1} \times \partial X_i \times X_{i+1} \times \ldots \times X_s$. For $r = 0$ both formulas can be found in 1.7; for $r > 0$, they are plain.

It is clear that when we extend the C^r-structure of the C^r-space X to a C^q-structure, the C^r-structures induced on the open subsets of X and on its boundary X also extend to C^q-structures, and that $C^q(X_1 \times \ldots \times X_s) = C^q X_1 \times \ldots \times C^q X_s$. In particular, the topology defined by the above induced C^r-structures coincide with the relative topology, and the product of the C^{r_i}-spaces X_i, $i = 1,\ldots,s$, considered as a topological space, is just the product of the topological spaces $X_1,\ldots,X_s$.

Smooth Maps

8. A continuous map f of a $C^{\geq r}$-space X into a $C^{\geq r}$-space Y is of <u>class</u> C^r, or a C^r-<u>map</u>, if for any chart φ on X and any chart ψ on Y, the composite map

$$\varphi(\mathrm{supp}\,\varphi \,\cap\, f^{-1}(\mathrm{supp}\,\psi)) \xrightarrow{\;\mathrm{ab}\,\varphi^{-1}\;} \mathrm{supp}\,\varphi \,\cap\, f^{-1}(\mathrm{supp}\,\psi) \xrightarrow{\;\mathrm{ab}\,f\;} \mathrm{supp}\,\psi \xrightarrow{\;\psi\;} \mathrm{Im}\,\psi$$

is of class C^r (see 1). Such a composite map is called a <u>local representative</u> of the map f; we use the notation $\mathrm{loc}(\varphi,\psi)f$. Obviously, a map $f: X \to Y$ of $C^{\geq r}$-spaces is of class C^r if and only if the local representatives of f constructed for all the charts of of some atlases of X and Y are of class C^r.

Note that this general definition of C^r-maps contains the definition given in 1. Now, as before, a C^0-map is just a continuous map. The maps of class C^1 are called <u>smooth</u>, and the maps of class C^a — <u>(real) analytic</u>.

The composition of two C^r-maps is obviously a C^r-map. If A

is an open subset or the boundary of a C^r-space X, then the inclusion $A \to X$ is a C^r-map. If A is open in X or $A = \partial X$, and B is open in Y or $B = \partial Y$, then the compression $A \to B$ of any C^r-map $X \to Y$ is a C^r-map.

9. A map f of a $C^{\geq 1}$-space X into a $C^{\geq 1}$-space Y is a <u>diffeomorphism</u> if it is invertible and both f and f^{-1} are smooth. The space Y is said to be <u>diffeomorphic</u> to the space X if there is a diffeomorphism $X \to Y$, and C^r-<u>diffeomorphic</u> to X, if there is a C^r-diffeomorphism $X \to Y$.

Of course, the identity map of a $C^{\geq r}$-space with $r \geqslant 1$ is a C^r-diffeomorphism. Also, the composition of two C^r-diffeomorphisms is a C^r-diffeomorphism, and the inverse of a C^r-diffeomorphism is a C^r-diffeomorphism itself, as we may easily see from 2 (iii). Therefore, the property of being C^r-diffeomorphic is an equivalence relation.

Using 2 (i), we conclude that nonempty diffeomorphic spaces have the same dimension.

10. Let $f_1 \colon X_1 \to Y_1, \ldots, f_m \colon X_m \to Y_m$ be C^r-maps with $r \geqslant 1$, where no more than one of the spaces $X_1, \ldots, X_m$, and no more than one of the spaces $Y_1, \ldots, Y_m$, has a boundary. Then

$$f_1 \times \ldots \times f_m \colon X_1 \times \ldots \times X_m \longrightarrow Y_1 \times \ldots \times Y_m$$

is obviously a C^r-map. If $f_1, \ldots, f_m$ are diffeomorphisms, then so is $f_1 \times \ldots \times f_m$.

The canonical homeomorphism $X_1 \times X_2 \to X_2 \times X_1$ is a C^r-diffeomorphism for any two C^r-spaces X_1 and X_2 with $r \geqslant 1$, such that one of them has no boundary. The canonical homeomorphisms

$(X_1 \times \ldots \times X_{m-1}) \times X_m \to X_1 \times \ldots \times X_m$ and $X_1 \times (X_2 \times \ldots \times X_m) \to$ $\to X_1 \times \ldots \times X_m$ are C^r-diffeomorphisms for any C^r-spaces $X_1, \ldots, X_m$ with $r \geqslant 1$, such that no more that one of them has a boundary.

Subspaces

11. A subset A of an n-dimensional C^r-space X with $r \geqslant 1$ is a <u>k-dimensional subspace of</u> X if each point of A is covered by a chart φ on X such that the pair $(\mathrm{Im}\,\varphi, \varphi(A \cap \mathrm{supp}\,\varphi))$ coincides with one of the pairs $(\mathbb{R}^n, \mathbb{R}^k)$, $(\mathbb{R}^n_-, \mathbb{R}^k)$, or $(\mathbb{R}^n_-, \mathbb{R}^k_-)$. If $k > 0$, then this condition is obviously equivalent to the following one: each point of A is covered by a chart φ of the space X such that $\varphi(A \cap \mathrm{supp}\,\varphi) = \mathrm{Im}\,\varphi \cap \mathbb{R}^k$ or $\mathrm{Im}\,\varphi \cap \mathbb{R}^k_-$.

Suppose A is a k-dimensional subspace of an n-dimensional

C^r-space X, and consider the maps ab $\varphi : A \cap \text{supp } \varphi \to \varphi(A \cap \text{supp } \varphi)$
corresponding to all charts φ on X such that $\varphi(A \cap \text{supp } \varphi)$ =
= $\text{Im } \varphi \cap \mathbb{R}^k$ or $\text{Im } \varphi \cap \mathbb{R}^k_-$. Each such map is a k-dimensional chart on
A, and together they yield a k-dimensional C^r-atlas of A. The
C^r-structure defined by this atlas transforms A into a k-dimensional
C^r-space. The topology of this C^r-space obviously coincides with the
relative topology.

When we extend the C^r-structure of a space to a C^q-structure
with $q \geqslant 1$, its subspaces remain subspaces. The subspaces of $C^q X$
are called C^q-subspaces of the original C^r-space X.

The codimension of a subspace is the difference between the
dimension of the ambient space and that of the subspace.

It is evident that the open subsets of a C^r-space (with
$r \geqslant 1$) are among its subspaces; in particular, we cite its interior
and its components. Warning: if not empty, the boundary of a C^r-space
is not a subspace.

It is readily seen that if A is a subspace of codimension
0 of a C^r-space X, then $\text{int } A = \text{Int } A \cap \text{int } X$.

A subspace of a C^r-space is neat if it is closed as a subset
and its boundary is contained in the boundary of the space. Note that
every neat subspace of codimension 0 is made of whole components of
the ambient space.

12. The definition of a subspace given in 11 contains
implicitly the generally used method of defining subspaces through
equations and inequalities. Namely, according to the second variant
of our definition, a subset A of an n-dimensional C^r-space X $(r \geqslant 1)$
is a subspace of X of positive dimension k if and only if each point
of A has a neighborhood U with coordinates $\varphi_1 , \ldots , \varphi_n$, such that
the intersection $A \cap U$ is defined in U either by the equations
$\varphi_{k+1}(x) = 0 , \ldots , \varphi_n(x) = 0$, or by the equations $\varphi_{k+1}(x) = 0 , \ldots ,$
$\varphi_n(x) = 0$ and the inequality $\varphi_1(x) \leqslant 0$.

As a procedure for defining subspaces, this formulation has
an obvious disatvantage; namely, we must assume from the beginning that
$\varphi_1 , \ldots , \varphi_n$ are local coordinates. We make use the implicit function
theorem to make it more efficient.

Let X be a C^r-space $(r \geqslant 1)$, $x_0 \in X$, and let $f_1 , \ldots , f_m$
be real C^1-functions defined on a neighborhood U_0 of x_0. We say that
$f_1 , \ldots , f_m$ are independent at the point x_0 if there is a chart φ on
X with $x_0 \in \text{supp } \varphi$ and such that the functions
$g_1 , \ldots , g_m : \varphi(U_0 \cap \text{supp } \varphi) \to \mathbb{R}$ defined by $g_i(y) = f_i(\varphi^{-1}(y))$ have
linearly independent gradients at the point $\varphi(x_0)$. The following

statements are consequences of the implicit function theorem:

(i)   if the C^r-functions $f_1,\ldots,f_m$ are independent at the interior point x_0 of the C^r-space X, then they can be completed to a system of coordinates in a neighborhood of x_0;

(ii)   if the C^r-functions $f_1,\ldots,f_m$ are independent at the boundary point x_0 of the C^r-space X, and the function f_1 is negative at the interior points and zero on the boundary, then $f_1,\ldots,f_m$ can be completed to a system of coordinates in a neighborhood of x_0 having f_1 as the first coordinate.

Comparing with the previous, coordinate description of the subspaces of a C^r-space, we see that a subset A of the n-dimensional C^r-space X is a subspace of positive dimension k of X if:

(i)   each point $x_0 \in A$ which is an interior point for X has a neighborhood U where the intersection $A \cap U$ is defined either by the equations $\varphi_{k+1}(x) = 0,\ldots,\varphi_n(x) = 0$, where $\varphi_{k+1},\ldots,\varphi_n$ are C^r-functions independent at x_0, or by the equations $\varphi_{k+1}(x) = 0,\ldots,$ $\varphi_n(x) = 0$ and the inequality $\varphi_1(x) \leqslant 0$, where $\varphi_1,\varphi_{k+1},\ldots,\varphi_n$ are C^r-functions independent at x_0;

(ii)   each point $x_0 \in A$ which lies on the boundary of X has a neighborhood U such that the intersections $\mathrm{int}\,X \cap U$, $\partial X \cap U$, and $A \cap U$ are defined in U by the inequality $\varphi_1(x) < 0$, by the equation $\varphi_1(x) = 0$, and by the equations $\varphi_{k+1}(x) = 0,\ldots,\varphi_n(x) = 0$, respectively, where $\varphi_1,\varphi_{k+1},\ldots,\varphi_n$ are C^r-functions independent at x_0.

13.   An obvious consequence of the definition of a subspace is that the interior $\mathrm{int}\,A$ of a subspace A of the C^r-space X is contained in $\mathrm{int}\,X$, and that the intersections $\partial A \cap \mathrm{int}\,X$ and $\partial A \cap \partial X$ are open in ∂A, i.e., they consist of whole components of ∂A. Moreover, it is clear that $\mathrm{int}\,A$ is a subspace of both X and $\mathrm{int}\,X$, while $\partial A \cap \mathrm{int}\,X$ is a subspace of X. In particular, if A is a neat subspace of X, then $\partial A = A \cap \partial X$ and $\mathrm{int}\,A$ is a neat subspace of $\mathrm{int}\,X$, while ∂A is a neat subspace of ∂X.

The inclusion $A \to X$ of a subspace A of a C^r-space X is obviously a C^r-map. The compression $A \to B$ of any C^q-map $X \to Y$, where A (B) is a subspace of X (respectively, Y) is a C^q-map.

Let A_i be a subspace of the C^r-space X_i, $i = 1,\ldots,s$, and assume that at most one of the spaces $X_1,\ldots,X_s$ has a boundary, the same being true for the subspaces $A_1,\ldots,A_s$. Then $A_1 \times \ldots \times A_s$ is a subspace of $X_1 \times \ldots \times X_s$, and is a neat subspace if each A_i is neat. For example, the fibers of the product $X_1 \times \ldots \times X_s$ are neat subspaces.

Let A be a subspace of the C^r-space X, and let B be a subspace of A. Then, using the description of subspaces in 12, we see that B is a subspace of X; in particular, if B is a neat subspace of A and A is a neat subspace of X, then B is also a neat subspace of X. In a similar fashion we conclude that a neat subspace B of the C^r-space X which is contained in a neat subspace A of X is also a neat subspace of A.

C^r-manifolds

14. A C^r-space is a C^r-manifold, or a manifold of class C^r, if it is a topological manifold, i.e., a second-countable, Hausdorff space. A manifold of class C^0 is simply a topological manifold; see 5. The manifolds of class C^r with $r \geqslant 1$ are called smooth, or differentiable. The manifolds of class C^a are called (real) analytic.

Since any C^r-structure is defined by one of its atlases, it is interesting to discover those properties of an atlas of a C^r-space which guarantee that the space is Hausdorff and second-countable. Here we formulate only two obvious conditions: if each pair of points is covered by a chart of the atlas or by two disjoint charts of the atlas, then the space is Hausdorff; if the atlas is countable, then the space is second-countable.

There is no need to check that a space is Hausdorff and second-countable if the differentiable structure is introduced on a set which is already a topological manifold, and the topology defined by the differentiable structure coincides with the initial one. If the differentiable structure is defined by an atlas $\{\varphi_\alpha\}$, then the two topologies agree if and only if $\operatorname{supp}\varphi_\alpha$ are open and φ_α are homeomorphisms; see 5.

A smooth manifold is closed if it is compact and has no boundary; cf. 1.10. Warning: at the present time the equality $\partial X = \partial(C^0 X)$ is not proven (see 6). Therefore, we should be careful to distinguish between the smooth manifold X being closed and the topological manifold $C^0 X$ being closed.

16. Reconsidering the statements made in 7 and 11 in the light of the corresponding properties of Hausdorff spaces, second-countable spaces, and compact spaces, we may deduce the following: every open subset of an n-dimensional C^r-manifold is an n-dimensional C^r-manifold; the interior of an n-dimensional C^r-manifold is an n-dimensional C^r-manifold without boundary; the boundary of an

n-dimensional C^r-manifold is an (n-1)-dimensional C^r-manifold without boundary; the boundary of a compact C^r-manifold is a closed C^r-manifold; the product of the C^r-manifolds $X_1,\ldots,X_s$ of dimensions $n_1,\ldots,n_s$, such that no more than one of them has a boundary, is an $(n_1+\ldots+n_s)$-dimensional C^r-manifold; a subspace of a C^r-manifold is a C^r-manifold.

The subspaces of smooth manifolds are called <u>submanifolds</u>, and the neat subspaces - <u>neat submanifolds</u>.

16. The basic examples of n-dimensional C^r-manifolds with $r \geqslant 1$ are again $\mathbb{R}^n$ and $\mathbb{R}^n_-$ (see 4). The submanifolds of these spaces provide an unlimited supply of examples of C^r-manifolds. The simplest are the submanifolds of $\mathbb{R}^n$ defined by one system of equations $\varphi_{k+1}(x) = 0,\ldots,\varphi_n(x) = 0$, where $\varphi_{k+1},\ldots,\varphi_n$ are C^r-functions defined on an open subset of $\mathbb{R}^n$ and having linearly independent gradients on the set of their common zeros. We may add to such a system the inequality $\varphi_1(x) \leqslant 0$, where φ_1 is any C^r-function defined in a neighborhood of the set of common zeros of $\varphi_{k+1},\ldots,\varphi_n$, zero on this set, and such that the gradients of $\varphi_1,\varphi_{k+1},\ldots,\varphi_n$ are linearly independent on the same set. For example, the sphere S^{n-1} is defined, in standard coordinates, by the equation $x_1^2 + \ldots + x_n^2 - 1 = 0$, and the ball D^n - by the inequality $x_1^2 + \ldots + x_n^2 - 1 \leqslant 0$. Hence S^{n-1} and D^n are submanifolds of $\mathbb{R}^n$ and, in particular, C^a-manifolds.

The following facts are clear: $\mathbb{R}^k$, for $k \leqslant n$, and S^k, for $k < n$, are neat submanifolds of $\mathbb{R}^n$; $\mathbb{R}^k_-$ and D^k are not neat submanifolds of $\mathbb{R}^n$ for $k \leqslant n$; $\mathbb{R}^k_-$ is a neat submanifold of $\mathbb{R}^n_-$ for $k \leqslant n$; S^k is a neat submanifold of S^n for $k \leqslant n$; and D^k is a neat submanifold of D^n for $k \leqslant n$.

17. Finally, we note that every real, n-dimensional vector space has a natural C^r-structure for any r $(0 \leqslant r \leqslant a)$, which makes it into an n-dimensional C^r-manifold. This structure is defined by the linear charts i.e., by the linear maps onto $\mathbb{R}^n$.

3. Orientations

1. Consider a C^r-manifold X, and let $\mathrm{Catl}\,X$ denote the atlas of X consisting of all the charts with connected support. If φ and ψ are two charts on the smooth manifold X, we denote by $J(\varphi,\psi)$ the Jacobian of the map $\mathrm{loc}(\varphi,\psi)\mathrm{id}$, i.e., of the composite map

$$\varphi(\mathrm{supp}\,\varphi \,\cap\, \mathrm{supp}\,\psi) \xrightarrow{\mathrm{ab}\,\varphi^{-1}} \mathrm{supp}\,\varphi \cap \mathrm{supp}\,\psi \xrightarrow{\mathrm{ab}\,\psi} \psi(\mathrm{supp}\,\varphi \cap \mathrm{supp}\,\psi).$$

An _orientation_ of the smooth manifold X is a function
ω: Catl X $\to$ S^0 such that $\omega(\varphi)$ = [sign J(φ,ψ)](y)$\omega(\psi)$ for each two
charts $\varphi,\psi \in$ Catl X , where y is an arbitrary point of
$\varphi($supp $\varphi \cap$ supp $\psi)$ [therefore, the function sign J(φ,ψ) must be
constant on $\varphi($supp $\varphi \cap$ supp $\psi)$]. A smooth manifold endowed with an
orientation is said to be _oriented._ A smooth manifold which can be
oriented is _orientable._

Clearly, an orientation Catl X $\to$ S^0 is determined by its
values on any subatlas of Catl X . Moreover, every function which
carries a subatlas of Catl X   into S^0 and satisfies the previous
compatibility condition, i.e., its values on two charts φ,ψ of the
given subatlas are obtained one from another multiplying by
[sign J(φ,ψ)](y) with y $\in \varphi($supp $\varphi \cap$ supp $\psi)$, extends to an
orientation Catl X $\to$ S^0.

When one extends the C^r-structure of a C^r-manifold X to a
C^q structure with q < r, Catl X becomes a subatlas of Catl C^qX . If
q $\geqslant$ 1, this establishes a one-to-one correspondence between the
orientations of the manifolds X and C^qX.

2. For each orientation ω: Catl X $\to$ S^0 there exists the
opposite orientation $-\omega$, and thus every nonempty orientable manifold
has at least two orientations.

Given two arbitary orientations of the manifold X, the set
covered by the charts on which they agree, and the set covered by the
charts on which they do not agree are open, disjoint, and together they
cover X; hence, each of them is a union of components of the manifold
X. Consequently, the orientation of a connected manifold is uniquely
determined by its value at one chart. Moreover, every smooth, connected,
orientable manifold has exactly two orientations, whereas an orientable
smooth manifold with s components has 2^s orientations.

The standard way of describing an orientation of a smooth
connected manifold is to indicate the charts where it is positive.
For example, $\mathbb{R}^n$ has a natural orientation, positive on the chart
id $\mathbb{R}^n$.

3. Since any point of a zero-dimensional manifold X is
covered by only one chart of Catl X , it follows that every zero-
dimensional manifold is orientable, and has actually a natural
orientation, identically equal to +1. We shall see later (in 5.3.1
and 5.6.3.4) that all one-dimensional manifolds are orientable,
whereas the manifolds of dimension $\geqslant$ 2 are not necessarily so.

In Chapter 5 we shall give effective sufficient conditions
for the orientability of a manifold of arbitrary dimension (see 5.6.3).

The crudest of them is that the manifold be simply connected.

4. Let A be an open subset of a smooth manifold X. Then $\text{Catl}\,A \subset \text{Catl}\,X$, and every orientation of X induces an orientation of A; in particular, A is orientable whenever X is. If A intersects all the components of X, then the orientation of X is determined by the orientation induced on the submanifold A.

In the case $A = \text{int}\,X$, we may say more: not only does each orientation of X restrict to an orientation of the manifold $\text{int}\,X$, but also each orientation of $\text{int}\,X$ extends to an orientation of X. Indeed, for each connected subset U of X, $U \cap \text{int}\,X$ is connected (see 2.6 and 2.7) and so the compression $\text{ab}\,\varphi : \text{supp}\,\varphi \cap \text{int}\,X \to \varphi(\text{supp}\,\varphi \cap \text{int}\,X)$ belongs to $\text{Catl}(\text{int}\,X)$ for any $\varphi \in \text{Catl}\,X$. This enables us to extend any orientation $\omega : \text{Catl}(\text{int}\,X) \to S^0$ to an orientation $\text{Catl}\,X \to S^0$ through the formula $\varphi \mapsto \omega(\text{ab}\,\varphi)$. Thus, if we associate to each orientation $\text{Catl}\,X \to S^0$ its restriction $\text{Catl}(\text{int}\,X) \to S^0$, we obtain a one-to-one correspondence between the orientations of the manifolds X and $\text{int}\,X$. In particular, X is orientable whenever $\text{int}\,X$ is such.

According to 2.7, for each chart φ on the smooth manifold X we have the corresponding chart $\text{ab}\,\varphi : \text{supp}\,\varphi \cap \partial X \to \varphi(\text{supp}\,\varphi \cap \partial X)$ on its boundary ∂X. It is clear that every chart in $\text{Catl}\,\partial X$ is of the form $\text{ab}\,\varphi$, where $\varphi \in \text{Catl}\,X$, and that for each orientation $\omega : \text{Catl}\,X \to S^0$ of the manifold X we have an orientation of its boundary, defined by the rule $\text{ab}\,\varphi \mapsto \omega(\varphi)$. In particular, ∂X is orientable whenever X is such. If all the components of X have a boundary, then the orientation of X is uniquely determined by the orientation it induces on ∂X.

The only submanifolds of a smooth manifold X which inherit a natural orientation when X is oriented are those of codimension zero. This orientation has been implied in the discussion of the open subsets of X. If A is an arbitrary submanifold of X of codimension zero, then the induced orientation is defined by the orientation of its interior $\text{int}\,A$, which is open in X. In particular, A is orientable if X is such. If A intersects all the components of X, then the orientation of X is uniquely determined by the orientation it induces on A.

Since the manifold $\mathbb{R}^n$ has a natural orientation, all its n-dimensional submanifolds inherit a natural orientation too. In particular, D^n carries a natural a natural orientation, which in turn induces an orientation of its boundary S^{n-1}. Warning: the orientation of S^0 induced by the orientation of D^1 does not coincide with the canonical orientation of S^0, considered as a zero-dimensional manifold (see 3).

Orientations and Diffeomorphisms

5. Every diffeomorphism $f: X \to Y$ establishes a one-to-one correspondence between the orientations of the manifolds X and Y. If both manifolds are oriented and f transforms the orientation of X into the orientation of Y (into the opposite orientation of Y), then we say that f _preserves_ (_reverses_) _the orientation_ or that f is _orientation preserving_ (respectively, _orientation reversing_). To determine whether a diffeomorphism $f: X \to Y$ is orientation preserving or not, we may look at its local representatives; if X and Y are connected, it is enough to analyze only one local representative. Namely, suppose that ω_X and ω_Y are orientations of the connected manifolds X and Y. Let $\varphi \in \text{Catl } X$ and $\psi \in \text{Catl } Y$ be two charts and pick $x \in \text{supp}\,\varphi \cap f^{-1}(\text{supp}\,\psi)$. If the sign of the Jacobian of $\text{loc}(\varphi,\psi)f$ at the point $\varphi(x)$ coincides with (is opposite to) the sign of the product $\omega_X(\varphi)\omega_Y(\psi)$, then f is clearly orientation preserving (respectively, reversing).

6. Of special interest is the case where $X = Y$ is a connected manifold. It is readily seen that in this situation a diffeomorphism which preserves (reverses) one orientation of X, will preserve (respectively, reverse) all orientations of X. Therefore, in this case one can talk about an orientation preserving (reversing) diffeomorphism without fixing an orientation. In particular, every (auto)diffeomorphism of a smooth, connected, orientable manifold is either orientation preserving or orientation reversing.

As an example, consider a nonsingular linear transformation $f: \mathbb{R}^n \to \mathbb{R}^n$. It is clear that f is a diffeomorphism and that f is orientation preserving (reversing) if $\det f > 0$ (respectively, $\det f < 0$). If the transformation f is orthogonal, then its compressions $\text{ab}\,f: D^n \to D^n$ and $\text{ab}\,f: S^{n-1} \to S^{n-1}$ are meaningful, are diffeomorphisms, and preserve (reverse) orientation if $\det f = 1$ (respectively, $\det f = -1$) [here S^0 is oriented as ∂D^1]. If f is given by $f(x) = -x$, then $\det f = (-1)^n$. Hence this diffeomorphism and the induced antipodal map $\text{ab}\,f: S^{n-1} \to S^{n-1}$ are orientation preserving for n even and orientation reversing for n odd.

Orientations and Products of Manifolds

7. We have shown in 2.7 that every product $\varphi_1 \times \ldots \times \varphi_n$ of charts $\varphi_1,\ldots,\varphi_s$ on the smooth manifolds $X_1,\ldots,X_s$ <u>without boundary</u> of dimensions $n_1,\ldots,n_s$, may be regarded, using the canonical identification $\mathbb{R}^{n_1} \times \ldots \times \mathbb{R}^{n_s} = \mathbb{R}^{n_1+\ldots+n_s}$ [defined by the formula $((x_{11},\ldots,x_{1n_1}),\ldots,(x_{s1},\ldots,x_{sn_s})) \mapsto (x_{11},\ldots,x_{1n_1},\ldots,x_{s1},\ldots,x_{sn_s})]$, as a chart on the product $X_1 \times \ldots \times X_s$. Given orientations $\omega_1,\ldots,\omega_s$ of the manifolds $X_1,\ldots,X_s$, let us define a mapping from the collection of all charts $\varphi_1 \times \ldots \times \varphi_s$ with $\varphi_1 \in \text{Catl} X_1, \ldots, \varphi_s \in \text{Catl} X_s$, into S^0 by the formula

$$\varphi_1 \times \ldots \times \varphi_s \mapsto \omega_1(\varphi_1) \ldots \omega_s(\varphi_s).$$

It is obvious that the above collection of charts is a subatlas of $\text{Catl}(X_1 \times \ldots \times X_s)$, and that this mapping satisfies the compatibility condition introduced in 1. Therefore, it extends to an orientation of the manifold $X_1 \times \ldots \times X_s$, called the <u>product of the orientations</u> $\omega_1,\ldots,\omega_s$. The latter can be defined even if one of the manifolds $X_1,\ldots,X_s$ has a boundary: it is the orientation induced by the orientation of the interior $\text{int} X_1 \times \ldots \times \text{int} X_s$. Thus, a product of smooth, oriented manifolds (such that no more than one of them has a boundary) is oriented, and a product of smooth, orientable manifolds is orientable.

We note that the orientability of $X_1 \times \ldots \times X_s$ implies the orientability of each factor $X_1,\ldots,X_s$. In fact, if ω is an orientation of the product and $\varphi_1,\ldots,\varphi_{i-1},\varphi_{i+1},\ldots,\varphi_s$ are fixed charts of $\text{Catl}(\text{int} X_1),\ldots,\text{Catl}(\text{int} X_{i-1}),\text{Catl}(\text{int} X_{i+1}),\ldots,\text{Catl}(\text{int} X_s)$, then the mapping $\text{Catl}(\text{int} X_i) \to S^0$ given by $\varphi \mapsto \omega(\varphi_1 \times \ldots \times \varphi_{i-1} \times \varphi \times \varphi_{i+1} \times \ldots \times \varphi_s)$ is an orientation of the manifold $\text{int} X_i$.

8. <u>Suppose X_1 and X_2 are smooth oriented manifolds of dimensions n_1 and n_2, such that one of them has no boundary. The canonical diffeomorphism $X_1 \times X_2 \to X_2 \times X_1$ preserves orientation if $n_1 n_2$ is even and reverses orientation if $n_1 n_2$ is odd.</u>

PROOF. Let $\varphi_1 \in \text{Catl}(\text{int} X_1)$ and $\varphi_2 \in \text{Catl}(\text{int} X_2)$ be charts with $\text{Im}\,\varphi_1 = \mathbb{R}^{n_1}$ and $\text{Im}\,\varphi_2 = \mathbb{R}^{n_2}$, respectively. Clearly,

$\varphi_1 \times \varphi_2 \in \text{Catl}(\text{int}(X_1 \times X_2))$, $\varphi_2 \times \varphi_1 \in \text{Catl}(\text{int}(X_2 \times X_1))$, and the local representative $\mathbb{R}^{n_1+n_2} \to \mathbb{R}^{n_1+n_2}$ of the canonical diffeomorphism $X_1 \times X_2 \to X_2 \times X_1$ relative to these charts is given by $(x_1,\ldots,x_{n_1+n_2}) \mapsto (x_{n_1+1},\ldots,x_{n_1+n_2},x_1,\ldots,x_{n_1})$. Therefore, the Jacobian of this local representative is $(-1)^{n_1 n_2}$.

9. <u>Suppose</u> $X_1,\ldots,X_s$ <u>are smooth oriented manifolds of dimensions</u> $n_1,\ldots,n_s$, <u>such that only one, say</u> X_i, <u>has a boundary. Then the canonical orientation of the product</u>

$$X_1 \times \ldots \times X_{i-1} \times \partial X_i \times X_{i+1} \times \ldots \times X_s$$

<u>and the natural orientation that this product receives as</u>

$\partial(X_1 \times \ldots \times X_s)$ <u>differ by a factor of</u> $(-1)^{n_1+\ldots+n_{i-1}}$.

PROOF. Obviously, the above orientations agree if $i = 1$. The case $i > 1$ reduces to $i = 1$ with the aid of the diffeomorphisms

$$X_1 \times \ldots \times X_s \to X_i \times X_1 \times \ldots \times X_{i-1} \times X_{i+1} \times \ldots \times X_s$$

and

$$X_1 \times \ldots \times X_{i-1} \times \partial X_i \times X_{i+1} \times \ldots \times X_s \to$$
$$\to \partial X_i \times X_1 \times \ldots \times X_{i-1} \times X_{i+1} \times \ldots \times X_s.$$

The first of them is the product of the canonical diffeomorphism

$$(X_1 \times \ldots \times X_{i-1}) \times X_i \to X_i \times (X_1 \times \ldots \times X_{i-1})$$

with $\text{id}(X_{i+1} \times \ldots \times X_s)$, while the second is the product of the canonical diffeomorphism

$$(X_1 \times \ldots \times X_{i-1}) \times \partial X_i \to \partial X_i \times (X_1 \times \ldots \times X_{i-1})$$

with $\text{id}(X_{i+1} \times \ldots \times X_s)$. The first diffeomorphism preserves (reverses) the orientation if the product $(n_1+\ldots+n_{i-1})n_i$ is even (respectively, odd), while the second has the same property if the product $(n_1+\ldots+n_{i-1})(n_i-1)$ is even (respectively, odd); see 8. This explains the factor $(-1)^{n_1+\ldots+n_{i-1}}$.

Orientations of Vector Spaces

10. Real vector spaces are smooth manifolds (see 2.17) and hence the definition of orientation given in 1 applies. On the other

hand, there is a well-known, purely vectorial definition of an orientation of a real vector space: an orientation is a mapping from the set of all bases of the space into S^0, which takes the same value on two bases if and only if the matrix transforming one basis into the other has a positive determinant. This vectorial definition clearly agrees with the definition given in 1 and is often more convenient.

In particular, the vectorial definition makes obvious the following remark. Let f be a linear map of a real vector space V into another real vector space. Fix a subspace $V' \subset V$ that is mapped isomorphically by f onto $\mathrm{Im}\, f$. Then we may represent V as the direct sum of the spaces $\mathrm{Im}\, f$ and $\mathrm{Ker}\, f$. Consequently, the orientations of any two of the three spaces V, $\mathrm{Im}\, f$, and $\mathrm{Ker}\, f$ determine the orientation of the third. As an immediate result of our definition, we see that this connection between the orientations of V, $\mathrm{Im}\, f$, and $\mathrm{Ker}\, f$ does not depend upon the choice of V', but only on the map f.

4. The Manifold of Tangent Vectors

1. Suppose that X is an n-dimensional C^r-manifold, $r \geqslant 1$. For each point $x \in X$, let $\mathrm{Atl}_x X$ be the collection of all charts $\varphi \in \mathrm{Atl}\, X$ such that $x \in \mathrm{supp}\,\varphi$; recall that $\mathrm{Atl}\, X$ denotes the complete atlas of X. If $\varphi, \psi \in \mathrm{Atl}_x X$, then at the point $\varphi(x)$ the differential of the diffeomorphism

$$\mathrm{loc}\,(\varphi,\psi)\,\mathrm{id}\colon \varphi(\mathrm{supp}\,\varphi \cap \mathrm{supp}\,\psi) \to \psi(\mathrm{supp}\,\varphi \cap \mathrm{supp}\,\psi)$$

is meaningful [and is the linear map $\mathbb{R}^n \to \mathbb{R}^n$ whose matrix is the Jacobi matrix of the map $\mathrm{loc}\,(\varphi,\psi)\,\mathrm{id}$ at the point $\varphi(x)$]. We denote this differential by $d_x(\varphi,\psi)$.

Now consider the real vector space of all the maps $\mathrm{Atl}_x X \to \mathbb{R}^n$ (with the natural operations). It is clear that those maps $v\colon \mathrm{Atl}_x X \to \mathbb{R}^n$ which satisfy the relation $v(\psi) = d_x(\varphi,\psi)\,v(\varphi)$ for any two charts $\varphi, \psi \in \mathrm{Atl}_x X$ yield a subspace of this vector space, i.e., they form a real vector space. The latter is called the <u>tangent space to X at the point</u> x and is denoted by $\mathrm{Tang}_x X$. The maps in $\mathrm{Tang}_x X$ are called <u>tangent vectors at the point</u> x (<u>on</u> X), or <u>vectors tangent to X at</u> x.

Obviously, a tangent vector is completely defined by its value on an arbitrary chart of $\mathrm{Atl}_x X$, and for each chart $\varphi \in \mathrm{Atl}_x X$ and each vector $u \in \mathbb{R}^n$ there is a vector $v \in \mathrm{Tang}_x X$ with $v(\varphi) = u$. Therefore, the mapping $\varphi_{\#}\colon \mathrm{Tang}_x X \to \mathbb{R}^n$, $\varphi_{\#}(v) = v(\varphi)$, defined for a chart $\varphi \in \text{Atl}_x X$, is invertible. Moreover, $\varphi_{\#}$ being linear, it is an isomorphism; in particular, $\dim \text{Tang}_x X = n$. The isomorphism $\varphi_{\#}^{-1}$ takes the canonical basis $\text{ort}_1,\ldots,\text{ort}_n$ of $\mathbb{R}^n$ into a basis of $\text{Tang}_x X$, which we shall call the φ-basis. The coordinates of a vector $v \in \text{Tang}_x X$ relative to the φ-basis coincide with the usual coordinates of the vector $v(\varphi)$ and are called the φ-coordinates of v.

2. We denote the union $\bigcup_{x\in X}\text{Tang}_x X$, i.e., the space of all vectors tangent to X, by $\text{Tang}\,X$. The map $\text{Tang}\,X \rightarrow X$ transforming $\text{Tang}_x X$ into x is called projection and is denoted by pr. Thus $\text{pr}^{-1}(x) = \text{Tang}_x X$.

The set $\text{Tang}\,X$ has a natural topology which makes it a topological manifold. Furthermore, for $r \geqslant 2$, $\text{Tang}\,X$ has a natural differentiable structure. In order to describe these structures, we define, for each chart $\varphi \in \text{Atl}\,X$, the map $\text{tn}\,\varphi : \text{pr}^{-1}(\text{supp}\,\varphi) \rightarrow$ $\rightarrow \text{Im}\,\varphi \times \mathbb{R}^n$, $\text{tn}\,\varphi(v) = (\varphi \circ \text{pr}(v), \varphi_{\#}(v))$. If we identify $\mathbb{R}^n \times \mathbb{R}^n$ with $\mathbb{R}^{2n}$, and consider $\text{Im}\,\varphi \times \mathbb{R}^n$ to be an open subset of $\mathbb{R}^{2n}$ or $\mathbb{R}^{2n}_-$, then we may interpret $\text{tn}\,\varphi$ as a 2n-dimensional chart on $\text{Tang}\,X$. Obviously, if $\psi \in \text{Atl}\,X$ is another chart, the composition

$$\varphi(\text{supp}\,\varphi \cap \text{supp}\,\psi) \times \mathbb{R}^n =$$

$$= \text{tn}\,\varphi\,(\text{pr}^{-1}(\text{supp}\,\varphi) \cap \text{pr}^{-1}(\text{supp}\,\psi)) \xrightarrow{\text{ab}\,(\text{tn}\,\varphi)^{-1}}$$

$$\text{pr}^{-1}(\text{supp}\,\varphi) \cap \text{pr}^{-1}(\text{supp}\,\psi) \xrightarrow{\text{ab}\,\text{tn}\,\psi}$$

$$\text{tn}\,\psi\,(\text{pr}^{-1}(\text{supp}\,\varphi) \cap \text{pr}^{-1}(\text{supp}\,\psi)) = \psi(\text{supp}\,\varphi \cap \text{supp}\,\psi) \times \mathbb{R}^n$$

$$\tag{1}$$

is given by the formula $(a,u) \mapsto (\text{loc}(\varphi,\psi)\,\text{id}(a), d_{\varphi^{-1}(a)}(\varphi,\psi)u)$. Equivalently, writing this composite map in coordinates, we have $(a_1,\ldots,a_n;u_1,\ldots,u_n) \mapsto (b_1,\ldots,b_n;v_1,\ldots,v_n)$, where

$$\left.\begin{aligned} b_j &= \ell_j(a_1,\ldots,a_n), \\ v_j &= \sum_{i=1}^m D_i\ell_j(a_1,\ldots,a_n)u_i \end{aligned}\right\} \quad j = 1,\ldots,n. \tag{2}$$

and $\ell_1,\ldots,\ell_n$ are the coordinate functions of the map $\text{loc}(\varphi,\psi)\,\text{id}$ (D_i denotes the partial derivative with respect to the i-th coordinate). Formula (2) shows that the charts $\text{tn}\,\varphi$ and tn are C^{r-1} compatible (we set $C^{\infty-1} = C^{\infty}$ and $C^{a-1} = C^a$). Moreover, the charts $\text{tn}\,\varphi$, $\varphi \in \text{Atl}\,X$, cover $\text{Tang}\,X$, and thus yield a C^{r-1}-atlas of the set $\text{Tang}\,X$. This atlas has a countable subatlas (since $\text{Atl}\,X$ has such a subatlas). Furthermore, for any two vectors of $\text{Tang}\,X$, it has either

a chart which contains both of them, or a pair of disjoint charts, each
containing one of the vectors (indeed, recall that $\text{Atl}\,X$ contains,
for any two points of X, either a chart containing both of them, or
a pair of disjoint charts, each containing one of the points).
Therefore, it makes $\text{Tang}\,X$ into a 2n-dimensional C^{r-1}-manifold, which
we call the <u>total manifold of vectors tangent to the manifold</u> X.

Clearly, the projection $\text{Tang}\,X \to X$, the inclusions
$\text{Tang}_x X \to \text{Tang}\,X$, and the natural map $X \to \text{Tang}\,X$ which takes each
point x into the zero vector of the space $\text{Tang}_x X$, are all C^{r-1} maps.

Formula (2) shows that for $r \geqslant 2$ the Jacobian of the
composite map (1) at the point (a,u) is equal to the square of the
Jacobian of the map $\text{loc}\,(\varphi,\psi)\,\text{id}$ at the point a. We deduce that for
$r \geqslant 2$ the manifold $\text{Tang}\,X$ is always orientable and even carries a
canonical orientation, namely the one which is positive on the charts
$\text{tn}\,\varphi$ with $\varphi \in \text{Catl}\,X$.

One more remark: let X_1 and X_2 be two arbitrary smooth
manifolds such that $\partial X_2 = \emptyset$. Then for any two points $x_1 \in X_1$ and
$x_2 \in X_2$, $\text{Tang}_{(x_1,x_2)}(X_1 \times X_2)$ and $\text{Tang}_{x_1} X_1 \oplus \text{Tang}_{x_2} X_2$ are
isomorphic as vector spaces, and the isomorphism is natural. In
addition, the isomorphisms corresponding to all pairs (x_1,x_2) yield
a diffeomorphism of $\text{Tang}(X_1 \times X_2)$ onto $\text{Tang}\,X_1 \times \text{Tang}\,X_2$.

The Differential of a Smooth Map

3. Let f be a C^r-map of an m-dimensional $C^{\geqslant r}$-manifold X
into an n-dimensional $C^{\geqslant r}$-manifold Y, $r \geqslant 1$. For a point $x \in X$ and
two charts $\varphi \in \text{Atl}_x X$ and $\psi \in \text{Atl}_{f(x)} Y$, we let $d_x(f;\varphi,\psi)$ denote
the differential of the map $\text{loc}\,(\varphi,\psi)\,f$ at the point $\varphi(x)$, regarded
as the linear map $\mathbb{R}^m \to \mathbb{R}^n$ whose matrix is the Jacobi matrix of
$\text{loc}\,(\varphi,\psi)\,f$ at $\varphi(x)$. If $\varphi' \in \text{Atl}_x X$ and $\psi' \in \text{Atl}_{f(x)} Y$ are two
other charts, then $d_x(f;\varphi',\psi') = d_{f(x)}(\psi,\psi') \circ d_x(f;\varphi,\psi) \circ d_x(\varphi',\varphi)$.
Combining this relation with the equalities $d_x(\varphi',\varphi) = \varphi_\# \circ (\varphi'_\#)^{-1}$
and $d_{f(x)}(\psi,\psi') = \psi'_\# \circ \psi^{-1}$, we see that

$$(\psi'_\#)^{-1} \circ d_x(f;\varphi',\psi') \circ \varphi'_\# = \psi_\#^{-1} \circ d_x(f;\varphi,\psi) \circ \varphi_\# \,,$$

i.e., the linear map $\psi_\#^{-1} \circ d_x(f;\varphi,\psi) \circ \varphi_\#: \text{Tang}_x X \to \text{Tang}_{f(x)} Y$ does
not depend upon the choice of the charts φ and ψ. This linear
map is called the <u>differential of the map</u> f <u>at the point</u> x and is

denoted by $d_x f$. The map $\text{Tang } X \to \text{Tang } Y$ which equals $d_x f$ on $\text{Tang}_x X$ for all $x \in X$ is called the <u>differential of the map</u> f and is denoted by df. The resulting diagram

$$
\begin{array}{ccc}
\text{Tang } X & \xrightarrow{\ df\ } & \text{Tang } Y \\
{\scriptstyle pr}\downarrow & & \downarrow{\scriptstyle pr} \\
X & \xrightarrow{\ f\ } & Y
\end{array}
$$

is clearly commutative.

Let $\ell_1,\ldots,\ell_n$ be the coordinate functions of the map $\text{loc}(\varphi,\psi)f$. Then the local representative $\text{loc}(\text{tn }\varphi,\text{tn }\psi)df$ of the differential df is given in coordinates by the formula $(a_1,\ldots,a_m;u_1,\ldots,u_m) \mapsto (b_1,\ldots,b_n;v_1,\ldots,v_n)$, where

$$
\left.
\begin{aligned}
b_j &= \ell_j(a_1,\ldots,a_m), \\[2mm]
v_j &= \textstyle\sum_{i=1}^{m} D_i\ell_j(a_1,\ldots,a_m)u_i
\end{aligned}
\right\} \quad j = 1,\ldots,n.
$$

Therefore, df is of class C^{r-1}.

Of course, $d(h \circ f) = dh \circ df$ for any smooth map h of Y into a third smooth manifold, and $df = \text{id}(\text{Tang } X)$ when $X = Y$ and $f = \text{id } X$. Also, if f is a diffeomorphism, then df is a diffeomorphism for $r \geqslant 2$ and a homeomorphism for $r = 1$.

In the special case when X is an open subset of $\mathbb{R}^m$ or of $\mathbb{R}^m_-$, and Y is an open subset of $\mathbb{R}^n$ or of $\mathbb{R}^n_-$, in addition to the differential $d_x f\colon \text{Tang}_x X \to \text{Tang}_{f(x)} Y$ one has the classical differential of the map f, i.e., the linear map $\mathbb{R}^m \to \mathbb{R}^n$ whose matrix is the Jacobi matrix of f at the point x. Let us identify the spaces $\text{Tang}_x X$ and $\mathbb{R}^m$ ($\text{Tang}_{f(x)} Y$ and $\mathbb{R}^n$) via the linear isomorphism $(\text{id } X)_\#\colon \text{Tang}_x X \to \mathbb{R}^m$ (respectively, $(\text{id } Y)_\#\colon \text{Tang}_{f(x)} Y \to \mathbb{R}^n$). Then it is clear that $d_x f$ becomes the classical differential.

4. If A is a submanifold or the boundary of a smooth manifold X, then the differential $d_x \text{in}\colon \text{Tang}_x A \to \text{Tang}_x X$ of the inclusion $\text{in}\colon A \to X$ is a monomorphism for each point $x \in A$. Thus, we can identify the tangent space $\text{Tang}_x A$ with the subspace $d_x \text{in}(\text{Tang}_x A)$ of $\text{Tang}_x X$, and $\text{Tang } A$ - with $d\,\text{in}(\text{Tang } A)$.

If A is a submanifold of $\mathbb{R}^n$, then along with the identification $\text{Tang}_x A = d_x \text{in}(\text{Tang}_x A)$ one has the identification $\text{Tang}_x \mathbb{R}^n = \mathbb{R}^n$ via the canonical (linear) isomorphism $(\text{id } \mathbb{R}^n)_\#\colon \text{Tang}_x \mathbb{R}^n \to \mathbb{R}^n$, and so $\text{Tang}_x A$ becomes a subspace of $\mathbb{R}^n$.

It is easy to describe this subspace explicitly when A is defined in a neighborhood of x by the independent functions $\varphi_{k+1}, \ldots, \varphi_n$, as in 2.12. In this situation, $\mathrm{Tang}_x A$ consists of all the vectors of $\mathbb{R}^n$ which are orthogonal to the $n-k$ vectors $\mathrm{grad}\,\varphi_{k+1}(x), \ldots, \mathrm{grad}\,\varphi_n(x)$. For example, $\mathrm{Tang}_x S^{n-1}$ is the subspace of $\mathbb{R}^n$ composed of all vectors orthogonal to the vector x.

Vector Fields

5. A <u>vector field</u> on a smooth manifold X is a continuous map $X \to \mathrm{Tang}\,X$ which takes each point $x \in X$ into a vector tangent to X at x. A trivial example is the <u>zero</u> vector field, whose value at each point $x \in X$ is the zero vector of the space $\mathrm{Tang}_x X$ (see 2).

A smooth, n-dimensional, $C^{\geq r+1}$-manifold X is <u>C^r-parallelizable</u> if there exist n C^r-vector fields $f_1, \ldots, f_n \colon X \to \mathrm{Tang}\,X$ such that, at each point $x \in X$, the vectors $f_1(x), \ldots, f_n(x)$ yield a basis of the space $\mathrm{Tang}_x X$. For example, $\mathbb{R}^n$ (regarded as a C^a-manifold) is C^a-parallelizable: a parallelization is given by the vector fields which associate to each point $x \in \mathbb{R}^n$ the $(\mathrm{id}\,\mathbb{R}^n)$-basis of the space $\mathrm{Tang}_x \mathbb{R}^n$.

C^0-parallelizability is simply called <u>parallelizability.</u> It will be shown in Chapter 4 (see 4.6.4.3) that the parallelizabililty of a compact $C^{\geq r+1}$-manifold with $r \leqslant \infty$ implies its C^r-paralleliza-bility.

INFORMATION. The C^r-parallelizability is a consequence of the parallelizability of a $C^{\geq r+1}$-manifold even if the manifold is not compact, or if $r = a$.

6. Suppose that the smooth manifold X is parallelized by the C^r-vector fields $f_1, \ldots, f_n \colon X \to \mathrm{Tang}\,X$. Then the formula $(x, (y_1, \ldots, y_n)) \mapsto y_1 f_1(x) + \ldots + y_n f_n(x)$ defines a C^r-diffeomorphism of the product $X \times \mathbb{R}^n$ onto $\mathrm{Tang}\,X$. Indeed, the formula $v \mapsto (\mathrm{pr}(v), (y_1, \ldots, y_n))$, where $y_1, \ldots, y_n$ are the coordinates of the vector v relative to the basis $f_1(\mathrm{pr}(v)), \ldots, f_n(\mathrm{pr}(v))$ of $\mathrm{Tang}_{\mathrm{pr}(v)} X$, defines the inverse map $\mathrm{Tang}\,X \to X \times \mathbb{R}^n$, and obviously both maps are of class C^r. Thus <u>the total manifold of vectors tangent to a C^r-parallelizable n-dimensional smooth manifold X is C^r-diffeo-morphic to the product $X \times \mathbb{R}^n$.</u>

In particular, we deduce that <u>every C^1-parallelizable smooth</u>

manifold is orientable. In fact, $\mathrm{Tang}\, X$ is orientable for any C^r-manifold X with $r \geqslant 2$ (see 2); hence, in our case, the product $X \times \mathbb{R}^n$ is orientable, which in turn implies the orientability of X (see 3.7).

7 (EXAMPLES). For each odd n, there is a vector field with no zeros on the sphere S^n. For example, such a vector field $S^n \to \mathbb{R}^{n+1}$ is given by $x = (x_1,\ldots,x_{n+1}) \mapsto (y_1,\ldots,y_{n+1}) \in \mathbb{R}^{n+1}$, $x \in S^{n+1}$, where $y_{2k-1} = -x_{2k}$, $y_{2k} = x_{2k-1}$ $(k = 1,\ldots,(n+1)/2)$; here we consider $(y_1,\ldots,y_{n+1})$ as a vector in $\mathrm{Tang}_x S^n$ (see 4).

Note that the same vector field can be defined in a more concise fashion as the map $x \mapsto xi$, regarding $\mathbb{R}^{n+1}$ as $\mathbb{C}^{(n+1)/2}$. If $n+1$ is divisible by 4, $\mathbb{R}^{n+1}$ may be regarded as $\mathbb{H}^{(n+1)/4}$, and the formulas $x \mapsto x\,\mathrm{ort}_2$, $x \mapsto x\,\mathrm{ort}_3$, and $x \mapsto x\,\mathrm{ort}_4$ (ort_2, ort_3, and ort_4 are considered here as the imaginary quaternion units) define three vector fields on S^n which are linearly independent at each point. If $n+1$ is divisible by 8, $\mathbb{R}^{n+1}$ may be regarded as $\mathbb{C}\mathrm{a}^{(n+1)/8}$, and the formulas $x \mapsto x\,\mathrm{ort}_2,\ldots$, $x \mapsto x\,\mathrm{ort}_8$ ($\mathrm{ort}_2,\ldots,\mathrm{ort}_8$ are considered here as the imaginary Cayley units) define seven vector fields on S^n which are linearly independent at each point. Since all the above vector fields are analytic, this construction shows, in particular, that S^1, S^3, and S^7 are C^a-parallelizable.

INFORMATION. For $n \neq 0,1,3,7$ the sphere S^n is not parallelizable. For a proof, see [1] and [2].

5. Embeddings, Immersions, and Submersions

1. A map $f\colon X \to Y$ of smooth manifolds is a $\underline{C^r\text{-embedding}}$ if $f(X)$ is a C^r-submanifold of Y and $\mathrm{ab}\, f\colon X \to f(X)$ is a C^r-diffeomorphism. For example, the inclusion of a submanifold into its ambient $C^{\geqslant r}$-manifold is a C^r-embedding. Since every map $f\colon X \to Y$ can be written as the composition of its compression $\mathrm{ab}\, f\colon X \to f(X)$ with the inclusion $f(X) \to Y$, a C^r-embedding is really a map of class C^r.

The C^1-embeddings are also termed $\underline{\text{differentiable embeddings.}}$ Using Theorem 3 we shall prove below, we can see that a differentiable embedding which is of class C^r is a C^r-embedding. Moreover, it is evident that differentiable embeddings are topological embeddings. The latter are sometimes called C^0-embeddings.

A differentiable embedding $f\colon X \to Y$ is $\underline{\text{neat}}$ if $f(X)$ is a

neat submanifold of Y . For example, the inclusion of a neat submanifold into its smooth ambient manifold is such an embedding. Clearly,
if $\dim X = \dim Y$, Y is connected, and $X \neq \emptyset$, then every neat
differentiable embedding $X \to Y$ is a diffeomorphism.

It is obvious that for each C^r-embedding $f \colon X \to Y$ and each
point $x \in X$, there are charts $\varphi \in \mathrm{Atl}_x C^r X$ and $\psi \in \mathrm{Atl}_{f(x)} C^r Y$ such
that $\mathrm{loc}(\varphi,\psi)f$ coincides with one of the inclusions

$$\mathbb{R}^m \to \mathbb{R}^n, \quad \mathbb{R}^m_- \to \mathbb{R}^n_- , \quad \text{or} \quad \mathbb{R}^m_- \to \mathbb{R}^n_- ,$$

where $m = \dim X$ and $n = \dim Y$. If f is neat, then the second case
must be excluded.

Immersions

2. A smooth map $f \colon X \to Y$ of smooth manifolds is an
immersion if: (i) $d_x f \colon \mathrm{Tang}_x X \to \mathrm{Tang}_{f(x)} Y$ is a monomorphism for any
$x \in X$; (ii) $(df)^{-1}(\mathrm{Tang}\, \partial Y) \subset \mathrm{Tang}\, \partial X$.

We remark that condition (i) implies that $\dim X \leqslant \dim Y$,
and condition (ii) - that $f(\mathrm{int}\, X) \subset \mathrm{int}\, Y$. If $\partial Y = \emptyset$, then (ii) is
automatically fulfilled.

Trivially, the composition of two immersions is an immersion.
The differentiable embeddings are examples of immersions.

3. _If $f \colon X \to Y$ is an immersion of class C^r, then each
point of the manifold X has a neighborhood N such that the
restriction of f to N is a C^r-embedding._

PROOF. Let $x_0 \in X$ be an arbitrary point, and put $m = \dim X$,
$n = \dim Y$. Now pick some charts $\varphi \in \mathrm{Atl}_{x_0} X$, $\psi \in \mathrm{Atl}_{f(x_0)} Y$, and denote
by $\ell_1,\ldots,\ell_n$ the coordinate functions of the map $\mathrm{loc}(\varphi,\psi)f$.
According to condition 2(i), the Jacobi matrix of $\mathrm{loc}(\varphi,\psi)f$ at
$\varphi(x_0)$ has rank m. Using condition 2(ii), one can assume that the
minor M of this matrix, constructed from its first m rows, is not
zero. [If $f(x_0) \in \mathrm{int}\, Y$, 2(ii) is not necessary: one can achieve
$M \neq 0$ by reindexing the local coordinates $\psi_1,\ldots,\psi_n$ of the chart ψ,
i.e., by permuting the rows of the matrix. When $f(x_0) \in \partial Y$, 2(ii)
ensures that all the elements of the elements of the first row of the
Jacobi matrix, starting with the second, are zero; this allows to
achieve $M \neq 0$ by reindexing the coordinates $\psi_2,\ldots,\psi_n$.] Next apply
the implicit function theorem to deduce the existence of a neighborhood
W of the point $(\psi_1(f(x_0)),\ldots,\psi_m(f(x_0)))$ in $\mathbb{R}^m$ and of a

C^r-embedding $h: W \to \mathrm{Im}\,\varphi$, such that, for $i = 1,\ldots,m$, one has:

$$h_i(\psi_1(f(x_0)),\ldots,\psi_m(f(x_0))) = \varphi_i(x_0)$$

and

$$\ell_i(h_1(y_1,\ldots,y_m),\ldots,h_m(y_1,\ldots,y_m)) = y_i \qquad [(y_1,\ldots,y_m) \in W],$$

where $h_1,\ldots,h_m$ are the coordinate functions of h. Let
$N = \varphi^{-1}(h(W))$. Evidently, N is a neighborhood of x_0. Its image
$f(N)$ in $\varphi^{-1}(W \times \mathbb{R}^{n-m})$ is defined by the equations
$\psi_j - \ell_j(h_1(\psi_1,\ldots,\psi_m),\ldots,h_m(\psi_1,\ldots,\psi_m)) = 0$, and, possibly, by the
additional inequality $h_1(\psi_1,\ldots,\psi_m) \leqslant 0$. Therefore, $f(N)$ is a
C^r-submanifold of Y. The composite map

$$f(N) \xrightarrow{\ \mathrm{ab}\,\psi\ } \psi(f(N)) \longrightarrow W \xrightarrow{\ h\ } h(W) \xrightarrow{\ (\mathrm{ab}\,\varphi)^{-1}\ } N\ ,$$

(where the second map is the compression of the orthogonal projection
$\mathbb{R}^n \to \mathbb{R}^m$) is of class C^r and is the inverse of the map $\mathrm{ab}\, f: N \to f(N)$.
We conclude that $f|_N$ is a C^r-embedding.

4 (COROLLARY). <u>If an immersion of class C^r is a topological</u>
<u>embedding, then it is a C^r-embedding. In particular, every injective</u>
<u>C^r-immersion of a compact manifold is a C^r-embedding.</u>

5. <u>Let $f: X \to Y$ be a smooth map such that</u>
<u>$(df)^{-1}(\mathrm{Tang}\,\partial Y) \subset \mathrm{Tang}\,\partial X$, and let A be a compact subset of the</u>
<u>manifold X. If $f|_A$ is injective and the differential $d_x f$ is</u>
<u>nondegenerate for each point $x \in A$, then f is a differentiable embed-</u>
<u>ding on some neighborhood of A. In particular, if we require, in</u>
<u>addition to the previous conditions, that $\dim X = \dim Y$ and</u>
<u>$f(\partial X) \subset \partial Y$, then f carries a neighborhood of A diffeomorphically</u>
<u>onto a neighborhood of $f(A)$.</u>

PROOF. Fix for each point $x \in A$ a neighborhood U_x such
that $f|_{U_x}$ is a C^r-embedding (see 3), and then cover A by a finite
number of such neighborhoods, say $U_{x_1},\ldots,U_{x_s}$. Since the set
$Y \times Y \smallsetminus \mathrm{diag}\,Y$ is open in $Y \times Y$ (see 1.2.2.4), its preimage W
under the map $f \times f: X \times X \to Y \times Y$ is also open. But f is injective
on A; whence $W \cup [\cup_{i=1}^{s}(U_{x_i} \times U_{x_i})]$ contains $A \times A$, and is actually
a neighborhood of $A \times A$. Next introduce a metric on X (see 1.1.12),
get the corresponding metric on $X \times X$ (see 1.2.2.9), and then set

$$B = \{x \in X \mid \mathrm{Dist}(A,x) < \mathrm{Dist}((X \times X) \smallsetminus W, A \times A)/2 \}.$$

Since B is open and contains A (see 1.1.7.15), the intersection

$B \cap (\cup_{i=1}^{s} U_{x_i})$ is also open and contains A. This intersection contains a relatively compact neighborhood U of A, because A is compact. Moreover, $B \times B \subset W$, and so f is injective on B. We conclude that $f\big|_{Cl\ U}$ is a topological embedding and $f\big|_{U}$ - a differentiable embedding (see 4).

Submersions

6. A smooth map $f: X \to Y$ of smooth manifolds is a <u>submersion</u> if: (i) $d_x f: Tang_x X \to Tang_{f(x)} Y$ is an epimorphism for any $x \in X$; (ii) $d_x f(Tang_x \partial X) = Tang_{f(x)} Y$ for any point $x \in \partial X \cap f^{-1}(int\ X)$; (iii) $\partial X \cap f^{-1}(int\ Y)$ is a union of whole components of the manifold ∂X.

We remark that condition (i) implies $dim\ X \geqslant dim\ Y$, and that conditions (ii) and (iii) are automatically fulfilled whenever $\partial X = \emptyset$.

As examples, consider smooth real functions $f: X \to \mathbb{R}$. As in the classical calculus, a point $x \in X$ and the corresponding value $f(x)$ are said to be <u>critical</u> for f if $d_x f = 0$. It it clear that f is a submersion if and only if the functions f and $f\big|_{\partial X}$ have no critical points.

As additional examples of submersions we cite the projections of the product of two smooth manifolds (one of them being without boundary) onto its factors.

7. <u>A map</u> f <u>of a smooth m-dimensional manifold</u> X <u>into a smooth n-dimensional manifold</u> Y <u>is a submersion of class</u> C^r <u>if and only if for each point</u> $x \in X$ <u>there are charts</u> $\varphi \in Atl_x C^r X$ <u>and</u> $\psi \in Atl_{f(x)} C^r Y$ <u>such that the following holds</u>: $f(supp\ \varphi) \subset supp\ \psi$; <u>the pair</u> $(Im\ \varphi, Im\ \psi)$ <u>coincides with one of the pairs</u> $(\mathbb{R}^m, \mathbb{R}^n)$, $(\mathbb{R}^m_-, \mathbb{R}^n)$, <u>or</u> $(\mathbb{R}^m_-, \mathbb{R}^n_-)$, <u>and in each case</u> $\varphi(x) = 0$ <u>and</u> $\psi(f(x)) = 0$; <u>the corresponding local representative</u> $loc(\varphi, \psi)f$ <u>can be described, in the first case, as the projection of the product</u> $\mathbb{R}^n \times \mathbb{R}^{m-n}$ <u>onto its first factor, in the</u> second <u>case - as the projection of the product</u> $\mathbb{R}^{m-n}_- \times \mathbb{R}^n$ <u>onto its second factor, and in the third case - as the projection of the product</u> $\mathbb{R}^n_- \times \mathbb{R}^{m-n}$ <u>onto its first factor.</u>

PROOF. The sufficiency of this condition is obvious. Let us prove its necessity. Let $\varphi^1 \in Atl_x X$ and $\psi^1 \in Atl_{f(x)} Y$ be arbitrary charts such that $f(supp\ \varphi^1) = supp\ \psi^1$, $\varphi^1(x) = 0$, and $\psi^1(f(x)) = 0$. We denote by $\varphi^1_1, \ldots, \varphi^1_m$ and $\psi^1_1, \ldots, \psi^1_n$ the corresponding coordinate functions, and by $\ell_1,\ldots,\ell_n$ the coordinate functions of the map $\mathrm{loc}(\varphi^1,\psi^1)f$. We consider three distinct cases: a) $x \in \mathrm{int}\,X$, $f(x) \in \mathrm{int}\,Y$; b) $x \in \partial X$, $f(x) \in \mathrm{int}\,Y$; and c) $f(x) \in \partial Y$. Condition 6(i) says that the Jacobi matrix of the map $\mathrm{loc}(\varphi^1,\psi^1)f$, computed at the point 0, has rank n in each of the three cases. Condition 6(ii) says that in case b) this rank does not decrease when we remove from the matrix the first column. Therefore, in cases a) and c) we may assume that the minor constructed from the first n columns does not vanish, while in case b) the same is true for the minor constructed from the last n columns. Moreover, in case c), 0 is a boundary point of $\mathrm{Im}\,\varphi^1$ in $\mathbb{R}^m$, and the function ℓ_1 vanishes on $\varphi^1(\partial X \cap \mathrm{Im}\,\varphi^1)$ in a neighborhood of this point. Indeed, the first part of the last assertion follows from the fact that ℓ_1 is a non-positive function, vanishes at 0, and has nonzero gradient at 0; now the second part of the assertion is seen to be a consequence of 6(iii).

We pass from φ^1,ψ^1 to the required charts φ,ψ through the intermediary charts $\varphi^2 \in \mathrm{Atl}_x\,C^r X$ and $\psi^2 \in \mathrm{Atl}_{f(x)}\,C^r Y$. In cases a) and c), φ^2 is the chart whose local coordinates are the restrictions of the functions $\ell_1 \circ \varphi^1,\ldots,\ell_n \circ \varphi^1,\varphi^1_{n+1},\ldots,\varphi^1_m$ to a small enough neighborhood U_2 of the point x, while ψ^2 is the chart $\mathrm{ab}\,\psi^1 : V_2 \to \psi^1(V_2)$, where $V_2 = f(U_2)$. In case b), φ^2 and ψ^2 are similarly defined, but one replaces $\ell_1 \circ \varphi^1,\ldots,\ell_n \circ \varphi^1,\varphi^1_{n+1},\ldots,\varphi^1_m$ by the functions $\varphi^1_1,\ldots,\varphi^1_{m-n},\ell_1 \circ \varphi^1,\ldots,\ell_n \circ \varphi^1$. In all cases $f(U_2) = V_2$, and obviously, in cases a) and c), the map $\mathrm{loc}(\varphi^2,\psi^2)f$ is given in the new coordinates $\varphi^2_1,\ldots,\varphi^2_m$ and $\psi^2_1,\ldots,\psi^2_n$ by the formulas $\psi^2_1 = \varphi^2_1,\ldots,\psi^2_n = \varphi^2_n$, whereas in case b) the corresponding formulas are $\psi^2_1 = \varphi^2_{m-n+1},\ldots,\psi^2_n = \varphi^2_m$. Fix a positive ε and define the subsets U and V of U_2 and V_2 by the inequalities $|\varphi^2_i| < \varepsilon$ ($i = 1,\ldots,m$) and $|\psi^2_j| < \varepsilon$ ($j = 1,\ldots,n$), respectively. It is clear that for ε small enough, the charts φ,ψ with $\mathrm{supp}\,\varphi = U$, $\mathrm{supp}\,\psi = V$, and local coordinates

$$\varphi_i(y) = \frac{\varphi^2_i(y)}{\varepsilon - |\varphi^2_i(y)|} \qquad (i = 1,\ldots,m)$$

and

$$\psi_j(z) = \frac{\psi^2_j(z)}{\varepsilon - |\psi^2_j(z)|} \qquad (j = 1,\ldots,n)$$

have the desired properties.

8 (COROLLARIES). _If_ $f: X \to Y$ _is a submersion, then_
$f(\text{int } X) \subset \text{int } Y$, $f^{-1}(\partial Y) \subset \partial X$, _and the maps_ $\text{ab } f : \text{int } X \to \text{int } Y$
and $\text{ab } f : f^{-1}(\partial Y) \to \partial X$ _are submersions._

_If_ $f: X \to Y$ _is a submersion of class_ C^r, _then_ $f^{-1}(y)$
is a neat C^r-_submanifold of_ X _for_ $y \in \text{int } Y$, _and a neat_ C^r-_submanifold of_ ∂X _for_ $y \in \partial Y$.

Every submersion is an open map.

The composition of two submersions is a submersion.

9. _A_ C^r-_map_ $f: X \to Y$ _satisfying condition 6(iii) is a submersion if and only if for each point_ $x_0 \in X$ _there is a neighborhood_ V _of the point_ $f(x_0)$ _and a_ C^r-_map_ $g: V \to X$, _such that_ $f(g(y)) = y$ _for all_ $y \in V$ _and_ $g(f(x_0)) = x_0$.

The sufficiency of this condition is clear, its necessity results from 7.

10. _Let_ $f: X \to Y$ _be a submersion of class_ C^r _such that_ $f(X) = Y$, _and let_ h _be a map of_ Y _into a third manifold. If the composition_ $h \circ f$ _is of class_ C^r, _then_ h _is of class_ C^r _too._

PROOF. According to 9, one can find for each point of Y a neighborhood V and a C^r-map $g: V \to X$ such that $f \circ g = [\text{in}: V \to Y]$, and hence $h\big|_V = (h \circ f) \circ g$.

6. Complex Structures

1. Recall that a map of an open subset of $\mathbb{C}^m$ into a subset of $\mathbb{C}^n$ is _holomorphic_ if its coordinate functions are holomorphic, and _biholomorphic_ if it is invertible and both the map and its inverse are holomorphic. Obviously, in the last case we must have $m = n$; cf. 2.2.

If we regard $\mathbb{C}^m$ and $\mathbb{C}^n$ as $\mathbb{R}^{2m}$ and $\mathbb{R}^{2n}$, respectively, then the holomorphic maps become C^a-maps, and the biholomorphic ones — C^a-diffeomorphisms. A smooth map of an open subset of $\mathbb{R}^{2m}$ into a subset of $\mathbb{R}^{2n}$ is holomorphic relative to the complex structures on $\mathbb{R}^{2m}$ and $\mathbb{R}^{2n}$ resulting from the identifications $\mathbb{R}^{2m} = \mathbb{C}^m$ and $\mathbb{R}^{2n} = \mathbb{C}^n$ if and only if it satisfies the Cauchy-Riemann conditions. Evidently, a map which is the inverse of a diffeomorphism satisfying the Cauchy-Riemann conditions also satisfies these conditions. Consequently, every holomorphic diffeomorphism is a biholomorphic map.

2. Suppose that a holomorphic map between subsets of $\mathbb{C}^n$ has a nondegenerate differential at some point. Then it retains the same property when considered as a C^a-map, and hence it maps a neighborhood of the given point diffeomorphically onto its image. Thus, a holomorphic map whose Jacobian does not vanish at a point maps a neighborhood of the point biholomorphically onto its image.

This statement is the exact analogue, and also the result of the theorem concerning local inversion of a smooth map in the real case. In a similar fashion, one can translate a more general theorem from the real calculus into the complex language - the implicit function theorem. Namely, let f be a holomorphic map of an open subset A of $\mathbb{C}^m \times \mathbb{C}^n$ into $\mathbb{C}^n$, and let $(z^0, w^0) \in A$ be such that $f(z^0, w^0) = 0$. Suppose that the Jacobian of f with respect to the second variable does not vanish at (z^0, w^0). The complex implicit function theorem states the existence of a neighborhood U of the point z^0 in $\mathbb{C}^m$, of a neighborhood V of the point w^0 in $\mathbb{C}^n$, and of a holomorphic map $g: U \to V$, such that $U \times V \subset A$ and the preimage of 0 under $f|_{U \times V}$ is precisely the graph of g.

3. A linear transformation of $\mathbb{C}^n$ is also linear as a transformation of $\mathbb{R}^{2n}$; hence to each complex $n \times n$-matrix C one can associate a real $2n \times 2n$-matrix R. If $C = A + iB$ and $z = x + iy$ are the decompositions of the matrix C and the vector $z \in \mathbb{C}^n$ into real and imaginary parts, then $Cz = (Ax - By) + i(Bx + Ay)$, and

$$R = \begin{pmatrix} A & -B \\ B & A \end{pmatrix}.$$

In particular, $\det R = |\det C|^2$. Indeed, if we add to the first block-row of R the second, multiplied by i, and then add to the second block-column the first, multiplied by $-i$, we obtain the matrix $\begin{pmatrix} C & 0 \\ B & \overline{C} \end{pmatrix}$, which has the same determinat as R.

Complex Manifolds

4. An n-dimensional <u>complex chart</u> on the set X is an invertible mapping of a subset of X onto an open subset of $\mathbb{C}^n$. Two complex charts, φ and ψ, are <u>compatible</u> if the set $\varphi(\operatorname{supp} \varphi \cap \operatorname{supp} \psi)$ is open in $\operatorname{Im} \varphi$, the set $\psi(\operatorname{supp} \varphi \cap \operatorname{supp} \psi)$ is open in $\operatorname{Im} \psi$, and the composite maps

$$\varphi(\mathrm{supp}\,\varphi \cap \mathrm{supp}\,\psi) \xrightarrow{\mathrm{ab}\,\varphi^{-1}} \mathrm{supp}\,\varphi \cap \mathrm{supp}\,\psi \xrightarrow{\mathrm{ab}\,\psi} \psi(\mathrm{supp}\,\varphi \cap \mathrm{supp}\,\psi)$$

and

$$\psi(\mathrm{supp}\,\varphi \cap \mathrm{supp}\,\psi) \xrightarrow{\mathrm{ab}\,\psi^{-1}} \mathrm{supp}\,\varphi \cap \mathrm{supp}\,\psi \xrightarrow{\mathrm{ab}\,\varphi} \varphi(\mathrm{supp}\,\varphi \cap \mathrm{supp}\,\psi),$$

which are inverses of each other, are holomorphic (here supp is
defined as in the real case). If two overlapping charts of dimensions
m and n are compatible, then m = n.

A collection of complex charts is an <u>n-dimensional holomorphic
atlas of the set</u> X if these charts cover X, are n-dimensional, and
are pairwise compatible. Two holomorphic atlases are <u>holomorphically
equivalent</u> if their union is again an atlas. The family of n-dimensional
holomorphic atlases of X is divided into disjoint classes of holomorphically equivalent atlases. These classes are called <u>n-dimensional
complex structures on</u> X.

5. Any n-dimensional complex chart may be regarded as a
2n-dimensional real chart, i.e., a 2n-dimensional chart in the sense of
2.3. Furthermore, compatible complex charts yield C^a-compatible charts,
holomorphic atlases yield C^a-atlases, and holomorphically equivalent
atlases yield C^a-equivalent atlases. Therefore, an n-dimensional complex
structure on the set X induces a 2n-dimensional C^a-structure on X.
For us, the most important case occurs when this C^a-structure makes X
into a manifold, i.e., when it defines a Hausdorff, second countable
topology. A set X equipped with an n-dimensional complex structure
enjoying this property is an <u>n-dimensional complex manifold.</u>

We let Atl X denote the complete atlas of the complex
manifold X, i.e., the collection of all charts of all atlases of its
complex structure.

6. A continuous map f: X → Y between complex manifolds
is <u>holomorphic</u> if all its local representatives are holomorphic, i.e.,
the maps $\varphi(\mathrm{supp}\,\varphi \cap f^{-1}(\mathrm{supp}\,\psi)) \to \mathrm{Im}\,\psi$, $x \to \psi(f(\varphi^{-1}(x)))$,
constructed by means of the charts φ ∈ Atl X and ψ ∈ Atl Y, are
holomorphic. A map f: X → Y is <u>biholomorphic</u> if it is holomorphic,
invertible, and its inverse is also holomorphic. Two complex manifolds
which can be transformed one into another by a biholomorphic map are
said to be <u>biholomorphically equivalent.</u>

If complex manifolds are considered as C^a-manifolds, the
holomorphic maps become C^a-maps, and the biholomorphic maps - C^a-diffeo-
morphisms.

7. Let A be a subset of an n-dimensional complex manifold
X. A is a <u>k-dimensional submanifold of</u> X if for each point x ∈ A

there is a chart $\varphi \in \text{Atl}\, X$ such that $x \in \text{supp}\,\varphi$ and $\varphi(\text{supp}\,\varphi \cap A) = \text{Im}\,\varphi \cap \mathbb{C}^k$. The charts $\text{ab}\,\varphi : \text{supp}\,\varphi \cap A \to \text{Im}\,\varphi \cap \mathbb{C}^k$ derived from the charts $\varphi \in \text{Atl}\, X$ form a k-dimensional holomorphic atlas of the set A, thus transforming A into a complex manifold.

The notion of independent functions defined in 2.12 makes sense for the complex case too. Therefore, using the implicit function theorem, we deduce that a subset A of the complex n-dimensional manifold X is a k-dimensional submanifold of X if and only if for each point $x_0 \in A$ there are a neighborhood U of x_0 in X and holomorphic functions $\varphi_{k+1}, \ldots, \varphi_n : U \to \mathbb{C}$, independent at x_0, and such that the intersection $A \cap U$ is defined in U by the equations $\varphi_{k+1}(x) = 0, \ldots, \varphi_n(x) = 0$; cf. 2.12.

If the complex manifold X is regarded as a C^a-manifold, then a submanifold remains a submanifold and its C^a-structure induced from the C^a-structure of X is identical to the C^a-structure induced by its own complex structure.

If A is a submanifold of the complex manifold X, then the inclusion $A \to X$ is holomorphic.

A map $f: X \to Y$ between complex manifolds is a <u>holomorphic embedding</u> if $\text{ab}\, f : X \to f(X)$ is a biholomorphic map of X onto a submanifold of Y. In this case f is the composition of the biholomorphic map $X \to f(X)$ and the inclusion $f(X) \to Y$. We conclude that every holomorphic embedding is a holomorphic map.

8. Suppose $X_1, \ldots, X_s$ are complex manifolds of dimensions $n_1, \ldots, n_s$. The products $\varphi_1 \times \ldots \times \varphi_s$ of all charts $\varphi_i \in \text{Atl}\, X_i$ form an $(n_1 + \ldots + n_s)$-dimensional holomorphic atlas of the set $X_1 \times \ldots \times X_s$, transforming it into an $(n_1 + \ldots + n_s)$-dimensional complex manifold. Considered as a C^a-manifold, the latter is just the product of the C^a-manifolds $X_1, \ldots, X_s$.

9. The same definitions of tangent vectors, tangent vector spaces, total manifold of tangent vectors, and differential of a map (see Subsection 4) apply in the complex case. The space $\text{Tang}_x X$ tangent to the n-dimensional complex manifold X at the point x is an n-dimensional complex vector space; the total manifold $\text{Tang}\, X$ is a 2n-dimensional complex manifold, and the projection $\text{Tang}\, X \to X$ is holomorphic. The differential $d_x f$ of a holomorphic map $f: X \to Y$ at the point $x \in X$ is a linear mapping $\text{Tang}_x X \to \text{Tang}_{f(x)} Y$ of complex vector spaces, and the differential df is a holomorphic mapping $\text{Tang}\, X \to \text{Tang}\, Y$.

Again, the tangent spaces $\text{Tang}_x X$ and the manifold $\text{Tang}\, X$

may be considered as real vector spaces and as a C^a-manifold, respectively. As one may guess, they coincide with the tangent space and the total manifold of tangent vectors to X, regarded as a C^a-manifold. The differential of a holomorphic map f, regarded as a C^a-map, coincides with the differential of the C^a-map f.

10. The simplest examples of complex manifolds are the spaces $\mathbb{C}^n$ themselves. We obtain an unlimited supply of additional examples by defining submanifolds of $\mathbb{C}^n$ through systems of equations; cf. 2.16. However, this method will never produce compact manifolds of positive dimension: <u>every compact submanifold of $\mathbb{C}^n$ has dimension zero.</u>

To convince ourselves that this is true, it is enough to show that on a compact manifold the only holomorphic functions are the constants. This is a straightforward consequence of the well-known theorem stating that a function holomorphic on an open subset of the complex line $\mathbb{C}$ which attains its maximum modulus is constant. Now suppose X is a compact, connected, complex manifold, and $f: X \to \mathbb{C}$ is holomorphic. Let c be a value of f such that $|c| = \max |f(w)|$, and let $x \in X$ be such that $f(x) = c$. Then for each chart $\varphi \in$ Atl X with $x \in \operatorname{supp} \varphi$ and $\operatorname{Im} \varphi = \operatorname{Int} D^{2 \dim X}$, and for each point $y \in \operatorname{supp}$, the set of complex numbers z such that $(1 - z)\varphi(x) + z\varphi(y) \in \operatorname{Im} \varphi$ is an open disc with center 0 and radius greater than 1. The formula $z \to f(\varphi^{-1}((1 - z)\varphi(x) + z\varphi(y)))$ defines a holomorphic function on this disc, which attains its maximum modulus as 0. Since such a function is necessarily a constant, we see that $f(y) = c$ by setting $z = 0$ and $z = 1$. Therefore, the set $f^{-1}(c)$ is open. Because $f^{-1}(c)$ is also closed and nonempty, it is all of X.

Examples of compact complex manifolds will appear in § 2.

Manifolds of Complex Origin

11. Every complex manifold gives rise to a C^a-manifold when we pass from complex to real numbers, as described in 5. A C^a-manifold arising in this way is called a <u>manifold of complex origin.</u>

Clearly, the manifolds of complex origin are even-dimensional and have no boundary. They are orientable, and if the original complex structure is known, they receive a canonical orientation, namely that orientation which is positive on the real connected charts which arise from the charts of the complete atlas of the complex structure when we pass to the reals. (Using 3 we see that the compatibility condition required in 3.1 is satisfied.) It is also clear that the products of manifolds of complex origin are again manifolds of complex origin.

7. Exercises

1. Show that a nonempty, closed, smooth manifold of dimension $n > 0$ cannot be immersed in $\mathbb{R}^n$.

2. Show that the equation $z_1^2 + \ldots + z_n^2 = 1$ defines a submanifold of $\mathbb{C}^n$ which is C^a-diffeomorphic to $\mathrm{Tang}\, S^{n-1}$.

3. Show that the map $f: S^2 \to \mathbb{R}^4$, defined by the formula $f(x_1,x_2,x_3) = (x_1^2 - x_2^2, x_1 x_2, x_1 x_3, x_2 x_3)$, is an imerssion and that $f(S^2)$ is a submanifold of $\mathbb{R}^4$ diffeomorphic to $\mathbb{R}P^2$.

4. Show that if one of the numbers $n_1,\ldots,n_s$ is odd and $s > 1$, then the manifold $S^{n_1} \times \ldots \times S^{n_s}$ is parallelizable.

§2. STIEFEL AND GRASSMAN MANIFOLDS

1. Stiefel Manifolds

1. We denote by $\mathbb{R}V(n,k)$, or simply by $V(n,k)$ $(0 \leqslant k \leqslant n)$, the set of linear isometric maps $\mathbb{R}^k \to \mathbb{R}^n$. Such a map is uniquely determined by the images of the vectors $\mathrm{ort}_1,\ldots,\mathrm{ort}_k \in \mathbb{R}^k$, i.e., by an orthonormal k-frame in $\mathbb{R}^n$. The coordinates of the vectors of this frame form the matrix of the map, which has n rows and k columns. In this way, $V(n,k)$ can be interpreted as the set of orthonormal k-frames in $\mathbb{R}^n$, or as the set of the n×k-matrices $\|v_{si}\|$ such that

$$\sum_{s=1}^{n} v_{si} v_{sj} = \delta_{ij} \qquad (1 \leqslant i \leqslant j \leqslant k). \tag{1}$$

We may regard a matrix $\|v_{si}\|$ as a point of $\mathbb{R}^{nk}$, if we index its entries v_{si} in dictionary order. Thus $V(n,k)$ becomes the subset of $\mathbb{R}^{nk}$ defined by the equations (1). An easy computation shows that the gradients of the left-hand sides of (1) do not vanish and are pairwise orthogonal on this subset. Hence $V(n,k)$ is an

$[nk-k(k+1)/2]$-dimensional C^a-submanifold without boundary of $\mathbb{R}^{nk}$ (see 1.2.12). $V(n,k)$ is called the <u>Stiefel manifold.</u>

Clearly, $V(n,0)$ reduces to a point, $V(n,1)$ is just the sphere S^{n-1}, and $V(n,2)$ is the submanifold of all vectors of unit length in $\text{Tang } S^{n-1}$.

2. The points of $V(n,n)$ are orthogonal transformations of $\mathbb{R}^n$, or orthogonal matrices of order n, and $V(n,n)$ is usually denoted by $O(n)$. The composition of transformation (multiplication of matrices) induces a group structure on $O(n)$. The subgroup of $O(n)$ consisting of all matrices with determinant $+1$ is denoted by $SO(n)$. The two sets $SO(n)$ and $O(n) \smallsetminus SO(n)$ are open in $O(n)$, and hence are C^a-manifolds. They are actually C^a-diffeomorphic: multiplication by an arbitrary matrix from $O(n) \smallsetminus SO(n)$ establishes a diffeomorphism. Moreover, the manifold $SO(n)$ is canonically C^a-diffeomorphic to $V(n,n-1)$: a matrix from $V(n,n-1)$ is carried by this diffeomorphism into a matrix from $SO(n)$ through the addition of a column; that is to say, we complete each orthonormal $(n-1)$-frame in $\mathbb{R}^n$ to a positive orthonormal n-frame.

We further note that $SO(2) = V(2,1) = S^1$, and the group structure on $SO(2)$ agrees with the group structure on the circle S^1, considered as the multiplicative group of complex numbers of modulus 1.

3. The inclusion $\mathbb{R}^n \to \mathbb{R}^{n+q}$ induces a C^a-embedding $V(n,k) \to V(n+q,k)$, which transforms each map $\varphi \colon \mathbb{R}^k \to \mathbb{R}^n$ into the composite map $\mathbb{R}^k \xrightarrow{\varphi} \mathbb{R}^n \xrightarrow{\text{in}} \mathbb{R}^{n+q}$. There is also a canonical C^a-embedding $V(n,k) \to V(n+q,k+q)$, which transforms $\varphi \colon \mathbb{R}^k \to \mathbb{R}^n$ into the map

$$\mathbb{R}^{k+q} = \mathbb{R}^k \times \mathbb{R}^q \xrightarrow{\varphi \times \text{id}} \mathbb{R}^n \times \mathbb{R}^q = \mathbb{R}^{n+q}.$$

Finally, the inclusion $\mathbb{R}^{k-q} \to \mathbb{R}^k$ induces a C^a-submersion $V(n,k) \to V(n,k-q)$, which transforms each map $\varphi \colon \mathbb{R}^k \to \mathbb{R}^n$ into the composite map $\mathbb{R}^{k-q} \xrightarrow{\text{in}} \mathbb{R}^k \xrightarrow{\varphi} \mathbb{R}^n$, i.e., each frame $(v_1,\ldots,v_k)$ is taken into the frame $(v_1,\ldots,v_{k-q})$. Using Theorem 1.5.9, it is readily seen that this is indeed a submersion, because given any frame $(v_1^0,\ldots,v_k^0) \in V(n,k)$ we have an explicit construction of the neighborhood V of the frame $(v_1^0,\ldots,v_{k-q}^0)$ in $V(n,k-q)$ and of the C^a-map $g \colon V \to V(n,k)$ which are required by this theorem. In fact, one can take as V the set of frames $(v_1,\ldots,v_{k-q})$ in $V(n,k-q)$ such that the vectors $v_1,\ldots,v_{k-q},v_{k-q+1}^0,\ldots,v_k^0$ are linearly independent and then, for $(v_1,\ldots,v_{k-q}) \in V$, define $g(v_1,\ldots,v_{k-q})$ to be the frame obtained from $(v_1,\ldots,v_{k-q},v_{k-q+1}^0,\ldots,v_k^0)$ through standard orthogonalization. Let us add that the preimage of an arbitrary frame $(v_1^0, \ldots, v_{k-q}^0) \in V(n,k-q)$ under this submersion is the submanifold of $V(n,k)$ consisting of all the frames $(v_1^0, \ldots, v_{k-q}^0, v_{k-q+1}, \ldots, v_k)$, where $(v_{k-q+1}, \ldots, v_k)$ is an orthonormal q-frame of the $(n-k+q)$-dimensional subspace of $\mathbb{R}^n$ which is orthogonal to the vectors $v_1^0, \ldots, v_{k-q}^0$; in particular, this submanifold is diffeomorphic to $V(n-k+q,q)$.

4. We see from equations (1) that the set $V(n,k)$ is bounded and closed in $\mathbb{R}^{nk}$. Therefore, $V(n,k)$ is a closed manifold.

The manifold $V(n,n-1) = SO(n)$ is connected: each matrix of $SO(n)$ can be expressed as $cu(\phi_1, \ldots, \phi_r)c^{-1}$, where

$$
u(\phi_1, \ldots, \phi_r) = \begin{pmatrix}
\cos\phi_1 & -\sin\phi_1 & & & & & & \\
\sin\phi_1 & \cos\phi_1 & & & & & 0 & \\
& & \ddots & & & & & \\
& & & \cos\phi_r & -\sin\phi_r & & \\
& & & \sin\phi_r & \cos\phi_r & & \\
& & & & & 1 & & \\
& 0 & & & & & \ddots & \\
& & & & & & & 1
\end{pmatrix}
$$

(with $\phi_1, \ldots, \phi_r \in \mathbb{R}$) and c is an orthogonal matrix. Therefore, each matrix can be joined to the identity matrix by the path $t \mapsto cu((1 - t)\phi_1, \ldots, (1 - t)\phi_r)c^{-1}$. Since the manifolds $V(n,k)$ with $k < n-1$ are the images under continuous maps of $V(n,n-1)$ (see 3), they are also connected. For $n > 0$, $V(n,n)$ has two connected components: $SO(n)$ and $O(n) \smallsetminus SO(n)$.

The Complex Case

5. Let $\mathbb{C}V(n,k)$, $0 \leqslant k \leqslant n$, be the set of linear isometric maps $\mathbb{C}^k \to \mathbb{C}^n$. In other words, $\mathbb{C}V(n,k)$ consists of the orthonormal k-frames in $\mathbb{C}^n$ or, equivalently, of the complex $n \times k$-matrices $\|v_{si}\|$ such that

$$
\sum_{s=1}^{n} v_{si}\bar{v}_{sj} = \delta_{ij} \qquad (1 \leqslant i \leqslant j \leqslant k).
$$

These equations show that $\mathbb{C}V(n,k)$ is a subset of $\mathbb{C}^{nk} = \mathbb{R}^{2nk}$. Now we separate their real and imaginary parts and obtain k^2 real equations such that the gradients of their left-hand sides do not vanish and are pairwise orthogonal on $\mathbb{C}V(n,k)$. Thus $\mathbb{C}V(n,k)$ is a $(2nk-k^2)$-dimensional C^a-submanifold without boundary of $\mathbb{R}^{2nk}$, called the <u>complex Stiefel manifold</u>.

Warning: $\mathbb{C}V(n,k)$ is not a complex manifold in the sense of 1.6.5. The present definition does not equip it with a complex structure; in fact, such a structure does not exist in general, since $\mathbb{C}V(n,k)$ is odd-dimensional for k odd.

Clearly, $\mathbb{C}V(n,0)$ reduces to a point, and $\mathbb{C}V(n,1)$ is just the sphere S^{2n-1}. The points of the manifold $\mathbb{C}V(n,n)$ are unitary transformations of $\mathbb{C}^n$, and $\mathbb{C}V(n,n)$ is usually denoted by $U(n)$. Like $O(n)$, $U(n)$ is a group under the composition operation $\circ$. The subgroup of $U(n)$ consisting of all matrices with determinant 1 is denoted by $SU(n)$ and is a C^a-submanifold of $U(n)$, canonically diffeomorphic to $\mathbb{C}V(n,n-1)$ (cf. 2).

The manifold $U(n)$ is canonically diffeomorphic to $SU(n) \times S^1$: this diffeomorphism takes each pair $(u,z) \in SU(n) \times S^1$ into the matrix obtained from u by multiplying its first row by z. Warning: this diffeomorphism is not a group isomorphism between the direct product of groups $SU(n) \times S^1$ and $U(n)$.

The C^a-embeddings $\mathbb{C}V(n,k) \to \mathbb{C}V(n+q,k)$ and $\mathbb{C}V(n,k) \to \mathbb{C}V(n+q,k+q)$, and the C^a-submersion $\mathbb{C}V(n,k) \to \mathbb{C}V(n,k-q)$ are defined exactly as in the real case. Moreover, since each linear isometric map $\mathbb{C}^k \to \mathbb{C}^n$ may be regarded as a linear isometric map $\mathbb{R}^{2k} \to \mathbb{R}^{2n}$, there is a canonical C^a-embedding $\mathbb{C}V(n,k) \to \mathbb{R}V(2n,2k)$.

The manifolds $\mathbb{C}V(n,k)$ are compact. Moreover, they are all connected. Indeed, $\mathbb{C}V(n,n) = U(n)$ is connected, because every unitary matrix can be expressed as $cu(\phi_1,\ldots,\phi_n)c^{-1}$, where $u(\phi_1,\ldots,\phi_n)$ is the diagonal matrix with entries $\exp(i\phi_1),\ldots,\exp(i\phi_n)$ $(\phi_1,\ldots,\phi_n \in \mathbb{R})$ and c is a unitary matrix, and then joined to the identity matrix by the path $t \mapsto cu((1-t)\phi_1,\ldots,(1-t)\phi_n)$. The manifolds $\mathbb{C}V(n,k)$ with $k < n$ are connected as continuous images of $\mathbb{C}V(n,n)$.

The Quaternionic Case

6. The discussion in 5 can be repeated almost word for word if one replaces the field of complex numbers by the field of quaternions. (Recall that $\mathbb{H}^n$ is considered as a left vector space; see 1.2.5.4; consequently, a linear map $\mathbb{H}^k \to \mathbb{H}^n$ is left-linear here, and the scalar product of the vectors $(u_1,\ldots,u_n)$ and $(v_1,\ldots,v_n)$ is defined as

$\sum_{i=1}^{n} u_i \bar{v}_i$.) In this way, we obtain: the C^a-manifold without boundary $\mathbb{H}V(n,k)$ $(0 \leqslant k \leqslant n)$ of dimension $4nk-(2k^2-k)$, called the quaternionic Stiefel manifold; the C^a-embeddings $\mathbb{H}V(n,k) \to \mathbb{H}V(n+q,k)$ $\mathbb{H}V(n,k) \to \mathbb{H}V(n+q,k+q)$, and $\mathbb{H}V(n,k) \to \mathbb{C}V(2n,2k)$; and the C^a-submersion $\mathbb{H}V(n,k) \to \mathbb{H}V(n,k-q)$

Clearly, $\mathbb{H}V(n,0)$ reduces to a point, and $\mathbb{H}V(n,1)$ is just S^{4n-1}. The manifold $\mathbb{H}V(n,n)$ is usually denoted by $Sp(n)$; its points are the linear isometric transformations of $\mathbb{H}^n$, and the composition of transformations makes $Sp(n)$ into a group.

All the manifolds $\mathbb{H}V(n,k)$ are compact and connected. Since the proof of connectedness along the lines in 4 or 5 requires the normal form of a matrix from $Sp(n)$, which is less known than the normal forms for $SO(n)$ and $U(n)$, we remark that a different proof of connectedness is given in Chapter 5 (see 5.2.7.4).

Noncompact Stiefel Manifolds

7. Let $\mathbb{R}V'(n,k)$, or simply $V'(n,k)$, $0 \leqslant k \leqslant n$, denote the set of linear monomorphisms $\mathbb{R}^k \to \mathbb{R}^n$. Alternatively, one may describe $V'(n,k)$ as the set of nondegenerate k-frames in $\mathbb{R}^n$, or as the set of the real $n \times k$-matrices of rank k. Clearly, this set is open in the space $\mathbb{R}^{nk}$ of all real $n \times k$-matrices, and hence $V'(n,k)$ is an nk-dimensional C^a-manifold containing $V(n,k)$ as a submanifold.

Let $T(k,\mathbb{R})$, or simply $T(k)$, be the set of real upper triangular matrices of order k with positive diagonal entries. Trivially, $T(k)$ is open in the space of all upper triangular matrices of order k, $\mathbb{R}^{k(k+1)/2}$, and is actually diffeomorphic to $\mathbb{R}^{k(k+1)/2}$. If we orthogonalize a given nondegenerate k-frame in $\mathbb{R}^n$ via the standard procedure, then the matrix corresponding to this frame takes the form ut, where $u \in V(n,k)$ and $t \in T(k)$. Moreover, it is obvious that this representation is unique and defines a diffeomorphism $V'(n,k) \to V(n,k) \times T(k)$, transforming $V(n,k)$ into the fiber $V(n,k) \times E$, where E is the identity matrix. Therefore, the manifold $V'(n,k)$ is C^a-diffeomorphic to $V(n,k) \times \mathbb{R}^{k(k+1)/2}$, and $V(n,k)$ is its strong deformation retract. In particular, $V'(n,k)$ is connected for $k < n$, and $V'(n,n)$ has two components.

Again, $V'(n,0)$ reduces to a point, while $V'(n,1)$ coincides with $\mathbb{R}^n \smallsetminus 0$. The manifold $V'(n,n)$ is usually denoted by $GL(n,\mathbb{R})$ and its points are the nondegenerate linear transformations of $\mathbb{R}^n$ or, equivalently, the nondegenerate matrices of order n. The composition of transformations (multiplication of matrices) defines a group structure on $GL(n,\mathbb{R})$. The subgroup of $GL(n,\mathbb{R})$ consisting of all matrices with positive determinant is denoted by $GL_+(n,\mathbb{R})$. The sets $GL_+(n,\mathbb{R})$ and $GL(n,\mathbb{R}) \smallsetminus GL_+(n,\mathbb{R})$ are open in $GL(n,\mathbb{R})$ and C^a-diffeomorphic to $SO(n) \times \mathbb{R}^{n(n+1)/2}$; in fact, they are the components of the manifold $GL(n,\mathbb{R})$.

Corresponding to each monomorphism $\varphi: \mathbb{R}^k \to \mathbb{R}^n$ we have the composite maps $\mathbb{R}^k \overset{\varphi}{\to} \mathbb{R}^n \overset{in}{\longrightarrow} \mathbb{R}^{n+q}$, $\mathbb{R}^{k+q} \overset{id}{\longrightarrow} \mathbb{R}^k \times \mathbb{R}^q \overset{\varphi \times id}{\longrightarrow} \mathbb{R}^n \times \mathbb{R}^q \overset{id}{\to} \mathbb{R}^{n+q}$ and $\mathbb{R}^{k-q} \overset{in}{\longrightarrow} \mathbb{R}^k \overset{\varphi}{\to} \mathbb{R}^n$, and so we obtain the C^a-embeddings $V'(n,k) \to V'(n+q,k)$, $V'(n,k) \to V'(n+q,k+q)$, and the C^a-submersion $V'(n,k) \to V'(n,k-q)$, respectively; cf. 3.

8. We let $\mathbb{C}V'(n,k)$ $(0 \leqslant k \leqslant n)$ denote the set of linear monomorphisms $\mathbb{C}^k \to \mathbb{C}^n$. Alternatively, $\mathbb{C}V(n,k)$ is the set of non-degenerate k-frames in $\mathbb{C}^n$, or the set of complex $n \times k$-matrices of rank k. Again, it it clear that this is an open set in the space $\mathbb{C}^{nk}$ of all complex $n \times k$-matrices. Thus $\mathbb{C}V'(n,k)$ is a 2nk-dimensional C^a-manifold containing $\mathbb{C}V(n,k)$ as a submanifold.

Repeating what was said in 7 (with obvious modifications), we obtain a C^a-diffeomorphism $\mathbb{C}V'(n,k) \to \mathbb{C}V(n,k) \times T(k,\mathbb{C})$, where $T(k,\mathbb{C})$ is the manifold of complex upper triangular matrices of order k with positive diagonal elements. Clearly, $T(k,\mathbb{C})$ is C^a-diffeomorphic to $\mathbb{R}^{k^2}$, and hence $\mathbb{C}V'(n,k)$ is diffeomorphic to $\mathbb{C}V(n,k) \times \mathbb{R}^{k^2}$. In particular, the manifolds $\mathbb{C}V'(n,k)$ are all connected. $\mathbb{C}V'(n,0)$ reduces to a point, while $\mathbb{C}V'(n,1)$ is just $\mathbb{C}^n \smallsetminus 0$. Usually the manifold $\mathbb{C}V'(n,n)$ is denoted by $GL(n,\mathbb{C})$; it consists of all non-degenerate linear transformations of $\mathbb{C}^n$, and is a group with group operation $\circ$. There are also the natural C^a-embeddings $\mathbb{C}V'(n,k) \to \mathbb{C}V'(n+q,k)$ and $\mathbb{C}V'(n,k) \to \mathbb{C}V'(n+q,k+q)$, as well as the natural C^a-submersion $\mathbb{C}V'(n,k) \to \mathbb{C}V'(n,k-q)$.

9. If we replace the commutative fields $\mathbb{R}$ and $\mathbb{C}$ by the field $\mathbb{H}$ in the previous definitions, we obtain: the 4nk-dimensional C^a-manifold $\mathbb{H}V'(n,k)$ of all linear monomorphisms $\mathbb{H}^k \to \mathbb{H}^n$; the C^a-diffeomorphism $\mathbb{H}V'(n,k) \to \mathbb{H}V(n,k) \times \mathbb{R}^{2k^2-k}$; the C^a-embeddings $\mathbb{H}V'(n,k) \to \mathbb{H}V'(n+q,k)$ and $\mathbb{H}V'(n,k) \to \mathbb{H}V'(n+q,k+q)$; and finally, the C^a-submersion $\mathbb{H}V'(n,k) \to \mathbb{H}V'(n,k-q)$. As before, $\mathbb{H}V'(n,0)$ reduces to a point, while $\mathbb{H}V'(n,1)$ is just $\mathbb{H}^n \smallsetminus 0$. The manifold $\mathbb{H}V'(n,n)$ consists of all nondegenerate linear transformations of $\mathbb{H}^n$, is a group with group operation $\circ$, and is usually denoted by $GL(n,\mathbb{H})$.

2. <u>Grassman Manifolds</u>

1. Let $\mathbb{R}G(n,k)$, or simply $G(n,k)$, $0 \leqslant k \leqslant n$, be the set of k-dimensional linear subspaces (or k-planes passing through 0) of $\mathbb{R}^n$. If γ is such a plane, we let U_γ denote the collection of planes in $G(n,k)$ whose projection on γ is nondegenerate. Let us fix an orthonormal basis $e = \{e_1,\ldots,e_k\}$ of γ and complete it to an ortho-normal basis of $\mathbb{R}^n$, adding a frame $\varepsilon = \{\varepsilon_1,\ldots,\varepsilon_{n-k}\}$. For each $\gamma' \in U_\gamma$ there exists a unique frame $u_1,\ldots,u_k$ in γ' which projects onto e. Expressing the vectors $u_1,\ldots,u_k$ in terms of the basis $e_1,\ldots,e_k,\varepsilon_1,\ldots,\varepsilon_{n-k}$, we obtain

$$u_i = e_i + \sum_{s=1}^{n-k} c_{is}\varepsilon_s \qquad (i = 1,\ldots,k; \quad c_{is} \in \mathbb{R}).$$

This construction yields a map $\varphi(e,\varepsilon)$ from the set U_γ into the space $\mathbb{R}^{k(n-k)}$ of real $k\times(n-k)$-matrices. Obviously, $\varphi(e,\varepsilon)$ is invertible, and the collection of maps $\varphi(e,\varepsilon)$ obtained from all the pairs e,ε is a C^a-atlas of the set $G(n,k)$. Since any two planes in $G(n,k)$ are contained in some set U_γ, the topology defined by the atlas $\{\varphi(e,\varepsilon)\}$ is Hausdorff. Moreover, this atlas contains finite subatlases (for example, the subatlas consisting of the charts $\varphi(e,\varepsilon)$ where e and ε are subframes of the standard frame $ort_1,\ldots,ort_n$ of $\mathbb{R}^n$), and hence the above topology has a countable base. Thus, the atlas $\{\varphi(e,\varepsilon)\}$ transforms $G(n,k)$ into a $k(n-k)$-dimensional C^a-manifold, called the <u>Grassman manifold</u>.

Clearly, $G(n,0)$ and $G(n,n)$ both reduce to points, and $G(n,1)$, as a set and a topological space, is identical to $\mathbb{R}P^{n-1}$. So we see that the space $\mathbb{R}P^{n-1}$ has a C^a-structure compatible with its topology, and hence $\mathbb{R}P^{n-1}$ is an $(n-1)$-dimensional C^a-manifold. We remark that this C^a-structure may be described directly and conveniently as follows: for $k = 1$, the finite atlas of the manifold $G(n,k)$ given above consists of n charts $\varphi_1\colon U_1 \to \mathbb{R}^{n-1},\ldots,$ $\varphi_n\colon U_n \to \mathbb{R}^{n-1}$, defined in homogeneous coordinates by the formulas

$$U_i = \{(x_1:\ldots:x_n) \mid x_i \neq 0\},$$

$$\varphi_i((x_1:\ldots:x_n)) = (x_1/x_i,\ldots,x_{i-1}/x_i,x_{i+1}/x_i,\ldots,x_n/x_i).$$

It is also clear that relative to this C^a-structure the canonical homeomorphism $\mathbb{R}P^1 \to S^1$ (see 1.2.5.6) becomes a C^a-diffeomorphism.

2. Obviously, we may modify the definition of the manifold
$G(n,k)$ by replacing the nonoriented planes with oriented ones. More
precisely, $G(n,k)$ is replaced by the set $G_+(n,k)$ of oriented
k-dimensional planes (oriented k-planes, for short) of $\mathbb{R}^n$ passing
through 0. One has to modify the set U_γ accordingly and take it to
be the collection of all planes in $G_+(n,k)$ whose projections onto the
plane $\gamma \in G_+(n,k)$ are nondegenerate and orientation preserving. The
maps $\varphi(e,\varepsilon): U_\gamma \to \mathbb{R}^{n(n-k)}$ are defined as in 1 and again they form a
C^a-atlas possessing finite subatlases. Since any pair of points of
$G_+(n,k)$ is covered by a set of the form $U_\gamma \cup U_{\bar\gamma}$, where the plane $\bar\gamma$
differs from γ only by its orientation (obviously, $U_\gamma \cap U_{\bar\gamma} = \emptyset$),
the topology defined by the atlas $\{\varphi(e,\varepsilon)\}$ is Hausdorff. The
$k(n-k)$-dimensional C^a-manifold so obtained is termed the <u>upper Grassman</u>
<u>manifold.</u>

The manifolds $G_+(n,0)$ and $G_+(n,n)$ are canonically homeomorphic to S^0, while $G_+(n,1)$ is canonically diffeomorphic to S^{n-1}:
under this diffeomorphism each point $x \in S^{n-1}$ goes into the oriented
line defined by the pair of points $0,x$.

3. If one associates to each plane $\gamma \in G(n,k)$ its orthogonal complement $\gamma^\perp$, one obtains a mapping of $G(n,k)$ onto $G(n,n-k)$
which is clearly a C^a-diffeomorphism. A C^a-diffeomorphism
$G_+(n,k) \to G_+(n,n-k)$ is similarly defined, if the orientation given to
the orthogonal plane $\gamma^\perp$ and the orientation of γ behave in accordance
to the classical rule: let a basis of $\gamma^\perp$ which is compatible with this
orientation be written to the right of a basis of γ which is compatible
with its orientation; then the resulting basis of $\mathbb{R}^n$ should be
compatible with the standard orientation of $\mathbb{R}^n$ (see 1.3.10). In
particular, $G(n,n-1) = \mathbb{R}P^{n-1}$ and $G_+(n,n-1) = S^{n-1}$.

The inclusion $\mathbb{R}^n \to \mathbb{R}^{n+q}$ induces obvious C^a-embeddings
$G(n,k) \to G(n+q,k)$ and $G_+(n,k) \to G_+(n+q,k)$. Furthermore, the formula
$\gamma \mapsto \gamma \times \mathbb{R}^q$ defines C^a-embeddings $G(n,k) \to G(n+q,n+k)$ and
$G_+(n,k) \to G_+(n+q,n+k)$ (the orientation of the product $\gamma \times \mathbb{R}^q$ is
defined by the orientation of its factors - see 1.3.7). These
embeddings and the previous diffeomorphisms form the following
commutative diagrams:

$$
\begin{array}{ccc}
G(n,k) & \longrightarrow & G(n+q,k) \\
\downarrow & & \downarrow \\
G(n,n-k) & \longrightarrow & G(n+q,n-k+q)
\end{array}
\qquad
\begin{array}{ccc}
G_+(n,k) & \longrightarrow & G_+(n+q,k) \\
\downarrow & & \downarrow \\
G_+(n,n-k) & \longrightarrow & G_+(n+q,n-k+q).
\end{array}
$$

The map $G_+(n,k) \to G(n,k)$ which takes each oriented plane and "forgets" its orientation is clearly a C^a-submersion such that the preimage of each point of $G(n,k)$ consists of just two points. For $k = 1$ we recover the projection $S^{n-1} \to \mathbb{R}P^{n-1}$.

If to each frame from $V(n,k)$ we associate the oriented plane that it spans, we obtain a C^a-map $V(n,k) \to G_+(n,k)$. This is a submersion, a fact that can be readily checked with the aid of Theorem 1.5.9, if for a given frame $v^0 \in V(n,k)$ one explicitly indicates the neighborhood V of the oriented plane γ^0 spanned by v^0 and the map $g: V \to V(n,k)$ which are needed in this theorem. One can take $V = U_{\gamma^0}$ (see 3) and define $g(\gamma)$ for each $\gamma \in V$ to be the frame which is obtained from v^0 after projection on γ and standard orthogonalization. The preimage of an oriented plane $\gamma \in G_+(n,k)$ under this submersion is the set of all orthonormal positive k-frames of γ and, in particular, is diffeomorphic to $SO(k)$.

The map of $V(n,k)$ onto $G(n,k)$ which takes each frame into the nonoriented plane that it spans is also a submersion. In fact, it is exactly the composition of the two previous submersions. The preimage of a plane $\gamma \in G(n,k)$ consists of all the orthonormal frames of γ and, in particular, is diffeomorphic to $O(k)$.

Finally, the maps $V'(n,k) \to G_+(n,k)$ and $V'(n,k) \to G(n,k)$, which transform each frame into the plane that it spans are both submersions. The preimages of the points of $G_+(n,k)$ and $G(n,k)$ under these maps are diffeomorphic to $GL_+(k,\mathbb{R})$ and $GL(k,\mathbb{R})$, respectively.

4. Since the manifolds $G(n,k)$ and $G_+(n,k)$ are continuous images of $V(n,k)$, they are compact and, excepting $G_+(n,0)$ and $G_+(n,n)$, connected.

The family of manifolds $G(n,k)$ $(k \neq 0,n)$ contains both orientable and nonorientable manifolds. More precisely, <u>for</u> $k \neq 0$ $G(n,k)$ <u>is orientable if</u> n <u>is even and nonorientable if</u> n <u>is odd.</u> To see this, use the atlas made of the charts $\varphi(e,\varepsilon)$, where e and ε are complementary subframes of the standard basis of $\mathbb{R}^n$, with the order of vectors they inherit from this basis (cf. 1). Denote the indices of the vectors of e by $j_1(e),\ldots,j_k(e)$ $(1 \leqslant j_1(e) < \cdots < j_k(e) \leqslant n)$, and say that two charts $\varphi(e,\varepsilon)$ and $\varphi(e',\varepsilon')$ of the atlas are <u>contiguous</u> if there exists ℓ such that $j_\ell(e') = j_\ell(e) \pm 1$ and $j_p(e') = j_p(e)$ for $p \neq \ell$. Obviously, for $k \neq 0,n$ we can always exhibit contiguous charts, and actually any two charts in the atlas may be connected by a finite chain of charts such that each two neighboring charts in the chain are contiguous. Suppose that c_{is} and c'_{is} are the coordinate functions of two contiguous charts, $\varphi(e,\varepsilon)$ and

$\varphi(e',\varepsilon')$, such that $j_\ell(e') = j_\ell(e) + 1$. Then it is readily seen that the coordinate $c_{\ell m}$ with $m = j_\ell - \ell + 1$ does not vanish on the intersection of the supports of the two charts, but takes all the remaining real values on this intersection. A simple computation shows that

$$
c'_{is} = \begin{cases}
(c_{is}c_{\ell m} - c_{im}c_{\ell s})c_{\ell m}^{-1}, & \text{if } i \neq \ell, \; s \neq m, \\
c_{\ell s}c_{\ell m}^{-1}, & \text{if } i = \ell, \; s \neq m, \\
-c_{im}c_{\ell m}^{-1}, & \text{if } i \neq \ell, \; s = m, \\
c_{\ell m}^{-1}, & \text{if } i = \ell, \; s = m,
\end{cases}
$$

and thus the Jacobian $J(\varphi(e,\varepsilon),\varphi(e',\varepsilon'))$ [see 1.3.1] is equal to $(-1)^n c_{\ell m}^{-n}$. If n is odd, this Jacobian takes both positive and negative values; hence the manifold $G(n,k)$ is not orientable for such n. For n even, the formula $\varphi(e,\varepsilon) \mapsto (-1)^{k[j_1(e)+\dots+j_k(e)]}$ defines a map of the atlas under consideration into S^0, which satisfies the compatibility condition in 1.3.1: this condition obviously holds for contiguous charts, which in turn implies the compatibility for non-contiguous charts. Hence, for n even $G(n,k)$ is orientable.

5. The next constructions produce C^a-embeddings of $G(n,k)$ and $G_+(n,k)$ in Euclidean spaces; see also 4.8.

Let us start with $G_+(n,k)$. For a matrix $v \in V'(n,k)$, let $M_{i_1 \dots i_k}(v)$ be the minor constructed from the rows with indices $i_1,\dots,i_k$, and put $N_{i_1 \dots i_k}(v) = M_{i_1 \dots i_k}(v)/\mu(v)$, where $\mu(v)$ is the positive square root of the sum of the squares of all minors having maximal order. The functions $N_{i_1 \dots i_k} : V'(n,k) \to \mathbb{R}$ are clearly analytic. Moreover, if two frames in $V'(n,k)$, v^1 and v^2, span the same k-plane, then $N_{i_1 \dots i_k}(v^1) = N_{i_1 \dots i_k}(v^2)$. Therefore, the map $N: V'(n,k) \to \mathbb{R}^{\binom{n}{k}}$ with coordinate functions $N_{i_1 \dots i_k}$ is the composition of the canonical submersion $V'(n,k) \to G_+(n,k)$, defined in 3, with a C^a-map $g_+: G_+(n,k) \to \mathbb{R}^{\binom{n}{k}}$ (see 1.5.10). We next show that g_+ is a C^a-embedding.

Using Theorem 1.5.4, it suffices to verify that g_+ is an injective immersion. To demonstrate that g_+ is injective, one has to show that if $v^1,v^2 \in V'(n,k)$ and $N(v^1) = N(v^2)$, then the $n\times 2k$-matrix constructed by adjoining the matrices v^1 and v^2 has rank k. But this is plain, because any $(k+1)\times(k+1)$-minor of this

$n \times 2k$-matrix, constructed from k columns of v^1 and one column of v^2 is equal to zero (expand the minor with respect to the column of v^2). To verify that g_+ is an immersion, it is enough to show that at each point $\gamma_0 \in G_+(n,k)$, the rank of the differential $d_{\gamma_0} g_+$ is $k(n-k)$. This in turn will be true if we can prove that at each point $v_0 \in V'(n,k)$, the differential $d_{v_0} N$ has rank $\geqslant k(n-k)$. Finally, to obtain this property of $d_{v_0} N$, we prove that at each point $v_0 \in V'(n,k)$, the differential $d_{v_0} M$ has rank $\geqslant k(n-k) + 1$, where $M: V'(n,k) \to \mathbb{R}^{\binom{n}{k}}$ is the map with coordinate functions $M_{i_1 \ldots i_k}$. So let A be any $k \times k$-submatrix of the matrix $v \in V'(n,k)$ such that A is nondegenerate for $v = v_0$, and let $v_{i_0 j_0}$ be an element of A such that its cofactor, call it α, does not vanish for $v = v_0$. Next isolate those minors $M_{i_1 \ldots i_k}(v)$ of the matrix v which have at least $k-1$ rows in common with A, and then form a submatrix of the Jacobi matrix of the map M as follows. Take the derivatives of the chosen minors with respect to those elements of the matrix v which do not appear in A and the one derivative with respect to $v_{i_0 j_0}$. We obtain a square matrix of order $k(n-k) + 1$ which has, in a neighborhood of the point v_0 (and for a suitable arrangement of the rows and columns) the form

$$
\left(
\begin{array}{cccc|c}
(A^t)^{-1}\det A & 0 & \cdots & 0 & \beta_1 \\
0 & (A^t)^{-1}\det A & \cdots & 0 & \vdots \\
\cdot\ \cdot\ \cdot\ \cdot\ \cdot\ & \cdot\ \cdot\ \cdot\ \cdot & \cdot\ \cdot\ \cdot & \cdot\ \cdot & \vdots \\
0 & 0 & \cdots & (A^t)^{-1}\det A & \vdots \\
& & & & \beta_{k(n-k)} \\
\hline
0 & \cdots & & \cdots\ 0 & \alpha
\end{array}
\right),
$$

where t indicates transposition. The determinant of this matrix is equal to $\alpha (\det A)^{(n-k)(k-1)}$ and therefore does not vanish.

Now let us turn to $G(n,k)$ and compose the embedding g_+ with the map $q: \mathbb{R}^{\binom{n}{k}} \to \mathbb{R}^{\binom{n}{k}\left(\binom{n}{k}+1\right)/2}$, defined as $q(x_1, \ldots, x_{\binom{n}{k}}) =$

$$
= (x_1^2, x_1 x_2, \ldots, x_1 x_{\binom{n}{k}}, x_2^2, x_2 x_3, \ldots, x_2 x_{\binom{n}{k}}, \ldots, x_{\binom{n}{k}-1} x_{\binom{n}{k}}, x_{\binom{n}{k}}^2).
$$

Clearly, the restriction $q\big|_{\mathbb{R}^{\binom{n}{k}} \smallsetminus 0}$ is a C^a-immersion, and $q(x) = q(y)$ if and only if $x = \pm y$. Since the equality $g_+(\gamma') = -g_+(\gamma)$ holds for planes $\gamma, \gamma' \in G_+(n,k)$ which coincide geometrically (but have opposite orientations), we see that $q \circ g_+$ is a C^a-immersion of $G_+(n,k)$ into $\mathbb{R}^{\binom{n}{k}(\binom{n}{k}+1)/2}$. Moreover, $q \circ g_+(\gamma') = q \circ g_+(\gamma)$ if and only if γ and γ' coincide or differ only by their orientations. Therefore, the map $\mathrm{fact}(q \circ g_+)\colon G(n,k) \to \mathbb{R}^{\binom{n}{k}(\binom{n}{k}+1)/2}$ is well defined and is a C^a-embedding.

6 (INFORMATION). The coordinate functions $g^+_{i_1 \ldots i_k}$ of the map $g_+\colon G_+(n,k) \to \mathbb{R}^{\binom{n}{k}}$ satisfy the relations

$$g^+_{i_1 \ldots i_k} g^+_{j_1 \ldots j_k} - \sum_{s=1}^{k} g^+_{i_1 \ldots i_{k-1} j_s} g^+_{j_1 \ldots j_{s-1} i_k j_{s+1} \ldots j_k} = 0.$$

Considering these relations as equations relative to the coordinates in $\mathbb{R}^{\binom{n}{k}}$, they define a subset of $\mathbb{R}^{\binom{n}{k}}$ which is exactly $g_+(G_+(n,k))$. See [9] for details.

The numbers $g^+_{i_1 \ldots i_k}(\gamma)$ are known as the <u>Grassman-Plücker coordinates</u> of the oriented plane γ.

The Complex and Quaternionic Cases

7. To obtain the complex version of the manifold $\mathbb{R}G(n,k)$, one has to take k-dimensional planes of $\mathbb{C}^n$ passing through 0 instead of k-dimensional planes of $\mathbb{R}^n$ passing through 0. The result is a complex manifold $\mathbb{C}G(n,k)$ of dimension $k(n-k)$, called the <u>complex Grassman manifold.</u>

Obviously, $\mathbb{C}G(n,0)$ and $\mathbb{C}G(n,n)$ reduce to points, whereas $\mathbb{C}G(n,1)$ coincides, as a topological space, with $\mathbb{C}P^{n-1}$. Thus we equip the projective space $\mathbb{C}P^{n-1}$ with a complex structure compatible with its topology, which makes $\mathbb{C}P^{n-1}$ into an $(n-1)$-dimensional complex manifold. This structure may be given in homogeneous coordinates just as we described the C^a-structure of $\mathbb{R}P^{n-1}$. Relative to this structure the canonical homeomorphism $\mathbb{C}P^1 \to S^2$ (see 1.2.5.6) becomes a C^a-diffeomorphism.

The complex analogues $\mathbb{C}G(n,k) \to \mathbb{C}G(n,n-k)$, $\mathbb{C}G(n,k) \to \mathbb{C}G(n+q,k)$, $\mathbb{C}G(n,k) \to \mathbb{C}G(n+q,k+q)$, $\mathbb{C}V(n,k) \to \mathbb{C}G(n,k)$, and

$\mathbb{C}V'(n,k) \to \mathbb{C}G(n,k)$ of the maps $G(n,k) \to G(n,n-k)$, $G(n,k) \to G(n+q,k)$, $G(n,k) \to G(n+q,k+q)$, $V(n,k) \to G(n,k)$, and $V'(n,k) \to G(n,k)$, constructed in 3, are defined in an obvious way. The first is a biholomorphic map, the second and the third are holomorphic embeddings, and the fourth and the fifth are C^a-submersions. We mention also the C^a-embedding $\mathbb{C}G(n,k) \to G_+(2n,2k)$: when we transform $\mathbb{C}^n$ into $\mathbb{R}^{2n}$, every k-plane becomes an oriented 2k-plane.

The manifold $\mathbb{C}G(n,k)$ is the image of $\mathbb{C}V(n,k)$ under a continuous map, and as such it is compact and connected. Moreover, it readily seen that $\mathbb{C}G(n,k)$ can be analytically embedded in $\mathbb{R}^{\binom{2n}{2k}}$: the composition of the embedding $\mathbb{C}G(n,k) \to G_+(2n,2k)$ described above with the canonical embedding $G_+(2n,2k) \to \mathbb{R}^{\binom{2n}{2k}}$ is a C^a-embedding.

8. Substituting quaternions for complex numbers in these definitions, we obtain a connected $4k(n-k)$-dimensional C^a-manifold $\mathbb{H}G(n,k)$, called the <u>quaternionic Grassman manifold.</u> For $k = 0,n$, $\mathbb{H}G(n,k)$ reduces to a point. For $k = 1,n-1$, $\mathbb{H}G(n,k)$ is topologically the quaternionic projective space $\mathbb{H}P^{n-1}$. Thus $\mathbb{H}P^{n-1}$ becomes a C^a-manifold, and the canonical homeomorphism $\mathbb{H}P^1 \to S^4$ becomes a C^a-diffeomorphism.

The quaternionic analogues of the maps described in 3 and 7 are C^a-maps. In particular, there is a canonical C^a-embedding $\mathbb{H}G(n,k) \to G_+(4n,4k)$. Composing it with the canonical embedding $G_+(4n,4k) \to \mathbb{R}^{\binom{4n}{4k}}$, we obtain a C^a-embedding $\mathbb{H}G(n,k) \to \mathbb{R}^{\binom{4n}{4k}}$.

9. The maps

$$S^{2n-1} = \mathbb{C}V(n,1) \to \mathbb{C}G(n,1) = \mathbb{C}P^{n-1}$$

and

$$S^{4n-1} = \mathbb{H}V(n,1) \to \mathbb{H}G(n,1) = \mathbb{H}P^{n-1},$$

which are particular cases of the canonical maps in 7 and 8, obviously coincide with the corresponding Hopf maps (see 1.2.5.8). As a result, we see that these Hopf maps are C^a-submersions. The Hopf map $S^{15} \to S^8$ is a C^a-submersion too, a fact that follows directly from its definition.

Projective Cayley Plane

10. The projective Cayley plane is also equipped with a C^a-structure which transforms $\mathbb{C}aP^2$ into a C^a-manifold. We next

describe this structure.

Identify $\mathbb{R}^{16}$ with $\mathbb{C}a^2$, and $\mathbb{R}^{17}$ with $\mathbb{C}a^2 \times \mathbb{R}$, and then define three maps $D^{16} \to S^{16}$ by the formulas

$$(y_1,y_2) \mapsto (2(1 - \rho^2)^{1/2}y_1, 2(1 - \rho^2)^{1/2}y_2, 2\rho^2 - 1)$$

$$(y_1,y_2) \mapsto (2\bar{y}_1 y_2, 2(1 - \rho^2)^{1/2}\bar{y}_1, 1 - 2|y_1|^2),$$

$$(y_1,y_2) \mapsto (2(1 - \rho^2)^{1/2}\bar{y}_2, 2\bar{y}_2 y_1, 1 - 2|y_2|^2),$$

where $y_1, y_2 \in \mathbb{C}a$ and $\rho = (|y_1|^2 + |y_2|^2)^{1/2}$ (the first map is just DS, but we do not use this fact explicitly). These three maps yield a map $F: D^{16} \to S^{16} \times S^{16} \times S^{16}$, and one can easily verify that

$$\mathrm{zer}(F) = \mathrm{zer}[\mathrm{pr}: D^{16} \to \mathbb{C}aP^2].$$

Since the injective factor of F is a topological embedding of the quotient space $D^{16}/\mathrm{zer}(F)$ into $S^{16} \times S^{16} \times S^{16}$ (see 1.1.7.10), and since, according to the last equality, this quotient space coincides with $D^{16}/\mathrm{zer}(\mathrm{pr}) = \mathbb{C}aP^2$, we can identify $\mathbb{C}aP^2$ with $F(D^{16})$. To transform $\mathbb{C}aP^2$ into a C^a-manifold, it suffices to show that $F(D^{16})$ is a C^a-submanifold of $S^{16} \times S^{16} \times S^{16}$.

Define $f,g: S^{16} \smallsetminus \mathrm{ort}_{17} \to S^{16}$ as

$$f(u_1,u_2,t) = (\frac{u_2\bar{u}_1}{1 - t}, \bar{u}_1, \frac{|u_2|^2}{1 - t} - t)$$

and

$$g(u_1,u_2,t) = (\bar{u}_2, \frac{u_1\bar{u}_2}{1 - t}, \frac{|u_1|^2}{1 - t} - t), \qquad [u_1,u_2 \in \mathbb{C}a, \ t \in \mathbb{R}]$$

and consider the three maps $h_1, h_2, h_3: S^{16} \smallsetminus \mathrm{ort}_{17} \to S^{16} \times S^{16} \times S^{16}$ given by

$$h_1(x) = (x,f(x),g(x)), \quad h_2(x) = (g(x),x,f(x)),$$

$$h_3(x) = (f(x),g(x),x).$$

Since f and g are analytic, h_1, h_2 and h_3 are analytic embeddings. One can check directly that

$$F(D^{16}) \cap \mathrm{pr}_i^{-1}(S^{16} \smallsetminus \mathrm{ort}_{17}) = h_i(S^{16} \smallsetminus \mathrm{ort}_{17}).$$

Therefore, the left-hand side intersections are C^a-submanifolds of $S^{16} \times S^{16} \times S^{16}$, and it remains to note that

$$\cup_i \mathrm{pr}_i^{-1}(S^{16} \smallsetminus \mathrm{ort}_{17}) = (S^{16} \times S^{16} \times S^{16}) \smallsetminus (\mathrm{ort}_{17}, \mathrm{ort}_{17}, \mathrm{ort}_{17}),$$

and that $(ort_{17}, ort_{17}, ort_{17}) \notin F(D^{16})$.

Noncompact Grassman Manifolds

1. We denote by $\mathbb{R}G'(n,k)$, or simply by $G'(n,k)$ $(0 \leqslant k \leqslant n)$ the set of <u>all</u> k-dimensional planes of $\mathbb{R}^n$, i.e., the planes need not pass through 0. It is clear that given a plane $\gamma' \in G'(n,k)$, there is a unique (k+1)-plane γ in $\mathbb{R}^{n+1}$ that passes both through 0 and the k-plane which results by translating γ' by the vector ort_{n+1}. Moreover, the formula $\gamma' \mapsto \gamma$ defines an injective mapping $G'(n,k) \to G(n+1,k+1)$, which has an open image. So we may regard $G'(n,k)$ as an open subset of the manifold $G(n+1,k+1)$. In particular, $G'(n,k)$ is a (k+1)(n-k)-dimensional C^a manifold. We call it the <u>noncompact Grassman manifold</u>.

$G'(n,k)$ can be mapped naturally onto $G(n,k)$: for each plane $\gamma' \in G'(n,k)$, consider the parallel plane $\gamma \in G(n,k)$. It is a straightforward consequence of Theorem 1.5.9 that this map is a C^a-submersion (one can take the entire $G(n,k)$ for the required neighborhood V of γ in $G(n,k)$, and one can take the map which takes each plane from $G(n,k)$ into the parallel plane passing through an arbitrary, but fixed point of γ', for the required g). The preimage of a plane $\gamma \in G(n,k)$ under this submersion is the set of all k-planes of $\mathbb{R}^n$ which are parallel to γ, and is canonically diffeomorphic to the orthogonal complement $\gamma^{\perp}$ of γ (a unique k-plane parallel to γ passes through each point of $\gamma^{\perp}$).

The oriented, complex, and quaternionic versions of these definitions are immediate.

3. <u>Some Low-Dimensional Stiefel and Grassman Manifolds</u>

1. <u>$SO(3)$ is canonically C^a-diffeomorphic to $\mathbb{R}P^3$.</u>

We let shi denote the map of $\mathbb{R}^3$ into $\mathbb{R}_1^3$ which takes (y_1, y_2, y_3) into $(0, y_1, y_2, y_3)$. The canonical C^a-diffeomorphism $\mathbb{R}P^3 \to SO(3)$ takes the line of $\mathbb{R}P^3$ passing through the point $x \in \mathbb{R}^4 \smallsetminus 0$ into the orthogonal transformation $\varphi: \mathbb{R}^3 \to \mathbb{R}^3$ defined by the quaternionic formula $\varphi(y) = shi^{-1}(x\, shi(y)\, x^{-1})$. The inverse diffeomorphism $SO(3) \to \mathbb{R}P^3$ takes each transformation $\varphi: \mathbb{R}^3 \to \mathbb{R}^3$ into the line consisting of the quaternions of the form

$q - \text{shi}(\varphi(\text{ort}_1)) \, q \, \text{ort}_2 - \text{shi}(\varphi(\text{ort}_2)) \, q \, \text{ort}_3 - \text{shi}(\varphi(\text{ort}_3)) \, q \, \text{ort}_4,$

where q is an arbitrary quaternion, and ort_2, ort_3, and ort_4 are regarded as quaternion units. It is routine to check that quaternions of this form describe precisely a line and that the constructed maps $\mathbb{R}P^3 \to SO(3)$ and $SO(3) \to \mathbb{R}P^3$ are inverses of one another.

2. $\mathbb{R}V(4,2)$ <u>is canonically C^a-diffeomorphic to</u> $S^3 \times S^2$.

The canonical C^a-diffeomorphism $S^3 \times S^2 \to \mathbb{R}V(4,2)$ takes the pair $(x,y) \in S^3 \times S^2$ into the frame $\{x, x\,\text{shi}(y)\}$.

3. $SO(4)$ <u>is canonically C^a-diffeomorphic to</u> $S^3 \times SO(3)$.

The canonical C^a-diffeomorphism $S^3 \times SO(3) \to SO(4)$ is defined by the quaternion formula $(x,\{y,z\}) \mapsto \{x, x\,\text{shi}(y), x\,\text{shi}(z)\}$ (here the points of the manifolds $SO(3)$ and $SO(4)$ are interpreted as frames).

4. $G_+(4,2)$ <u>is canonically C^a-diffeomorphic to</u> $S^2 \times S^2$.

The canonical C^a-diffeomorphism $G_+(4,2) \to S^2 \times S^2$ takes the oriented plane spanned by the frame $\{x,y\} \in V(4,2)$ into the pair $(\text{shi}^{-1}(xy^{-1}), \text{shi}^{-1}(x^{-1}y))$. The inverse diffeomorphism transforms each pair $(u,v) \in S^2 \times S^2$ into the two-dimensional plane consisting of quaternions of the form $\text{shi}(u)q + q\,\text{shi}(v)$, where q is an arbitrary quaternion. Again, it is routine to check that the pair $(\text{shi}^{-1}(xy^{-1}), \text{shi}^{-1}(x^{-1}y))$ is uniquely determined by the oriented plane spanned by the frame $\{x,y\}$, that the quaternions $\text{shi}(u)q + q\,\text{shi}(v)$ fill exactly a two-dimensional plane, and that the maps $G_+(4,2) \to S^2 \times S^2$ and $S^2 \times S^2 \to G_+(4,2)$ constructed above are inverses of one another.

4. <u>Exercises</u>

1. A homogeneous polynomial in $n+1$ variables and with real (complex) coefficients is <u>nonsingular</u> if there are no points in in $\mathbb{R}^{n+1} \smallsetminus 0$ (respectively, in $\mathbb{C}^{n+1} \smallsetminus 0$) where all its partial derivatives vanish. Show that the projection $\mathbb{R}^{n+1} \smallsetminus 0 \to \mathbb{R}P^n$ (respectively, $\mathbb{C}^{n+1} \smallsetminus 0 \to \mathbb{C}P^n$) transforsm the set of zeros different from 0 of such a polynomial into a submanifold of $\mathbb{R}P^n$ (respectively, $\mathbb{C}P^n$).

2. Let $p(x_1, x_2, x_3)$ be a nonsingular homogeneous polynomial

of degree k with real coefficients. Show that the submanifold of the projective plane $\mathbb{R}P^2$ defined by the equation $p(x_1,x_2,x_3) = 0$ has an orientable neighborhood if and only if k is even.

3. Let $p(x_1,x_2,x_3)$ be a nonsingular homogeneous polynomial of degree 3 with real coefficients. Show that the submanifold of $\mathbb{R}P^2$ defined by the equation $p(x_1,x_2,x_3) = 0$ is homeomorphic to either S^1 or $S^1 \sqcup S^1$, and that both cases are realized.

4. Show that the equation $x_1^2 + x_2^2 + x_3^2 = 0$ defines in $\mathbb{C}P^2$ a submanifold homeomorphic to S^2.

5. Show that the equation $x_1^3 + x_2^3 + x_3^3 = 0$ defines in $\mathbb{C}P^2$ a submanifold homeomorphic to $S^1 \times S^1$.

6. Show that the equation $x_1^2 + x_2^2 + x_3^2 + x_4^2 = 0$ defines in $\mathbb{C}P^3$ a submanifold homeomorphic to $S^2 \times S^2$.

7. Show that $\mathbb{R}G(n,k)$ $(\mathbb{C}G(n,k))$ admits a C^a-embedding in $\mathbb{R}P^{\binom{n}{k}-1}$ (respectively, $\mathbb{C}P^{\binom{n}{k}-1}$).

8. Show that the map $\mathbb{R}G(n,k) \to R^{n^2}$ which takes each plane $\gamma \in \mathbb{R}G(n,k)$ into the matrix of the composite map $\mathbb{R}^n \xrightarrow{\text{pr}} \gamma \xrightarrow{\text{in}} \mathbb{R}^n$ (where pr is the orthogonal projection) is a C^a-embedding. Show that the same is true for the map $\mathbb{C}G(n,k) \to \mathbb{C}^{n^2}$ which takes each plane $\gamma \in \mathbb{C}G(n,k)$ into the matrix of the composite map $\mathbb{C}^n \xrightarrow{\text{pr}} \gamma \xrightarrow{\text{in}} \mathbb{C}^n$.

9. Show that $\mathbb{R}V(8,k)$ is C^a-diffeomorphic to $S^7 \times \mathbb{R}V(7,k-1)$ $(1 \leqslant k \leqslant 8)$. Show that $\mathbb{C}V(4,k)$ is C^a-diffeomorphic to $S^7 \times \mathbb{C}V(3,k-1)$ $(1 \leqslant k \leqslant 4)$.

§3. A DIGRESSION: THREE THEOREMS FROM CALCULUS

1. <u>Polynomial Approximation of Functions</u>

1. The purpose of this section is to state and prove three theorems from calculus, namely theorem 4, 2.3, and 3.3. They differ in character and we grouped them together here because all three are needed in this chapter, and none of them is included in the traditional calculus course.

The main theorem of this subsection, Theorem 4, is a corollary of Lemma 3, whose proof, in turn, requires Lemma 2.

2 (LEMMA). <u>For any positive</u> $\delta < 1$

$$\lim_{k \to \infty} \frac{1}{a_k} \int_{-\delta}^{\delta} (1 - t^2)^k dt = 1,$$

<u>where</u>

$$a_k = \int_{-1}^{1} (1 - t^2)^k dt . \tag{1}$$

This is an easy consequence of the inequalities

$$0 < 1 - \frac{1}{a_k} \int_{-\delta}^{\delta} (1 - t^2)^k dt < \frac{2(1 - \delta)}{\delta} \left(\frac{1 - \delta^2}{1 - \frac{\delta^2}{4}} \right)^k .$$

The left inequality is plain, while the right one follows from the estimates

$$a_k - \int_{-\delta}^{\delta} (1 - t^2)^k dt = 2 \int_{\delta}^{1} (1 - t^2)^k dt < 2(1 - \delta)(1 - \delta^2)^k$$

and

$$a_k > \int_{-\delta/2}^{\delta/2} (1 - t^2)^k dt > \delta (1 - \frac{\delta^2}{4})^k.$$

3 (LEMMA). <u>There exists a sequence of mappings</u>
$\{p_k : C(I^n, \mathbb{R}) \to C(\mathbb{R}^n, \mathbb{R})\}_{k=1}^{\infty}$ <u>such that:</u>

(i) $p_k(f)$ <u>is a polynomial for any</u> $f \in C(I^n, \mathbb{R})$ <u>and any</u> k;

(ii) <u>if</u> f <u>equals</u> 0 <u>on</u> $\mathrm{Fr}\, I^n$, <u>then the sequence</u>
$\{p_k(f)\big|_{I^n}\}$ <u>converges uniformly to</u> f;

(iii) <u>if</u> f <u>equals</u> 0 <u>on</u> $\mathrm{Fr}\, I^n$ <u>and has continuous</u>
<u>partial derivative</u> $D_i f$, <u>then</u>

$$p_k(D_i f) = D_i p_k(f) .$$

PROOF. For $f \in C(I^n, \mathbb{R})$ we set

$$[p_k(f)](x_1, \ldots, x_n) = \frac{1}{a_k^n} \int_{I^n} f(t_1, \ldots, t_n) \prod_{j=1}^{n} [1 - (t_j - x_j)^2]^k dt_1 \ldots dt_n, \tag{2}$$

where a_k is defined by (1). This clearly defines a polynomial and therefore we must check only properties (ii) and (iii).

To check (ii), we extend the function f to $\mathbb{R}^n$, setting $f(x) = 0$ for $x \in \mathbb{R}^n \smallsetminus I^n$, and denote by M the maximum of $|f(x)|$.

Given an arbitrary $\varepsilon > 0$, we can find δ, $0 < \delta < 1$, such that $|f(x_1', \ldots, x_n') - f(x_1, \ldots, x_n)| < \varepsilon/2$ for $|x_1' - x_1| < \delta, \ldots,$ $|x_n' - x_n| < \delta$. Moreover, we can find a number K such that

$$1 - \left[\frac{1}{a_k} \int_{-\delta}^{\delta} (1 - t^2)^k dt\right]^n < \frac{\varepsilon}{4M}$$

for all $k > K$ (see 2). We next show that $|[p_k(f)](x) - f(x)| < \varepsilon$ for $x \in I^n$ and $k > K$.

Write $[p_k(f)](x) - f(x)$ as

$$\frac{1}{a_k^n} \int_{[-1,1]^n} (f(t_1 + x_1, \ldots, t_n + x_n) - f(x_1, \ldots, x_n)) \prod_{j=1}^{n} (1 - t_j^2)^k dt_1 \ldots dt_n$$

and then replace the integrand by its absolute value. Now divide the new integral into two integrals, one over the cube $[-\delta, \delta]^n$ and one over its complement $[-1, 1]^n \smallsetminus [-\delta, \delta]^n$. We obtain

$$|[p_k(f)](x) - f(x)| \leqslant$$

$$\leqslant \frac{1}{a_k^n} \int_{[-\delta, \delta]^n} |f(t_1 + x_1, \ldots, t_n + x_n) - f(x_1, \ldots, x_n)| \prod_{j=1}^{n} (1 - t_j^2)^k dt_1 \ldots dt_n +$$

$$+ \frac{1}{a_k^n} \int_{[-1,1]^n \smallsetminus [-\delta, \delta]^n} |f(t_1 + x_1, \ldots, t_n + x_n) - f(x_1, \ldots, x_n)| \prod_{j=1}^{n} (1 - t_j^2)^k dt_1 \ldots dt_n .$$

The first term is smaller than $\varepsilon/2$, since

$$|f(t_1 + x_1, \ldots, t_n + x_n) - f(x_1, \ldots, x_n)| < \varepsilon/2$$

for $(t_1, \ldots, t_n) \in [-\delta, \delta]^n$ and

$$\frac{1}{a_k^n} \int_{[-\delta, \delta]^n} \prod_{j=1}^{n} (1 - t_j^2)^k dt_1 \ldots dt_n = \left[\frac{1}{a_k} \int_{-\delta}^{\delta} (1 - t^2)^k dt\right]^n < 1.$$

The second term is also smaller than $\varepsilon/2$, since

$$|f(t_1 + x_1, \ldots, t_n + x_n) - f(x_1, \ldots, x_n)| \leqslant 2M$$

and

$$\frac{1}{a_k^n} \int_{[-1,1]^n \smallsetminus [-\delta, \delta]^n} \prod_{j=1}^{n} (1 - t_j^2)^k dt_1 \ldots dt_n =$$

$$= \frac{1}{a_k^n} \int_{[-1,1]^n} \prod_{j=1}^{n} (1 - t_j^2)^k dt_1 \ldots dt_n - \frac{1}{a_k^n} \int_{[-\delta, \delta]^n} \prod_{j=1}^{n} (1 - t_j^2)^k dt_1 \ldots dt_n =$$

190

$$= 1 - \left[\frac{1}{a_k} \int_{-\delta}^{\delta} (1 - t^2)^k dt \right]^n < \frac{\varepsilon}{4M} .$$

Therefore, $|[p_k(f)](x) - f(x)| < \varepsilon$.

To check property (iii), replace f by $D_i f$ in definition (2) and then integrate by parts with respect to t_i in the right-hand side. We obtain

$$[p_k(D_i f)](x_1,\ldots,x_n) =$$

$$= - \frac{1}{a_k^n} \int_{I^n} f(t_1,\ldots,t_n) \frac{\partial}{\partial t_i} \prod_{j=1}^{n} (1 - (t_j - x_j)^2)^k dt_1 \ldots dt_n . \quad (3)$$

Since

$$\frac{\partial}{\partial t_i} \prod_{j=1}^{n} (1 - (t_j - x_j)^2)^k = - \frac{\partial}{\partial x_i} \prod_{j=1}^{n} (1 - (t_j - x_j)^2)^k ,$$

the right-hand side of (3) is equal to $[D_i p_k(f)](x_1,\ldots,x_n)$.

4. **Suppose** X **is a compact set in** $\mathbb{R}^n$ **and** f **is a real function defined and of class** C^r **in a neighborhood of** X. **If** $r \leqslant \infty$, **then for any** $\varepsilon > 0$ **and any nonnegative integer** $s \leqslant r$, **there is a polynomial** $g: \mathbb{R}^n \to \mathbb{R}$ **such that**

$$\max_{x \in X} |D_1^{s_1} \ldots D_n^{s_n} g(x) - D_1^{s_1} \ldots D_n^{s_n} f(x)| < \varepsilon$$

for any collection $s_1,\ldots,s_n$ **of nonnegative integers with** $s_1 + \ldots + s_n \leqslant s$.

PROOF. Clearly, one can assume that $X \subset \mathrm{Int}\, I^n$. Denote by U the neighborhood of X mentioned above, and let $\beta: \mathbb{R}^n \to \mathbb{R}$ be any C^r-function equal to 0 on D^n and equal to 1 outside the concentric ball of radius 2. For $y \in (I^n \smallsetminus U) \cup \mathrm{Fr}\, I^n$, we let $d(y)$ denote the ball of center y and radius $\mathrm{Dist}(y,X)/4$. Cover $(I^n \smallsetminus U) \cup \mathrm{Fr}\, I^n$ by a finite number of balls $\mathrm{Int}\, d(y)$, say $\mathrm{Int}\, d(y_1),\ldots,\mathrm{Int}\, d(y_p)$, and define $h: I^n \to \mathbb{R}$ as

$$h(y) = \begin{cases} f(y) \prod_{i=1}^{p} \beta(4(y - y_i)/\mathrm{Dist}(y_i,X)), & \text{if } y \in U, \\ \\ 0, & \text{if } y \notin U. \end{cases}$$

Then h is of class C^r, agrees with f on X, and vanishes on a neighborhood of $\mathrm{Fr}\, I^n$. This shows that one can take g to be $p_k(h)$, with p_k as in Lemma 3 and k large enough.

2. <u>Singular Values</u>

1. In the present subsection we assume that we are given an open subset U of $\mathbb{R}^n$ and a C^∞-map $f: U \to \mathbb{R}^q$ with $q \geqslant 1$. We let F denote the set of points in U where the rank of the Jacobi matrix of f is less than q. Our aim is to prove Theorem 3, which is needed in the next section (see 4.7.4).

We need two auxilliary notations: f_j for the j-th coordinate function of the map f $(j = 1,\ldots,n)$ and F_s for the set of points in U where all partial derivatives of order $1,\ldots,s$ of the functions f_j vanish. Clearly, F_s is closed in U and $F \supset F_1 \supset F_2 \supset \ldots$.

2 (LEMMA). <u>Suppose</u> $s > (n/q)-1$ <u>and</u> C <u>is any compact part of the set</u> F_s. <u>The image</u> $f(C)$ <u>is nowhere dense.</u>

It is enough to show that for any n-dimensional qube $Q \subset U$, the set $f(C \cap Q)$ is nowhere dense. Indeed, one can cover C by a finite number of such cubes, and use the fact that a finite union of nowhere dense sets is nowhere dense.

Let a be the edge length of Q. Consider the standard partition of Q into m^n small cubes of edge length a/m (with m an arbitrary positive integer). Let Q' be a small cube in this partition which intersects C. Now apply Taylor's theorem to the functions f_j and use the fact that their partial derivatives of order $s+1$ are bounded on Q to show that there is a constant b such that

$$\mathrm{dist}(f(x),f(y)) \leqslant b[\mathrm{dist}(x,y)]^{s+1}$$

for any $x \in F_s \cap Q, \in y \in Q$. Since Q' has diameter $a\sqrt{n}/m$, we see that $f(Q')$ is contained in a ball of radius $b(a\sqrt{n}/m)^{s+1}$. Therefore, $f(Q')$ is contained in a q-dimensional cube with edge length $2b(a\sqrt{n}/m)^{s+1}$, and $f(C \cap Q)$ - in a union of no more than m^n such cubes. The volume of each such cube is $[2b(a\sqrt{n}/m)^{s+1}]^q$, and so the sum of their volumes does not exceed

$$m^n[2b(a\sqrt{n}/m)^{s+1}]^q = cm^{n-q(s+1)},$$

where c is independent of m. By hypothesis, this sum goes to 0 as $m \to \infty$. This shows that the set $f(C \cap Q)$ can have no interior points. Since the latter is a closed set, it is also nowhere dense.

3. <u>The image</u> $f(C)$ <u>of any compact part of the set</u> F <u>is</u>

<u>nowhere dense.</u>

We proceed inductively on n. For $n = 0$ there is nothing to prove; hence it is enough to show that the theorem holds for $n = k+1$ if it holds for $n = k$.

We start with the special case $F_1 = \emptyset$. Since C is compact, it suffices to find, for each point $x \in C$, a neighborhood V in U such that $f(C \cap V)$ is nowhere dense. Assume that $D_{k+1}f_q(x) \neq 0$ (we can always achieve this by reindexing the coordinates in $\mathbb{R}^{k+1}$ and $\mathbb{R}^q$) and consider the map $g \colon U \to \mathbb{R}^{k+1}$ defined by the formula

$$y = (y_1, \ldots, y_{k+1}) \mapsto (y_1, \ldots, y_k, f_q(y)).$$

Its Jacobian at the point x does not vanish (being equal to $D_{k+1}f_q$); hence g is a C^∞-diffeomorphism of a neighborhood W of x onto a neighborhood of $g(x)$. Now one can take as V any neighborhood of x with compact closure included in W. To see this, set h to be the composite map

$$g(W) \xrightarrow{\ (ab\ g)^{-1}\ } W \xrightarrow{\ ab\ f\ } \mathbb{R}^q.$$

It transforms each point from $g(W)$ into a point which has the same last coordinate. In particular, for any real number u, h can be compressed to a map

$$g(W) \cap [\mathbb{R}^k \times u] \to \mathbb{R}^{q-1} \times u.$$

If we identify $g(W) \cap [\mathbb{R}^k \times u]$ with its orthogonal projection on $\mathbb{R}^k$ in the standard fashion, and $\mathbb{R}^{q-1} \times u$ – with its orthogonal projection on $\mathbb{R}^{q-1}$, we obtain a C^∞-map h_u of an open subset of $\mathbb{R}^k$ into $\mathbb{R}^{q-1}$. Clearly, the Jacobi matrix of h_u at the point $(y_1, \ldots, y_k)$ is obtained from the Jacobi matrix of the map h at the point $(y_1, \ldots, y_k, u)$ by deleting the last column and the last row, which has the form $0, \ldots, 0, 1$. Therefore, the rank of the first matrix is less than $q-1$ if and only if the rank of the second matrix is less than q, i.e., if $(y_1, \ldots, y_k, u) \in g(F \cap W)$. Applying the induction hypothesis to h_u, we deduce that the intersection of $h(g(C \cap Cl\,V))$ with each hyperplane $\mathbb{R}^{q-1} \times u$ is nowhere dense in $\mathbb{R}^{q-1} \times u$. But if this is the case, $h(g(C \cap Cl\,V))$ has no interior points in $\mathbb{R}^q$ and we need only note that this set is closed and coincides with $f(C \cap Cl\,V)$.

Now let us turn to a second special case: $C \subset F_s$ and $F_{s+1} = \emptyset$ (for some s). Again, it is enough to exhibit for each point $x \in C$ a neighborhood V in U such that the set $f(C \cap V)$ is nowhere dense. Let φ be a derivative of order s of one of the functions f_j

which satisfies the following condition: one of the derivatives $D_i\varphi$, say $D_{k+1}\varphi$, does not vanish at the point x. Consider the map $g: U \to \mathbb{R}^{q+1}$ defined as

$$y = (y_1, \ldots, y_{k+1}) \mapsto (y_1, \ldots, y_k, \varphi(y)).$$

Its Jacobian does not vanish at x (being equal to $D_{k+1}\varphi$); hence g yields a C^∞-diffeomorphism of a neighborhood W of x onto a neighborhood of $g(x)$. We show, with the aid of the composite map

$$g(W) \xrightarrow{\ (ab\ g)^{-1}\ } W \xrightarrow{\ ab\ f\ } \mathbb{R}^q \tag{4}$$

that one can take any neighborhood of x with compact closure included in W for V. To do this, note that $g(C) \subset \mathbb{R}^k$ and restrict the map (4) to a map $h: g(W) \cap \mathbb{R}^k \to \mathbb{R}^q$. Clearly, all the derivatives of order $\leqslant s$ of the coordinate functions of h vanish on $g(C \cap W)$. Using the induction hypothesis, it is evident that h carries the compact parts of the set $g(C \cap W)$ into nowhere dense sets. Finally, we observe that $g(C \cap \text{Cl } V)$ is a compact part of $g(C \cap W)$, and that $h(g(C \cap \text{Cl } V))$ is just $f(C \cap \text{Cl } V)$.

At last, we come to the general case. According to Lemma 2, there exists a number r such that the set $f(C \cap F_r)$ is nowhere dense. We shall prove statement 3 by induction on r, i.e., assuming that $f(C \cap F_r)$ is nowhere dense, we show that, for $r = 1$, $f(C)$ is nowhere dense, and for $r > 1$, $f(C \cap F_{r-1})$ is nowhere dense.

Let G be an open nonempty subset of $\mathbb{R}^q$. Since $C \cap F_r$ is compact and $f(C \cap F_r)$ is nowhere dense, the set $C \cap F_r$ has a neighborhood N in $\mathbb{R}^n$ such that $\text{Cl } N$ is compact and $\text{Cl } N \subset U$, $f(\text{Cl } N) \not\supseteq G$. Next replace the map f by its restriction to $U \smallsetminus F_r$ and C - by the set

$$C' = \begin{cases} C \smallsetminus N, & \text{if } r = 1 \\[2mm] (C \smallsetminus F_r) \smallsetminus N, & \text{if } r > 1. \end{cases}$$

Now we are back to one of the cases covered by the first part of the proof (namely, in the first case for $r = 1$, and in the second one for $r > 1$). Therefore, we conclude that $f(C')$ does not cover $G \smallsetminus f(\text{Cl } N)$. Consequently, if $r = 1$ the set $f(C)$ does not cover G, while if $r > 1$ the set $f(C \cap F_r)$ does not cover G. This completes the proof, because $f(C)$ and $f(C \cap F_{r-1})$ are closed.

4 (INFORMATION). In Theorem 3, the condition that f be C^∞-smooth is unnecessarily strong: in fact, the proof uses only the

fact that f is of class C^r, with $r = 2 + \max(n-q,0)$. A more precise analysis shows that this r can be decreased by 1 (see, for example, [21]), but no further (for $q = 1$, this is showed in [23], and the case $q > 1$ reduces easily to the case $q = 1$).

3. Nondegenerate Critical Points

1. Let f be a real C^2-function defined on an open subset of $\mathbb{R}^n$. A critical point y of f is <u>nondegenerate</u> if the second differential of f at y (considered as a quadratic form) has rank n. The index of the second differential of f at y (i.e., the number of negative squares in the diagonal representation of this form) is called the <u>index of the point</u> y and is denoted by $\mathrm{ind}_f y$.

We remark that if φ is a C^2-diffeomorphism of an open subset U of $\mathbb{R}^n$ onto another open subset of $\mathbb{R}^n$ and y is a non-degenerate critical point of $f: U \to \mathbb{R}$, then $\varphi(y)$ is a nondegenerate critical point of the function $f \circ \varphi^{-1}: \varphi(U) \to \mathbb{R}$, and $\mathrm{ind}_{f \circ \varphi^{-1}} \varphi(y) = \mathrm{ind}_f y$. Both conclusions remain true in the more general situation where φ is only of class C^1, but the function $f \circ \varphi^{-1}$ is of class C^2.

Now consider the function $\mathbb{R}^n \to \mathbb{R}$ defined as

$$(x_1, \ldots, x_n) \mapsto -x_1^2 - \ldots - x_k^2 + x_{k+1}^2 + \ldots + x_n^2 + c , \qquad (5)$$

where c is a real number $(0 \leqslant k \leqslant n)$. This function has a unique critical point, at 0, which clearly is nondegenerate and of index k. The main goal in the present subsection is to show that, in a suitably chosen system of coordinates, any sufficiently smooth function has the above form (5) in the vicinity of a nondegenerate critical point.

2 (LEMMA). <u>Let</u> V <u>be an open ball in</u> $\mathbb{R}^n$ <u>with center</u> 0, <u>and let</u> $f: V \to \mathbb{R}$ <u>be a</u> C^r-<u>function</u>, $r \geqslant 1$, <u>with</u> $f(0) = 0$. <u>There are</u> C^{r-1}-<u>functions</u> $f_1, \ldots, f_n: V \to \mathbb{R}$ <u>such that</u>

$$f(x) = \sum_{i=1}^{n} x_i f_i(x) \qquad (6)$$

<u>for all points</u> $x = (x_1, \ldots, x_n) \in V$.

To prove the lemma, it is enough to set

$$f_i(x) = \int_0^1 D_i f(tx)\, dt,$$

and then observe that (6) is an immediate consequence of the equality

$$\frac{\partial}{\partial t} f(tx) = \sum_{i=1}^{n} x_i D_i f(tx).$$

3. <u>Suppose that</u> y <u>is a nondegenerate critical point of a C^r-function</u> f <u>defined on an open subset of</u> $\mathbb{R}^n$. <u>If</u> $r \geqslant 3$, <u>then there exist a neighborhood</u> U <u>of</u> y <u>and a diffeomorphism</u> φ <u>of</u> U <u>onto a neighborhood</u> V <u>of</u> 0, <u>such that the restriction</u> $f|_U$ <u>coincides with the composite map</u> $U \overset{\varphi}{\to} V \overset{(5)}{\longrightarrow} \mathbb{R}$, <u>where</u> $k = \mathrm{ind}_f y$ <u>and</u> $c = f(y)$.

PROOF. Without loss of generality, we may assume that $y = 0$ and $f(y) = 0$. By Lemma 2,

$$f(x) = \sum_{i=1}^{n} x_i f_i(x)$$

in some neighborhood of 0, where f_i are C^{r-1}-functions. Differentiating, we obtain $D_i f(x) = \sum_{j=1}^{n} x_j D_i f_j(x)$, since $f_1(0) = \ldots = f_n(0)$. Again we apply Lemma 2 and write, in a neighborhood V_0 of 0,

$$f_i(x) = \sum_{j=1}^{n} x_j f_{ij}(x) \qquad (i = 1,\ldots,n),$$

with C^{r-2}-functions f_{ij}. Therefore, for $x \in V_0$,

$$f(x) = \sum_{i,j=1}^{n} g_{ij}(x) x_i x_j ,$$

where $g_{ij}(x) = \{f_{ij}(x) + f_{ji}(x)\}/2$. Clearly, $g_{ij}(0) = D_i D_j f(0)/2$.

The subsequent constructions mimic the standard reduction of a quadratic form to canonical form through linear transformations. For $p = 0,\ldots,n$, we construct neighborhoods V_p and W_p of the point 0 in $\mathbb{R}^n$, C^{r-2}-diffeomorphisms $\varphi_p: W_p \to V_p$, and also C^{r-2}-functions $g_{ij}^p: V_p \to \mathbb{R}$, $i,j = p+1,\ldots,n$, such that: (i) $W_p \subset V_{p-1}$; (ii) (ii) $\varphi_p(0) = 0$; (iii) the composite map

$$V_p \overset{\varphi_p^{-1}}{\longrightarrow} W_p \overset{in}{\longrightarrow} V_{p-1} \overset{\varphi_{p-1}^{-1}}{\longrightarrow} \ldots \overset{\varphi_1^{-1}}{\longrightarrow} W_1 \overset{in}{\longrightarrow} V_0 \overset{ab\ f}{\longrightarrow} \mathbb{R}$$

is represented by the formula

$$x \mapsto \pm x_1^2 \pm \ldots \pm x_p^2 + \sum_{i,j=p+1}^{n} g_{ij}^p(x) x_i x_j ; \qquad (7)$$

(iv) $g_{ij}^p = g_{ji}^p$. Then we will have finished, since one could take

$$\varphi_1^{-1}(\ldots(\varphi_n^{-1}(V_n))\ldots)$$

for V, define φ as the composition

$$V \xrightarrow{\text{ab } \varphi_1} \varphi_1(V) \xrightarrow{\text{ab } \varphi_2} \varphi_2(\varphi_1(V)) \xrightarrow{\text{ab } \varphi_3} \cdots \xrightarrow{\text{ab } \varphi_n} V_n \xrightarrow{\text{ab } \pi} \pi(V_n),$$

where π is a suitable permutation of the standard coordinates in $\mathbb{R}^n$, and set $U = \varphi^{-1}(V)$.

The neighborhood V_0 is already given. We let $W_0 = V_0$, $\varphi_0 = \text{id } V_0$, $g_{ij}^0 = g_{ij}$, and assume that we have constructed V_p, W_p, φ_p, and g_{ij}^p satisfying (i), (ii), (iii), and (iv) for $p \le q$. It is clear that 0 is a nondegenerate critical point of the function (7) with $p = q$. Hence the matrix $G = \|g_{ij}^p(0)\|_{i,j=q+1}^k$ is nondegenerate and there exists a nondegenerate $(n-q) \times (n-q)$-matrix A such that the left upper element of the matrix $A^t G A$ is not zero. Let ℓ denote the linear transformation of $\mathbb{R}^n$ having matrix $\begin{pmatrix} E & 0 \\ 0 & A \end{pmatrix}$, where E is the $q \times q$-identity matrix. The composition of the diffeomorphism ab $\ell : \ell^{-1}(V_q) \to V_q$ with the function (3) is given by

$$x \mapsto \pm x_1^2 \pm \cdots \pm x_q^2 + \sum_{i,j=q+1}^n h_{ij}(x) x_i x_j,$$

where $h_{ij} = h_{ji}$ and $h_{q+1,q+1}(0) \ne 0$. Now consider the subset L of $\ell^{-1}(V_q)$ consisting of all the points x where $h_{q+1,q+1}(x) \ne 0$ and has the same sign as $h_{q+1,q+1}(0)$, and then define $\psi : L \to \mathbb{R}^n$ as

$$\psi(x) = (x_1, \ldots, x_q, \xi | h_{q+1,q+1}(x) |^{1/2}, x_{q+2}, \ldots, x_n),$$

where

$$\xi = x_{q+1} + \sum_{s>q+1} x_s \frac{h_{s,q+1}(x)}{h_{q+q,q+1}(x)}.$$

A simple computation shows that the Jacobian of ψ at the point 0 does not vanish. Therefore, the compression of ψ to a neighborhood M of 0 and to its image $\psi(M)$ is a C^{r-2}-diffeomorphism. It is now readily verified that the sets $V_{q+1} = \psi(M)$ and $W_{q+1} = \ell(M)$, the map $\varphi_{q+1} : W_{q+1} \to V_{q+1}$ defined by $\varphi_{q+1}(x) = \psi(\ell^{-1}(x))$, and the functions $g_{ij}^{q+1} : V_{q+1} \to \mathbb{R}$, defined as

$$g_{ij}^{q+1}(x) = h_{ij}(x) - \frac{h_{i,q+1}(x) h_{j,q+1}(x)}{h_{q+1,q+1}(x)}$$

enjoy the properties (i)-(iv).

§4. EMBEDDINGS. IMMERSIONS. SMOOTHINGS.

APPROXIMATIONS

1. Spaces of Smooth Maps

1. Let X and X' be $C^{\geq r}$-manifolds $(0 \leq r \leq a)$. We denote by $C^r(X,X')$ the set of all C^r-maps $X \to X'$. If $r \leq \infty$, we equip $C^r(X,X')$ with the $\underline{C^r\text{-topology}}$ which makes $C^r(X,X')$ into a topological space, as follows. Given two arbitrary charts $\varphi \in \mathrm{Atl}\, X$ and $\varphi' \in \mathrm{Atl}\, X'$, a sequence of nonnegative integers $r_1,\ldots,r_n$ with $n = \dim X$ and $r_1+\ldots+r_n \leq r$, a compact subset A of $\mathrm{Im}\,\varphi$, and an open subset A' of $\mathbb{R}^{n'}$, where $n' = \dim X'$, consider the subset of $C^r(X,X')$ consisting of all maps f such that

$$[D_1^{r_1}\ldots D_n^{r_n}\mathrm{loc}(\varphi,\varphi')f](A) \subset A'.$$

These subsets form a prebase of the C^r-topology on $C^r(X,X')$.

Clearly, $C^0(X,X') = C(X,X')$, and the C^0-topology is simply the compact-open topology (see 1.2.7.1). Also, for $s < r$, the inclusion $C^r(X,X') \to C^s(X,X')$ is obviously continuous. Another direct consequence of the definition of the C^r-topology is that all the spaces $C^r(X,X')$ are regular. In addition, we note that the sets closed (open) in $C^\infty(X,X')$ are exactly those sets closed (respectively, open) in all the C^r-topologies with r finite.

For each pair of C^r-maps $f: Y \to X$ and $f': X' \to Y'$, there is a map $C^r(X,X') \to C^r(Y,Y')$ defined by the formula $g \to f' \circ g \circ f$, and denoted by $C^r(f,f')$. Obviously, $C^0(f,f') = C(f,f')$ (see 1.2.7.1), $C^r(f,f') = \mathrm{ab}\, C^s(f,f')$ for $s < r$, and $C^r(f,f')$ is continuous for all $r \leq \infty$.

We list some particular subsets of $C^r(X,X')$ which are important in the sequel. These are the sets of all C^r-embeddings, C^r-immersions, C^r-submersions, and C^r-diffeomeorphisms $X \to X'$, and they are denoted by $\mathrm{Emb}^r(X,X')$, $\mathrm{Imm}^r(X,X')$, $\mathrm{Subm}^r(X,X')$, and $\mathrm{Diff}^r(X,X')$, respectively $(1 \leq r \leq a)$. Moreover, we let $C_\partial^r(X,X')$ denote the set of all C^r-maps $f:X \to X'$ such that $f(\partial X) \subset \partial X'$ and $d_x f(\mathrm{Tang}_x X) \subset \mathrm{Tang}_{f(x)}(\partial X')$ for all $x \in \partial X$. Usually one writes

$\text{Diff}^r X$ instead of $\text{Diff}^r(X,X)$.

For $1 \leqslant r \leqslant \infty$, the map $C_\partial^r(X,X') \to C^r(\partial X, \partial X')$, defined as $f \mapsto \text{ab}\, f$ is continuous, and the map $C^r(X,X') \to C^{r-1}(\text{Tang}\, X, \text{Tang}\, X')$, defined as $f \mapsto df$ is a topological embedding.

2. <u>If</u> X <u>is compact, then the set</u> $\text{Imm}^r(X,X')$ <u>is open in</u> $C^r(X,X')$ $(1 \leqslant r \leqslant \infty)$.

We have to exhibit, for a given C^r-immersion $f_0 \colon X \to X'$, a neighborhood of f_0 in $C^r(X,X')$ consisting only of immersions. To do this, pick for each point $x \in X$ two charts, $\varphi_x \in \text{Atl}_x X$ and $\varphi_x' \in \text{Atl}\, X'$, such that $f_0(\text{supp}\, \varphi_x) \subset \text{supp}\, \varphi_x'$ and $\text{loc}(\varphi_x, \varphi_x')f$ equals one of the inclusions $\mathbb{R}^n \to \mathbb{R}^{n'}$, $\mathbb{R}_-^n \to \mathbb{R}^{n'}$, or $\mathbb{R}_-^n \to \mathbb{R}_-^{n'}$ (see 1.5.3 and 1.5.1; here $n = \dim X$ and $n' = \dim X'$). Now cover X with a finite number of sets $U_x = \varphi_x^{-1}(\text{Int}\, D^n)$, say $U_{x_1}, \ldots, U_{x_s}$, and denote by U_i the subset of $C^r(X,X')$ consisting of all the maps f such that $f(\text{Cl}\, U_{x_i}) \subset \text{supp}\, \varphi_{x_i}'$ and the upper $n \times n$-minor of the Jacobi matrix of the map $\text{loc}(\varphi_{x_i}, \varphi_{x_i}')f$ has no zeros on D^n. The intersection $U_1 \cap \ldots \cap U_s$ is the desired neighborhood of the map f_0.

3. <u>If</u> X <u>is compact and</u> X' <u>has no boundary, then the set</u> $\text{Subm}^r(X,X')$ <u>is open in</u> $C^r(X,X')$ $(1 \leqslant r \leqslant \infty)$.

We have to exhibit, for a given C^r-submersion $f_0 \colon X \to X'$, a neighborhood of f_0 in $C^r(X,X')$ consisting only of submersions. Again, for each point $x \in X$ we choose charts $\varphi_x \in \text{Atl}_x X$ and $\varphi_x' \in \text{Atl}\, X'$, such that $f_0(\text{supp}\, \varphi_x) \subset \text{supp}\, \varphi_x'$ and $\text{loc}(\varphi_x, \varphi_x')f$ equals one of the orthogonal projections $\mathbb{R}^n \to \mathbb{R}^{n'}$, $\mathbb{R}_-^n \to \mathbb{R}^{n'}$ (see 1.5.7). Now cover X with a finite number of sets $U_x = \varphi_x^{-1}(\text{Int}\, D^n)$, say $U_{x_1}, \ldots, U_{x_s}$, and denote by U_i the subset of $C^r(X,X')$ consisting of all maps f such that $f(\text{Cl}\, U_{x_i}) \subset \text{supp}\, \varphi_{x_i}'$ and the left $n' \times n'$-minor of the Jacobi matrix of the map $\text{loc}(\varphi_{x_i}, \varphi_{x_i}')f$ has no zeros on D^n. The intersection $U_1 \cap \ldots \cap U_s$ is the desired neighborhood of the map f_0.

4. <u>If</u> X <u>is compact, then the set</u> $\text{Emb}^r(X,X')$ <u>is open in</u> $C^r(X,X')$ $(1 \leqslant r \leqslant \infty)$.

PROOF. Given a C^r-embedding $f_0 \colon X \to X'$, theorems 2 and 1.5.4 show that it is enough to produce a neighborhood of f_0 in $C^r(X,X')$ consisting only of injective maps. For each point $x \in X$, choose two charts, $\varphi_x \in \text{Atl}_x X$ and $\varphi'_x \in \text{Atl} X'$, such that $f_0(\text{supp }\varphi_x) \subset \text{supp }\varphi'_x$ and $\text{loc}(\varphi_x,\varphi'_x)f$ coincides with one of the inclusions $\mathbb{R}^n \to \mathbb{R}^{n'}$, $\mathbb{R}^n_- \to \mathbb{R}^{n'}$, or $\mathbb{R}^n_- \to \mathbb{R}^{n'}_-$ (see 1.5.1). Now cover X with a finite number of sets $U_x = \varphi_x^{-1}(\text{Int } D^n)$, say $U_{x_1},\ldots,U_{x_s}$. Let U_i be the subset of $C^r(X,X')$ consisting of all maps f such that $f(\text{Cl } U_{x_i}) \subset \text{supp }\varphi'_{x_i}$ and, if we symmetrize the upper $n\times n$-part of the Jacobi matrix of $\text{loc}(\varphi_{x_i},\varphi'_{x_i})f$ and take all the principal minors, they are all positive on the ball D^n. (The principal minors are the left-upper minors; the symmetrized matrix is half the sum of the matrix with its transpose.) Finally, denote by U that part of $C^r(X,X')$ consisting of all maps such that the preimage of any point of X' lies in one of the sets U_{x_i}. Let us show that the intersection $V = U_1 \cap \ldots \cap U_s \cap U$ is a neighborhood of f_0 with the necessary property. It is clear that $f_0 \in V$ and that all the sets U_i are open. Hence it suffices to verify that: (i) U is open, and (ii) the maps in U_i are injective.

To prove (i), note that U is the preimage of the set

$$W = C(X \times X, (X \times X) \smallsetminus U_i(U_{x_i} \times U_{x_i});X' \times X',(X' \times X') \smallsetminus \text{diag}(X'))$$

under the continuous mapping $C^r(X,X') \to C(X \times X, X' \times X')$, given by $f \mapsto f \times f$. Since $(X \times X) \smallsetminus U_i(U_{x_i} \times U_{x_i})$ is compact and $(X' \times X') \smallsetminus \text{diag}(X')$ is open in $X' \times X'$ (see 1.2.2.4), W is open in $C(X \times X, X' \times X')$ and U is open in $C^r(X,X')$.

To prove (ii), given a map $f \in U_i$ and arbitrary distinct points $y,z \in U_{x_i}$, let $s: I \to \mathbb{R}$ be the function which takes each point $t \in I$ into the inner product of the vectors $v = \varphi_{x_i}(z) - \varphi_{x_i}(y)$ and

$$[\text{loc}(\varphi_{x_i},\varphi'_{x_i})f]((1-t)\varphi_{x_i}(y) + t\varphi_{x_i}(z)) - \text{loc}(\varphi_{x_i},\varphi'_{x_i})f(y),$$

computed in $\mathbb{R}^{n'}$. Next denote by $J(u)$ the symmetrized upper $n\times n$-part of the Jacobi matrix of the map $\text{loc}(\varphi_{x_i},\varphi'_{x_i})f$ at the point $u \in \text{Int } D^n$, and by q_t - the bilinear form $\mathbb{R}^n \times \mathbb{R}^n \to \mathbb{R}^n$ having the matrix $J((1-t)\varphi_{x_i}(y) + t\varphi_{x_i}(z))$. The function s is smooth and its derivative at the point t is precisely $q_t(v,v)$, and is therefore positive (as a consequence of the definition of U_i). Moreover, $s(0) = 0$, which implies that $s(1) > 0$, i.e., $f(y) \neq f(z)$.

5. **If** X **is compact, then the set of neat** C^r**-embeddings** $X \to X'$ **is open in** $C_\partial^r(X,X')$ $(1 \leqslant r \leqslant \infty)$.

This is an immediate corollary of 4 since the set in question is just $\text{Emb}^r(X,X') \cap C_\partial^r(X,X')$.

6. **If** X **is compact, then the set** $\text{Diff}^r(X,X')$ **is open in** $C_\partial^r(X,X')$ $(1 \leqslant r \leqslant \infty)$.

This is a corollary of 5 (see 1.5.1)

2. The Simplest Embedding Theorems

1. **Every compact** $C^{\geqslant r}$**-manifold,** $1 \leqslant r \leqslant \infty$, **admits a** C^r**-embedding in a Euclidean space of sufficiently high dimension.**

PROOF. Let n be the dimension of the given manifold X, and let $\alpha : \mathbb{R}^n \to I$ be a C^r-function equal to 1 on D^n, smaller than 1 outside D^n, and equal to 0 outside the concentric ball of radius 2. For each point $x \in X$, fix a chart $\varphi_x \in \text{Atl}_x X$ with $\text{Im}\,\varphi_x = \mathbb{R}^n$ or $\mathbb{R}^n_-$, and $\varphi_x(x) = 0$. Define a map $j_x : X \to \mathbb{R} \times \mathbb{R}^n$ by the formula

$$
j_x(y) = \begin{cases} (\alpha(\varphi_x(y)), \alpha(\varphi_x(y))\varphi_x(y), & \text{if } y \in \text{supp}\,\varphi_x, \\[2ex] (0,0), & \text{if } y \in X \smallsetminus \text{supp}\,\varphi_x. \end{cases}
$$

Now cover X by a finite number of sets $U_x = \varphi_x^{-1}(\text{Int}\,D^n)$, say $U_{x_1}, \ldots, U_{x_s}$, and define $j : X \to (\mathbb{R} \times \mathbb{R}^n) \times \ldots \times (\mathbb{R} \times \mathbb{R}^n) = \mathbb{R}^{s(n+1)}$ by $j(y) = (j_{x_1}(y), \ldots, j_{x_s}(y))$. The map j is of class C^r and injective: if $y \in U_{x_i}$ and $y' \neq y$, then $j_{x_i}(y') \neq j_{x_i}(y)$. [Indeed, if $y' \notin U_{x_i}$, then $\alpha(\varphi_{x_i}(y')) < 1$, whereas $\alpha(\varphi_{x_i}(y)) = 1$; if $y' \in U_{x_i}$, then $\alpha(\varphi_{x_i}(y'))\varphi_{x_i}(y') = \varphi_{x_i}(y') \neq \varphi_{x_i}(y) = \alpha(\varphi_{x_i}(y))\varphi_{x_i}(y)$.] Moreover, j is an immersion, since j_{x_i} is an immersion on U_{x_i} (the second component $X \to \mathbb{R}^n$ of the map j_{x_i} agrees with φ_{x_i} on U_{x_i}). Therefore, j is a C^r-embedding (see 1.5.4).

Supplement for the Case of Nonempty Boundary

2 (LEMMA). **On any compact** $C^{\geqslant r}$**-manifold,** $1 \leqslant r \leqslant \infty$, **there**

is a (real) C^r-function h, <u>equal to</u> 0 <u>on</u> ∂X, <u>positive on</u> int X, and having no critical points on ∂X.

PROOF. Let $\alpha: \mathbb{R}^n \to I$, $n = \dim X$, be a C^r-function equal to 1 on D^n and equal to 0 outside the concentric ball of radius 2. For each point $x \in \partial X$ fix a chart $\varphi_x \in \text{Atl}_x X$ such that $\text{Im}\,\varphi_x = \mathbb{R}^n_-$ and $\varphi_x(x) = 0$, and define two functions $f_x, g_x: X \to \mathbb{R}$ through the formulas

$$
f_x(y) = \begin{cases} 1 - \alpha(\varphi_x(y)), & \text{if } y \in \text{supp}\,\varphi_x, \\ \\ 1, & \text{if } y \in X \smallsetminus \text{supp}\,\varphi_x, \end{cases}
$$

and

$$
g_x(y) = \begin{cases} -\beta(\varphi_x(y)), & \text{if } y \in \text{supp}\,\varphi_x, \\ \\ 0, & \text{if } y \in X \smallsetminus \text{supp}\,\varphi_x. \end{cases}
$$

Here $\beta: \mathbb{R}^n \to \mathbb{R}$ is given by $\beta(t_1,\ldots,t_n) = t_1 \alpha(t_1,\ldots,t_n)$. Covering ∂X by a finite number of sets $U_x = \varphi_x^{-1}(\text{Int}\,D^n)$, say $U_{x_1},\ldots,U_{x_s}$, and setting

$$
h(y) = \prod_{i=1}^{s} f_{x_i}(y) + \sum_{i=1}^{s} g_{x_i}(y),
$$

we obtain the needed function h: $X \to \mathbb{R}$. In fact, h vanishes identically on ∂X, since f_{x_i} is equal to 0 on U_{x_i} and all the functions g_{x_i} vanish identically on ∂X; h is positive on int X, since all the functions f_{x_i}, g_{x_i} are nonnegative and g_{x_i} is positive at all points of int X, excepting the zeros of f_{x_i}. Finally, h has no critical points on ∂X, since $\sum g_{x_i}$ has no critical points on ∂X (the derivative with respect to the first coordinate of the local representative $\text{loc}(\varphi_{x_k}, \text{id}\,\mathbb{R}) g_{x_i}$, i.e., of the composition $(\varphi_{x_i}|\text{supp}\,\varphi_{x_k}) \circ \varphi_{x_k}^{-1}$, is negative on $D^n \cap \mathbb{R}^{n-1}_1$ for $k = i$ and nonpositive on $D^n \cap \mathbb{R}^{n-1}_1$ for all k), while $\prod f_{x_i}$ vanishes identically on $\cup U_{x_i}$.

3. <u>Every compact $C^{\geqslant r}$-manifold, $1 \leqslant r \leqslant \infty$, admits a neat</u> C^r-<u>embedding in a Euclidean space of sufficiently high dimension.</u>

The formula $x \mapsto (-h(x), j(x))$, where j is an arbitrary

C^r-embedding in $\mathbb{R}^n$ (see 1), and h is the function constructed in 2, defines a neat C^r-embedding in $\mathbb{R}_-^{q+1} = \mathbb{R}_-^1 \times \mathbb{R}^q$.

Information

4. The compactness assumption and the condition that $r \neq a$ may be eliminated from the formulations of 1 and 3. Any smooth manifold of class $C^{\geqslant r}$, with $r \leqslant \infty$ or $r = a$, compact or not, can be C^r-embedded in Euclidean space, and any smooth manifold of class $C^{\geqslant r}$, with $r \leqslant \infty$ or $r = a$, compact or not, admits a neat C^r-embedding in Euclidean space. For proofs see [22] and [8].

We should mention that the case $r = a$ in Theorems 1 and 3 is exceedingly difficult and this is the reason why we excluded it here. In the sequel we shall exclude it from other formulations too: cf., for example, 4.2, 5.3, 6.5, and 4.6. 2.7.

3. Transversalizations and Tubes

1. In this subsection, we consider the image in Euclidean space of a smooth manifold under a differentiable embedding and study the structure of a neighborhood of this image. The results are concentrated in theorems 4, 5, and 7, and serve as the technical basis for the remaining part of the present section.

2. Let j be a differentiable embedding of the smooth, closed, n-dimensional manifold X in $\mathbb{R}^q$. A _transversalization_ of j is a continuous map $\tau: X \to G(q,q-n)$ such that, for each point $x \in X$, the plane $\tau(x)$ is transverse to the plane $d_x j(\mathrm{Tang}_x X)$ (i.e., the two planes intersect at only one point). A basic example is the _normal transversalization_ which associates to each point $x \in X$ the corresponding normal plane (i.e., the orthogonal complement to $d_x j(\mathrm{Tang}_x X)$ in $\mathbb{R}^q$); if j is of class C^r, then its normal transversalization is obviously of class C^{r-1} (cf. 1.4.2).

Given an embedding $j: X \to \mathbb{R}^q$ and a transversalization $\tau: X \to G(q,q-n)$ of j, one can construct the natural map $\tilde{\tau}: X \to G'(q,q-n)$, which takes each point x into the plane $j(x) + \tau(x)$ (which is parallel to $\tau(x)$ and passes through $j(x)$). We denote the ball and the sphere with center $j(x)$ and radius ρ in $j(x) + \tau(x)$ by $d_\tau(x,\rho)$ and $s_\tau(x,\rho)$, respectively. The unions $\bigcup_{x \in X} d_\tau(x,\rho)$ and $\bigcup_{x \in X} [d_\tau(x,\rho) \smallsetminus s_\tau(x,\rho)]$ are called the _tube_ (or the <u>tubular neighborhood</u>) and the <u>open tube</u> (or the <u>open tubular neighborhood</u>) of radius ρ of the transversalization τ, and are denoted by $\text{Tub}_\tau\rho$ and $\text{tub}_\tau\rho$, respectively.

A tube $\text{Tub}_\tau\rho$ is said to be <u>neat</u> if there is a $\sigma > \rho$ such that: (i) the open balls $d_\tau(x,\sigma) \smallsetminus s_\tau(x,\sigma)$, $x \in X$, are pairwise disjoint and the open tube $\text{tub}_\tau\sigma$ they form is a neighborhood of $j(X)$ in $\mathbb{R}^q$; (ii) the map of this neighborhood onto X, which transforms all the points of $d_\tau(x,\sigma) \smallsetminus s_\tau(x,\sigma)$ into x, is smooth. The restrictions of the last map to $\text{Tub}_\tau\rho$ or $\text{tub}_\tau\rho$ (which obviously do not depend on the choice of σ) are called <u>projections</u> and are denoted by pr_τ.

If $\text{Tub}_\tau\rho$ is neat, then all the tubes $\text{Tub}_\tau\rho'$ with $\rho' < \rho$ are obviously neat.

Warning: it may happen that the normal transversalization does not have a neat tube, or even a tube such that the balls $d_\tau(x,\rho)$ are disjoint; see exercise 11.4.

3. The following construction enables us to represent a neat tubular neighborhood as the image of some ideal model of itself, and is required in the proofs of theorems 4 and 5.

Let Tu_τ be that subset of the product $X \times \mathbb{R}^q$ consisting of all the pairs (x,t) with $t \in \tau(x)$. For $\rho > 0$, we let $\text{Tu}_\tau\rho$ and $\text{tu}_\tau\rho$ denote the pieces of Tu_τ such that $\text{dist}(0,t) \leqslant \rho$ and $\text{dist}(0,t) < \rho$, respectively. We also denote by $\text{nat}: \text{Tu}_\tau \to \mathbb{R}^q$ the map given by $\text{nat}(x,t) = j(x) + t$. nat is obviously an isometry of each plane $x \times \tau(x)$ onto the corresponding plane $j(x) + \tau(x)$, and transforms $\text{Tu}_\tau\rho$ ($\text{tu}_\tau\rho$) exactly into $\text{Tub}_\tau\rho$ (respectively, $\text{tub}_\tau\rho$). It is also clear that nat is injective on $\text{Tu}_\tau\rho$ ($\text{tu}_\tau\rho$) if and only if the balls $d_\tau(x,\rho)$ (respectively, the open balls $d_\tau(x,\rho) \smallsetminus s_\tau(x,\rho)$) are pairwise disjoint. In this case nat transforms the restriction to $\text{Tu}_\tau\rho$ of the projection $\text{pr}_1: X \times \mathbb{R}^q \to X$ into the projection $\text{pr}_\tau: \text{Tub}_\tau\rho \to X$ (respectively, the restriction to $\text{tu}_\tau'\rho$ of pr_1 into $\text{pr}_\tau: \text{tub}_\tau\rho \to X$).

We are interested only in smooth transversalizations τ. If j and τ are C^r-maps with $r \geqslant 1$, then $X \times \mathbb{R}^q$ is a $C^{\geqslant r}$-manifold, Tu_τ is a neat q-dimensional submanifold of $X \times \mathbb{R}^q$ (without boundary), and nat is a C^r-map. Moreover, in this case $\text{Tu}_\tau\rho$ is a compact q-dimensional submanifold of Tu_τ such that $\text{int}(\text{Tu}_\tau\rho) = \text{tu}_\tau\rho$ and the restriction of the projection $\text{pr}_1: X \times \mathbb{R}^q \to X$ to each of the manifolds Tu_τ, $\text{Tu}_\tau\rho$, and $\text{tu}_\tau\rho$ is a C^r-submersion.

4. <u>Let the maps</u> j <u>and</u> τ <u>be of class</u> C^r, $r \geqslant 1$. <u>If</u>

$\mathrm{Tub}_\tau \rho$ __is a neat tube, then it is a__ C^r__-submanifold of__ $\mathbb{R}^q$ __with__
$\mathrm{int}(\mathrm{Tub}_\tau \rho) = \mathrm{tub}_\tau \rho$, __and__ $\mathrm{pr}_\tau : \mathrm{Tub}_\tau \rho \to X$ __is a__ C^r__-submersion.__

PROOF. Let $\sigma > \rho$ be such that the conditions (i) and (ii) in 2 are satisfied. As 3 shows, the map $\mathrm{ab\,nat} : \mathrm{tu}_\tau \sigma \to \mathrm{tub}_\tau \sigma$ is invertible and its inverse $\mathrm{ab\,nat}^{-1} : \mathrm{tub}_\tau \sigma \to \mathrm{tu}_\tau \sigma$ is obviously given by $y \mapsto (\mathrm{pr}_\tau(y), y - j \circ \mathrm{pr}_\tau(y))$. This formula shows that $\mathrm{ab\,nat}^{-1}$ is smooth and thus a C^r-diffeomorphism. Now it is evident that the properties of $\mathrm{Tub}_\tau \rho$ and the projection $\mathrm{pr}_\tau : \mathrm{Tub}_\tau \rho \to X$ which we have to verify are consequences of the properties established in 3 for their models $\mathrm{Tu}_\tau \rho$ and $\mathrm{ab\,pr}_1 : \mathrm{Tu}_\tau \rho \to X$.

5. __Every smooth transversalization has a neat tube.__

PROOF. Let τ be a smooth transversalization of the embedding $j : X \to \mathbb{R}^q$. Since the planes $\tau(x)$ and $d_x j(\mathrm{Tang}_x X)$ are transverse, the differential $d_{(x,t)}\mathrm{nat}$ is nondegenerate when $t = 0$. Hence, nat defines a diffeomorphism of a neighborhood U of $X \times 0$ in Tu_τ onto a neighborhood of $j(X)$ (see 1.5.5). It is clear that if $\mathrm{Tu}_\tau \rho \subset U$, then $\mathrm{Tub}_\tau \rho$ is a neat tube of the transversalization τ.

6 (LEMMA). __Suppose__ X __and__ X' __are closed__ $C^{\geq r}$__-manifolds,__ $1 \leqslant r \leqslant \infty$. __If there exists a__ C^r__-embedding of__ X' __in Euclidean space together with a__ C^r__-transversalization, then the set__ $C^r(X,X')$ __is dense in__ $C(X,X')$.

PROOF. Fix a C^r-embedding $j : X \to \mathbb{R}^q$, and a C^r-embedding $j' : X' \to \mathbb{R}^{q'}$ together with a C^r-transversalization τ'. It suffices to show that given an arbitrary continuous map $f : X \to X'$ and an arbitrary $\varepsilon > 0$, there is a C^r-map $g : X \to X'$ such

$$\max_{x \in X} \mathrm{dist}(j' \circ f(x), j' \circ g(x)) < \varepsilon$$

(see 1.2.7.3). We construct the neat tube $\mathrm{Tub}_{\tau'}\rho'$ with $\rho' \leqslant \varepsilon/2$ and choose δ to be the smallest of the two numbers $\varepsilon/2$ and $\mathrm{Dist}(j'(X'), \mathbb{R}^{q'} \smallsetminus \mathrm{tub}_{\tau'}\rho')$; note that $\delta > 0$ (see 1.1.7.15). Then, according to 1.1.5.12, the composition

$$j(X) \xrightarrow{\ (\mathrm{ab}\,j)^{-1}\ } X \xrightarrow{\ f\ } X' \xrightarrow{\ j'\ } \mathbb{R}^{q'}$$

extends to a continuous map $f_1 : \mathbb{R}^q \to \mathbb{R}^{q'}$. Now Theorem 3.1.4. yields a map $g_1 : \mathbb{R}^q \to \mathbb{R}^{q'}$ with polynomial components and such that

$$\max_{x \in X} \mathrm{dist}(f_1 \circ j(x), g_1 \circ j(x)) < \delta.$$

This in turn shows that $g_1 \circ j(X) \subset \mathrm{tub}_\tau, \rho'$, and it is clear that $g(x) = \mathrm{pr}_\tau, (g_1 \circ j(x))$ defines the desired map $g: X \to X'$.

      7. <u>Every C^r-embedding of a closed $C^{\geqslant r}$-manifold, $1 \leqslant r \leqslant a$,</u> <u>in Euclidean space admits a C^r-transversalization.</u>

      If $r = a, \infty$, the normal transversalization will suffice. If $1 \leqslant r < \infty$, the existence of the normal transversalization shows that the set of all transversalizations of a given C^r-embedding $j: X \to \mathbb{R}^q$ is not empty. Since the latter set is (trivially) open in $C(X, G(q, q\text{-dim}\, X))$, it is enough to show that $C^r(X, G(q, q\text{-dim}\, X))$ is dense in $C(X, G(q, q\text{-dim}\, X))$. But this is a consequence of Lemma 6, because $G(q, q\text{-dim}\, X)$ can be analytically embedded in Euclidean space.

4.   <u>Smoothing Maps in the Case of Closed Manifolds</u>

      1. Now we arrive at the main topic of the present section – approximating maps of one smooth manifold into another by maps which are more regular in a sense or another as, for example, maps of a higher differentiability class, or embeddings, or immersions.

      In this subsection we consider only approximations which raise the differentiability class of maps without improving their other properties, and we restrict ourselves to the simplest case – that of the closed manifolds.

      2. <u>If $r \leqslant \infty$, then for any closed $C^{\geqslant r}$-manifolds X and X'</u> <u>and any $s < r$, the set $C^r(X, X')$ is dense in $C^s(X, X')$. The same is</u> <u>true when $r = a$ provided that X and X' can be C^a-embedded in</u> <u>Euclidean spaces.</u>

      We have to show that given a map $f \in C^s(X, X')$ and a neighborhood U of f in $C^s(X, X')$, there is a C^r-map in U. Fix C^r-embeddings $j: X \to \mathbb{R}^q$ and $j': X' \to \mathbb{R}^{q'}$, corresponding transversalizations τ and τ', and corresponding neat tubes $\mathrm{Tub}_\tau \rho$ and $\mathrm{Tub}_{\tau'} \rho'$. Now consider the mapping

$$C^s(\mathrm{ab}\, j, \mathrm{pr}_{\tau'}) : C^s(\mathrm{Tub}_\tau(\rho/2), \mathrm{tub}_{\tau'} \rho') \to C^s(X, X'),$$

where $\mathrm{ab}\, j = [\mathrm{ab}\, j: X \to \mathrm{Tub}_\tau(\rho/2)]$ (see 1.1). Since it takes C^r-maps into C^r-maps, it suffices to prove that the preimage V of U under this mapping intersects $C^r(\mathrm{Tub}_\tau(\rho/2), \mathrm{tub}_\tau \rho')$. But V is open and contains the restriction to $\mathrm{Tub}_\tau(\rho/2)$ of the composition

$$\mathrm{tub}_\tau \rho \xrightarrow{\mathrm{pr}_\tau} X \xrightarrow{f} X' \xrightarrow{\mathrm{ab}\, j'} \mathrm{tub}_{\tau'} \rho'. \tag{1}$$

The image of this restriction is compact, and so it lies at a positive distance from $\mathbb{R}^{q'} \setminus \mathrm{tub}_\tau, \rho'$. Consequently, there exists $\varepsilon > 0$ such that if, at the points of $\mathrm{Tub}_\tau(\rho/2)$, the partial derivatives of order $0,1,\dots,s$ of the coordinate functions of a map $g \in C^s(\mathrm{Tub}_\tau(\rho/2),\mathbb{R}^{q'})$ differ from the corresponding partial derivatives of the map (1) by less than ε, then $g(\mathrm{Tub}_\tau(\rho/2)) \subset \mathrm{tub}_\tau,\rho'$ and

[ab $g : \mathrm{Tub}_\tau(\rho/2) \to \mathrm{tub}_\tau,\rho'] \in V$. Finally, apply Theorem 3.1.4 to the coordinate functions of (1) to deduce that there exists a map g with polynomial coordinate functions whose partial derivatives have the property above. Thus $V \cap C^r(\mathrm{Tub}_\tau(\rho/2),\mathrm{tub}_\tau,\rho') \neq \emptyset$.

3. Comparing Theorem 2 with the Theorems 1.2, 1.3, 1.4, and 1.6, we see that for $r \leqslant \infty$ and $1 \leqslant s < r$, and for any given closed $C^{\geqslant r}$-manifolds X and X', the following holds: $\mathrm{Imm}^r(X,X')$ is dense in $\mathrm{Imm}^s(X,X')$, $\mathrm{Subm}^r(X,X')$ is dense in $\mathrm{Subm}^s(X,X')$, $\mathrm{Emb}^r(X,X')$ is dense in $\mathrm{Emb}^s(X,X')$, and finally $\mathrm{Diff}^r(X,X')$ is dense in $\mathrm{Diff}^s(X,X')$. The same is true when $r = a$ provided that X and X' can be C^a-embedded in Euclidean spaces.

4 (COROLLARY). If two closed $C^{\geqslant r}$-manifolds, $1 \leqslant r \leqslant \infty$, are diffeomorphic, then they are C^r-diffeomorphic. The same is true for $r = a$ provided that X and X' can be C^a-embedded in Euclidean spaces.

5 (INFORMATION). Two closed homeomorphic C^a-manifolds are not necessarily diffeomorphic. Historically, the first such examples where C^a-manifolds which are homeomorphic, but not diffeomorphic to S^7; see [15].

Supplements to Theorem 2

6. If $r \leqslant \infty$, then $C^r(X,X')$ is dense in $C^s(X,X')$, $s < r$, in the following more general situation too: X is a closed $C^{\geqslant r}$-manifold and X' is an open subset of a closed $C^{\geqslant r}$-manifold Y. The same is true for $r = a$ if X and Y can be C^a-embedded in Euclidean spaces.

The proof reduces to observing that the mapping

$$C^s(\mathrm{id},\mathrm{in}): C^s(X,X') \to C^s(X,Y)$$

is a topological embedding with open image which carries $C^r(X,X')$ into the intersection of this image with $C^r(X,Y)$. Since $C^r(X,Y)$ is dense in $C^s(X,Y)$, the above intersection is dense in this image, and $C^r(X,X')$ is dense in $C^s(X,X')$.

      7 (LEMMA). <u>Every pair of disjoint closed subsets of a closed</u> C^r<u>-manifold with</u> $r \leqslant \infty$ <u>has a Urysohn function of class</u> C^r.

      PROOF. Let $\phi: X \to I$ be an arbitrary Urysohn function for the given pair of subsets A, B of X. According to Theorem 6, there is a C^r-function $\psi: X \to \mathbb{R}$ such that $\max_{x \in X} |\psi(x) - \phi(x)| < 1/3$. If now $\lambda: \mathbb{R} \to I$ is a C^r-function such that $\lambda(y) = 0$ for $y \leqslant 1/3$ and $\lambda(y) = 1$ for $y \geqslant 2/3$, then $\lambda \circ \psi$ is obviously a Urysohn C^r-function for the pair A,B.

      8. <u>Let</u> X <u>and</u> X' <u>be closed</u> $C^{\geqslant r}$<u>-manifolds and let</u> A <u>be a closed subset of</u> X. <u>If</u> $0 \leqslant s < r \leqslant \infty$, <u>then that part of</u> $C^s(X,X')$ <u>consisting of the</u> C^r<u>-extensions of a given map</u> $\phi: A \to X'$ <u>is dense in the part of</u> $C^s(X,X')$ <u>consisting of the extensions of</u> ϕ <u>which are of class</u> C^r <u>in a neighborhood of</u> A (<u>the neighborhood depends upon the extension</u>).

      Let $f \in C^s(X,X')$ be an extension of ϕ which is of class C^r in a neighborhood U of A. Given a neighborhood U of f in $C^s(X,X')$, we have to show that U contains a C^r-extension of ϕ. Fix a C^r-embedding $j': X' \to \mathbb{R}^{q'}$, a C^r-transversalization τ' of j', and a neat tube $\mathrm{Tub}_{\tau', \rho'}$, and denote by V the piece of $C^s(X,X')$ consisting of all the maps g such that

$$\max_{x \in X} \mathrm{dist}(j' \circ f(x), j' \circ g(x)) < \mathrm{Dist}(j'(X), \mathbb{R}^{q'} \smallsetminus \mathrm{tub}_{\tau', \rho'}).$$

It is obvious that V is open and that for any $g \in V$, $x \in X$, the segment with endpoints $j' \circ f(x)$, $j' \circ g(x)$ is contained in $\mathrm{tub}_{\tau', \rho'}$. Next construct a Urysohn C^r-function ψ for the pair $A, X \smallsetminus U$ (see 7) and consider the map $\Phi: V \to C^s(X,X')$ which transforms g into the map

$$x \mapsto \mathrm{pr}_{\tau'}((1-\psi(x))j' \circ f(x) + \psi(x)j' \circ g(x)).$$

One may check directly that Φ is continuous and $\Phi(f) = f$; hence the set $\Phi^{-1}(U)$ is open and nonempty. Now Theorem 2 shows that $\Phi^{-1}(U)$ contains a C^r-map. Finally, we note that Φ takes C^r-maps into C^r-extensions of ϕ.

      9. <u>Suppose that</u> X <u>and</u> X' <u>are closed</u> $C^{\geqslant r}$<u>-manifolds and</u> A <u>is a submanifold of</u> X <u>which is itself closed as a manifold. Let</u> $\phi: A \to X'$ <u>be a</u> C^r<u>-map. If</u> $0 \leqslant s < r \leqslant \infty$, <u>then that part of</u> $C^s(X,X')$ <u>consisting of the</u> C^r<u>-extensions of</u> ϕ <u>is dense in the part of</u> $C^s(X,X')$

<u>consisting of all the C^s-extensions of</u> ϕ.

      Given a C^s-extension of ϕ and a neighborhood U of this extension in $C^s(X,X')$, we have to show that U contains a C^r-extension of ϕ. Fix C^r-embeddings $j: X \to \mathbb{R}^q$ and $j': X' \to \mathbb{R}^{q'}$, a C^r-transversalization τ of the embedding $j|_A: A \to \mathbb{R}^q$ and a C^r-transversalization τ' of j', and corresponding neat tubes $\mathrm{Tub}_\tau \rho$ and $\mathrm{Tub}_{\tau'}\rho'$. Further, denote by V the piece of $C^s(X,X')$ consisting of all the maps g such that

$$\max_{x \in X} \mathrm{dist}(j' \circ \phi(x), j' \circ g(x)) < \mathrm{Dist}(j'(X'), \mathbb{R}^{q'} \smallsetminus \mathrm{tub}_{\tau'}\rho').$$

Obviously, V is open and contains all the C^s-extensions of ϕ to X. Now take any Urysohn C^r-function ψ for the pair $X \smallsetminus j^{-1}(\mathrm{tub}_\tau \rho), A$ and consider the mapping $\Phi: V \to C^s(X,X')$ which transforms each map g into the map

$$x \mapsto \begin{cases} \mathrm{pr}_{\tau'}(j \circ g(x) + \psi(x)[j' \circ \phi \circ \mathrm{pr}_\tau(j(x)) - j' \circ g \circ \mathrm{pr}_\tau(j(x))]), & \\ & \text{if } j(x) \in \mathrm{Tub}_\tau \rho, \\ g(x), & \text{if } j(x) \notin \mathrm{Tub}_\tau \rho. \end{cases}$$

It is clear that Φ is continuous and that $\Phi(g) = g$ whenever g extends ϕ. This implies that $\Phi^{-1}(U)$ is an open nonempty set which, according to Theorem 2, contains a C^r-map. Finally, note that Φ takes C^r-maps into C^r-extensions of ϕ.

5.   <u>Glueing Manifolds Smoothly</u>

    1. Our main task in this subsection is to make the necessary preparations for extending the basic approximation theorems given in the previous subsection, i.e., theorems 4.2-4.4, in their nonanalytic version, to include compact manifolds with boundary. The main tool used in the extension is that of smooth doubling of a compact manifold, an operation which transforms it into a closed manifold. However, we find it convenient to define and study a more general operation, which is useful for other purposes too - the smooth glueing of smooth compact manifolds. To begin with, we need to investigate the structure of a smooth compact manifold in the vicinity of its boundary.

Collars

2. A <u>collaring</u> of a compact C^r-manifold X $(0 \leqslant r \leqslant a)$ is a C^r-embedding of the cylinder $\partial X \times I$ into X, which takes the point $(x,0)$ into x, for each $x \in \partial X$. The image of $\partial X \times I$ under such an embedding is known as a <u>collar</u> (on X).

If X is a smooth manifold (i.e., $r \geqslant 1$), a collaring is a differentiable embedding and its image is a submanifold of codimension 0, whose boundary consists of ∂X and of a submanifold of $\mathrm{int}\, X$ diffeomorphic to ∂X.

3. <u>If</u> $1 \leqslant r \leqslant \infty$, <u>every compact C^r-manifold admits a collaring.</u>

PROOF. Let X be the given manifold. Pick a neat C^r-embedding $j \colon X \to \mathbb{R}^q_-$ (see 2.3), a C^r-transversalization τ of the composite embedding $X \xrightarrow{\ \mathrm{ab}\ j\ } \mathbb{R}^q_- \xrightarrow{\ \mathrm{in}\ } \mathbb{R}^q$, and a neat tube $\mathrm{Tub}_\tau \rho$. Consider the map $\phi \colon j^{-1}(\mathrm{tub}_\tau \rho) \to \partial X \times \mathbb{R}^1_-$, defined as $x \mapsto (\mathrm{pr}_\tau(j(x)), j_1(x))$, where j_1 is the first coordinate function of j. Since j is neat, the differential $d_x\phi$ is nondegenerate at each point $x \in \partial X$, so that ϕ realizes a diffeomorphism of a neighborhood of ∂X onto a neighborhood of $\partial X \times 0$ (see 1.5.5). Now let $\varepsilon > 0$ be small enough so that the product $\partial X \times [-\varepsilon, 0]$ is contained in the previous neighborhood. Then the formula $(x,t) \mapsto \phi^{-1}(x,-\varepsilon t)$ obviously defines a collaring of X.

4 (INFORMATION). The compact topological manifolds $(r = 0)$ and the compact analytic manifolds $(r = a)$ admit collarings too. The case $r = 0$ is considered in [4].

Glueing

5. Suppose that X and X' are compact n-dimensional C^r-manifolds with $r \geqslant 1$, and let C and C' be submanifolds of ∂X and $\partial X'$, respectively, consisting of whole components of these boundaries. Assuming that C and C' are diffeomorphic, pick a C^r-diffeomorphism $\phi \colon C \to C'$ and attach X to X' by the composite map $C \xrightarrow{\ \phi\ } C' \xrightarrow{\ \mathrm{in}\ } X'$ (see 1.2.4.8). The resulting space $Y = X' \cup_{\mathrm{in} \circ \phi} X$ is obviously a compact, n-dimensional, topological manifold. However, if X and X' have collars then it turns out that Y has a natural C^r-structure that makes it into a collared C^r-manifold.

The atlas that defines this C^r-structure consists of the charts of
Atl$(X \smallsetminus C)$ and Atl$(X' \smallsetminus C')$ (we regard X and X' as parts of Y),
as well as the charts Ψ constructed from both the charts $\psi \in$ Atl C
and the collarings $k: \partial X \times I \to X$ and $k': \partial X' \times I \to X'$ by the
formulas

$$\text{supp}\,\Psi = k(\text{supp}\,\psi \times [0,1)) \cup k'(\text{supp}\,\psi \times [0,1))$$

and

$$\left.\begin{array}{l} \Psi(k(z,t)) = (\psi(z),-t) \\[2mm] \Psi(k'(z,t)) = (\psi(z),t) \end{array}\right\} \quad z \in C, \quad t \in [0,1)$$

($\text{Im}\,\psi \subset \mathbb{R}^{n-1}$ and $\text{Im}\,\Psi = \text{Im}\,\psi \times (-1,1) \subset \mathbb{R}^{n-1} \times \mathbb{R} = \mathbb{R}^n$). It is readily
seen that these charts are all pairwise compatible. One constructs a
collar on the resulting C^r-manifold Y from those pieces of the
collarings k and k' which are preserved under the above procedure.
We say that the diffeomorphism ϕ __glues__ X and X' into Y. It is
clear that $\partial Y = (\partial X \smallsetminus C) \cup (\partial X' \smallsetminus C')$ and that X, X', and C are
submanifolds of Y. The pieces of the collarings k and k' which
are related to C and C' yield a __two-sided collaring of the manifold__
C __in__ Y, i.e., a C^r-embedding $C \times [-1,1] \to Y$ such that $(z,0) \mapsto z$;
its image is a __two-sided collar__ of C in Y.

In the particular situation $X' = X$, $C' = C = \partial X$, and
$\phi = \text{id}\,\partial X$, we use the term __doubling__ instead of glueing and denote Y
by $\text{dopp}\,X$. This definition agrees with 1.1.9, i.e., $C^0(\text{dopp}\,X) =$
$= \text{dopp}(C^0 X)$.

Now suppose that X and X' are oriented manifolds and C
and C' are equipped with the induced orientations (see 1.3.4). If
ϕ is orientation preserving, then Y is orientable and can be actually
oriented in a canonical way. This canonical orientation is that which
induces the original orientation on X and the orientation opposite to
the original one on X'. In particular, $\text{dopp}\,X$ is oriented for any
oriented manifold X.

The simplest examples show that the C^r-structure on Y
depends not only upon the C^r-structures of the manifolds X and X'
and the diffeomorphism ϕ, but also upon the collarings k and k'.
Our next objective is to demonstrate that for $r \neq a$ this last
dependence is eliminated if we regard Y as distinct up to
C^r-diffeomorphisms.

6 (LEMMA). <u>Let</u> X <u>and</u> X' <u>be closed</u> $C^{\geq r}$-<u>manifolds,</u>
$1 \leq r \leq \infty$, <u>and let</u> $f: X \times [-1,1] \to X' \times [-1,1]$ <u>be such that</u>
$f(X \times [-1,0]) \subset X' \times (-1,0]$ <u>and</u> $f(X \times [0,1]) \subset X' \times [0,1)$. <u>Assume</u>
<u>that</u> $ab\,f: X \times [-1,0] \to X' \times (-1,0]$ <u>and</u> $ab\,f: X \times [0,1] \to X' \times [0,1)$
<u>are differentiable embeddings of class</u> C^r, <u>while</u> $ab\,f: X \times 0 \to X' \times 0$
<u>is a diffeomorphism. Then there exists a</u> C^r-<u>embedding</u>
$g: X \times [-1,1] \to X' \times [-1,1]$ <u>which agrees with</u> f <u>on</u>
$(X \times [-1,-1/2]) \cup (X \times 0) \cup (X \times [1/2,1])$, <u>and such that</u>
$g(X \times [-1,0]) = f(X \times [-1,0])$ <u>and</u> $g(X \times [0,1]) = f(X \times [0,1])$.

In the proof that follows we let f_1 and f_2 denote the composite maps

$$X \times [-1,1] \xrightarrow{\ f\ } X' \times [-1,1] \xrightarrow{\ pr_1\ } X'$$

and

$$X \times [-1,1] \xrightarrow{\ f\ } X' \times [-1,1] \xrightarrow{\ pr_2\ } [-1,1] \ ,$$

respectively.

For a start, assume that for some positive ε the map f_1 is constant on each set $x \times [-\varepsilon,\varepsilon]$, $x \in X$. In this case, fix positive numbers δ and η, such that $\delta \leq \min(\varepsilon,1/2)$ and for any $x \in X$ the derivative of the function $t \mapsto f_2(x,t)$ is not less than η on the intervals $[-\delta,0)$ and $(0,\delta]$. [The existence of such δ and η results from the continuity of the functions $X \times [-1,0] \to \mathbb{R}$ and $X \times [0,1] \to \mathbb{R}$, given by the formula $(x,t) \mapsto \partial f_2(x,\tau)\partial\tau\big|_{\tau=t}$ (for $t = 0$ one takes the left derivative for the first function and the righ derivative for the second one), and the positivity of both functions on $X \times 0$. Since X is compact, these functions are bounded from below by a positive constant on $X \times 0$, and hence on $X \times [-\delta,\delta]$ for some $\delta > 0$.] To proceed further, pick a C^r-function $\alpha: [-\delta,\delta] \to I$ such that $\alpha(t) = 0$ for $|t| \leq \delta/4$ and $\alpha(t) = 1$ for $|t| \geq \delta/2$. It is not hard to verify that the formula

$$g(x,t) = \begin{cases} (f_1(x,t),(1-\alpha(t))\eta t + \alpha(t)f_2(x,t)), & \text{if } |t| \leq \delta, \\[2mm] f(x,t), & \text{if } |t| \geq \delta, \end{cases}$$

defines a map $g: X \times [-1,1] \to X' \times [-1,1]$ with the desired properties.

In the general case, choose $0 < \varepsilon_1 \leq 1/2$ such that the map $\phi_t: X \to X'$, defined as $\phi_t(x) = f_1(x,t)$, is a diffeomorphism for all $|t| \leq \varepsilon_1$. [The existence of such an ε_1 is a consequence of: the continuous dependence of ϕ_t on t in the C^r-topology, the fact that ϕ_t is a diffeomorphism (recall that $ab\,f: X \times 0 \to X' \times 0$ is a diffeomorphism), and the fact that $\mathrm{Diff}^r(X,X')$ is open in $C^r(X,X')$

(see 1.6). Let $\gamma: [-\varepsilon_1, \varepsilon_1] \to \mathbb{R}$ be a nondecreasing C^r-function such that $\gamma(t) = 0$ for $|t| \leqslant \varepsilon_1/4$ and $\gamma(t) = t$ for $|t| \geqslant \varepsilon_1/2$. Now define $\widetilde{f}: X \times [-1,1] \to X' \times [-1,1]$ by

$$\widetilde{f}(x,t) \;=\; \begin{cases} f(\phi_t^{-1} \circ \phi_{\gamma(t)}(x),t), & \text{if } |t| \leqslant \varepsilon_1, \\[2ex] f(x,t), & \text{if } |t| \leqslant \varepsilon_1. \end{cases}$$

Then $\widetilde{f}$ satisfies all the conditions imposed to f in the statement of the lemma. Moreover, $\widetilde{f}$ satisfies the extra conditions under which the lemma has already been proved, namely that the composite map

$$X \times [-1,1] \xrightarrow{\;\widetilde{f}\;} X' \times [-1,1] \xrightarrow{\;\mathrm{pr}_1\;} X'$$

is constant on the sets $x \times [-\varepsilon, \varepsilon]$, with $\varepsilon = \varepsilon_1/4$. The map g corresponding to $\widetilde{f}$ via the above procedure has the needed properties because $\widetilde{f}$ agrees with f on $(X \times [-1,-1/2]) \cup (X \times 0) \cup (X \times [1,1/2])$.

7. <u>Suppose that</u> X_1, X_1', X_2, <u>and</u> X_2' <u>are collared</u> C^r-<u>manifolds,</u> $1 \leqslant r \leqslant \infty$, C_1, C_1', C_2, <u>and</u> C_2' <u>are pieces of their boundaries consisting of whole components, and</u> $\phi_1: C_1 \to C_1'$ <u>and</u> $\phi_2: C_2 \to C_2'$ <u>are</u> C^r-<u>diffeomorphisms. Assume that there are</u> C^r-<u>diffeomorphisms</u> $F: X_1 \to X_2$ <u>and</u> $F': X_1' \to X_2'$, <u>such that</u> $F(C_1) = C_2$, $F'(C_1') = C_2'$, <u>and the diagram</u>

$$\begin{array}{ccc} C_1 & \xrightarrow{\;\phi_1\;} & C_1' \\[0.5ex] {\scriptstyle \mathrm{ab}\,F}\Big\downarrow & & \Big\downarrow{\scriptstyle \mathrm{ab}\,F'} \\[0.5ex] C_2 & \xrightarrow[\;\phi_2\;]{} & C_2' \end{array}$$

<u>is commutative. If</u> Y_1 <u>is the result of glueing</u> X_1 <u>and</u> X_1' <u>by</u> ϕ_1, <u>and</u> Y_2 <u>is the result of glueing</u> X_2 <u>and</u> X_2' <u>by</u> ϕ_2, <u>then the manifolds</u> Y_1 <u>and</u> Y_2 <u>are</u> C^r-<u>diffeomorphic. Moreover, there exists a</u> C^r-<u>diffeomorphism</u> $G: Y_1 \to Y_2$ <u>such that</u> $G(X_1) = X_2$, $G(X_1') = X_2'$, $G(C_1) = C_2$, <u>and</u> $[\mathrm{ab}\,G: C_1 \to C_2] = [\mathrm{ab}\,F: C_1 \to C_2]$.

PROOF. Let $\ell_1: C_1 \times [-1,1] \to Y_1$ and $\ell_2: C_2 \times [-1,1] \to Y_2$ be two-sided collarings. Denote by $H: Y_1 \to Y_2$ the map defined by the formulas $H|_{X_1} = [\mathrm{in}: X_2 \to Y_2] \circ F$ and $F|_{X_1'} = [\mathrm{in}: X_2' \to Y_2] \circ F'$, and choose $\varepsilon > 0$ so that $H \circ \ell_1(C_1 \times [-\varepsilon, \varepsilon]) \subset \ell_2(C_2 \times (-1,1))$. Now apply Lemma 6 to the map $f: C_1 \times [-1,1] \to C_2 \times [-1,1]$ given by $f(z,t) = \ell_2^{-1}(H \circ \ell_1(z,\varepsilon t))$. This lemma guarantees the existence of a C^r-embedding $g: C_1 \times [-1,1] \to C_2 \times [-1,1]$ which agrees with f on

$(C_1 \times [-1,-1/2]) \cup (C_1 \times 0) \cup (C_1 \times [1/2,1])$, and satisfies
$g(C_1 \times [-1,0]) = f(C_1 \times [-1,0])$, $g(C_1 \times [0,1]) = f(C_1 \times [0,1])$.
Clearly,

$$G(y) = \begin{cases} H(y), & \text{if } y \notin \ell_1(C_1 \times [-\varepsilon,\varepsilon]), \\[2ex] \ell_2 \circ g(z,t/\varepsilon), & \text{if } y = \ell_1(z,t) \text{ with } z_1 \in C_1, \\ & \qquad\qquad\qquad\qquad t \in [-\varepsilon,\varepsilon] \end{cases}$$

defines the required C^r-diffeomorphism $G: Y_1 \to Y_2$.

Cutting

8. <u>Let</u> Y <u>be a</u> C^r-<u>manifold</u>, $1 \leqslant r \leqslant \infty$, <u>and let</u> X <u>and</u> X'
<u>be compact submanifolds of</u> Y <u>such that</u> $\dim X = \dim X' = \dim Y$ <u>and</u>
$Y = X \cup X'$. <u>If</u> $C = X \cap X'$ <u>is a piece of both boundaries</u> ∂X <u>and</u> $\partial X'$
<u>consisting of whole components of</u> ∂X <u>and</u> $\partial X'$, <u>then there is a</u>
C^r-<u>embedding</u> $\ell: C \times [-1,1] \to Y$ <u>such that</u> $\ell(z,0) = z$ <u>for any point</u>
$z \in C$ <u>and</u> $\ell(C \times [-1,0)) \subset \text{int } X$, $\ell(C \times (0,1]) \subset \text{int } X'$.

PROOF. Fix a C^r-embedding $j: Y \to \mathbb{R}^q$, a C^r-transversalization
τ of the embedding $j|_C: C \to \mathbb{R}^q$, and a neat tube $\text{Tub}_\tau \rho$. Consider
the map $\phi: C \to S^{q-1}$ which takes each point $z \in C$ into the unit
vector tangent to $j(Y)$ at the point $j(z)$, contained in $\tau(z)$, and
pointing towards $j(X')$. ϕ is continuous (in fact, of class C^{r-1}),
and so Theorem 4.2 yields a C^r-map $\phi_1: C \to S^{q-1}$ such that the inner
product $<\phi(z),\phi_1(z)>$ is positive on C. Denote by $\psi: \text{tub}_\tau \rho \to C \times \mathbb{R}$
the map defined by the formula $\psi(z) = (\text{pr}_\tau(z),<z - j \circ \text{pr}_\tau(z),\phi_1(z)>)$.
Clearly, ψ is of class C^r and for $z \in C$ the differential

$$d_{j(z)}\psi: \text{Tang}_{j(z)}(\text{tub}_\tau \rho) \to \text{Tang}_{(z,0)}(C \times \mathbb{R}) = \text{Tang}_z C \oplus \mathbb{R}$$

induces an isomorphism of $\text{Tang}_{j(z)}j(C)$ onto $\text{Tang}_z C$ and carries the
vector $\phi(z)$ into $<\phi(z),\phi_1(z)> \in \mathbb{R}$. Moreover, both $\text{Tang}_{j(z)}j(C)$
and $\phi(z)$ are contained in $\text{Tang}_{j(z)}j(Y)$; hence $d_{j(z)}\psi$ takes
$\text{Tang}_{j(z)}j(Y)$ onto $\text{Tang}_{(z,0)}(C \times \mathbb{R})$. Since $\dim \text{Tang}_{j(z)} =$
$= \dim \text{Tang}_{(z,0)}(C \times \mathbb{R})$, we see that $d_{j(z)}\psi \text{Tang}_{j(z)}j(Y)$, i.e.,
the linear map $d_{j(z)}(\psi|_{j(Y) \cap \text{tub}_\tau \rho})$, is an isomorphism. By Theorem
1.5.5, $\psi|_{j(Y) \cap \text{tub}_\tau \rho}$ defines a diffeomorphism from a neighborhood of
$j(C)$ onto a neighborhood of $C \times 0$. Accordingly, $C \times [-\varepsilon,\varepsilon]$ will

lie in the previous neighborhood provided that $\varepsilon > 0$ is small enough. Now it is plain that $\ell(z,t) = j^{-1}(\psi^{-1}(z,\varepsilon t))$ defines the desired embedding $\ell\colon C \times [-1,1] \to Y$.

9 (COROLLARY). <u>Let</u> Y <u>be a</u> C^r<u>-manifold,</u> $1 \leqslant r \leqslant \infty$, <u>and let</u> X <u>and</u> X' <u>be compact submanifolds of</u> Y <u>such that</u> $\dim X = \dim X' = \dim Y$ <u>and</u> $Y = X \cup X'$. <u>If</u> $X \cap X'$ <u>is a piece of both boundaries</u> ∂X <u>and</u> $\partial X'$, <u>consisting of whole components of</u> ∂X <u>and</u> $\partial X'$, <u>then</u> $\mathrm{id}\,X$ <u>and</u> $\mathrm{id}\,X'$ <u>together define a</u> C^r<u>-diffeomorphism of</u> Y <u>onto the manifold obtained from the appropriately collared manifolds</u> X <u>and</u> X' <u>glueing</u> X <u>and</u> X' <u>by</u> $\mathrm{id}(X \cap X')$.

The Simplest Application

10. <u>Every smooth compact manifold is a CNRS.</u>

When the manifold is closed, this is a consequence of 2.1, 3.7, and 3.5, because the image of a smooth manifold under a differentiable embedding in Euclidean space is the retract of the interior of a neat tube corresponding to a smooth transversalization of the given embedding. Theorem 1.3.6.4 enables us to reduce the case of manifolds with boundary to the closed case; namely, any compact smooth manifold has a smooth closed double (see 5), and is obviously a retract of this double.

6. <u>Smoothing Maps in the Presence of a Boundary</u>

1. The main results of this subsection are Theorems 5 and 9, which generalize Theorem 4.2. Lemma 2 is necessary to the proof of Lemma 3, Lemma 3 — to the proof of Lemma 4, and Lemma 4 — to the proof of Theorem 5. Finally, lemmas 7 and 8 are necessary to the proof of Theorem 9.

2 (LEMMA). <u>Let</u> Y <u>be a</u> $C^{\geqslant r}$<u>-manifold with</u> $r < \infty$, <u>and let</u> $f\colon Y \times \mathbb{R}^1_- \to \mathbb{R}$ <u>be a</u> C^r<u>-function. Then the function</u> $F\colon Y \times \mathbb{R} \to \mathbb{R}$ <u>defined by</u>

$$F(y,t) = \begin{cases} f(y,t), & \text{if } t \leqslant 0, \\[2mm] \sum_{k=0}^{r} (-1)^k \binom{r+1}{k+1} f(y,-kt), & \text{if } t \geqslant 0 \end{cases}$$

<u>is also of class</u> C^r.

PROOF. All we must check is that the two expressions defining F, as well as their partial derivatives with respect to t and local coordinates on Y agree for $t = 0$. To see this, it suffices to note that the equality

$$\sum_{k=0}^{r} (-1)^{k+s} \binom{r+1}{k+1} k^s D^s \phi(0) = D^s \phi(0)$$

holds for $s \leqslant r$ and any C^s-function $\phi \colon \mathbb{R}_-^1 \to \mathbb{R}$. Indeed, this last equality is equivalent to

$$\sum_{k=-1}^{s} (-1)^k \binom{r+1}{k+1} k^s = 0 \qquad (s \leqslant r)$$

and this is valid if we interpret the sum as the "$(r+1)$-th difference" of the integral function $k \to k^s$, computed at $k = -1$.

3 (LEMMA). <u>Let X be a collared $C^{\geqslant r}$-manifold. If $1 \leqslant r < \infty$, then every function in $C^r(X,\mathbb{R})$ extends to a function in $C^r(\operatorname{dopp} X, \mathbb{R})$.</u>

PROOF. Fix a two-sided collaring of the manifold X in $\operatorname{dopp} X$, $\ell \colon \partial X \times [-1,1] \to \operatorname{dopp} X$ (see 5.5), and pick a C^r-function $\alpha \colon \mathbb{R} \to I$ such that $\alpha(t) = 1$ for $t \leqslant 0$ and $\alpha(t) = 0$ for $t \geqslant 1/r$. Let $\phi \in C^r(X,\mathbb{R})$. Consider the function $f \colon \partial X \to \mathbb{R}_-^1$ defined by the formula

$$f(y,t) = \begin{cases} \alpha(t)\phi(\ell(y,t)), & \text{if } t \leqslant 1/r, \\[1em] 0, & \text{if } t \geqslant 1/r. \end{cases}$$

Since f is C^r, Lemma 2 processes it into a C^r-function $F \colon \partial X \times \mathbb{R} \to \mathbb{R}$ (take $Y = \partial X$). Now it is plain that

$$\psi(x) = \begin{cases} 0, & \text{if } x \in \operatorname{dopp} X \smallsetminus (X \cup \ell(\partial X \times [0,1/r])), \\ \phi(x), & \text{if } x \in X \\ F(\ell^{-1}(x)), & \text{if } x \in \ell(\partial X \times [0,1/r]), \end{cases}$$

defines a C^r-function $\psi \colon \operatorname{dopp} X \to \mathbb{R}$ extending ϕ.

4 (LEMMA). <u>Let X be a collared $C^{\geqslant r}$-manifold, and let X' be a closed $C^{\geqslant r}$-manifold. If $0 \leqslant r < \infty$, then every map in $C^r(X,X')$ extends to a map in $C^r(\operatorname{dopp} X, X')$.</u>

PROOF. This is evident when $r = 0$. Suppose $r > 0$ and $f \in C^r(X,X')$. Fix a C^r-embedding $j' \colon X' \to \mathbb{R}^{q'}$, a C^r-transversalization τ' of j', and a neat tube $\operatorname{Tub}_{\tau',\rho'}$. Lemma 3 ensures that the coordinate functions of $j' \circ f$ extend to C^r-functions $\operatorname{dopp} X \to \mathbb{R}$, i.e., $j' \circ f$

extends to a C^r-map $g: \mathrm{dopp}\, X \to \mathbb{R}^{q'}$. Let U be the neighborhood of ∂X in X consisting of the points x such that

$$\mathrm{dist}(j' \circ f(x), g(\mathrm{cop}(x))) < \mathrm{Dist}(j'(X'), \mathbb{R}^{q'} \smallsetminus \mathrm{tub}_{\tau, \rho'}).$$

Now construct a Urysohn C^r-function $\phi: \mathrm{dopp}\, X \to I$ for the pair $X, \mathrm{cop}(X \smallsetminus U)$. It is clear that for any $x \in X$ the segment with end-points $j' \circ f(x)$ and $g(\mathrm{cop}(x))$ lies in $\mathrm{tub}_{\tau, \rho'}$. Moreover, we see that the formulas

$$h(x) = j' \circ f(x), \quad \text{if } x \in X,$$

$$h(\mathrm{cop}(x)) = j' \circ f(x), \quad \text{if } x \in X \smallsetminus U,$$

$$h(\mathrm{cop}(x)) = (1 - \phi(\mathrm{cop}(x)))g(\mathrm{cop}(x)) + \phi(\mathrm{cop}(x))j' \circ f(x),$$
$$\text{if } x \in \mathrm{Cl}\, U,$$

define a C^r-map $h: \mathrm{dopp}\, X \to \mathbb{R}^{q'}$ which extends $j' \circ f$ and satisfies $h(\mathrm{dopp}\, X) \subset \mathrm{tub}_{\tau, \rho'}$. Finally, the composite map

$$\mathrm{dopp}\, X \xrightarrow{\mathrm{ab}\, h} \mathrm{tub}_{\tau, \rho'} \xrightarrow{\mathrm{pr}_{\tau'}} X'$$

is the desired C^r-extension of f to $\mathrm{dopp}\, X$.

5. <u>Let X and X' be compact $C^{\geqslant r}$-manifolds with X' closed. If $0 \leqslant s < r \leqslant \infty$, then:</u>

(i) $C^r(X,X')$ <u>is dense in</u> $C^s(X,X')$;

(ii) <u>given an arbitrary C^r-map $\phi: \partial X \to X'$, that part of</u> $C^s(X,X')$ <u>consisting of the C^r-extensions of ϕ is dense in the part of $C^s(X,X')$ consisting of the C^s-extensions of ϕ.</u>

PROOF. The mapping

$$C^s(\mathrm{in}: X \to \mathrm{dopp}\, X, \mathrm{id}\, X'): C^s(\mathrm{dopp}\, X, X') \to C^s(X,X')$$

transforms C^r-maps into C^r-maps and, according to Lemma 4, its image is precisely $C^s(X,X')$. Hence, (i) is a consequence of the fact that $C^r(\mathrm{dopp}\, X, X')$ is dense in $C^s(\mathrm{dopp}\, X, X')$ (see 4.2), while (ii) follows from the fact that the set of all C^r-extensions of ϕ is dense in the part of $C^s(\mathrm{dopp}\, X, X')$ consisting of the C^s-extensions of ϕ (see 4.9).

6 (COROLLARY). <u>Let X and X' be closed $C^{\geqslant r}$-manifolds, $1 \leqslant r \leqslant \infty$. If two maps in $C^r(X,X')$ are homotopic, then they can be connected by a C^r-homotopy $X \times I \to X'$.</u>

7 (LEMMA). <u>Let X and X' be compact $C^{\geqslant r}$-manifolds,</u>

$1 \leqslant r \leqslant \infty$. <u>Then the mapping</u> $C_\partial^r(X,X') \to C^r(\partial X, \partial X')$, $f \mapsto ab\, f$, <u>is open.</u>

PROOF. We have already seen in 1.1 that this mapping is continuous. We presently show that it is open. Given a map $f \in C_\partial^r(X,X')$, it is enough to find a neighborhood U of the map $ab\, f : \partial X \to \partial X'$ and a continuous mapping $\Phi : U \to C_\partial^r(X,X')$ such that $\Phi(ab\, f) = f$ and $[ab(\Phi(g)) : \partial X \to \partial X'] = g$ for all $g \in U$ (see 1.1.4.4). Fix collarings $k : \partial X \times I \to X$ and $k' : \partial X' \times I \to X'$, a C^r-embedding $j' : \partial X' \to \mathbb{R}^{q'}$, a C^r-transveralization τ' of j', and a neat tube $\mathrm{Tub}_{\tau'}, \rho'$. Now construct a C^r-function $\alpha : I \to I$ such that $\alpha(t) = 1$ for $0 \leqslant t \leqslant 1/3$ and $\alpha(t) = 0$ for $2/3 \leqslant t \leqslant 1$, and choose $\varepsilon > 0$ with $f(k(\partial X \times [0,\varepsilon]) \subset k'(\partial X' \times I)$. Let f_1 and f_2 be the composite maps

$$\partial X \times [0,\varepsilon] \xrightarrow{\;ab\,k\;} k(\partial X \times [0,\varepsilon]) \xrightarrow{\;ab\,f\;} k'(\partial X' \times I) \xrightarrow{\;(ab\,k')^{-1}\;}$$

$$\longrightarrow \partial X' \times I \underset{pr_2}{\overset{pr_1}{\lessgtr}} \begin{matrix} \partial X', \\ I. \end{matrix}$$

We define U as the set of all $g \in C^r(\partial X, \partial X')$ such that

$$\max_{y \in \partial X} \mathrm{dist}(j' \circ f(y), j' \circ g(y)) < \mathrm{Dist}(j'(\partial X'), \mathbb{R}^{q'} \smallsetminus \mathrm{tub}_{\tau'}, \rho'),$$

and define Φ as

$$[\Phi(g)](x) = f(x), \qquad \text{if } x \in X \smallsetminus k(\partial X \times [0,\varepsilon]),$$

$$[\Phi(g)](k(y,t)) =$$

$$= k'(pr_{\tau'}, (j'(f_1(y,t)) + \alpha(t/\varepsilon)[j'(f(y)) - j'(g(y))]), f_2(y,t)),$$
$$\text{if } y \in \partial X, \; t \in [0,\varepsilon].$$

It is routine to check that U and Φ have the needed properties.

8 (LEMMA). <u>Let</u> X <u>and</u> X' <u>be compact</u> $C^{\geqslant r}$-<u>manifolds. If</u> $1 \leqslant r \leqslant \infty$, <u>then the set of all composite maps</u> $X \xrightarrow{f} X' \xrightarrow{in} \mathrm{dopp}\, X'$ <u>with</u> $f \in C_\partial^r(X,X')$ <u>is open in that part of</u> $C^r(X, \mathrm{dopp}\, X')$ <u>consisting of the extensions of all maps</u> $\partial X \to \partial X'$

PROOF. Let $f_0 \in C_\partial^r(X,X')$ and denote by g_0 the composite map $X \xrightarrow{f_0} X' \xrightarrow{in} \mathrm{dopp}\, X'$. Fix a collaring of X, $k : \partial X \times I \to X$ and a two-sided collaring of $\partial X'$ in $\mathrm{dopp}\, X$, $\ell' : \partial X' \times [-1,1] \to \mathrm{dopp}\, X'$. Further, take positive δ, η such that: (i)

(i) $g_0(k(\partial X \times [0,\delta]) \subset \ell'(\partial X' \times (-1,1))$;

(ii) for any $y \in \partial X$ the derivative of the function

$$t \mapsto [pr_2 \colon \partial X \times [-1,1] \to [-1,1]](\ell'^{-1}(g_0 \circ k(y,t)), \qquad (2)$$

$[0,\delta] \to [-1,1]$, is everywhere less than $-\eta$. [The function $\partial X \times [0,1] \to \mathbb{R}$, which carries each point (y,t) to the derivative of the function (2) at the point t, is continuous on $\partial X \times [0,1]$ and negative on $\partial X \times 0$. Since ∂X is compact, this function is bounded above by a negative constant on $\partial X \times 0$, and hence on $\partial X \times [0,\delta]$, for some positive δ. This ensures the existence of δ and η as above.] Clearly, the C^r-maps $g \colon X \to \mathrm{dopp}\, X'$ which fulfil conditions (i) and (ii) (writing g instead of g_0) and satisfy $g(X \smallsetminus k(\partial X \times [0,\delta])) \subset \mathrm{int}\, X'$ form an open set in $C^r(X, \mathrm{dopp}\, X')$. It remains to observe that the intersection of this set with that part of $C^r(X, \mathrm{dopp}\, X')$ consisting of the extensions of all maps $\partial X \to \partial X'$ is a neighborhood of g_0 in the set of all composite maps $X \xrightarrow{\ f\ } X' \xrightarrow{\ \mathrm{in}\ } \mathrm{dopp}\, X'$ with $f \in C^r_\partial(X,X')$.

9. <u>Let</u> X <u>and</u> X' <u>be compact</u> $C^{\geqslant r}$<u>-manifolds. If</u> $1 \leqslant s < r \leqslant \infty$, <u>then</u>:

(i) $C^r_\partial(X,X')$ <u>is dense in</u> $C^s_\partial(X,X')$;

(ii) <u>that part of</u> $C^s_\partial(X,X')$ <u>consisting of all</u> C^r-<u>extensions</u> <u>of a given map</u> $\phi \in C^r(\partial X, \partial X')$ <u>is dense in the part of</u> $C^s_\partial(X,X')$ <u>consisting of all</u> C^s-<u>extensions of</u> ϕ.

PROOF. Lemma 7 and Theorem 4.2 (the latter applied to the manifolds ∂X and $\partial X'$) show that the set of all maps $g \in C^s_\partial(X,X')$ such that $[\mathrm{ab}\, g \colon \partial X \to \partial X'] \in C^r(\partial X, \partial X')$ is dense in $C^s_\partial(X,X')$. Therefore, (i) is a consequence of (ii).

To prove (ii), suppose that $f \in C^s_\partial(X,X')$ is an extension of ϕ, and let U be a neighborhood of f in $C^s_\partial(X,X')$. We have to show that U contains a C^r-extension of ϕ. Consider the composite map $X \xrightarrow{\ f\ } X' \xrightarrow{\ \mathrm{in}\ } \mathrm{dopp}\, X$. By virtue of 8, this map has a neighborhood V in $C^s(X, \mathrm{dopp}\, X')$ such that for any $g \in V$ with $g(\partial X) \subset \partial X'$ one has $g(X) \subset X'$ and $[\mathrm{ab}\, g \colon \partial X \to \partial X'] \in U$. Thus, it suffices to find a C^r-extension of ϕ in V; but such an extension is provided by 5.

10. Comparing Theorems 5 and 9 with Theorems 1.2 - 1.6, we arrive at the following statements for $1 \leqslant s < r \leqslant \infty$. Given any compact $C^{\geqslant r}$-manifold X and any closed $C^{\geqslant r}$-manifold X', $\mathrm{Imm}^r(X,X')$ is dense in $\mathrm{Imm}^s(X,X')$, $\mathrm{Subm}^r(X,X')$ is dense in $\mathrm{Subm}^s(X,X')$, and

$\mathrm{Emb}^r(X,X')$ is dense in $\mathrm{Emb}^s(X,X')$. For any compact $C^{\geqslant r}$-manifolds X and X', the set of neat embeddings in $C^r(X,X')$ is dense in the set of neat embeddings in $C^s(X,X')$, and $\mathrm{Diff}^r(X,X')$ is dense in $\mathrm{Diff}^s(X,X')$.

11 (COROLLARY). **Two compact $C^{\geqslant r}$-manifolds, $1 \leqslant r \leqslant \infty$, which are diffeomorphic are C^r-diffeomorphic.**

12 (INFORMATION). As with Theorem 4.2, Theorems 5 and 9 remain valid for $r = a$ too (cf. 2.4). We excluded this case in view of its difficulty.

Theorems 4.2, 5, and 9 (as well as their C^a-variants) also hold for noncompact X and X'. However, this generalization is of limited interest. For example, it does not suffice if one wants to eliminate the compactness assumption in Corollary 11 (which is actually possible). The appropriate extensions of Theorems 4.2, 5, and 9 to the noncompact case are related to topologies which are stronger than those defined in 1.1, and require analytic tools stronger than Theorem 3.1.4.

7. General Position

1. The main result of this subsection is the final Theorem 7, which constitutes the basis of a large part of the material below. We emphasize that this theorem is formulated and proved only in the C^∞-case. In Subsection 9 we add a statement covering the case of C^r-maps with r finite (see 9.10).

The technical part of the subsection is concentrated in Theorem 2, which establishes the fundamental topological property of the spaces $C^r(X,X')$, and Theorem 4, which represents the only corollary of Theorem 3.2.3 that we need.

To simplify the formulations of Theorems 2 and 3, we give a special name to those topological spaces where the intersection of any countable collection of dense open sets is dense: we call them _Baire spaces_.

Technicalities

2. $C^r(X,X')$ **is a Baire space for any $C^{\geqslant r}$-manifolds X and X' with $0 \leqslant r \leqslant \infty$.**

We have to show that given arbitrary open dense subsets

220

$u_1, u_2, \ldots$ of $C^r(X,X')$, and an arbitrary open subset W of $C^r(X,X')$, the intersection $W \cap (\cap_{i=1}^\infty u_i)$ is not empty. Let $\{\varphi_i\}_{i=1}^\infty$ and $\{\psi_i\}_{i=1}^\infty$ be atlases of the manifold X, indexed so that the set $K_i = \mathrm{Cl}\,\mathrm{supp}\,\psi_i$ is compact and contained in $\mathrm{supp}\,\varphi_i$ and $\psi_i = \mathrm{ab}\,\varphi_i$ for all i. Similarly, let $\{\varphi_j'\}_{j=1}^\infty$ and $\{\psi_j'\}_{j=1}^\infty$ be atlases of X', indexed so that $K_j' = \mathrm{Cl}\,\mathrm{supp}\,\psi_j' \subset \mathrm{supp}\,\varphi_j'$ and $\psi_j' = \mathrm{ab}\,\varphi_j'$ for all j. We construct a sequence of C^r-maps $f_1: X \to X', f_2: X \to X', \ldots,$ a sequence $V_1, V_2, \ldots$ of open subsets of $C^r(X,X')$, and a sequence of positive integers $n(1), n(2), \ldots,$ such that:

     (i) $f_i \in V_i$;

     (ii) if $i \geq 2$, then $V_i \subset V_{i-1}$;

     (iii) $\mathrm{Cl}\,V_i \subset W \cap u_i$;

     (iv) $f_i(K_i) \subset \cup_{j=1}^{n(i)} \mathrm{supp}\,\psi_j'$;

     (v) if $s \leq i$ and $t \leq n(s)$, then $f_i(K_s \cap f_s^{-1}(K_t')) \subset \mathrm{supp}\,\varphi_t'$;

     (vi) if $i \geq 2$, $s \leq i$, and $t \leq n(s)$, then the partial derivatives of order $\leq \min(r, i-2)$ of the coordinate functions of the local representatives $\mathrm{loc}(\varphi_s, \varphi_t')f_i$ and $\mathrm{loc}(\varphi_s, \varphi_t')f_{i-1}$ differ by less than 2^{-i} at the points of $\varphi_s(K_s \cap f_s^{-1}(K_t'))$.

     Then we will have finished the proof. Indeed, (v) and (vi) imply that for any s, t such that $t \leq n(s)$ the sequence

$$\{\mathrm{loc}(\varphi_s, \varphi_t')f_i \,\big|_{\varphi_s(\mathrm{supp}\,\psi_s \cap f_s^{-1}(\mathrm{supp}\,\psi_t'))}\}_{i=s}^\infty \;,$$

together with all its partial derivatives of order $\leq r$ converges uniformly on $\varphi_s(\mathrm{supp}\,\psi_s \cap f_s^{-1}(\mathrm{supp}\,\psi_t'))$ to a C^r-map

$$g_{st}: \varphi_s(\mathrm{supp}\,\psi_s \cap f_s^{-1}(\mathrm{supp}\,\psi_t')) \to \mathrm{Im}\,\psi_t' \;.$$

Moreover, the composite maps

$$\mathrm{supp}\,\psi_s \cap f_s^{-1}(\mathrm{supp}\,\psi_t') \xrightarrow{\;\mathrm{ab}\,\varphi_s\;} \varphi_s(\mathrm{supp}\,\psi_s \cap f_s^{-1}(\mathrm{supp}\,\psi_t')) \longrightarrow$$

$$\xrightarrow{\;g_{st}\;} \mathrm{Im}\,\varphi_t' \xrightarrow{\;(\varphi_t')^{-1}\;} \mathrm{supp}\,\varphi_t'$$

together define a C^r-map $g: X \to X'$ [(iv) shows that the sets $\mathrm{supp}\,\psi_s \cap f_s^{-1}(\mathrm{supp}\,\psi_t')$ cover X and clearly the maps g_{st} are compatible on the intersections of these sets]. g is the limit (in $C^r(X,X')$ of the sequence $f_1, f_2, \ldots,$ and (i)-(iii) show that

$g \in W \cap (\cap_{i=1}^{\infty} U_i)$.

We proceed by induction. Take f_1 to be any element of $W \cap U_1$, and take V_1 to be any neighborhood of f_1 (in $C^r(X,X')$) such that $\text{Cl } V_1 \subset W \cap U_1$ (recall that $C^r(X,X')$ is a regular space – see 1.1). Assume that for some $k \geqslant 2$, the maps $f_i \in C^r(X,X')$, the open sets $V_i \subset C^r(X,X')$, and the positive integers $n(i)$ with $i < k$ have been defined and satisfy (i)-(vi). We let G denote the set of all C^r-maps $g: X \to X'$ such that for $s \leqslant k-1$ and $t \leqslant n(s)$

$$g(K_s \cap f_s^{-1}(K_t')) \subset \text{supp } \varphi_t' ,$$

and all partial derivative of order $\leqslant \min(r,k-2)$ of the coordinate functions of the local representatives $\text{loc}(\varphi_s,\varphi_t')g$ and $\text{loc}(\varphi_s,\varphi_t')f_{k-1}$ differ by less than 2^{-k} at the points of $\varphi_s(K_s \cap f_s^{-1}(K_t'))$. Obviously, G is open and $f_{k-1} \in G$. Therefore, $G \cap V_{k-1} \neq \emptyset$ and, since U_k is dense, $G \cap (V_{k-1} \cap U_k) \neq \emptyset$. Choose V_k to be any nonempty open set with $\text{Cl } V_k \subset G \cap V_{k-1} \cap U_k$, f_k to be any element of V_k, and $n(k)$ to be any positive integer such that $f_k(K_k) \subset \cup_{t=1}^{n(k)} \text{supp } \psi_t'$. It is readily seen that the objects V_k, f_k, and $n(k)$ satisfy conditions (i)-(vi) for $i = k$.

3. <u>Every topological manifold is a Baire space.</u>

This is a special case of Theorem 2: in fact, the topological manifold X may be regarded as the space $C^0(D^0,X)$.

4. <u>Let X and X' be manifolds of class C^∞ or C^a. If $f: X \to X'$ is a C^∞-map and F is the set of all points $x \in X$ such that $d_x f(\text{Tang}_x X) \neq \text{Tang}_{f(x)} X'$, then $f(F)$ is the union of a countable family of nowhere dense sets.</u>

PROOF. Let Φ and Φ' be arbitrary countable atlases of the manifolds X and X'. For each pair $(\varphi,\varphi') \in \Phi \times \Phi'$, write $\text{supp } \varphi \cap f^{-1}(\text{supp } \varphi')$ as the union of a sequence of compact sets $K_1(\varphi,\varphi'), K_2(\varphi,\varphi'),\ldots$, and let $C_i(\varphi,\varphi')$ be the set of all points $x \in K_i(\varphi,\varphi')$ such that the rank of the Jacobi matrix of $\text{loc}(\varphi,\varphi')f$ at $\varphi(x)$ is less than $\dim X'$. Theorem 3.2.3 shows that the sets $f(C_i(\varphi,\varphi'))$ are all nowhere dense, and obviously $f(F) = \cup_i f(C_i(\varphi,\varphi'))$.

The Basic Theorem

5. Let X_1, X_2, and X' be smooth manifolds, and let A_1 and A_2 be subsets of X_1 and X_2. Two smooth maps $f_1: X_1 \to X'$ and $f_2: X_2 \to X'$ are said to be <u>transverse (one to the other)</u> <u>on</u> A_1, A_2 if for any $x_1 \in A_1$, $x_2 \in A_2$ with $f_1(x_1) = f_2(x_2)$, the vector space $\mathrm{Tang}_{f(x_1)} X'$ is spanned by its subspaces $d_{x_1} f_1(\mathrm{Tang}_{x_1} X_1)$ and $d_{x_2} f_2(\mathrm{Tang}_{x_2} X_2)$, and the following holds: if $x_1 \in \partial X_1$, then $\mathrm{Tang}_{f_1(x_1)} X'$ is already spanned by its subspaces $d_{x_1} f_1(\mathrm{Tang}_{x_1} \partial X_1)$ and $d_{x_2} f_2(\mathrm{Tang}_{x_2} X_2)$; if $x_2 \in \partial X_2$, then $\mathrm{Tang}_{f_1(x_1)} X'$ is already spanned by $d_{x_1} f_1(\mathrm{Tang}_{x_1} X_1)$ and $d_{x_2} f_2(\mathrm{Tang}_{x_2} \partial X_2)$; and if $x_1 \in \partial X_1$ and $x_2 \in \partial X_2$, then $\mathrm{Tang}_{f_1(x_1)} X'$ is already spanned by $d_{x_1} f_1(\mathrm{Tang}_{x_1} \partial X_1)$ and $d_{x_2} f_2(\mathrm{Tang}_{x_2} \partial X_2)$. Two maps $f_1: X_1 \to X'$ and $f_2: X_2 \to X'$ which are transverse on X_1, X_2 are simply referred to as transverse.

Let us make three obvious remarks. First, if $\dim X_1 + \dim X_2 < \dim X'$, then the fact that f_1 and f_2 are transverse on A_1, A_2 implies that $f_1(A_1) \cap f_2(A_2) = \emptyset$. Secondly, if f_1 and f_2 are transverse on A_1, A_2, then they are transverse on some neighborhoods of A_1 and A_2. And thirdly, if X_1, X_2, and X' are of class C^r with $1 \leqslant r \leqslant \infty$, and A_1, A_2 are compact, then given $f_2 \in C^r(X_2, X')$, the set of all $f_1 \in C^r(X_1, X')$ such that f_1 and f_2 are transverse on A_1, A_2 is open in $C^r(X_1, X')$.

6 (LEMMA). <u>Suppose</u> X_1 <u>and</u> X_2 <u>are</u> C^∞- <u>or</u> C^a-<u>manifolds</u> <u>and</u> $f_1: X_1 \to \mathbb{R}^{q'}$ <u>and</u> $f_2: X_2 \to \mathbb{R}^{q'}$ <u>are</u> C^∞-<u>maps.</u> <u>Then there is a</u> <u>dense set</u> V <u>in</u> $\mathbb{R}^{q'}$ <u>such that for each vector</u> $v \in V$, <u>the map</u> $X_1 \to \mathbb{R}^{q'}$ <u>defined by</u> $x_1 \to f_1(x_1) + v$ <u>is transverse to</u> f_2.

PROOF. We may take V to be the set of all $v \in \mathbb{R}^{q'}$ which satisfy the following condition. Consider the four maps $\mathrm{int}\, X_1 \times \mathrm{int}\, X_2 \to \mathbb{R}^{q'}$, $\mathrm{int}\, X_1 \times \partial X_2 \to \mathbb{R}^{q'}$, $\partial X_1 \times \mathrm{int}\, X_2 \to \mathbb{R}^{q'}$, and $\partial X_1 \times \partial X_2 \to \mathbb{R}^{q'}$, given by $(x_1, x_2) \mapsto f_2(x_2) - f_1(x_1)$. Then given any of these maps, v is not the image of a point where the differential of that map has rank less than q'. The map $x_1 \mapsto f_1(x_1) + v$ with $v \in V$ is obviously transverse to f_2, and theorems 4 and 3 show that V is dense in $\mathbb{R}^{q'}$.

7. <u>Let</u> X_1, X_2, <u>and</u> X' <u>be</u> C^∞- <u>or</u> C^a-<u>manifolds with</u> $\partial X' = \emptyset$. <u>If</u> $f_2: X_2 \to X'$ <u>is a</u> C^∞-<u>map, then the subset of</u> $C^\infty(X_1, X')$

<u>consisting of all maps transverse to f_2 is the intersection of a</u>
<u>countable collection of dense open sets.</u>

PROOF. For two sets $A_1 \subset X_1$ and $A_2 \subset X_2$, we let $F(A_1,A_2)$ denote the set of all C^∞-maps $f_1: X_1 \to X'$ such that f_1 and f_2 are transverse on A_1,A_2. If A_1 and A_2 are compact, then $F(A_1,A_2)$ is obviously open in $C^\infty(X_1,X')$, and we shall presently show that if A_1 and A_2 are compact, then $F(A_1,A_2)$ is dense in $C^\infty(X_1,X')$. These two facts are enough: express X_1 as the union of a sequence of compact sets $K_{11},K_{12},\ldots,$ express X_2 as the union of a sequence of compact sets $K_{21},K_{22},\ldots,$ and then observe that the part in which we are interested (i.e, $F(X_1,X_2)$) can be written as $\cap_{i,j} F(K_{1i},K_{2j})$.

Thus, suppose that A_1 and A_2 are compact, and let U be a neighborhood (in $C^\infty(X_1,X')$) of an arbitrarily given map $g_1: X_1 \to X'$. We have to produce a map contained in both $F(A_1,A_2)$ and U. We do this in two steps: first, we compress U to a a neighborhood V of g_1 having a more special form, and then construct a map belonging to $F(A_1,A_2)$ and to V.

Let $\dim X_1 = q_1$, $\dim X_2 = q_2$, and $\dim X' = q'$. To begin building the neighborhood V, fix for each point $x' \in f_2(A_2)$ a chart $\varphi'_{x'} \in \mathrm{Atl}_{x'} X'$ with $\mathrm{Im}\,\varphi'_{x'} = \mathbb{R}^{q'}$ and $\varphi'_{x'}(x') = 0$. Further, fix for each point $x_2 \in A_2$ a chart $\varphi_{2x_2} \in \mathrm{Atl}_{x_2} X_2$ such that $\mathrm{Im}\,\varphi_{2x_2} = \mathbb{R}^{q_2}$ or $\mathbb{R}^{q_2}_-$, $\varphi_{2x_2}(x_2) = 0$, $\mathrm{Cl\,supp}\,\varphi_{2x_2}$ is compact, and $f_2(\mathrm{Cl\,supp}\,\varphi_{2x_2}) \subset$ $\subset [\varphi'_{f_2(x_2)}]^{-1}(\mathrm{Int}\,D^{q'})$. Now cover A_2 by a finite number of sets $\varphi^{-1}_{2x_2}(\mathrm{Int}\,D^{q_2})$, say $\varphi^{-1}_{2x_{21}}(\mathrm{Int}\,D^{q_2}),\ldots,\varphi^{-1}_{2x_{21}}(\mathrm{Int}\,D^{q_2})$, and denote the chart $\varphi'_{f_2(x_{2j})}$ simply by ψ'_j, and the chart $\varphi_{2x_{2j}}$ — by ψ_{2j}.

Continuing, for each point $x_1 \in A_1$ choose a chart $\varphi_{1x_1} \in \mathrm{Atl}_{x_1} X_1$ such that $\mathrm{Im}\,\varphi_{1x_1} = \mathbb{R}^{q_1}$ or $\mathbb{R}^{q_1}_-$, $\varphi_{1x_1}(x_1) = 0$, $\mathrm{Cl\,supp}\,\varphi_{1x_1}$ is compact, and $g_1(\mathrm{Cl\,supp}\,\varphi_{1x_1})$ is contained in one of the sets $X' \smallsetminus f_2(\mathrm{Cl\,supp}\,\psi_{2j})$, $[\psi'_j]^{-1}(\mathrm{Int}\,D^{q'})$, for any given $j = 1,\ldots,1$. Finally, cover A_1 by a finite number of the sets $\varphi^{-1}_{1x_1}(\mathrm{Int}\,D^{q_1})$, say $\varphi^{-1}_{1x_{11}}(\mathrm{Int}\,D^{q_1}),\ldots,\varphi^{-1}_{1x_{1k}}(\mathrm{Int}\,D^{q_1})$, and denote the chart $\varphi_{1x_{1i}}$ by ψ_{1i}. At last, we may define V as the subset of U consisting of all maps $h_1: X_1 \to X'$ such that: $h_1(\mathrm{Cl}\ \mathrm{supp}\ \psi_{1i}) \subset X' \smallsetminus f_2(\mathrm{Cl}\ \mathrm{supp}\ \psi_{2j})$ whenever $g_1(\mathrm{Cl}\ \mathrm{supp}\ \psi_{1i}) \subset X' \smallsetminus f_2(\mathrm{Cl}\ \mathrm{supp}\ \psi_{2j})$, and $h_1(\mathrm{Cl}\ \mathrm{supp}\ \psi_{1i}) \subset X' \smallsetminus (\psi'_j)^{-1}(\mathrm{Int}\ D^{q'})$ whenever $g_1(\mathrm{Cl}\ \mathrm{supp}\ \psi_{1i}) \subset X' \smallsetminus (\psi'_j)^{-1}(\mathrm{Int}\ D^{q'})$. That V is a neighborhood of g_1 in $C^\infty(X_1,X')$ is plain.

Turning now to the final part of the construction, arrange the pairs (i,j), $i = 1,\ldots,k$, $j = 1,\ldots,l$, in a sequence $(i_1,j_1),\ldots,(i_m,j_m)$, with $m = k\cdot l$. Next construct inductively maps $h_1^0,\ldots,h_1^m: X_1 \to X'$ with the following properties: (i) $h_1^s \in V$; (ii) if $r \leqslant s$, then the maps h_1^r and f_2 are transverse on $\psi_{1i_r}^{-1}(D^{q_1})$, $\psi_{2j_r}^{-1}(D^{q_2})$. Then h_1^m will belong to $F(A_1,A_2) \cap V$.

Put $h_1^0 = f_1$ and assume that the maps h_1^s satisfying (i) and (ii) are already defined for $s < t \leqslant m$. If the maps h_1^{t-1} and f_2 are transverse on $\psi_{1i_t}^{-1}(D^{q_1})$, $\psi_{2j_t}^{-1}(D^{q_2})$, put $h_1^t = h_1^{t-1}$. Otherwise, $h_1^{t-1}(\psi_{1i_t}^{-1}(D^{q_1})) \subset (\psi'_{j_t})^{-1}(D^{q'})$, and Lemma 6 guarantees the existence of a set V dense in $\mathbb{R}^{q'}$ such that the composite map

$$\mathrm{Im}\ \psi_{1i_t} \xrightarrow{\ \mathrm{loc}(\psi_{1i_t},\psi'_{j_t})h_1^{t-1}\ } \mathbb{R}^{q'} \xrightarrow{\ x \to x+v\ } \mathbb{R}^{q'}$$

is transverse to $\mathrm{loc}(\psi_{2j_t},\psi'_{j_t})f_2: \mathrm{Im}\ \psi_{2j_t} \to \mathbb{R}^{q'}$, for each $v \in V$. Pick a C^∞-function $\alpha: \mathbb{R}^{q'} \to \mathbb{R}$, equal to 1 on $D^{q'}$ and equal to 0 outside the concentric ball $D^{q'}$ of radius 2. Now for $v \in V$ define $g_v: X_1 \to X'$ by

$$g_v(x_1) = \begin{cases} h_1^{t-1}(x_1), & \text{if } x_1 \in X_1 \smallsetminus (h_1^{t-1})^{-1}[(\psi'_{j_t})^{-1}(2D^{q'})], \\[2mm] (\psi'_{j_t})^{-1}(\psi'_{j_t}(h_1^{t-1}(x_1)) + \alpha(\psi'_{j_t}(h_1^{t-1}(x_1)))v), \\[2mm] \qquad\qquad \text{if } x_1 \in (h_1^{t-1})^{-1}(\mathrm{supp}\ \psi'_{j_t}). \end{cases}$$

We easily see that $g_0 = h_1^{t-1}$, that g_v is C^∞, and that the map $\mathbb{R}^{q'} \mapsto C^\infty(X_1,X')$, $v \to g_v$, is continuous. Consequently, there is an open set $U \subset \mathbb{R}^{q'}$, such that $0 \in U$ and for $v \in U$ the map $g_v \in V$, and g_v and f_2 are transverse on $\psi_{1i_r}^{-1}(D^{q_1})$, $\psi_{2j_t}^{-1}(D^{q_2})$ for $r < t$. On the other hand, the definition of V shows that for $v \in V$, g_v

and f_2 are transverse on $\psi_{1i_t}^{-1}(D^{q_1})$, $\psi_{2j_t}^{-1}(D^{q_2})$, and hence one may take $h_1^t = g_v$ for any $v \in U \cap V$.

8. Maps Transverse to a Submanifold

1. Theorem 7.7 is used mainly when X_2 is a submanifold of X' and f_2 is the corresponding inclusion. In such a situation we use a simpler terminology; namely, instead of saying that the map $f_1: X_1 \to X'$ is transverse to the inclusion $\mathrm{in}: X_2 \to X'$, we say that f_1 _is transverse to_ X_2. Then Theorem 7.7 states that if X_1 and X' are of class C^∞ or C^a and $\partial X' = \emptyset$, then the set of all maps in $C^\infty(X_1, X')$ which are transverse to X_2 is dense in $C^\infty(X_1, X')$.

The more special case, when both X_1 and X_2 are submanifolds of a manifold X', and f_1 and f_2 are the corresponding inclusions, deserves particular attention. If the maps $\mathrm{in}: X_1 \to X'$ and $\mathrm{in}: X_2 \to X'$ are transverse, we say that the submanifolds X_1 and X_2 themselves are transverse. Comparing Theorems 7.7 and 1.4, we see that given arbitrary submanifolds X_1 and X_2 of a closed, C^∞- or C^a-manifold X', every neighborhood of the inclusion $\mathrm{in}: X_1 \to X'$ in $C^\infty(X_1, X')$ contains embeddings transverse to X_2. The last statement is frequently formulated in a more geometric and less formal fashion: two submanifolds can be made transverse through an arbitrarily small displacement of either of them.

Of course, the $C^{<\infty}$-complement to Theorem 7.7 that was mentioned in 7.1 (Theorem 9.10) applies to these special cases too. However, note that in order to bring two submanifolds into general position in this way one must displace both of them, and not only one.

2. _Suppose_ X_1 _and_ X' _are_ $C^{\geq r}$_-manifolds_ $(1 \leq r \leq a)$ _with_ $\dim X_1 = q_1$ _and_ $\dim X' = q'$, _and_ X_2 _is a submanifold of_ X' _with_ $\dim X_2 = q_2$. _Let_ $f_1: X_1 \to X'$ _be a_ C^r_-map transverse to_ X_2 _and such that_ $f_1(\partial X_1) \cap \partial X_2 \subset \partial X'$. _Then_ $X_{12} = f_1^{-1}(X_2)$ _is a_ $(q_1 + q_2 - q')$_-dimensional_ C^r_-submanifold of_ X_1 _with_ $\partial X_{12} = f_1^{-1}(\partial X_2)$. _Moreover, this submanifold is neat whenever_ X_2 _is neat. Finally, the linear map_

$$\text{fact } d_{x_1} f_1 : \mathrm{Tang}_{x_1} X_1 / \mathrm{Tang}_{x_1} X_{12} \to \mathrm{Tang}_{f_1(x_1)} X' / \mathrm{Tang}_{f_1(x_1)} X_2 \tag{3}$$

is an isomorphism for any point $x_1 \in X_1$.

(It is understood that a submanifold of negative dimension is void.)

PROOF. Note that the second assertion is a straightforward supplement to the first, while the third assertion becomes evident if one observes that both quotient spaces appearing in (3) have the same dimension. Therefore, we have to prove only the first assertion, and in order to do this, we follow the scheme in 1.2.12.

To begin with, consider, given $x_1 \in X_{12}$, all possible positions of the point $f_1(x_1)$. The case $f_1(x_1) \in \text{int}\, X_2 \cap \partial X'$ is excluded, because $\text{int}\, X_2 \cap \partial X' = \emptyset$ (X_2 is a submanifold of X'). If $x_1 \in \text{int}\, X_1$, the case $f_1(x_1) \in \partial X_2 \cap \partial X'$ is excluded since it would contradict the assumption that f_1 is transverse to X_2 (if $x_1 \in \text{int}\, X_1$ and $f_1(x_1) \in \partial X_2 \cap \partial X'$, then the spaces $\text{Im}\, d_{x_1} f_1$ and $\text{Tang}_{f_1(x_1)} \partial X_2$ are both contained in $\text{Tang}_{f_1(x_1)} \partial X'$, and so they cannot span $\text{Tang}_{f_1(x_1)} X'$)). When $x_1 \in \partial X_1$, the relation $f_1(x_1) \in \partial X_2 \cap \text{int}\, X'$ is impossible since $f_1(\partial X_1) \cap \partial X_2 \subset \partial X'$. Therefore, we are left with four possibilities: (i) $x_1 \in \text{int}\, X_1$, $f_1(x_1) \in \text{int}\, X_2 \cap \text{int}\, X'$; (ii) $x_1 \in \text{int}\, X_1$, $f_1(x_1) \in \partial X_2 \cap \text{int}\, X'$; (iii) $x_1 \in \partial X_1$, $f_1(x_1) \in \text{int}\, X_2 \cap \text{int}\, X'$; and (iv) $x_1 \in \partial X_1$, $f_1(x_1) \in \partial X_2 \cap \partial X'$.

Fix a chart $\varphi' \in \text{Atl}_{f_1(x_1)}^r C^r X'$ which transforms the triple $(\text{supp}\, \varphi', X_2 \cap \text{supp}\, \varphi', f_1(x_1))$ into one of the triples $(\mathbb{R}^{q'}, \mathbb{R}^{q_2}, 0)$, $(\mathbb{R}^{q'}, \mathbb{R}^{q_2}_-, 0)$, or $(\mathbb{R}^{q'}_-, \mathbb{R}^{q_2}_-, 0)$, with corresponding local coordinates $\varphi'_1, \ldots, \varphi'_{q'}$. Let $\psi_1, \ldots, \psi_{q'}\colon f_1^{-1}(\text{supp}\, \varphi') \to \mathbb{R}$ be the functions $\psi_i(x) = \varphi'_i(f_1(x))$. In cases (i), (iii), and (iv) above, the intersection $X_{12} \cap f_1^{-1}(\text{supp}\, \varphi')$ is defined by the equations $\psi_{q_2+1}(x) = 0, \ldots,$ $\psi_{q'}(x) = 0$, and in the case (ii) - by the same equations and the inequality $\psi_1(x) < 0$. Let us verify that for all cases (i)-(iv), the above equations and inequality satisfy the independence conditions displayed in 1.2.12.

In case (i), we have to show that $\psi_{q_2+1}, \ldots, \psi_{q'}$ are independent at the point x_1. This follows from the equality $\dim \cap_{j=q_2+1}^{q'} \text{Ker}(d_{x_1} \psi_j) = q'-q_2$, which in turn follows from the trivial inclusion $d_{x_1} f_1(\cap_{j=q_2+1}^{q'} \text{Ker}(d_{x_1} \psi_j)) \subset \text{Tang}_{f_1(x_1)} X_2$ and the equality $\text{Tang}_{f_1(x_1)} X' = \text{Im}\, d_{x_1} f_1 + \text{Tang}_{f_1(x_1)} X_2$ (which is part of the definition of transversality).

In case (ii), we have to show that $\psi_1, \psi_{q_2+1}, \ldots, \psi_{q'}$ are independent at x_1. The proof is a repeat of the previous one, except that one must replace the equality $\mathrm{Tang}_{f_1(x_1)} X' = \mathrm{Im}\, d_{x_1} f_1 + \mathrm{Tang}_{f_1(x_1)} X_2$

by $\mathrm{Tang}_{f_1(x_1)} X' = \mathrm{Im}\, d_{x_1} f_1 + \mathrm{Tang}_{f_1(x_1)} \partial X_2$.

Finally, in cases (iii) and (iv), we have to produce a C^r-function $\psi: f_1^{-1}(\mathrm{supp}\,\varphi') \to \mathbb{R}$ which is zero on $\partial X_1 \cap f_1^{-1}(\mathrm{supp}\,\varphi')$, negative on $\mathrm{Int}\, X_1 \cap f_1^{-1}(\mathrm{supp}\,\varphi')$, and is such that $\psi, \psi_{q_2+1}, \ldots, \psi_{q'}$ are independent at x_1. The existence of such a function is equivalent to the restrictions of $\psi_{q_2+1}, \ldots, \psi_{q'}$ to $\partial X_1 \cap f_1^{-1}(\mathrm{supp}\,\varphi')$ being independent at x_1. The latter can be proved as in (i), employing the equality $\mathrm{Tang}_{f_1(x_1)} X' = \mathrm{Im}\, d_{x_1}(f_1|\partial X_1) + \mathrm{Tang}_{f_1(x_1)} X_2$ rather than

$\mathrm{Tang}_{f_1(x_1)} X' = \mathrm{Im}\, d_{x_1} f_1 + \mathrm{Tang}_{f_1(x_1)} X_2$.

3 (COROLLARY). <u>Let</u> X_1 <u>and</u> X_2 <u>be transverse submanifolds of a smooth manifold</u> X', <u>and assume that</u> X_1 <u>is neat. Then</u> $X_1 \cap X_2$ <u>is a</u> $(\dim X_1 + \dim X_2 - \dim X')$-<u>dimensional submanifold of</u> X', <u>and is neat whenever</u> X_2 <u>is neat.</u>

The Simplest Applications

4. <u>Let</u> A <u>be a closed subset of a closed</u> C^r-<u>manifold</u> X, <u>and let</u> U <u>be a neighborhood of</u> A. <u>If</u> $1 \leqslant r \leqslant \infty$, <u>then there is in</u> U <u>a compact submanifold</u> B <u>of codimension</u> 0 <u>such that</u> $A \subset \mathrm{int}\, B$.

PROOF. Let $\phi: X \to I$ be a Urysohn C^r-function for the pair A, $X \smallsetminus U$ (see 4.7), and suppose that $c \in (0,1)$ is not a critical value of ϕ (see 1). Set $B = \phi^{-1}([0,c])$. Then B is the preimage of the submanifold $(-\infty, c]$ of $\mathbb{R}$ under the composite C^r-map $X \xrightarrow{\phi} I \xrightarrow{\mathrm{in}} \mathbb{R}$. Since the latter is transverse to $(-\infty, c]$, B is a submanifold of codimension 0. It is immediate from the construction that B is closed as a subset, that $A \subset \mathrm{int}\, B$, and that $B \subset U$.

5. <u>Every CNRS is homeomorphic to a retract of a closed, orientable,</u> C^∞-<u>manifold.</u>

PROOF. Let j be an embedding of the CNRS X in S^q, with q large enough, and let U be a neighborhood of $j(X)$ which retracts on $j(X)$. Theorem 4 provides a compact submanifold $B \subset U$

such that $B \supset j(X)$, and clearly $j(X)$ is a retract of the double of B.

9. Raising the Smoothness Class of a Manifold

1. The main results of this subsection are Theorems 6 and 8. Lemmas 2-5 are needed for the proof of Theorem 6, while Lemma 7 in conjunction with Theorem 6 yield Theorem 8.

2 (LEMMA). _Let_ X _and_ X' _be_ $C^{\geqslant r}$-_manifolds,_ $1 \leqslant r \leqslant \infty$, X _compact, and_ $\partial X' = \emptyset$. _Suppose_ A' _is a submanifold of_ X', f: X → X' _is a_ C^r-_map transverse to_ A' _such that_ $f(\partial X) \subset X' \smallsetminus A'$, _and_ $\rho: X \to f^{-1}(A')$ _is both a retraction and a_ C^r-_submersion. Then there is a neighborhood_ U _of the map_ f _in_ $C^r(X,X')$ _such that every_ $g \in U$ _satisfies:_

(i)   g _is transverse to_ A';

(ii) $g(\partial X) \subset X' \smallsetminus A'$;

(iii) $\rho\big|_{g^{-1}(A')}: g^{-1}(A') \to f^{-1}(A')$ _is a submersion._

PROOF. We obtain U as the intersection of three open sets, U_1, U_2, and U_3. U_1 is the set of all maps in $C^r(X,X')$ which are transverse to A'. U_2 is the set of all $g \in C^r(X,X')$ with $g(\partial X) \subset X' \smallsetminus A'$. Finally, U_3 is the set of all $g \in C^r(X,X')$ such that the intersection of the subspaces $\mathrm{Ker}\, d_x\rho$ and $(d_x g)^{-1}(\mathrm{Tang}_{g(x)} A')$ of $\mathrm{Tang}_x X$ reduces to 0 for all points $x \in g^{-1}(A')$.

We already know that U_1 is open (see 7.5). The openness of U_2 is a consequence of the compactness of ∂X and the openness of $X' \smallsetminus A'$. To prove that U_3 is open, we shall describe it in a different way. Fix a C^r-embedding $j: X \to \mathbb{R}^q$ and let $C \subset \mathrm{Tang}\, X$ be the subset of all vectors u such that $d\rho(u) = 0$ and $<dj(u),dj(u)> = 1$. Then clearly U_3 is just the set of all $g \in C^r(X,X')$ such that $dg(C) \subset \mathrm{Tang}\, X' \smallsetminus \mathrm{Tang}\, A'$. The openness of this last set is a consequence of the following facts: C is compact, $\mathrm{Tang}\, X' \smallsetminus \mathrm{Tang}\, A'$ is open (in $\mathrm{Tang}\, X'$), and the mapping $C^r(X,X') \to C^{r-1}(\mathrm{Tang}\, X, \mathrm{Tang}\, X')$, which takes each $g \in C^r(X,X')$ into dg, is continuous (see 1.1).

Therefore, U is open, and we see at once that $f \in U$ and that any map $g \in U$ satisfies (i) and (ii). It is easily checked that (iii) also holds: indeed, $g \in U_1$ implies that $g^{-1}(A')$ is a neat submanifold of X (see 8.2); since $g \in U_2$, $g^{-1}(A') \subset \mathrm{int}\, X$, and as

such it is closed as an independent manifold. Finally, $g \in U_3$
implies that the differential $d_x(\rho\big|_{g^{-1}(A')})$ is nondegenerate at all
points $x \in g^{-1}(A')$.

3 (LEMMA). _Let_ X _and_ X' _be smooth closed manifolds of
equal dimensions, and let_ $f: X \to X'$ _be a submersion. If_ X' _is
connected and the preimage under_ f _of one of its points reduces to a
point, then_ f _is a diffeomorphism._

It suffices to show that f is invertible. Let $A' = f(X)$,
and denote by B' the set of all $x' \in X'$ such that $f^{-1}(x')$ consists
of more than one point. Being a submersion, the map f is open (see
1.5.8), so A' is an open set. Also, B' is open: if $f(x_1) = f(x_2) =$
$= x'$ and $x_1 \neq x_2$, then x_1 and x_2 have disjoint neighborhoods,
U_1 and U_2 , in X such that $f\big|_{U_1}$ and $f\big|_{U_2}$ are differentiable
embeddings, and $f(U_1) \cap f(U_2)$ is a neighborhood of x' contained in
B'. On the other hand, A' is closed because X is compact. Also,
B' is closed, because if $U_1,\ldots,U_s$ are open sets covering X and
such that the restrictions $f\big|_{U_i}$ are differentiable embeddings, then

$$B' = \bigcap_{i=1}^{s} f(X \smallsetminus U_i),$$

and the sets $f(X \smallsetminus U_i)$ are closed. Finally, since X' is connected,
$A' \neq \emptyset$, and $B' \neq X'$, we have $A' = X'$ and $B' = \emptyset$, i.e., f is
invertible.

4 (LEMMA). _Suppose_ X _is a compact_ C^r-_submanifold of_ $\mathbb{R}^q$,
$1 \leqslant r \leqslant \infty$, U _is a neighborhood of_ X _in_ $\mathbb{R}^q$, _and_ X' _is an open
subset of a closed_ C^a-_manifold admitting a_ C^a-_embedding in Euclidean
space. If_ $f \in C^r(U,X')$, _then every neighborhood in_ $C^r(X,X')$ _of
the restriction_ $f\big|_X$ _contains the restriction of some map belonging to_
$C^a(U,X')$.

PROOF. For a start, suppose that X' is itself a closed
C^a-manifold. Denote by $\mathcal{U}$ the given neighborhood of $f\big|_X$ in $C^r(X,X')$,
and fix a C^a-embedding $j': X' \to \mathbb{R}^{q'}$, a C^a-transversalization τ' of
j' , and a neat tube $\mathrm{Tub}_{\tau'}\rho'$. Since

$$C^r(\mathrm{id},\mathrm{pr}_{\tau'}): C^r(X,\mathrm{tub}_{\tau'}\rho') \to C^r(X,X')$$

is continuous, the preimage of $\mathcal{U}$ under this mapping is open; moreover,
it is not empty because it contains the composite map
$[\mathrm{ab}\, j': X' \to \mathrm{tub}_{\tau'}\rho'] \circ f$. Therefore, Theorem 3.1.4 yields a map
$g: U \to \mathrm{tub}_{\tau'}\rho'$ with polynomial components, such that $g\big|_X$ belongs to
the above preimage of $\mathcal{U}$. Clearly, the composition $\mathrm{pr}_{\tau'} \circ g$ is the

230

desired map belonging to $C^a(U,X')$.

One can reduce the general case to the above situation: if X' is an open subset of the closed C^a-manifold Y, then

$$C^r(\text{id},\text{in}): \ C^r(X,X') \to C^r(X,Y)$$

is a topological embedding with open image, and transforms $C^a(X,X')$ into the intersection of this image with $C^a(X,Y)$ (cf. 4.6).

5 (LEMMA). <u>For each compact, q-dimensional C^r-submanifold X of $\mathbb{R}^q$, $1 \leqslant r \leqslant a$, and each compact subset A of $\mathbb{R}^q$, the set of all C^r-embeddings $f: X \to \mathbb{R}^q$ with $f(\text{Int}\,X) \supset A$ is open in</u> $C^r(X, \mathbb{R}^q)$.

(In the present subsection we shall apply Lemma 5 only in the case where A is a point. However, we shall need it in full generality in the next subsection.)

PROOF. For a start, suppose that A is the ball in $\mathbb{R}^q$ with center c and radius ρ. Let $f: X \to \mathbb{R}^q$ be an embedding of class C^r such that $f(\text{Int}\,X) \supset A$, and denote by U the set of all C^r-embeddings $g: X \to \mathbb{R}^q$ satisfying for any $x \in X$ the inequality

$$\text{dist}(f(x),g(x)) < \min(\rho,\text{Dist}(f(\text{Fr}\,X),A)) .$$

Since U is open in $C^r(X,\mathbb{R}^q)$ (see 1.4) and $f \in U$, it is enough to show that $g(\text{Int}\,X) \supset A$ for any $g \in U$. But if $g \in U$, then $g(\text{Fr}\,X) \cap A = \emptyset$ and thus the two sets $g(X) \cap A$ and $g(\text{Int}\,X) \cap A$ are equal, while the first is closed in A and the second is open in A. Moreover, $g(\text{Int}\,X) \cap A \supset g(f^{-1}(c))$ (since $\text{dist}(g(f^{-1}(c)),c) < \rho$), so that $g(\text{Int}\,X) \cap A \neq \emptyset$. Consequently, $g(\text{Int}\,X) \cap A = A$, i.e., $g(\text{Int}\,X) \supset A$.

The more general situation where A is the union of a finite number of balls reduces to the case already considered. To prove the theorem in the most general case, it remains to observe that for any C^1-embedding $f: X \to \mathbb{R}^q$ with $f(\text{Int}\,X) \supset A$, there is a finite number of balls whose union contains A and is contained in $f(\text{Int}\,X)$.

6. <u>Every closed C^r-manifold X with $1 \leqslant r \leqslant \infty$ is C^r-diffeomorphic to a C^a-submanifold of Euclidean space.</u>

It is sufficient to consider a connected manifold X. Fix a C^r-embedding $j: X \to \mathbb{R}^q$, a C^r-transversalization τ of j, and a neat tube $\text{Tub}_\tau\rho$. Consider the map $f: \text{Tub}_\tau\rho \to G'(q,n{=}\dim X)$ which takes each point $y \in \text{Tub}_\tau\rho$ into the plane $y - j \circ \text{pr}_\tau(y) + (\tau \circ \text{pr}_\tau(y))^\perp$ (which passes through the point $y - j \circ \text{pr}_\tau(y)$ and is orthogonal to

$\tau \circ \mathrm{pr}_\tau(y))$. It is clear that f is transverse to $G(q,n)$ (even the maps $f|_{\mathrm{pr}_\tau^{-1}(x)}$, $x \in X$, are transverse to $G(q,n)$) and that $f^{-1}(G(q,n)) = j(X)$. Pick some point $x_0 \in X$ and let $V \subset G'(q,n)$ be the set of all planes transverse to $\tau(x_0)$ (i.e., intersecting $\tau(x_0)$ at only one point). Finally, let π be the map $V \to \tau(x_0)$ which takes each plane belonging to V into its intersection with $\tau(x_0)$. According to Lemma 2, $f|_{\mathrm{Tub}_\tau(\rho/2)}$ has in $C^r(\mathrm{Tub}_\tau(\rho/2), G'(q,n))$ a

neighborhood U such that if $g \in U$, then:

(i) g is transverse to $G(q,n)$;

(ii) $g(\partial\mathrm{Tub}_\tau(\rho/2)) \subset G'(q,n) \smallsetminus G(q,n)$;

(iii) ab $\mathrm{pr}_\tau : g^{-1}(G'(q,n)) \to X$ is a submersion.

Moreover, as Lemma 5 shows, the set of all C^r-embeddings $\phi: d_\tau(x_0,\rho/2) \to \tau(x_0)$ with $\phi(\mathrm{Int}\, d_\tau(x_0,\rho/2)) \ni 0$ is open in $C^r(d_\tau(x_0,\rho/2),\tau(x_0))$. Hence $f|_{\mathrm{Tub}_\tau(\rho/2)}$ has in $C^r(\mathrm{Tub}_\tau(\rho/2),G'(q,n))$

a neighborhood V such that for each $g \in V$ one has $g(d_\tau(x_0,\rho/2)) \subset V$ and $\pi \circ [\mathrm{ab}\, g : d_\tau(x_0,\rho/2) \to V]$ is a C^r-embedding whose image contains 0. Since $G'(q,n)$ is an open subset of a closed C^a-manifold (see 2.2.10), Lemma 4 guarantees the existence of a C^a-map $h: \mathrm{tub}_\tau\rho \to G'(q,n)$ whose restriction to $\mathrm{Tub}_\tau(\rho/2)$ belongs to $U \cap V$. Set $Y = [h|_{\mathrm{Tub}_\tau(\rho/2)}]^{-1}(G(q,n))$. By virtue of Theorem 8.2, Y is a

C^a-manifold, and we have only to verify that $\mathrm{pr}_\tau|_Y : Y \to X$ is a

diffeomorphism. But this claim follows from Lemma 3. Indeed, since $h|_{\mathrm{Tub}_\tau(\rho/2)} \in U \cap V$, the conditions of this lemma are satisfied: the fact that $h|_{\mathrm{Tub}_\tau(\rho/2)} \in U$ shows that $\mathrm{pr}_\tau|_Y$ is a submersion, while $h|_{\mathrm{Tub}_\tau(\rho/2)} \in V$ implies that the preimage $(\mathrm{pr}_\tau|_Y)^{-1}(x_0)$ reduces to a point.

7 (LEMMA). <u>Let</u> X <u>be a closed C^r-manifold with</u> $1 \leqslant r \leqslant \infty$, <u>and let</u> A <u>and</u> B <u>be compact submanifolds of</u> X <u>such that</u> $A \subset \mathrm{int}\, B$ <u>and</u> $\dim B = \dim X$. <u>If</u> B <u>is endowed with the C^∞-structure which is the restriction of its C^r-structure, then for any closed C^∞-manifold</u> X', <u>that part of the space</u> $C^r(X,X')$ <u>consisting of all extensions of maps from</u> $C^\infty(A,X')$ <u>is dense in</u> $C^r(X,X')$.

PROOF. Suppose $f \in C^r(X,X')$ and U is a neighborhood of f in $C^r(X,X')$. We have to produce in U a map extending a map from $C^\infty(A,X')$. Fix a C^∞-embedding $j': X' \to \mathbb{R}^{q'}$, a C^∞-transversalization

τ' of j', and a neat tube $\text{Tub}_{\tau',\rho'}$. Let $V \subset C^r(B,X')$ be the set of all maps g such that

$$\max_{x \in B} \; \text{dist}(j' \circ f(x), j'(g(x))) < \text{Dist}(j'(X'), \mathbb{R}^{q'} \smallsetminus \text{tub}_{\tau',\rho'}).$$

Clearly, V is open and the segment with endpoints $j' \circ f(x)$ and $j'(g(x))$ lies in $\text{tub}_{\tau',\rho'}$, for any $g \in V$ and $x \in B$. Now choose a neighborhood U of ∂B in B such that $\text{Cl}\, U \cap A = \emptyset$, construct a Urysohn C^∞-function $\phi: B \to I$ for the pair $\text{Cl}\, U, A$, and consider the mapping $\Phi: V \to C^r(X,X')$ which takes $g \in V$ into the map

$$x \mapsto \begin{cases} f(x), & \text{if } x \in X \smallsetminus B. \\[2ex] \text{pr}_{\tau'}((1-\phi(x))j' \circ f(x) + \phi(x)j'(g(x))), & \text{if } x \in B. \end{cases}$$

Obviously, Φ is continuous, and $\Phi(f\big|_B) = f$. These properties of Φ show that the set $\Phi^{-1}(U)$ is open and nonempty. Applying Theorem 6.5, we deduce that $\Phi^{-1}(U)$ contains a C^∞-map. Finally, note that Φ transforms C^∞-maps into extensions of maps from $C^\infty(A,X')$.

   8. <u>Every compact C^r-manifold with</u> $1 \leqslant r < \infty$ <u>is</u> C^r-<u>diffeomorphic to a</u> C^∞-<u>manifold. Moreover, if</u> X <u>is a compact</u> C^r-<u>manifold with</u> $1 \leqslant r < \infty$, Y <u>is a</u> C^∞-<u>manifold, and</u> $\psi: Y \to \partial X$ <u>is a</u> C^r-<u>diffeomorphism, then there exists a</u> C^∞-<u>manifold</u> X' <u>together</u> <u>with a</u> C^r-<u>diffeomorphism</u> $\phi: X \to X'$, <u>such that the composite map</u>

$$Y \xrightarrow{\ \psi\ } \partial X \xrightarrow{\ \text{ab}\ \phi\ } \partial X'$$

<u>is a</u> C^∞-<u>diffeomorphism.</u>

   PROOF. Composing the C^r-diffeomorphism $\psi \times \text{id}: Y \times [-1,1] \to \partial X \times [-1,1]$ with an arbitrary two-sided C^r-collaring $\partial X \times [-1,1] \to \text{dopp}\, X$, we obtain a C^r-embedding $Y \times [-1,1] \to \text{dopp}\, X$. Since $Y \times [-1,1]$ is a C^∞-manifold, the image B of this embedding inherits a C^∞-structure which is the restriction of the induced C^r-structure, and obviously $\partial X \subset \text{int}\, B$. Theorem 6 provides a closed C^a-manifold Z together with a C^r-diffeomorphism $\text{dopp}\, X \to Z$. In addition, Lemma 7 and Theorem 1.6 imply that this C^r-diffeomorphism may be taken of class C^∞ on ∂X. Now set X' to be the image of X under the diffeomorphism chosen in this way, and take ϕ to be the compression of this diffeomorphism to a diffeomorphism $X \to X'$. It is immediate that X' and ϕ have the desired properties.

9 (INFORMATION). Theorems 6 and 8 can be substantially strengthened: they are valid for noncompact manifolds too, and in Theorem 8 one may replace ∞ by a. However, one cannot eliminate the hypothesis that $r \geqslant 1$ from these theorems: there exist closed topological manifolds which are not homeomorphic to smooth manifolds. The first example of this kind appeared in [12].

Application: A Supplement to Theorem 7.7

10. <u>Let</u> X_1, X_2, <u>and</u> X' <u>be compact $C^{\geqslant r}$-manifolds with</u> $1 \leqslant r \leqslant \infty$. <u>If</u> X' <u>is closed, then the pairs</u> (f_1, f_2) <u>of transverse maps form a dense set in</u> $C^r(X_1, X') \times C^r(X_2, X')$.

PROOF. By Theorem 8, there exist C^∞-manifolds Y_1, Y_2, and Y', together with C^r-diffeomorphisms $Y_1 \to X_1$, $Y_2 \to X_2$, and $Y' \to X'$. Hence it is enough to prove that the pairs of transverse maps are dense in $C^r(Y_1, Y') \times C^r(Y_2, Y')$. By Theorem 7.7, these pairs are dense in $C^\infty(Y_1, Y') \times C^\infty(Y_2, Y')$, and according to Theorem 6.5 this product is dense in $C^r(Y_1, Y') \times C^r(Y_2, Y')$.

10. <u>Approximation of Maps by Embeddings and Immersions</u>

1. In this subsection we complete the program outlined in 4.1. The main results are Theorems 4 and 5. Lemma 3 is our basic tool, and this same lemma can be used to derive many other corollaries; see, in particular, 11.11, 11.12, and 11.13.

Auxiliary Manifolds

2. Suppose X is a closed n-dimensional $C^{\geqslant r}$-manifold with $1 \leqslant r \leqslant a$, and let $j: X \to \mathbb{R}^q$ be an embedding of class C^r. Also, let m be a positive integer such that $0 < m \leqslant q$. We shall need two constructions.

The first construction: we denote by Aux_1 or, more specifically by $\mathrm{Aux}_1(j;m)$, that subset of $\mathrm{Tang}\,X \times G(q,m)$ consisting of the pairs (u,γ) such that $\langle dj(u), dj(u) \rangle = 1$ and $dj(u) \in \gamma$. Further, let $\mathrm{aux}_1: \mathrm{Aux}_1 \to G'(q,m)$ be the map defined by $\mathrm{aux}_1(u,\gamma) = j(\mathrm{pr}(u)) + \gamma$, where $\mathrm{pr} = [\mathrm{pr}: \mathrm{Tang}\,X \to X]$. Aux_1 is a

[2n-1+(m-1)(q-m)]-dimensional submanifold of Tang X × G(q,m) [indeed, it is the preimage of the submanifold S^{q-1} × G(q,m) of $\mathbb{R}^q$ × G'(q,m) under the mapping Tang X × G(q,m) → $\mathbb{R}^q$ × G'(q,m) given by (u,γ) ↦ (dj(u),dj(u) + γ); this mapping is transverse to S^{q-1} × G(q,m)]. aux_1 is of class C^{r-1} and its image consists of those m-planes of $\mathbb{R}^q$ which contain lines tangent to j(X).

The second construction: we denote by Aux_2 or, more specifically by $Aux_2(j;m)$, the subset of X × X × G(q,m) consisting of the triples (x,x',γ) such that x ≠ x' and j(x') - j(x) ∈ γ. Further, let aux_2: Aux_2 → G'(q,m) be the map defined by $aux_2(x,x',γ)$ = j(x) + γ. Aux_2 is a [2n+(m-1)(q-m)]-dimensional submanifold of X × X × G(q,m) [indeed, it is the preimage of the submanifold G(q,m) of G'(q,m) under the mapping ((X × X) ∖ diag X) × G(q,m) → G'(q,m) given by (x,x',γ) → j(x') - j(x) + γ; this mapping is transverse to G(q,m)]. aux_2 is of class C^r and its image consists of those m-planes of $\mathbb{R}^q$ which intersect j(X) at more than one point.

The Basic Theorems

3 (LEMMA). _Let X be a closed n-dimensional $C^{\geq r}$-manifold, 1 ≤ r ≤ ∞, and let j: X → $\mathbb{R}^q$ be an embedding of class C^r together with a C^r-transversalization τ: X → G(q,q-n) and a neat tube $Tub_\tau \rho$. Then there exist a neighborhood U of the map $\tilde{\tau}$: X → G'(q,q-n) (see 3.2) in $C^r(X,G'(q,q-n))$ and a continuous mapping Φ: U → $Diff^r X$ such that, for each map g ∈ U:_

(i) [j ∘ Φ(g)](x) ∈ g(x) _for all_ x ∈ X;

(ii) _the map_ $τ_g$: X → G(q,q-n), $τ_g(x)$ = g(x) - [j ∘ Φ(g)](x) _is a transversalization of the embedding_ j ∘ Φ(g): X → $\mathbb{R}^q$;

(iii) _some neat tube of this transversalization contains_ $Tub_\tau(\rho/2)$.

PROOF. Given g ∈ $C^r(X,G'(q,q-n))$, let h_g: $Tub_\tau \rho$ → $\mathbb{R}^q$ denote the map which carries each point y ∈ $Tub_\tau \rho$ into its image under the orthogonal projection onto the plane g(pr_τ(y)). Obviously, g ↦ h_g is a continuous mapping $C^r(X,G'(q,q-n))$ → $C^r(Tub_\tau \rho,\mathbb{R}^q)$, and if g = $\tilde{τ}$, then h_g = [in: $Tub_\tau \rho$ → $\mathbb{R}^q$]. Consequently, $\tilde{τ}$ has a neighborhood V in $C^r(X,G'(q,q-n))$ such that, for any g ∈ V, h_g

is a C^r-embedding and $h_g(\text{tub}_\tau \rho) \supset \text{Tub}_\tau(\rho/2)$ (see 9.5). For $g \in V$, let i_g denote the composition

$$X \xrightarrow{\text{ab } j} \text{Tub}_\tau(\rho/2) \xrightarrow{(\text{ab } h_g)^{-1}} h_g^{-1}(\text{Tub}_\tau(\rho/2)) \xrightarrow{\text{in}} \text{Tub}_\tau\rho \xrightarrow{\text{pr}_\tau} X$$

An obvious verification shows that $g \mapsto i_g$ is a continuous mapping $V \to C^r(X,X)$, and that $i_{\tilde\tau} = \text{id}$. Therefore, $\tilde\tau$ has a neighborhood U in V such that i_g is a diffeomorphism for all $g \in U$. Set $\Phi(g) = i_g^{-1}$ for $g \in U$. It is immediate that Φ is continuous and that U and Φ satisfy the conditions (i)-(iii).

4. <u>Suppose</u> X <u>is a compact n-dimensional $C^{\geqslant r}$-manifold,</u> $1 \leqslant r \leqslant \infty$, <u>and</u> X' <u>is a closed n'-dimensional $C^{\geqslant r}$-manifold. Then for</u> $n' \geqslant 2n$, $\text{Imm}^r(X,X')$ <u>is dense in</u> $C^r(X,X')$, <u>and for</u> $n' \geqslant 2n+1$, $\text{Emb}^r(X,X')$ <u>is dense in</u> $C^r(X,X')$.

Without loss of generality, we shall prove these statements in the case $r = \infty$; when $r < \infty$, we simply apply Theorems 9.6 and 6.5 to reduce to the first case.

Let $f \in C^\infty(X,X')$, and let U be a neighborhood of f. We have to show that, for $n' \geqslant 2n$, U contains an immersion, and that for $n' \geqslant 2n+1$, U contains an embedding.

Fix C^∞-embeddings $j: X \to \mathbb{R}^q$ and $j': X' \to \mathbb{R}^{q'}$, and define an embedding $J': X' \to \mathbb{R}^{q+q'} = \mathbb{R}^q \times \mathbb{R}^{q'}$ by the formula $J'(x') = (j'(x'),0)$. Further, pick a C^∞-transversalization τ' of J' and a neat tube $\text{Tub}_{\tau'}\rho'$. Then $J_t(x) = (j'(f(x)),tj(x))$ defines a C^∞-embedding $J_t: X \to \mathbb{R}^{q+q'}$, for any fixed $t > 0$. Pick ε small enough so that $J_\varepsilon(X) \subset \text{Tub}_{\tau'}(\rho'/2)$ and denote J_ε simply by J. Applying Lemma 3 (to the manifold X', the embedding J', the transversalization τ' and the tube $\text{Tub}_{\tau'}\rho'$), we conclude that there are a neighborhood U' of the map $(\tau')^{\tilde{}}$ in $C^\infty(X',G'(q'+q,q'+q-n'))$ and a continuous mapping $\Phi': U' \to \text{Diff}^\infty X'$ such that, for each $g' \in U'$:

    (i) $[J' \circ \Phi'(g')](x') \in g'(x')$ for all $x' \in X'$;

    (ii) the map $\tau'_{g'}: X' \to G(q'+q,q'+q-n')$, defined as

$\tau'_{g'}(x') = g'(x') - [J' \circ \Phi'(g')](x')$ is a transversalization of the embedding $J' \circ \Phi'(g')$;

    (iii) some neat tube of this transversalization contains $\text{Tub}_{\tau'}(\rho'/2)$.

Now consider the mapping $\Psi': U' \to C^\infty(X,X')$ given by $[\Psi'(g')](x) = \text{pr}_{\tau'_{g'}}(J(x))$. We see that Ψ' is continuous and

$\Psi'(\tau') = f$. Hence, $(\Psi')^{-1}(\mathcal{U})$ is open in $C^\infty(X',G'(q'+q,q'+q-n'))$ and nonempty. This fact together with Theorem 7.7 show that $(\Psi')^{-1}(\mathcal{U})$ contains a map h transverse to each of the maps

$$aux_1 : Aux_1(J;q'+q-n') \to G'(q'+q,q'+q-n')$$

and

$$aux_2 : Aux_2(J;q'+q-n') \to G'(q'+q,q'+q-n').$$

We shall presently show that $\Psi'(h)$ is an immersion for $n' \geqslant 2n$, and a differentiable embedding for $n' \geqslant 2n+1$. This will complete the proof, since $\Psi'(h) \in \mathcal{U}$.

The inequality $n' \geqslant 2n$ is equivalent to

$$\dim X' + \dim Aux_1(J;q'+q-n') < \dim G'(q'+q,q'+q-n') \; ;$$

hence for $n' \geqslant 2n$ the fact that aux_1 and h are transverse means that $h(X')$ does not intersect $\mathrm{Im}\, aux_1$, i.e., none of the planes $h(x')$, $x' \in X'$, contains a line tangent to $J(X)$. The latter, in turn, means that the differential $d_y(pr_{\tau_h'}|_{J(X)})$ of the restriction $pr_{\tau_h'}|_{J(X)}$ is a monomorphism for any point $y \in J(X)$. Thus $pr_{\tau_h'}|_{J(X)}$, and also $\Psi'(h)$, are immersions.

Similarly, the inequality $n' \geqslant 2n+1$ is equivalent to

$$\dim X' + \dim Aux_2(J;q'+q-n') < \dim G'(q'+q,q'+q-n') \; ;$$

hence for $n' \geqslant 2n+1$ the fact that aux_2 and h are transverse means that $h(X')$ does not intersect $\mathrm{Im}\, aux_2$, i.e., none of the planes $h(x')$, $x' \in X'$, intersects $J(X)$ at more than one point. This says that the restriction $pr_{\tau_h'}|_{J(X)}$ is injective. We see at once that $\Psi'(h)$ is also an injective map, and since it is an immersion, $\Psi'(h)$ is a differentiable embedding (see 1.5.4).

Embeddings and Immersions in Euclidean Space

5. <u>Every compact n-dimensional $C^{\geqslant r}$-manifold with</u> $1 \leqslant r \leqslant \infty$ <u>can be C^r-immersed in</u> $\mathbb{R}^{2n}$ <u>and C^r-embedded in</u> $\mathbb{R}^{2n+1}$.

This is a consequence of Theorem 4 (where we take X to be the given manifold, and set first $X' = S^{2n}$, and then $X' = S^{2n+1}$; we disregard the trivial case $n = 0$).

6 (INFORMATION). As a matter of fact, every n-dimensional $C^{\geqslant r}$-manifold with $r \geqslant 1$ admits a C^r-embedding in $\mathbb{R}^{2n}$ whenever

$n \geqslant 1$, and a C^r-immersion in $\mathbb{R}^{2n-1}$ whenever $n \geqslant 2$. If $n > 0$ and is not a power of 2, then every $C^{\geqslant r}$-manifold with $r \geqslant 1$ admits a C^r-embedding in $\mathbb{R}^{2n-1}$. However, for each $n = 2^s$, $s \geqslant 0$, there are smooth, closed, n-dimensional manifolds which cannot be even topologically embedded in $\mathbb{R}^{2n-1}$ (such an example is $\mathbb{R}P^n$). Every n-dimensional C^r-manifold, $r \geqslant 1$, without closed components, can be C^r-embedded in $\mathbb{R}^{2n-1}$. Every orientable n-dimensional $C^{\geqslant r}$-manifold with $r \geqslant 1$ and $n \neq 1,4$ can be C^r-embedded in $\mathbb{R}^{2n-1}$. [It is not known whether a smooth, closed, orientable, four-dimensional manifold admits a differentiable embedding in $\mathbb{R}^7$; a topological embedding always exists.]

More information, details, and references can be found in [20] and [18].

11. Exercises

1. Show that for any $C^{\geqslant r}$-manifolds X and X', with $0 \leqslant r \leqslant \infty$, the space $C^r(X,X')$ has a countable base.

2. Let X be a compact $C^{\geqslant r}$-manifold, $1 \leqslant r \leqslant \infty$, and let X' be an arbitrary $C^{\geqslant r}$-manifold. Show that the set $\mathrm{Subm}^r(X,X') \cap C^r_\partial(X,X')$ is open in $C^r_\partial(X,X')$.

3. Show that every compact topological manifold is a CNRS. [cf. 5.10.]

4. Suppose X is the smooth one-dimensional submanifold of $\mathbb{R}^2$, closed as an independent manifold, and containing the graph of the function $x \mapsto x^2 \sin(1/x)$ defined on the interval $(0,1)$. Let τ be the normal transversalization of the inclusion $X \to \mathbb{R}^2$. Show that for any ρ the segment $d_\tau((0,0),\rho)$ intersects a segment $d_\tau(x,\rho)$ for some $x \neq (0,0)$. [cf. 3.2.]

5. Suppose X is a compact C^r-manifold with $1 \leqslant r \leqslant \infty$, and A is a submanifold of X. Show that for a suitable collaring of X, $A \cup \mathrm{cop}(A)$ is a submanifold of $\mathrm{dopp}\, X$.

6. Let X and Y be compact C^r-manifolds with $1 \leqslant r \leqslant \infty$. Check that: (i) the product $X \times Y$ has a C^r-structure which induces the usual C^r-structures on $\mathrm{int}\, X \times Y$ and $X \times \mathrm{int}\, Y$; (ii) the C^r-manifold obtained by equipping $X \times Y$ with a C^r-structure having these properties is unique up to a diffeomorphism.

7. Let X_1, X_2, and X be compact C^∞-manifolds, and let

$f_2 \in C_\partial^\infty(X_2, X')$. Show that the subset of $C_\partial^\infty(X_1, X')$ consisting of all maps transverse to f_2 is dense in $C_\partial^\infty(X_1, X')$.

8. Let X_1, X_2, X', and f_2 be as in exercise 7. Show that for any C^∞-map $\phi: \partial X_1 \to \partial X'$ transverse to ab $f_2: \partial X_2 \to \partial X'$, the subset of $C_\partial^\infty(X_1, X')$ consisting of all the extensions of ϕ which are transverse to f_2 is dense in the subspace of $C_\partial^\infty(X_1, X')$ consisting of all the extensions of ϕ.

9. Let X and X' be compact $C^{\geqslant r}$-manifolds with $1 \leqslant r \leqslant \infty$, and let $\phi: \partial X \to \partial X'$ be a C^r-immersion. If $\dim X' \geqslant 2 \dim X$, show that the subset of $C_\partial^r(X, X')$ consisting of all C^r-immersions that extend ϕ is dense in the subspace of $C_\partial^r(X, X')$ consisting of all extensions of ϕ.

10. Let X and X' be compact $C^{\geqslant r}$-manifolds with $1 \leqslant r \leqslant \infty$, and let $\phi: \partial X \to \partial X'$ be a C^r-embedding. If $\dim X' \geqslant 2 \dim X + 1$, show that the subset of $C_\partial^r(X, X')$ consisting of all C^r-embeddings that extend ϕ is dense in the subspace of $C_\partial^r(X, X')$ consisting of all extensions of ϕ. [In particular, every compact n-dimensional $C^{\geqslant r}$-manifold admits a neat embedding in D^{2n+1}.]

11. Let X and X' be closed $C^{\geqslant r}$-manifolds with $1 \leqslant r \leqslant \infty$. Show that the set of all C^r-maps $f: X \to X'$ such that $\mathrm{Tang}_{f(x_1)} X' = \mathrm{Im}\, d_{x_1} f + \mathrm{Im}\, d_{x_2} f$ for any two distinct points $x_1, x_2 \in X$ with $f(x_1) = f(x_2)$, is dense in $C^r(X, X')$.

12. Let X and X' be closed $C^{\geqslant r}$-manifolds with $1 \leqslant r \leqslant \infty$. If $\dim X' = 2 \dim X - 1$, show that the set of all C^r-maps $f: X \to X'$ such that $\mathrm{rank}\, d_x f = \dim X$ for all but a finite number of points $x \in X$, where $\mathrm{rank}\, d_x f = \dim X - 1$, is dense in $C^r(X, X')$.

13. Let X and X' be closed $C^{\geqslant r}$-manifolds with $1 \leqslant r \leqslant \infty$. If $2 \dim X' > 3 \dim X$, show that the set of all C^r-maps $f: X \to X'$ such that the preimage of each point of X' under f contains at most two points, is dense in $C^r(X, X')$.

§5. THE SIMPLEST STRUCTURE THEOREMS

1. Morse Functions

1. The central result of this section is Theorem 2.10, whose main conclusion is that every compact n-dimensional C^∞-manifold can be obtained from an empty n-dimensional manifold through a finite number of fairly standard operations, namely, by attaching handles. The entire present subsection and that part of Subsection 2 preceding Theorem 2.10 are essentially devoted to the preparation of its formulation and proof. The remaining part of Subsection 2 contains corollaries of Theorem 2.10. In Subsection 3 this theorem is used to effectively classify the compact smooth two-dimensional manifolds.

It should come as no surprise that, in contrast to the previous sections, here we consider, in general, only the C^∞-case: the theorems concerning smoothing of diffeomorphisms and manifolds (i.e., Theorems 4.4.4, 4.6.11, and 4.9.6, 4.9.8) show that we may replace the class C^∞ by any class C^r, $1 \leqslant r < \infty$, without affecting the theory discussed here.

Cobordisms and Morse Functions

2. A compact C^∞-manifold X is called a <u>cobordism</u> if its boundary ∂X is the disjoint union of two parts, $\partial_0 X$ and $\partial_1 X$, each consisting of whole components of ∂X. Those two parts are termed the <u>beginning</u> and the <u>end of the cobordism</u> X. Each of them may be empty; when both are empty, X is closed. In general, given a compact C^∞-manifold, one can transform it into a cobordism in 2^1 ways, where 1 is the number of components of ∂X. Among these cobordisms, there is one without beginning $(\partial_0 X = \emptyset,\ \partial_1 X = \partial X)$ and one without end $(\partial_0 X = \partial X,\ \partial_1 X = \emptyset)$.

Two cobordisms, X and X', are said to be <u>diffeomorphic</u> if there is a diffeomorphism (and hence a C^∞-diffeomorphism) $f: X \to X'$ such that $f(\partial_0 X) = \partial_0 X'$ and $f(\partial_1 X) = \partial_1 X'$.

Suppose that X and X' are two cobordisms such that $\partial_1 X$ and $\partial_0 X'$ are diffeomorphic, and let $\varphi: \partial_1 X \to \partial_0 X'$ be a C^∞-diffeomorphism. Then one can form a manifold Y by glueing the somehow collared manifolds X and X' with the aid of φ. Now Y naturally becomes a cobordism if we set $\partial_0 Y = \partial_0 X$ and $\partial_1 Y = \partial_1 X'$. We say that _the cobordism_ Y _is the result of glueing the cobordisms_ X _and_ X' _by_ φ. If the cobordisms X and X' are oriented and φ is orientation _reversing_ (here the orientations of $\partial_1 X$ and $\partial_0 X'$ are those induced by the orientations of X and X'; see 1.3.4), then one can orient Y in such a manner that both embeddings, $X \to Y$ and $X' \to Y$, become orientation preserving. Warning: this definition of the orientation of the glued cobordism is not in accordance with the definition of the orientation of a glued manifold, given in 4.5.5.

Two smooth closed manifolds, V_0 and V_1, are _cobordant_ if there is a cobordism with the beginning and the end diffeomorphic to V_0 and V_1, respectively. If, in addition, V_0 and V_1 are oriented, and there is an oriented cobordism X such that one of the diffeomorphisms $V_0 \to \partial_0 X$ and $V_1 \to \partial_1 X$ preserves orientation, whereas the other reverses it, then we say that V_0 and V_1 are _oriented cobordant._ Clearly, the cobordism and oriented cobordism relations are reflexive and symmetric, and since cobordisms can be glued, they are also transitive, i.e., they are genuine equivalence relations.

3. A critical point x of a C^2-function $f: X \to \mathbb{R}$, where X is a $C^{\geqslant 2}$-manifold is _nondegenerate_ if for some chart $\varphi \in \mathrm{Atl}_x C^2 X$ (and hence for any such chart) $\varphi(x)$ is a nondegenerate critical point of the function $(f|_{\mathrm{supp}\,\varphi}) \circ \varphi^{-1}: \mathrm{Im}\,\varphi \to \mathbb{R}$ (see 3.3.1). The corresponding index is independent of the choice of the chart φ (see 3.3.1), and is called the _index of the point_ x _relative to_ f.

Suppose X is a cobordism, and let $f: X \to \mathbb{R}$ be a C^∞-function; f is a _Morse function_ if: $\mathrm{Im}\,f \subset I$; $f^{-1}(0) = \partial_0 X$, and $f^{-1}(1) = \partial_1 X$; and all critical points of f lie in $\mathrm{int}\,X$ and are nondegenerate.

We say that a Morse function is _proper_ if its values at distinct critical points are distinct.

The Local Structure of Morse Functions

4. _Suppose that_ X _is an n-dimensional cobordism and_ $f: X \to \mathbb{R}$ _is a Morse function. Then for every point_ $x \in X$ _there is a_

chart $\varphi \in \text{Atl}_x X$ <u>with</u> $\varphi(x) = 0$, <u>such that the restriction</u> $f\big|_{\text{supp}\,\varphi}$ coincides with the composite map

$$\text{supp}\,\varphi \xrightarrow{\ \varphi\ } \text{Im}\,\varphi \longrightarrow \mathbb{R}\,,$$

<u>where the second arrow denotes one of the following functions:</u>

$$(t_1,\dots,t_n) \mapsto -t_1, \quad \underline{if} \quad x \in \partial_0 X;$$

$$(t_1,\dots,t_n) \mapsto 1 + t_1, \quad \underline{if} \quad x \in \partial_1 X;$$

$$(t_1,\dots,t_n) \mapsto f(x) + t_1, \quad \underline{if} \quad x \in \text{int}\,X \quad \underline{\text{is not a critical}}$$

<u>point of</u> f;

$$(t_1,\dots,t_n) \mapsto f(x) - t_1^2 - \dots - t_k^2 + t_{k+1}^2 + \dots + t_n^2, \quad \underline{if} \quad x$$

<u>is a critical point of index</u> k.

To prove the first three cases we need only remark that the function $X \to \mathbb{R}$, defined by $y \mapsto f(y) - x$ for $x \in \text{int}\,X \cup \partial_1 X$ (respectively, by $y \mapsto -f(y)$ for $x \in \partial_0 X$) can be completed, in a neighborhood of x, to a system of coordinates (see 1.2.12). For the fourth case, we refer to Theorem 3.3.3.

5 (COROLLARY). <u>A Morse function has only a finite number of</u> <u>critical points.</u>

An Existence Theorem

6 (LEMMA). <u>Given any</u> C^∞<u>-function</u> $f: D^n \to \mathbb{R}$, <u>there exists</u> <u>an open dense subset</u> A <u>of</u> $\mathbb{R}^n$ <u>such that for</u> $a \in A$ <u>the function</u> $D^n \to \mathbb{R}$ <u>defined by</u>

$$x \mapsto f(x) - \langle a,x\rangle \tag{1}$$

<u>has no degenerate critical points.</u>

PROOF. Consider the map $\text{grad}\,f : D^n \to \mathbb{R}^n$ with coordinate functions $D_1 f,\dots,D_n f$. One may take A to be $\mathbb{R}^n \smallsetminus [\text{grad}\,f](F)$, where $F = \{x \in D^n \mid \text{rank}\,d_x \text{grad}\,f < n\}$. F is clearly open, and Theorem 4.7.4 implies that it is also dense in $\mathbb{R}^n$. Moreover, it is evident that if $x \in D^n$ is a critical point of the function (1), then $[\text{grad}\,f](x) = a$, and the matrix of the second-order partial derivatives of (1) at x is precisely the matrix of the differential $d_x \text{grad}\,f$

relative to the standard coordinates in $\text{Tang}_x D^n$ and $\text{Tang}_a \mathbb{R}^n$. Therefore, if $a \in A$, then this matrix of second-order partial derivatives is nonsingular.

7. <u>On every cobordism there is a proper Morse function.</u>

First, let us show that if there exists some Morse function on the cobordism X, then there exists a proper Morse function on X. Let $x_1,\dots,x_m$ be the critical points of the Morse function $f\colon X \to \mathbb{R}$, and let $U_1,\dots,U_m$ be pairwise disjoint neighborhoods of these points in $\text{int}\,X$. Further, let $V_1,\dots,V_m$ be neighborhoods of $x_1,\dots,x_m$ such that $\text{Cl}\,V_1 \subset U_1,\dots,\text{Cl}\,V_m \subset U_m$, and let $\phi_1,\dots,\phi_m$ be Urysohn C^∞-functions for the pairs $(X \smallsetminus U_1,\text{Cl}\,V_1),\dots,(X \smallsetminus U_m,\text{Cl}\,V_m)$. (The existence of such Urysohn functions results from Theorem 4.7, applied here to the double of the manifold X.) Clearly, if the numbers $\varepsilon_1,\dots,\varepsilon_m$ are small enough, then the function $X \to \mathbb{R}$ defined by $x \mapsto f(x) + \varepsilon_1 \phi_1(x) + \dots + \varepsilon_m \phi_m(x)$ is, together with f, a Morse function, and its only critical points are $x_1,\dots,x_m$. Moreover, its values at the points $x_1,\dots,x_m$ are $f(x_1) + \varepsilon_1,\dots,f(x_m) + \varepsilon_m$, and so, by choosing suitable $\varepsilon_1,\dots,\varepsilon_m$, one can force these values to be distinct.

Now we show that there exists Morse functions on X. Fix a collaring $k\colon \partial X \times I \to X$ and a C^∞-function $\beta\colon I \to I$ which equals 1 on the segment $[0,1/2]$ and vanishes on a neighborhood of 1. Next define $g\colon X \to \mathbb{R}$ by the formulas

$$g(x) = \frac{1}{2}, \quad \text{for} \quad x \in X \smallsetminus k(\partial X \times [0,1)),$$

$$g(k(z,t)) = \frac{1}{2} + \frac{1}{2}(t - 1)\beta(t), \quad \text{for} \quad z \in \partial_0 X,\ t \in I,$$

$$g(k(z,t)) = \frac{1}{2} + \frac{1}{2}(1 - t)\beta(t), \quad \text{for} \quad z \in \partial_1 X,\ t \in I.$$

It is clear that g is C^∞, takes X into I, $\partial_0 X$ into 0, and $\partial_1 X$ into 1, and has no critical points in $k(\partial X \times [0,1/2])$. Pick charts $\varphi_1,\dots,\varphi_s \in \text{Atl}\,X$ such that $\text{Im}\,\varphi_1 = \dots = \text{Im}\,\varphi_s = \mathbb{R}^n$, $n = \dim X$, and the sets $\varphi_1^{-1}(\text{Int}\,D^n),\dots,\varphi_s^{-1}(\text{Int}\,D^n)$ cover $X \smallsetminus k(\partial X \times [0,1/2))$. Further, choose a C^∞-function $\alpha\colon \mathbb{R}^n \to I$ which equals 1 on D^n and 0 ouside the concentric ball $2D^n$ of radius 2. Using induction, we shall build functions $g_0,\dots,g_s\colon X \to \mathbb{R}$, such that g_i coincides with g in a neighborhood of ∂X, maps $\text{int}\,X$ into $\text{int}\,I$, has no critical points in $k(\partial X \times [0,1/2])$, and has no degenerate critical points in $\cup_{j=1}^{i}\varphi_j^{-1}(D^n)$. Then g_s will be a Morse function on X.

Let $g_0 = g$ and assume that for some $k \geqslant 1$ functions g_i enjoying the required properties are already constructed for $i < k$. For each point $a \in \mathbb{R}^n$, the formula

$$
h_a(x) = \begin{cases} g_{k-1}(x), & \text{if } x \in X \smallsetminus \varphi_k^{-1}(2D^n), \\[2ex] g_{k-1} - <a, \varphi_k(x) > \alpha \circ \varphi_k(x), & \text{if } x \in \operatorname{supp} \varphi_k, \end{cases}
$$

defines a C^∞-function $h_a : X \to \mathbb{R}$ which agrees with g_{k-1} in a neighborhood of ∂X, and for $a = 0$ coincides with g_{k-1} on all of X. Clearly, the point 0 has a neighborhood U in $\mathbb{R}^n$ such that for $a \in U$ the function h_a has no critical points in $k(\partial X \times [0,1/2])$, has no degenerate critical points in $\bigcup_{j=1}^{k-1} \varphi_j^{-1}(D^n)$, and maps $\operatorname{int} X$ into $\operatorname{int} I$. Thus, by Lemma 6, we can find $a \in U$ such that h_a also has no degenerate critical points in $\varphi_k^{-1}(D^n)$, and hence we can take $g_k = h_a$ for such a value of a.

2. <u>Cobordisms and Surgery</u>

1. This subsection is devoted to two types of special operations on cobordisms, called <u>attaching of handles</u> and <u>spherical modifications.</u> The former were already mentioned in 1.1. Spherical modifications are simpler that the attaching of handles, but their applications are more limited: we shall define them solely for the closed case, and starting with a closed C^∞-manifold they are capable of producing only manifolds cobordant with the given one.

Standard Cobordisms

2. We define <u>standard cobordisms</u> of two kinds: the <u>standard trivial cobordisms</u> and the <u>standard elementary cobordisms of index</u> k. On every standard cobordism there is a standard Morse function.

A standard trivial cobordism is constructed by taking an arbitrary closed C^∞-manifold V and simply forming the cylinder $V \times I$, with $\partial_0(V \times I) = V \times 0$ and $\partial_1(V \times I) = V \times 1$. The corresponding standard Morse function is defined as $(v,t) \mapsto t$ and has no critical points.

The standard elementary cobordism of index k is defined for an arbitrary closed $(n-1)$-dimensional C^∞-manifold V with $n \geq k$, and an arbitrary C^∞-embedding $\varphi\colon S^{k-1} \times D^{n-k} \to V$, and is denoted by $El(V,\varphi)$. Up to a canonical homeomorphism, $El(V,\varphi)$ is simply $(V \times I) \cup_{\varphi_1} (D^k \times D^{n-k})$, where the map $\varphi_1\colon S^{k-1} \times D^{n-k} \to V \times I$ is defined by $\varphi_1(x,y) = (\varphi(x,y),1)$. To define $El(V,\varphi)$ as a C^∞-manifold, we let $E(n,k)$ denote the set

$$\{(t_1,\ldots,t_n) \in \mathbb{R}^n \mid \textstyle\sum_{k+1}^{n} t_i^2 \leq 1,\ |\sum_1^k t_i^2 - \sum_{k+1}^n t_i^2| \leq \tfrac{1}{16}\},$$

and let el denote the homeomorphism of the intersection

$$E(n,k) \cap [\mathbb{R}^n \smallsetminus (\mathbb{R}^k \times \mathrm{Int}(\tfrac{1}{2}D^{n-k}))]$$

onto the cylinder

$$\varphi(S^{k-1} \times (D^{n-k} \smallsetminus \mathrm{Int}(\tfrac{1}{2}D^{n-k}))) \times I,$$

given by the formula

$$(t_1,\ldots,t_n) \mapsto (\varphi((t_1,\ldots,t_k)/(t_1^2+\ldots+t_k^2)^{1/2}, (t_{k+1},\ldots,t_n)),$$

$$8(\tfrac{1}{16} - t_1^2 - \ldots - t_k^2 + t_{k+1}^2 + \ldots + t_n^2)).$$

As a topological space, $El(V,\varphi)$ is the result of glueing the cylinder $[V \smallsetminus \varphi(S^{k-1} \times \mathrm{Int}(\tfrac{1}{2}D^{n-k}))] \times I$ and $E(n,k)$ by the homeomorphism el (see Fig. 5).

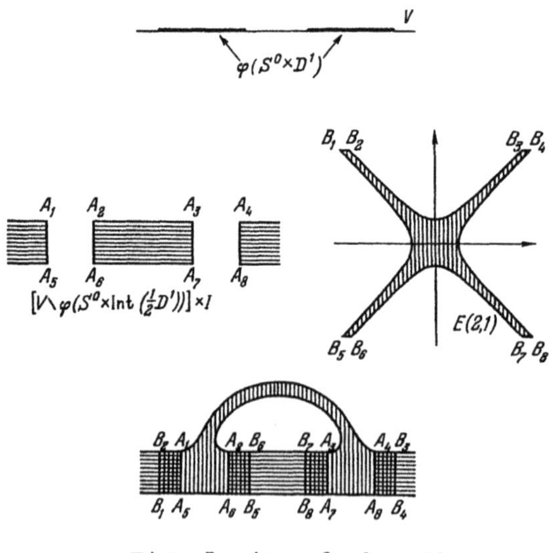

Fig. 5 $(n = 2, k = 1)$

The C^∞-structure on $El(V,\varphi)$ is fixed by an atlas consisting of the charts of arbitrary atlases of the C^∞-manifolds

$$[V \smallsetminus \varphi(S^{k-1} \times (\tfrac{1}{2}D^{n-k}))] \times I \quad \text{and} \quad E(n,k) \cap (\mathbb{R}^k \times \operatorname{Int} D^{n-k})$$

(these manifolds are regarded as parts of the space $El(V,\varphi)$). The beginning of $\partial_0 El(V,\varphi)$ is formed by the two sets

$$[V \smallsetminus \varphi(S^{k-1} \times \operatorname{Int}(\tfrac{1}{2}D^{n-k}))] \times 0$$

and

$$\{(t_1,\ldots,t_n) \in E(n,k) \mid -t_1^2 - \ldots - t_k^2 + t_{k+1}^2 + \ldots + t_n^2 = -\tfrac{1}{16}\}.$$

Similarly, the sets

$$[V \smallsetminus \varphi(S^{k-1} \times \operatorname{Int}(\tfrac{1}{2}D^{n-k}))] \times 1$$

and

$$\{(t_1,\ldots,t_n) \in E(n,k) \mid -t_1 - \ldots - t_k + t_{k+1} + \ldots + t_n = \tfrac{1}{16}\}$$

constitute the end $\partial_1 El(V,\varphi)$. The standard Morse function on $El(V,\varphi)$ is given by the two functions

$$[V \smallsetminus \varphi(S^{k-1} \times \operatorname{Int}(\tfrac{1}{2}D^{n-k}))] \times I \to \mathbb{R} \quad \text{and} \quad E(n,k) \to \mathbb{R},$$

defined by the formulas

$$(v,t) \mapsto t$$

and

$$(t_1,\ldots,t_n) \mapsto 8(\tfrac{1}{16} - t_1^2 - \ldots - t_k^2 + t_{k+1}^2 + \ldots + t_n^2),$$

respectively. It has a unique critical point $0 \in E(n,k)$, of index k. We use the symbol mo to denote the standard Morse function.

Using Fig. 5 as a guide, it is readily seen that the manifold $El(V,\varphi)$ is homeomorphic to $(V \times I) \cup_{\varphi_1} (D^k \times D^{n-k})$. The formulas describing this canonical homeomorphism are cumbersome, and we shall not burden the reader with their precise form.

The basic property shared by all standard cobordisms constructed from a given manifold V is that the beginning of each is canonically C^∞-diffeomorphic to V. For the standard trivial cobordism, this diffeomorphism is $v \mapsto (v,0)$. In the case of the cobordism $El(V,\varphi)$, the diffeomorphism is defined as $v \mapsto (v,0)$ on the part

$$V \smallsetminus \varphi(S^{k-1} \times \operatorname{Int}(\tfrac{1}{2}D^{n-k})) \to [V \smallsetminus \varphi(S^{k-1} \times \operatorname{Int}(\tfrac{1}{2}D^{n-k}))] \times 0,$$

and as

$$\varphi((t_1,\ldots,t_k),(t_{k+1},\ldots,t_n)) \mapsto (t_1(\tfrac{1}{16} + t_{k+1}^2 + \ldots + t_n^2)^{1/2},$$

$$\ldots,t_k(\tfrac{1}{16} + t_{k+1}^2 + \ldots + t_n^2)^{1/2},t_{k+1},\ldots,t_n)$$

on the part

$$\varphi(S^{k-1} \times D^{n-k}) \rightarrow$$

$$\rightarrow \{(t_1,\ldots,t_n) \in E(n,k) \mid - t_1^2 - \ldots - t_k^2 + t_{k+1}^2 + \ldots + t_n^2 = - \tfrac{1}{16}\}.$$

3. The construction of the standard elementary cobordism is the simplest when the index k is 0 or n.

If $k = 0$, then φ embeds the empty set into V, $E(n,k)$ is the ball $\tfrac{1}{4}D^n$, and el is a homeomorphism of empty sets. In this case the similarity dilatation (with coefficient 4) of the ball $E(n,k)$ transforms $El(V,\varphi)$ into the sum $(V \times I) \bigsqcup D^n$, with $\partial_0[(V \times I) \bigsqcup D^n] = in_1(V \times 0)$ and $\partial_1[(V \times I) \bigsqcup D^n] = (V \times 1) \bigsqcup S^{n-1}$; it also transforms the standard Morse function mo into the function

$$\begin{cases} V \times I \rightarrow \mathbb{R}, \quad (v,t) \mapsto t, \\[2ex] D^n \rightarrow \mathbb{R}, \quad (t_1,\ldots,t_n) \mapsto (1 + t_1^2 + \ldots + t_n^2)/2. \end{cases}$$

If $k = n$, then φ embeds S^{n-1} into V, the image of this embedding being one of the components of V (see 1.5.1). Again, $E(n,k)$ is the ball $\tfrac{1}{4}D^n$ and el is a homeomorphism of empty sets. Here the similarity dilatation of the ball $E(n,k)$ transforms $El(V,\varphi)$ into the sum $([V \smallsetminus \varphi(S^{n-1})] \times I) \bigsqcup D^n$, with

$$\partial_0\{([V \smallsetminus \varphi(S^{n-1})] \times I) \bigsqcup D^n\} = ([V \smallsetminus \varphi(S^{n-1})] \times 0) \bigsqcup S^{n-1}$$

and

$$\partial_1\{([V \smallsetminus \varphi(S^{n-1})] \times I) \bigsqcup D^n\} = in_1([V \smallsetminus \varphi(S^{n-1})] \times 1),$$

while the standard Morse function mo is taken into the function

$$\begin{cases} [V \smallsetminus \varphi(S^{n-1})] \times I \rightarrow \mathbb{R}, \quad (v,t) \mapsto t, \\[2ex] D^n \rightarrow \mathbb{R}, \quad (t_1,\ldots,t_n) \mapsto (1 - t_1^2 - \ldots - t_n^2)/2. \end{cases}$$

We also remark that $k = 0$ and $k = n$ are the only values of the index for which the standard elementary cobordism can have a connected boundary: for $k = 0$ the boundary is connected if and only if the initial manifold V is empty, while for $k = n$ the boundary is connected if and only if φ is a diffeomorphism. We have seen that in both cases the cobordism is diffeomorphic to D^n.

Trivial Cobordisms and Elementary Cobordisms

4. A cobordism is said to be _trivial_ if it is diffeomorphic to a standard trivial cobordism. A cobordism is an _elementary cobordism of index_ k if it is diffeomorphic to a standard elementary cobordism of index k.

From these definitions it follows that if X is a trivial cobordism, then the manifolds $\partial_0 X$ and $\partial_1 X$ are diffeomorphic, and that on a trivial cobordism there is a Morse function without critical points, whereas on an elementary cobordism of index k there is a Morse function with a single critical point of index k and no other critical points.

5 (LEMMA). _Suppose_ X _and_ X' _are two cobordisms such that the manifolds_ $\partial_1 X$ _and_ $\partial_0 X'$ _are diffeomorphic. If_ X' _is trivial, the the cobordism obtained by glueing_ X _and_ X' _by an arbitrary_ C^∞-_diffeomorophism_ $\partial_1 X \to \partial_0 X'$ _(and using arbitrary collarings of_ X _and_ X'_) is diffeomorphic to_ X.

Theorem 4.5.9 shows that given a C^∞-diffeomorphism $\varphi : \partial_1 X \to \partial_0 X'$, it suffices to produce two C^∞-embeddings, $j : X \to X$ and $j' : X' \to X$, such that $j(X) \cup j'(X') = X$, $j(X) \cap j'(X') = j(\partial_1 X) = j'(\partial_0 X')$, and the composite diffeomorphism

$$\partial_0 X' \xrightarrow{\text{ab } j'} j'(\partial_0 X') \xrightarrow{\text{ab } j^{-1}} \partial_1 X$$

coincides with φ^{-1}. In order to accomplish this, let us fix: a collaring $k : \partial X \times I \to X$; a C^∞-diffeomorphism $f : X' \to \partial_0 X' \times I$ such that $f(x') = (x',0)$ for all $x' \in \partial_0 X'$; and an increasing C^∞-function $\alpha : I \to I$, such that $\alpha(t) = 1/2 + t/3$ for $t \leqslant 3/4$ and $\alpha(t) = t$ for $t \geqslant 7/8$. Further, using the function $\beta : I \to I$, $\beta(t) = (1 - t)/2$, set

$$\begin{cases} j(x) = x, & \text{if } x \in X \smallsetminus k(\partial_1 X \times [0,1)), \\[2ex] j(k(z,t)) = k(z,\alpha(t)), & \text{if } z \in \partial_1 X, \ t \in I, \end{cases}$$

and

$$j'(x') = k \circ (\varphi^{-1} \times \beta) \circ f(x'), \quad \text{if } x' \in X'.$$

We can verify directly that j and j' have the required properties.

6. _If on a cobordism there is a Morse function without critical point, then the cobordism is trivial._

PROOF. Let $f: X \to \mathbb{R}$ be a Morse function with no critical points. According to 1.5.8 (or, if one prefers, to 4.8.2), the preimage $f^{-1}(t)$ of any point $t \in (0,1)$ is a neat submanifold of X; moreover, $f^{-1}(t)$ is obviously closed as an independent manifold (the preimages $f^{-1}(0) = \partial_0 X$ and $f^{-1}(1) = \partial_1 X$ are also closed manifolds). By Theorems 4.5.3 and 4.5.8, the manifold $f^{-1}(t)$ has a neighborhood U_t together with a C^∞-submersion $\pi_t: U_t \to f^{-1}(t)$ which is the identity map on $f^{-1}(t)$, for each fixed $t \in I$. Define, for $t \in I$, a C^∞-map $F_t: U_t \to f^{-1}(t) \times I$ by $F_t(x) = (\pi_t(x), f(x))$. Obviously, the differential $d_x F_t$ is nondegenerate for $x \in f^{-1}(t)$, and F_t induces a diffeomorphism of $f^{-1}(t)$ onto $f^{-1}(t) \times t$. Consequently, F_t is a C^∞-embedding on a neighborhood of $f^{-1}(t)$ (see 1.5.5), i.e., there is a neighborhood Δ_t of the point t in I such that F_t induces a diffeomorphism of $F_t^{-1}(\Delta_t)$ onto $f^{-1}(t) \times \Delta_t$. Let m be large enough so that if we divide I into m intervals of length $1/m$, then each such interval is contained in one of the sets Δ_t. Then all the cobordisms $f^{-1}([(i-1)/m, i/m])$, with $\partial_0 f^{-1}([(i-1)/m, i/m]) =$

$= f^{-1}((i-1)/m)$ and $\partial_1 f^{-1}([(i-1)/m, i/m]) = f^{-1}(i/m)$, are trivial, and now Lemma 5 shows that the entire cobordism X is trivial.

7. _If on a cobordism X there is a Morse function with a single critical point of index k and no other critical points, then X is an elementary cobordism of index k._

The proof is quite long and we shall begin by constructing an auxiliary cobordism Y. Fix a Morse function $f: X \to I$ with a single critical point of index k, say x, and no other critical points. By Theorem 1.4, there is a chart $\varphi \in \mathrm{Atl}_x X$, $\varphi(x) = 0$, such that $f|_{\mathrm{supp}\,\varphi}$ equals the composition of φ with the function $\mathrm{Im}\,\varphi \to \mathbb{R}$ given by

$$(t_1, \ldots, t_n) \mapsto f(x) - t_1^2 - \ldots - t_k^2 + t_{k+1}^2 + \ldots + t_n^2 .$$

Obviously, $\mathrm{Im}\,\varphi$ contains the subset A of $\mathbb{R}^n$ determined (in the standard coordinates) by the inequalities:

$$t_{k+1}^2 + \ldots + t_n^2 \leqslant 4\varepsilon^2 ,$$

and

$$-\frac{\varepsilon^2}{16} \leqslant -t_1^2 - \ldots - t_k^2 + t_{k+1}^2 + \ldots + t_n^2 \leqslant \frac{\varepsilon^2}{16} \, ,$$

for some $\varepsilon > 0$. Let

$$B = A \cap (\, \mathbb{R}^k \times (\, \mathbb{R}^{n-k} \smallsetminus \mathrm{Int}(\tfrac{1}{2}D^{n-k})))\,.$$

As a topological space, Y is the result of glueing the subset

$$f^{-1}([\,f(x) - \frac{\varepsilon^2}{16}, f(x) + \frac{\varepsilon^2}{16}]) \smallsetminus \varphi^{-1}(A \smallsetminus B)$$

of X and the product $S^{k-1} \times 2D^{n-k} \times I$ by the homeomorphism

$$\varphi^{-1}(B) \to S^{k-1} \times [\,2D^{n-k} \smallsetminus \mathrm{Int}(\tfrac{1}{2}D^{n-k})] \times I$$

given by

$$\varphi^{-1}(t_1,\ldots,t_n) \mapsto \left(\frac{(t_1,\ldots,t_k)}{(t_1^2+\ldots+t_k^2)^{1/2}} \, , (t_{k+1},\ldots,t_n) \, , \right.$$
$$\left. \frac{8}{\varepsilon^2}(\frac{\varepsilon^2}{16} - t_1^2 - \ldots - t_k^2 + t_{k+1}^2 + \ldots + t_n^2) \right).$$

[Here we use the inequalities $\frac{\varepsilon^2}{16} < f(x) < 1 - \frac{\varepsilon^2}{16}$, which are immediate consequences of the inclusion $A \subset \mathrm{Im}\,\varphi$.]

The C^∞-structure of Y is fixed by the atlas consisting of arbitrary atlases of the C^∞-manifolds

$$f^{-1}([\,f(x) - \frac{\varepsilon^2}{16}, f(x) + \frac{\varepsilon^2}{16}]) \smallsetminus \mathrm{Cl}\,\varphi^{-1}(A \smallsetminus B)$$

and

$$S^{k-1} \times \mathrm{Int}(2D^{n-k}) \times I.$$

The beginning $\partial_0 Y$ is the union of the two sets

$$f^{-1}(f(x) - \frac{\varepsilon^2}{16}) \smallsetminus [\varphi^{-1}(A \smallsetminus B) \cap f^{-1}(f(x) - \frac{\varepsilon^2}{16})]$$

and

$$S^{k-1} \times 2D^{n-k} \times 0.$$

Similarly, the end $\partial_1 Y$ is the union of the two sets

$$f^{-1}(f(x) + \frac{\varepsilon^2}{16}) \smallsetminus [\varphi^{-1}(A \smallsetminus B) \cap f^{-1}(f(x) + \frac{\varepsilon^2}{16})]$$

and

$$S^{k-1} \times 2D^{n-k} \times 1.$$

The functions

$$f^{-1}([f(x) - \frac{\epsilon^2}{16}, f(x) + \frac{\epsilon^2}{16}]) \smallsetminus \varphi^{-1}(A \smallsetminus B) \to \mathbb{R},$$

$$y \mapsto \frac{8}{\epsilon^2}(\frac{\epsilon^2}{16} + f(y) - f(x)),$$

and

$$S^{k-1} \times \mathrm{Int}(2D^{n-k}) \times I \to \mathbb{R},$$

$$(u,v,t) \mapsto t \quad [u \in S^{k-1}, \quad v \in \mathrm{Int}(2D^{n-k}), \quad t \in I],$$

yield together a function $g: Y \to \mathbb{R}$. It is readily seen that g is a Morse function with no critical points. (In virtue of Theorem 6, this implies that Y is a trivial cobordism; however, in what follows this property of Y is not used directly.) Now consider the composite embedding

$$S^{k-1} \times D^{n-k} \xrightarrow{(z_1,z_2) \mapsto (z_1,z_2,1/2)} S^{k-1} \times D^{n-k} \times I \xrightarrow{\ \mathrm{in}\ }$$

$$\xrightarrow{\ \mathrm{in}\ } S^{k-1} \times \mathrm{Int}(2D^{n-k}) \times I \xrightarrow{\ \mathrm{in}\ } Y$$

and its compression $\psi: S^{k-1} \times D^{n-k} \to g^{-1}(1/2)$. To complete the proof of the theorem we shall presently verify that the cobordism X is diffeomorphic to $\mathrm{El}(g^{-1}(1/2), \psi)$.

As a preliminary step, we find a number δ, $0 < \delta < 1/2$, and a C^∞-diffeomorphism

$$H: \ g^{-1}(\tfrac{1}{2}) \ \times \ [\tfrac{1}{2} - \delta, \tfrac{1}{2} + \delta] \to g^{-1}([\tfrac{1}{2} - \delta, \tfrac{1}{2} + \delta]),$$

such that

$$H((u,v,\tfrac{1}{2}),t) = (u,v,t)$$

for all $u \in S^{k-1}$, $v \in D^{n-k}$, and $t \in [\tfrac{1}{2} - \delta, \tfrac{1}{2} + \delta]$.

By Theorem 4.5.8, one can find η, $0 < \eta < 1/2$, such that there is a C^∞-submersion $\pi: g^{-1}((\tfrac{1}{2} - \eta, \tfrac{1}{2} + \eta)) \to g^{-1}(\tfrac{1}{2})$ which is the identity map on $g^{-1}(\tfrac{1}{2})$. Fix a C^∞-function $\alpha: \mathbb{R}^{n-k} \to \mathbb{R}$, equal to 1 on D^{n-k} and to 0 outside $2D^{n-k}$, and define a new C^∞-submersion $\rho: g^{-1}((\tfrac{1}{2} - \eta, \tfrac{1}{2} + \eta)) \to g^{-1}(\tfrac{1}{2})$ by the formula

$$\rho(y) = \begin{cases} \pi(y), & \text{if } y \notin S^{k-1} \times \mathrm{Int}(2D^{n-k}) \times I, \\[2mm] \pi(u,v,\tfrac{1}{2} + (t - \tfrac{1}{2})(1 - \alpha(v))), & \text{if } y = (u,v,t), \\[2mm] \qquad\qquad \text{where } u \in S^{k-1}, \ v \in 2D^{n-k}, \ t \in I. \end{cases}$$

Further, define

$$G: g^{-1}((\tfrac{1}{2} - \eta, \tfrac{1}{2} + \eta)) \to g^{-1}(\tfrac{1}{2}) \times (\tfrac{1}{2} - \eta, \tfrac{1}{2} + \eta)$$

by $G(y) = (\rho(y), g(y))$. Since G induces a diffeomorphism of $g^{-1}(\tfrac{1}{2})$ onto $g^{-1}(\tfrac{1}{2}) \times \tfrac{1}{2}$ and the differential $d_y G$ is nondegenerate as long as $y \in g^{-1}(\tfrac{1}{2})$, there is δ, $0 < \delta < \eta$, such that the restriction of G to $g^{-1}([\tfrac{1}{2} - \delta, \tfrac{1}{2} - \delta])$ is a C^∞-embedding. Now it is clear that one can take for H the map

$$(ab\, G)^{-1} : g^{-1}(\tfrac{1}{2}) \times [\tfrac{1}{2} - \delta, \tfrac{1}{2} + \delta] \to g^{-1}([\tfrac{1}{2} - \delta, \tfrac{1}{2} + \delta]).$$

Finally, to show that the cobordisms X and $El(g^{-1}(\tfrac{1}{2}), \psi)$ are diffeomorphic, cut each of them into three cobordisms: X into the cobordisms $f^{-1}([0, f(x) - \tfrac{\delta\varepsilon^2}{8}])$, $f^{-1}([f(x) - \tfrac{\delta\varepsilon^2}{8}, f(x) + \tfrac{\delta\varepsilon^2}{8}])$, and $f^{-1}([f(x) + \tfrac{\delta\varepsilon^2}{8}, 1])$, and $El(g^{-1}(\tfrac{1}{2}), \psi)$ into the cobordisms $mo^{-1}([0, \tfrac{1}{2} - \delta])$, $mo^{-1}([\tfrac{1}{2} - \delta, \tfrac{1}{2} + \delta])$, and $mo^{-1}([\tfrac{1}{2} + \delta, 1])$. By Theorem 6, the first and the third cobordisms in each of these triples are trivial, and hence the second cobordism is diffeomorphic to the entire cobordism. Therefore, it is enough to exhibit a diffeomorphism

$$mo^{-1}([\tfrac{1}{2} - \delta, \tfrac{1}{2} + \delta]) \to f^{-1}([f(x) - \tfrac{\delta\varepsilon^2}{8}, f(x) + \tfrac{\delta\varepsilon^2}{8}]).$$

To do this, consider the composite map

$$(g^{-1}(\tfrac{1}{2}) \smallsetminus \psi(S^{k-1} \times \text{Int}\, D^{n-k})) \times [\tfrac{1}{2} - \delta, \tfrac{1}{2} + \delta] \xrightarrow{\ ab\, H\ }$$

$$(Y \smallsetminus (S^{k-1} \times \text{Int}\, D^{n-k} \times I)) \cap g^{-1}([\tfrac{1}{2} - \delta, \tfrac{1}{2} + \delta]) \xrightarrow{\ id\ }$$

$$(X \smallsetminus \varphi^{-1}(\mathbb{R}^k \times \text{Int}\, D^{n-k})) \cap f^{-1}([f(x) - \tfrac{\delta\varepsilon^2}{8}, f(x) + \tfrac{\delta\varepsilon^2}{8}]),$$

and the map

$$E(n,k) \cap mo^{-1}([\tfrac{1}{2} - \delta, \tfrac{1}{2} + \delta]) \longrightarrow$$

$$\varphi^{-1}(\text{Im}\,\varphi \cap (\mathbb{R}^k \times \varepsilon D^{n-k})) \cap f^{-1}([f(x) - \tfrac{\delta\varepsilon^2}{8}, f(x) + \tfrac{\delta\varepsilon^2}{8}])$$

given by the formula

$$(t_1, \ldots, t_n) \mapsto \varphi^{-1}(\varepsilon t_1, \ldots, \varepsilon t_n),$$

which together provide the desired diffeomorphism.

8 (COROLLARY). <u>Suppose that on a given n-dimensional</u> <u>cobordism</u> X <u>with connected boundary there exists a Morse function</u> <u>with a unique critical point. Then</u> X <u>is diffeomorphic to</u> D^n.

Attaching Handles

9. Let X be an n-dimensional cobordism and let $\varphi: S^{k-1} \times D^{n-k} \to \partial_1 X$ be a C^∞-embedding. The result of glueing X and the elementary cobordism $El(\partial_1 X, \varphi)$ by the canonical diffeomorphism $\partial_1 X \to \partial_0 El(\partial_1 X, \varphi)$ (see 2) is said to be obtained from X by <u>attaching</u> <u>a handle of index</u> k.

By attaching a handle of index 0 to X we replace, up to a diffeomorphism, X by $X \sqcup D^n$; the new component of the boundary, i.e., $in_2(S^{n-1})$, is added to $\partial_1 X$. To attach a handle of index n, we actually glue X and D^n by a diffeomorphism of one of the components of $\partial_1 X$ onto S^{n-1}.

10. <u>Every cobordism</u> X <u>can be obtained, up to a diffeomor-</u> <u>phism, from the standard trivial cobordism</u> $\partial_0 X \times I$, <u>by attaching a</u> <u>finite number of handles. Moreover, given any proper Morse function</u> $f: X \to \mathbb{R}$, <u>one may choose these handles so that their number will not</u> <u>exceed the number of critical points of</u> f.

We prove this statement by induction on the number of critical points of f. If f has no critical points, Theorem 6 suffices. If f has $m \geqslant 1$ critical points, then there is $c \in (0,1)$ such that one of the critical values of f is greater than c, while the remaining ones are smaller than c. We cut X into two cobordisms: $f^{-1}([0,c])$ and $f^{-1}([c,1])$. On $f^{-1}([0,c])$ there is a proper Morse function with m-1 critical points, for example, $x \mapsto f(x)/c$. On $f^{-1}([c,1])$ there is a Morse function with a unique critical point, for example, $x \mapsto (f(x) - c)/(1 - c)$. Finally, note that by Theorem 7 the second cobordism is elementary.

11. <u>A closed n-dimensional</u> C^∞<u>-manifold</u> X <u>on which there is</u> <u>Morse function having only two critical points is homeomorphic to</u> S^n.

PROOF. We remark (leaving the trivial case n = 0 aside) that every Morse function with only two critical points is proper (the two points are necessarily a maximum and a minimum). Thus Theorem 10 shows that X can be obtained from an empty manifold by attaching two handles. Obviously, the first handle has index 0, and the second index n, and hence X actually results from glueing two copies of

D^n by a diffeomorphism of S^{n-1}.

A Homotopy Corollary

12 (LEMMA). _The cobordism $El(V,\varphi)$ is homotopy equivalent to $V \cup_f D^k$, where $f: S^{k-1} \to V$ is given by $f(y) = \varphi(y,0)$. Moreover, there is a homotopy equivalence $V \cup_f D^k \to El(V,\varphi)$ which agrees on V with the inclusion $V[= \partial_0 El(V,\varphi)] \to El(V,\varphi)$._

One can assemble such a homotopy equivalence from the above inclusion $V \to El(V,\varphi)$ and the embedding $D^k \to El(V,\varphi)$ which takes each point $x \in D^k$ into the point $x/4 \in E(n,k)$. To complete the proof, it is enough to remark that the constructed mapping $V \cup_f D^k \to El(V,\varphi)$ is a topological embedding whose image is a strong deformation retract of $El(V,\varphi)$.

13. _Every compact n-dimensional smooth manifold is homotopy equivalent to a finite cellular space of dimension $\leqslant n$._

PROOF. The discussion in 1.2 implies that one may assume that the given manifold is a cobordism with an empty beginning. Therefore, all we have to show is that if an n-dimensional cobordism is homotopy equivalent to a finite cellular space of dimension $\leqslant n$, then it retains this property after we attach to it an arbitrary handle; see 10. But from Lemma 12 it follows that attaching a handle of index k to a cobordism X has the same homotopy effect as attaching D^k to X by some embedding $f: S^{k-1} \to X$. Now replace X by a finite cellular space Y of dimension $\leqslant n$ with the same homotopy type, replace the map f by its composition with a homotopy equivalence $X \to Y$, and subsequently replace this composition by a homotopic cellular map $g: S^{k-1} \to Y$ (see 2.3.2.4). By Theorem 1.3.7.8, the cobordism which results by attaching a handle of index k to X is homotopy equivalent to the space $Y \cup_g D^k$; according to 2.1.5.5, $Y \cup_g D^k$ is a finite cellular space of dimension $\leqslant n$.

Spherical Modifications

14. Let V be a closed n-dimensional C^∞-manifold, and let $\varphi: S^{k-1} \times D^{n-k+1} \to V$ be a C^∞-embedding. Fix arbitrary collars on $V \smallsetminus \varphi(S^{k-1} \times \operatorname{Int} D^{n-k+1})$ and $D^k \times S^{n-k}$, and then glue these manifolds by the diffeomorphism $ab\,\varphi: S^{k-1} \times S^{n-k} \to \varphi(S^{k-1} \times S^{n-k})$ of the

boundary of the second onto the boundary of the first. We say that the glued manifold is obtained from V by a <u>spherical modification along the embedding</u> φ.* The number k is the <u>index of the modification</u>.

15. <u>If the cobordism X' is obtained from the cobordism X by attaching a handle using an embedding</u> $\varphi: S^{k-1} \times D^{n-k} \to \partial_1 X$, <u>then $\partial_1 X'$ is obtained from $\partial_1 X$ by a spherical modification along the same embedding</u> $\varphi: S^{k-1} \times D^{n-k} \to \partial_1 X$.

PROOF. Since $\partial_1 X' = \partial_1 \mathrm{El}(\partial_1 X, \varphi)$, we actually claim that $\partial_1 \mathrm{El}(\partial_1 X, \varphi)$ is obtained from $\partial_1 X$ by a spherical modification along φ. Recall that $\mathrm{El}(\partial_1 X, \varphi)$ is the result of glueing the spaces $[\partial_1 X \smallsetminus \varphi(S^{k-1} \times \mathrm{Int}(\tfrac{1}{2}D^{n-k}))] \times I$ and $E(n,k)$ by el (see 2). Obviously, $[\partial_1 X \smallsetminus \varphi(S^{k-1} \times \mathrm{Int}\, D^{n-k})] \times 1$ and $E(n,k) \cap \partial_1 \mathrm{El}(\partial_1 X, \varphi)$ are compact $(n-1)$-dimensional submanifolds of the manifold $\partial_1 \mathrm{El}(\partial_1 X, \varphi)$, which they cover, and they intersect along their common boundary. Consider the mappings

$$\mathrm{pr}_1: [\partial_1 X \smallsetminus \varphi(S^{k-1} \times \mathrm{Int}\, D^{n-k})] \times 1 \to \partial_1 X \smallsetminus \varphi(S^{k-1} \times \mathrm{Int}\, D^{n-k})$$

and

$$\psi: E(n,k) \cap \partial_1 \mathrm{El}(\partial_1 X, \varphi) \to D^k \times S^{n-k-1},$$

$$\psi(t_1, \ldots, t_n) = \left(\frac{16}{15}(t_1, \ldots, t_k), \frac{(t_{k+1}, \ldots, t_n)}{(t_{k+1}^2 + \ldots + t_n^2)^{1/2}} \right).$$

It is clear that ψ is a C^∞-diffeomorphism such that $\mathrm{pr}_1(\psi^{-1}(z_1, z_2)) = \varphi(z_1, z_2)$ for all $z_1 \in S^{k-1}$ and $z_2 \in S^{n-k-1}$. Now applying Theorem 4.5.9, we see that $\partial_1 \mathrm{El}(\partial_1 X, \varphi)$ is really obtained by glueing the manifolds $\partial_1 X \smallsetminus \varphi(S^{k-1} \times \mathrm{Int}\, D^{n-k})$ and $D^k \times S^{n-k-1}$ by ab φ.

16 (COROLLARY). <u>Given an arbitrary cobordism X, the manifold $\partial_1 X$ can be obtained from $\partial_0 X$ through a finite number of spherical modifications. In particular, the boundary of an arbitrary compact C^∞-manifold can be obtained from an empty manifold through a finite number of spherical modifications.</u>

---

* Translator's note: or <u>by doing a surgery on</u> V <u>using</u> φ.

3. <u>Two-dimensional Manifolds</u>

1. The theory presented in the two previous subsections
represents, first of all, an attempt to somehow visualize, survey, and
classify smooth compact manifolds of a given dimension. To apply this
theory in complex situations, one needs to develop it further, but in
the simplest cases Theorem 2.10 emerges as a sufficiently effective
tool. For example, in the one-dimensional case every standard cobordism
is a sum of segments, and hence Theorem 2.10 implies that every smooth
compact one-dimensional manifold is diffeomorphic to the sum of a finite
number of segments and circles. Warning: although this differentiable
classification coincides with the topological classification given in
1.1.17, it has an entirely different meaning. [We add that also in the
noncompact case (which, due to its relative complexity, is not
considered in our book in any systematic way), the differentiable classification of one-dimensional manifolds is identical with the topological one; see 4.1.]

The present subsection is devoted to the differentiable classification of the smooth, compact, two-dimensional manifolds, and again
Theorem 2.10 is enough. We begin by drawing up a list of model manifolds
(which play the same role as the segments and circles in the one-dimensional case). Next we show that every smooth, compact, connected, two-
dimensional manifold is diffeomorphic to one of the models. No two of
the model manifolds are diffeomorphic or homeomorphic. However, we
prove this fact only in Chapter 5 (see 5.3.4.3); here we merely prepare
the geometric part of the proof, decribing for the closed model manifolds
canonical rigged cellular decomposition, and for the nonclosed model
manifolds - the bouquets of circles to which they are homotopy
equivalent. As we did in the classification of one-dimensional manifolds,
we shall not worry about the differentiability class (cf. 1.1).

Model Surfaces

2. We begin with the <u>elementary</u> model surfaces: the <u>spheres</u>
<u>with holes</u> and the <u>Möbius strips.</u>

A sphere with l holes is S^2 with the interiors of l
pairwise disjoint sperical caps (segments) removed; its boundary is a
sum of l circles and inherits a well-defined orientation. A sphere
with one hole is diffeomorphic to D^2, while a sphere with two holes

is diffeomorphic to the cylinder $S^1 \times D^1$; we call the latter a <u>handle</u> (do not confuse with the "handles" in Subsection 2 !).

A Möbius strip is a submanifold of $\mathbb{R}^3$ produced by the motion of a segment of length 1 whose middle glides along a circle S^1 in such a manner that the segment remains normal to the circle and turns uniformly through a total angle π (see Fig. 6). Every Möbius strip is a nonorientable compact submanifold of $\mathbb{R}^3$ with boundary diffeomorphic to S^1.

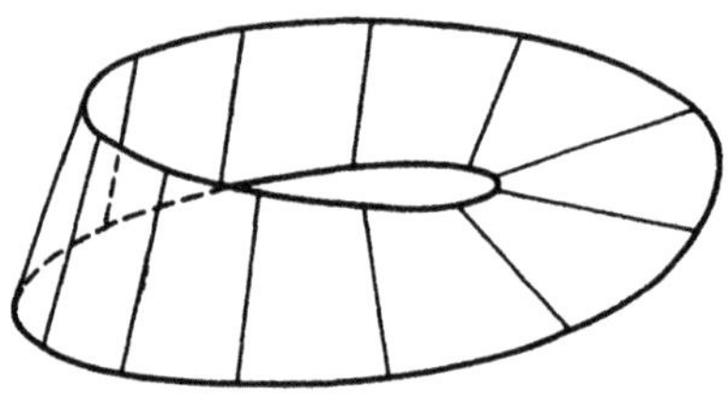

Fig. 6

The list of all <u>model surfaces</u> is: the empty two-dimensional manifold, the <u>spheres with handles and holes,</u> and the <u>spheres with crosscaps.</u>

A sphere with g handles and l holes ($g \geqslant 0$, $l \geqslant 0$) is defined as the result of glueing a sphere with $2g+l$ holes and a sum of g handles by an orientation preserving diffeomorphism from the boundary of the sum of handles onto the union of $2g$ components of the sphere with holes. This object is a smooth, compact, orientable, two-dimensional manifold, uniquely defined up to an orientation preserving diffeomorphism by the numbers g and l, and whose boundary is diffeomorphic to a sum of l circles. This manifold coincides with S^2 if $g = 0, l = 0$; it is diffeomorphic to D^2 if $g = 0, l = 1$, and to $S^1 \times S^1$ if $g = 1, l = 0$, and then it is known as a <u>torus;</u> for $g = 2$ and $l = 0$, it is called a <u>double torus</u> or a <u>pretzel.</u> For any g and l, a sphere with g handles and l holes can be differentiably embedded in $\mathbb{R}^3$; for $g = 3$ and $l = 2$, the standard embedding is depicted in Fig. 7.

A sphere with h crosscaps and l holes ($h \geqslant 0$, $l \geqslant 0$) is defined as the result of glueing a sphere with $h+l$ holes and a sum of h Möbius strips by a diffeomorphism of the boundary of the sum of strips onto the union of h components of the sphere with holes. One obtains a smooth, compact, two-dimensional manifold, uniquely defined up to a diffeomorphism by the numbers h and l, and whose boundary is diffeomorphic to a sum of l circles. This manifold is orientable

only for h = 0. For h = 1, l = 0, it is diffeomorphic to $\mathbb{R}P^2$, and
for h = 1, l = 1, to a Möbius strip. For h = 2, l = 0, this
manifold is called the <u>Klein bottle</u> and is well known because of its
immersion in $\mathbb{R}^3$, depicted in Fig. 8; for h = 2, l = 1, it is called
a <u>disc with an inverted handle</u> and is depicted in Fig. 9, on the left
(the drawing on the right in Fig. 9 demonstrates that by glueing a disc
with an inverted handle and a usual disc by a diffeomorphism of their
boundaries we actually get a Klein bottle).

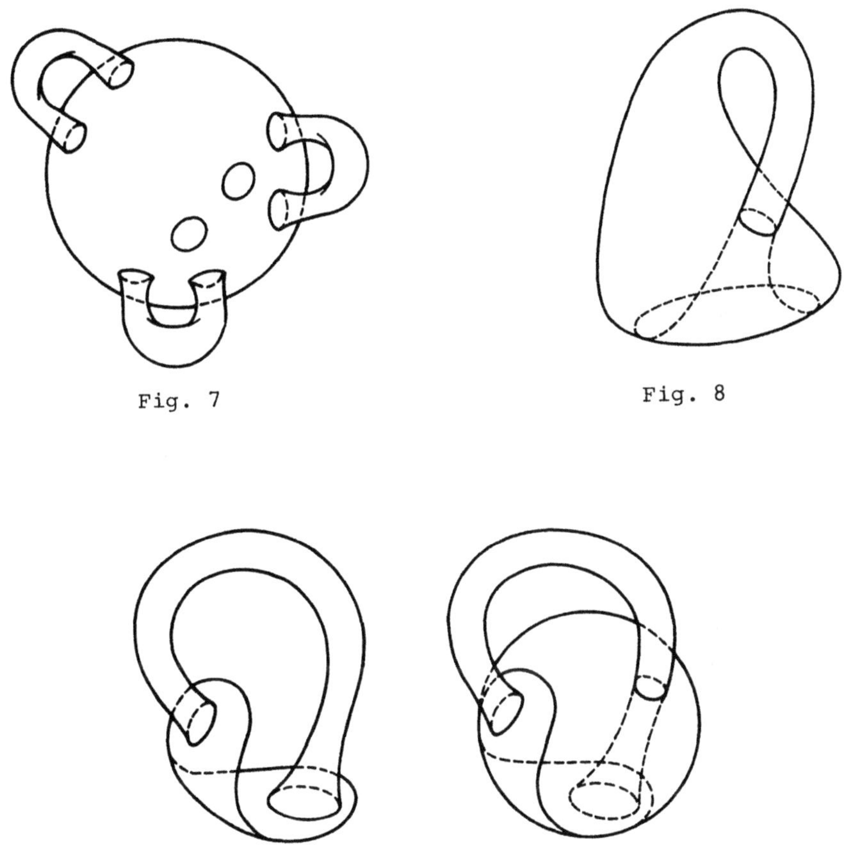

Fig. 7   Fig. 8

Fig. 9

We emphasize that in our list of model manifolds, the
crosscaps do not meet with the handles, i.e., we excluded the case of
a sphere with holes to which we attach both handles and Möbius strips.
We show in 3 that for h ≥ 1, every sphere with g handles, h crosscaps, and l holes is diffeomorphic to a sphere with 2g+h crosscaps
and l holes.

Auxiliary Propositions

3 (LEMMA). <u>Let X_1 and X_2 be smooth, compact, two-dimensional manifolds, and let φ be a diffeomorphism of a component of ∂X_1 onto a component of ∂X_2. Denote by X the manifold obtained by glueing X_1 and X_2 by φ. Then:</u>

(i) <u>if X_1 and X_2 are diffeomorphic to a sphere with g_1 handles and l_1 holes, and to a sphere with g_2 handles and l_2 holes, respectively, then X is diffeomorphic to a sphere with g_1+g_2 handles and l_1+l_2-2 holes;</u>

(ii) <u>if X_1 and X_2 are diffeomorphic to a sphere with h_1 crosscaps and l_1 holes, and to a sphere with h_2 crosscaps and l_2 holes, respectively, then X is diffeomorphic to a sphere with h_1+h_2 crosscaps and l_1+l_2-2 holes;</u>

(iii) <u>if X_1 and X_2 are diffeomorphic to a sphere with g_1 handles and l_1 holes, and to a sphere with $h_2 > 0$ crosscaps and l_2 holes, respectively, then X is diffeomorphic to a sphere with $2g_1+h_2$ crosscaps and l_1+l_2-2 holes.</u>

The only case which needs proof is (iii), where we see immediately that X is diffeomorphic to a sphere with g_1 handles, h_2 crosscaps, and l_1+l_2-2 holes.

To begin with, suppose that $g_1 = 1$, $h_2 = 1$, and $l_1+l_2-2 = 1$. Then X is diffeomorphic to a Möbius strip with a handle attached; see the left drawing in Fig. 10. The last manifold is obviously diffeomorphic to a Möbius strip with an inverted handle, as that in the right drawing in Fig. 10, and hence can be cut into a sphere with one crosscap and a disc with an inverted handle. According to (ii), this implies that X is diffeomorphic to a sphere with three crosscaps and one hole.

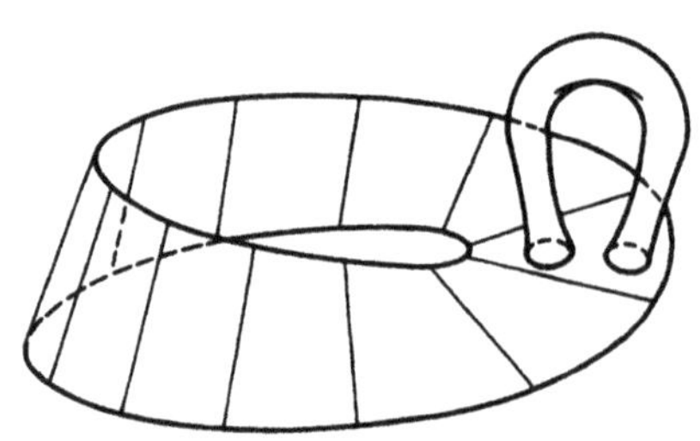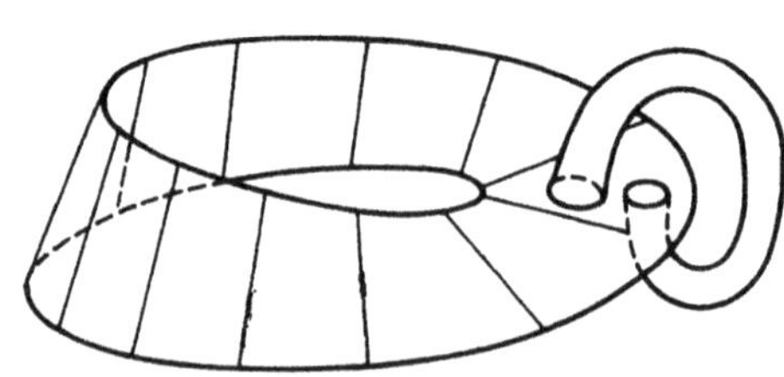

Fig. 10

In the general case, we use induction on g_1. If $g_1 = 0$, there is nothing to prove. If $g_1 > 0$, then X can be glued from a sphere with g_1-1 handles, h_2-1 crosscaps, and l_1+l_2-1 holes and a sphere with one handle, one crosscap, and one hole. But we already proved that this second manifold is diffeomorphic to a sphere with three crosscaps and one hole. Consequently, X is diffeomorphic to a sphere with g_1-1 handles, h_2+2 crosscaps, and l_1+l_2-2 holes, and so diffeomorphic to a sphere with $2g_1+h_2$ crosscaps and l_1+l_2-2 holes.

4 (LEMMA). <u>Let</u> X <u>be a smooth, compact, connected, two-dimensional manifold, and let</u> X' <u>be the result of attaching a handle to</u> X <u>by a diffeomorphism of the boundary of the handle onto the union of two components of</u> ∂X. <u>Then:</u>

(i) <u>if</u> X <u>is diffeomorphic to a sphere with</u> g <u>handles and</u> l <u>holes, then</u> X' <u>is diffeomorphic to a sphere with</u> $g+1$ <u>handles and</u> $l-2$ <u>holes, or to a sphere with</u> $2g+2$ <u>crosscaps and</u> $l-2$ <u>holes;</u>

(ii) <u>if</u> X <u>is diffeomorphic to a sphere with</u> h <u>crosscaps and</u> l <u>holes, then</u> X' <u>is diffeomorphic to a sphere with</u> $h+2$ <u>crosscaps and</u> $l-2$ <u>holes.</u>

This is a corollary of Lemma 3: indeed, in both cases one can cut X' into two manifolds, such that the first one differs from X by having one hole less, while the second is diffeomorphic to a sphere with one handle and one hole, or to a disc with an inverted handle.

5. <u>Let</u> X_1 <u>and</u> X_2 <u>be smooth, compact, two-dimensional manifolds, and let</u> φ <u>be a diffeomorphism of a nonempty union of whole components of</u> ∂X_2 <u>onto a nonempty union of whole components of</u> ∂X_1. <u>Denote by</u> X <u>the manifold glued from</u> X_1 <u>and</u> X_2 <u>by means of</u> φ. <u>If both</u> X_1 <u>and</u> X_2 <u>are diffeomorphic to model surfaces, then</u> X <u>is also diffeomorphic to one of the model surfaces.</u>

This is a consequence of Lemmas 3 and 4, because glueing by means of φ is equivalent to first glueing by the diffeomorphism of one of the components of ∂X_2 onto the corresponding component of ∂X_1, obtained by compressing φ, and subsequently attaching a number of handles equal to half the number of the components of ∂X_1 and ∂X_2 which remain to be identified.

The Main Theorem

6. <u>Every smooth, connected, compact, two-dimensional manifold is diffeomorphic to one of the model surfaces.</u>

Applying Theorems 2.10 and 5, all we need to show is that the components of the elementary two-dimensional cobordisms are diffeomorphic to model surfaces. And this is not hard to check directly by examining all possible cases, if we recall that every smooth, closed, one-dimensional manifold is diffeomorphic to a sum of circles. To spell it out, every elementary cobordism of index 0 constructed from a sum of m circles is diffeomorphic to a sum of m spheres with two holes and a sphere with one hole. Next, every elementary cobordism of index 2 constructed from a sum of m circles (and a differentiable embedding of a circle in this sum) is diffeomorphic to a sum of m-1 spheres with two holes and a sphere with one hole. And finally, every elementary cobordism of index 1 constructed from a sum of m circles and a differentiable embedding of $S^0 \times D^1$ in this sum is diffeomorphic to one of the following three manifolds: a sum of m-2 spheres with two holes and a sphere with three holes; a sum of m-1 spheres with two holes and a sphere with three holes; a sum of m-1 spheres with two holes and a sphere with one crosscap and two holes (for m = 2, one see the three cases in Fig. 11).

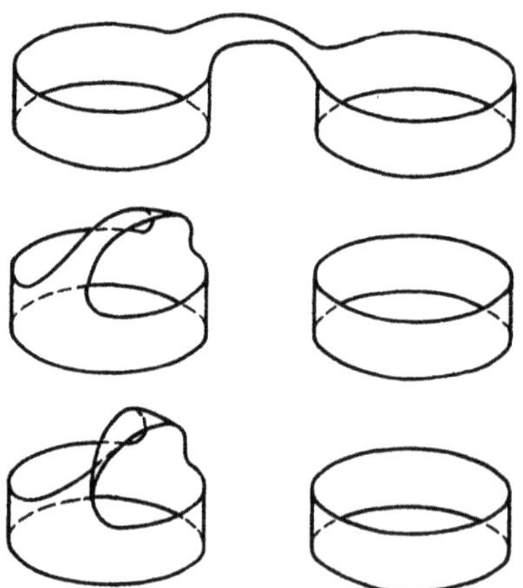

Fig.11

7 (INFORMATION). Every compact, connected, two-dimensional topological manifold is homeomorphic to one of the model surfaces.

Cellular Decompositions
of the Closed Model Surfaces

8. The closed model surfaces possess standard rigged cellular decompositions, which generalize the canonical decompositions of the sphere S^2, the complex projective space $\mathbb{C}P^2$, and the torus $S^1 \times S^1$ into two cells, three cells, and four cells, respectively. Each of these standard decompositions, except the no-cell decomposition of the empty model surface, contains only one 0-cell and only one 2-cell, while the number of 1-cells is $2g$ for a sphere with g handles, and h for a sphere with h crosscaps. Therefore, the 1-skeleton of a sphere with g handles is a bouquet of $2g$ circles, while the 1-skeleton of a sphere with h crosscaps is a bouquet of h circles. Moreover, the description of the entire rigged cellular decomposition reduces to the characterization of the attaching map for the 2-cell, i.e., of a certain map of S^1 into the aforementioned bouquet.

We disregard the values $g = 0,1$ and $h = 0,1$, already considered, and for the case of a sphere with g handles, we represent S^1 as the contour of a regular polygon with first vertex ort_1 and $4g$ edges, arranged successively as

$$a_1, \ b_1, \ a_1', \ b_1', \ \ldots , \ a_g, \ b_g, \ a_g', \ b_g' .$$

In the case of a sphere with h crosscaps, we represent S^1 as the contour of a regular polygon with first vertex ort_1 and $2h$ edges, which are arranged successively as

$$c_1, \ c_1', \ \ldots, \ c_h, \ c_h' .$$

In both cases we form a quotient space of S^1 by identifying each edge with the correspondig "primed" edge, as follows: a_i is identified with a_i', and b_i with b_i', through a reflection with respect to a line (relative to which these edges are symmetric), while c_i is identified with c_i' through a rotation of the polygon (around its center). In either of cases the quotient space is a bouquet of circles: in the first case the number of circles is $2g$, and in the second h. The projection of S^1 onto this quotient space is the required attaching map.

Of course, we still have to convince the reader that the cellular spaces produced in this manner are homeomorphic to the model surfaces. To this end, let us divide our 4g-gon into a g-gon and g pentagons, by drawing diagonals which cut out quadruplets a_i, b_i, a_i', b_i'. Similarly, we divide our 2h-gon into a h-gon and h triangles, by drawing diagonals which cut out pairs c_i, c_i' (see Fig. 12).

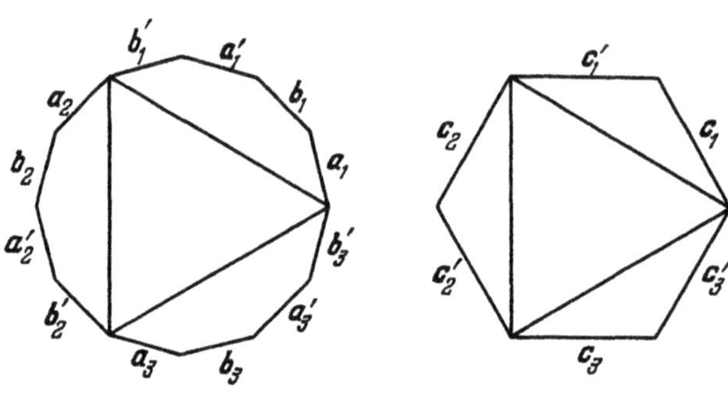

Fig. 12   (g = 3, h = 3)

Identifying the edges a_i, a_i' and b_i, b_i' in the way described above, all the vertices of the remaining g-gon become one and the same point, thus transforming the g-gon into a sphere with g circular apertures; at the same time, the edges of the pentagons are identified in such a manner that each pentagon becomes a torus with a circular aperture. Attaching these holed tori to the holed sphere in such a way as to restore all that was destroyed by the auxiliary (diagonal) cuts, we obtain, up to a homeomorphism, a sphere with g handles. Similarly, the prescribed identifications of the edges c_i, c_i' take all the vertices of the h-gon into one and the same point, thus transforming the h-gon into a sphere with h circular apertures; at the same time, each triangle becomes a Möbius strip, since two of its edges are identified. Attaching these strips to the holed sphere, we obtain, up to a homeomorphism, a sphere with h crosscaps. [Warning: the boundaries of the previous circular apertures (in the sphere) have a common point, and for g = 2 or h = 2, they even coincide.]

The Homotopy Structure
of the Nonclosed Model Surfaces

9. A sphere with g handles and l holes is homotopy equivalent to a bouquet of $2g+l-1$ circles. A sphere with h crosscaps and l holes is homotopy equivalent to a bouquet of $h+l-1$ circles.

To prove these assertions, we first note that by attaching the $4g$-gon to the bouquet of $2g$ circles as in 8, we produce, up to a homeomorphism, a sphere with g handles and l holes, provided we first remove the interiors of l pairwise disjoint discs from the interior of the $4g$-gon. Now let us arrange these discs such that every line passing through the first vertex, ort_1, intersects no more than one of them. Denote by A the set consisting of:

(i) the contour of the $4g$-gon;

(ii) the $2l-2$ segments tangent to $l-1$ of the removed discs and passing through ort_1;

(iii) the outer arcs of the boundaries of these discs having as endpoints the tangency points.

Then A is obviously a strong deformation retract of the holed $4g$-gon. If we now project this $4g$-gon onto the sphere with g handles, the above strong deformation retraction is transformed into a strong deformation retraction of the holed sphere with g handles onto the image of A under the projection. Finally, note that this image is manifestly homeomorphic to a bouquet of $2g+l-1$ circles.

The proof for a sphere with h crosscaps and l holes is a verbatim repetition of the previous argument, with the $4g$-gon replaced by the $2h$-gon.

4. Exercises

1. Show that every smooth, connected, noncompact, one-dimensional manifold is homeomorphic to a line or a half-line (see 3.1).

2. Define a submanifold of $\mathbb{C}P^2$ in homogeneous coordinates by the equation $z_1^m + z_2^m + z_3^m = 0$, and show that it is diffeomorphic to a sphere with $(m-1)(m-2)/2$ handles (see 2.4.4 and 2.4.5).

3. Show that the subset of $\mathbb{C}P^1 \times \mathbb{C}P^1$ consisting of the

points $((z_1:z_2),(w_1:w_2))$ such that $z_1^q(w_1^p + w_2^p) = z_2^2(w_1^p - w_2^p)$ is a manifold diffeomorphic to a sphere with $(p-1)(q-1)$ handles.

4. Show that every smooth, closed, connected, orientable, three-dimensional manifold can be obtained by glueing two copies of a handle-body by a diffeomorphism of its boundary. (A handle-body is a part of $\mathbb{R}^3$ bounded by a sphere with handles which is standardly embedded in $\mathbb{R}^3$).

5. Consider the manifold obtained by glueing two copies of the solid torus $S^1 \times D^2$ by a diffeomorphism of its boundary $S^1 \times S^1$, given by the formula $(z_1,z_2) \mapsto (z_1^a z_2^b, z_1^c z_2^d)$, where a,b,c,d are integers satisfying $ad - bc = \pm 1$. Show that this manifold is diffeomorphic to S^3 for $a = 0$, to $S^2 \times S^1$ for $a = \pm 1$, and to $\mathbb{R}P^3$ for $a = \pm 2$.

6. Show that on every connected closed C^∞-manifold there is a Morse function with a unique local minimum and a unique local maximum.

7. Show that on every connected cobordism X with nonempty $\partial_0 X$ and $\partial_1 X$ there is a proper Morse function with no local maxima and minima lying in $\operatorname{int} X$.

8. Show that on every cobordism there is a proper Morse function such that, for any of its critical points x_1 and x_2, of indices k_1 and k_2, $k_1 < k_2$ implies $f(x_1) < f(x_2)$.

9. Suppose that on a cobordism X there is a Morse function with no critical points of index 1 and that the manifold $\partial_0 X$ is orientable. Show that the cobordism X is orientable and that every orientation of $\partial_0 X$ is induced by some orientation of X.

Chapter 4. Bundles

§1. BUNDLES WITHOUT GROUP STRUCTURE

1. General Definitions

1. A <u>bundle</u> is a triple (T,p,B), where T and B are topological spaces and $p\colon T \to B$ is a continuous map. The spaces T and B are called the <u>total space</u> and the <u>base</u> of the bundle (T,p,B), respectively, and the map p is its <u>projection.</u> For a bundle ξ, we denote its total space, its base, and its projection by $\mathrm{tl}\,\xi$, $\mathrm{bs}\,\xi$, and $\mathrm{pr}\,\xi$, respectively: $\xi = (\mathrm{tl}\,\xi, \mathrm{pr}\,\xi, \mathrm{bs}\,\xi)$.

The preimage $\mathrm{pr}\,\xi^{-1}(b)$ of a point $b \in \mathrm{bs}\,\xi$ is called the <u>fiber of the bundle</u> ξ <u>over the point</u> b.

A <u>section</u> of the bundle ξ is a continuous map $s\colon \mathrm{bs}\,\xi \to \mathrm{tl}\,\xi$ such that $\mathrm{pr}\,\xi \circ s = \mathrm{id}\,\mathrm{bs}\,\xi$. Two sections of ξ are <u>homotopic</u> if they can be connected by a homotopy consisting only of sections, i.e., by a homotopy $h\colon \mathrm{bs}\,\xi \times I \to \mathrm{tl}\,\xi$ such that $\mathrm{pr}\,\xi \circ h$ equals $\mathrm{pr}_1\colon \mathrm{bs}\,\xi \times I \to \mathrm{bs}\,\xi$.

The <u>restriction of the bundle</u> ξ <u>to a subspace</u> $B \subset \mathrm{bs}\,\xi$ is the bundle $(\mathrm{pr}\,\xi^{-1}(B), \mathrm{ab}\,\mathrm{pr}\,\xi, B)$, denoted by $\xi\big|_B$.

The <u>product of the bundles</u> ξ_1 <u>and</u> ξ_2 is the bundle $(\mathrm{tl}\,\xi_1 \times \mathrm{tl}\,\xi_2, \mathrm{pr}\,\xi_1 \times \mathrm{pr}\,\xi_2, \mathrm{bs}\,\xi_1 \times \mathrm{bs}\,\xi_2)$, denoted by $\xi_1 \times \xi_2$. The fiber of $\xi_1 \times \xi_2$ over a point $(b_1,b_2) \in \mathrm{bs}(\xi_1 \times \xi_2)$ is precisely the product of the fibers $\mathrm{pr}\,\xi_1^{-1}(b_1)$ and $\mathrm{pr}\,\xi_2^{-1}(b_2)$.

2. A <u>map of the bundle</u> ξ' <u>into the bundle</u> ξ is a pair of continuous maps $F\colon \mathrm{tl}\,\xi' \to \mathrm{tl}\,\xi$, $f\colon \mathrm{bs}\,\xi' \to \mathrm{bs}\,\xi$, such that the diagram

$$\begin{array}{ccc} \mathrm{tl}\;\xi' & \xrightarrow{\;\;F\;\;} & \mathrm{tl}\;\xi \\ \mathrm{pr}\;\xi' \downarrow & & \downarrow \mathrm{pr}\;\xi \\ \mathrm{bs}\;\xi' & \xrightarrow{\;\;f\;\;} & \mathrm{bs}\;\xi \end{array} \qquad (1)$$

is commutative. If $\phi = (F,f)$ is such a pair, we write $\phi: \xi' \to \xi$, $F = \mathrm{tl}\,\phi$, $f = \mathrm{bs}\,\phi$.

A map $\phi: \xi' \to \xi$ is said to be an <u>isomorphism</u> if $\mathrm{tl}\,\phi$ and $\mathrm{bs}\,\phi$ are homeomorphisms, and an <u>equivalence</u> if, in addition, $\mathrm{bs}\,\xi' =$ $= \mathrm{bs}\,\xi$ and $\mathrm{bs}\,\phi = \mathrm{id}\,\mathrm{bs}\,\xi$. If there is an isomorphism (equivalence) between ξ' and ξ, then the bundles ξ' and ξ are said to be <u>isomorphic</u> (respectively, <u>equivalent</u>).

A map $\phi: \xi' \to \xi$ is called an <u>inclusion</u> if $\mathrm{tl}\,\phi$ and $\mathrm{bs}\,\phi$ are inclusions. For example, given any subset B of $\mathrm{bs}\,\xi$, the inclusions $\mathrm{in}: \mathrm{pr}\,\xi^{-1}(B) \to \mathrm{tl}\,\xi$ and $\mathrm{in}: B \to \mathrm{bs}\,\xi$ form the inclusion of the bundle $\xi|_B$ in ξ.

3. The commutativity of the diagram (1) implies that F is a <u>fiber preserving map</u> (or a <u>fibered map</u>), i.e., it takes each fiber of ξ' into a fiber of ξ. Obviously, if $\mathrm{pr}\,\xi'(\mathrm{tl}\,\xi') = \mathrm{bs}\,\xi'$, then given an arbitrary fiber preserving map $F: \mathrm{tl}\,\xi' \to \mathrm{tl}\,\xi$ there is a unique map $f: \mathrm{bs}\,\xi' \to \mathrm{bs}\,\xi$ which makes diagram (1) commutative. Moreover, if the map $\mathrm{pr}\,\xi'$ is factorial, then the continuity of F implies the continuity of f. Therefore, <u>if</u> ξ' <u>is a bundle with factorial</u> <u>projection, then given any fiber preserving map</u> $F: \mathrm{tl}\,\xi' \to \mathrm{tl}\,\xi$ <u>there</u> <u>is one and only one continuous map</u> $\phi: \xi' \to \xi$ <u>such that</u> $\mathrm{tl}\,\phi = F$.

4. Let f be a continuous map of a topological space B into the base of a bundle ξ. We may define a new bundle having base B, total space $\{(b,x) \in B \times \mathrm{tl}\,\xi \mid f(b) = \mathrm{pr}\,\xi(x)\}$, and projection – the restriction of the projection $\mathrm{pr}_1: B \times \mathrm{tl}\,\xi \to B$ to the last space. This new bundle is called the <u>bundle induced from</u> ξ <u>by</u> f, and is denoted by $f^!\xi$.

It is clear that the restriction of the projection $\mathrm{pr}_2: B \times \mathrm{tl}\,\xi \to \mathrm{tl}\,\xi$ to $\mathrm{tl}(f^!\xi)$ defines for each $b \in \mathrm{bs}(f^!\xi)$ a homeomorphism of the fiber of $f^!\xi$ over the point b onto the fiber of ξ over the point $f(b) \in \mathrm{bs}\,\xi$, and determines, together with f, a map $f^!\xi \to \xi$. This map is called the <u>adjoint of</u> f and is denoted by $\mathrm{ad}\,f$.

The following observations also need no proofs or explanations. If f is a homeomorphism, then the adjoint map $\mathrm{ad}\,f: f^!\xi \to \xi$ is an isomorphism; if, in addition, $f = \mathrm{id}\,\mathrm{bs}\,\xi$, then $\mathrm{ad}\,f$ is an equivalence.

If f is an inclusion, then $\operatorname{ad} f : f^! \xi \to \xi$ establishes an equivalence between $f^! \xi$ and $\xi|_B$. Finally, given arbitrary continuous maps $f : B \to \operatorname{bs} \xi$ and $g : B' \to B$, the bundles $(f \circ g)^! \xi$ and $g^! (f^! \xi)$ are canonically equivalent.

    5. If $\phi : \xi' \to \xi$ is a map of bundles, then the formula $x \mapsto (\operatorname{pr} \xi'(x), \operatorname{tl} \phi(x))$ defines a continuous map $\operatorname{tl} \xi' \to \operatorname{tl}((\operatorname{bs} \phi)^! \xi)$. This map defines, together with $\operatorname{id} \operatorname{bs} \xi'$, a map of ξ' into the bundle $(\operatorname{bs} \phi)^! \xi$, which we denote by $\operatorname{corr} \phi$; we say that $\operatorname{corr} \phi$ <u>corrects</u> the map ϕ. Obviously, $\operatorname{ad}(\operatorname{bs} \phi) \circ \operatorname{corr} \phi = \phi$.

2. <u>Locally Trivial Bundles</u>

    1. The obvious example of a bundle having a given base B and fibers homeomorphic to a given space F is the <u>standard trivial bundle</u> (or the <u>product bundle</u>) $(B \times F, \operatorname{pr}_1, B)$. Its fibers are the fibers $b \times F$ of the product $B \times F$, and are obviously canonically homeomorphic to F.

    Notice that the sections $B \to B \times F$ of the standard trivial bundle $(B \times F, \operatorname{pr}_1, B)$ are in a one-to-one correspondence with the continuous functions $B \to F$: for each function $f : B \to F$ there is the corresponding section $s : B \to B \times F$, $s(b) = (b, f(b))$; we say that f and s are <u>associated.</u>

    2. A bundle ξ is <u>trivial</u> or, more specifically, <u>topologically trivial,</u> if it is equivalent to a standard trivial bundle. Any equivalence between a standard trivial bundle and ξ is referred to as a <u>trivialization</u> of ξ.

    A bundle ξ is <u>locally trivial</u> or, more specifically, <u>topologically locally trivial,</u> if every point of $\operatorname{bs} \xi$ has a neighborhood U such that the bundle $\xi|_U$ is trivial.

    Since the projection of a product of topological spaces onto one of its factors is an open map, the projection of a trivial bundle is open, and hence so is the projection of a locally trivial bundle.

    It is immediate that the product of two trivial (locally trivial) bundles is a trivial (respectively, locally trivial) bundle. Furthermore, any bundle induced from a trivial (locally trivial) bundle is trivial (respectively, locally trivial). If $f : B \to \operatorname{bs} \xi$ is constant, then $f^! \xi$ is a trivial bundle, for any ξ.

    3. The fibers of a trivial bundle are, as those of the standard trivial bundle, homeomorphic to each other. However, in a

trivial, but not standard trivial bundle, these homeomorphisms are not canonical any longer. If the base of a locally trivial bundle is connected, then its fibers are also mutually homeomorphic; indeed, the set of the points of the base having fibers homeomorphic to a given fiber is open, and the sets of this type form a partition of the base (see 1.3.3.5). On the other hand, the example of the locally trivial bundle $((B \times F) \bigsqcup (B' \times F'), pr_1 \bigsqcup pr_1, B \bigsqcup B')$, where B, F, B', and F' are arbitrary topological spaces, demonstrates that in a locally trivial bundle the fibers over points situated in different components of the base are not necessarily homeomorphic. Moreover, we see that there are locally trivial bundles which are not trivial.

A nontrivial, locally trivial bundle may have a connected base; see 5 and 6.

Coverings

4. A locally trivial bundle is a <u>covering in the broad sense</u> if all its fibers are discrete spaces. In this case the total space and the projection are usually called a <u>covering space</u> and a <u>covering projection,</u> respectively.* Clearly, every point of a covering space has a neighborhood such that the restriction of the projection to this neighborhood is a homeomorphism onto its image in the base.

A covering in the broad sense is said to be a <u>covering in the narrow sense</u> or, simply, a <u>covering,</u> if both the covering and base spaces are connected and nonempty. According to 3, all the fibers of a covering have the same cardinality, called the <u>number of sheets</u> (or the <u>multiplicity</u>) of the given covering.

5. <u>A covering whose number of sheets is greater than one cannot be trivial.</u>

Indeed, the total space of a trivial bundle is homeomorphic to the product of the base and a fiber, and hence it cannot be connected when the fiber is discrete and has more than one point.

6 (EXAMPLES). The bundle (S^1, hel_m, S^1), where $hel_m \colon S^1 \to S^1$ is defined as $hel_m(z) = z^m$, is an m-sheeted covering for any $m \neq 0$. The bundle $(\mathbb{R}^1, hel, S^1)$, where $hel \colon \mathbb{R} \to S^1$ is defined as $hel(x) = e^{2\pi i x}$, is a countably-sheeted covering.

If $k \neq 0, n$, then the bundle having total space $G_+(n,k)$,

---

* Translator's note: Frequently, the terms covering space and covering projection are themselves used to designate the whole covering.

base G(n,k), and projection equal to the submersion exhibited in 3.2.2.3, is a two-sheeted covering. In particular, so is $(S^n, pr, \mathbb{R}P^n)$ for $n \geqslant 1$.

Finally, let us show that every sphere with h crosscaps admits as a two-sheeted covering space a sphere with h-1 handles (see Subsection 3.5.3); here h is an arbitrary positive integer. For h = 1, we already encountered such a covering, namely $(S^2, pr, \mathbb{R}P^2)$. Generally, one may construct it starting with h copies of $(S^2, pr, \mathbb{R}P^2)$. To do this, restrict one copy of $(S^2, pr, \mathbb{R}P^2)$ to a covering over the projective plane with h-1 holes, and restrict the remaining h-1 copies to coverings over the projective plane with one hole (i.e., over the Möbius strip). Now glue the bases of these h restricted coverings into a sphere with h crosscaps by diffeomorphisms of the boundaries of the holes. The resulting glued space is the base of a new covering, whose total space is obtained by glueing the h total spaces of the above restricted coverings: one of these total spaces is a sphere with h-1 pairs of antipodal holes, while the remaining h-1 total spaces are spheres with two antipodal holes, i.e., cylinders over circles. (Each of these h-1 cylinders has two possible covering attaching maps, and we may use either one of them.) Since "sealing" a pair of holes by a cylinder results in replacing this pair by a handle, what we actually obtain is a two-sheeted covering having a sphere with h crosscaps as the base and a sphere with h-1 handles as the total space.

3. Serre Bundles

1. A bundle ξ is a _Serre bundle_* if it satisfies _Serre's condition_: for any positive integer r or r = 0, and every continuous maps $f: I^r \to bs\, \xi$ and $f_0^{\sim}: I^{r-1} \to tl\, \xi$, related by $pr\, \xi \circ f_0^{\sim} = f\big|_{I^{r-1}}$, there is a continuous map $f^{\sim}: I^r \to tl\, \xi$ such that $pr\, \xi \circ f^{\sim} = f$ and $f^{\sim}\big|_{I^{r-1}} = f_0^{\sim}$. (We identify the cube I^{r-1} with that face of I^r whose points have the last coordinate equal to zero; see 1.2.5.7.)

The requirement that $pr\, \xi \circ f = f$ appearing in Serre's condition is fundamental in the theory of bundles, and is encountered also when f and f are defined on spaces other than cubes. If two maps, $f^{\sim}: X \to tl\, \xi$ and $f: X \to bs\, \xi$, satisfy this last requirement,

---

* Translator's note: Frequently called a _Serre fiber space_ or a _weak fibration_.

we say that $\tilde{f}$ <u>covers</u> f (or that $\tilde{f}$ is a <u>lifting</u> or <u>lift</u> of f; we also say that f can be lifted to $\text{tl}\,\xi$); this terminology is valid for arbitrary ξ and X.

Obviously, the product of two Serre bundles and a bundle induced from a Serre bundle are again Serre bundles.

2. Examples of bundles which do not satisfy Serre's condition are (I,p,I), where $p(x) = x/2$ or $p(x) = 4x(1-x)$. In the first case, take $r = 1$, $f = \text{id}\,I$, $\tilde{f}_0(0) = 0$; in the second case, take $r = 2$, $f(x_1,x_2) = 4x_1(1-x_1)(1-x_2)$, $\tilde{f}_0 = \text{id}\,I$. Then there is no continuous map $\tilde{f}$ such that $p \circ \tilde{f} = f$ and $\tilde{f}\big|_{I^{r-1}} = \tilde{f}_0$.

We remark that the first bundle has both empty and nonempty fibers, while the second has a single connected fiber, the others being not connected. As we shall see later (see 5.4.3.6), such features of a bundle are not compatible with Serre's condition when the base is connected.

Serre's Condition is Local

3. <u>If every point of the base</u> $\text{bs}\,\xi$ <u>of a bundle</u> ξ <u>has a neighborhood</u> U <u>such that</u> $\xi\big|_U$ <u>is a Serre bundle, then</u> ξ <u>itself is a Serre bundle.</u>

PROOF. Let $f: I^n \to \text{bs}\,\xi$ and $\tilde{f}_0: I^{n-1} \to \text{tl}\,\xi$ be continuous maps satisfying $\text{pr}\,\xi \circ \tilde{f}_0 = f\big|_{I^{n-1}}$. Since $\text{bs}\,\xi$ can be covered by open sets such that the restriction of ξ to each of these sets satisfies Serre's condition, Theorem 1.1.7.16 yields a positive integer N such that every cube of edge $1/N$ contained in I^n is taken by f into one of these open sets. Divide I^n into N^n cubes of edge $1/N$, arrange these cubes in dictionary order $Q_1,\ldots,Q_{N^n}$, and set $W_i = I^{n-1} \cup (\cup_{j=1}^{i} Q_j)$ and $Q_i' = Q_i \cap W_{i-1}$. It is clear that $I^{n-1} = W_0 \subset W_1 \subset \ldots \subset W_{N^n} = I^n$, and examining all possible cases, we see that each pair (Q_i,Q_i') is homeomorphic to (I^n,I^{n-1}). Now assume that for some $i \leqslant N^n$ there is a map $\tilde{f}_{i-1}: W_{i-1} \to \text{tl}\,\xi$ such that $\text{pr}\,\xi \circ \tilde{f}_{i-1} = f\big|_{W_{i-1}}$ and $\tilde{f}_{i-1}\big|_{I^{n-1}} = \tilde{f}_0$. Set $g_i = f\big|_{Q_i}$ and $\tilde{g}_{0i} = \tilde{f}_{i-1}\big|_{Q_i'}$. Since the restriction $\xi\big|_{f(Q_i)}$ satisfies Serre's condition, there is a continuous map $\tilde{g}_i: Q_i \to \text{tl}\,\xi$ such that

$pr \, \xi \circ \tilde{g_i} = g_i$ and $\tilde{g_i} \Big|_{Q_i^!} = \tilde{g_{0i}}$. Furthermore, because $\tilde{f_{i-1}}$ and $\tilde{g_i}$ agree on $W_{i-1} \cap Q_i = Q_i^!$, together they form a continuous map $\tilde{f_i}: W_i = W_{i-1} \cup Q_i \to tl \, \xi$, and obviously $pr \, \xi \circ \tilde{f_i} = f\Big|_{W_i}$ and $\tilde{f_i}\Big|_{I^{n-1}} = \tilde{f_0}$. This shows that induction on i works, starting with $i = 0$, and the result is a map $\tilde{f} = \tilde{f}_{N^n}$ such that $pr \, \xi \circ \tilde{f} = f$ and $\tilde{f}\Big|_{I^{n-1}} = \tilde{f_0}$.

Serre's Condition and Local Triviality

4. **Every locally trivial bundle is a Serre bundle.**

By Theorem 3, one need only consider the case $(B \times F, pr_1, B)$, where B and F are arbitrary topological spaces. Let $f: I^n \to B$ and $\tilde{f_0}: I^{n-1} \to B \times F$ be continuous maps such that $pr_1 \circ \tilde{f_0} = f\Big|_{I^{n-1}}$. Define $\tilde{f}: I^n \to B \times F$ as

$$\tilde{f}(x_1, \ldots, x_n) = (f(x_1, \ldots, x_n), pr_2 \circ f_0(x_1, \ldots, x_{n-1})).$$

Clearly, $pr_1 \circ \tilde{f} = f$ and $\tilde{f}\Big|_{I^{n-1}} = \tilde{f_0}$.

5. The following example demonstrates that there are Serre bundles which are not locally trivial. Let T be the triangle in $\mathbb{R}^2$ with vertices $(0,0)$, $(0,1)$, and $(1,0)$, and let $p_1, p_2: T \to I$ be defined by $p_1(x_1, x_2) = x_1$, $p_2(x_1, x_2) = x_2$. The bundle (T, p_1, I) is not locally trivial; indeed, the fibers over the points 0 and 1 are not homeomorphic. However, (T, p_1, I) does satisfy Serre's condition: if $f: I^n \to I$ and $\tilde{f_0}: I^{n-1} \to T$ are continuous, and $p_1 \circ \tilde{f_0} = f\Big|_{I^{n-1}}$, then the map $\tilde{f}: I^n \to T$ defined by

$$\tilde{f}(x_1, \ldots, x_n) = (f(x_1, \ldots, x_n), \min(1 - f(x_1, \ldots, x_n), p_2 \circ f_0(x_1, \ldots, x_{n-1})))$$

is continuous, covers f, and equals $\tilde{f_0}$ on I^{n-1}.

This example shows also that in a Serre bundle with connected base there can be nonhomeomorphic fibers. Actually, there are Serre bundles with connected base and in which some fibers are not even homotopy equivalent, being instead equivalent in a certain weaker sense (see 5.4.4.3 and 5.4.3.6).

The Covering Homotopy Theorem

6. *Suppose that* ξ *is a Serre bundle and* (X,A) *is a cellular pair. Then for every continuous map* $\tilde{f}: X \to tl\,\xi$, *every homotopy* $F: X \times I \to bs\,\xi$ *of* $pr\,\xi \circ \tilde{f}$, *and every homotopy* $G: A \times I \to tl\,\xi$ *of* $\tilde{f}\big|_A$ *covering* $F\big|_{A \times I}$, *there is a homotopy of* $\tilde{f}$ *which covers* F *and extends* G.

PROOF. Assume that X is rigged, and that for some $r \geqslant 0$ there is a homotopy $\tilde{F}_{r-1}: (A \cup ske_{r-1}X) \times I \to tl\,\xi$ of $\tilde{f}\big|_{A \cup ske_{r-1}X}$ covering $F\big|_{(A \cup ske_{r-1}X) \times I}$. If e is an r-cell from $X \smallsetminus A$, then $\phi_e(x,t) = F(cha_e(x),t)$ defines a continuous map $\phi_e: D^r \times I \to bs\,\xi$, while the formula

$$\tilde{\phi}_{0,e}(x,t) = \begin{cases} \tilde{f}(cha_e(x)), & \text{if } t = 0, \\[2mm] \tilde{F}_{r-1}(cha_e(x),t), & \text{if } x \in S^{r-1}, \end{cases}$$

defines a continuous map $\tilde{\phi}_{0,e}: (D^r \times 0) \cup (S^{r-1} \times I) \to tl\,\xi$. Obviously, the pairs $(D^r \times I, (D^r \times 0) \cup (S^{r-1} \times I))$ and (I^{r+1}, I^r) are homeomorphic, and $pr\,\xi \circ \tilde{\phi}_{0,e} = \phi_e\big|_{(D^r \times 0) \cup (S^{r-1} \times I)}$. Consequently, there is a continuous map $\tilde{\phi}_e: D^r \times I \to tl\,\xi$ covering ϕ_e and extending $\tilde{\phi}_{0,e}$. Since $\tilde{\phi}_e(x,t) = \tilde{F}_{r-1}(cha_e(x),t)$ for all $x \in S^{r-1}$, the maps $\tilde{\phi}_e$ corresponding to all r-cells from $X \smallsetminus A$ together with $\tilde{F}_{r-1}$ yield a continuous map $\tilde{F}_r: (A \cup ske_r X) \times I \to tl\,\xi$, and it is evident that $pr\,\xi \circ \tilde{F}_r = F\big|_{(A \cup ske_r X) \times I}$ and $\tilde{F}_r\big|_{(A \cup ske_{r-1}X) \times I} = \tilde{F}_{r-1}$. Hence, we may use induction on r, setting $\tilde{F}_{-1} = G$, to produce a sequence $\{\tilde{F}_r: (A \cup ske_r X) \times I \to tl\,\xi\}_{r=-1}^{\infty}$ of homotopies which extend each other. These homotopies define a homotopy of $\tilde{f}$ covering F and extending G.

7. *Let* X *be a cellular space and let* $\tilde{f}: X \to tl\,\xi$ *be continuous. If* ξ *is a Serre bundle, then every homotopy of* $pr\,\xi \circ \tilde{f}$ *is covered by a homotopy of* $\tilde{f}$.

This is precisely Theorem 6 for the absolute case $A = \emptyset$, and hence the proof is immediate. We note that for $X = I^n$, Theorem 7 reduces to Serre's condition for $r = n+1$.

The Case of Coverings

8. **Suppose that** ξ **is a covering in the broad sense,** X **is a connected topological space, and** $f,g: X \to \mathrm{tl}\,\xi$ **are continuous maps. If** $\mathrm{pr}\,\xi \circ f = \mathrm{pr}\,\xi \circ g$ **and** f **equals** g **at some point, then** $f = g$.

PROOF. Since the set $\{x \in X \mid f(x) \neq g(x)\}$ is open and, by assumption, its complement is not empty, it suffices to show that this complement is also open. In other words, let us verify that if $f(x_0) = g(x_0)$, then x_0 has a neighborhood U such that $f(x) = g(x)$ for all $x \in U$. Let V be a neighborhood of $f(x_0)$ such that $\mathrm{pr}\,\xi|_V : V \to \mathrm{pr}\,\xi(V)$ is a homeomorphism (see 2.4), and take U to be any neighborhood of x_0 with $f(U) \subset V$ and $g(U) \subset V$. Since $\mathrm{pr}\,\xi(f(x)) = \mathrm{pr}\,\xi(g(x))$ for all $x \in X$, we have $f(x) = g(x)$ for all $x \in U$.

9. **Suppose that** ξ **is a covering in the broad sense,** X **is a connected cellular space with a distinguished 0-cell** x_0, **and** $f,g: X \to \mathrm{tl}\,\xi$ **are continuous. If the maps** $\mathrm{pr}\,\xi \circ f$ **and** $\mathrm{pr}\,\xi \circ g$ **are** x_0**-homotopic and** $f(x_0) = g(x_0)$, **then** f **and** g **are** x_0**-homotopic.**

PROOF. By Theorem 6, any x_0-homotopy from $\mathrm{pr}\,\xi \circ f$ to $\mathrm{pr}\,\xi \circ g$ is covered by an x_0-homotopy from f to some map h. Since $h(x_0) = f(x_0) = g(x_0)$ and $\mathrm{pr}\,\xi \circ h = \mathrm{pr}\,\xi \circ g$, Proposition 8 yields $h = g$.

4. Bundles with Map Spaces as Total Spaces

1. We say that a bundle ξ satisfies the **strong Serre condition*** if for every topological space X, every continuous map $\tilde{f}: X \to \mathrm{tl}\,\xi$, and every homotopy F of $\mathrm{pr}\,\xi \circ \tilde{f}$, there is a homotopy of $\tilde{f}$ which covers F.

If we replace X by a cube of arbitrary dimension, then this becomes the simple Serre condition; moreover, when X is restricted to be an arbitrary cellular space, we obtain again a condition equivalent to the simple Serre condition; see 3.7.

2. **Let** (X,A) **and** Y **be a Borsuk pair and a topological space, respectively. If** X **is Hausdorff and locally compact, then the bundle** $(C(X,Y), C(\mathrm{in},\mathrm{id}), C(A,Y))$ **satisfies the strong Serre condition.**

---

* Translator's note: Such a bundle is frequently called a Hurewicz fiber space or a fibration.

PROOF. Consider a topological space Z, a continuous map $\tilde{f}: Z \to C(X,Y)$, and a homotopy $F: Z \times I \to C(A,Y)$ of $C(in,id) \circ \tilde{f}$. Since X is Hausdorff and locally compact, the maps $\tilde{g}: Z \times X \to Y$ and $G: Z \times A \times I \to Y$ given by $\tilde{g}(z,x) = (\tilde{f})^{\wedge}(z,x) = [\tilde{f}(z)](x)$ and $G(z,x,t) = [F(z,t)](x)$ are continuous (see 1.2.7.6). It is clear that G is a homotopy of $\tilde{g}|_{Z \times A}$. Now $(Z \times X, Z \times A)$ is a Borsuk pair (see 1.3.5.5. and 1.3.5.3), and hence G extends to a homotopy $\tilde{G}$ of $\tilde{g}$. Finally, the formula $[\tilde{F}(z,t)](x) = \tilde{G}(z,x,t)$ defines a homotopy $\tilde{F}: Z \times I \to C(X,Y)$ of $\tilde{f}$ which covers F.

3. <u>In a bundle with connected base and satisfying the strong Serre condition, the fibers are pairwise homotopy equivalent.</u>

Let ξ be the given bundle, and let s be a path joining two given points of $\mathrm{bs}\,\xi$. Set $F_0 = \mathrm{pr}\,\xi^{-1}(s(0))$ and $F_1 = \mathrm{pr}\,\xi^{-1}(s(1))$. Now consider two homotopies $J_0: F_0 \times I \to \mathrm{bs}\,\xi$ and $J_1: F_1 \times I \to \mathrm{bs}\,\xi$ of the composite maps $F_0 \xrightarrow{\ \mathrm{in}\ } \mathrm{tl}\,\xi \xrightarrow{\ \mathrm{pr}\,\xi\ } \mathrm{bs}\,\xi$ and $F_1 \xrightarrow{\ \mathrm{in}\ } \mathrm{tl}\,\xi \xrightarrow{\ \mathrm{pr}\,\xi\ } \mathrm{bs}\,\xi$, respectively, given by $J_0(x,t) = s(t)$ and $J_1(x,t) = s^{-1}(t)$. Since ξ satisfies the strong Serre condition, J_0 and J_1 are covered by two homotopies, $\tilde{J}_0: F_0 \times I \to \mathrm{tl}\,\xi$ and $\tilde{J}_1: F_1 \times I \to \mathrm{tl}\,\xi$, of the maps $\mathrm{in}: F_0 \to \mathrm{tl}\,\xi$ and $\mathrm{in}: F_1 \to \mathrm{tl}\,\xi$, respectively. Now $\tilde{J}_0(F_0 \times 1) \subset F_1$, $\tilde{J}_1(F_1 \times 1) \subset F_0$, and hence there are well-defined maps $f_0: F_0 \to F_1$, $f_0(x) = \tilde{J}_0(x,1)$, and $f_1: F_1 \to F_0$, $f_1(x) = \tilde{J}_1(x,1)$. We next show that $f_1 \circ f_0$ is homotopic to $\mathrm{id}\,F_0$, and since the construction is symmetric, $f_0 \circ f_1$ will be homotopic to $\mathrm{id}\,F_1$.

The formulas

$$j(x,t) = \begin{cases} \tilde{J}_0(x,2t), & \text{if } t \leqslant 1/2, \\[2ex] \tilde{J}_1(f_0(x),2t-1), & \text{if } t \geqslant 1/2, \end{cases}$$

and

$$H(x,t,\tau) = s((1-\tau)(1 - |1-2t|))$$

define a map $j: F_0 \times I \to \mathrm{tl}\,\xi$ and a homotopy $H: (F_0 \times I) \times I \to \mathrm{bs}\,\xi$ of $\mathrm{pr}\,\xi \circ j$. Again, using Serre's strong condition, H can be lifted to a homotopy $\tilde{H}: (F_0 \times I) \times I \to \mathrm{tl}\,\xi$ of j. Since $(1-\tau)(1 - |1-2t|) = 0$ for $\tau = 1$ or $t = 0,1$, we see that $\tilde{H}((F_0 \times (0 \cup 1)) \times I) \cup \tilde{H}((F_0 \times I) \times 1) \subset F_0$. Therefore, the formula

$$K(x,t) = \begin{cases} \widetilde{H}((x,0),3t), & \text{if } t \leqslant 1/3, \\ \widetilde{H}((x,3t-1),1), & \text{if } 1/3 \leqslant t \leqslant 2/3, \\ \widetilde{H}((x,1),3-3t), & \text{if } t \geqslant 2/3, \end{cases}$$

defines a homotopy $K: F_0 \times I \to F_0$. Since $K(x,0) = \widetilde{H}((x,0),0) =$ $= j(x,0) = \widetilde{J}_0(x,0) = x$ and $K(x,1) = \widetilde{H}((x,1),0) = j(x,1) =$ $= \widetilde{J}_1(f_0(x),1) = f_1(f_0(x))$, it follows that K is a homotopy from $\operatorname{id} F_0$ to $f_1 \circ f_0$.

4. <u>Given arbitrary points</u> x_0, x_1, x_0', x_1' <u>of a connected topological space</u> X, <u>the spaces</u> $C(I,0,1;X,x_0,x_1)$ <u>and</u> $C(I,0,1;X,x_0',x_1')$ <u>have the same homotopy type.</u>

PROOF. $C(I,0,1;X,x_0,x_1)$ and $C(I,0,1;X,x_0',x_1')$ are the fibers of the bundle $(C(I,X),C(\operatorname{in},\operatorname{id}),C((0 \cup 1),X))$ over the points (x_0,x_1) and (x_0',x_1') of its base $C((0 \cup 1),X) = X \times X$; hence by theorems 2 and 3, they have the same homotopy type.

The Adjoint Serre Bundle

5. Given an arbitrary bundle ξ, we let $\operatorname{ad} \xi$ denote the bundle with the same base, total space

$$\{ (x,s) \in \operatorname{tl} \xi \times C(I,\operatorname{bs} \xi) \mid s(0) = \operatorname{pr} \xi(x) \},$$

and projection $(x,s) \to s(1)$. We call $\operatorname{ad} \xi$ the <u>bundle adjoint to</u> ξ.

Notice that the total spaces $\operatorname{tl} \operatorname{ad} \xi$ and $\operatorname{tl} \xi$ have the same homotopy type: the formulas $x \mapsto (x,u_x)$ and $(x,s) \mapsto x$, where u_x is the constant path in $\operatorname{bs} \xi$ with $u_x(0) = \operatorname{pr} \xi(x)$, define homotopy equivalences $\operatorname{tl} \xi \to \operatorname{tl} \operatorname{ad} \xi$ and $\operatorname{tl} \operatorname{ad} \xi \to \operatorname{tl} \xi$ which are inverses to one another. Indeed, the composition of the first map with the second one is $\operatorname{id} \operatorname{tl} \xi$, while the composition of the second map with the first one is homotopic to $\operatorname{id}(\operatorname{tl} \operatorname{ad} \xi)$ via the homotopy $((x,s),t) \mapsto (x,s_t)$, where s_t is the path in $\operatorname{bs} \xi$ defined by $s_t(\tau) = s(t\tau)$.

6. <u>The bundle</u> $\operatorname{ad} \xi$ <u>satisfies the strong Serre condition, for any bundle</u> ξ.

PROOF. Consider a topological space Z, a continuous map $\widetilde{f}: Z \to \operatorname{tl} \operatorname{ad} \xi$, and a homotopy $F: Z \times I \to \operatorname{bs} \operatorname{ad} \xi \ (= \operatorname{bs} \xi)$ of $\operatorname{pr} \operatorname{ad} \xi \circ \widetilde{f}$. Denote by g_1 and g_2 the composite maps

$$z \xrightarrow{\ \widetilde{f}\ } tl\ ad\ \xi \xrightarrow{\ in\ } tl\ \xi\ \times\ \mathcal{C}(I,bs\ \xi) \begin{cases} \xrightarrow{\ pr_1\ } tl\ \xi \\[2ex] \xrightarrow{\ pr_2\ } \mathcal{C}(I,bs\ \xi), \end{cases}$$

and define a homotopy $g:\ Z\ \times\ I\ \to\ (I,bs\ \xi)$ by

$$[g(z,t)]\,(\tau) = \begin{cases} [g_2(z)](\tau(1+t)), & \text{if}\quad \tau \leqslant 1/(1+t), \\[2ex] F(z,\tau(1+t)\ -\ 1), & \text{if}\quad \tau \geqslant 1/(1+t). \end{cases}$$

It is readily verified that $\widetilde{F}(z,t) = (g_1(z),g(z,t))$ defines a homotopy $\widetilde{F}:\ Z\ \times\ I\ \to\ tl\ ad\ \xi$ of $\widetilde{f}$ and that $\widetilde{F}$ covers F.

5. Exercises

1. Show that for any $g \geqslant 1$ a sphere with g handles admits a sphere with $2g-1$ handles as a covering space.

2. Show that for any $h \geqslant 1$ a sphere with h crosscaps admits a sphere with $2h-2$ crosscaps as a covering space.

3. Show that the spaces $\mathcal{C}(S^1,ort_1;\ \mathbb{RP}^n,(1:0:\ldots:0))$ and $\mathcal{C}(S^1,ort_1;S^n,ort_1)\ \times\ S^0$ are homeomorphic for any $n \geqslant 1$.

4. Show that the bundle with total space $\mathcal{C}(I,0;S^n,ort_1)$, base S^n, and projection $s \mapsto s(1)$, is locally trivial (see 1.2.9.4).

§2. A DIGRESSION:
TOPOLOGICAL GROUPS AND TRANFORMATION GROUPS

1. Topological Groups

1. A set G is a __topological group__ or a __group space__ if it is endowed with both a topology and a group structure such that the group operations, i.e., the maps $G\ \times\ G\ \to\ G,$ $(g,h) \mapsto gh,$ and $G\ \to\ G,$ $g \mapsto g^{-1}$, are continuous. Obviously, the continuity of these two maps is equivalent to the continuity of the single map $G\ \times\ G\ \to\ G,$ $(g,h) \mapsto g^{-1}h.$

By the definition of the (product) topology on $G\ \times\ G,$ the continuity of the map $(g,h) \mapsto gh$ at the point (g_0, h_0) means that for every neighborhood W of the point $g_0 h_0$ one can find neighborhoods U and V of g_0 and h_0 such that $UV \subset W$. Similarly, the continuity of the map $(g,h) \mapsto g^{-1}h$ means that for every neighborhood W of the point $g_0^{-1} h_0$ there are neighborhoods U and V of g_0 and h_0 such that $U^{-1}V \subset W$.

Clearly, every group becomes a topological group if it is equipped with the discrete topology.

2. The continuity of the group operations implies that the left and right translations by group elements (i.e., the maps $G \to G$ given by $g \mapsto ag$ and $g \mapsto ga$), and the map $g \mapsto g^{-1}$ are homeomorphisms of the space G. In particular, if $B \subset G$ is open or closed, then so are the sets B^{-1}, aB and Ba, for all $a \in G$.

We note also that if B is open and A is arbitrary, then AB and BA are open. Indeed, $AB = \bigcup_{a \in A} aB$ and $BA = \bigcup_{a \in A} Ba$.

Subgroups and Quotients

3. Any subset H of the topological group G which is a subgroup in the algebraic sense inherits both a group structure and a topology from G, and it is immediate that the group operations in H are continuous.

A subgroup of a topological group is _normal_ if it is normal in the algebraic sense. As in ordinary group theory, normal subgroups are termed also _normal divisors_ or _invariant subgroups._

An example of a normal subgroup of a topological group G, which has no analog in ordinary group theory, is the _component of the identity,_ i.e., that component of the space G which contains the identity element, e_G, of G. This component is obviously a subgroup: if u and v are paths joining e_G with g and h, respectively, then the path $t \mapsto u(t)^{-1} v(t)$ joins e_g with $g^{-1}h$. Since the inner automorphisms of G are continuous and take e_G into itself, this subgroup is normal. It is also clear that the cosets of this subgroup in G are exactly the components of the space G, and the corresponding quotient group coincides, as a set, with $\operatorname{comp} G$.

4. _Every open subgroup of a topological group is also closed._

In fact, the complement of an open subgroup is a union of left cosets and by 2 each of these cosets is open. Therefore, the complement is also open.

5. The partition of a topological group G into the left cosets of a subgroup H is denoted by $\text{zer}(G,H)$, and the corresponding quotient space is called the <u>space of left cosets of</u> H <u>in</u> G and is denoted by G/H. We shall not need here right coset spaces.

The basic topological property of the partition $\text{zer}(G,H)$ and of the projection $G \to G/H$ is their openness. Indeed, the saturation of a set B relative to $\text{zer}(G,H)$ equals BH, which is an open set whenever B is open (see 2).

6. <u>Given a topological group, the space of left cosets of a closed subgroup is regular. In particular, a topological group whose identity element is closed is regular.</u>

PROOF. Since the cosets gH are closed (see 2), G/H satisfies Axiom T_1. To show that G/H additionally satisfies Axiom T_3, it suffices to produce, given a coset $g_0 H$ and a neighborhood U of $g_0 H$ which is saturated relative to $\text{zer}(G,H)$, a saturated neighborhood V of $g_0 H$ such that $\text{Cl}\, V \subset U$. To see this, suppose that we have such a neighborhood. Then for every point $\text{pr}(g_0) \in G/H$ and every neighborhood $\text{pr}(U)$ of $\text{pr}(g_0)$, there is a neighborhood, $\text{pr}(V)$, of $\text{pr}(g_0)$, such that $\text{Cl}\, \text{pr}(V) \subset \text{pr}(U)$. The last inclusion follows from the inclusion $\text{pr}(V) \subset \text{pr}(\text{Cl}\, V)$ together with the fact that $\text{pr}(\text{Cl}\, V)$ is closed (which follows from the fact that $\text{Cl}\, V$ is closed and saturated; $\text{Cl}\, V$ is saturated because the partition $\text{zer}(G,H)$ is open; see 1.2.3.10).

Now to produce the desired neighborhood V, note that $e_G^{-1} g_0 = g_0$ and so the points e_G and g_0 have neighborhoods W and W_0 such that $W^{-1} W_0 \subset U$. Set $V = W_0 H$. If $g \in \text{Cl}\, V$, then Wg, being a neighborhood of g, intersects V, i.e., there exist $w \in W$, $w_0 \in W_0$, and $h \in H$ such that $wg = w_0 h$. We have $g = w^{-1} w_0 h$, and thus $g = w^{-1} w_0 h \in W^{-1} W_0 H \subset UH = U$. Therefore, $\text{Cl}\, V \subset U$.

7. Let H be a normal subgroup of the topological group G. According to ordinary group theory, the set of cosets G/H is endowed with a group structure. Let us show that the map $G/H \times G/H \to G/H$, $(x,y) \mapsto x^{-1} y$, is continuous.

First, note that the composition $\psi \colon G \times G \to G/H$ of the map $G \times G \to G$, $(g,h) \mapsto g^{-1} h$, with the projection $G \to G/H$ is constant on the elements of the partition $\text{zer}(G,H) \times \text{zer}(G,H)$. Secondly, the map $(x,y) \mapsto x^{-1} y$ equals $\text{fact}\, \psi \circ \alpha^{-1}$, where

$$\alpha^{-1} \colon G/H \times G/H \to (G \times G)/(\text{zer}(G,H) \times \text{zer}(G,H))$$

is the inverse of the injective factor, α, of the map

$$\text{pr} \times \text{pr}: G \times G \to G/H \times G/H,$$

and

$$\text{fact } \psi: (G \times G)/(\text{zer}(G,H) \to \text{zer}(G,H)) \to G/H.$$

Since the partition $\text{zer}(G,H)$ is open, α is a homeomorphism (see 1.2.3.11), and this implies that the map $(x,y) \mapsto x^{-1}y$ is continuous.

We conclude that G/H is a topological group; G/H is called the <u>quotient</u> of the <u>factor group</u> of the topological group G by H.

Homomorphisms

8. A map $f: G \to G'$, where G and G' are topological groups, is a <u>homomorphism</u> if it is an algebraic homomorphism as well as continuous.

As in ordinary group theory, the kernel $\text{Ker } f$ of f is defined as the preimage of the identity of G'. A homomorphism f is a <u>monomorphism</u> if it is injective, i.e., if its kernel $\text{Ker } f$ is the identity element of G, and an <u>epimorphism</u> if its image $\text{Im } f = f(G)$ is all of G'. An example of monomorphism (epimorphism) is the inclusion of a subgroup in a topological group (respectively, the projection of a topological group onto a factor group).

An invertible homomorphism whose inverse is also a homomorphism is called an <u>isomorphism.</u> In other words, an isomorphism of topological groups is a map which is both an algebraic homomorphism and a homemorphism.

9. <u>Let</u> $f: G \to G'$ <u>be a homomorphism of topological groups.</u> <u>Then:</u>

(i) $\text{Im } f$ <u>is a subgroup of</u> G', <u>and the compression</u> $\text{ab } f: G \to \text{Im } f$ <u>is an epimorphism;</u>

(ii) $\text{Ker } f$ <u>is a normal subgroup of</u> G, <u>and the injective factor of</u> f, $\text{fact } f: G/\text{Ker } f \to G'$, <u>is a monomorphism.</u>

In addition to recognizing that this copies a well-know statement of ordinary group theory, one has to check that the maps $\text{ab } f$ and $\text{fact } f$ are continuous, which is trivial.

10. <u>An epimorphism</u> $f: G \to G'$ <u>is open if and only if its injective factor,</u> $\text{fact } f: G/\text{Ker } f \to G'$, <u>is an isomorphism.</u>

The necessity of this condition is obvious. The oppenness of the projection $G \to G/\mathrm{Ker}\, f$ shows that the condition is also sufficient.

11. <u>An epimorphism of a compact topological group onto a topological group with closed identity element is open.</u>

PROOF. Let $f: G \to G'$ be the given epimorphism. Since G is compact, $G/\mathrm{Ker}\, f$ is compact. Moreover, as the identity element of G' is closed, G' is Hausdorff (see 6). Finally, $\mathrm{fact}\, f: G/\mathrm{Ker}\, f \to G'$ is invertible and continuous, and hence in our case a homeomorphism (see 1.1.7.10). Applying 10, f is open.

Direct Products

12. Let G_1 and G_2 be topological groups. In ordinary group theory, the product $G_1 \times G_2$ is given a group structure, and in topology it is given a topology (see 1.2.2.1), and it is clear that these two structures are compatible in the sense of 1, i.e., the group operations in $G_1 \times G_2$ are continuous. The result is a topological group $G_1 \times G_2$, called the <u>direct product</u> of the topological groups G_1 and G_2.

This product operation is both commutative and associative: there are obvious canonical isomorphisms $G_1 \times G_2 \to G_2 \times G_1$ and $(G_1 \times G_2) \times G_3 \to G_1 \times (G_2 \times G_3)$.

We remark that the inclusions $\mathrm{in}_1: G_1 \to G_1 \times G_2$, $x_1 \mapsto (x_1, e_{G_2})$, and $\mathrm{in}_2: G_2 \to G_1 \times G_2$, $x_2 \mapsto (e_{G_1}, x_2)$, are monomorphisms (of topological groups), while the projections $\mathrm{pr}_1: G_1 \times G_2 \to G_1$ and $\mathrm{pr}_2: G_1 \times G_2 \to G_2$ are open epimorphisms such that $\mathrm{Ker}\, \mathrm{pr}_1 = \mathrm{in}_2(G_2)$ and $\mathrm{Ker}\, \mathrm{pr}_2 = \mathrm{in}_1(G_1)$. The last observation together with Theorem 10 imply that $\mathrm{fact}\, \mathrm{pr}_1: (G_1 \times G_2)/\mathrm{in}_2(G_2) \to G_1$ and $\mathrm{fact}\, \mathrm{pr}_2: (G_1 \times G_2)/\mathrm{in}_1(G_1) \to G_2$ are isomorphisms.

13. We say that <u>the topological group</u> G <u>decomposes into the direct product of its subgroups</u> G_1 <u>and</u> G_2 if the map $G_1 \times G_2 \to G$, $(g_1, g_2) \mapsto g_1 g_2$, is an isomorphism of topological groups. If this is the case, the groups G and $G_1 \times G_2$ are usually identified via this isomorphism.

Recall that a similar definition exists in ordinary group theory, only there the isomorphism is simply an algebraic isomorphism. Moreover, in that theory, G decomposes into the direct product of its subgroups G_1 and G_2 if and only if G_1 and G_2 generate G, are normal subgroups of G, and $G_1 \cap G_2 = e_G$. Consequently, if these conditions are satisfied in our case, then $(g_1, g_2) \mapsto g_1 g_2$ is an algebraic isomorphism. This map is obviously continuous; however, there are obvious examples where the algebraic inverse isomorphism is not continuous. But the algebraic inverse isomorphism is continuous whenever the space G is compact and Hausdorff. Therefore, every compact Hausdorff topological group which decomposes algebraically into the direct product of two subgroups, decomposes also into the direct product of these subgroups in the sense of our topological definition.

The Simplest Examples

14. The real line $\mathbb{R}$ with addition as the group operation is a topological group, as is the space $\mathbb{R}^n$. Obviously, $\mathbb{R}^n = \mathbb{R} \times \ldots \times \mathbb{R}$ (n factors; the product is understood as in 12).

15. The punctured real line $\mathbb{R}^* = \mathbb{R} \smallsetminus 0$, with multiplication as group operation, is a topological group. Its subgroup $\mathbb{R}^*_+$ consisting of the positive reals, is isomorphic to $\mathbb{R}$: an isomorphism $\mathbb{R} \to \mathbb{R}^*_+$ is provided by the exponential function $x \mapsto a^x$, with arbitrary $a \neq 1$. Another evident subgroup of $\mathbb{R}^*$ is S^0, and obviously $\mathbb{R}^* = S^0 \times \mathbb{R}^*_+$.

The punctured complex line $\mathbb{C}^* = \mathbb{C} \smallsetminus 0$ and the punctured quaternionic line $\mathbb{H}^* = \mathbb{H} \smallsetminus 0$ are also topological groups under multiplication. Here S^1 is a subgroup of $\mathbb{C}^*$, S^3 is a subgroup of $\mathbb{H}^*$, and $\mathbb{C}^* = S^1 \times \mathbb{R}^*_+$, $\mathbb{H}^* = S^3 \times \mathbb{R}^*_+$.

16. The map $\mathrm{hel} \colon \mathbb{R} \to S^1$ (see 1.2.6) is an open epimorphism. Its kernel is the subgroup of integers, $\mathbb{Z}$, of $\mathbb{R}$, and hence the factor group $\mathbb{R}/\mathbb{Z}$ is isomorphic to S^1.

The map $\mathrm{hel} \times \ldots \times \mathrm{hel} \colon \mathbb{R}^n \to (S^1)^n = S^1 \times \ldots \times S^1$ is also an open epimorphism. Its kernel is the integer lattice, $\mathbb{Z} \times \ldots \times \mathbb{Z} = \mathbb{Z}^n$, of $\mathbb{R}^n$, and hence the factor group $\mathbb{R}^n/\mathbb{Z}^n$ is isomorphic to $(S^1)^n$.

17. The subgroup of S^3 consisting of the real quaternions (i.e., of the quaternions (x_1, x_2, x_3, x_4) such that $x_2 = x_3 = x_4 = 0$) is simply S^0. This is a normal subgroup, and the factor group S^3/S^0 is, as a topological space, the same as $\mathbb{R}P^3$.

The sugbroup of S^3 consisting of the complex quaternions (i.e., of the quaternions (x_1, x_2, x_3, x_4) such that $x_3 = x_4 = 0$) is simply S^1. However, this is not a normal subgroup. The coset space S^3/S^1 is canonically homeomorphic to S^2: this canonical homeomorphism is provided by the injective factor of the Hopf map $S^3 \to S^2$ (obviously,

$$\mathrm{zer}(S^3,S^1) = \mathrm{zer}(S^3 \to S^2)).$$

2. <u>Groups of Homeomorphisms</u>

1. By 1.1.4.7, the homeomorphisms of a topological space X are a subgroup of the group $\mathrm{Symm}\,X$ of all invertible transformations $X \to X$, i.e., they form a group under the $\circ$ (composition) operation. We denote the group of homeomorphisms of X by $\mathrm{Top}\,X$.

We may define two topologies on $\mathrm{Top}\,X$. The first one is induced by the inclusion $\mathrm{Top}\,X \subset C(X,X)$ (see 1.2.7.1), i.e., is defined by the prebase consisting of the sets $\mathrm{Nb}(K,O) =$ $= C(X,K;X,O) \cap \mathrm{Top}\,X$, with K compact and O open. The second topology is defined by the prebase consisting of the sets U, U^{-1}, where U is open in the first topology. Equivalently, the second topology is generated by the prebase consisting of the sets $\mathrm{Nb}(K,O)$, $[\mathrm{Nb}(K,O)]^{-1}$.

2 (LEMMA). <u>If X is a locally compact Hausdorff space, then the map</u> $\mathrm{Top}\,X \times \mathrm{Top}\,X \to \mathrm{Top}\,X$, $(g,h) \mapsto gh\ (= g \circ h)$, <u>is continuous in either of the above topologies.</u>

PROOF. If $gh \in \mathrm{Nb}(K,O)$, then $h(K) \subset g^{-1}(O)$, and by 1.1.7.22, every point of $h(K)$ has a neighborhood whose closure is compact and contained in $g^{-1}(O)$. Let O' denote the union of a finite collection of such neighborhoods which cover $h(K)$. Clearly, $\mathrm{Cl}\,O'$ is compact, and $g \in \mathrm{Nb}(\mathrm{Cl}\,O',O)$, $h \in \mathrm{Nb}(K,O')$, $\mathrm{Nb}(\mathrm{Cl}\,O',O)\mathrm{Nb}(K,O') \subset$ $\subset \mathrm{Nb}(K,O)$.

Now if $gh \in [\mathrm{Nb}(K,O)]^{-1}$, then $h^{-1}g^{-1} \in \mathrm{Nb}(K,O)$, and the above argument yields two sets, $U,V \subset \mathrm{Top}\,X$, open in the first topology, and such that $h^{-1} \in V$, $g^{-1} \in U$, $VU \subset \mathrm{Nb}(K,O)$. Clearly, $g \in U^{-1}$, $h \in V^{-1}$, and $U^{-1}V^{-1} \subset [\mathrm{Nb}(K,O)]^{-1}$.

3. <u>If X is a locally compact Hausdorff space, then</u> $\mathrm{Top}\,X$, <u>equipped with the second topology, is a topological group.</u>

This is a corollary of 2 and of the obvious continuity of the map $g \mapsto g^{-1}$ in the second topology.

4. <u>If X is a compact Hausdorff space, then the first and the second topologies on</u> $\mathrm{Top}\,X$ <u>are identical.</u>

This is an immediate corollary of the relation

$$[\mathrm{Nb}(K,O)]^{-1} = \mathrm{Nb}(X \smallsetminus O, X \smallsetminus K).$$

5 (LEMMA). <u>Let X be a locally compact, locally connected, Hausdorff space. Then for a prebase of the second topology on $\text{Top } X$ it suffice to take the sets of the form $\text{Nb}(K,O)$, where K is the closure of a connected open set and O is open.</u>

PROOF. Given a compact K, an open O, and a homeomorphism $f \in \text{Nb}(K,O)$, it suffices to produce connected open sets $U_1,\ldots,U_s$ with compact closures, such that $f \in \cap_1^s \text{Nb}(\text{Cl } U_i,O) \subset \text{Nb}(K,O)$. For each point $x \in K$, fix a connected neighborhood of x, V_x, with $\text{Cl } V_x$ compact and $\text{Cl } V_x \subset f^{-1}(O)$ (see 1.1.7.22 and 1.3.4.3). Now cover K by a finite number of the V_x's, say $V_{x_1},\ldots,V_{x_s}$. It is clear that the sets $U_i = V_{x_i}$ have the required properties.

6. <u>If X is a locally compact, locally connected, Hausdorff space, then the two topologies on $\text{Top } X$ are identical.</u>

By Lemma 5, given an open connected U with compact $\text{Cl } U$, an open O, and a homeomorphism $f \in \text{Nb}(\text{Cl } U,O)$, it is enough to find a subset $\mathcal{U} \subset \text{Top } X$, open in the first topology, and such that $f \in \mathcal{U}^{-1} \subset \text{Nb}(\text{Cl } U,O)$. Leaving aside the trivial case $U = \emptyset$, we fix a point $x_0 \in f(U)$ and find a set W with compact closure contained in O, such that $f(\text{Cl } U) \subset W$, and then take an open set V satisfying $f(\text{Cl } U) \subset V \subset \text{Cl } V \subset W$. Now set

$$\mathcal{U} = \text{Nb}(x_0,U) \cap \text{Nb}(\text{Cl } W \smallsetminus U, f^{-1}(O) \smallsetminus \text{Cl } U).$$

The inclusion $f \in \mathcal{U}^{-1}$ is trivial, and all that remains is to show that $\mathcal{U}^{-1} \subset \text{Nb}(\text{Cl } U,O)$, i.e., that $g^{-1}(\text{Cl } U) \subset O$ for any $g \in \mathcal{U}$. But $g \in \mathcal{U}$ implies $g(\text{Cl } W \smallsetminus U) \subset f^{-1}(O) \smallsetminus \text{Cl } U$, whence $U \subset g(V) \cup g(X \smallsetminus \text{Cl } W)$. Since $g(V)$ and $g(X \smallsetminus \text{Cl } W)$ are open and disjoint, and U is connected, we have only two possibilities: either $U \cap g(U) = \emptyset$ or $U \subset g(V)$. Since $g \in \mathcal{U}$, $g(x_0) \in U$; since $x_0 \in f(U) \subset V$, $g(x_0) \in g(V)$. Consequently, $U \cap g(V) \neq \emptyset$, and thus $U \subset g(V)$, i.e., $g^{-1}(U) \subset V$. Finally, $g^{-1}(\text{Cl } U) \subset \text{Cl } V \subset O$.

Groups of Diffeomorphisms

7. Let X be a $C^{\geq r}$-manifold, $1 \leq r \leq \infty$. By 3.1.2.9, the set of its C^r-diffeomorphisms, $\text{Diff}^r X$, is a group under the composition operation $\circ$. By 3.4.1.1, $\text{Diff}^r X$ can be endowed with the C^r-topology. We show that these two structures are compatible and conclude that $\text{Diff}^r X$ <u>is a topological group.</u>

Obviously, the case $r = \infty$ reduces to $r < \infty$, and so we may assume from the beginning that r is finite. Consider the mapping

$$\underbrace{d \circ \ldots \circ d}_{r}\colon \mathrm{Diff}^{r} X \to \mathrm{Top}(\underbrace{\mathrm{Tang} \ldots \mathrm{Tang}\, X}_{r})\,.$$

This is clearly a group monomorphism. Moreover, by 3.4.1.1, $d \circ \ldots \circ d$ is a topological embedding when the group $\mathrm{Top}(\mathrm{Tang} \ldots \mathrm{Tang}\, X)$ is equipped with the second topology. However, the first and the second topologies on $\mathrm{Top}(\mathrm{Tang} \ldots \mathrm{Tang}\, X)$ coincide (see 6), and hence the operations $\circ$ and $f \mapsto f^{-1}$ are continuous in these topologies. Consequently, both operations are continuous in $\mathrm{Diff}^{r} X$ also.

Let us add that the inclusion $\mathrm{Diff}^{r} X \to \mathrm{Top}\, X$ is a monomorphism of topological groups, and that the same is true for the inclusions $\mathrm{Diff}^{r} X \to \mathrm{Diff}^{s} X$ with $s < r$.

The Classical Groups

10. The analytic manifolds $O(n)$, $SO(n)$, $U(n)$, $SU(n)$, $Sp(n)$, and also $GL(n, \mathbb{R})$, $GL_{+}(n, \mathbb{R})$, $GL(n, \mathbb{C})$, $GL(n, \mathbb{H})$, defined in Subsection 3.2.1 (see 3.1.1.2 and 3.2.1.5 – 3.2.1.9) and endowed there with group structures, are obviously topological groups. $O(n)$ is called the <u>orthogonal group</u>, $SO(n)$ – the <u>special orthogonal group,</u> $U(n)$ – the <u>unitary group,</u> $SU(n)$ – the <u>special unitary group</u>, and $Sp(n)$ – the <u>symplectic group.</u> $GL(n, \mathbb{R})$, $GL(n, \mathbb{C})$, and $GL(n, \mathbb{H})$ are known as the <u>general linear groups.</u> It is immediate that $SO(n)$ is the component of the identity of $O(n)$, while $GL_{+}(n, \mathbb{R})$ is the component of the identity of $GL(n, \mathbb{R})$.

11. The topological group $GL(n, \mathbb{R})$ is manifestly a subgroup of the topological group $\mathrm{Top}\, \mathbb{R}^{n}$ (in the sense of 1.3). In the same sense, $O(n)$, $SO(n)$, and $GL_{+}(n, \mathbb{R})$ are subgroups of $\mathrm{Top}\, \mathbb{R}^{n}$.

Similarly, $GL(n, \mathbb{C})$ and its subgroups $U(n)$, $SU(n)$, are subgroups of $\mathrm{Top}\, \mathbb{C}^{n}$, while $GL(n, \mathbb{H})$ and its subgroup $Sp(n)$ are subgroups of $\mathrm{Top}\, \mathbb{H}^{n}$.

We also note that the inclusions $U(n) \subset SO(2n)$, $Sp(n) \subset SU(2n)$, $GL(n, \mathbb{C}) \subset GL_{+}(2n, \mathbb{R})$, and $GL(n, \mathbb{H}) \subset GL(2n, \mathbb{C})$ are inclusions of a subgroup into a group in the sense of 1.3.

3. Actions

1. An <u>action of the group</u> G <u>on a set</u> X is a map
$G \times X \to X$ with the following two properties:

  (i) $(e_G, x) \mapsto x$;

  (ii) if $(g_1, x) \mapsto x_1$ and $(g_2, x_1) \mapsto x_2$, then $(g_2 g_1, x) \mapsto x_2$.

The image of (g, x) under the given action is usually denoted
by gx, and so one may write conditions (i),(ii) in the form: $e_G x = x$,
$g_2(g_1 x) = (g_2 g_1)x$.

Every element $g \in G$ defines a map $X \to X$, $x \mapsto gx$, called
the <u>transformation induced by the element</u> g. We see from (i) and (ii)
that this map is invertible (its inverse is the transformation induced
by g^{-1}) and that the map $G \to \text{Symm}\, X$ which takes each g into the
corresponding transformation is a homomorphism. We call it the <u>adjoint</u>
<u>homomorphism of the given action.</u> Actually, this homomorphism uniquely
determines the action, and it is clear that every homomorphism
$h: G \to \text{Symm}\, X$ is the adjoint homomorphism of a certain action, namely,
of $(g, x) \mapsto (h(g))x$. Therefore, <u>an action of a group</u> G <u>on</u> X <u>can be</u>
<u>interpreted as a homomorphism</u> $G \to \text{Symm}\, X$.

We are mainly interested in the case where the adjoint
homomorphism is a monomorphism. An action with this property is said
to be <u>effective.</u> Generally, the kernel of the adjoint homomorphism will
be referred to as the <u>noneffectiveness kernel</u> of the given action. If
K is this kernel, then we can write the adjoint homomorphism as the
composition of the projection $G \to G/K$ with the monomorphism
$G/K \to \text{Symm}\, X$. Moreover, the action itself may be expressed as the
composition of the map $\text{pr} \times \text{id}\, X: G \times X \to G/K \times X$ with the effective
action $G/K \times X \to X$. We call the action $G/K \times X \to X$ the <u>effective</u>
<u>factor</u> of the action $G \times X \to X$.

The image of the set $G \times x$ under a given action $G \times X \to X$
is a subset of X called the <u>orbit</u> of the point x. Obviously, the
orbits of two points are either identical or disjoint, and hence the
orbits partition X. An action with only one orbit is said to be
<u>transitive.</u> In general, we denote the space of orbits by X/G.

If $h: G_1 \to G$ is a group homomorphism, then by composing the
mapping $h \times \text{id}\, X: G_1 \times X \to G \times X$ with an action $G \times X \to X$ of G on
X, we obtain an action of G_1 on X. We say that this new action is
<u>induced</u> by the initial action via the homomorphism h. The adjoint
homomorphism of the induced action is simply the composition of h with

the adjoint homomorphism of the initial action. If h is a monomorphism (epimorphism), then an effective (respectively, transitive) action induces an effective (respectively, a transitive) one.

If G_1 is a subgroup of G and h is the inclusion of G_1 in G, then we say that the induced action is obtained by <u>restricting</u> (or by <u>reducing</u>) <u>the group</u> G <u>to</u> G_1; we also say that the initial action is obtained by <u>extending</u> (or by <u>prolonging</u>) <u>the group</u> G_1 <u>to</u> G. The discussion above shows that when one restricts the group, an effective action remains efective. Also, when one extends the group, a transitive action remains transitive.

A subset X_1 of X is <u>invariant under the action</u> $G \times X \to X$ if it is saturated with respect to the partition of X into orbits. If this is the case, then we have an action $G \times X_1 \to X_1$, and clearly this is effective whenever the initial action $G \times X \to X$ is effective.

Given two actions, $G_1 \times X_1 \to X_1$ and $G_2 \times X_2 \to X_2$, their <u>product</u> is the action $(G_1 \times G_2) \times (X_1 \times X_2) \to X_1 \times X_2$ defined by $(g_1, g_2)(x_1, x_2) = (g_1 x_1, g_2 x_2)$. A product of effective (transitive) actions is again effective (respectively, transitive).

Given two actions, $G \times X \to X$ and $G \times X' \to X'$, of the same group, a map $f: X \to X'$ is a G-<u>map</u> (or a G-<u>equivariant map</u>) if $f(gx) = gf(x)$ for all $x \in X$ and $g \in G$. We can describe the more general situation when we are given two actions, $G \times X \to X$ and $G' \times X' \to X'$, of different groups, and a homomorphism $\gamma: G \to G'$; then $f: X \to X'$ is called a γ-<u>map</u> if $f(gx) = \gamma(g)f(x)$ for all $x \in X$ and $g \in G$.

Two actions, $G \times X \to X$ and $G \times X' \to X'$, are <u>equivalent</u> if there is an invertible G-map $X \to X'$.

2. The action defined in 1 should actually be called a <u>left action</u>, to distinguish it from a <u>right action</u>, which is defined as a map $X \times G \to X$ with the following two properties: (i) $(x, e_G) \mapsto x$; (ii) if $(x, g_1) \mapsto x_1$ and $(x_1, g_2) \mapsto x_2$, then $(x, g_1 g_2) \mapsto x_2$. For a right action, we write xg instead of gx, and properties (i),(ii) can be reformulated as: $xe_G = x$, $(xg_1)g_2 = x(g_1 g_2)$. Furthermore, the adjoint homomorphism becomes the adjoint antihomomorphism, and the rest of the discussion in 1 can be repeated word for word for a right action.

It is clear that the formula $xg = g^{-1}x$ transforms a left action into a right one, and that the formula $gx = xg^{-1}$ yields the inverse transformation. We say that the actions thus related are <u>conjugate</u>.

Henceforth, by <u>action</u> we shall mean a left action, unless we mention explicitly that we are dealing with a right action.

3. If the transformations induced by the elements of the group G, acting on X from the left (right), are elements of a subgroup H of $\mathrm{Symm}\, X$, then the action (respectively, right action) of G can be thought of as a homomorphism (respectively, antihomomorphism) of G into H. In this case we say that G acts on X (respectively, acts from the right on X) <u>by transformations from</u> H.

The group H is not always indicated explicitly. For example, if X is a topological space and $H = \mathrm{Top}\, X$, then one simply says that G acts on X (acts on X from the right) by homeomorphisms. Similarly, if X is a group and H is a group of automorphisms of X, then one says that G acts on X (respectively, acts on X from the right) by automorphisms; in this case, the action itself will be referred to as a <u>group-action.</u>

Important special examples are the actions $G \times G \to G$ given by $(g,x) \mapsto gx$ or $(g,x) \mapsto gxg^{-1}$, and the right actions $G \times G \to G$ given by $(x,g) \mapsto xg$ and $(x,g) \mapsto g^{-1}xg$ (here all the products are taken in G). These are called, in order: the <u>left canonical action,</u> the <u>left inner action,</u> the <u>right canonical action,</u> and the <u>right inner action.</u> The canonical actions are effective and transitive, while the inner actions are group-actions.

Let us remark that by restricting the left canonical action $G \times G \to G$ to the action $G_1 \times G \to G$, where G_1 is a subgroup of G, the orbits become the (right) cosets of G_1. Thus, the two interpretations (the usual group-theoretic one, and that given in 1) of the notation G/G_1 agree (if we denote the right coset space also by G/G_1).

The following generalization of the left canonical action is already of general importance. Let G_1 be a subgroup of G. Since every left translation takes left cosets into left cosets, the map $(g,x) \mapsto gx$ induces a map $G \times G/G_1 \to G/G_1$, and this is clearly an action, called the <u>canonical action of the group</u> G <u>on</u> G/G_1. It is transitive, and its noneffectiveness kernel is the intersection of all subgroups of G which are conjugate to G_1. The projection $G \to G/G_1$ is a G-map with respect to the left canonical action of G on G and the canonical action of G on G/G_1.

It turns out that every transitive action of G is equivalent to the canonical action on some quotient G/G_1. Specifically, let $G \times X \to X$ be a transitive action, and let $x_1 \in X$ be an arbitrarily chosen point. Consider the map $f: G \to X$, $f(g) = gx_1$. The preimage of x_1 under f is a subgroup G_1 of G, while the preimages under f of points $x \in X$, $x \neq x_1$, are left cosets of G_1. It is routine to check that the injective factor of f, $\mathrm{fact}\, f : G/G_1 \to X$, is a G-map.

The subgroup G_1 has a special name: it is known as the _isotropy_ (or _stability_, or _stationary_) _subgroup of the action_ $G \times X \to X$, or _of the group_ G, _at the point_ x_1. Obviously, the isotropy subgroup at the point gx_1 is gG_1g^{-1}, and so the isotropy subgroups of a transitive action of G constitute exactly one of its classes of conjugate subgroups.

Continuous Actions

5.   A _continuous action of a topological group_ G _on a topological space_ X is a continuous map $G \times X \to X$ which is an action in the sense of 1.

For a continuous action, the transformations induced by the elements of the group are manifestly homeomorphisms. Therefore, the adjoint homomorphism of a continuous action $G \times X \to X$ can be compressed to an algebraic isomorphism $G \to \mathrm{Top}\, X$. By 1.2.7.6, this last homomorphism is continuous in the first topology of the group $\mathrm{Top}\, X$ (see 2.1). If, in addition, X is Hausdorff and locally compact, then the existence of a continuous (in the first topology on $\mathrm{Top}\, X$) compression $G \to \mathrm{Top}\, X$ of the adjoint homomorphism of the given action is equivalent to the continuity of the action. It is clear that an algebraic homomorphism $G \to \mathrm{Top}\, X$ which is continuous in one of the two topologies of $\mathrm{Top}\, X$ is continuous also in the other one, and we know that if X is locally compact and Hausdorff, then $\mathrm{Top}\, X$ with the second topology is a topological group (see 2.3). Therefore, _if_ X _is a locally compact Hausdorff space, then a continuous action of_ G _on_ X _may be defined as a homomorphism_ $G \to \mathrm{Top}\, X$ _of topological groups_ (see 1.8).

A discrete group which acts by homeomorphisms always acts continuously. Thus, we may regard the actions of nontopologized groups which act by homeomorphisms as continuous actions.

6.   A G-_space_ is a topological space endowed with a continuous action of the group G. A G-space is called _effective_ if the action of G is effective. In the general case, by shifting to the effective factor of the action of G, the given G-space becomes an effective (G/K)-space, where K is the noneffectiveness kernel.

When the action $G \times X \to X$ is continuous, X/G is a topological space (a quotient space of X), known as an _orbit space_.

Since the saturation of any subset $A \subset X$ with respect to the partition of X into orbits is the union of the sets gA, $g \in G$, this partition is always open. When the group G is finite, the partition into orbits is also closed. In particular, X/G is second countable together with X, and when G is finite, X/G is normal together with X; see 1.2.3.10 and 1.2.3.9. Let us add that the partition into orbits is again closed whenever X is compact and Hausdorff and G is compact. Indeed, in this case the action $G \times X \to X$ is a closed map, and since it transforms every product $G \times A$ into the saturation of the set A, this saturation is closed whenever A is closed.

By restricting the topological group G to a subgroup G_1, we transform every G-space into a G_1-space. Any invariant subspace of a G-space is obviously a G-space; the G-spaces of this type are termed <u>subspaces</u> of the initial G-space. The product of two continuous actions is continuous, and hence the product of a G_1-space with a G_2-space is a $(G_1 \times G_2)$-space.

One can define the notions of G-map and γ-map for the case of continuous actions. To spell it out, a G-<u>map</u> (or a G-<u>equivariant map</u>) <u>of a G-space into another G-space</u> is any continuous map which is a G-map in the sense of 1; similarly, a γ-<u>map of a G-space into a G'-space</u> is any continuous map which is also a γ-map in the sense of 1 (here $\gamma \colon G \to G'$ is a homomorphism of topological groups). Two continuous actions of G are <u>equivalent</u> if the corresponding G-spaces are G-homeomorphic.

When G is a topological group, the special actions introduced in 3, i.e., the left canonical and left inner actions, are continuous.

As in 3, the notation G/G_1 can be interpreted in two ways (as the space of cosets of G_1 in G and as an orbit space), but again the two interpretations agree.

7. The canonical action of a group G on the space G/G_1 of cosets of a subgroup G_1 is also continuous. To see this, consider the composition $\psi \colon G \times G \to G/G_1$ of the map $G \times G \to G$, $(g,h) \mapsto gh$, with the projection $G \to G/G_1$. Clearly, ψ is constant on the elements of the partition $\mathrm{zer}(G,e_G) \times \mathrm{zer}(G,G_1)$. Furthermore, the action $G \times G/G_1 \to G/G_1$ in which we are interested is the composition of the map $G \times G/G_1 \to (G \times G)/(\mathrm{zer}(G,e_G) \times \mathrm{zer}(G,G_1))$, given by the inverse of the injective factor of $\mathrm{id}\,G \times \mathrm{pr} \colon G \times G \to G \times G/G_1$, with $\mathrm{fact}\,\psi \colon (G \times G)/(\mathrm{zer}(G,e_G) \times \mathrm{zer}(G,G_1)) \to G/G_1$. Since the partitions $\mathrm{zer}(G,e_G)$ and $\mathrm{zer}(G,G_1)$ are open, the above injective factor is a homeomorphism (see 1.2.3.11), which in turn implies the continuity of

the action of G on G/G_1.

Therefore, G/G_1 becomes a G-space with respect to the canonical action $G \times G/G_1 \to G/G_1$, and is called a <u>homogeneous space.</u>

The equivalence between an arbitrary transitive action $G \times X \to X$ and the canonical action of G on G/G_1, where G_1 is an isotropy subgroup, is not complete in the case of continuous actions. More precisely, the map $f: G \to X$, $f(g) = gx_1$, and its injective factor, $\text{fact}\, f : G/G_1 \to X$, are indeed continuous, but obvious examples* show that $(\text{fact}\, f)^{-1}$ is not necessarily continuous. However, $(\text{fact}\, f)^{-1}$ is continuous provided that G is compact and X is Hausdorff, i.e., <u>every transitive continuous action of a compact topological group on a Hausdorff topological space with a distinguished point is canonically equivalent to the canonical action of the group on the space of cosets of the isotropy group at the distinguished point.</u>

8. A continuous action $G \times X \to X$ is <u>free</u> if for every point $x \in X$ the map $G \to X$ given by $g \mapsto gx$ is an embedding.

Every free action is clearly effective. Moreover, if the action $G \times X \to X$ is free, then by restricting G to one of its subgroups, or by restricting X to one of its G-subspaces, the action remains free. A product of free actions is free. The canonical action $G \times G \to G$ is free, while the canonical action $G \times G/G_1 \to G/G_1$ is not free unless $G_1 = e_G$.

Given a free action $G \times X \to X$, consider the bundle $(X, pr, X/G)$. If every point $x \in X$ has a neighborhood U such that $gU \cap g'U = \emptyset$ for all $g, g' \in G$ with $g \neq g'$ (which happens, in particular, when G is finite and X is Hausdorff), then the image of U under the projection $pr: X \to X/G$ is open, and the restriction of the bundle $(X, pr, X/G)$ to $pr(U)$ is a trivial bundle with discrete fibers. Therefore, in this case $(X, pr, X/G)$ is a covering in the broad sense.

9. A continuous <u>right</u> action of a topological group G on a topological space X is a continuous map $X \times G \to X$ which is also a right action in the sense of 2.

All definitions and facts discussed in 5, 6, and 8 can be adapted immediately to the case of right actions. In particular, a topological space endowed with a right continuous action is called a <u>right G-space.</u> We keep the simple term G-space <u>only</u> for <u>left</u> G-spaces.

---

* Translator's note: Think of S^1 as the set of complex numbers of modulus 1 and let α be irrational. Take $G = \mathbb{R}$, $X = S^1 \times S^1$, and $G \times X \to X$, $(t, (e^{2\pi i x}, e^{2\pi i y})) \mapsto (e^{2\pi i (x+t)}, e^{2\pi i (y+\alpha t)})$ (an "irrational flow" on the torus).

Examples

10. Let X be a topological space. The identity homomorphism $\text{Top}\, X \to \text{Top}\, X$ defines an effective action of $\text{Top}\, X$ on X, and hence an effective action of any subgroup of $\text{Top}\, X$ on X. If X is Hausdorff and locally compact, then $\text{Top}\, X$ is a topological group and all these actions are continuous. In particular, $GL(n, \mathbb{R})$, $GL_+(n, \mathbb{R})$, $O(n)$, and $SO(n)$ act effectively and continuously on $\mathbb{R}^n$, while $GL(n, \mathbb{C})$, $U(n)$, and $SU(n)$ act in the same manner on $\mathbb{C}^n$, and $GL(n, \mathbb{H})$ and $Sp(n)$ - on $\mathbb{H}^n$.

Given an arbitrary $C^{\geqslant r}$-manifold X with $r \leqslant \infty$, the inclusion $\text{Diff}^r X \to \text{Top}\, X$ defines an effective and continuous action of $\text{Diff}^r X$ on X.

11. Since S^{n-1} is invariant under the action of $O(n)$ on $\mathbb{R}^n$, $O(n)$ and its subgroup $SO(n)$ act continuously on S^{n-1}. Similarly, $U(n)$ and $SU(n)$, being subgroups of $O(2n)$, act continuously on S^{2n-1}, while $Sp(n)$, being a subgroup of $O(4n)$, acts continuously on S^{4n-1}. All these actions are effective, and, if we exclude the trivial cases $SO(1) \times S^0 \to S^0$ and $SU(1) \times S^1 \to S^1$, transitive.

The isotropy subgroups of the actions $O(n) \times S^{n-1} \to S^{n-1}$ and $SO(n) \times S^{n-1} \to S^{n-1}$ at ort_n are exactly $O(n-1)$ and $SO(n-1)$. Similarly, the isotropy subgroups of the actions $U(n) \times S^{2n-1} \to S^{2n-1}$ and $SU(n) \times S^{2n-1} \to S^{2n-1}$ at ort_{2n} are $U(n-1)$ and $SU(n-1)$, while the isotropy subgroup of the action $Sp(n) \times S^{4n-1} \to S^{4n-1}$ at ort_{4n} is $Sp(n-1)$. The corresponding homeomorphisms, $O(n)/O(n-1) \to S^{n-1}$, $SO(n)/SO(n-1) \to S^{n-1}$, $U(n)/U(n-1) \to S^{2n-1}$, $SU(n)/SU(n-1) \to S^{2n-1}$, and $Sp(n)/Sp(n-1) \to S^{4n-1}$ (see 7) equal the injective factors of the submersions $V(n,n)\ [= O(n)] \to V(n,1)\ [= S^{n-1}]$, $V(n,n-1) \to V(n,1)$, $\mathbb{C}V(n,n) \to \mathbb{C}V(n,1)$, $\mathbb{C}V(n,n-1) \to \mathbb{C}V(n,1)$, and $\mathbb{H}V(n,n) \to \mathbb{H}V(n,1)$, defined in Subsection 3.2.1 (see 3.2.1.3, 3.2.1.5, and 3.2.1.6).

If we restrict $O(n)$, $U(n)$, and $Sp(n)$ $(n \geqslant 1)$ to their subgroups which consists of scalar multiples of the identity matrix, and which are usually identified with S^0, S^1, and S^3, respectively, we obtain continuous actions $S^0 \times S^{n-1} \to S^{n-1}$, $S^1 \times S^{2n-1} \to S^{2n-1}$, and $S^3 \times S^{4n-1} \to S^{4n-1}$. These are free actions, and the corresponding orbit spaces S^{n-1}/S^0, S^{2n-1}/S^1, and S^{4n-1}/S^3 are simply $\mathbb{R}P^{n-1}$, $\mathbb{C}P^{n-1}$, and $\mathbb{H}P^{n-1}$.

We remark also that D^n, D^{2n}, and D^{4n} are invariant under the actions of $O(n)$, $U(n)$, and $Sp(n)$ on $\mathbb{R}^n$, $\mathbb{C}^n$, and $\mathbb{H}^n$. Hence

O(n) and SO(n) act continuously on D^n, U(n) and SU(n) act continuously on D^{2n}, and Sp(n) acts continuously on D^{4n}. All these are effective actions.

12. The groups O(n) and SO(n) (O(k) and SO(k)) act continuously from the left (right) on the Stiefel manifolds V(n,k): the left actions are defined by $(g,v) \mapsto g \circ v$ [$g \in$ O(n) or SO(n), $v \in$ V(n,k); g and v are regarded as linear maps]; the right actions are given by $(v,g) \mapsto v \circ g$. Similarly, U(n) and SU(n) (U(k) and SU(k)) act continuously from the left (respectively, from the right) on $\mathbb{C}$V(n,k), and Sp(n) (Sp(k)) acts continuously from the left (respectively, from the right) on $\mathbb{H}$V(n,k).

For $k \neq 0$, all the left actions are effective, and the only intransitive ones are SO(n) $\times$ V(n,n) $\to$ V(n,n) and SU(n) $\times$ $\mathbb{C}$V(n,n) $\to$ $\to$ $\mathbb{C}$V(n,n), $n \geqslant 1$. The isotropy subgroups of O(n) and SO(n) at the point $[(x_1,\dots,x_k) \mapsto (0,\dots,0,x_1,\dots,x_k)] \in$ V(n,k) (the elements of V(n,k) are considered as linear isometric maps $\mathbb{R}^k \to \mathbb{R}^n$) coincide with O(n-k) and SO(n-k). Similarly, the isotropy subgroups of U(n), SU(n), and Sp(n) at the points $[(x_1,\dots,x_k) \mapsto (0,\dots,0,x_1,\dots,x_k)]$ of $\mathbb{C}$V(n,k) and $\mathbb{H}$V(n,k) coincide with U(n-k), SU(n-k), and Sp(n-k), respectively. The corresponding homeomorphisms

$$O(n)/O(n-k) \to V(n,k), \quad SO(n)/SO(n-k) \to V(n,k),$$

$$U(n)/U(n-k) \to \mathbb{C}V(n,k), \quad SU(n)/SU(n-k) \to \mathbb{C}V(n,k),$$

and

$$Sp(n)/Sp(n-k) \to \mathbb{H}V(n,k),$$

are precisely the injective factors of the maps O(n) $\to$ V(n,k), SO(n) $\to$ V(n,k), U(n) $\to$ $\mathbb{C}$V(n,k), SU(n) $\to$ $\mathbb{C}$V(n,k), and Sp(n) $\to$ $\mathbb{H}$V(n,k), defined in Subsection 3.2.1 (see 3.2.1.3, 3.2.1.5, and 3.2.1.6). When $k = 1$, these actions reduce to those discussed in 11.

All the right actions are free. The corresponding orbit spaces, V(n,k)/O(k), V(n,k)/SO(k), $\mathbb{C}$V(n,k)/U(k), $\mathbb{C}$V(n,k)/SU(k), and $\mathbb{H}$V(n,k)/Sp(k), are canonically homeomorphic to the Grassman manifolds G(n,k), G_+(n,k), $\mathbb{C}$G(n,k), and $\mathbb{H}$G(n,k), respectively; the corresponding canonical homeomorphisms are the injective factors of the maps V(n,k) $\to$ G(n,k), V(n,k) $\to$ G_+(n,k), $\mathbb{C}$V(n,k) $\to$ $\mathbb{C}$G(n,k), and $\mathbb{H}$V(n,k) $\to$ $\mathbb{H}$G(n,k), defined in Subsection 3.2.2 (see 3.2.2.3, 3.2.2.7, and 3.2.2.8).

13. The same formulas, i.e., $(g,v) \mapsto g \circ v$, $(g,v) \mapsto v \circ g$, define left actions of GL(n, $\mathbb{R}$) and GL_+(n, $\mathbb{R}$) on V'(n,k), of GL(n,$\mathbb{C}$) on $\mathbb{C}$V'(n,k), and of GL(n, $\mathbb{H}$) on $\mathbb{H}$V'(n,k), and right

actions of GL(k, IR) and GL$_+$(k, IR) on V'(n,k), of GL(k,$\mathbb{C}$) on
$\mathbb{C}$V'(n,k), and of GL(k, IH) on IHV'(n,k).

All the left actions are effective and, excepting the action
GL$_+$(n, IR) × V'(n,n) → V'(n,n), transitive. The isotropy subgroups of
GL(n, IR), GL(n,$\mathbb{C}$), and GL(n, IH) at the points [(x$_1$,...,x$_k$) ↦
→ (0,...,0,x$_1$,...,x$_k$)] of V'(n,k), $\mathbb{C}$V'(n,k), and IHV'(n,k) are
GL(n-k, IR), GL(n-k,$\mathbb{C}$), and GL(n-k, IH), respectively. The
corresponding homeomorphisms

GL(n, IR)/GL(n-k, IR) → V'(n,k), GL(n,$\mathbb{C}$)/GL(n-k,$\mathbb{C}$) → $\mathbb{C}$V'(n,k),

and

GL(n, IH)/GL(n-k, IH) → IHV'(n,k),

are the injective factors of the maps GL(n, IR) → V'(n,k),
GL(n,$\mathbb{C}$) → $\mathbb{C}$V'(n,k), and GL(n, IH) → IHV'(n,k), defined in Subsection
3.2.1 (see 3.2.1.7, 3.2.1.8, and 3.2.1.9). The isotropy subgroup of
GL$_+$(n, IR) at the point [(x$_1$,...,x$_k$) ↦ (0,...,0,x$_1$,...,x$_k$)] ∈ V'(n,k)
is GL$_+$(n-k, IR).

All the right actions are free. The corresponding orbit
spaces, V'(n,k)/GL(k, IR), V'(n,k)/GL$_+$(k, IR), $\mathbb{C}$V'(n,k)/GL(k,$\mathbb{C}$), and
IHV'(n,k)/GL(k, IH), are canonically homeomorphic to the Grassman
manifolds G(n,k), G$_+$(n,k), $\mathbb{C}$G(n,k), and IHG(n,k); the corresponding
canonical homeomorphisms are the injective factors of the maps
V'(n,k) → G(n,k), V'(n,k) → G$_+$(n,k), $\mathbb{C}$V'(n,k) → $\mathbb{C}$G(n,k), and
IHV'(n,k) → IHG(n,k), defined in Subsection 3.2.2 (see 3.2.2.3, 3.2.2.7,
and 3.2.2.8).

14. GL(n, IR) and its subgroups GL$_+$(n, IR), O(n), and
SO(n) obviously act continuously from the left on the Grassman
manifolds G(n,k), G$_+$(n,k). Similarly, GL(n,$\mathbb{C}$) and its subgroups
U(n) and SU(n) act continuously from the left on $\mathbb{C}$G(n,k), while
GL(n, IH) and Sp(n) act continuously from the left on IHG(n,k).
For k odd, the actions of O(n) and SO(n) on G$_+$(n,k) are effective.
The noneffectiveness kernels of the actions GL(n, IR) × G$_+$(n,k) → G$_+$(n,k)
and GL$_+$(n,k) × G$_+$(n,k) → G$_+$(n,k) for k odd consist of scalar
matrices with positive diagonal elements. If we exclude the trivial
cases k = 0 and k = n, the noneffectivennes kernels of the remaining
actions consists of all scalar matrices contained in the corresponding
group. The only intransitive actions are GL(n, IR) × G$_+$(n,0) → G$_+$(n,0),
GL$_+$(n, IR) × G$_+$(n,0) → G$_+$(n,0), O(n) × G$_+$(n,0) → G$_+$(n,0),
SO(n) × G$_+$(n,0) → G$_+$(n,0), GL$_+$(n, IR) × G$_+$(n,n) → G$_+$(n,n), and
SO(n) × G$_+$(n,n) → G$_+$(n,n). Take the plane x$_1$ = 0,...,x$_{n-k}$ = 0

(oriented in the case of $G_+(n,k)$) as a distinguished point in the
manifolds $G(n,k)$, $G_+(n,k)$, $\mathbb{C}G(n,k)$, and $\mathbb{H}G(n,k)$. Then the isotropy
subgroups of the actions $GL(n,\mathbb{R}) \times G(n,k) \to G(n,k)$,
$GL(n,\mathbb{R}) \times G_+(n,k) \to G_+(n,k)$, $GL(n,\mathbb{C}) \times \mathbb{C}G(n,k) \to \mathbb{C}G(n,k)$, and
$GL(n,\mathbb{H}) \times \mathbb{H}G(n,k) \to \mathbb{H}G(n,k)$ at these distinguished points are the
subgroups of all matrices of the form

$$\begin{pmatrix} A & C \\ 0 & B \end{pmatrix},$$

where A and B are nonsingular matrices of order $n-k$ and k,
respectively, and C is an arbitrary $(n-k) \times k$ matrix (and $B \in GL_+(k,\mathbb{R})$
in the case of $G_+(n,k)$).

If we restrict the acting group to a subgroup, then the new
isotropy subgroup is the intersection of the original isotropy subgroup
with the new acting group. In particular, for the actions of $O(n)$ on
$G(n,k)$ and $G_+(n,k)$, the action of $SO(n)$ on $G_+(n,k)$, the action of
$U(n)$ on $\mathbb{C}G(n,k)$, and the action of $Sp(n)$ on $\mathbb{H}G(n,k)$, the
corresponding isotropy subgroups are the images of the monomorphisms
$O(n-k) \times O(k) \to O(n)$, $O(n-k) \times SO(k) \to O(n)$, $SO(n-k) \times SO(k) \to SO(n)$,
$U(n-k) \times U(k) \to U(n)$, and $Sp(n-k) \times Sp(k) \to Sp(n)$, all defined by
the matrix formula

$$(A,B) \mapsto \begin{pmatrix} A & 0 \\ 0 & B \end{pmatrix}.$$

If we identify these product with their images, we obtain canonical
homeomorphisms

$$O(n)/[O(n-k) \times O(k)] \to G(n,k), \quad O(n)/[O(n-k) \times SO(k)] \to G_+(n,k),$$

$$SO(n)/[SO(n-k) \times SO(k)] \to G_+(n,k),$$

$$U(n)/[U(n-k) \times U(k)] \to \mathbb{C}G(n,k),$$

$$Sp(n)/[Sp(n-k) \times Sp(k)] \to \mathbb{H}G(n,k).$$

15. Let $m, \ell_1, \ldots, \ell_n$ be relatively prime positive integers.
The complex-number formula

$$(k,(z_1,\ldots,z_n)) \mapsto (z_1 e^{2\pi i k \ell_1/m}, \ldots, z_n e^{2\pi i k \ell_n/m}),$$

where $k \in \mathbb{Z}$, $(z_1,\ldots,z_n) \in S^{2n-1}$, defines an action $\mathbb{Z} \times S^{2n-1} \to$
$\to S^{2n-1}$ with noneffectiveness kernel $m\mathbb{Z}$, which becomes, by shifting
to the effective factor, a free action of the group $\mathbb{Z}_m = \mathbb{Z}/m\mathbb{Z}$ on

S^{2n-1}. The orbit space $S^{2n-1}/\mathbb{Z}_m$ is denoted by $L(m;\ell_1,\ldots,\ell_n)$ and is called a <u>lens</u> (or a <u>lens space</u>).

There are also <u>infinite lenses</u> $L(m;\ell_1,\ell_2,\ldots)$, with $m,\ell_1,\ell_2,\ldots$ relatively prime positive integers. The lens $L(m;\ell_1,\ell_2,\ldots)$ is defined as the orbit space of the free action resulting from passing to the effective factor of the action

$$(k,(z_1,z_2,\ldots)) \mapsto (z_1 e^{2\pi ik\ell_1/m}, z_2 e^{2\pi ik\ell_2/m}, \ldots)$$

of $\mathbb{Z}$ on S^∞. An equivalent decription: $L(m;\ell_1,\ell_2,\ldots) =$
$= \lim(L(m;\ell_1,\ldots,\ell_n),\text{in}: L(m;\ell_1,\ldots,\ell_n) \to L(m;\ell_1,\ldots,\ell_{n+1}))$. The infinite lens $L(m;1,1,\ldots)$ is denoted simply by $L(m)$.

According to 8, the triples $(S^{2n-1},\mathrm{pr},L(m;\ell_1,\ldots,\ell_n))$ and $(S^\infty,\mathrm{pr},L(m;\ell_1,\ell_2,\ldots))$ are coverings.

16. The formula $(y,x) \mapsto yxy^{-1}$, where x and y are quaternions and y has norm 1, defines a continuous action $S^3 \times \mathbb{R}^4 \to \mathbb{R}^4$. The space $\mathbb{R}^3_1$ of imaginary quaternions is invariant under this action, and hence $\mathbb{R}^3_1$, and also $\mathbb{R}^3$, are S^3-spaces. [We identify $\mathbb{R}^3_1$ with $\mathbb{R}^3$ via the map $\mathrm{shi}: \mathbb{R}^3 \to \mathbb{R}^3_1$; see 3.2.3.1.] The noneffectiveness kernel of the action $S^3 \times \mathbb{R}^3 \to \mathbb{R}^3$ is obviously S^0, and now it is clear that the effective action of the factor group $S^3/S^0 = \mathbb{R}P^3$ on $\mathbb{R}^3$ becomes the standard action of $SO(3)$ on $\mathbb{R}^3$ (see 11) under the canonical identification of the spaces $\mathbb{R}P^3$ and $SO(3)$ (see 3.2.3.1).

17. Let P be a convex regular polyhedron in $\mathbb{R}^3$ (a tetrahedron, cube, octahedron, dodecahedron, or icosahedron) with center 0. Let GP be the subgroup of $SO(3)$ consisting of those rotations which take P into itself, and let $\tilde{G}P$ be the preimage of GP under the projection $S^3 \to SO(3)$ (see 16). Obviously, GP and $\tilde{G}P$ do not change if we replace P by the dual polyhedron, while they are transformed into conjugate subgroups of $SO(3)$ and S^3 if we replace P by any convex regular polyhedron with the same number of faces and center 0. Therefore, in $SO(3)$ (S^3) there are exactly three classes of conjugate subgroups GP (respectively, $\tilde{G}P$). The groups in the first class are called tetrahedral groups (respectively, binary tetrahedral groups), those in the second class - cube or octahedral groups (respectively, binary cube or octahedral groups), and those in the third class - dodecahedral or icosahedral groups

(respectively, binary dodecahedral and icosahedral groups).

To every rotation in GP we may associate the image of a marked oriented edge of the polyhedron P, and in this way define an invertible mapping of the group GP onto the set of oriented edges of P. Consequently, the order of the group GP is twice the number of edges of P, i.e., 12 when P is a tetrahedron, 24 when P is a cube or octahedron, and 60 when P is a dodecahedron or an icosahedron. The corresponding binary groups $\tilde{G}P$ have order 24, 48, and 120.

The coset spaces $SO(3)/GP$ and $S^3/\tilde{G}P$ are orbit spaces of the free actions induced by the left canonical actions of $SO(3)$ and S^3 under the inclusions $GP \to SO(3)$ and $\tilde{G}P \to S^3$. Therefore, the triples $(SO(3),pr,SO(3)/GP)$ and $(S^3,pr,S^3/\tilde{G}P)$ are coverings (see 8). Obviously, we can write $SO(3)/GP = S^3/\tilde{G}P$.

4. Exercises

1. Show that for any smooth manifold X the first and the second topologies on $\text{Top}\,X$ coincide.

2. Let X denote the subset of $\mathbb{R}$ consisting of the points 0 and 2^n, all $n \in \mathbb{Z}$. Show that the first and the second topologies on $\text{Top}\,X$ are distinct.

3. Show that the canonical diffeomorphism $SU(2) \to S^3$ (see 3.2.1.5) is a group isomorphism.

4. Show that the lenses $L(m;\ell_1,\ldots,\ell_k)$ and $L(m;\ell_1',\ldots,\ell_k')$ are homeomorphic whenever for each i the sum $\ell_i + \ell_i'$ or the difference $\ell_i - \ell_i'$ is a multiple of m.

5. Show that the submanifold $\text{Tang}_1\,\mathbb{R}P^2$ of $\text{Tang}\,\mathbb{R}P^2$ consisting of the unit tangent vectors (i.e., of the images under the map $d\,pr : \text{Tang}\,S^2 \to \text{Tang}\,\mathbb{R}P^2$ of the unit tangent vectors) is homeomorphic to the lens $L(4;1,1)$.

6. Consider the action of $\mathbb{Z}_2$ on the manifold $V(3,2)$ of unit vectors tangent to S^2, where the nonzero element of $\mathbb{Z}_2$ takes each vector v into $-v$. Show that the orbit space $V(3,2)/\mathbb{Z}_2$ is homeomorphic to $L(4;1,1)$.

7. Consider the action $\mathbb{Z}_2 \times \text{Tang}_1\,\mathbb{R}P^2 \to \text{Tang}_1\,\mathbb{R}P^2$ (see 5), where the nonzero element of $\mathbb{Z}_2$ takes each vector v into $-v$. Show that the orbit space $\text{Tang}_1\,\mathbb{R}P^2/\mathbb{Z}_2$ is homeomorphic to the coset

space S^3/H, where H is the subgroup of S^3 consisting of the quaternions $\pm\mathrm{ort}_1$, $\pm\mathrm{ort}_2$, $\pm\mathrm{ort}_3$, $\pm\mathrm{ort}_4$.

8. Consider the action $\mathbb{Z}_2 \times \mathbb{C}P^2 \to \mathbb{C}P^2$, where the nonzero element of $\mathbb{Z}_2$ takes each point $(z_1:z_2:z_3)$ into $(\bar{z}_1:\bar{z}_2:\bar{z}_3)$. Show that the orbit space $\mathbb{C}P^2/\mathbb{Z}_2$ is homeomorphic to S^4.

9. Consider the action of $\mathbb{Z}_2$ on $\mathbb{C}P^1 \times \mathbb{C}P^1$, where the nonzero element of $\mathbb{Z}_2$ takes each point $((z_1:z_2),(w_1:w_2))$ into $((\bar{z}_1:\bar{z}_2),(\bar{w}_1:\bar{w}_2))$. Show that the orbit space $(\mathbb{C}P^1 \times \mathbb{C}P^1)/\mathbb{Z}_2$ is homeomorphic to S^4.

10. Consider the action of $\mathbb{Z}_2$ on $S^2 \times S^2$, where the nonzero element of $\mathbb{Z}_2$ takes each point (x,y) into (y,x). Show that the orbit space $(S^2 \times S^2)/\mathbb{Z}_2$ is homeomorphic to $\mathbb{C}P^2$.

§3. BUNDLES WITH A GROUP STRUCTURE

1. Spaces With F-Structure

1. The bundles which we encouter most frequently have fibers that besides being merely topological spaces, carry some additional structure: for example, they may be vector, Euclidean, or Hermitian spaces. In the present section we shall introduce this concept of additional structure into the theory of bundles.

We begin by giving an exact description of the necessary type of structures and then fit them systematically into the basic definitions of the theory, given in § 1 (see Subsections 1.1 and 1.2).

2. Let G be a topological group, and let F be an effective G-space. We say that the topological space W is endowed with an F-structure if there is given a nonempty set A of homeomorphism $F \to W$ such that, for an arbitrarily fixed homeomorphism $\alpha \in A$, a given homeomorphism $\beta: F \to W$ belongs to A if and only if $\beta^{-1} \circ \alpha$ is the transformation induced by one of the elements of G. The homeomorphisms of A are called marked.

Every marked homeomorphism naturally carries the action of G from F to W. If G is commutative, then the resulting action

$G \times W \to W$ does not depend upon the choice of the marked homeomorphism, and hence in this case the F-structure reduces to the action of G. If G is not commutative, then an F-structure does not define a canonical action of G on W.

We remark that F itself has a canonical F-structure, namely that whose marked homeomorphisms are the transformations induced by the elements of G.

In the simplest case when G is the trivial group, a space with an F-structure is simply a topological space canonically homeomorphic to F.

3 (EXAMPLES). If $G = GL(n, \mathbb{R})$ and $F = \mathbb{R}^n$ with the usual action of this group, then a space with an F-structure is nothing else but an n-dimensional vector space, and fixing a marked homeomorphism is simply fixing a basis of the space.

If G is one of the groups $GL_+(n, \mathbb{R})$, $O(n)$, or $SO(n)$, and F is $\mathbb{R}^n$ with the usual action of these groups, then a space with an F-structure is an oriented n-dimensional real vector space, an n-dimensional Euclidean space, or an oriented n-dimensional Euclidean space, respectively. When G is $GL(n,\mathbb{C})$ or $U(n)$, and F is $\mathbb{C}^n$ with the usual action of G, then a space with an F-structure is an n-dimensional complex vector space, or an n-dimensional Hermitian space, respectively.

If $G = \mathrm{Diff}^r X$, where $X = F$ is a C^r-manifold $(1 \leqslant r \leqslant a)$ and G acts as usual, then a space with an F-structure is a C^r-manifold which is C^r-diffeomorphic to X.

If $G = \mathrm{Top}\, X$, where X is a locally compact Hausdorff space, and $F = X$ with the usual action of $\mathrm{Top}\, X$, then a space with an F-structure is simply a topological space homeomorphic to X.

If G is the group of all simplicial autohomeomorphisms of the unit simplex T^n, and $F = T^n$ with the standard action of this group, then a space with an F-structure is simply an n-dimensional topological simplex.

4. A homeomorphism $W \to W'$, where W and W' are spaces with F-structure, which takes the set of marked homeomorphisms of W into the set of marked homeomorphisms of W', is called an isomorphism or, more specifically, an F-isomorphism.

In each of the previous examples, the F-isomorphisms form a well-known class of maps: in the first and the fifth cases they are the linear isomorphism, in the second - the orientation preserving linear isomorphisms, in the third and sixth - the linear isometric isomorphisms, in the fourth - the orientation preserving linear isomorphisms, in the seventh - the C^r-diffeomorphisms, in the eighth - the homeomorphisms, and in the ninth - the simplicial homeomorphisms.

5. Given a space W with an F-structure and a space W' with an F'-structure, the product $W \times W'$ is obviously a space with an $F \times F'$-structure (see 2.3.6); the marked homeomorphisms $F \times F' \to W \times W'$ are those of the form $\alpha \times \alpha'$, where α and α' are marked homeomorphisms.

If the G_1-space F_1 is obtained from the G-space F by reducing the group G to G_1, then by returning to F from F_1 every space W_1 with an F_1-structure becomes a space W with an F-structure: topologically, W is the same as W_1, while the new marked homeomorphisms are defined as the compositions of the transformations induced by the elements of G with the old marked homeomorphisms. We say that W is obtained from W_1 by <u>extending</u> (or <u>prolonging</u>) the group G_1 to G.

2. <u>Steenrod Bundles</u>

1. Let G and F be a topological group and an effective G-space, respectively. A bundle ξ is a <u>weak F-bundle</u>, or a <u>W-F-bundle</u>, if each of its fibers is endowed with an F-structure. In this case, F and G are called the <u>standard fiber</u> and the <u>structure group of</u> ξ, respectively. The set of all marked homeomorphisms from F onto the fibers of ξ is denoted by $MH(\xi)$. The group G acts naturally from the right on $MH(\xi)$ by the rule: $[\alpha g](y) = \alpha(gy)$ $[\alpha \in MH(\xi), \; g \in G, \; y \in F]$.

If ξ is a W-F-bundle and $f : B \to bs\,\xi$ is continuous, then clearly the induced bundle $f^!\xi$ is a W-F-bundle: the F-structures on its fibers are defined via the homeomorphisms $ab\,tl\,ad\,f : pr\,f^!\xi^{-1}(b) \to$ $\to pr\,\xi^{-1}(f(b))$ $[b \in B]$.

Given two W-F-bundles, ξ and η, a map f of ξ into η is called a <u>W-F-map</u> if the maps $ab\,tl\,f$ from the fibers of ξ into the fibers of η are isomorphisms (see 1.4). A W-F-map which is an isomorphism (respectively, equivalence) in the pure topological sense, i.e., in the sense of 1.1.2, is called a <u>W-F-isomorphism</u> (respectively, a W-F-equivalence). Two W-F-bundles which can be mapped into each other by a W-F-isomorphism (W-F-equivalence) are said to be <u>W-F-isomorphic</u> (respectively, W-F-<u>equivalent</u>).

To each W-F-map $f : \xi \to \eta$ corresponds the map

$MH(f): MH(\xi) \to MH(\eta)$, which takes each marked homeomorphism $\alpha: F \to \operatorname{pr} \xi^{-1}(b)$ into the composite homeomorphism $F \xrightarrow{\alpha} \operatorname{pr} \xi^{-1}(b) \xrightarrow{\text{ab tl } f} \operatorname{pr} \eta^{-1}(\operatorname{bs} f(b))$. Moreover, we see that $MH(f)$ is a G-map with respect to the natural right actions of G on $MH(\xi)$ and $MH(\eta)$.

The standard trivial bundle, $(B \times F, \operatorname{pr}_1, B)$, with B an arbitrary topological space, is obviously a W-F-bundle: the F-structures on its fibers are defined by the homeomorphisms $F \to b \times F$, $y \mapsto (b, y)$. As in Subsection 1.2, every W-F-bundle which is W-F-equivalent to a standard trivial W-F-bundle is called a <u>W-F-trivial W-F-bundle.</u>

2. A bundle ξ is a <u>strong F-bundle</u> or, simply, <u>an F-bundle</u> if it is a W-F-bundle and $MH(\xi)$ is endowed with a topology.

If ξ is an F-bundle and $f: B \to \operatorname{bs} \xi$ is continuous, then the induced bundle $f^!\xi$ is also an F-bundle: to introduce a topology on $MH(f^!\xi)$, we use the injective mapping $MH(f^!\xi) \to B \times MH(\xi)$ given by $\alpha \mapsto (\operatorname{pr} f^!\xi(\alpha(F)), [MH(\operatorname{ad} f)](\alpha))$.

A map $f: \xi \to \eta$, where ξ and η are F-bundles, is said to be an F-<u>map</u> if it is a W-F-map and $MH(f)$ is continuous. An F-map f is an F-<u>isomorphism</u> (F-<u>equivalence</u>) if it is an isomorphism (respectively, equivalence) in the pure topological sense and $MH(f)$ is a homeomorphism.

The standard trivial bundle $(B \times F, \operatorname{pr}_1, B)$, with B an arbitrary topological space, is obviously an F-bundle: the F-structures of its fibers were already introduced in 1, and one can introduce a topology on $MH((B \times F, \operatorname{pr}_1, B))$ by means of the invertible mapping $B \times G \to MH((B \times F, \operatorname{pr}_1, B))$, which takes each pair (b, g) into the homeomorphism $F \to b \times F$, $y \mapsto (b, gy)$. An F-bundle which is F-equivalent to a standard trivial bundle is called F-<u>trivial</u>, and every such equivalence is an F-<u>trivialization.</u>

3. The F-bundle ξ is <u>locally F-trivial</u> if every point of $\operatorname{bs} \xi$ has a neighborhood U such that the restriction $\xi\big|_U$ is F-trivial. The locally F-trivial bundles are called <u>Steenrod F-bundles.</u>

Steenrod bundles play a major role in what follows, which accounts also for the importance of the F-bundles. The weak F-bundles are only auxiliary.

We remark that for Steenrod bundles the canonical right action of the structure group on the space of marked homeomorphisms is continuous and free. This is plainly true in the standard trivial case, to which the general case reduces.

4. If ξ is a Steenrod F-bundle and $f: B \to \operatorname{bs} \xi$ is continuous, then the induced bundle $f^!\xi$ is again a Steenrod F-bundle: by 2, $f^!\xi$ is an F-bundle, and the obvious fact that $f^!\xi$ is F-trivial

if ξ is so implies the local F-triviality of $f^!\xi$. Clearly, the map $\mathrm{ad}\, f : f^!\xi \to \xi$ is an F-map, the canonical equivalence $(\mathrm{id}\,\mathrm{bs}\,\xi)^!\xi \to \xi$, and the canonical equivalences of the form $g^!(f^!\xi) \to (f \circ g)^!\xi$ (see 1.1.4), are F-equivalences. Moreover, given any F-map h of ξ into another Steenrod F-bundle, the correcting map, $\mathrm{corr}\, h$, is an F-map.

The product of a Steenrod F-bundle ξ with a Steenrod F'-bundle ξ' is a Steenrod $F \times F'$-bundle: the $F \times F'$-structures on its fibers is defined as in 1.5; the topology on $MH(\xi \times \xi')$ is introduced by means of the invertible mapping $MH(\xi) \times MH(\xi') \to MH(\xi \times \xi')$ given by $(\alpha,\alpha') \mapsto \alpha \times \alpha'$; the local $F \times F'$-triviality of the resulting $F \times F'$-bundle follows from the fact that it is $F \times F'$-trivial whenever ξ is F-trivial and ξ' is F'-trivial.

If the G_1-space F_1 comes from the effective G-space F by reducing the group G to G_1, then by returning to F from F_1, every Steenrod F_1-bundle ξ_1 becomes a Steenrod F-bundle ξ: topologically, ξ is the same as ξ_1; the F-structures on the fibers of ξ are those described in 1.5; further, to define a topology on $MH(\xi)$, consider the action of G_1 on $G \times MH(\xi_1)$, defined as $g_1(g,\alpha) \mapsto (g_1 g, \alpha^{-1} g_1)$, where G_1 acts canonically from the right on $MH(\xi_1)$ (see 1), and then use the invertible mapping $(G \times MH(\xi_1))/G_1 \to MH(\xi)$, which takes the orbit of the pair (g,α) into the homeomorphism $y \mapsto \alpha(gy)$, to transfer the topology of $(G \times MH(\xi_1))/G_1$ to $MH(\xi)$; finally, the local F-triviality of the resulting F-bundle is a consequence of its F-triviality in the case when ξ_1 is F_1-trivial. This transformation of F_1-bundles into F-bundles is known as the <u>extension</u> (or <u>prolongation</u>) <u>of the structure group.</u> It takes F_1-maps into F-maps, and F_1-equivalences into F-equivalences. It is also clear that the extension of the structure group commutes with the induction operation; that is to say, if ξ is obtained from ξ_1 by extension of the structure group and $f: B \to \mathrm{bs}\,\xi$ is an arbitrary continuous map, then $f^!\xi$ is obtained from $f^!\xi_1$ by extension of the structure group.

5. <u>Every Steenrod F-bundle with trivial structure group is F-trivial.</u>

PROOF. Let ξ be a Steenrod bundle with standard fiber F and trivial structure group. Let Γ be an open cover of $\mathrm{bs}\,\xi$ such that the bundle $\xi|_U$ is F-trivial for any $U \in \Gamma$. Set $\eta = (\mathrm{bs}\,\xi \times F, \mathrm{pr}_1, \mathrm{bs}\,\xi)$. Clearly, the F-trivialization $\eta|_U \to \xi|_U$ is unique for any $U \in \Gamma$, and these F-trivializations together yield an F-trivialization $\eta \to \xi$.

Theorems About F-maps

6. <u>Suppose that</u> ξ <u>and</u> ξ' <u>are Steenrod bundles with standard fiber</u> F, B <u>is a topological space, and</u> p: B $\to$ bs ξ <u>is a factorial map. If</u> τ: tl ξ $\to$ tl ξ' <u>and</u> β: bs ξ $\to$ bs ξ' <u>are maps such that</u> $(\tau \circ \text{tl ad } p, \beta \circ p)$ <u>is an F-map</u> $p^!\xi \to \xi'$, <u>then</u> (τ,β) <u>is an F-map</u> $\xi \to \xi'$.

We need only check the continuity of β, τ, and MH(τ,β). The continuity of β is an immediate consequence of the continuity of the composition $\beta \circ p$ and the fact that p is factorial (see 1.2.3.4). As for τ and MH(τ,β), it is enough to verify their continuity when ξ and ξ' are standard trivial F-bundles. In this situation, τ is given by $\tau(b,y) = (\beta(b),\phi(b)y)$, where ϕ is some map from bs ξ into the structure group G. Moreover, if we use the homeomorphisms B $\times$ G $\to$ MH$(p^!\xi)$, bs ξ $\times$ G $\to$ MH(ξ), and bs ξ' $\times$ G $\to$ MH(ξ') (which define the topologies on MH$(p^!\xi)$, MH(ξ), and MH(ξ'), respectively; see 2), then the maps

$$\text{MH}(\tau \circ \text{tl ad } p, \beta \circ p): \text{MH}(p^!\xi) \to \text{MH}(\xi')$$

and

$$\text{MH}(\tau,\beta): \text{MH}(\xi) \to \text{MH}(\xi')$$

are transformed into the maps

$$B \times G \to \text{bs } \xi' \times G, \quad (b,g) \mapsto (\beta \circ p(b),(\phi \circ p(b))g),$$

and

$$\text{bs } \xi \times G \to \text{bs } \xi' \times G, \quad (b,g) \mapsto (\beta(b),\phi(b)g),$$

respectively. The first formula shows that $\phi \circ p$ is continuous, and since p is factorial, ϕ is continuous. Finally, the continuity of ϕ implies the continuity of τ and MH(τ,β).

7. <u>Suppose that</u> ξ <u>and</u> ξ' <u>are Steenrod bundles with standard fiber</u> F <u>and</u> β: bs ξ $\to$ bs ξ' <u>is continuous. If</u> τ: tl ξ $\to$ tl ξ' <u>is a map such that the restrictions</u> $\tau\big|_{\text{pr } \xi^{-1}(U)}$, $\beta\big|_U$ <u>form an F-map</u> $\xi\big|_U \to \xi'$, <u>for each element</u> U <u>of some fundamental cover of</u> bs ξ, <u>then</u> (τ,β) <u>is an F-map</u> $\xi \to \xi'$.

This is a corollary of 6: take p to be the map pr: $\bigsqcup_{U \in \Gamma} U \to$ bs ξ, where Γ is the given fundamental cover of bs ξ.

8. <u>If the Steenrod F-bundles</u> ξ <u>and</u> ξ' <u>have the same</u>

base, then every F-map f: $\xi \to \xi'$ with bs f = id bs ξ is an
F-equivalence.

All we need to prove is that tl f^{-1} and MH(f)$^{-1}$ are
continuous, and it suffices to examine the case when $\xi' = \xi$ and ξ
is a standard trivial F-bundle. Then tl f ,tl f^{-1}: bs ξ × F → bs ξ → F
are given by

$$(b,y) \mapsto (b,\phi(b)y), \quad (b,y) \mapsto (b,\psi(b)y) \quad [b \in bs\,\xi, \; y \in F],$$

where ϕ and ψ are some maps from bs ξ into the structure group G.
Moreover, if we use the topologizing homeomorphism bs ξ × G → MH(ξ),
then MH(f) and MH(f)$^{-1}$ become the maps bs ξ × G → bs ξ × G given
by

$$(b,g) \mapsto (b,\phi(b)g) \quad \text{and} \quad (b,g) \mapsto (b,\psi(b)g) \quad [b \in bs\,\xi, \; y \in F].$$

Obviously, $\psi(b) = [\phi(b)]^{-1}$, and thus the continuity of MH(f) implies
first the continuity of ϕ and ψ, and then the continuity of tl f^{-1}
and MH(f)$^{-1}$.

9 (COROLLARY). <u>The correcting map,</u> corr f , <u>is an
F-equivalence for every F-map</u> f <u>between Steenrod F-bundles.</u>

Principal Bundles

10. A Steenrod bundle is called <u>principal</u> if its standard
fiber is the structure group G which acts canonically from the left
on itself (see 2.3.6). We take the liberty to denote the last G-space
simply by G and, accordingly, the principal bundles with structure
group G will be referred to as <u>Steenrod G-bundles.</u>

A fundamental property of the principal bundles is that their
spaces of marked homeomorphisms can be identified with their total
spaces. More precisely, given a principal G-bundle ξ, the formula
$\alpha \mapsto \alpha(e_G)$ defines a homeomorphism MH(ξ) → tl ξ. For a standard trivial
bundle, this is evident, and the general case is readily reduced to the
standard trivial one.

If we identify MH(ξ) and tl ξ via the homeomorphism
$\alpha \mapsto \alpha(e_G)$, then the natural right action of G on MH(ξ) (see 1)
becomes the free right action of G on tl ξ. This free action,
tl ξ × G → tl ξ, can be also described directly: its orbits are exactly
the fibers of ξ, and on each fiber the action is simply the right
canonical action, transferred from G to the fiber by means of marked

homeomorphisms.

11. This construction of the free right action of G on the
total space of a principal bundle with structure group G can be
partially reversed. Assume that the topological group G acts
continuously and freely from the right on the topological space X, and
consider the bundle (X,pr,X/G). Its fibers (orbits) carry natural
G-structures: the marked homeomorphisms $G \to pr^{-1}(b)$ (b ∈ X/G) are
given by $g \mapsto xg$, $x \in pr^{-1}(b)$. Since to distinct points x correspond
distinct homeomorphisms $g \mapsto xg$, we obtain also an invertible map of X
onto the set of marked homeomorphisms, and thus we get a topology on the
last set. Therefore, (X,pr,X/G) is a G-bundle.

To explain why we called this last construction a partial
inversion of the original one, apply it now to the right action
tl ξ × G → tl ξ described in 10; the resulting bundle is exactly ξ.
More precisely, the injective factor of the projection pr ξ maps
tl ξ/G onto bs ξ, and together with id(tl ξ) forms a G-isomorphism
(tl ξ,pr,tl ξ/G) → ξ.

12. <u>If the G-bundle (X,pr,X/G), defined by a free right
action of G, has a section, then it is G-trivial. In particular,
every Steenrod G-bundle having a section is G-trivial.</u>

Indeed, if s: X/G → X is a section, then the map
f: ((X/G) × G,pr_1,X/G) → (X,pr,X/G), given by tl f(b,g) = s(b)g, is
a G-trivialization of the bundle (X,pr,X/G).

13 (COROLLARY). <u>If the G-bundle (X,pr,X/G), defined by a
free right action of G, is topologically trivial, then it is G-trivial.
If (X,pr,G/X) is locally topologically trivial, then it is locally
G-trivial, i.e., it is a Steenrod G-bundle.</u>

3. Associated Bundles

1. Let G be a topological group, and let F and F' be
effective G-spaces. The construction below associates to each Steenrod
F-bundle ξ a certain Steenrod F'-bundle having the same base.

The formula $g(\alpha,y) = (\alpha g, g^{-1}y)$, where g ∈ G, α ∈ MH(ξ),
and y ∈ F', defines a right action of G on MH(ξ) × F' (here G
acts canonically from the right on MH(ξ); see 2.3). Let ξ' denote
the bundle with total space (MH(ξ) × F')/G, base bs ξ, and whose
projection takes the orbit of a pair (α,y) ∈ MH(ξ) × F' into the point
pr ξ(α(F)). The fibers of this bundle carry a natural F'-structure:

the marked homeomorphisms $F' \to (\mathrm{pr}\,\xi')^{-1}(b)$ are given by $y \mapsto \mathrm{pr}(\alpha,y)$, where $\alpha \in MH(\xi)$ is such that $\alpha(F) = \mathrm{pr}\,\xi^{-1}(b)$. Since distinct homeomorphisms α yield distinct homeomorphisms $y \mapsto \mathrm{pr}(\alpha,y)$, we obtain at the same time an invertible map $MH(\xi) \to MH(\xi')$, which we use to topologize $MH(\xi')$, and thus make from ξ' an F'-bundle. Finally, $\xi'|_U$ is F'-trivial for each set U such that $\xi|_U$ is F-trivial. Consequently, ξ' is locally F'-trivial, i.e., it is a Steenrod F'-bundle. We say that ξ' is <u>the F'-bundle associated with</u> ξ and denote it by $\mathrm{asso}(\xi,F')$.

2. We add four remarks to the above description of the asso construction:

(i) The map $\mathrm{tl}\,\xi \to (MH(\xi) \times F)/G$ which takes each point $x \in \mathrm{tl}\,\xi$ into the orbit consisting of the pairs $(\alpha,y) \in MH(\xi) \times F$ with $\alpha(y) = x$, is obviously a homeomorphism; together with $\mathrm{id}\,\mathrm{bs}\,\xi$, this map defines an F-equivalence $\xi \to \mathrm{asso}(\xi,F)$. Therefore, <u>the bundle</u> $\mathrm{asso}(\xi,F)$ <u>is canonically F-equivalent to</u> ξ.

(ii) The invertible map $MH(\xi) \to MH(\mathrm{asso}(\xi,F'))$ that we used to topologize $MH(\mathrm{asso}(\xi,F'))$, is a G-map with respect to the right canonical actions of G on $MH(\xi)$ and $MH(\mathrm{asso}(\xi,F'))$. As a corollary, we may state that, given an arbitrary effective G-space F'', the product of the above invertible G-map with $\mathrm{id}\,F''$ is a G-map $MH(\xi) \times F'' \to$ $\to MH(\mathrm{asso}(\xi,F')) \times F''$, where G acts from the right on $MH(\xi) \times F''$ and $MH(\mathrm{asso}(\xi,F')) \times F''$ by $g(\alpha,y) = (\alpha g, g^{-1}y)$. The resulting homeomorphism $(MH(\xi) \times F'')/G \to (MH(\mathrm{asso}(\xi,F')) \times F'')/G$, together with $\mathrm{id}\,\mathrm{bs}\,\xi$, define an F''-equivalence $\mathrm{asso}(\xi,F'') \to \mathrm{asso}(\mathrm{asso}(\xi,F'),F'')$. Therefore, <u>the bundles</u> $\mathrm{asso}(\mathrm{asso}(\xi,F'),F'')$ <u>and</u> $\mathrm{asso}(\xi,F'')$ are <u>canonically F''-equivalent.</u>

(iii) The bundle $\mathrm{asso}(\xi,G)$, i.e., the principal bundle associated with ξ, is canonically G-isomorphic to the G-bundle $(MH(\xi),\mathrm{pr},MH(\xi)/G)$ defined by the canonical right action of G on $MH(\xi)$ (see 2.11). The canonical G-isomorphism $(MH(\xi),\mathrm{pr},MH(\xi)/G) \to$ $\to \mathrm{asso}(\xi,G)$ is given by the homeomorphism $MH(\xi) \to (MH(\xi) \times G)/G$ which takes each $\alpha \in MH(\xi)$ into the orbit of (α,e_G).

(iv) If F' is a subspace of the G-space F'' (see 2.3.6), then $(MH(\xi) \times F')/G$ is a subset of $(MH(\xi) \times F'')/G$, and the inclusion $(MH(\xi) \times F')/G \to (MH(\xi) \times F'')/G$ together with $\mathrm{id}\,\mathrm{bs}\,\xi$ yield an inclusion of the bundle $\mathrm{asso}(\xi,F')$ into $\mathrm{asso}(\xi,F'')$. Moreover, $MH(\mathrm{asso}(\xi,F'))$ is exactly the set of maps $\mathrm{ab}\,\alpha : F' \to \alpha(F')$ with $\alpha \in MH(\mathrm{asso}(\xi,F''))$.

Behavior With Respect to Maps

3. Let F and F' again be effective G-spaces, and suppose that ξ and η are Steenrod bundles with standard fiber F, and f: $\xi \to \eta$ is an arbitrary F-map. Define the map

$$\text{asso}(f,F'): \text{asso}(\xi,F') \to \text{asso}(\eta,F')$$

by the formulas bs asso(f,F') = bs f   and   tl asso(f,F') =

$$= [\text{fact}(MH(f) \times \text{id} F') : (MH(\xi) \times F')/G \to (MH(\eta) \times F')/G].$$

It is clear that asso(f,F') is an F'-map. Moreover, asso(f,F') is an F'-isomorphism (F'-equivalence) whenever f is an F-isomorphism (respectively, F-equivalence). Next, consider the diagrams

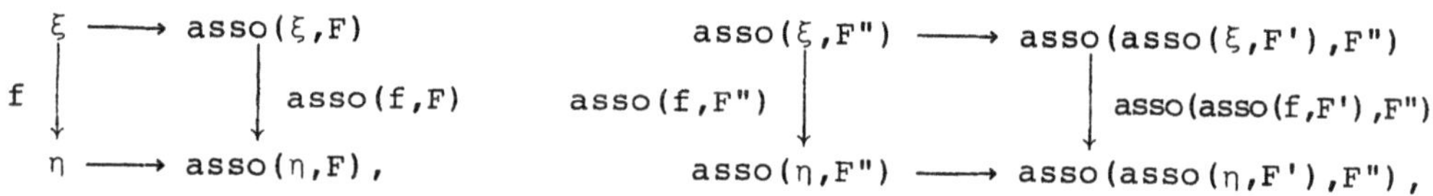

and

$$
\begin{array}{ccc}
(MH(\xi),\text{pr},MH(\xi)/G) & \longrightarrow & \text{asso}(\xi,G) \\
(MH(f),\text{fact}\,MH(f)) \downarrow & & \downarrow \text{asso}(f,G) \\
(MH(\eta),\text{pr},MH(\eta)/G) & \longrightarrow & \text{asso}(\eta,G),
\end{array}
$$

where F" is any effective G-space, and the horizontal arrows denote successively the canonical F-equivalences described in 2(i) and 2(ii), and the canonical F-isomorphisms from 2(iii) These diagrams clearly commute.

4. The asso and induction operations commute. Namely, the map

$$\text{corr}[\text{asso}(\text{ad}\,h,F')] : \text{asso}(h^!\xi,F') \to h^!\text{asso}(\xi,F')$$

is an F'-equivalence, for any Steenrod F-bundle ξ and any continuous map h: B $\to$ bs ξ; see 2.9.

Furthermore, the asso operation commutes with the extension of the structure group. That is to say, let F_1 and F_1' be the effective G_1-spaces obtained from F and F' by reducing the group G to a subgroup G_1. If the Steenrod F_1-bundle ξ_1 is taken into ξ by the extension of the group G_1 to G, then the map

$$\text{fact}(\text{in} \times \text{id}\,F') : (MH(\xi_1) \times F'_1)/G_1 \to (MH(\xi) \times F')/G,$$

where $\text{in} = [\text{in}: MH(\xi_1) \to MH(\xi)]$, defines an F'-equivalence between the bundle obtained from $\text{asso}(\xi_1, F'_1)$ by extending G_1 to G, and $\text{asso}(\xi, F')$.

Weakly Associated Bundles

5. The construction described in 1 can be generalized to the situation where the action of G on F' is not effective: we need only shift, as a preliminary step, to the effective factor of this action, and thus transform F' into an effective G/K-space, $\underset{\sim}{F'}$, where K is the noneffectiveness kernel. Therefore, $\text{asso}(\xi, F')$ may be defined for any Steenrod F-bundle ξ and any G-space F' and is, in the general case, a Steenrod bundle with structure group G/K and standard fiber $\underset{\sim}{F'}$. We say that $\text{asso}(\xi, F')$ is <u>weakly associated with</u> ξ.

The map $\text{asso}(f, F')$ defined in 3 remains viable under this extension of the asso construction, and becomes an $\underset{\sim}{F'}$-map. The properties of asso discussed in 2, 3, and 4 must be modified in an obvious manner; for example, when F'' is not effective (but F' is effective), the canonical F''-equivalence $\text{asso}(\xi, F'') \to \text{asso}(\text{asso}(\xi, F'), F'')$ becomes an $\underset{\sim}{F''}$-equivalence.

Sections Associated with F-Maps

6. Let ξ and ξ' be Steenrod bundles with structure group G and standard fiber F, and let $f: \text{bs}\,\xi \to \text{bs}\,\xi'$ be continuous. The construction below establishes a one-to-one correspondence between the F-maps $h: \xi \to \xi'$ with $\text{bs}\,h = f$ and the sections of a specially constructed bundle, $\text{Fibr}(\xi, \xi'; f)$.

Let $G^{\times}$ denote the group G endowed with the action of the group $G \times G$ given by $(g_1, g_2)g = g_1 g g_2^{-1}$ (generally speaking, this is not an effective action). Set

$$\text{Fibr}(\xi, \xi'; f) = \text{asso}(\text{diag}^!(\xi \times f^!\xi'), G^{\times}),$$

where $\text{diag} = [\text{diag}: \text{bs}\,\xi \to \text{bs}\,\xi \times \text{bs}\,\xi]$, and asso is taken in the weak sense of 5. It is clear that for every F-map h such that $\text{bs}\,h = f$ and every point $b \in \text{bs}\,\xi$, the composite homeomorphism

$$F \xrightarrow{\;\alpha\;} \text{pr}\,\xi^{-1}(b) \xrightarrow{\;\text{ab tl}\,h\;} (\text{pr}\,\xi')^{-1}(f(b)) \xrightarrow{\;(\text{ab tl ad}\,f)^{-1}\;}$$

$$\longrightarrow (\text{pr}\,f^!\xi')^{-1}(\beta(F)) \xrightarrow{\;\beta^{-1}\;} F\,,$$

where $\alpha \in MH(\xi)$, $\text{pr}\,\xi(\alpha(F)) = b$, and $\beta \in MH(f^!\xi')$, $\text{pr}\,f^!\xi'(\beta(F)) =$
$= b$, is simply one of the transformations $F \to F$ induced by the
elements of G. We denote the corresponding element by $g(\alpha,\beta)$, and
note that $g(\alpha g_1, \beta g_2) = g_1 g(\alpha,\beta) g_2^{-1}$ for any $g_1, g_2 \in G$ (here G acts
canonically from the right on $MH(\xi)$ and $MH(f^!\xi')$). This shows that
the orbit of the pair $(\alpha \times \beta, g(\alpha,\beta)) \in MH(\xi \times f^!\xi') \times G^{\times}$ under the
right action of $G \times G$ on $MH(\xi \times f^!\xi') \times G^{\times}$, constructed as in 1,
does not depend upon the choice of α and β, provided h and b
are fixed. When h is fixed, the map $\text{bs}\,\xi \to (MH(\xi \times f^!\xi') \times G^{\times})/(G \times G)$
taking $b \in \text{bs}\,\xi$ into this orbit is continuous and manifestly a section
of the bundle $\text{Fibr}(\xi,\xi';f)$. We call it the <u>section associated with</u> h
and denote it by $h^{\times}$. The correspondence $h \mapsto h^{\times}$ defines an invertible
map from the set of all F-maps $h\colon \xi \to \xi'$ with $\text{bs}\,h = f$ onto the set
of sections of $\text{Fibr}(\xi,\xi';f)$: the inverse map takes each section
$s\colon \text{bs}\,\xi \to \text{tl}\,\text{Fibr}(\xi,\xi';f)$ into the F-map $h\colon \xi \to \xi'$ given by

$$\text{tl}\,h(x) = \text{tl}\,\text{ad}\,f(\beta_x(g_x(\alpha_x^{-1}(x))))\,,$$

where $x \in \text{tl}\,\xi$, $\alpha_x \in MH(\xi)$, $\beta_x \in MH(f^!\xi')$, $g_x \in G$, and
$(\alpha_x \times \beta_x, g_x) \in s \circ \text{pr}\,\xi(x)$.

Let us remark that when $\text{bs}\,\xi' = \text{bs}\,\xi$ and $f = \text{id}\,\text{bs}\,\xi$, an
F-map $h\colon \xi \to \xi'$ with $\text{bs}\,h = f$ is simply an F-equivalence (see 2.8).
Therefore, when ξ and ξ' have the same base, the above construction
yields a one-to-one correspondence between the F-equivalences $\xi \to \xi'$
and the sections of the bundle $\text{Fibr}(\xi,\xi';\text{id}\,\text{bs}\,\xi)$.

4. Ehresmann-Feldbau Bundles

1. An <u>Ehresmann-Feldbau bundle</u> is a W-F-bundle which is
locally W-F-trivial; the last means that every point of the base has a
neighborhood such that the restriction of the bundle to this neighborhood
is W-F-trivial.

The theory of Ehresmann-Feldbau bundles is a variant of the
theory of bundles with a group structure; it is simpler than the theory
of Steenrod bundles (there are fewer structures), but also less pithy
(there are no associated bundles). This relative poverty nearly deprives

it of any independent value; however, the fact that it is equivalent, for a large class of standard fibers which includes the most important cases, to the theory of Steenrod bundles, makes it useful, as it enables us to simplify the latter.

The Case of Topologically Effective Actions

2. A continuous effective action $G \times X \to X$ is said to be _topologically effective_ if given any topological space Y and any map $f: Y \to G,$ the continuity of the composite map

$$Y \times X \xrightarrow{\;f \times id\;} X \times G \to X \tag{1}$$

implies the continuity of f. In this case we also say that _the G-space_ X _is topologically effective._

Clearly, if we reduce the group, a topologically effective action remains so, and every G-space which has a topologically effective subspace is itself topologically effective.

The free actions are immediate examples of topologically effective actions; in particular, the left canonical action of a topological group (on itself) is always topologically effective. Also, the usual actions of $GL(n, \mathbb{R})$ and of its subgroups on $\mathbb{R}^n$ are all topologically effective.

In order that a G-space be topologically effective, it is necessary that the map $c: G \to C(X,X)$, which takes each $g \in G$ into the transformation induced by g, be a topological embedding; if X is locally compact and Hausdorff, then this condition is also sufficient. The necessity is plain (take $Y = c(G)$ and $f = [(ab\,c)^{-1}: c(G) \to G]$). Now assume that the condition is satisfied. Then the continuity of $f: Y \to G$ is equivalent to the continuity of the composite map $Y \xrightarrow{\;f\;} G \xrightarrow{\;c\;} C(X,X)$. By Theorem 1.2.7.6, for X locally compact and Hausdorff the last map is continuous because so is the map (1). In particular, for G compact, every effective, locally compact Hausdorff G-space is topologically effective (see 1.1.7.10 and 1.2.7.2). If X is Hausdorff, locally compact, and locally connected, then the usual action of the group $Top\,X$ on X is topologically effective, and the same holds when X is Hausdorff and compact (see 2.2.5 and 2.2.6).

3. _If the standard fiber_ F _is topologically effective, then every W-F-map (respectively, W-F-isomorphism, W-F-equivalence) between Steenrod bundles is an F-map (respectively, F-isomorphism,_

<u>F-equivalence</u>).

In fact, let ξ and ξ' be Steenrod F-bundles. We need only prove the continuity of the map $MH(f)$ corresponding to the given W-F-map $f: \xi \to \xi'$, and we may assume that ξ and ξ' are standard trivial bundles. In this case, $\mathrm{tl}\,\xi = \mathrm{bs}\,\xi \times F$, $\mathrm{tl}\,\xi' = \mathrm{bs}\,\xi' \times F$, and $\mathrm{tl}\,f$ is given by $(b,y) \mapsto (\mathrm{bs}\,f(b), \phi(b)y)$, where ϕ is some map of $\mathrm{bs}\,\xi$ into the structure group G. At the same time, if we use the canonical homeomorphisms $\mathrm{bs}\,\xi \times G \to MH(\xi)$ and $\mathrm{bs}\,\xi' \times G \to MH(\xi')$ (see 2.2), then $MH(f)$ becomes the map $\mathrm{bs}\,\xi \times G \to \mathrm{bs}\,\xi' \times G$, $(b,g) \mapsto (\mathrm{bs}\,f(b), \phi(b)g)$. The continuity of $\mathrm{tl}\,f$ implies the continuity of the composition $\mathrm{pr}_2 \circ \mathrm{tl}\,f : \mathrm{bs}\,\xi \times F \to F$, which equals the composition $\mathrm{bs}\,\xi \times F \xrightarrow{\;\phi \times \mathrm{id}\;} G \times F \longrightarrow F$, where the last arrow denotes the action. Since this action is topologically effective, ϕ is continuous, and so is $MH(f)$.

4. <u>Given an Ehresmann-Feldbau bundle with a topologically effective standard fiber, there is a unique topology on the set of its marked homeomorphisms which transforms this bundle into a Steenrod bundle.</u>

The uniqueness of this topology is a consequence of Theorem 3. Let us prove its existence. Let ξ be an Ehresmann-Feldbau bundle with topologically effective fiber F. Cover $\mathrm{bs}\,\xi$ by open sets U such that $\xi|_U$ is W-F-trivial, and fix W-F-equivalences $h_U: (U \times F, \mathrm{pr}_1, U) \to \xi|_U$. Now topologize the sets $MH(\xi|_U)$ with the aid of the maps $MH(h_U): MH((U \times F, \mathrm{pr}_1, U)) \to MH(\xi|_U)$. We obtain a cover of $MH(\xi)$ by topological spaces $MH(\xi|_U)$, and Theorem 3 shows that these spaces induce the same topologies on their intersections, as required for the construction in 1.2.4.3. The topology on $MH(\xi)$ produced by this construction transforms ξ into a Steenrod bundle.

Locally Trivial Bundles as
Ehresmann-Feldbau Bundles

5. If F is a locally compact Hausdorff space endowed with the usual action of the group $\mathrm{Top}\,F$, then an Ehresmann-Feldbau W-F-bundle is simply a locally trivial bundle with fibers homeomorphic to F. Therefore, any ordinary locally trivial bundle whose fibers are locally compact Hausdorff spaces homeomorphic one to another, may be regarded as an Ehresmann-Feldbau bundle, and as such it has an implicit group structure. If, in addition, the fibers are locally connected or compact, then such bundles can be also regarded as Steenrod bundles.

We remark that the last assertion is also true for all coverings in the broad sense with connected bases.

5. Exercises

1. Show that all the effective actions listed in 2.3.11, 2.3.12, 2.3.13, and 2.3.14. are topologically effective.

2. Let X be an arbitrary C^r-manifold $(r \geqslant 1)$ of positive dimension. Show that the usual action of $\mathrm{Diff}^r X$ on X is topologically effective.

3. Consider $\mathbb{R}$ as a $\mathbb{Z}$-space with the action $(n,t) \mapsto t + n$ $(n \in \mathbb{Z}, \ t \in \mathbb{R})$, and using the same formula, extend this action to an action of the additive group $\mathbb{R}$, equipped with the discrete topology. Show that this extension of the structure group takes $\mathrm{asso}(\mathbb{R},\mathrm{hel},S^1),\mathbb{R})$ into a bundle which is not trivial as a Steenrod bundle, but is trivial as an Ehresmann-Feldbau bundle. (Cf. 4.3.)

4. Suppose that G is a connected topological group, F is an effective G-space, and ξ is a nontrivial Steenrod F-bundle with simply connected base. Denote G^δ, F^δ, and ξ^δ the group G equipped with the discrete topology, the space F, regarded as a G^δ-space, and the bundle ξ, regarded as a W-F^δ-bundle, respectively. Show that there is no topology on the set of marked homeomorphisms of ξ^δ which makes from ξ^δ a Steenrod F^δ-bundle. (Cf. 4.4.)

§ 4. THE CLASSIFICATION OF STEENROD BUNDLES

1. Steenrod Bundles and Homotopies

1. We now turn to the problem of classifying the Steenrod bundles with a given standard fiber F and a given cellular base B with respect to F-equivalence. Our main achievement in this section is to establish a canonical one-to-one correspondence between the classes of F-equivalent bundles over B and the homotopy classes of maps from B into a specially constructed space that depends only upon the structure group. This correspondence reduces the given classifi-

cation problem to a problem in ordinary homotopy theory.

Lemmas About F-Trivial Bundles

2. **Let ξ be a Steenrod bundle with standard fiber F, and let B_1 and B_2 be closed subspaces of $\mathrm{bs}\,\xi$, such that $B_1 \cup B_2 = \mathrm{bs}\,\xi$ and $B_1 \cap B_2$ is a retract of B_2. If the restrictions $\xi\big|_{B_1}$ and $\xi\big|_{B_2}$ are F-trivial, then ξ is also F-trivial.**

PROOF. Choose a retraction $\rho: B_2 \to B_1 \cap B_2$ and two F-trivializations,

$$h_1 : \eta_1 = (B_1 \times F, \mathrm{pr}_1, B_1) \to \xi\big|_{B_1}$$

and

$$h_2 : \eta_2 = (B_2 \times F, \mathrm{pr}_1, B_2) \to \xi\big|_{B_2},$$

and denote by f the composite F-equivalence

$$\eta_2\big|_{B_1 \cap B_2} \xrightarrow{\ \mathrm{id}\ } \eta_1\big|_{B_1 \cap B_2} \xrightarrow{\ \mathrm{ab}\,h_1\ } \xi\big|_{B_1 \cap B_2} \xrightarrow{\ (\mathrm{ab}\,h_2)^{-1}\ } \eta_2\big|_{B_1 \cap B_2}.$$

Obviously, $\mathrm{tl}\,f$ is given by $(b,y) \mapsto (b,\phi(b)y)$ $[b \in B_1 \cap B_2,\ y \in F]$, where ϕ is some map of $B_1 \cap B_2$ into the structure group G. Moreover, if we use the canonical homeomorphism $(B_1 \cap B_2) \times G \to \mathrm{MH}(\xi\big|_{B_1 \cap B_2})$, then $\mathrm{MH}(f)$ becomes the map $(B_1 \cap B_2) \times G \to (B_1 \cap B_2) \times G$, given by $(b,g) \mapsto (b,\phi(b)g)$ $[b \in B_1 \cap B_2,\ g \in G]$. Therefore, the continuity of $\mathrm{MH}(f)$ implies the continuity of ϕ, which in turn yields the continuity of the map $B_2 \times F \to B_2 \times F$, $(b,y) \mapsto (b,\phi \circ \rho(b)y)$. But this last map, together with $\mathrm{id}\,B_2$, form an F-equivalence $f' : \eta_2 \to \eta_2$, and the obvious equality

$$f = [\mathrm{ab}\,f' : \eta_2\big|_{B_1 \cap B_2} \to \eta_2\big|_{B_1 \cap B_2}]$$

shows that the composite F-maps

$$\eta_1 \xrightarrow{\ h_1\ } \xi\big|_{B_1} \xrightarrow{\ \mathrm{in}\ } \xi \quad \text{and} \quad \eta_2 \xrightarrow{\ f'\ } \eta_2 \xrightarrow{\ h_2\ } \xi\big|_{B_2} \xrightarrow{\ \mathrm{in}\ } \xi$$

coincide on $B_1 \cap B_2$. By 3.2.7, from this it follows that these two maps yield an F-map $(B \times F, \mathrm{pr}_1, B) \to \xi$. Finnally, by 3.2.8, the last map is an F-equivalence.

3. **Every Steenrod F-bundle with base I^n is F-trivial.**

PROOF. Let η be an arbitrary Steenrod F-bundle with

bs $\eta = I^n$. Find a positive integer N such that η is F-trivial over any cube of edge $1/N$ contained in I^n. Now divide I^n, as usual, into N^n such cubes, arrange them in dictionary order $Q_1, \ldots, Q_{N^n}$, and set $W_i = \cup_{j=1}^{i} Q_j$. Induction shows that η is F-trivial over each of the sets $W_1, \ldots, W_{N^n}$: to go from W_i to W_{i+1}, apply Lemma 2 to $\xi = \eta\big|_{W_{i+1}}$, $B_1 = W_i$, and $B_2 = Q_{i+1}$. We conclude that η is F-trivial over $W_{N^n} = I^n$.

4. <u>Let ξ_1 and ξ_2 be Steenrod bundles with common standard fiber F and common base B, and let A be a retract of B. If ξ_1 and ξ_2 are F-trivial, then for every F-equivalence $h: \xi_1\big|_A \to \xi_2\big|_A$ there is an F-equivalence $h': \xi_1 \to \xi_2$ such that $[ab\, h': \xi_1\big|_A \to \xi_2\big|_A] = h$.</u>

It is enough to prove this assertion for the case where ξ_1 is the standard trivial F-bundle $(B \times F, pr_1, B)$, and $\xi_2 = \xi_1$. Let $\rho: B \to A$ be a retraction. Obviously, $tl\, h$ is given by $(a,y) \mapsto (a, \phi(a)y)$ $[a \in A,\ y \in F]$, where ϕ is some map of A into the structure group G. Moreover, via the canonical homeomorphism $A \times G \to MH(\xi_1\big|_A)$ $[= MH(\xi_2\big|_A)]$, $MH(h)$ becomes the map $A \times G \to A \times G$, $(a,g) \mapsto (a, \phi(a)g)$ $[a \in A,\ g \in G]$. Therefore, the continuity of $MH(h)$ implies the continuity of ϕ, which in turn implies the continuity of the map $B \times F \to B \times F$, $(b,y) \mapsto (b, \phi \circ \rho(b)y)$. The latter and $id\, B$ yield an F-equivalence $h': \xi_1 \to \xi_2$ which extends h.

The Homotopy Invariance of the Induced Bundle

5. <u>Let ξ be a Steenrod bundle with standard fiber F, and let f_1 and f_2 be continuous maps of a cellular space X into bs ξ. If f_1 and f_2 are homotopic, then the bundles $f_1^! \xi$ and $f_2^! \xi$ are F-equivalent. Moreover, if f_1 and f_2 are A-homotopic, where A is a cellular subspace of X, then there is an F-equivalence $f_1^! \xi \to f_2^! \xi$ which is the identity on $f_1^! \xi\big|_A$.</u>

PROOF. Pick an A-homotopy, $H: X \times I \to bs\, \xi$, from f_1 to f_2, and set $\xi_1 = (f_1^! \xi) \times (I, id\, I, I)$, $\xi_2 = H^! \xi$. It is clear that $\xi_1\big|_{(X \times 0) \cup (A \times I)} = \xi_2\big|_{(X \times 0) \cup (A \times I)}$ and that the canonical homeomorphism $X \to X \times 1$ transforms $\xi_1\big|_{X \times 1}$ and $\xi_2\big|_{X \times 1}$ into $f_1^! \xi$ and $f_2^! \xi$, respectively. Therefore, it suffices to find an F-equivalence $\xi_1 \to \xi_2$

which is the identity over $(X \times 0) \cup (A \times I)$.

We produce such an F-equivalence by taking the limit of a sequence of F-equivalences, $h_i : \xi_1|_{C_i} \to \xi_2|_{C_i}$, where where $C_i = (X \times 0) \cup (A \times I) \cup (\mathrm{ske}_i X \times I)$, such that each map h_i extends the preceding one. Take h_{-1} to be the identity map, and assume that the F-equivalence h_i is already constructed. To get h_{i+1}, suppose that X is rigged, and for each cell $e \in \mathrm{cell}_{i+1} X \smallsetminus \mathrm{cell}_{i+1} A$ consider the bundles

$$[\mathrm{ab}(\mathrm{cha}_e \times \mathrm{id}\, I)]^!(\xi_1|_{C_i}) = [(\mathrm{cha}_e \times \mathrm{id}\, I)^!\xi_1]\Big|_{(D^{i+1}\times 0)\cup(S^i\times I)}$$

and

$$[\mathrm{ab}(\mathrm{cha}_e \times \mathrm{id}\, I)]^!(\xi_2|_{C_i}) = [(\mathrm{cha}_e \times \mathrm{id}\, I)^!\xi_2]\Big|_{(D^{i+1}\times 0)\cup(S^i\times I)},$$

where

$$\mathrm{ab}(\mathrm{cha}_e \times \mathrm{id}\, I) = [\mathrm{ab}(\mathrm{cha}_e \times \mathrm{id}\, I) : (D^{i+1} \times 0) \cup (S^i \times I) \to C_i].$$

Let g_e denote the F-equivalence of these bundles defined by h_i. By Lemma 3, $(\mathrm{cha}_e \times \mathrm{id}\, I)^!\xi_1$ and $(\mathrm{cha}_e \times \mathrm{id}\, I)^!\xi_2$ are F-trivial, and since $(D^{i+1} \times 0) \cup (S^i \times I)$ is a retract of $D^{i+1} \times I$, Lemma 4 shows that g_e extends to an F-equivalence

$$\tilde{g_e} : (\mathrm{cha}_e \times \mathrm{id}\, I)^!\xi_1 \to (\mathrm{cha}_e \times \mathrm{id}\, I)^!\xi_2 .$$

Further, note that the map $\mathrm{tl}\, \tilde{g_e}$ is constant on the elements of the partition $\mathrm{zer}(\mathrm{tl}\, \mathrm{ad}(\mathrm{cha}_e \times \mathrm{id}\, I))$ and apply Theorem 3.2.6 (with $B = D^{i+1} \times I$ and $p = [\mathrm{ab}(\mathrm{cha}_e \times \mathrm{id}\, I) : D^{i+1} \times I \to \mathrm{Cl}\, e \times I]$) to conclude that the composite map

$$(\mathrm{cha}_e \times \mathrm{id}\, I)^!\xi_1 \xrightarrow{\tilde{g_e}} (\mathrm{cha}_e \times \mathrm{id}\, I)^!\xi_2 \xrightarrow{\mathrm{ab}\, \mathrm{ad}(\mathrm{cha}_e \times \mathrm{id}\, I)} \xi_2|_{\mathrm{Cl}\, e \times I}$$

defines an F-map $\xi_1|_{\mathrm{Cl}\, e \times I} \to \xi_2|_{\mathrm{Cl}\, e \times I}$. By Theorem 3.2.8, this is an F-equivalence, which we denote by h_e. Now note that for any cells $e_1, e_2 \in \mathrm{cell}_{i+1} X \smallsetminus \mathrm{cell}_{i+1} A$, $\mathrm{tl}\, h_{e_1}$ and $\mathrm{tl}\, h_{e_2}$ agree over $(\mathrm{Cl}\, e_1 \times I) \cap (\mathrm{Cl}\, e_2 \times I)$, and that for each cell $e \in \mathrm{cell}_{i+1} X \smallsetminus \mathrm{cell}_{i+1} A$, $\mathrm{tl}\, h_e$ and $\mathrm{tl}\, h_i$ agree over $(\mathrm{Cl}\, e \times I) \cap C_i$. Since the sets C_i and $\mathrm{Cl}\, e \times I$, $e \in \mathrm{cell}_{i+1} X \smallsetminus \mathrm{cell}_{i+1} A$, constitute a fundamental cover of C_{i+1}, we use 3.2.7 to conclude that h_i and h_e, $e \in \mathrm{cell}_{i+1} X \smallsetminus \mathrm{cell}_{i+1} A$, form together an F-map $\xi_1|_{C_{i+1}} \to \xi_2|_{C_{i+1}}$. We take this map for h_{i+1} and note that it obviously extends h_i; moreover, by 3.2.7, h_{i+1}^{-1} is also an F-map.

To check the rest, i.e., that the sequence $\{h_i : \xi_1|_{C_i} \to \xi_2|_{C_i}\}$ converges to an F-equivalence $\xi_1 \to \xi_2$, it is enough to remark that the sets C_i constitute a fundamental cover of $X \times I$, and then apply Theorem 3.2.7 to the sequences $\{h_i\}$ and $\{h_i^{-1}\}$.

The Sets Stee(B,F)

6. We let Stee(B,F) denote the set of F-equivalence classes of Steenrod F-bundles over B. Below we shall study the mappings of this set into itself, defined by the induced bundle construction, by the extension of the structure group, and by the associated bundle construction.

For any continuous map $f: B' \to B$, the rule $\xi \mapsto f^!\xi$ defines a mapping

$$f^! : \text{Stee}(B,F) \to \text{Stee}(B',F).$$

If B' is a cellular space, Theorem 5 shows that $f^!$ depends only on the homotopy class of f. In particular, if B and B' are both cellular and f is a homotopy equivalence, then $f^!$ is invertible.

The extension of the structure group, which transforms the G_1-space F_1 into the effective G-space F, defines a mapping

$$\text{ext} : \text{Stee}(B,F_1) \to \text{Stee}(B,F),$$

for any topological space B. This mapping is natural: that is to say, the diagram

$$
\begin{array}{ccc}
\text{Stee}(B,F_1) & \xrightarrow{\text{ext}} & \text{Stee}(B,F) \\
f^! \downarrow & & \downarrow f^! \\
\text{Stee}(B',F_1) & \xrightarrow{\text{ext}} & \text{Stee}(B',F)
\end{array}
$$

commutes for any continuous map $f: B' \to B$; see 3.2.4.

Given another effective G-space, F', the rule $\xi \mapsto \text{asso}(\xi,F')$ defines for any topological space B a mapping

$$\text{asso} : \text{Stee}(B,F) \to \text{Stee}(B,F').$$

This mapping is invertible [its inverse is $\text{asso}: \text{Stee}(B,F') \to \text{Stee}(B,F)$] and also natural, i.e., the diagram

$$\begin{array}{ccc}
\text{Stee}(B,F) & \xrightarrow{\text{asso}} & \text{Stee}(B,F') \\
f^! \downarrow & & \downarrow f^! \\
\text{Stee}(B',F) & \xrightarrow{\text{asso}} & \text{Stee}(B',F')
\end{array}$$

commutes for any continuous map $f: B' \to B$; see 3.3.4.
Moreover, the diagram

$$\begin{array}{ccc}
\text{Stee}(B,F_1) & \xrightarrow{\text{asso}} & \text{Stee}(B,F_1') \\
f^! \downarrow & & \downarrow f^! \\
\text{Stee}(B,F) & \xrightarrow{\text{asso}} & \text{Stee}(B,F')
\end{array}$$

commutes for any topological space B, any effective G_1-spaces F_1 and
F_1', and any effective G-spaces, F and F', obtained from F_1 and
F_1' by extension of the structure group; see 3.3.4 and 3.3.5.

2. Universal Bundles

1. Let F be an effective G-space. By Theorem 1.5, given
any Steenrod F-bundle ξ, any cellular space B, and any continuous
map $f: B \to \text{bs}\,\xi$, we may consider the bundle $f^!\xi$. This defines a
mapping $\pi(B,\text{bs}\,\xi) \to \text{Stee}(B,F)$, which we denote by $\text{induz}(B,\xi)$:
$\text{induz}(B,\xi)$ (the homotopy class of f) = the F-equivalence class of $f^!\xi$.
The following diagram is obviously commutative for any
topological space C and any continuous map $g: C \to \text{bs}\,\xi$

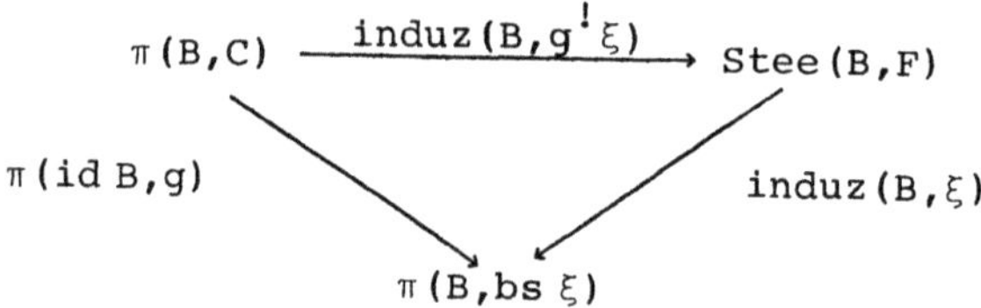

Similarly, the diagram

$$\begin{array}{ccc}
\pi(B,\text{bs}\,\xi) & \xrightarrow{\text{induz}(B,\text{asso}(\xi,F'))} & \text{Stee}(B,F') \\
& \text{induz}(B,\xi) \searrow \quad \swarrow \text{asso} & \\
& \text{Stee}(B,F) &
\end{array}$$

commutes for any effective G-space F'.

2. A Steenrod F-bundle ζ is called <u>universal</u> if the map $induz(B,\zeta)$ is invertible for any cellular space B. In other words, a Steenrod F-bundle ζ is universal if: (i) given any Steenrod F-bundle ξ with cellular base, there is a continuous map $f: bs\,\xi \to bs\,\zeta$ such that the bundle $f^!\zeta$ is F-equivalent to ξ, and (ii) if for two arbitrary continuous maps, f_1 and f_2, of a cellular space into $bs\,\zeta$ the bundles $f_1^!\zeta$ and $f_2^!\zeta$ are F-equivalent, then f_1 and f_2 are homotopic.

Condition (i) has an equivalent formulation (i'): given any Steenrod F-bundle ξ with cellular base, there is an F-map $\xi \to \zeta$. Indeed, if $f: bs\,\xi \to bs\,\zeta$ is continuous and $g: \xi \to f^!\zeta$ is an F-equivalence, then $ad\,f \circ g: \xi \to \zeta$ is an F-map. Conversely, if $h: \xi \to \zeta$ is an F-map, then $corr\,h: \xi \to (bs\,h)^!\zeta$ is an F-equivalence (see 3.2.9).

Similarly, condition (ii) is equivalent to (ii'): given any Steenrod F-bundle ξ with cellular base and any F-maps $h_0,h_1: \xi \to \zeta$, $bs\,h_0$ and $bs\,h_1$ are homotopic. Indeed, suppose that ξ, h_0, h_1 have these properties; then both bundles, $(bs\,h_0)^!\zeta$ and $(bs\,h_1)^!\zeta$, are F-equivalent to ξ, and so, by (ii), $bs\,h_0$ and $bs\,h_1$ are homotopic. Conversely, if f_0 and f_1 are continuous maps of a cellular space into $bs\,\zeta$ and $h: f_0^!\zeta \to f_1^!\zeta$ is an F-equivalence, then $f_0 = bs\,ad\,f_0$, $f_1 = bs(ad\,f_1 \circ h)$ and, by (ii'), f_0 and f_1 are homotopic.

We remark also that both (i') and (ii') (and hence (i) and (ii)) are consequences of the following condition: given an arbitrary F-bundle ξ with cellular base and an arbitrary subspace A of $bs\,\xi$, every F-map $\xi|_A \to \zeta$ extends to an F-map $\xi \to \zeta$. To see that this condition implies (i'), it suffices to take $A = \emptyset$. To see that it implies (ii'), take the F-bundle $\xi \times (I,id\,I,I)$, the subspace $A = (bs\,\xi \times 0) \cup (bs\,\xi \times 1)$ of $bs(\xi \times (I,id\,I,I)) = bs\,\xi \times I$, and take as the F-map that must be extended

$$g: \xi \times (I,id\,I,I)\,\Big|_{(bs\,\xi \times 0) \cup (bs\,\xi \times 1)} \to \zeta,$$

with

$$bs\,g(b,0) = bs\,h_0(b), \quad bs\,g(b,1) = bs\,h_1(b) \quad [b \in bs\,\xi],$$
$$tl\,g(x,0) = tl\,h_0(x), \quad tl\,g(x,1) = tl\,h_1(x) \quad [x \in tl\,\xi].$$

3. <u>Every bundle induced from a universal bundle by a homotopy equivalence is universal.</u>

Indeed, if ζ is a universal F-bundle and $f: B \to bs\,\zeta$ is a homotopy equivalence, then the map $induz(B,f^!\zeta)$ coincides with $induz(B,\zeta) \circ \pi(f,id\,B)$ for any cellular space (see 1), and hence it is invertible.

4. <u>If the F-bundles</u> ζ <u>and</u> ζ' <u>are universal and</u> $\mathrm{bs}\,\zeta$ <u>and</u> $\mathrm{bs}\,\zeta'$ <u>are cellular spaces, then</u> $\mathrm{bs}\,g : \mathrm{bs}\,\zeta \to \mathrm{bs}\,\zeta'$ <u>is a homotopy equivalence for any F-map</u> $g : \zeta \to \zeta'$.

PROOF. Pick an F-map $g' : \zeta' \to \zeta$. Since $g' \circ g : \zeta \to \zeta$ and $g \circ g' : \zeta' \to \zeta'$ are F-maps, condition 2(ii') implies that the map $\mathrm{bs}\,g' \circ \mathrm{bs}\,g = \mathrm{bs}(g' \circ g)$ (respectively, $\mathrm{bs}\,g \circ \mathrm{bs}\,g' = \mathrm{bs}(g \circ g')$) is homotopic to $\mathrm{id}(\mathrm{bs}\,\zeta)$ (respectively, to $\mathrm{id}(\mathrm{bs}\,\zeta')$).

5. <u>A bundle associated with a universal bundle is itself universal.</u>

Indeed, let ζ be a universal F-bundle and let F' be another effective G-space. Then for any cellular space B, $\mathrm{induz}(B,\mathrm{asso}(\zeta,F'))$ is precisely the composition of the maps $\mathrm{induz}(B,\zeta)$ and $\mathrm{asso} : \mathrm{Stee}(B,F) \to \mathrm{Stee}(B,F')$ (see 1), and hence is invertible.

Classifying Spaces

6. As Theorem 5 shows, the base of a universal bundle with structure group G is simultaneously the base of all universal bundles with structure group G and all possible standard fibers, and so we may say that it does not depend upon the choice of the standard fiber. This base is called a <u>classifying space</u> of the group G.

From Theorem 3 it follows that every space which has the same homotopy type as a classifying space of G is itself a classifying space of G. Moreover, it results from Theorem 4 that any two cellular classifying spaces of G have the same homotopy type.

k-Universal Bundles

7. A Steenrod F-bundle ζ is called k-<u>universal</u> if $\mathrm{induz}(B,\zeta) : \pi(B,\mathrm{bs}\,\zeta) \to \mathrm{Stee}(B,F)$ is surjective for each cellular space B of dimension $\leqslant k$ and injective for each cellular space B of dimension $\leqslant k-1$. In other words, a Steenrod F-bundle ζ is k-universal if: a) given any Steenrod F-bundle ξ with cellular base of dimension $\leqslant k$, there is a continuous map $f : \mathrm{bs}\,\xi \to \mathrm{bs}\,\zeta$ such that the bundle $f^{!}\zeta$ is F-equivalent to ξ, and b) any two continuous maps, f_0 and f_1, from a cellular space of dimension $\leqslant k-1$ into $\mathrm{bs}\,\zeta$, such that

the bundles $f_0^!\zeta$ and $f_1^!\zeta$ are F-equivalent, are homotopic. Equivalent conditions are: a') given any Steenrod F-bundle ξ with cellular base of dimension $\leqslant k$, there is an F-map $\xi \to \zeta$ and b') for any Steenrod F-bundle ξ with cellular base of dimension $\leqslant k-1$ and any F-maps $g_0, g_1 : \xi \to \zeta$, bs g_0 and bs g_1 are homotopic.

In all these formulations k is a positive integer, and the universal bundles are sometimes termed ∞-<u>universal.</u> Every k-universal bundle is obviously l-universal for $l \leqslant k$. Moreover, Theorems 2.3.2.4 and 2.3.2.5 show that the restriction of a k-universal bundle with cellular base to a subspace of the base which contains its l-skeleton is l-universal, for any $l \leqslant k$.

Theorems 3 and 5 have immediate analogs for k-universal bundles: every bundle induced from a k-universal bundle by a homotopy equivalence is k-universal; a bundle aasociated with a k-universal bundle is itself k-universal.

3. The Milnor Bundles

1. Below we shall construct for any topological group G a principal bundle with structure group G, called the <u>Milnor G-bundle</u> and denoted by $\mathrm{Mi}\,G$. In 2-6 we shall prove that $\mathrm{Mi}\,G$ is a universal G-bundle.

Let $TG(k)$ denote the join of k copies of the group G. Then $TG(k)$ embeds naturally in $TG(k+1)$ (as a base of the join $TG(k) * G = TG(k+1)$), and so the limit $TG = \lim TG(k)$ is meaningful. The right action $G \times G \to G$, $(g_1, g) \mapsto g^{-1}g_1$, extends to a free, continuous, right action of G on $TG(k)$. Since the inclusions $TG(k) \to TG(k+1)$ are G-maps with respect to this action, G acts also on TG. Therefore, a G-bundle $(TG, pr, TG/G)$ results, and this is $\mathrm{Mi}\,G$.

If G_1 is a subgroup of G, then there exist the inclusions in $* \ldots *$ in: $TG_1(k) \to TG(k)$ $(k = 1, 2, \ldots)$, where in $= [\mathrm{in}: G_1 \to G]$, and all of them are in-maps. Moreover, they are compatible with the embeddings $TG(k) \to TG(k+1)$ and $TG_1(k) \to TG_1(k+1)$, and hence the limit map $TG_1 \to TG$ is meaningful (see 1.2.4.4). This map is also an in-map, and together with the map $TG_1/G_1 \to TG/G$ that it induces, it clearly yields an (in,in)-map $\mathrm{Mi}\,G_1 \to \mathrm{Mi}\,G$. Therefore, to each inclusion $G_1 \to G$ corresponds an (in,in)-map $\mathrm{Mi}\,G_1 \to \mathrm{Mi}\,G$.

Mi G is Locally Trivial

2. It is convenient to identify $TG(k)$ with that subspace of the product $con\,G \times \ldots \times con\,G$ (k factors) which consists of the points $\{pr(g_i,t_i)\}_{i=1}^{k}$ such that $t_1 + \ldots + t_k = 1$, and is canonically homeomorphic to $TG(k)$ (see 1.2.6.4; here $pr = [\mathbf{pr}\colon G \times I \to con\,G]$) . After this identification, the points of TG may be represented as sequences $\{pr(g_i,t_i)\}_{i=1}^{\infty}$ such that $\sum t_i = 1$ and only a finite number of the t_i's are not zero. Now the right action of G on TG (see 1) is described as $\{pr(g_i,t_i)\}g = \{pr(g^{-1}g_i,t_i)\}$.

3. Mi G <u>is locally G-trivial.</u>

Let U_s be the collection of sequences $\{pr(g_i,t_i)\}$ with $t_s \neq 0$, and consider the sets $pr\,Mi\,G(U_1)$, $pr\,Mi\,G(U_2),\ldots$. These sets are open, cover TG/G, and over each of them the bundle Mi G is G-trivial: the G-trivialization $pr\,Mi\,G(U_s) \times G \to (pr\,Mi\,G)^{-1}(pr\,Mi\,G(U_s))$ $[= U_s]$ takes each point (x,g) into the sequence $\{pr(g_i,t_i)\}$ determined by the conditions: $pr\,Mi\,G(\{pr(g_i,t_i)\}) = x$ and $g_s = g$.

Mi G is Universal

4. Mi G <u>is universal.</u>

According to 2.2, it suffices to show that given any G-bundle ξ with cellular base and any subspace $A \subset bs\,\xi$, every G-map $f\colon \xi|_A \to Mi\,G$ extends to a G-map $\xi \to Mi\,G$.

We consider first the case $bs\,\xi = D^{r+1}$, $A = S^r$, for some r. Then ξ is G-trivial (see 1.3), and we may actually assume that ξ is the standard trivial G-bundle $(D^{r+1} \times G, pr_1, D^{r+1})$. The desired extension $h\colon \xi \to Mi\,G$ has an explicit description: let k be the smallest number s such that $TG(s) \supset tl\,f(S^r \times e_G)$, and let ϕ_i denote the composite map

$$S^r \times G \xrightarrow{\;ab\,tl\,f\;} TG(k) \xrightarrow{\;in\;} \underbrace{con\,G \times \ldots \times con\,G}_{k} \xrightarrow{\;pr_i\;} con\,G$$

$(i = 1,\ldots,k)$; further, define $\psi\colon D^{r+1} \times G \to TG(k+1)$ by

$$\psi(ty,g) = (t\phi_1(y),\ldots,t\phi_k(y),pr(g,1-t)),$$

where $y \in S^r$, $t \in I$, and $pr = [\mathbf{pr}\colon G \times I \to con\,G]$. Now set $tl\,h = [in\colon TG(k+1) \to TG] \circ \psi$.

The general case reduces to this special one. Indeed, assume that the space $bs\,\xi$ is rigged and that a G-map $h_r\colon \xi\big|_{A\cup ske_r bs\,\xi} \to Mi\,G$ extending f is already available. The above argument shows that for each cell $e \in cell_{r+1}bs\,\xi \smallsetminus cell_{r+1}A$ the G-map

$$f_e = h_r \circ [ab\,cha_e \colon S^r \to A \cup ske_r bs\,\xi] \colon cha_e^! \xi\big|_{S^r} \to Mi\,G$$

extends to a G-map $g_e \colon cha_e^! \xi \to Mi\,G$, and it is clear that $tl\,g_e$ is constant on the elements of the partition $zer(tl\,ad\,cha_e)$. Applying Theorem 3.2.6 (with $B = D^{r+1}$ and $p = [ab\,cha_e \colon D^{r+1} \to Cl\,e]$), we see that g_e defines a G-map $\xi\big|_{Cl\,e} \to Mi\,G$, which we denote by h_e.

Further, note that for any cells $e_1, e_2 \in cell_{r+1}bs\,\xi \smallsetminus cell_{r+1}A$, $tl\,h_{e_1}$ and $tl\,h_{e_2}$ agree over $Cl\,e_1 \cap Cl\,e_2$, and that for any cell $e \in cell_{r+1}bs\,\xi \smallsetminus cell_{r+1}A$, $tl\,h_e$ and $tl\,h_r$ agree over $Cl\,e \cap (A \cup ske_r bs\,\xi)$. This implies that h_r and h_e, $e \in cell_{r+1}bs\,\xi \smallsetminus cell_{r+1}A$, yield together a G-map $h_{r+1}\colon \xi\big|_{A\cup ske_{r+1}bs\,\xi} \to Mi\,G$ extending h_r (see 3.2.7). Therefore, using induction, we can produce a sequence $\{h_s \colon \xi\big|_{A \cup ske_s bs\,\xi} \to Mi\,G\}_{s=-1}^{\infty}$ of G-maps with $h_{-1} = f$, such that h_s extends h_{s-1} for all $s > 0$. Since the sets $A \cup ske_s bs\,\xi$ constitute a fundamental cover of $bs\,\xi$, using again 3.2.7 we conclude that the h_s's yield a G-map $\xi \to Mi\,G$ extending f.

A Promise

5. The base of the bundle $Mi\,G$ is not a cellular space. However, we shall see in the next chapter that for any topological group G there are also universal G-bundles with cellular base; see 5.6.1.4. By 2.7, this will imply the existence of k-universal G-bundles with cellular base of dimension $\leqslant k$, for any given topological group G and any positive integer k.

4. Reducings of the Structure Group

1. We say that the Steenrod F_1-bundle ξ_1 with structure group G_1 is obtained from the Steenrod F-bundle ξ with structure group G by <u>reducing the group</u> G <u>to</u> G_1 if ξ is obtained from ξ_1 by

extending the group G_1 to G.

While the extension of the structure group of a Steenrod bundle is a well-defined operation, the reduction of the structure group cannot be carried out for every Steenrod bundle, and even when it is possible, it may produce bundles which are not equivalent with respect to the reduced group. In other words, the mapping

$$\text{ext: Stee}(B,F_1) \to \text{Stee}(B,F) \tag{2}$$

defined in 1.6 may be both nonsurjective and noninjective.

We remark that the set-theoretic properties of the mapping (2) are uniquely determined by the triple B, G, G_1, i.e., they are preserved when we replace F and F_1 by other effective G-spaces and their corresponding G_1-spaces, while keeping B, G, and G_1 the same; this is plain from diagram (1).

2. Recall that given a cellular space B, $\text{Stee}(B,F)$ can be interpreted as the set of homotopy classes of continuous maps from B into a classifying space of the structure group. Below we describe (2) in the same homotopy terms.

Let G, G_1, F, F_1 be as in 1, and let ζ and ζ_1 be universal bundles with standard fibers F and F_1. A continuous map $\psi: \text{bs }\zeta_1 \to \text{bs }\zeta$ is called <u>classifying</u> if $\psi^!\zeta$ is F-equivalent to the bundle obtained from ζ_1 by extending the structure group G_1 to G. By the definition of a universal bundle, such a map exists whenever $\text{bs }\zeta_1$ is a cellular space, and so it certainly exists when $\zeta = \text{Mi }G$ and $\zeta_1 = \text{Mi }G_1$ (see 3.1). Our main claim is that <u>the diagram</u>

$$
\begin{array}{ccc}
\text{Stee}(B,F_1) & \xrightarrow{\;\text{ext}\;} & \text{Stee}(B,F) \\[2mm]
{\scriptstyle \text{induz}(B,\zeta_1)}\Big\uparrow & & \Big\uparrow{\scriptstyle \text{induz}(B,\zeta)} \\[2mm]
\pi(B,\text{bs }\zeta_1) & \xrightarrow{\;\pi(\text{id }B,\psi)\;} & \pi(B,\text{bs }\zeta)
\end{array}
\tag{3}
$$

<u>commutes for any classifying map</u> ψ <u>and any cellular space</u> B.

PROOF. The composition $\text{induz}(B,\zeta) \circ \pi(\text{id }B,\zeta)$ takes the homotopy class of $f_1: B \to \text{bs }\zeta_1$ into the class of the bundle $(\psi \circ f_1)^!\zeta$, while the same homotopy class is taken by $\text{ext} \circ \text{induz}(B,\zeta_1)$ into the class of the bundle obtained from $f_1^!\zeta_1$ by extending the structure group G_1 to G. Since the extension of the structure group and the induction construction commute (see 1.6), the last class contains $f_1^!(\psi^!\zeta)$, and it remains to observe that $f_1^!(\psi^!\zeta) = (\psi \circ f_1)^!\zeta$.

3. The commutativity of the diagram (3) and the invertibility

of its vertical mappings imply that $\text{induz}(B, \zeta_1)$ is an injective mapping from the set of homotopy classes of maps $g: B \to \text{bs}\,\zeta_1$ such that $\psi \circ g$ is homotopic to a given map $f: B \to \text{bs}\,\zeta$, onto the set of classes of F_1-equivalent F_1-bundles which are obtained from $f^!\zeta$ by reducing the structure group G to G_1. In particular, a Steenrod F-bundle ξ with cellular base admits the reduction of the group G to G_1 if and only if any continuous map $f: \text{bs}\,\xi \to \text{bs}\,\zeta$ such that $f^!\zeta$ is F-equivalent to ξ is homotopic to the composition of some continuous map $\text{bs}\,\xi \to \text{bs}\,\zeta_1$ with ψ.

5. Exercises

1. Given a topological group G and a positive integer k, denote by $Mi(G,k)$ the restriction of the bundle MiG to $TG(k)/G$, i.e., the bundle $(TG(k), \text{pr}, TG(k)/G)$. Show that $Mi(G,k)$ is a $(k-1)$-universal G-bundle.

2. Show that $Mi\,\mathbb{Z}_2$ is isomorphic to $(S^\infty, \text{pr}, \mathbb{R}P^\infty)$, while $Mi(\mathbb{Z}_2, k)$ (see Exercise 1) is isomorphic to $(S^{k-1}, \text{pr}, \mathbb{R}P^{k-1})$.

3. Show that $Mi\,S^1$ is isomorphic to $(S^\infty, \text{pr}, \mathbb{C}P^\infty)$, while $Mi(S^1, k)$ is isomorphic to $(S^{2k-1}, \text{pr}, \mathbb{C}P^{k-1})$.

4. Let X be a compact n-dimensional C^r-manifold, $1 \leqslant r \leqslant \infty$. Consider the right action of $\text{Diff}^r X$ on $\text{Emb}^r(X, \mathbb{R}^q)$, given by $(j, \phi) \mapsto j \circ \phi$ $[j \in \text{Emb}^r(X, \mathbb{R}^q),\ \phi \in \text{Diff}^r X]$, and the limit right action of $\text{Diff}^r X$ on

$$\lim(\text{Emb}^r(X, \mathbb{R}^q), \text{ab}\,C^r(\text{id}\,X, \text{in}): \text{Emb}^r(X, \mathbb{R}^q) \to \text{Emb}^r(X, \mathbb{R}^{q+1})).$$

Show that $(\lim \text{Emb}^r(X, \mathbb{R}^q), \text{pr}, [\lim \text{Emb}^r(X, \mathbb{R}^q)]/\text{Diff}^r X)$ is a universal $\text{Diff}^r X$-bundle, while $(\text{Emb}^r(X, \mathbb{R}^q), \text{pr}, \text{Emb}^r(X, \mathbb{R}^q)/\text{Diff}^r X)$ is a $(q-2n-1)$-universal $\text{Diff}^r X$-bundle, for any $q \geqslant 2n+1$.

§5. VECTOR BUNDLES

1. General Definitions

1. The main objective of this section is to study those Steenrod bundles whose standard fiber is either $\mathbb{R}^n$ with the usual action of one of the groups $GL(n, \mathbb{R})$, $GL_+(n, \mathbb{R})$, $O(n)$, or $SO(n)$, or $\mathbb{C}^n$ with the usual action of $GL(n,\mathbb{C})$ or $U(n)$.

Since all the standard fibers listed above are topologically effective, the corresponding bundles may be also regarded as Ehresmann- -Feldbau bundles with the same standard fibers (see Subsection 3.4). We shall proceed in this way and ignore completely the topology on the set of marked homeomorphisms in the course of the entire section.

To simplify the discussion, we introduce a special notation for the above standard fibers: $GL\,\mathbb{R}^n$, $GL_+\mathbb{R}^n$, $O\mathbb{R}^n$, $SO\mathbb{R}^n$, and $GL\mathbb{C}^n$, $U\mathbb{C}^n$.

Standard Fiber $GL\mathbb{R}^n$

2. A Steenrod bundle with standard fiber $GL\mathbb{R}^n$ is called an <u>n-dimensional real vector bundle.</u>

Since a space with a $GL\mathbb{R}^n$-structure is simply an n-dimensional real vector space (see 3.1.3), a W-$GL\mathbb{R}^n$-bundle is simply a bundle whose fibers are n-dimensional real vector spaces. Moreover, a W-$GL\mathbb{R}^n$-equivalence of W-$GL\mathbb{R}^n$-bundles is an equivalence that is linear on fibers. Therefore, an n-dimensional real vector bundle is a bundle whose fibers are n-dimensional real vector spaces, and which is locally trivial in the natural vector sense: every point of the base has a neighborhood U over which the given bundle is equivalent to $(U \times \mathbb{R}^n, pr_1, U)$ via an equivalence which is linear on each fiber.

3. <u>A bundle</u> ξ <u>whose fibers are n-dimensional real vector spaces is an n-dimensional real vector bundle (i.e.,</u> ξ <u>is locally trivial in the previous vector sense) if and only if:</u>

(i) ξ <u>is topologically locally trivial;</u>

(ii) <u>the partial vector operations in</u> $tl\,\xi$, <u>i.e., the maps</u>

$$\mathbb{R} \times tl\,\xi \to tl\,\xi, \quad (\lambda,x) \mapsto \lambda x,$$

<u>and</u>

$$\{(x_1,x_2) \in \mathrm{tl}\,\xi \times \mathrm{tl}\,\xi \mid \mathrm{pr}\,\xi(x_1) = \mathrm{pr}\,\xi(x_2)\} \to \mathrm{tl}\,\xi, \quad (x_1,x_2) \mapsto x_1 + x_2$$

<u>are continuous.</u>

The necessity of these conditions is obvious. Let us verify their sufficiency. Let $b_0 \in \mathrm{bs}\,\xi$. Fix an arbitrary basis, $v_1,\ldots,v_n$, of the vector space $\mathrm{pr}\,\xi^{-1}(b_0)$, a neighborhood U of b_0 such that $\xi|_U$ is topologically trivial, and a trivialization $h: (U \times \mathbb{R}^n, \mathrm{pr}_1, U) \to$ $\to \xi|_U$. Define a map, $h': (U \times \mathbb{R}^n, \mathrm{pr}_1, U) \to \xi$, linear on fibers, by the formula $\mathrm{tl}\,h'(b,\mathrm{ort}_i) = \mathrm{tl}\,h(b, \mathrm{pr}_2 \circ \mathrm{tl}\,h^{-1}(v_i))$, where $\mathrm{pr}_2 = [\mathrm{pr}_2: U \times \mathbb{R}^n \to \mathbb{R}^n]$. Now pick disjoint neighborhoods, K and N, of the set $\mathrm{pr}_2 \circ \mathrm{tl}\,h^{-1} \circ \mathrm{tl}\,h'(b_0 \times S^{n-1})$ and of the point $\mathrm{pr}_2 \circ \mathrm{tl}\,h^{-1} \circ \mathrm{tl}\,h'(b_0,0)$ in $\mathbb{R}^n$, respectively, and denote by V the neighborhood of b_0 consisting of all $b \in U$ such that $\mathrm{tl}\,h'(b \times S^{n-1}) \subset \mathrm{tl}\,h(b \times K)$ and $\mathrm{tl}\,h'(b,0) \in \mathrm{tl}\,h(b \times N)$. It is clear that $\mathrm{tl}\,h'(b,0) \notin \mathrm{tl}\,h'(b \times S^{n-1})$ when $b \in V$, and thus the map $\mathrm{ab}\,h' : (V \times \mathbb{R}^n, \mathrm{pr}_1, V) \to \xi|_V$ is nondegenerate on each fiber. Consequently, we can apply Theorem 3.2.8 to $\mathrm{ab}\,h'$, taking F to be $\mathbb{R}^n$, regarded as a Top $\mathbb{R}^n$-space ($(V \times \mathbb{R}^n, \mathrm{pr}_1, V)$ and $\xi|_V$ are thought of as Steenrod F-bundles; see 3.4.5). We conclude that $\mathrm{ab}\,h'$ is an equivalence in the topological sense, and since $\mathrm{ab}\,h'$ is also linear on fibers, the proof is complete.

Standard Fiber $O\mathbb{R}^n$

4. A Steenrod bundle with standard fiber $O\mathbb{R}^n$ is called an <u>n-dimensional Euclidean bundle.</u>

Since a space with an $O\mathbb{R}^n$-structure is an n-dimensional Euclidean space, a W-$O\mathbb{R}^n$-bundle is simply a bundle whose fibers are n-dimensional Euclidean spaces. Moreover, it is clear that any W-$O\mathbb{R}^n$-equivalence of W-$O\mathbb{R}^n$-bundles is an equivalence which is an orthogonal map on each fiber. Therefore, an n-dimensional Euclidean bundle is a bundle whose fibers are n-dimensional Euclidean spaces and which is locally trivial in the natural Euclidean sense: every point of the base has a neighborhood U over which the bundle is equivalent to $(U \times \mathbb{R}^n, \mathrm{pr}_1, U)$ via an equivalence which is an orthogonal map on each fiber.

5. <u>A bundle</u> ξ <u>whose fibers are n-dimensional Euclidean spaces is an n-dimensional Euclidean bundle</u> (<u>i.e.,</u> ξ <u>is locally trivial in the Euclidean sense</u>) <u>if and only if it satisfies</u> (i),(ii) <u>of</u> 3 <u>and the following condition:</u>

      (iii) <u>the function</u> $\mathrm{tl}\,\xi \to \mathbb{R}$, <u>which takes each vector into its length, is continuous.</u>

These conditions are obviously necessary. Let us verify that they are also sufficient. By Theorem 3, (i) and (ii) imply that every point of $\mathrm{bs}\,\xi$ has a neighborhood U together with a trivialization linear on fibers, $h: (U \times \mathbb{R}^n, \mathrm{pr}_1, U) \to \xi\big|_U$. Let $v_1(b),\ldots,v_n(b)$ be the basis of the vector space $\mathrm{pr}\,\xi^{-1}(b)$, $b \in U$, resulting from the standard orthogonalization of the basis $\mathrm{tl}\,h(b,\mathrm{ort}_1),\ldots,\mathrm{tl}\,h(b,\mathrm{ort}_n)$. Now (iii) shows that the vectors $v_1(b),\ldots,v_n(b)$ depend continuously on b, and it is clear that the map linear on fibers, $h': (U \times \mathbb{R}^n, \mathrm{pr}_1, U) \to \xi\big|_U$, given by $\mathrm{tl}\,h'(b,\mathrm{ort}_i) = v_i(b)$ $(i = 1,\ldots,n)$, is a trivialization, orthogonal on each fiber, of the bundle $\xi\big|_U$.

6. Since $O(n) \subset GL(n,\mathbb{R}^n)$, every n-dimensional Euclidean bundle ξ determines a unique n-dimensional real vector bundle ξ', through extension of the structure group. One may use Theorem 5 to interpret the reduction of the structure group transforming ξ' into ξ as enriching the bundle ξ' with an additional structure: namely, a Euclidean metric on each fiber, such that the corresponding length function $\mathrm{tl}\,\xi' \to \mathbb{R}$ is continuous. This additional structure is termed a <u>Euclidean metric on</u> ξ'.

Standard Fibers $GL_+\mathbb{R}^n$ and $SO\mathbb{R}^n$

7. A Steenrod bundle with standard fiber $GL_+\mathbb{R}^n$ (respectively, $SO\mathbb{R}^n$) is called an <u>n-dimensional oriented vector bundle</u> (respectively, an <u>n-dimensional oriented Euclidean bundle</u>).

Since a space with $GL_+\mathbb{R}^n$-structure ($SO\mathbb{R}^n$-structure) is simply an n-dimensional oriented vector space (respectively, an n-dimensional oriented Euclidean space), a W-$GL_+\mathbb{R}^n$-bundle (a W-$SO\mathbb{R}^n$-bundle) is simply a bundle whose fibers are n-dimensional oriented vector (respectively, Euclidean) spaces. It is also plain that a W-$GL_+\mathbb{R}^n$-equivalence of W-$GL_+\mathbb{R}^n$-bundles (a W-$SO\mathbb{R}^n$-equivalence of W-$SO\mathbb{R}^n$-bundles) is simply an equivalence which is orientation preserving and linear (respectively, orthogonal) on fibers.

Consequently, an n-dimensional oriented vector bundle (Euclidean bundle) is a bundle whose fibers are n-dimensional oriented vector spaces (respectively, Euclidean spaces), and which is locally trivial in the following sense: every point of the base has a neighborhood over which the bundle has a trivialization that is orientation preserving and linear (respectively, orthogonal) on fibers.

8. To obtain a version of Theorem 3 which is suitable for the oriented case, note that the orientation existing on each fiber of an n-dimensional oriented vector bundle ξ maps the set of nondegenerate n-frames of the given fiber into S^0. Furthermore, the set of all nondegenerate n-frames of the fibers of ξ is the total space of the associated bundle $asso(\xi,V'(n,n))$ [where $GL_+(n,\mathbb{R})$ acts on $V'(n,n)$ as usual; see 2.3.1.3], and the orientations of the fibers combine to define a map $tl\,asso(\xi,V'(n,n)) \to S^0$. The "oriented" version of Theorem 3 asserts that a bundle ξ whose fibers are n-dimensional oriented real vector spaces is an n-dimensional oriented vector bundle if and only it satisfies conditions (i) and (ii) of 3 and the condition (iv): the function $tl\,asso(\xi,V'(n,n)) \to S^0$, defined by the orientations of the fibers of ξ, is continuous.

Theorem 5 must be modified in a similar fashion. Namely, given an n-dimensional oriented Euclidean bundle ξ, the orientation of each fiber of ξ maps the set of orthonormal n-frames of the given fiber into S^0. Since the set of all orthonormal n-frames of all fibers of ξ equals the total space of the associated bundle $asso(\xi,V(n,n))$, we obtain a function $tl\,asso(\xi,V(n,n)) \to S^0$. The "oriented" version of Theorem 5 asserts that a bundle ξ whose fibers are n-dimensional oriented Euclidean spaces is an n-dimensional oriented Euclidean bundle if and only if it satisfies conditions (i),(ii) of 3, condition (iii) of 5, and condition (v): the function $tl\,asso(\xi,V(n,n)) \to S^0$, defined by the orientations of the fibers of ξ, is continuous.

9. Since $GL_+(n,\mathbb{R}) \subset GL(n,\mathbb{R})$, every n-dimensional oriented real vector bundle ξ determines a unique n-dimensional real vector bundle ξ', obtained from ξ by extending the structure group. As it follows from the discussion in 8, when we reduce the structure group and produce ξ from ξ', we are endowing the fibers of ξ' with orientations which combine to define a continuous map $asso(\xi',V'(n,n)=GL(n,\mathbb{R})) \to S^0$. This additional structure is termed an <u>orientation of the bundle</u> ξ'.

Similarly, the inclusion $SO(n) \subset O(n)$ associates with every n-dimensional oriented Euclidean bundle ξ a unique n-dimensional

Euclidean bundle ξ', obtained from ξ by extension of the structure group. Again, the reduction of the structure group which transforms ξ' into ξ is seen to provide ξ' with an orientation, i.e., as orienting its fibers in such a manner that the corresponding function tl asso$(\xi',V(n,n) = O(n)) \to S^0$ is continuous.

The real vector and Euclidean bundles possessing orientations are reffered to as <u>orientable.</u> Since every orientation may be replaced by the <u>opposite</u> one, as a result of multiplication by -1, every orientable bundle has at least two orientations.

Standard Fibers $GL\mathbb{C}^n$ and $U\mathbb{C}^n$

10. A Steenrod bundle with standard fiber $GL\mathbb{C}^n$ $(U\mathbb{C}^n)$ is called an <u>n-dimensional complex vector</u> (respectively, <u>Hermitian</u>) <u>bundle.</u>

Our discussion of real vector bundles in 2 and 3 carries over, word-for-word, for complex vector bundles, and the same is true for the discussion in 4, 5, and 6 of Euclidean bundles and the Hermitian bundles. In particular, a W-$GL\mathbb{C}^n$-bundle (W-$U\mathbb{C}^n$-bundle) is simply a bundle whose fibers are n-dimensional complex vector spaces (respectively, n-dimensional Hermitian spaces); a W-$GL\mathbb{C}^n$-equivalence (W-$U\mathbb{C}^n$-equivalence) of W-$GL\mathbb{C}^n$-bundles (respectively, W-$U\mathbb{C}^n$-bundles) is simply an equivalence linear (respectively, Hermitian) on fibers; a W-$GL\mathbb{C}^n$-bundle (W-$U\mathbb{C}^n$-bundle) is locally trivial if and only if it is topologically locally trivial and, in addition, the vector operations (respectively, the vector operations and the length function) are continuous; the reduction of the structure group which turns a given complex vector bundle, ξ', into a Hermitian bundle ξ, may be interpreted as endowing ξ' with a Hermitian metric, i.e., as supplying a Hermitian metric (inner product) on each fiber of ξ', in such a manner that the resulting length function is continuous on tl ξ'.

11. Given an arbitrary n-dimensional complex vector bundle ξ, we may construct an n-dimensional complex vector bundle, conj ξ, by replacing each marked homeomorphism α with the composition $\alpha \circ$ conj, where conj: $\mathbb{C}^n \to \mathbb{C}^n$ is the usual complex conjugation; conj ξ is called the <u>bundle conjugate to</u> ξ.

This construction carries over to Hermitian bundles and produces again Hermitian bundles, i.e., to every n-dimensional Hermitian bundle ξ there corresponds the conjugate Hermitian bundle conj ξ.

12. The extension of structure group defined by the inclusion

$GL(n,\mathbb{C}) \subset GL(2n,\mathbb{R})$ turns n-dimensional complex vector bundles into 2n-dimensional real vector bundles. Similarly, the extension defined by the inclusion $U(n) \subset O(2n)$ turns n-dimensional Hermitian bundles into 2n-dimensional Euclidean bundles. In both cases we call the extension of the structure group <u>realification</u>, and we denote by $\mathbb{R}\xi$ the bundle obtained from ξ by realification.

We recall that the additional structure which turns a given 2n-dimensional real vector space into an n-dimensional complex space may be described as a linear transformation whose square equals $-id$ (a "multiplication by $-i$"). Similarly, the additional structure which turns a given 2n-dimensional Euclidean space into an n-dimensional Hermitian space may be described as an orthogonal transformation whose square equals $-id$. Accordingly, the additional structure which distinguishes between the n-dimensional complex bundle ξ and $\mathbb{R}\xi$ may be regarded as a $GL\mathbb{R}^{2n}$-equivalence, $I: \mathbb{R}\xi \to \mathbb{R}\xi$, such that $I^2 = -(id\,\mathbb{R}\xi)$ (the minus sign is defined fiberwise), and that which distinguishes between the n-dimensional Hermitian bundle ξ and $\mathbb{R}\xi$ - - as an $O\mathbb{R}^{2n}$-equivalence, $I: \mathbb{R}\xi \to \mathbb{R}\xi$, such that $I^2 = -id(\mathbb{R}\xi)$. Moreover, we may think of the reduction of structure group which turns $\mathbb{R}\xi$ into ξ as endowing $\mathbb{R}\xi$ with one of the last two equivalences.

Obviously, $\mathbb{R}(conj\,\xi) = \mathbb{R}\xi$ for any complex vector or Hermitian bundle ξ, and the shift $\xi \mapsto conj\,\xi$ may be described in the previous language as the shift $I \mapsto -I$.

13. Since we have not only $GL(n,\mathbb{C}) \subset GL(2n,\mathbb{R})$ and $U(n) \subset O(2n)$, but also $GL(n,\mathbb{C}) \subset GL_+(2n,\mathbb{R})$ and $U(n) \subset SO(2n)$, the realification of a given n-dimensional complex vector or Hermitian bundle may be effected in two steps: one may first extend $GL(n,\mathbb{C})$ to $GL_+(2n,\mathbb{R})$ ($U(n)$ to $SO(2n)$), and then extend $GL_+(2n,\mathbb{R})$ to $GL(2n,\mathbb{R})$ (respectively, $SO(2n)$ to $O(2n)$). Therefore, every complex vector or Hermitian bundle ξ provides $\mathbb{R}\xi$ with a canonical orientation.

Maps

14. A map of a real or complex vector bundle into another is said to be <u>linear</u> if it is linear on fibers. Those linear maps which are nondegenerate (injective) on fibers are called <u>linear monomorphisms.</u> A map between Euclidean bundles which is both isometric and linear on fibers is called an <u>orthogonal monomorphism.</u> A map between Hermitian bundles which is both isometric and linear on fibers is called a <u>unitary monomorphism.</u>

Note that a linear monomorphism between n-dimensional vector bundles is nothing else but a $GL\mathbb{R}^n$-map in the real case and a $GL\mathbb{C}^n$-map in the complex case. Similarly, an orthogonal (unitary) monomorphism between two n-dimensional Euclidean (respectively, Hermitian) bundles is simply an $O\mathbb{R}^n$-map (respectively, a $U\mathbb{C}^n$-map).

Vector Fields

15. Sections of vector bundles are called <u>vector fields.</u> This term is applied equally to the real and complex vector bundles, to the Euclidean and Hermitian bundles, and to the oriented bundles of both types.

A sequence of k vector fields is termed a <u>field of k-frames.</u> A point of the base where the corresponding frame is degenerate is a <u>singularity</u> of the given field. A field of k-frames with no singularities in an n-dimensional vector bundle may be regarded as a section of the associated bundle with fiber $\mathbb{R}V'(n,k)$ or $\mathbb{C}V'(n,k)$. Similarly, a field of orthonormal k-frames with no singularities in an n-dimensional Euclidean or Hermitian bundle may be thought of as a section of the associated bundle with fiber $\mathbb{R}V(n,k)$ or $\mathbb{C}V(n,k)$ [in all cases, the structure group acts as usual].

For every vector bundle there is the <u>zero</u> vector field, which takes each point of the base into the zero vector of the fiber over the given point. As we shall see, there are vector bundles having no sections without zeros. An n-dimensional real (complex) vector bundle ξ admits a field of n-frames with no singularities if and only if ξ is $GL\mathbb{R}^n$-trivial (respectively, $GL\mathbb{C}^n$-trivial); conversely, any such field yields a $GL\mathbb{R}^n$-trivialization (respectively, a $GL\mathbb{C}^n$-trivialization) of ξ. Similarly, for an n-dimensional Euclidean (Hermitian) bundle ξ, giving a field of orthonormal n-frames is equivalent to giving an $O\mathbb{R}^n$-trivialization (respectively, a $U\mathbb{C}^n$-trivialization) of ξ. Moreover, given an n-dimensional oriented real vector (Euclidean) bundle ξ, a field of n-frames without singularities (respectively, a field of orthonormal n-frames) such that the orientations of the fibers are positive on the frames of the field, yields a $GL_+\mathbb{R}^n$-trivialization (respectively, an $SO\mathbb{R}^n$-trivialization) of ξ.

2. Constructions

1. In this subsection we discuss a number of constructions which shall be applied afterwards to vector, Euclidean, and Hermitian bundles, and which were not covered by the general theory of Steenrod bundles from §3.

Subbundles

2. Let ξ be an n-dimensional real vector bundle, and let g be a section of the bundle $\mathrm{asso}(\xi,\mathbb{R}G(n,k))$ weakly associated with ξ [here $GL(n,\mathbb{R})$ acts on $\mathbb{R}G(n,k)$ in the usual way], i.e., a continuous function which takes each point $b \in \mathrm{bs}\,\xi$ into a k-dimensional subspace of the fiber $\mathrm{pr}\,\xi^{-1}(b)$. Denote by T the union of all these k-dimensional subspaces, and let $\xi|_g$ be the bundle $(T,\mathrm{pr}\,\xi|_T,\mathrm{bs}\,\xi)$. Since the fibers of $\xi|_g$ are the subspaces singled out by g, $\xi|_g$ is a $GL\mathbb{R}^k$-bundle, and it is clear that $\xi|_g$ is locally $GL\mathbb{R}^k$-trivial. [In fact, let ξ be the standard trivial bundle $(B \times \mathbb{R}^n,\mathrm{pr}_1,B)$. Given $b_0 \in B$, choose a linear isomorphism $\ell:\mathbb{R}^k \to g(b_0)$. Then for a small enough neighborhood U of b_0, the restriction of $\xi|_g$ to U admits even a canonical $GL\mathbb{R}^n$-trivialization, $h:(U \times \mathbb{R}^k,\mathrm{pr}_1,U) \to (\xi|_g)|_U$, given by $\mathrm{tl}\,h(b,v) = \mathrm{pr}_b(b,\ell(v))$; here pr_b denotes the projection of the fiber $b \times \mathbb{R}^n$ onto its subspace $g(b)$]. Therefore, $\xi|_g$ is a k-dimensional real vector bundle. We call it the <u>subbundle of</u> ξ <u>associated with</u> g.

3. The subbundles of Euclidean, complex vector, or Hermitian bundles are defined similarly. In the Euclidean case, $\mathbb{R}G(n,k)$ is regarded as an $O(n)$-space, g remains a section of $\mathrm{asso}(\xi,\mathbb{R}G(n,k))$, and the resulting subbundle, $\xi|_g$, turns out to be Euclidean (its local $O\mathbb{R}^k$-triviality is a consequence of its local triviality linear on fibers, established in 2, and of the continuity of the length; see 1.5). The discussion in 2 carries over, word-for word, to the complex case: all we have to do is to replace $\mathbb{R}$ by $\mathbb{C}$ [in particular, the $GL(n,\mathbb{R})$-space $\mathbb{R}G(n,k)$ must be replaced by the $GL(n,\mathbb{C})$-space $\mathbb{C}G(n,k)$]. The resulting subbundle, $\xi|_g$, is a complex vector bundle. Finally, in the Hermitian case, $\mathbb{C}G(n,k)$ is considered as a $U(n)$-space and the subbundle $\xi|_g$ is Hermitian ($\xi|_g$ is locally $U\mathbb{C}^k$-trivial because it is

locally trivial, linearly on fibers, and the length function is
continuous; see 1.10).

4. It is clear that the inclusion $\xi|_g \to \xi$ is a linear mono-
morphism in both vector cases, an orthogonal monomorphism in the
Euclidean case, and a unitary monomorphism in the Hermitian case.
Conversely, to every linear, orthogonal, or unitary monomorphism,
$f: \xi_1 \to \xi$, with $bs\,\xi_1 = bs\,\xi$ and $bs\,f = id$, there corresponds a
subbundle of ξ, namely the subbundle associated with the section
$b \mapsto tl\,f(pr\,\xi_1^{-1}(b))$ of $asso(\xi, \mathbb{R}G(n, \dim\xi_1))$ or $asso(\xi, \mathbb{C}G(n, \dim\xi_1))$.
This subbundle is the <u>image</u> of the monomorphism f, denoted $Im\,f$.
By Theorem 3.2.8, $ab\,f : \xi_1 \to Im\,f$ is a $GL\mathbb{R}^n$-, $GL\mathbb{C}^n$-, $O\mathbb{R}^n$-, or
$U\mathbb{C}^n$-equivalence, depending on whether we are in the real vector, complex
vector, Euclidean, or Hermitian case.

We note that the correcting map, $corr\,f : \xi_1 \to (bs\,f)^!\xi$,
which is a linear, orthogonal, or unitary monomorphism together with f
(see 3.2.4), always satisfies $bs\,corr\,f = id$. Therefore, $Im\,corr\,f$ is
meaningful for any linear, orthogonal, or unitary monomorphism f.

Orthogonal Complements and Quotient Bundles

5. If ξ is an n-dimensional Euclidean (Hermitian) bundle,
then to each section g of $asso(\xi, \mathbb{R}G(n,k))$ (respectively,
of $asso(\xi, \mathbb{C}G(n,k))$) there corresponds the section $g^\perp$ <u>orthogonal to</u>
g: $g^\perp$ is the section of $asso(\xi, \mathbb{R}G(n,n-k))$ (respectively, of
$asso(\xi, \mathbb{C}G(n,n-k))$) which takes each $b \in bs\,\xi$ into the orthogonal
complement of the subspace $g(b)$ in the fiber $pr\,\xi^{-1}(b)$. Therefore,
to every k-dimensional subbundle $\eta = \xi|_g$ of the n-dimensional
Euclidean or Hermitian bundle ξ there corresponds an (n-k)-dimensional
subbundle, $\xi|_{g^\perp}$, called the <u>orthogonal complement</u> of the subbundle
η, and denoted $\eta^\perp$.

6. A subbundle of a vector bundle has no orthogonal
complement, but a corresponding <u>quotient bundle</u> is well defined. Namely,
let η be a k-dimensional subbundle of the n-dimensional real or complex
vector bundle ξ. Consider the bundle $(T, fact\,pr\,\xi, bs\,\xi)$, where T is
the quotient space of $tl\,\xi$ by its partition into the sets $x + pr\,\xi^{-1}(b)$,
with $x \in pr\,\xi^{-1}(b)$. The fibers of this bundle are the quotient spaces
$pr\,\xi^{-1}(b)/pr\,\eta^{-1}(b)$, and so it is a $GL\mathbb{R}^{n-k}$- or a $GL\mathbb{C}^{n-k}$-bundle.
$(T, fact\,pr\,\xi, bs\,\xi)$ is called the <u>quotient bundle of</u> ξ <u>by</u> η, denoted
by ξ/η.

This construction and the previous one are related: indeed, if we apply the quotient bundle construction to a Euclidean or Hermitian bundle, we obtain the result of the orthogonal complement construction. More precisely, in the Euclidean case each quotient $\operatorname{pr}\xi^{-1}(b)/\operatorname{pr}\eta^{-1}(b)$ is an $(n-k)$-dimensional Euclidean space; hence, ξ/η is an $O\mathbb{R}^{n-k}$-bundle, and the map $h\colon \eta \to \xi/\eta$, given by $tl\,h(x) = \operatorname{pr}(x)$, where $\operatorname{pr} = [\operatorname{pr}\colon tl\,\xi \to tl(\xi/\eta)]$, is an $O\mathbb{R}^{n}$-equivalence. Similarly, in the Hermitian case, $\operatorname{pr}\xi^{-1}(b)/\operatorname{pr}\eta^{-1}(b)$ are $(n-k)$-dimensional Hermitian spaces, and hence ξ/η is a $U\mathbb{C}^{n-k}$-bundle, and the same h is a $U\mathbb{C}^{n-k}$-equivalence. Therefore, the quotient bundle of a Euclidean (Hermitian) bundle ξ by a subbundle η is a Euclidean (respectively, Hermitian) bundle of dimension $\dim \xi - \dim \eta$.

From this it is readily seen that the quotient of a vector bundle ξ by a subbundle η is a vector bundle of dimension $\dim \xi - \dim \eta$. All we have to check is that ξ/η is locally trivial linearly on fibers whenever ξ is a standard trivial bundle. But this is immediate from the previous discussion if we note that such a ξ may be assumed to be Euclidean in the real case, and Hermitian in the complex case.

7. For real vector or Euclidean bundles, the orientation of each of the three spaces, $\operatorname{pr}\xi^{-1}(b)$, $\operatorname{pr}\eta^{-1}(b)$, $\operatorname{pr}\xi^{-1}(b)/\operatorname{pr}\eta^{-1}(b)$, is uniquely determined by the orientations of the other two (see 3.1.3.10). Consequently, the orientability of two out of the three bundles, ξ, η, ξ/η, implies the orientability of the third, and given orientations of two of them canonically determine the orientation of the third.

Sums

8. The vector bundle ξ is said to **decompose into the (direct) sum** of its subbundles ξ_1 and ξ_2 if each fiber $\operatorname{pr}\xi^{-1}(b)$ is the direct sum of its subspaces $\operatorname{pr}\xi_1^{-1}(b)$ and $\operatorname{pr}\xi_2^{-1}(b)$. A Euclidean or Hermitian bundle ξ **decomposes into the (orthogonal) sum** of its subbundles ξ_1 and ξ_2 if each fiber $\operatorname{pr}\xi^{-1}(b)$ is the orthogonal sum of its subspaces $\operatorname{pr}\xi_1^{-1}(b)$ and $\operatorname{pr}\xi_2^{-1}(b)$.

In the Euclidean and Hermitian cases, every subbundle ξ_1 of the given bundle ξ splits ξ into the sum of its subbundles ξ_1 and $\xi_1^{\perp}$. We show in Subsection 4 that given any vector bundle ξ with cellular base and any subbundle ξ_1 of ξ, there is a subbundle ξ_2 of

ξ such that ξ decomposes into the direct sum of ξ_1 and ξ_2; see 4.2.

If the vector bundle ξ decomposes into the sum of its sub-bundles ξ_1 and ξ_2, then the quotient bundle ξ/ξ_1 is canonically $GL\mathbb{R}^{\dim\xi_2}$ - or $GL\mathbb{C}^{\dim\xi_2}$-equivalent to ξ_2: this canonical equivalence $h\colon \xi_2 \to \xi/\xi_1$ is given by $tl\,h(x) = pr(x)$, where $pr = [pr\colon tl\,\xi \to tl(\xi/\xi_1)]$. Such an equivalence exists also in the Euclidean and Hermitian cases, when O or U replaces GL (in fact, we have already established this in 6).

9. We now introduce a construction which reverses the above process and, in particular, allows us to recover a bundle which decomposes into a sum of subbundles from its summands.

Let ξ_1 and ξ_2 be real vector bundles of dimensions n_1 and n_2, and with common base. Then $\xi_1 \times \xi_2$ is a Steenrod bundle with base $bs\,\xi_1 \times bs\,\xi_2$ and structure group $GL(n_1,\mathbb{R}) \times GL(n_2,\mathbb{R})$. The bundle $diag^!(\xi_1 \times \xi_2)$, where $diag\colon bs\,\xi_1 \to bs\,\xi_1 \times bs\,\xi_1$, has the same structure group (and base $bs\,\xi_1$). Extending this group to $GL(n_1+n_2,\mathbb{R})$, we turn $diag^!(\xi_1 \times \xi_2)$ into an (n_1+n_2)-dimensional real vector bundle, called the (direct) sum of the bundles ξ_1 and ξ_2, and denoted $\xi_1 \oplus \xi_2$. In the Euclidean, complex vector, and Hermitian cases, the definition and notation of a sum of two bundles are the same [in the Euclidean case, $O(n_1) \times O(n_2)$ is extended to $O(n_1+n_2)$, in the complex vector case, $GL(n_1,\mathbb{C}) \times GL(n_2,\mathbb{C})$ is extended to $GL(n_1+n_2,\mathbb{C})$, and in the Hermitian case, $U(n_1) \times U(n_2)$ is extended to $U(n_1+n_2)$; moreover, a sum of Euclidean, complex vector, or Hermitian bundles is again a Euclidean, complex vector, or Hermitian bundle, respectively]. In all cases, $bs(\xi_1 \oplus \xi_2) = bs\,\xi_1 (= bs\,\xi_2)$ and $pr(\xi_1 \oplus \xi_2)^{-1}(b) = pr\,\xi_1^{-1}(b) \oplus pr\,\xi_2^{-1}(b)$; the last sum is orthogonal in the Euclidean and Hermitian situations. These equalities define linear, orthogonal, or unitary monomorphisms, $\xi_1 \to \xi_1 \oplus \xi_2$ and $\xi_2 \to \xi_1 \oplus \xi_2$, which act as the identity on the base. These monomorphisms identify ξ_1 and ξ_2 with subbundles of $\xi_1 \oplus \xi_2$ and split $\xi_1 \oplus \xi_2$ into the sum of these subbundles.

Note that the same identifications allow us to take the quotients $(\xi_1 \oplus \xi_2)/\xi_1$ and $(\xi_1 \oplus \xi_2)/\xi_2$, and using the canonical equivalences defined in 8, we may actually write $(\xi_1 \oplus \xi_2)/\xi_1 = \xi_2$ and $(\xi_1 \oplus \xi_2)/\xi_2 = \xi_1$. In particular, in the real vector and Euclidean cases, the orientability of two out of the three bundles, ξ_1, ξ_2, and $\xi_1 \oplus \xi_2$, implies the orientability of the third, and the orientations of any two of them define canonically an orientation of the third.

Let us add that $\xi \oplus \xi$ is always orientable and has a canonical orientation, for any real vector or Euclidean bundle ξ. This canonical orientation is determined on each fiber by arbitrary orientations of its summands, provided that we take identical orientations for both summands.

10. The sum of a real vector or Euclidean bundle ξ and the one-dimensional trivial $GL\mathbb{R}^1$- or $O\mathbb{R}^1$-bundle $(bs\,\xi \times \mathbb{R}, pr_1, bs\,\xi)$ is called the __suspension of__ ξ, and is denoted by $su\,\xi$. Similarly, for any complex vector or Hermitian bundle ξ, the suspension of ξ, $su\,\xi$, is the sum of ξ and the one-dimensional standard trivial $GL\mathbb{C}^1$- or $U\mathbb{C}^1$-bundle $(bs\,\xi \times \mathbb{C}, pr_1, bs\,\xi)$.

Two real vector bundles, ξ_1 and ξ_2, with $bs\,\xi_1 = bs\,\xi_2$, are said to be __stably equivalent__ if there exist k_1 and k_2 such that $\dim \xi_1 + k_1 = \dim \xi_2 + k_2$ and the bundles $su^{k_1}\xi_1$ and $su^{k_2}\xi_2$ are $GL\mathbb{R}^{\dim \xi_1 + k_1}$-equivalent. Stable equivalence of Euclidean, complex vector, and Hermitian bundles is similarly defined (replacing $GL\mathbb{R}^{\dim \xi_1 + k_1}$ by $O\mathbb{R}^{\dim \xi_1 + k_1}$, $GL\mathbb{C}^{\dim \xi_1 + k_1}$, and $U\mathbb{C}^{\dim \xi_1 + k_1}$, respectively). A bundle which is stably equivalent to a standard trivial bundle is called __stably trivial.__

Finally, we note that for real vector or Euclidean bundles, the orientability of one of the bundles ξ and $su\,\xi$ implies the orientability of the other, and any orientation of one of them canonically defines an orientation of the other; see 9.

Complexification

11. Given an n-dimensional real vector bundle ξ, consider the map $I: \xi \oplus \xi \to \xi \oplus \xi$, given by $tl\,I(x,y) = (-y,x)$ [x and y sit in the same fiber of ξ]. Obviously, I is a $GL\mathbb{R}^{2n}$-equivalence satisfying $I^2 = -id$. Therefore, I turns $\xi \oplus \xi$ into an n-dimensional complex vector bundle (see 1.12), called the __complexification of__ ξ, and denoted $\mathbb{C}\xi$.

If ξ is an n-dimensional Euclidean bundle, the same construction turns $\xi \oplus \xi$ into an n-dimensional Hermitian bundle, which is also called the complexification of ξ and is denoted $\mathbb{C}\xi$.

The operation $\xi \to \mathbb{C}\xi$ is called __complexification__ in both cases. Note that $\mathbb{R}\mathbb{C}\xi = \xi \oplus \xi$ in both cases.

12. Each of the bundles $\mathbb{RC}\xi$ and $\xi \oplus \xi$, appearing in the last equality, carries a canonical orientation; see 1.13 and 9. It may be shown that these orientations coincide when $n \equiv 0,1 \bmod 4$, and are opposite when $n \equiv 2,3 \bmod 4$.

Indeed, pick an arbitrary fiber $\mathrm{pr}\,\xi^{-1}(b)$ of ξ and an arbitrary basis (respectively, orthonormal basis) $v_1,\ldots,v_n$ in $\mathrm{pr}\,\xi^{-1}(b)$. The canonical orientation of the fiber $\mathrm{pr}\,\xi^{-1}(b) \times \mathrm{pr}\,\xi^{-1}(b)$ of $\mathbb{RC}$ takes the value $+1$ on the basis $(v_1,0),(0,v_1),\ldots,(v_n,0),(0,v_n)$ of this fiber, and the same holds for the canonical orientation of the fiber $\mathrm{pr}\,\xi^{-1}(b) \times \mathrm{pr}\,\xi^{-1}(b)$ of $\xi \oplus \xi$ and its basis $(v_1,0),\ldots,(v_n,0),(0,v_1),\ldots,(0,v_n)$. It remains to note that, in order to shift from one basis to the other, we have to perform $n(n-1)/2$ permutations of adjacent vectors, and that this number is even for $n \equiv 0,1 \bmod 4$, and odd for $n \equiv 2,3 \bmod 4$.

13. <u>The map</u> $\mathrm{conj}\colon \mathbb{C}\xi \to \mathrm{conj}\,\mathbb{C}\xi$, <u>given by</u> $\mathrm{tl}\,\mathrm{conj}(x,y) = (x,-y)$, <u>is a</u> $\mathrm{GL}\mathbb{C}^n$-<u>equivalence</u> ($\mathrm{U}\mathbb{C}^n$-<u>equivalence</u>) <u>for any n-dimensional real vector (respectively, Euclidean) bundle</u> ξ.

Indeed, the equivalences $I_1,I_2\colon \xi \oplus \xi \to \xi \oplus \xi$, which turn $\xi \oplus \xi$ into $\mathbb{C}\xi$ and $\mathrm{conj}\,\mathbb{C}\xi$, are given by $\mathrm{tl}\,I_1(x,y) = (-y,x)$ and $I_2 = -I_1$, and hence $I_2 \circ \mathrm{conj} = \mathrm{conj} \circ I_1$.

14. <u>The map</u> $K\colon \xi \oplus \mathrm{conj}\,\xi \to \mathbb{CR}\,\xi$, <u>given by</u> $\mathrm{tl}\,K(x,y) = (\frac{1}{2}(x+y),\frac{1}{2}(\mathrm{tl}\,I(y) - \mathrm{tl}\,I(x)))$, <u>where</u> I <u>is the equivalence which turns</u> $\mathbb{R}\xi$ <u>into</u> ξ, <u>is a</u> $\mathrm{GL}\mathbb{C}^n$-<u>equivalence</u> ($\mathrm{U}\mathbb{C}^n$-<u>equivalence</u>) <u>for any n-dimensional complex vector (respectively, Hermitian) bundle.</u>

The equivalences $I_1,I_2\colon \mathbb{R}\xi \oplus \mathbb{R}\xi \to \mathbb{R}\xi \oplus \mathbb{R}\xi$, which turn $\mathbb{R}\xi \oplus \mathbb{R}\xi$ into $\xi \oplus \mathrm{conj}\,\xi$ and $\mathbb{CR}\,\xi$, are given by $\mathrm{tl}\,I_1(x,y) = (\mathrm{tl}\,I(x),-\mathrm{tl}\,I(y))$ and $\mathrm{tl}\,I_2(x,y) = (-y,x)$, and hence $I_2 \circ K = K \circ I_1$.

3. <u>The Classical Universal Vector Bundles</u>

1. The main value of the construction in Subsection 4.3 is that it establishes the existence of a universal G-bundle for a completely arbitrary topological group G. However, for the groups $\mathrm{GL}(n,\mathbb{R})$, $\mathrm{GL}_+(n,\mathbb{R})$, $\mathrm{O}(n)$, $\mathrm{SO}(n)$, $\mathrm{GL}(n,\mathbb{C})$, and $\mathrm{U}(n)$, there are more convenient, classical constructions, which are described in the present subsection.

The Grassman Spaces

2. Set

$$G(\infty,n) = \lim(G(m,n),\text{in}: G(m,n) \to G(m+1,n)),$$

$$G_+(\infty,n) = \lim(G_+(m,n),\text{in}: G_+(m,n) \to G_+(m+1,n)),$$

and

$$\mathbb{C}G(\infty,n) = \lim(\mathbb{C}G(m,n),\text{in}: \mathbb{C}G(m,n) \to \mathbb{C}G(m+1,n)).$$

$G(\infty,n)$ consists of all n-dimensional planes of $\mathbb{R}^\infty$ passing through 0, and is called the n-th (real) Grassman space. Similarly, the n-th upper Grassman space, $G_+(\infty,n)$, consists of all oriented n-dimensional planes in $\mathbb{R}^\infty$ passing through 0. Finally, the n-th complex Grassman space, $\mathbb{C}G(\infty,n)$, consists of all n-dimensional planes in $\mathbb{C}^\infty$ passing through 0.

The canonical maps: $G_+(m,n) \to G(m,n)$, $\mathbb{C}G(m,n) \to G(2m,2n)$, $\mathbb{C}G(m,n) \to G_+(2m,2n)$, $G(m,n) \to G(m+q,n+q)$, $G_+(m,n) \to G_+(m+q,n+q)$, and $\mathbb{C}G(m,n) \to \mathbb{C}G(m+q,n+q)$ (see 3.2.2.3 and 3.2.2.7) define for any n the following maps:

$$G_+(\infty,n) \to G(\infty,n), \quad \mathbb{C}G(\infty,n) \to G(\infty,2n), \quad \mathbb{C}G(\infty,n) \to G_+(\infty,2n),$$

$$G(\infty,n) \to G(\infty,n+q), \quad G_+(\infty,n) \to G_+(\infty,n+q),$$

and

$$\mathbb{C}G(\infty,n) \to \mathbb{C}G(\infty,n+q).$$

The first of these maps (as the canonical map $G_+(m,n) \to G(m,n)$ with $m < \infty$) is a two-sheeted covering projection; the second is the composition of the first and the third; and all these maps, except the first, are embeddings.

3. The Grassman spaces possess natural cellular decompositions which we shall presently describe.

First consider $G(\infty,n)$. Let Ω_n denote the set of all sequences of integers, $\omega = \{\omega(1),\ldots,\omega(n)\}$, with $0 \leqslant \omega(1) \leqslant \ldots \leqslant \omega(n)$, and let us agree to add the term $\omega(0) = 0$ to each sequence $\omega \in \Omega_n$. Further, let $e(\omega)$ denote the subset of $G(\infty,n)$ consisting of those n-planes γ in $\mathbb{R}^\infty$ (passing through 0) such that

$$\dim(\gamma \cap \mathbb{R}^m) = \max\{s \mid \omega(s) + s \leqslant m\},$$

for all m. We show that the sets $e(\omega)$, $\omega \in \Omega_n$, yield a cellular decomposition of $G(\infty,n)$, with $\dim e(\omega) = d(\omega)$, where $d(\omega) =$

$= \omega(1) + \ldots + \omega(n)$.

We have to produce a characteristic map, $\mathrm{cha}_{e(\omega)} : D^{d(\omega)} \to G(\infty,n)$. Fix ω and, for points $u,v \in S^{\omega(n)+n-1}$ such that $u + v \neq 0$, denote by $r(u,v) \in SO(\omega(n)+n)$ the orthogonal transformation which takes u into v and keeps fixed all the vectors of $\mathbb{R}^{\omega(n)+n}$ which are orthogonal to u and v. Further, let H_i denote the $\omega(i)$-dimensional hemisphere consisting of all points $(x_1,\ldots,x_{\omega(i)+i}) \in S^{\omega(i)+i-1}$ such that $x_{\omega(j)+j} = 0$ for $j = 1,\ldots,i+1$, and $x_{\omega(i)+i} \geq 0$. Consider the map $\phi : H_1 \times \ldots \times H_n \to G(\infty,n)$ which takes each sequence $(u_1,\ldots,u_n)$ into the plane spanned by the n-frame:

$$u_1, \quad [r(\mathrm{ort}_{\omega(1)+1},u_1)](u_2),\ldots,$$

$$[r(\mathrm{ort}_{\omega(1)+1},u_1) \circ \ldots \circ r(\mathrm{ort}_{\omega(n-1)+n-1},u_{n-1})](u_n) \tag{1}$$

(which is obviously orthonormal). Then ϕ is continuous and maps $\mathrm{int}(H_1 \times \ldots \times H_n)$ into $e(\omega)$, and $\partial(H_1 \times \ldots \times H_n)$ into a union of sets $e(\omega')$ with $d(\omega') < d(\omega)$. Moreover, its compression $\mathrm{ab}\,\phi : \mathrm{int}(H_1 \times \ldots \times H_n) \to e(\omega)$ is a homeomorphism: its inverse takes each plane $\gamma \in e(\omega)$ into the sequence $(u_1,\ldots,u_n)$, with

$$u_1 = v_1, \quad u_2 = [r(\mathrm{ort}_{\omega(1)+1},u_1)]^{-1}(v_2),\ldots,$$

$$u_n = [r(\mathrm{ort}_{\omega(1)+1},u_1) \circ \ldots \circ r(\mathrm{ort}_{\omega(n-1)+n-1},u_{n-1})]^{-1}(v_n),$$

where $v_1,\ldots,v_n$ is an orthogonal basis of γ, selected in such a way that v_i sits in the hemisphere $x_{\omega(i)+i} > 0$ of $S^{\omega(i)+i-1}$ (the vectors $v_1,\ldots,v_n$ are uniquely determined by these requirements). Therefore, one may take $\mathrm{cha}_{e(\omega)}$ to be the composition

$$D^{d(\omega)} \to D^{\omega(1)} \times \ldots \times D^{\omega(n)} \to H_1 \times \ldots \times H_n \overset{\phi}{\to} G(\infty,n),$$

where the left map is the inverse of the homeomorphism indicated in 1.2.6.9, and the middle map is the product of the homeomorphisms $D^{\omega(i)} \to H_i$ given by $(x_1,\ldots,x_{\omega(i)}) \mapsto$

$$\mapsto (x_1,\ldots,x_{\omega(1)},0,x_{\omega(1)+1},\ldots,x_{\omega(2)},0,\ldots,0,x_{\omega(i-1)+1},\ldots,$$

$$x_{\omega(i)},(1 - \textstyle\sum_1^{\omega(i)} x_j^2)^{1/2}).$$

The cellular decomposition of $G_+(\infty,n)$ has twice as many cells as that of $G(\infty,n)$. Namely, over each cell $e(\omega)$ sit two cells, $e_+(\omega)$ and $e_-(\omega)$, of $G_+(\infty,n)$, which are homeomorphically mapped onto $e(\omega)$ by the projection $G_+(\infty,n) \to G(\infty,n)$: $e_+(\omega)$ is made up of planes oriented in such a way that the orientation is positive on the basis $v_1,\ldots,v_n$ described above, while $e_-(\omega)$ is made up of the same planes, but with the opposite orientations. The characteristic maps for $e_+(\omega)$ and $e_-(\omega)$ are constructed in the same manner as $\mathrm{cha}_{e(\omega)}$, but now the plane spanned by the frame (1) is oriented [its orientation is positive (negative) on (1) for $e_+(\omega)$ (respectively, for $e_-(\omega)$)].

The cellular decomposition of $\mathbb{C}G(\infty,n)$ is given by the cells $\mathbb{C}e(\omega)$, $\omega \in \Omega_n$, which are defined precisely as the $e(\omega)$'s if one replaces $\mathbb{R}^\infty$ and $\mathbb{R}^m$ by $\mathbb{C}^\infty$ and $\mathbb{C}^m$; $\dim \mathbb{C}e(\omega) = 2d(\omega)$, and $\mathrm{cha}_{\mathbb{C}e(\omega)}$ is the exact complex-Hermitian analog of the map $\mathrm{cha}_{e(\omega)}$.

4. The above cellular decompositions contain only a finite number of cells of a given dimension, and hence satisfy property (C). Since each of the manifolds $G(m,n)$, $G_+(m,n)$, and $\mathbb{C}G(m,n)$ is covered by a finite number of cells, and since these manifolds constitute fundamental covers of the spaces $G(\infty,n)$, $G_+(\infty,n)$, and $\mathbb{C}G(\infty,n)$, our cellular decompositions satisfy also property (W). Finally, from Theorem 1.2.4.6 it follows that $G(\infty,n)$, $G_+(\infty,n)$, and $\mathbb{C}G(\infty,n)$ are normal. Therefore, the above cellular decompositions turn $G(\infty,n)$, $G_+(\infty,n)$, and $\mathbb{C}G(\infty,n)$ into cellular spaces.

It is clear that $G(m,n)$, $G_+(m,n)$, and $\mathbb{C}G(m,n)$ are subspaces of $G(\infty,n)$, $G_+(\infty,n)$, and $\mathbb{C}G(\infty,n)$ in the cellular sense, and so they are cellular spaces too.

For $n = 1$, $G(m,n)$, $\mathbb{C}G(m,n)$, $G(\infty,n)$, and $\mathbb{C}G(\infty,n)$ are simply $\mathbb{R}P^{m-1}$, $\mathbb{C}P^{m-1}$, $\mathbb{R}P^\infty$, and $\mathbb{C}P^\infty$, respectively, and the above cellular decompositions are identical with those described in 2.1.3.4 and 2.1.3.5.

5. Note that the canonical embeddings

$$G(\infty,n) \to G(\infty,n+q), \quad G_+(\infty,n) \to G_+(\infty,n+q), \quad \mathbb{C}G(\infty,n) \to \mathbb{C}G(\infty,n+q)$$

(see 2) are cellular. The image of the first contains $\mathrm{ske}_n G(\infty,n+q)$, the image of the second - $\mathrm{ske}_n G_+(\infty,n+q)$, and the image of the third - $\mathrm{ske}_{2n+1} \mathbb{C}G(\infty,n+q)$.

The inclusions $G(m,n) \supset \mathrm{ske}_{m-n} G(\infty,n)$, $G_+(m,n) \supset \mathrm{ske}_{m-n} G_+(\infty,n)$, and $\mathbb{C}G(m,n) \supset \mathrm{ske}_{2m-2n+1} \mathbb{C}G(\infty,n)$ are equally evident.

The Grassman Bundles

6. Let $0 \leqslant n \leqslant m \leqslant \infty$ and $n < \infty$. We let $T(m,n)$, $T_+(m,n)$, and $\mathbb{C}T(m,n)$, denote those subsets of the respective products $G(m,n) \times \mathbb{R}^m$, $G_+(m,n) \times \mathbb{R}^m$, and $\mathbb{C}G(m,n) \times \mathbb{C}^m$, consiting of all pairs (γ,x) such that $x \in \gamma$. Consider the bundles $(T(m,n),pr,G(m,n))$, $(T_+(m,n),pr,G_+(m,n))$, and $(\mathbb{C}T(m,n),pr,\mathbb{C}G(m,n))$, where $pr(\gamma,x) = \gamma$. The fibers of the first (second; third) bundle are Euclidean spaces (respectively, oriented Euclidean spaces; Hermitian spaces). Moreover, the first (second; third) bundle is locally W-$O\mathbb{R}^n$-trivial (respectively, locally W-$SO\mathbb{R}^n$-trivial; locally W-$U\mathbb{C}^n$-trivial), and hence it is a Euclidean (respectively, oriented Euclidean; Hermitian) bundle. We denote these bundles by $\mathrm{Gra}(m,O(n))$, $\mathrm{Gra}(m,SO(n))$, and $\mathrm{Gra}(m,U(n))$. In addition, by extending the respective structure groups, $O(n)$, $SO(n)$, and $U(n)$, to $GL(n,\mathbb{R})$, $GL_+(n,\mathbb{R})$, and $GL(n,\mathbb{C})$, we obtain bundles denoted by $\mathrm{Gra}(m,GL(n,\mathbb{R}))$, $\mathrm{Gra}(m,GL_+(n,\mathbb{R}))$, and $\mathrm{Gra}(m,GL(n,\mathbb{C}))$. The bundles of these six series are called <u>Grassman bundles.</u> For $m = \infty$, we use the simpler notations $\mathrm{Gra}\,O(n)$, $\mathrm{Gra}\,SO(n)$, $\mathrm{Gra}\,U(n)$, $\mathrm{Gra}\,GL(n,\mathbb{R})$, $\mathrm{Gra}\,GL_+(n,\mathbb{R})$, and $\mathrm{Gra}\,GL(n,\mathbb{C})$.

Note that for $m < \infty$, $\mathrm{Gra}(m,O(n))$ is nothing else but the subbundle of the standard trivial bundle $(G(m,n) \times \mathbb{R}^m,pr_1,G(m,n))$ (viewed as a Euclidean bundle) associated with the diagonal section, $\gamma \mapsto (\gamma,\gamma)$, of the bundle $(G(m,n) \times G(m,n),pr_1,G(m,n))$. The same holds, with obvious modifications, for the remaining five series.

7. <u>The bundles</u> $\mathrm{Gra}\,G$ with $G = GL(n,\mathbb{R})$, $GL_+(n,\mathbb{R})$, $O(n)$, $SO(n)$, $GL(n,\mathbb{C})$, $U(n)$ <u>are universal.</u>

The proofs for the different groups G differ only in some obvious details, and are all very similar to the proof of Theorem 4.3.4. We shall treat here only the group $GL(n,\mathbb{R})$. According to 4.2.2, it suffices to show that given any n-dimensional real vector bundle ξ with cellular base and any subspace $A \subset bs\,\xi$, every $GL\mathbb{R}^n$-map $g\colon \xi\big|_A \to \mathrm{Gra}\,GL(n,\mathbb{R})$ extends to a $GL\mathbb{R}^n$-map $\xi \to \mathrm{Gra}\,GL(n,\mathbb{R})$.

Assume first that $bs\,\xi = D^{r+1}$ (for some r) and $A = S^r$. In this case ξ is $GL\mathbb{R}^n$-trivial, and so we may actually assume that ξ is the standard trivial bundle $(D^{r+1} \times \mathbb{R}^n,pr_1,D^{r+1})$. The desired extension of g, $f\colon \xi \to \mathrm{Gra}\,GL(n,\mathbb{R})$, can be described explicitly: let g_1 be the composite map

$$S^r \times \mathbb{R}^n \xrightarrow{\;tl\,g\;} tl\,\mathrm{Gra}\,GL(n,\mathbb{R}) \xrightarrow{\;in\;} G(\infty,n) \times \mathbb{R}^\infty \xrightarrow{\;pr_2\;} \mathbb{R}^\infty$$

and define $f_1 : D^{r+1} \times \mathbb{R}^n \to \mathbb{R}^\infty$ by $f_1(ty,(x_1,\ldots,x_n)) =$

$$= tg_1(y,(x_1,\ldots,x_n)) + (1-t^2)^{1/2}(\underbrace{0,\ldots,0}_{m},x_1,\ldots,x_n,0,\ldots),$$

where $y \in S^r$, $t \in I$, and m is the smallest number s such that $\mathbb{R}^s \supset g_1(S^r \times S^{n-1})$; finally, set $\operatorname{tl} f(y,x) = (f_1(y \times \mathbb{R}^n), f_1(y,x))$.

The general case reduces to this special situation. Assume that the cellular space $\operatorname{bs} \xi$ is rigged and that we already have a $\mathrm{GL}\mathbb{R}^n$-map $f_r : \xi\big|_{A\cup\operatorname{ske}_r\operatorname{bs}\xi} \to \operatorname{Gra} \mathrm{GL}(n,\mathbb{R})$ which extends g. The above argument shows that for every cell $e \in \operatorname{cell}_{r+1}\operatorname{bs}\xi \smallsetminus \operatorname{cell}_{r+1}A$ the $\mathrm{GL}\mathbb{R}^n$-map

$$g_e = f_r \circ \operatorname{ad}[\operatorname{ab} \operatorname{cha}_e : S^r \to A \cup \operatorname{ske}_r \operatorname{bs}\xi] : \operatorname{cha}_e^! \xi\big|_{S^r} \to \operatorname{Gra} \mathrm{GL}(n,\mathbb{R})$$

extends to a $\mathrm{GL}\mathbb{R}^n$-map $h_e : \operatorname{cha}_e^! \xi \to \operatorname{Gra} \mathrm{GL}(n,\mathbb{R})$, and it is clear that $\operatorname{tl} h_e$ is constant on the elements of the partition $\operatorname{zer}(\operatorname{tl} \operatorname{ad} \operatorname{cha}_e)$. Applying Theorem 3.2.6 (with $B = D^{r+1}$ and $p = [\operatorname{ab}\operatorname{cha}_e : D^{r+1} \to \operatorname{Cl} e]$), we conclude that h_e defines a $\mathrm{GL}\mathbb{R}^n$-map $\xi\big|_{\operatorname{Cl} e} \to \operatorname{Gra} \mathrm{GL}(n,\mathbb{R})$, which we denote by f_e. Now note that for any two cells, $e_1,e_2 \in \operatorname{cell}_{r+1}\operatorname{bs}\xi \smallsetminus \operatorname{cell}_{r+1}A$, $\operatorname{tl} f_{e_1}$ and $\operatorname{tl} f_{e_2}$ agree over $\operatorname{Cl} e_1 \cap \operatorname{Cl} e_2$, and that for any cell $e \in \operatorname{cell}_{r+1}\operatorname{bs}\xi \smallsetminus \operatorname{cell}_{r+1}A$, $\operatorname{tl} f_e$ and $\operatorname{tl} f_r$ agree over $\operatorname{Cl} e \cap (A \cup \operatorname{ske}_r\operatorname{bs}\xi)$. From this compatibility it follows that the maps f_r and f_e, $e \in \operatorname{cell}_{r+1}\operatorname{bs}\xi \smallsetminus \operatorname{cell}_{r+1}A$, combine to define a $\mathrm{GL}\mathbb{R}^n$-map $f_{r+1} : \xi\big|_{A\cup\operatorname{ske}_{r+1}\operatorname{bs}\xi} \to \operatorname{Gra} \mathrm{GL}(n,\mathbb{R})$ which extends f_r; see 3.2.7. Therefore, using induction, we can produce a sequence of $\mathrm{GL}\mathbb{R}^n$-maps,

$$\{f_s : \xi\big|_{A\cup\operatorname{ske}_s\operatorname{bs}\xi} \to \operatorname{Gra} \mathrm{GL}(n,\mathbb{R})\}_{s=-1}^{\infty},$$

such that $f_{-1} = g$ and f_s extends f_{s-1}, $s \geqslant 0$. Finally, the maps f_s define a $\mathrm{GL}\mathbb{R}^n$-map $\xi \to \operatorname{Gra} \mathrm{GL}(n,\mathbb{R})$ extending g.

8. <u>The bundles</u> $\operatorname{Gra}(m,\mathrm{GL}(n,\mathbb{R}))$, $\operatorname{Gra}(m,\mathrm{GL}_+(n,\mathbb{R}))$, $\operatorname{Gra}(m,\mathrm{O}(n))$, <u>and</u> $\operatorname{Gra}(m,\mathrm{SO}(n))$ <u>are $(m-n)$-universal. The bundles</u> $\operatorname{Gra}(m,\mathrm{GL}(n,\mathbb{C}))$ <u>and</u> $\operatorname{Gra}(m,\mathrm{U}(n))$ <u>are $(2m-2n+1)$-universal.</u>

This is a corollary of 7 (see 5 and 4.2.7).

Associated Principal Bundles

9. When $m < \infty$, the total spaces of the principal bundles associated with the Grassman bundles

$$\text{Gra}(m,\text{GL}(n,\mathbb{R})), \quad \text{Gra}(m,\text{GL}_+(n,\mathbb{R})), \quad \text{Gra}(m,\text{GL}(n,\mathbb{C})),$$

and

$$\text{Gra}(m,O(n)), \quad \text{Gra}(m,SO(n)), \quad \text{Gra}(m,U(n)),$$

are obviously $V'(m,n)$, $V'(m,n)$, $\mathbb{C}V'(m,n)$, and $V(m,n)$, $V(m,n)$, $\mathbb{C}V(m,n)$. The corresponding projections are the maps described in 3.2.2.3 and 3.2.2.7:

and

$$V'(m,n) \to G(m,n), \quad V'(m,n) \to G_+(m,n), \quad \mathbb{C}V'(m,n) \to \mathbb{C}G(m,n) \quad (2)$$

$$V(m,n) \to G(m,n), \quad V(m,n) \to G_+(m,n), \quad \mathbb{C}V(m,n) \to \mathbb{C}G(m,n). \quad (3)$$

The same is true for $m = \infty$, if $V'(\infty,n)$, $\mathbb{C}V'(\infty,n)$, $V(\infty,n)$, and $\mathbb{C}V(\infty,n)$ are understood as $\lim(V'(m,n),in)$, $\lim(\mathbb{C}V'(m,n),in)$, $\lim(V(m,n),in)$, and $\lim(\mathbb{C}V(m,n),in)$, and the projections $(2),(3)$ with $m = \infty$ as the limits of the projections $(2),(3)$, $m < \infty$. $V'(\infty,n)$, $\mathbb{C}V'(\infty,n)$, $V(\infty,n)$, and $\mathbb{C}V(\infty,n)$ are called <u>Stiefel spaces.</u>

It is clear that for $m < \infty$ the canonical right actions of the structure groups on the above total spaces (see 3.2.10) are exactly the right actions described in 2.3.12 and 2.3.13, while for $m = \infty$ they are the limits of the latter.

The Bundles
$\text{asso}(\text{Gra}\,O(1),O(1))$ and $\text{asso}(\text{Gra}\,U(1),U(1))$

10. <u>The principal bundle associated with</u> $\text{Gra}\,O(1)$ <u>is</u> <u>$O(1)$-isomorphic to</u> $\text{Mi}\,O(1)$. <u>The principal bundle associated with</u> $\text{Gra}\,U(1)$ <u>is $U(1)$-isomorphic to</u> $\text{Mi}\,U(1)$.

PROOF. It suffices to find an $O(1)$-homeomorphism $\text{tl}\,\text{Mi}\,O(1) \to \text{tl}\,\text{asso}(\text{Gra}\,O(1),O(1))$, when we regard $\text{tl}\,\text{Mi}\,O(1)$ and $\text{tl}\,\text{asso}(\text{Gra}\,O(1),O(1)) = V(\infty,1)\,[= S^\infty]$ as right $O(1)$-spaces; similarly, viewing $\text{tl}\,\text{Mi}\,U(1)$ and $\text{tl}\,\text{asso}(\text{Gra}\,U(1),U(1)) = \mathbb{C}V(\infty,1)\,[= S^\infty]$ as right $U(1)$-spaces, we need only exhibit a $U(1)$-homeomorphism $\text{tl}\,\text{Mi}\,U(1) \to \text{tl}\,\text{asso}(\text{Gra}\,U(1),U(1))$; see 3.1.9 and 3.2.10. In both cases such a homeomorphism is given by the formula

$$\{\text{pr}(g_i,t_i)\}_{i=1}^{\infty} \to \{g_i\sqrt{t_i}\}_{i=1}^{\infty}\,.$$

The meaning of the left-hand side was explained in 4.3.2, while in the right-hand side the elements g_i of $O(1)$ or $U(1)$ are thought of as numbers (the following inclusions are used: $V(\infty,1) \subset \mathbb{R}^\infty$, $\mathbb{C}V(\infty,1) \subset \mathbb{C}^\infty$, $O(1) = S^0 \subset \mathbb{R}$, and $U(1) = S^1 \subset \mathbb{C}$).

4. The Most Important
Reductions of the Structure Group

1. The use of Grassman bundles enables us to apply the scheme presented in Subsection 4.4 to the problems raised in Subsection 1 concerning reductions of the structure group. This is the subject of the present subsection.

Recall that the reductions corresponding to the inclusions

$$O(n) \subset GL(n,\mathbb{R}) , \quad SO(n) \subset GL_+(n,\mathbb{R}) , \quad U(n) \subset GL(n,\mathbb{C}), \qquad (4)$$

are equivalent to the introduction of a Euclidean or Hermitian metric, while the reductions resulting from the inclusions

$$GL_+(n,\mathbb{R}) \subset GL(n,\mathbb{R}) , \quad SO(n) \subset O(n), \qquad (5)$$

are equivalent to the introduction of an orientation. Finally, the reductions resulting from the inclusions

$$GL(n,\mathbb{C}) \subset GL(2n,\mathbb{R}) , \quad U(n) \subset O(2n), \qquad (6)$$

mean the introduction of a complex structure.

For each of the inclusions (4), (5), and (6), we shall exhibit a canonical classifying map, and then list the most obvious consequences of these constructions. Moreover, we shall carry out the same program for the inclusions

$$GL(n-s,\mathbb{R}) \subset GL(n,\mathbb{R}) , \quad GL_+(n-s,\mathbb{R}) \subset GL_+(n,\mathbb{R}) ,$$
$$GL(n-s,\mathbb{C}) \subset GL(n,\mathbb{C}), \qquad (7)$$

and

$$O(n-s) \subset O(n), \quad SO(n-s) \subset SO(n), \quad U(n-s) \subset U(n). \qquad (8)$$

The reductions of the structure group corresponding to the six inclusions (7) and (8) may be interpreted as the representation of the given n-dimensional bundle as the s-fold suspension of an (n-s)dimensional bundle.

2. The outlined program is simple to carry out for inclusions

(4). Indeed, the bundles $Gra\,GL(n,\mathbb{R})$ and $Gra\,O(n)$ have the same base, $G(\infty,n)$, and the same is true for $Gra\,GL_+(n,\mathbb{R})$ and $Gra\,SO(n)$, with the base $G_+(\infty,n)$, and for $Gra\,GL(n,\mathbb{C})$ and $Gra\,U(n)$, with the base $\mathbb{C}G(\infty,n)$. It is obvious that in all three cases the identity map of the base is classifying. Therefore, the mappings

$$ext:\ Stee(B,O\mathbb{R}^n)\ \to\ Stee(B,GL\mathbb{R}^n)\ ,$$

$$ext:\ Stee(B,SO\mathbb{R}^n)\ \to\ Stee(B,GL_+\mathbb{R}^n)\ ,$$

and

$$ext:\ Stee(B,U\mathbb{C}^n)\ \to\ Stee(B,GL\mathbb{C}^n)$$

are invertible for any cellular space B; see 4.4.2. In particular, every real (complex) vector bundle with cellular base admits a Euclidean (respectively, Hermitian) metric.

As a corollary, we obtain the theorem already formulated in 2.8: given any vector bundle ξ with cellular base and any subbundle ξ_1, there exists a subbundle ξ_2 of ξ such that ξ decomposes into the sum of ξ_1 and ξ_2.

3. Similarly, the projection $G_+(\infty,n) \to G(\infty,n)$ is classifying for both inclusions (5), while the inclusion $\mathbb{C}G(\infty,n) \to G(\infty,2n)$ is classifying for both inclusions (6). However, a study of the homotopy properties of these classifying maps is already a quite difficult task. We shall return to the first of them in §5.6, armed with more sophisticated tools.

4. For the inclusions (7) and (8) there are also obvious classifying maps: for both left inclusions, such a map is the canonical embedding $G(\infty,n-s) \to G(\infty,n)$, for both middle inclusions - the canonical embedding $G_+(\infty,n-s) \to G_+(\infty,n)$, and for both last inclusions - the canonical embedding $\mathbb{C}G(\infty,n-s) \to \mathbb{C}G(\infty,n)$ (see 3.2). Identifying $G(\infty,n-s)$, $G_+(\infty,n-s)$, and $\mathbb{C}G(\infty,n-s)$ with their images under these embeddings, and using 3.5, we can write:

$$G(\infty,n-s) \supset ske_{n-s}G(\infty,n),\quad G_+(\infty,n-s) \supset ske_{n-s}G_+(\infty,n)$$

and

$$\mathbb{C}G(\infty,n-s) \supset ske_{2n-2s+1}\mathbb{C}G(\infty,n).$$

From the first inclusion it follows that the pair $(G(\infty,n),G(\infty,n-s))$ is $(n-s)$-connected (see 2.3.2.2), which in turn implies (by Theorems 2.3.2.4 and 2.3.2.5) that the map

$$\pi(id,in):\ \pi(B,G(\infty,n-s))\ \to\ \pi(B,G(\infty,n))$$

is invertible for any cellular space B with $\dim B \leqslant n-s$, and
surjective for any cellular space B with $\dim B = n-s$. Consequently,

$$\text{ext: } \text{Stee}(B, GL\mathbb{R}^{n-s}) \to \text{Stee}(B, GL\mathbb{R}^{n})$$

and

$$\text{ext: } \text{Stee}(B, O\mathbb{R}^{n-s}) \to \text{Stee}(B, O\mathbb{R}^{n})$$

are invertible for any cellular B with $\dim B \leqslant n-s$, and surjective
for any cellular B with $\dim B = n-s$. In exactly the same manner the
inclusion $G_{+}(\infty, n-s) \supset \text{ske}_{n-s} G_{+}(\infty, n)$ leads to the invertibility
(surjectivity) of the mappings

$$\text{ext: } \text{Stee}(B, GL_{+}\mathbb{R}^{n-s}) \to \text{Stee}(B, GL_{+}\mathbb{R}^{n})$$

and

$$\text{ext: } \text{Stee}(B, SO\mathbb{R}^{n-s}) \to \text{Stee}(B, SO\mathbb{R}^{n})$$

for any cellular B with $\dim B \leqslant n-s$ (respectively, $\dim B = n-s$),
while the inclusion $\mathbb{C}G(\infty, n-s) \supset \text{ske}_{2n-2s+1} \mathbb{C}G(\infty, n)$ implies the
invertibility (surjectivity) of the mappings

$$\text{ext: } \text{Stee}(B, GL\mathbb{C}^{n-s}) \to \text{Stee}(B, GL\mathbb{C}^{n})$$

and

$$\text{ext: } \text{Stee}(B, U\mathbb{C}^{n-s}) \to \text{Stee}(B, U\mathbb{C}^{n})$$

for any cellular B with $\dim B \leqslant 2(n-s)$ (respectively, $\dim B =
= 2n-2s+1$). Therefore, every n-dimensional real (complex) vector bundle
with cellular base of dimension $\leqslant n-s$ (respectively, $\leqslant 2n-2s+1$) is
$GL\mathbb{R}^{n}$-equivalent (respectively, $GL\mathbb{C}^{n}$-equivalent) to the s-fold suspension
of an (n-s)-dimensional bundle; furthermore, if given two (n-s)-dimensio-
nal real (complex) vector bundles with cellular base of dimension $< n-s$
(respectively, $< 2n-2s+1$) their s-fold suspensions are $GL\mathbb{R}^{n}$-equivalent
(respectively, $GL\mathbb{C}^{n}$-equivalent), then the bundles themselves are
$GL\mathbb{R}^{n-s}$-equivalent (respectively, $GL\mathbb{C}^{n-s}$-equivalent).

5. Exercises

1. Let ξ be an n-dimensional real vector bundle. Show that
$\text{asso}(\xi, \mathbb{R}^{n} \smallsetminus 0)$ is equivalent (in the sense of 1.1.2) to the bundle
with total space $\{x \in \text{tl}\,\xi \mid x \neq 0\}$ and whose projection is the
restriction of $\text{pr}\,\xi$ to this subspace of $\text{tl}\,\xi$.

Let ξ be an n-dimensional Euclidean bundle. Show that
$\text{asso}(\xi, D^{n})$ and $\text{asso}(\xi, S^{n-1})$ are equivalent to the bundles whose total

spaces are the subspaces of $tl\,\xi$ consisting of the vectors of length $\leqslant 1$ and $= 1$, respectively, and whose projections are the appropriate restrictions of $pr\,\xi$.

2. Let ξ be an n-dimensional real (complex) vector bundle. Show that $asso(\xi,V'(n,k))$ (respectively, $asso(\xi,\mathbb{C}V'(n,k))$ is equivalent with the bundle with total space

$$\{(x_1,\ldots,x_k) \in \underbrace{tl\,\xi \times \ldots \times tl\,\xi}_{k} \mid pr\,\xi(x_1) = \ldots = pr\,\xi(x_k),\ x_1,\ldots,x_k \text{ linearly independent}\},$$

and whose projection is the restriction of the composite map

$$tl\,\xi \times \ldots \times tl\,\xi \xrightarrow{\ pr_1\ } tl\,\xi \xrightarrow{\ pr\,\xi\ } bs\,\xi.$$

Let ξ be an n-dimensional Euclidean (Hermitian) bundle. Show that $asso(\xi,V(n,k))$ (respectively, $assso(\xi,\mathbb{C}V(n,k))$ is equivalent to the bundle with total space

$$\{(x_1,\ldots,x_k) \in \underbrace{tl\,\xi \times \ldots \times tl\,\xi}_{k} \mid pr\,\xi(x_1) = \ldots = pr\,\xi(x_k),\ x_1,\ldots,x_k \text{ is an orthonormal frame}\},$$

and whose projection is the restriction of the composite map

$$tl\,\xi \times \ldots \times tl\,\xi \xrightarrow{\ pr_1\ } tl\,\xi \xrightarrow{\ pr\,\xi\ } bs\,\xi.$$

3. Consider the spaces T and S introduced in Exercise 1.2.9.5. Now $GL_+(1,\mathbb{R})$ acts on $T \smallsetminus 0$ from the right by $(\{x_i\},t) \to \{tx_i\}$. Clearly $(T \smallsetminus 0)/GL_+(1,\mathbb{R}) = S$. Show that the $GL_+(1,\mathbb{R})$-bundle defined by this action is locally trivial, but not trivial. Show that the associated oriented one-dimensional real vector bundle does not admit a Euclidean metric.

§6. SMOOTH BUNDLES

1. <u>Fundamental Concepts</u>

1. Let $1 \leqslant r \leqslant a$. A bundle ξ is called a <u>bundle of class</u> C^r, or a C^r-<u>bundle</u> if $tl\,\xi$ and $bs\,\xi$ are C^r-manifolds, and for each point $b_0 \in bs\,\xi$ there are a neighborhood U of b_0, a C^r-manifold

F with $\partial F = \emptyset$ if $\partial U \neq \emptyset$, and a C^r-diffeomorphism $h: U \times F \to$ $\to \mathrm{pr}\,\xi^{-1}(U)$, such that $\mathrm{pr}\,\xi(h(b,x)) = b$ for all $b \in U$ and $x \in F$. The C^s-bundles with $s \geqslant r$ will be referred to as bundles of class $C^{\geqslant r}$, or $C^{\geqslant r}$-bundles. The $C^{\geqslant 1}$-bundles are called <u>smooth.</u>

If ξ is a C^r-bundle, then $\mathrm{pr}\,\xi$ is obviously a C^r-submersion. In particular, the fibers of a smooth bundle are neat submanifolds of the total manifold $\mathrm{tl}\,\xi$ (see 3.1.5.8). Moreover, the fibers over points belonging to the same component of the base of a C^r-bundle are pairwise C^r-diffeomorphic. If $\mathrm{bs}\,\xi$ is connected and $\partial\mathrm{bs}\,\xi \neq \emptyset$, then the fibers have no boundary, and $\partial\mathrm{tl}\,\xi = \mathrm{pr}\,\xi^{-1}(\partial\mathrm{bs}\,\xi)$, whereas if $\partial\mathrm{bs}\,\xi = \emptyset$, then $\partial\mathrm{tl}\,\xi = \cup_{b\in\mathrm{bs}\,\xi}\,\partial\mathrm{pr}\,\xi^{-1}(b)$; in the first situation, the restriction $(\partial\mathrm{tl}\,\xi, \mathrm{ab}\,\mathrm{pr}\,\xi, \partial\mathrm{bs}\,\xi)$ of the bundle ξ to $\partial\mathrm{bs}\,\xi$ is a C^r-bundle, whereas in the second $(\partial\mathrm{tl}\,\xi, \mathrm{ab}\,\mathrm{pr}\,\xi, \mathrm{bs}\,\xi)$ is a C^r-bundle.

Given two C^r-bundles, ξ_1 and ξ_2, such that $\partial\mathrm{bs}\,\xi_i = \emptyset$ and $\partial\mathrm{tl}\,\xi_i = \emptyset$, the product $\xi_1 \times \xi_2$ is a C^r-bundle.

The restriction of a C^r-bundle to a neat submanifold of its base is clearly a C^r-bundle.

Suppose that ξ is a C^r-bundle, B is a C^r-manifold, and $f: B \to \mathrm{bs}\,\xi$ is a C^r-map such that the fiber $\mathrm{pr}\,\xi^{-1}(f(b))$ has no boundary for all $b \in \partial B$. Then $f^!\xi$ is a C^r-bundle, and we say that the bundle $f^!\xi$ is <u>neatly induced.</u> For example, given a C^r-bundle ξ, $\mathrm{in}^!\xi$ is always neatly induced when in is either the inclusion of a neat submanifold in $\mathrm{bs}\,\xi$, or the inclusion $\partial\mathrm{bs}\,\xi \to \mathrm{bs}\,\xi$; obviously, $\mathrm{in}^!\xi$ coincides, as a C^r-bundle, with the corresponding restriction of ξ.

2. Let $0 \leqslant s \leqslant r$. A map ϕ from one $C^{\geqslant r}$-bundle into another is said to be a C^s-<u>map</u> (a C^s-<u>isomorphism</u>) if $\mathrm{tl}\,\xi$ and $\mathrm{bs}\,\xi$ are C^s-maps (respectively, C^s-diffeomorphisms for $s \geqslant 1$, and homeomorphisms for $s = 0$). A C^s-isomorphism which is also an equivalence is called a C^s-<u>equivalence.</u>

A $C^{\geqslant r}$-bundle ξ is said to be C^s-<u>trivial</u> if it is C^s-equivalent to a standard trivial bundle $(\mathrm{bs}\,\xi \times F, \mathrm{pr}_1, \mathrm{bs}\,\xi)$, where F is a $C^{\geqslant r}$-manifold (such that $\partial F = \emptyset$ if $\partial\mathrm{bs}\,\xi \neq \emptyset$). Every C^r-bundle is obviously locally C^r-trivial, meaning that each point of $\mathrm{bs}\,\xi$ has a neighborhood U such that $\xi|_U$ is C^r-trivial; in particular, every smooth bundle is topologically locally trivial.

If $f^!\xi$ is neatly induced from the $C^{\geqslant r}$-bundle ξ by a C^r-map f, then $\mathrm{ad}\,f: f^!\xi \to \xi$ is a C^r-map. Furthermore, if $\phi: \xi' \to \xi$ is a C^r-map, where ξ and ξ' are $C^{\geqslant r}$-bundles, and $(\mathrm{bs}\,\phi)^!\xi$ is neatly induced, then corr ϕ : $\xi' \to$ (bs ϕ)$^!\xi$ is also a C^r-map.

Smooth Bundles and Submersions

3. **Let** $r \leqslant \infty$, **and let** $f: X \to Y$ **be a** C^r**-submersion, where** X **and** Y **are** C^r**-manifolds,** X **is compact, and** $f^{-1}(\partial Y) = \partial X$. **Then** (X,f,Y) **is a** C^r**-bundle. The same holds true when** $r = a$, **provided that** X **admits a** C^a**-embedding in Euclidean space.**

(See 6.1 for a supplement to this theorem.)

PROOF. It suffices to examine the case $f(X) = Y$: indeed, in the general case the set $f(X)$ is both open (see 3.1.5.8) and closed (see 1.1.7.9), and hence is a union of whole components of Y. We show that for each point $y_0 \in Y$ there are a neighborhood U of y_0, a closed C^r-manifold F, and a C^r-diffeomorphism $h: U \times F \to f^{-1}(U)$, such that $f(h(y,x)) = y$ for all $y \in U$ and $x \in F$.

From 3.1.5.8 (or, if it is more convenient, from 3.4.8.2), it follows that $f^{-1}(y_0)$ is a neat submanifold of X for $y_0 \in$ int Y, and a neat submanifold of ∂X for $y_0 \in \partial Y$, and in both cases $f^{-1}(y_0)$ is closed as an independent manifold. Set $F = f^{-1}(y_0)$, and pick a C^r-embedding $j: X \to \mathbb{R}^q$, a C^r-transversalization τ of the embedding $j|_F: F \to \mathbb{R}^q$, and a neat tube $\mathrm{Tub}_\tau \rho$. Now consider the map

$$\phi: j^{-1}(\mathrm{tub}_\tau \rho) \to Y \times F, \quad \phi(x) = (f(x), \mathrm{pr}_\tau(j(x))).$$

The following properties of ϕ are immediate: ϕ is of class C^r; $\phi(\partial(j^{-1}(\mathrm{tub}_\tau \rho))) \subset \partial(Y \times F)$; ϕ is injective on F; and the differential $d_x\phi$ is nondegenerate for all $x \in F$. Since F is compact, we conclude that ϕ defines a diffeomorphism of a neighborhood of F onto a neighborhood of $\phi(F) = y_0 \times F$ (see 3.1.5.5). Using once more the compactness of F, we see that the last neighborhood contains a set of the form $V \times F$, where V is a neighborhood of y_0 (in Y). Let U be a smaller neighborhood of y_0, such that $j(f^{-1}(U)) \subset \mathrm{tub}_\tau \rho$. Then $\phi^{-1}(U \times F) = f^{-1}(U)$, and we can finally set $h = (ab \phi)^{-1}: U \times F \to f^{-1}(U)$.

4 (EXAMPLES). The bundles $(V(n,k),\mathrm{pr},G(n,k))$, $(V(n,k),\mathrm{pr},G_+(n,k))$, $(\mathbb{C}V(n,k),\mathrm{pr},\mathbb{C}G(n,k))$, and $(\mathbb{H}V(n,k),\mathrm{pr},\mathbb{H}G(n,k))$, whose projections are the submersions defined in Subsection 3.2.2, are principal C^a-bundles with structure groups O(k), SO(k), U(k), and Sp(k), respectively. Similarly, $(V(n,k),\mathrm{pr},V(n,k-q))$, $(\mathbb{C}V(n,k),\mathrm{pr},\mathbb{C}V(n,k-q))$, and $(\mathbb{H}V(n,k),\mathrm{pr},\mathbb{H}V(n,k-q))$, whose projections

are the submersions defined in Subsection 3.2.1, are Steenrod C^a-bundles with structure group $O(n-k+q)$, $U(n-k+q)$, and $Sp(n-k+q)$, and standard fibers $V(n-k+q,q)$, $\mathbb{C}V(n-k+q,q)$, and $\mathbb{H}V(n-k+q,q)$ (on which the above groups act canonically; see 2.3.12). The coverings $(\mathbb{R},\mathrm{hel},S^1)$, (S^1,hel_m,S^1), and $(G_+(n,k),\mathrm{pr},G(n,k))$, defined in 1.2.6, are also principal C^a-bundles.

Among the previously listed principal C^a-bundles we find (S^3,pr,S^2) and (S^7,pr,S^4), whose projections are the Hopf submersions (see 3.2.2.9); they are called the <u>Hopf bundles.</u> The Hopf submersion $S^{15} \to S^8$ defines a C^a-bundle, which is also known as a Hopf bundle; its fibers are diffeomorphic to S^7 (but this bundle is not endowed with any special structure group).

The Smooth Bundles as Steenrod Bundles

5. Let F be a C^r-manifold with $r \geq 1$. According to 2.3.10, F is an effective $\mathrm{Diff}^r F$-space, and thus every C^r-bundle ξ whose fibers are C^r-diffeomorphic to F is a W-F-bundle (see 3.1.3 and 3.2.1). Moreover, ξ is clearly locally W-F-trivial, i.e., it is a Ehresmann--Feldbau bundle. However, the procedure that enabled us in Subsection 3.4 to turn Ehresmann-Feldbau bundles into Steenrod bundles does not work here: as we already had the ocassion to note (see 3.5.2), when $\dim F > 0$, the natural action of $\mathrm{Diff}^r F$ on F is not topologically effective. Nevertheless, the set $MH(\xi)$ carries a natural topology, transferred from $C^r(F,\mathrm{tl}\,\xi)$ with the aid of the injective mapping $MH(\xi) \to C^r(F,\mathrm{tl}\,\xi)$ which takes each diffeomorphism $\alpha \in MH(\xi)$ into the map $[\mathrm{in}\colon \alpha(F) \to \mathrm{tl}\,\xi] \circ \alpha$. It is clear that with this topology on $MH(\xi)$, ξ becomes a Steenrod F-bundle.

It is instructive to compare the implicit group structures described above for smooth bundles with the implicit group structures of locally trivial bundles (see 3.4.5). Here we merely mention two difficulties encountered in the differential situation. Firstly, not every Steenrod F-bundle can be smoothed - the fact that its base need not be a manifold is already an obstruction. Secondly, simple examples show that there are F-maps of C^r-bundles with fibers diffeomorphic to F, which are not C^r-maps.

2. <u>Smoothings and Approximations</u>

1. This subsection is similar in character with §3.4: here we generalize some of the results obtained there for smooth manifolds to smooth bundles. Although part of these results are indeed rather important, some problems are not touched upon at all. For the sake of brevity, we shall consider below only the closed case; the reader can find some additional information concerning the (more general) compact case in Subsection 6 (see 6.2-6.5).

We shall need two notations, for $0 \leqslant s \leqslant r$: if ξ is a C^r-bundle, we let $\mathrm{Sec}^s\xi$ denote the set of all C^s-sections of ξ; and if ξ and ξ' are C^r-bundles, we let $C^s(\xi,\xi')$ denote the space of all C^s-maps $\xi \to \xi'$. If $s \neq a$, both sets carry natural topologies: $\mathrm{Sec}^s\xi$ is a subspace of $C^s(\mathrm{bs}\,\xi,\mathrm{tl}\,\xi)$, while $C^s(\xi,\xi')$ is a subspace of the product $C^s(\mathrm{tl}\,\xi,\mathrm{tl}\,\xi') \times C^s(\mathrm{bs}\,\xi,\mathrm{bs}\,\xi')$.

ξ-Transversalizations and Tubes

2. Our immediate task is to adapt the definitions and theorems of Subsection 3.4.3 for use in the more general setting of this subsection.

We start with the transversalizations. Let ξ be a smooth bundle such that $\mathrm{tl}\,\xi$ is closed, and let $j: \mathrm{tl}\,\xi \to \mathbb{R}^q$ be a differentiable embedding. A continuous map $\tau : \mathrm{tl}\,\xi \to G(q,q-n)$, where $n = \dim \mathrm{tl}\,\xi - \dim \mathrm{bs}\,\xi$, is called a ξ-<u>transversalization</u> of the embedding j if the restriction $\tau\big|_{\mathrm{pr}\,\xi^{-1}(y)}$ is a transversalization of the embedding $j\big|_{\mathrm{pr}\,\xi^{-1}(y)}$ for all points $y \in \mathrm{bs}\,\xi$. A fundamental example is the <u>normal ξ-transversalization</u>, which takes each point $x \in \mathrm{tl}\,\xi$ into the orthogonal complement of the plane $d_x j(\mathrm{Tang}_x(\mathrm{pr}\,\xi^{-1}(\mathrm{pr}\,\xi(x))))$ in $\mathbb{R}^q$. Our ξ-version of Theorem 3.4.3.7 asserts that if ξ and j are of class C^r $(1 \leqslant r \leqslant a)$, then j admits a ξ-transversalization of class C^r. The proof is an obvious modification of that of Theorem 3.4.3.7.

Now we move on to tubes. Let τ be an arbitrary ξ-transversalization of the embedding j. We define the tube $\mathrm{Tub}_\tau\rho$ and the open tube $\mathrm{tub}_\tau\rho$ as the following subsets of $\mathrm{bs}\,\xi \times \mathbb{R}^q$:

$$\mathrm{Tub}_\tau\rho = \bigcup_{x\in\mathrm{tl}\,\xi} (\mathrm{pr}\,\xi(x) \times d_\tau(x,\rho))$$

and

$$\text{tub}_\tau \rho \;=\; \textstyle\bigcup_{x \in \text{tl}\,\xi} \, (\text{pr}\,\xi\,(x) \;\times\; (d_\tau(x,\rho) \smallsetminus s_\tau(x,\rho))),$$

where $d_\tau(x,\rho)$ and $s_\tau(x,\rho)$ are the ball, and respectively the sphere, with center $j(x)$ and radius ρ in the plane $j(x) + \tau(x)$. Equivalently,

$$\text{Tub}_\tau \rho \;=\; \textstyle\bigcup_{y \in \text{bs}\,\xi} \, (y \;\times\; \text{Tub}_\tau\big|_{\text{pr}\,\xi^{-1}(y)}\, \rho)$$

and

$$\text{tub}_\tau \rho \;=\; \textstyle\bigcup_{y \in \text{bs}\,\xi} \, (y \;\times\; \text{tub}_\tau\big|_{\text{pr}\,\xi^{-1}(y)}\, \rho).$$

The tube $\text{Tub}_\tau \rho$ is <u>neat</u> if there is a $\sigma > \rho$ such that: the sets $\text{pr}\,\xi\,(x) \times (d_\tau(x,\sigma) \smallsetminus s_\tau(x,\sigma))$ are pairwise disjoint, $\text{tub}_\tau \sigma$ is open in $\text{bs}\,\xi \times \mathbb{R}^q$, and the map $\text{tub}_\tau \sigma \to \text{tl}\,\xi$, which takes $\text{pr}\,\xi\,(x) \times (d_\tau(x,\sigma) \smallsetminus s_\tau(x,\sigma))$ into x, is smooth. The restrictions of this last map to $\text{Tub}_\tau \rho$ and $\text{tub}_\tau \rho$ are called projections and are denoted by pr_τ (they clearly do not depend upon the choice of σ). If ξ, j, and τ are of class C^r, $r \geqslant 1$, then the following are true: there exists a neat tube; every neat tube $\text{Tub}_\tau \rho$ is a submanifold of $\text{bs}\,\xi \times \mathbb{R}^q$, with $\text{int}\,\text{Tub}_\tau \rho = \text{tub}_\tau \rho$; and $\text{pr}_\tau \colon \text{Tub}_\tau \rho \to \text{tl}\,\xi$ is a C^r-submersion. Again, the proof is an obvious modification of the proofs of Theorems 3.4.3.4 and 3.4.3.5. We must also modify appropriately the construction in 3.4.3.3: now the model Tu_τ is defined as the subset $\{(x,t) \in \text{tl}\,\xi \times \mathbb{R}^q \mid t \in \tau(x)\}$ of $\text{tl}\,\xi \times \mathbb{R}^q$, while $\text{nat}\colon \text{Tu}_\tau \to \text{bs}\,\xi \times \mathbb{R}^q$ is given by $\text{nat}(x,t) = (\text{pr}\,\xi\,(x), j(x)+t)$.

The Basic Theorems

3. <u>Let</u> $r \leqslant \infty$, <u>and let</u> ξ <u>and</u> ξ' <u>be</u> C^r-<u>bundles with closed total spaces</u> $\text{tl}\,\xi$ <u>and</u> $\text{tl}\,\xi'$, <u>and closed bases</u> $\text{bs}\,\xi$ <u>and</u> $\text{bs}\,\xi'$. <u>Then</u> $C^r(\xi,\xi')$ <u>is dense in</u> $C^s(\xi,\xi')$ <u>for any</u> $s < r$. <u>The same holds true for</u> $r = a$, <u>provided that</u> $\text{tl}\,\xi$, $\text{tl}\,\xi'$, $\text{bs}\,\xi$, <u>and</u> $\text{bs}\,\xi'$ <u>admit</u> C^a-<u>embeddings in Euclidean spaces.</u>

PROOF. Pick a C^r-embedding $j'\colon \text{tl}\,\xi' \to \mathbb{R}^{q'}$, a ξ'-transversalization τ' of j' of class C^r, and a neat tube $\text{Tub}_{\tau'}\rho'$. Let $\mathcal{U}$ denote the subset of $C^s(\text{tl}\,\xi, \text{tl}\,\xi') \times C^s(\text{bs}\,\xi, \text{bs}\,\xi')$ consisting of all the pairs $(F\colon \text{tl}\,\xi \to \text{tl}\,\xi', f\colon \text{bs}\,\xi \to \text{bs}\,\xi')$ such that $(f(\text{pr}\,\xi\,(x)), j'(F(x))) \in \text{tub}_{\tau'}\rho'$ for all $x \in \text{tl}\,\xi$. Then $\mathcal{U}$ is open

352

and contains $c^s(\xi,\xi')$. For $(F,f) \in U$, define $\Phi(F,f)\colon tl\,\xi \to tl\,\xi'$ by

$$x \mapsto pr_\tau,(f(pr\,\xi(x)),j'(F(x))) \,.$$

Obviously, $(\Phi(F,f),f) \in c^s(\xi,\xi')$ and the map $\Psi\colon U \to c^s(\xi,\xi')$, $\Psi(F,f) = (\Phi(F,f),f)$, is a retraction which takes $U \cap (c^r(tl\,\xi,tl\,\xi') \times c^r(bs\,\xi,bs\,\xi'))$ into $c^r(\xi,\xi')$. Since $c^r(tl\,\xi,tl\,\xi') \times c^r(bs\,\xi,bs\,\xi')$ is dense in $c^s(tl\,\xi,tl\,\xi') \times c^s(bs\,\xi,bs\,\xi')$ (see 3.4.4.2), the existence of such a retraction implies that $c^r(\xi,\xi')$ is dense in $c^s(\xi,\xi')$.

4. Let $s < r \leqslant \infty$, let ξ and ξ' be arbitrary $c^{\geqslant r}$-bundles such that $tl\,\xi$, $tl\,\xi'$, $bs\,\xi$, and $bs\,\xi'$ are closed manifolds, and let $f\colon bs\,\xi \to bs\,\xi'$ be a c^r-map. Then the set $\{\phi \in c^r(\xi,\xi') \mid bs\,\phi = f\}$ is dense in $\{\phi \in c^s(\xi,\xi') \mid bs\,\phi = f\}$. The same holds true for $r = a$, provided that $tl\,\xi$, $tl\,\xi'$ admit c^a-embeddings in Euclidean spaces.

PROOF. Let F denote the subspace of $c^s(tl\,\xi,tl\,\xi')$ consisting of the maps $tl\,\phi$ such that $\phi \in c^s(\xi,\xi')$ and $bs\,\phi = f$. We show that $F \cap c^r(\xi,\xi')$ is dense in F.

Pick a c^r-embedding $j'\colon tl\,\xi' \to \mathbb{R}^{q'}$, a ξ'-transversalization τ' of j' of class c^r, and a neat tube Tub_τ,ρ'. Consider the subset $U \subset c^s(tl\,\xi,tl\,\xi')$ consisting of all maps $F\colon tl\,\xi \to tl\,\xi'$ such that $(f(pr\,\xi(x)),j'(F(x))) \in \mathrm{tub}_\tau,\rho'$ for all $x \in tl\,\xi$. It is clear that U is open and contains F. Moreover, the mapping $U \to F$, transforming each $F \in U$ into the map

$$x \mapsto pr_\tau,(f(pr\,\xi(x)),j'(F(x))) \,,$$

is a retraction which takes $U \cap c^r(tl\,\xi,tl\,\xi')$ into $F \cap c^r(tl\,\xi,tl\,\xi')$. Since $c^r(tl\,\xi,tl\,\xi')$ is dense in $c^s(tl\,\xi,tl\,\xi')$, the existence of such a retraction shows that $F \cap c^r(\xi,\xi')$ is dense in F, as claimed.

5. Let $r \leqslant \infty$, and let ξ and ξ' be arbitrary $c^{\geqslant r}$-bundles such that the manifolds $tl\,\xi$, $tl\,\xi'$, $bs\,\xi$, and $bs\,\xi'$ are closed. If $0 < s < r$, then the set of all c^r-isomorphisms (c^r-equivalences) $\xi \to \xi'$ is dense in the subspace of all c^s-isomorphisms (respectively, c^s-equivalences) of $c^s(\xi,\xi')$. The same holds true for $r = a$, provided that $tl\,\xi$, $tl\,\xi'$, $bs\,\xi$, and $bs\,\xi'$ (respectively, $tl\,\xi$ and $tl\,\xi'$) admit c^a-embeddings in Euclidean spaces.

This is a consequence of 3, 4, and 3.4.1.6.

6 (COROLLARIES). If two $c^{\geqslant r}$-bundles with closed total manifolds and bases are c^1-isomorphic and $r < \infty$, then they are c^r-isomorphic. The same holds true for $r = a$, provided that the total manifolds and the bases admit c^a-embeddings in Euclidean spaces.

<u>If two $C^{\geq r}$-bundles with closed total manifolds are</u> C^1-<u>equivalent and</u> $r \leq \infty$, <u>then they are</u> C^r-<u>equivalent. The same holds true for</u> $r = a$, <u>provided that the total manifolds admit</u> C^a-<u>embeddings in Euclidean spaces.</u>

7. <u>If</u> $r \leq \infty$, <u>then given any</u> $C^{\geq r}$-<u>bundle</u> ξ <u>with closed</u> $\mathrm{tl}\,\xi$ <u>and</u> $\mathrm{bs}\,\xi$, <u>the space</u> $\mathrm{Sec}^r\xi$ <u>is dense in</u> $\mathrm{Sec}^s\xi$, <u>for any</u> $s < r$. <u>The same holds true for</u> $r = a$, <u>provided that</u> $\mathrm{tl}\,\xi$ <u>and</u> $\mathrm{bs}\,\xi$ <u>admit</u> C^a-<u>embeddings in Euclidean spaces.</u>

This is a result of Theorem 4, applied to the bundles ξ and $(\mathrm{bs}\,\xi, \mathrm{id}\,\mathrm{bs}\,\xi, \mathrm{bs}\,\xi)$ and to the map $\mathrm{id}\,\mathrm{bs}\,\xi$.

8. <u>Every</u> $C^{\geq r}$-<u>bundle</u> ξ <u>such that</u> $\mathrm{tl}\,\xi$ <u>and</u> $\mathrm{bs}\,\xi$ <u>are closed manifolds is</u> C^r-<u>isomorphic to a</u> C^a-<u>bundle</u> η <u>with the property that</u> $\mathrm{tl}\,\eta$ <u>and</u> $\mathrm{bs}\,\eta$ <u>can be</u> C^a-<u>embedded in Euclidean spaces.</u>

PROOF. By 3.4.9.6, there exist C^a-manifolds, T and B, admitting C^a-embeddings in Euclidean spaces, together with C^r-diffeomorphisms $F: T \to \mathrm{tl}\,\xi$ and $f: B \to \mathrm{bs}\,\xi$. Pick a C^r-embedding $j: \mathrm{tl}\,\xi \to \mathbb{R}^q$, a ξ-transversalization τ of j of class C^r, and a neat tube $\mathrm{Tub}_\tau\rho$. Let U denote the subset of $C^r(T,B)$ consisting of all the submersions $p: T \to B$ such that the image of the composite map

$$T \xrightarrow{\ \mathrm{diag}\ } T \times T \xrightarrow{\ (f\circ p)\times(j\circ F)\ } \mathrm{bs}\,\xi \times \mathbb{R}^q$$

is contained in $\mathrm{tub}_\tau\rho$. Obviously: U is open; $f^{-1} \circ \mathrm{pr}\,\xi \circ F \in U$; the mapping $\Phi: U \to C^r(T, \mathrm{tl}\,\xi)$, which takes each $g \in U$ into the map $x \mapsto \mathrm{pr}_\tau(f\circ g(x), j\circ F(x))$, is continuous and $\Phi(f^{-1} \circ \mathrm{pr}\,\xi \circ F) = F$. Since F is a C^r-diffeomorphism, there is a neighborhood V of $f^{-1} \circ \mathrm{pr}\,\xi \circ F$ in $C^r(T,B)$ with the property that $V \subset U$ and $\Phi(g)$ is a C^r-diffeomorphism for all $g \in V$. Now pick some C^a-map h and set $\eta = (T,h,B)$. By Theorem 1.3, η is a C^a-bundle, and it is plain that $\Phi(h)$ and f yield a C^r-isomorphism $\eta \to \xi$.

3. <u>Smooth Vector Bundles</u>

1. Usually, when we encounter a bundle which is smooth, it carries some additional structures, most frequently a group structure of Steenrod type. In such cases, smoothness plays the same role as does the topology in the theory of Steenrod bundles discussed earlier (§§ 3, 4), and it is natural to try developing this analogy into a weighty theory of smooth Steenrod bundles.

Unfortunately, such a program is beyond the scope of our book. Therefore, we shall restrict ourselves to the basic facts concerning smooth <u>vector</u> bundles, which are of main interest to this study, and can be derived in a less cumbersome manner that the general theory.

We remark that because the fibers of a vector bundle of positive dimension are not compact, it is not possible to deduce the smoothing and approximation theorems below (Theorems 8-12) from the results of the previous subsection without resorting to additional devices. However, we prefer to give simple, straightforward proofs of these theorems, so that this subsection becomes independent of the previous one.

Fundamental Concepts

2. ξ is an <u>n-dimensional real vector C^r-bundle</u> $(1 \leqslant r \leqslant a)$ if it is both an n-dimensional real vector bundle and a C^r-bundle, and these two structures are compatible, meaning that the restriction of ξ over a small enough neighborhood of an arbitrary point of bs ξ is C^r-GL$\mathbb{R}^n$-trivial (i.e., is C^r-GL$\mathbb{R}^n$-equivalent to a standard trivial bundle, where, of course, a C^r-GL$\mathbb{R}^n$-equivalence is just a GL$\mathbb{R}^n$-equivalence which is simultaneously a C^r-equivalence). The <u>Euclidean</u>, <u>complex vector</u>, and <u>Hermitian C^r-bundles</u> are similarly defined.

Products of vector, Euclidean, or Hermitian $C^{\geqslant r}$-bundles, as well as bundles induced (in particular, obtained by reducing) such bundles, are again $C^{\geqslant r}$-bundles of the same kind, provided the conditions imposed by the corresponding definitions from Subsection 1 (see 1.2) are fulfilled.

We add that the statements and proofs of Theorems 3.2.8 and 3.2.9 carry over, with obvious modifications, to vector, Euclidean, and Hermitian $C^{\geqslant r}$-bundles. Here we formulate only the C^r-GL$\mathbb{R}^n$-version of Theorem 3.2.9: let $f : \xi \to \eta$ be a C^r-GL$\mathbb{R}^n$-map, where ξ and η are $C^{\geqslant r}$-GL$\mathbb{R}^n$-bundles, and suppose that the bundle $(\text{bs } f)^! \eta$ is neatly induced (see 1.1.); then corr f is a C^r-GL$\mathbb{R}^n$-equivalence.

3. The explicit descriptions of the vector, Euclidean, and Hermitian bundles, given in Subsection 5.1, have obvious C^r-analogs $(1 \leqslant r \leqslant a)$. The C^r-analog of Theorem 5.1.3 asserts that a C^r-bundle whose fibers are n-dimensional real vector spaces is an n-dimensional real vector C^r-bundle if and only if the partial vector operations indicated in 5.1.3 are C^r-maps. Similarly, the C^r-analog of Theorem 5.1.5 asserts that a C^r-bundle whose fibers are n-dimensional Euclidean

is an n-dimensional Euclidean C^r-bundle if and only if the partial
vector operations and the metric (equivalently, the square of the length
of vectors, considered as a function on the total space) are of class
C^r. In particular, in order to turn a real vector C^r-bundle into a
Euclidean C^r-bundle, one has to equip it with a Euclidean C^r-metric.
The corresponding complex formulations (i.e., the C^r-analogs of the
theorems in 5.1.10) are obtained by replacing Euclidean bundles and
Euclidean metrics by Hermitian bundles and Hermitian metrics.

Smooth Grassman Bundles

4. The Grassman bundles defined in 5.3.6 provide (for $m < \infty$)
fundamental examples of smooth vector, Euclidean, and Hermitian bundles.
Namely, if $0 \leqslant n \leqslant m < \infty$, then $\mathrm{Gra}(m,GL(n,\mathbb{R}))$, $\mathrm{Gra}(m,GL(n,\mathbb{C}))$,
$\mathrm{Gra}(m,O(n))$, and $\mathrm{Gra}(m,U(n))$ are obviously real vector, complex
vector, Euclidean, and Hermitian C^a-bundles, respectively, all of them
n-dimensional. The third (fourth) bundle differs from the first
(respectively, second) by having a Euclidean C^a-metric (respectively, a
Hermitian C^a-metric).

Theorem 5 below may be thought of as a weakened C^r-analog
(for $r \neq a$) of that part of Theorem 5.3.8 concerning $\mathrm{Gra}(m,GL(n,\mathbb{R}))$
and $\mathrm{Gra}(m,GL(n,\mathbb{C}))$.

5. <u>Let ξ be an n-dimensional real (complex) vector C^r-bundle
with compact base. If $r \neq a$, then there are a number m and a C^r-map</u>
$f: \mathrm{bs}\,\xi \to G(m,n)$ <u>(respectively,</u> $f: \mathrm{bs}\,\xi \to \mathbb{C}G(m,n))$ <u>such that ξ is</u>
$C^r\text{-}GL\mathbb{R}^n$<u>-equivalent to the bundle</u> $f^!\mathrm{Gra}(m,GL(n,\mathbb{R}))$ <u>(respectively,</u>
$C^r\text{-}GL\mathbb{C}^n$<u>-equivalent to the bundle</u> $f^!\mathrm{Gra}(m,GL(n,\mathbb{C}))$.

Since the proofs of the real and complex cases differ only in
some obvious details, we shall prove only the former. Choose, for every
point $b \in \mathrm{bs}\,\xi$, a chart $\psi_b \in \mathrm{Atl}_b\,\mathrm{bs}\,\xi$ such that

$$\psi_b(\mathrm{supp}\,\psi_b, b) = (\mathbb{R}^q, 0) \quad \text{or} \quad (\mathbb{R}^q_-, 0) \quad [q = \dim \mathrm{bs}\,\xi]$$

and $\xi\big|_{\mathrm{supp}\,\psi_b}$ is $C^r\text{-}GL\mathbb{R}^n$-trivial, and then fix a $C^r\text{-}GL\mathbb{R}^n$-triviali-
zation

$$\tau_b: (\mathrm{supp}\,\psi_b \times \mathbb{R}^n, \mathrm{pr}_1, \mathrm{supp}\,\psi_b) \to \xi\big|_{\mathrm{supp}\,\psi_b}\ .$$

Now cover $\mathrm{bs}\,\xi$ by a finite number of sets $\psi_b^{-1}(D^q)$, say $\psi_{b_1}^{-1}(D^q), \ldots,$
$\psi_{b_s}^{-1}(D^q)$, and pick a C^r-function $\alpha: \mathbb{R}^q \to \mathbb{R}$ which equals 1 on D^q

356

and 0 outside $2D^q$. Finally, define $H_1,\ldots,H_s\colon \text{tl}\,\xi \to \mathbb{R}^n$ by

$$H_i(x) = \begin{cases} \alpha(\psi_{b_i}(\text{pr}\,\xi(x)))\,\text{pr}_2\circ\text{tl}\,\tau_{b_i}^{-1}(x), & \text{if } x \in \text{pr}\,\xi^{-1}(\text{supp}\,\psi_{b_i}), \\[2ex] 0, & \text{if } x \notin \text{pr}\,\xi^{-1}(\text{supp}\,\psi_{b_i}), \end{cases}$$

where $\text{pr}_2 = [\text{pr}_2\colon \text{supp}\,\psi_{b_i} \times \mathbb{R}^n \to \mathbb{R}^n]$, and let $H\colon \text{tl}\,\xi \to \mathbb{R}^n\times\ldots\times\mathbb{R}^n = \mathbb{R}^{sn}$ be given by $H(x) = (H_1(x),\ldots,H_s(x))$. Clearly, H is a C^r-map and its restrictions $H\big|_{\text{pr}\,\xi^{-1}(b)}$ are linear monomorphisms for all $b \in \text{bs}\,\xi$. Set $m = sn$ and $f\colon \text{bs}\,\xi \to G(m,n)$, $f(b) = H(\text{pr}\,\xi^{-1}(b))$. To verify that the bundles ξ and $f^!\text{Gra}(m,GL(n,\mathbb{R}))$ are C^r-$GL\mathbb{R}^n$-equivalent, it is enough to produce a C^r-$GL\mathbb{R}^n$-map $\phi\colon \xi \to \text{Gra}(m,GL(n,\mathbb{R}))$ with $\text{bs}\,\phi = f$. Such a ϕ is defined by $\text{tl}\,\phi\colon \text{tl}\,\xi \to \text{tl}\,\text{Gra}(m,GL(n,\mathbb{R}))$, $\text{tl}\,\phi(x) = (f(\text{pr}\,\xi(x)),H(x))$ (recall that $\text{tl}\,\text{Gra}(m,GL(n,\mathbb{R})) = \{(\gamma,y) \in G(m,n) \times \mathbb{R}^n \mid y \in \gamma\}$).

An Application

6. <u>If</u> $1 \leqslant r \leqslant \infty$, <u>then every real</u> (<u>complex</u>) <u>vector</u> C^r-<u>bundle</u> <u>with compact base has a Euclidean</u> (<u>respectively, Hermitian</u>) C^r-<u>metric.</u>

Since $\text{Gra}(m,GL(n,\mathbb{R}))$ $(\text{Gra}(m,GL(n,\mathbb{C})))$ has a Euclidean (respectively, Hermitian) C^a-metric, 6 is a consequence of 5.

Smoothings and Approximations

7. Given two real or complex vector $C^{\geqslant r}$-bundles, ξ and ξ', and a C^s-map $f\colon \text{bs}\,\xi' \to \text{bs}\,\xi$ with $r \geqslant s \geqslant 0$, we let $L^s(\xi,\xi';f)$ denote, in Theorems 8 and 10 below, the set of all linear C^s-maps $\phi\colon \xi' \to \xi$ such that $\text{bs}\,\phi = f$. If $s \neq a$, then $L^s(\xi,\xi';f)$ inherits a natural topology as a subset of $C^s(\xi',\xi)$.

When $\dim\xi = \dim\xi' = n$, $\text{bs}\,\xi = \text{bs}\,\xi'$, and $f = \text{id}$, $L^s(\xi,\xi';f)$ contains the set of all C^s-$GL\mathbb{R}^n$-equivalences $\xi' \to \xi$ in the real case, and the set of all C^s-$GL\mathbb{C}^n$-equivalences $\xi' \to \xi$ in the complex case. In both cases this subset is open for any $s \neq a$.

Notice that among the spaces $L^s(\xi,\xi';f)$ we find $\text{Sec}^s\xi$, $0 \leqslant s \leqslant r$ (see 2.1). More precisely, $\text{Sec}^s\xi$ is canonically homeomorphic to $L^s(\xi,\xi';f)$, where ξ' is the standard trivial bundle $(\text{bs}\,\xi \times \mathbb{R},\text{pr}_1,\text{bs}\,\xi)$, and $f = \text{id}\,\text{bs}\,\xi$; the canonical homeomorphism

$L^s(\xi,\xi';f) \to Sec^s\xi$ takes each map $\phi: \xi' \to \xi$ into the section $b \mapsto tl\,\phi(b,1)$.

8. <u>Let ξ and ξ' be real or complex vector $C^{\geqslant r}$-bundles with compact bases. If $0 \leqslant s < r \leqslant \infty$, then $L^r(\xi,\xi';f)$ is dense in $L^s(\xi,\xi';f)$ for any C^r-map $f: bs\,\xi' \to bs\,\xi$.</u>

Since the proofs of the real and complex cases differ only in some obvious details, we shall again prove only the former. By Theorem 5, we may assume that $\xi = g^! Gra(m,GL(n,\mathbb{R}))$ and $\xi' = g'^! Gra(m',GL(n',\mathbb{R}))$, where $g: bs\,\xi \to G(m,n)$ and $g': bs\,\xi' \to G(m',n')$ are some C^r-maps. Then we can identify $tl\,\xi$ with the C^r-submanifold $\{(b,y) \in bs\,\xi \times \mathbb{R}^m | y = g(b)\}$ of $bs\,\xi \times \mathbb{R}^n$ and, similarly, $tl\,\xi' = \{(b',y') \in bs\,\xi' \times \mathbb{R}^{m'} | y' = g(b')\}$. The orthogonal projections of $\mathbb{R}^m$ onto its subspaces $g(b)$ with $b \in bs\,\xi$ combine to define a C^r-map $p: bs\,\xi \times \mathbb{R}^m \to tl\,\xi$, and a C^r-map $p': bs\,\xi' \times \mathbb{R}^{m'} \to \xi'$ is similarly defined. Let A be the Euclidean space of all linear maps $\mathbb{R}^{m'} \to \mathbb{R}^m$ (which is the same as the space of all real $(m\times m')$-matrices; cf. 3.2.1.1. or 3.2.1.7), and consider the mappings

$$\Phi: L^s(\xi,\xi';f) \to C^s(bs\,\xi',A)$$

and

$$\Psi: C^s(bs\,\xi',A) \to L^s(\xi,\xi';f),$$

given by

$$\{[\Phi(\phi)](b')\}(y') = [pr_2: bs\,\xi \times \mathbb{R}^m \to \mathbb{R}^m](tl\,\phi \circ p'(b',y'))$$

and

$$[tl(\Psi(h))](b',y') = p(f(b'),[h(b')](y'))$$

(where $\phi \in L^s(\xi,\xi';f)$, $b' \in bs\,\xi'$, $y' \in \mathbb{R}^{m'}$, and $h \in C^s(bs\,\xi',A)$) . Clearly, Ψ is continuous (and so is Φ) and takes C^r-maps into C^r-maps. Moreover, $\Psi \circ \Phi = id\,L^s(\xi,\xi';f)$, and hence Ψ is surjective. Since $C^r(bs\,\xi',A)$ is dense in $C^s(bs\,\xi',A)$ (see 3.4.6.5), we conclude that $L^r(\xi,\xi';f)$ is dense in $L^s(\xi,\xi';f)$. [Explanation: Theorem 3.4.6.5 is applied after we have completed the space A to a sphere by adding a point; cf. 3.4.4.2 and 3.4.4.6.]

9. <u>Let ξ be a real or complex vector $C^{\geqslant r}$-bundle with compact base. If $0 \leqslant s < r \leqslant \infty$, then $Sec^r\xi$ is dense in $Sec^s\xi$.</u>

This is a consequence of 8: $Sec^r\xi = L^r(\xi,\xi';id\,bs\,\xi)$ and $Sec^s\xi = L^s(\xi,\xi';id\,bs\,\xi)$, where $\xi' = (bs\,\xi \times \mathbb{R},pr_1,bs\,\xi)$; see 7.

10. <u>Let ξ and ξ' be n-dimensional real (complex) vector $C^{\geqslant r}$-bundles with common compact base. If $0 \leqslant s < r \leqslant \infty$, then the set</u>

of C^r-GL$\mathbb{R}^n$-equivalences (respectively, C^r-GL$\mathbb{C}^n$-equivalences) $\xi' \to \xi$ is dense in the subset of $L^s(\xi,\xi';id)$ consisting of all C^s-GL$\mathbb{R}^n$-equivalences (respectively, C^s-GL$\mathbb{C}^n$-equivalences).

Since the last subset is open in $L^s(\xi,\xi';id)$, 10 is a consequence of 8.

11 (COROLLARY). If two n-dimensional real (complex) vector $C^{\geq r}$-bundles with compact base are GL$\mathbb{R}^n$-equivalent (respectively, GL$\mathbb{C}^n$-equivalent) and $r \neq a$, then they are C^r-GL$\mathbb{R}^n$-equivalent (respectively, C^r-GL$\mathbb{C}^n$-equivalent).

12. If the base of the n-dimensional real (complex) vector bundle ξ is a compact $C^{\geq r}$-manifold with $1 \leq r \leq \infty$, then ξ is GL$\mathbb{R}^n$-equivalent (GL$\mathbb{C}^n$-equivalent) to a real (complex) vector $C^{\geq r}$-bundle. If the base of the n-dimensional real (complex) vector C^s-bundle ξ is a compact $C^{\geq r}$-manifold, where $1 \leq s < r \leq \infty$, then ξ is C^s-GL$\mathbb{R}^n$-equivalent (C^s-GL$\mathbb{C}^n$-equivalent) to a real (complex) vector $C^{\geq r}$-bundle.

We shall prove again only the real case. Let ξ be an n-dimensional real vector bundle with bs ξ a compact $C^{\geq r}$-manifold. By 5.3.8 and 3.5.2.13, ξ is GL$\mathbb{R}^n$-equivalent to $f^!$Gra$(m,GL(n,\mathbb{R}))$, where m is large enough and f is some continuous map bs $\xi \to G(m,n)$. If ξ is an n-dimensional real vector C^s-bundle such that bs ξ is a compact $C^{\geq r}$-manifold, then by Theorem 5 ξ is C^s-GL$\mathbb{R}^n$-equivalent to $f^!$Gra$(m,GL(n,\mathbb{R}))$, where m is large enough and f is some C^s-map bs $\xi \to G(m,n)$. In both cases f is homotopic to a $C^{\geq r}$-map $g\colon$ bs $\xi \to G(m,n)$ (see 3.4.6.5, 1.3.6.6, and 3.4.5.10), so that ξ is GL$\mathbb{R}^n$-equivalent to $g^!$Gra$(m,GL(n,\mathbb{R}))$ (see 4.1.5). This completes the proof of the first claim; as for the second, we need only add that, by 11, ξ is C^s-GL$\mathbb{R}^n$-equivalent to $g^!$Gra$(m,GL(n,\mathbb{R}))$.

Constructions

13. We conclude this subsection with a short review of the constructions described in § 5.

By definition, a C^s-subbundle of a (real or complex) vector C^r-bundle ξ is a subbundle of ξ in the sense of 5.2.2 or 5.2.3, whose total space is a C^s-submanifold of tl ξ. A C^s-subbundle is clearly a vector C^s-bundle. The C^s-subbundles of Euclidean or Hermitian C^r-bundles are similarly defined. The C^r-bundles of C^r-bundles will be simply referred to as subbundles.

According to 5.2.5, every subbundle η of a Euclidean or Hermitian bundle ξ has an orthogonal complement $\eta^\perp$, and it is clear that: $\eta^\perp$ is a C^s-subbundle of ξ together with $\cdot\eta$; the canonical equivalence $\eta^\perp \to \xi/\eta$ (see 5.2.6) turns ξ/η into a Euclidean or Hermitian C^s-bundle (and thus becomes a C^s-equivalence). We see that in the (real or complex) vector C^s-case, ξ/η becomes a vector C^s-bundle by introducing on ξ a (Euclidean or Hermitian) C^s-metric. Recall, however, that we have established the existence of such a metric only under the assumptions that the base is compact and $s \neq a$ (see 6).

Let ξ_1 and ξ_2 be real vector C^r-bundles with a common boundaryless base. Then the construction of $\xi_1 \oplus \xi_2$ (see 5.2.9) shows that this sum is again a real vector C^r-bundle. The difficulty occuring when the base has a boundary (i.e., the fact that the product $\xi_1 \times \xi_2$ is not defined as a C^r-bundle) can be circumvented with the aid of the formulas

$$\mathrm{tl}(\xi_1 \oplus \xi_2) = \mathrm{tl}((\mathrm{pr}\,\xi_1)^!\xi_2)$$

and

$$\mathrm{pr}(\xi_1 \oplus \xi_2) = \mathrm{pr}\,\xi_2 \circ \mathrm{pr}((\mathrm{pr}\,\xi_1)^!\xi_2).$$

If the conditions in 5.2.9. are satisfied, then these formulas are equivalent to the definition of $\xi_1 \oplus \xi_2$, and this remains valid under our present circumstances, provided that the base has no boundary; the same formulas are now taken as the definition of the sum when the boundary is present. One can repeat the argument for complex vector, Euclidean, and Hermitian C^r-bundles. In particular, we can define the suspension (see 5.2.10) of a C^r-bundle.

The C^r-variants of the other constructions described in §5 and their mutual relations are already evident. In particular, the conjugate of a complex vector (Hermitian) C^r-bundle is a complex (respectively, Hermitian) C^r-bundle; the realification (see 5.1.12) of a complex vector (Hermitian) C^r-bundle is a real vector (respectively, Euclidean) C^r-bundle; the complexification (see 5.2.11) of a real vector (Euclidean) C^r-bundle is a complex vector (respectively, Hermitian) C^r-bundle; and in the C^r-versions of Theorems 5.2.13 and 5.2.14, the equivalences conj and K become C^r-equivalences.

4. Tangent and Normal Bundles

1. The basic notions of tangent and normal bundles have actually already been introduced and used in Chapter 3. However, only

now, that we have acquired the ideea of a smooth vector bundle, can we
present the full-fledged definitions of tangent and normal bundles and
give them the general, correct treatment that they deserve.

Tangent Bundles

2. Recall that in Chapter 3 we defined, for an arbitrarily
given C^r-manifold with $r \geqslant 1$, the real vector spaces $\mathrm{Tang}_x X$ $(x \in X)$,
the C^{r-1}-manifold $\mathrm{Tang}\, X$, and the projection $\mathrm{pr}: \mathrm{Tang}\, X \to X$ (see
3.1.4.1 and 3.1.4.2). Comparing these objects with the general
definitions given in 5.1.2 and 3.2, we readily see that $(\mathrm{Tang}\, X, \mathrm{pr}, X)$
is a real vector bundle of dimension $\dim X$ and, for $r > 2$, a real
vector C^{r-1}-bundle of dimension $\dim X$, called the <u>tangent bundle of
the manifold</u> X, and is denoted by $\mathrm{tang}\, X$.

Similarly, confronting the definition of the differential
$df: \mathrm{Tang}\, X \to \mathrm{Tang}\, Y$ of a C^r-map $f: X \to Y$ (see 3.1.4.3) with the
general definitions given in 5.1.14 and 3.2, we conclude that (df, f)
is a linear C^{r-1}-map $\mathrm{tang}\, X \to \mathrm{tang}\, Y$. If f is a C^r-diffeomorphism
then (df, f) is a linear C^{r-1}-isomorphism.

3. The notion of vector field has been defined twice: once
for smooth manifolds (see 3.1.4.5), and once for vector, Euclidean, and
Hermitian bundles (see 5.1.15). Now it is plain that the second
definition generalizes the first one: a vector field on a smooth
manifold X is simply a vector field in its tangent bundle $\mathrm{tang}\, X$.
In particular, the parallelizability (C^s-parallelizability) of an
n-dimensional smooth manifold X is equivalent to the $\mathrm{GL}\mathbb{R}^n$-triviality
(respectively, C^s-$\mathrm{GL}\mathbb{R}^n$-triviality) of the bundle $\mathrm{tang}\, X$. Comparing
this with Theorem 3.11, we see that a parallelizable compact C^r-manifold
with $r \leqslant \infty$ is C^{r-1}-parallelizable.

A smooth manifold is <u>stably parallelizable</u> if its tangent
bundle is stably trivial. The discussion in 5.4.4 and Theorem 3.5.2.13
show that if a compact manifold is stably parallelizable, then the
stabilization occurs already at the first step, i.e., the bundle
$\mathrm{su\, tang}\, X$ is $\mathrm{GL}\mathbb{R}^{n+1}$-trivial for any stably parallelizable n-dimensional
manifold X.

4. Recall that, given a point x of the smooth manifold X,
each chart $\phi \in \mathrm{Atl}_x X$ defines a ϕ-basis for the tangent space $\mathrm{Tang}_x X$,
and the matrix of the transformation from the ϕ-basis to the ψ-basis
is just the Jacobi matrix of the map $\mathrm{loc}(\phi, \psi)\mathrm{id}$, computed at $\phi(x)$.

From this it follows that the values of any orientation of X (assumed on the charts of $\mathrm{Catl}\,X$) are correctly transferred to the bases of the spaces $\mathrm{Tang}_x X$, and in this manner an orientation is defined on the bundle $\mathrm{tang}\,X$. This procedure is clearly reversible, and hence there is a one-to-one correspondence between the orientations of the smooth manifold X and the orientations of its tangent bundle $\mathrm{tang}\,X$. In particular, X is orientable if and only if $\mathrm{tang}\,X$ is orientable, and $\mathrm{tang}\,X$ is a $GL_+\mathbb{R}^n$-bundle for any oriented smooth n-dimensional manifold X.

We already know that every parallelizable manifold is orientable (see 3.1.4.6 and 3). Now we can add that every stably parallelizable manifold is also orientable.

5. One attractive feature of tangent bundles is that they can naturally be induced from Grassman bundles. Namely, if $j: X \to \mathbb{R}^q$ is a C^r-immersion (for example, a C^r-embedding), then dj maps each tangent space $\mathrm{Tang}_x X$ onto an n-dimensional plane of $\mathbb{R}^q$ passing through 0. Thus a map $t: X \to G(q,n)$ is defined, and it is clear that $\mathrm{tang}\,X$ is nothing else but $t^!\mathrm{Gra}(q,GL(n,\mathbb{R}))$ (up to a correcting C^{r-1}-$GL\mathbb{R}^n$-equivalence). [In fact, this obvious observation is older that the theory of bundles and was one of the factors which stimulated its creation.] If X is orientable, then one can generalize this observation and replace $G(q,n)$, the standard fiber $GL\mathbb{R}^n$, and the bundle $\mathrm{Gra}(q,GL(n,\mathbb{R}))$, by $G_+(q,n)$, $GL_+\mathbb{R}^n$, and $\mathrm{Gra}(q,GL_+(n,\mathbb{R}))$, respectively. In all cases t is known as a _tangential map_. If j is an embedding and no orientation is involved, then t coincides with the composition of the normal transversalization $X \to G(q,q-n)$ of j with the canonical diffeomorphism $G(q,q-n) \to G(q,n)$ (see 3.2.2.3).

6. A smooth manifold whose tangent bundle is equipped with a Euclidean metric is called a _Riemannian manifold._ In this case the metric is usually termed a _Riemannian metric._ If X is of class C^r, then the metric can be at most of class C^{r-1}. In fact, Theorem 3.6 shows that there is a Riemannian C^{r-1}-metric on every compact C^r-manifold with $1 \leqslant r \leqslant \infty$. Incidentally, the same result may be extracted from Theorem 3.4.2.1.

It is customary to look upon the tangent bundle of a Riemannian manifold as a Euclidean bundle. If the metric is of class C^s, $s \geqslant 1$, then this bundle is a Euclidean C^s-bundle.

Normal Bundles

7. The initial data involved in the concept of <u>normal bundle</u> are two smooth manifolds, X and X', and an immersion $j: X \to X'$. The most important case is that of an embedding j. In the definitions below $n = \dim X$ and $n' = \dim X'$.

Let us examine first the simplest case: X is a submanifold of the Riemannian manifold X', and j is the inclusion $X \to X'$. One may naturally view $\operatorname{tang} X$ as a subbundle of $\operatorname{tang} X'|_X$. By 5.2.5, $\operatorname{tang} X$ has in $\operatorname{tang} X'|_X$ an orthogonal complement, $(\operatorname{tang} X)^\perp$. This is an $(n'-n)$-dimensional Euclidean bundle with base X, called the <u>normal bundle</u> of X, and denoted $\operatorname{norm} X$. If X and X' are of class $C^{\geq r}$ with $r \geq 2$, and the Riemannian metric is of class C^{r-1}, then $\operatorname{norm} X$ is a Euclidean $C^{\geq r-1}$-bundle; see 3.13. In all cases, the sum $\operatorname{tang} X \oplus \operatorname{norm} X$ is canonically $O\mathbb{R}^{n'}$-equivalent to $\operatorname{tang} X'|_X$; when X and X' are of class $C^{\geq r}$ with $r \geq 2$, this equivalence is of class C^{r-1}.

To define the normal bundle for an arbitrary immersion $j: X \to X'$ and without recourse to a Riemannian metric, we have to replace the restriction $\operatorname{tang} X'|_X$ by the induced bundle $j^! \operatorname{tang} X'$ and then pass to a quotient instead of taking an orthogonal complement. More exactly, the normal bundle, $\operatorname{norm} j$, of the immersion $j: X \to X'$, with X and X' smooth manifolds, is defined as

$$\operatorname{norm} j = j^! \operatorname{tang} X'/\operatorname{Im} \operatorname{corr}(dj, j).$$

This formula defines an $(n'-n)$-dimensional real vector bundle over X. The sum $\operatorname{tang} X \oplus \operatorname{norm} X$ is canonically $GL\mathbb{R}^{n'}$-equivalent to the induced bundle $j^! \operatorname{tang} X'$. Unfortunately, there is less to say about the differentiability class of $\operatorname{norm} j$ and of the canonical equivalence $j^! \operatorname{tang} X' \to \operatorname{tang} X \oplus \operatorname{norm} j$: if X' is either compact or diffeomorphic to an open subset of a compact manifold, and j is C^r with $2 \leq r \leq \infty$, then $\operatorname{norm} j$ is a real vector $C^{\geq r-1}$-bundle, while the above canonical equivalence is C^{r-1}. These limitations are obviously due to the fact that we have had no theorems which guarantee the existence of a Riemannian metric in the noncompact and analytic situations.

If j is an inclusion, we may simply write $\operatorname{norm} X$ instead of $\operatorname{norm} j$.

Notice that, as the definition of the normal bundle $\operatorname{norm} j$ shows, the orientability of two of the three bundles $j^! \operatorname{tang} X'$, $\operatorname{tang} X$, and $\operatorname{norm} j$ implies the orientability of the third, and given orientations on two of them canonically define the orientation of the third; see 5.2.7. Comparing this with the discussion in 4, we see that if X' is orientable, then the orientability of $\operatorname{norm} j$ is equivalent to the orientability of the manifold X; moreover, if X' is oriented, then the orientations of X' and of $\operatorname{norm} j$ canonically determine each other.

8. The introduction of tangent and normal bundles allows us to better formulate and essentially complete the main result of Subsection 3.4.8, i.e., Theorem 3.4.8.2.

We may sharpen the formulation of the third part of this theorem, which asserts that the linear homomorphisms $\operatorname{fact} d_{x_1} f_1$ are actually isomorphisms. Now we can say that the maps $\operatorname{fact} d_{x_1} f_1$ combine with $\operatorname{ab} f_1 : X_{12} \to X_2$ to define a linear isomorphism from the normal bundle of the manifold X_{12} (taken in X_1) onto the normal bundle of the manifold X_2 (taken in X'). When X' is compact and $2 \leqslant r \leqslant \infty$, this is a linear C^{r-1}-isomorphism $\operatorname{norm} X_{12} \to \operatorname{norm} X_2$.

To complete Theorem 3.4.8.2, we consider orientations. Namely, suppose that X_1, X', and X_2 are orientable (oriented). Then, as an immediate consequence of 4 and 7, the manifold X_{12} is orientable (respectively, canonically oriented). In particular, under the assumptions of 3.4.8.3, the orientability of X', X_1, and X_2 implies the orientability of $X_1 \cap X_2$, and given orientations of X', X_1, and X_2 canonically orient $X_1 \cap X_2$.

9. <u>An n-dimensional smooth compact manifold is stably parallelizable if and only if it admits a differentiable embedding in some $\mathbb{R}^q$ having a $\operatorname{GL}\mathbb{R}^{q-n}$-trivial normal bundle.</u>

Let $j: X \to \mathbb{R}^q$ be an embedding enjoying the above property. Then $\operatorname{tang} X \oplus \operatorname{norm} j$ is $\operatorname{GL}\mathbb{R}^q$-equivalent to $j^! \operatorname{tang} \mathbb{R}^q$, and hence the condition is sufficient. To prove its necessity, consider an arbitrary differentiable embedding $j: X \to \mathbb{R}^q$ and the composite embedding $X \xrightarrow{\ j\ } \mathbb{R}^q \xrightarrow{\ \operatorname{in}\ } \mathbb{R}^{q+n+k}$, where k is large enough so that the suspension $\operatorname{su}^k \operatorname{tang} X$ is $\operatorname{GL}\mathbb{R}^{n+k}$-trivial (actually, it suffices to take $k = 1$; see 3). The normal bundle of this composite embedding is just $\operatorname{su}^{n+k} \operatorname{norm} j$, and it is $\operatorname{GL}\mathbb{R}^{q+k}$-trivial, being $\operatorname{GL}\mathbb{R}^{q+k}$-equivalent to the bundle

$$\operatorname{norm} j \oplus \operatorname{su}^k \operatorname{tang} X = \operatorname{su}^k (\operatorname{norm} j \oplus \operatorname{tang} X) = \operatorname{su}^k (j^! \operatorname{tang} \mathbb{R}^q).$$

The Complex Case

10. The basic definitions of this subsection, i.e., those of
the tangent bundle, $\operatorname{tang} X$, Riemannian metric, and normal bundles,
$\operatorname{norm} X$ and $\operatorname{norm} j$, carry over, word-for-word, to complex manifolds.
The bundles become complex bundles, the Riemannian metric is replaced
by a Hermitian one, while j is assumed to be a holomorphic immersion
(i.e., to be locally a holomorphic embedding). If $f: X \to X'$ is holo-
morphic, then (df,f) is a linear map $\operatorname{tang} X \to \operatorname{tang} X'$. The definition
and properties of tangential maps are preserved, but they become less
universal (see 3.1.6.10). Finally, the realification of a complex
manifold (see 3.1.6.9) leads to the realification of its tangent bundle
(see 5.1.12 and 5.13), and turns Hermitian metrics into Riemannian ones.

It is impossible not to notice the eclectic character of these
definitions. The reason for this inconsistency is that, whereas the
notion of differentiable structure, which lies at the heart of the
theory of smooth vector bundles, is specifically a real notion even when
we pass to complex vector bundles, in the complex case the tangent and
normal bundles carry an additional, complex-differentiable structure -
the so-called <u>holomorphic structure.</u> Unfortunately, the theory of
holomorphic vector bundles is beyond the scope of this book.

5. <u>Degree</u>

1. In this subsection we shall apply some of the simplest
results of differential topology to homotopy theory. Namely, given any
oriented, compact, smooth manifold X, any oriented, compact, connected,
smooth manifold Y with $\dim Y = \dim X$, and any continuous map
$f: (X,\partial X) \to (Y,\partial Y)$, we define an integer which depends only upon the
homotopy class of f. This number is called the <u>degree</u> of the map f
and is denoted by $\deg f$.

Although the degree $\deg f$ is a global characteristic of f,
and is actually defined for maps which are merely continuous, we shall
approach this notion by infinitesimal methods: we start by assuming that
$f \in C(X,\partial X;Y,\partial Y) \cap C^1(X,Y)$ and choose a point $y \in \operatorname{int} Y$ such that
f is transverse to y. Consider $f^{-1}(y)$. It consists of a finite
number of points, each of them having a neighborhood which is mapped
diffeomorphically by f onto a neighborhood of y (see 3.4.8.2 and
3.1.5.5), and each of these diffeomorphisms is either orientation

preserving or orientation reversing, where the neighborhoods are oriented
in agreement with the orientations of X and Y (see 4.8). The <u>degree
of the map</u> f <u>at the point</u> y, denoted $\deg_y f$, is the number of the
points of $f^{-1}(y)$ where the orientation is preserved, minus the number
of points of $f^{-1}(y)$ where the orientation is reversed. A popular
shorter version of this definition is: $\deg_y f$ <u>is the algebraic number
of the preimages of the point</u> y.

In this definition of $\deg_y f$, the assumption that $y \in \operatorname{int} Y$
is essential. However, one may repeat the definition for a boundary
point y, provided that $f \in C^1_\partial(X,Y) \cap C(X, \operatorname{int} X; Y, \operatorname{int} Y)$ and
$\operatorname{ab} f : \partial X \to \partial Y$ is transverse to y. The degree $\deg_y f$ thus defined
is obviously the same as $\deg_y [\operatorname{ab} f : f^{-1}(Z) \to Z]$, where Z is the
component of ∂Y containing y.

We add that if the degree is defined for a point y [i.e.,
either $y \in \operatorname{int} Y$, $f \in C(X, \partial X; Y, \partial Y) \cap C^1(X,Y)$, and f is transverse
to y, or $y \in \partial Y$, $f \in C^1_\partial(X,Y) \cap C(X, \operatorname{int} X; Y, \operatorname{int} Y)$, and $\operatorname{ab} f : \partial X \to \partial Y$
is transverse to y], then it is defined for any point y' in a
neighborhood of y (in Y), and $\deg_{y'} f = \deg_y f$.

2 (LEMMA). <u>Let</u> X <u>and</u> Y <u>be oriented, compact,</u> C^∞<u>-manifolds,</u>
<u>with</u> $\dim X = \dim Y$ <u>and</u> Y <u>connected, and let</u>

$$g, h \in C^\infty_\partial(X,Y) \cap C(X, \operatorname{int} X; Y, \operatorname{int} Y).$$

<u>Further, let</u> $y, z \in Y$ <u>be such that</u> $\deg_y g$ <u>and</u> $\deg_z h$ <u>are defined</u>
(<u>see</u> 1). <u>If the maps</u>

$$\operatorname{rel} g, \operatorname{rel} h : (X, \partial X) \to (Y, \partial Y)$$

<u>are homotopic, then</u> $\deg_z h = \deg_y g$.

PROOF. We disregard the trivial case $\dim Y = 0$ and assume
for a start that $h = g$. Using 1, we find neighborhoods U and V of
y and z, such that $\deg_{y'} g = \deg_y g$ and $\deg_{z'} h = \deg_z h$ for all
$y' \in U$ and $z' \in V$. It is clear that one can join y to z by a
path which is a C^∞-embedding $I \to Y$. By Theorems 3.4.1.4 and 3.4.7.7,
there is a path $s : I \to \operatorname{int} Y$ such that $s(0) \in U$, $s(1) \in V$, and
s is a C^∞-embedding transverse to g. The preimage $g^{-1}(s(I))$ is an
oriented, compact, one-dimensional C^∞-submanifold of X, and
$\partial g^{-1}(s(I)) = g^{-1}(s(0)) \cup g^{-1}(s(1))$ (see 3.4.8.2 and 4.8). Obviously,

$$\deg_{s(0)} g = \deg_{s(0)} \operatorname{ab} g, \quad \deg_{s(1)} g = \deg_{s(1)} \operatorname{ab} g,$$

where $\operatorname{ab} g = [\operatorname{ab} g : g^{-1}(s(I)) \to s(I)]$. But the contribution of a point from $g^{-1}(s(1))$ to $\deg_{s(1)} ab\, g$ equals the value of the orientation which this point inherits as a component of $\partial g^{-1}(s(I))$. Similarly, the contribution of a point from $g^{-1}(s(0))$ to $\deg_{s(0)} ab\, g$ is opposite to the value of the orientation which this point inherits as a component of $\partial g^{-1}(s(I))$. Consequently, $\deg_{s(1)} g - \deg_{s(0)} g$ is the number of points of $\partial g^{-1}(s(I))$ which inherit the orientation $+1$ from $g^{-1}(s(I))$, minus the number of points of $\partial g^{-1}(s(I))$ which inherit the orientation -1 from $g^{-1}(s(I))$. However, this difference must be 0, because every component of $g^{-1}(s(I))$ is diffeomorphic to either S^1 or D^1 (see 3.5.3.1).

Now drop the assumption that $h = g$, and suppose that Y is closed. Then X is also closed, and from the fact that g and h are homotopic it follows that there exists a C^∞-homotopy F from g to h (see 3.4.6.6). The map $\Phi: X \times I \to Y \times I$, $\Phi(x,t) = (F(x,t),t)$, obviously belongs to the intersection

$$C^\infty_\partial(X \times I, Y \times I) \cap C(X \times I, \mathrm{int}(X \times I); Y \times I, \mathrm{int}(Y \times I)).$$

Furthermore, if we identify $X \times 0$ and $X \times 1$ with X, and $Y \times 0$ and $Y \times 1$ with Y, then $ab\,\Phi: X \times 0 \to Y \times 0$ and $ab\,\Phi: X \times 1 \to Y \times 1$ become g and h, an we can write $\deg_{(y,0)} \Phi = \deg_y g$ and $\deg_{(z,1)} \Phi = \deg_z h$. By the argument above, $\deg_{(z,1)} \Phi = \deg_{(y,0)} \Phi$, and thus $\deg_z h = \deg_y g$.

Finally, if $\partial Y \neq \emptyset$, we find a component Z of ∂Y and points $y',z' \in Z$ such that $ab\,g: \partial X \to \partial Y$ is transverse to y', while $ab\,h: \partial X \to \partial Y$ is transverse to z'. We already know that

$$\deg_y g = \deg_{y'} g \quad \text{and} \quad \deg_z h = \deg_{z'} h,$$

while, according to 1,

$$\deg_{y'} g = \deg_{y'}[ab\,g : g^{-1}(Z) \to Z]$$

and

$$\deg_{z'} h = \deg_{z'}[ab\,h : h^{-1}(Z) \to Z].$$

But

$$\deg_{y'}[ab\,g : g^{-1}(Z) \to Z] = \deg_{z'}[ab\,h : h^{-1}(Z) \to Z]. \tag{1}$$

Indeed, since $rel\,g$ and $rel\,h$ are homotopic, $h^{-1}(Z) = g^{-1}(Z)$, and thus (1) follows from that part of the lemma which we have already proved.

3 (LEMMA). <u>For any compact C^∞-manifolds</u> X <u>and</u> Y, <u>the set</u>
$C_\partial^\infty(X,Y) \cap C(X,\text{int } X;Y,\text{int } Y)$ <u>is dense in</u> $C(X,\partial X;Y,\partial Y)$.

PROOF. Construct the doubles, dopp X and dopp Y, together
with two-sided collarings $k: \partial X \times D^1 \to$ dopp X and $1: \partial Y \times D^1 \to$ dopp Y.
Pick a C^∞-embedding $j:$ dopp $Y \to \mathbb{R}^q$, a C^∞-transversalization τ of
$j|_{\partial Y}$, and a neat tube $\text{Tub}_\tau \rho$. Then it suffices, given a map
$f \in C(X,\partial X;Y,\partial Y)$ and $\varepsilon > 0$, to find $g \in C_\partial^\infty(X,Y) \cap C(X,\text{int } X;Y,\text{int } Y)$
such that $\text{dist}(j(f(x)),j(g(x))) < \varepsilon$ for all $x \in X$. To produce such
a g, we shall construct successively three auxiliary maps,
$h_1,h_2,h_3:$ dopp X $\to$ dopp Y.

The map h_1 is very simply defined by

$$h_1(x) = f(x), \quad h_1(\text{cop}(x)) = \text{cop}(f(x)) \quad [x \in X].$$

To construct h_2, fix δ, $0 < \delta < 1$, such that:

(i) $\text{dist}(j \circ 1(z,t),j \circ 1(z,t')) < \varepsilon/4$ for $|t-t'| < \delta$ $[z \in \partial Y,$
$t,t' \in D^1]$;

(ii) $\text{dist}(j(h_1(k(z,t))),j(h_1(k(z,t')))) < \varepsilon/4$ for
for $|t-t'| < \delta$ $[z \in \partial Y, \ t,t' \in D^1]$;

(iii) for any $z \in \partial Y$, the ball with center $j(z)$ and
radius δ lies in $\text{Tub}_\tau \rho$, while its image unde the map
$j \circ \text{pr}_\tau: \text{Tub}_\tau \rho \to \mathbb{R}^q$ lies in the ball with center $j(z)$ and radius
$\varepsilon/4$. [The existence of such a δ is a consequence of the continuity
of j, k, 1, h_1, and pr_τ.] Further, pick a C^∞-map $\phi: \partial X \to \partial Y$ such
that $\text{dist}(j(\phi(z)),j \circ f(z)) < \delta$ for all $z \in \partial X$ (3.4.4.2 guarantees
that such a map ϕ exists), and define, for each $t \in I$, the map
$\phi_t: \partial X \to \partial Y$ by $\phi_t(x) = \text{pr}_\tau(tj(\phi(x)) + (1-t)j \circ f(x))$. Next, pick a
C^∞-map $\alpha: \mathbb{R} \to \mathbb{R}$ such that $\alpha(t) = 0$ for $|t| \leqslant 1/3$, and $\alpha(t) = 1$
for $|t| \geqslant 2/3$, and define $k_1: X \to X$ and $1_1: Y \to Y$ by

$$\begin{cases} k_1(k(z,t)) = k(z,(1-\delta)t + \delta), & \text{if } z \in \partial X, \ t \in I, \\ \\ k_1(x) = x, & \text{if } x \in X \smallsetminus k(\partial X \times I), \end{cases}$$

and

$$\begin{cases} 1_1(1(z,t)) = 1(z,(1-\delta)t + \delta), & \text{if } z \in \partial Y, \ t \in I, \\ \\ 1_1(y) = y, & \text{if } y \in Y \smallsetminus 1(\partial Y \times I). \end{cases}$$

It is clear that k_1 and 1_1 are topological embeddings.
Now define h_2 by

$$\begin{cases} h_2(k(z,t)) = 1(\phi_{\alpha(\delta t)}(z),t), & \text{if } z \in \partial X, \ |t| < \delta, \\ h_2(x) = 1_1(f(k_1^{-1}(x))), & \\ h_2(\text{cop}(x)) = \text{cop}(1_1(f(k_1^{-1}(x)))) & \end{cases} \Bigg\}, \quad \text{if } x \in X \smallsetminus k(\partial X \times [0,\delta)).$$

Then the following facts are evident:
$h_2 \in C(\text{dopp}\, X, \text{int}\, X, \partial X; \text{dopp}\, Y, \text{int}\, Y, \partial Y)$; $\text{dist}(j \circ h_1(x), j \circ h_2(x)) < \varepsilon/2$
for all $x \in \text{dopp}\, X$; the restriction of h_2 to $k(\partial X \times [-\delta/3, \delta/3])$
is of class C^∞; and if $z \in \partial X$, then

$$\text{Im}\, d_z(h_2\big|_{k(\partial X \times [-\delta/3, \delta/3])}) \not\subset \text{Tang}_{h_2(z)}\, \partial Y.$$

Finally, take h_3 to be any C^∞-map $\text{dopp}\, X \to \text{dopp}\, Y$ which
equals h_2 on $k(\partial X \times [-\delta/6, \delta/6]$ and enjoys the properties:

$$h_3(X \smallsetminus k(\partial X \times [0, \delta/6)) \subset \text{int}\, Y\,;$$

$$h_3(\text{cop}(X \smallsetminus k(\partial X \times [0, \delta/6)))) \subset \text{cop}(\text{int}\, Y)\,;$$

and

$$\text{dist}(j \circ h_3(x), j \circ h_2(x)) < \varepsilon/2 \quad \text{for all } x \in X.$$

[Theorem 3.4.4.8 guarantees that such a h_3 exists.]

Clearly, $h_3(X) \subset Y$. Now define the desired map g by
$g = [\text{ab}\, h_3 : X \to Y]$, and check directly that it has all the necessary
properties.

4. Let X and Y be (as in 1) oriented, compact, smooth
manifolds with $\dim X = \dim Y$, and let $f \in C(X, \partial X; Y, \partial Y)$. Restrict
the differentiable structures of the manifolds X and Y to
C^∞-structures (see 3.4.9.8), and find a map $g \in C_\partial^\infty(X,Y) \cap$
$\cap\, C(X, \text{int}\, X; Y, \text{int}\, Y)$ which is close enough to f in the C^0-topology,
and such that the maps $f, g: X \to Y$, as well as the maps
$\text{ab}\, f, \text{ab}\, g : \partial X \to \partial Y$ are homotopic (see 3, 3.4.5.10, and 1.3.6.6.). Now
compute $\deg_y g$ at some point $y \in \text{int}\, Y$ such that g is transverse
to y. By Lemma 2, $\deg_y g$ does not depend upon the choice of g or
y, while 3.4.6.10 and 3.4.1.6. show that $\deg_y g$ does not depend upon
the modality of restricting the differentiable structures of X and Y
to C^∞-structures (according to the aforementioned theorems, the
C^∞-manifolds resulting from the restriction of the differentiable
structure of a given compact, smooth manifold, effected in two distinct
ways, are C^∞-diffeomorphic via a diffeomorphism which can be as C^0-close
to the identity diffeomorphism as we choose). We call $\deg_y g$ the
degree of the map f, denoted $\deg f$.

The main properties of the degree are immediate consequences of its definition and of Lemmas 2 and 3. We list here some of them:

(i) if $f, f': (X, \partial X) \to (Y, \partial Y)$ are homotopic, then $\deg f = \deg f'$ (by Lemma 2);

(ii) the degree of the composite map $(X, \partial X) \overset{f}{\to} (Y, \partial Y) \overset{g}{\to} (Z, \partial Z)$ is $\deg f \cdot \deg g$ (by the definition of $\deg$);

(iii) the degree of the identity map is 1 (trivial);

(iv) if $f: (X, \partial X) \to (Y, \partial Y)$ is a homotopy equivalence, then $\deg f = 1$ (proof: if g is a homotopy inverse of f, then $\deg f \cdot \deg g = \deg(g \circ f) = \deg \mathrm{id}(X, \partial X) = 1$);

(v) if $f: (X, \partial X) \to (Y, \partial Y)$ is such that $f(X) \neq Y$, then $\deg f = 0$ (indeed, one can approximate f as closely as desired by a map from $C_\partial^\infty(X, Y) \cap C(X, \mathrm{int}\, X; Y, \mathrm{int}\, Y)$ enjoying the same properties);

(vi) if Z is any component of Y, then

$$\deg[f: (X, \partial X) \to (Y, \partial Y)] = \deg[\mathrm{ab}\, f: f^{-1}(Z) \to Z]$$

(a result of the discussion in 1) and, in particular, $\deg f = 0$ whenever X is closed but Y is not.

As examples, consider the maps $f: (D^n, S^{n-1}) \to (D^n, S^{n-1})$ and $\mathrm{ab}\, f: S^{n-1} \to S^{n-1}$, defined by an orthogonal $(n \times n)$-matrix V ($n \geqslant 2$). Obviously, $\deg f = \deg \mathrm{ab}\, f = \det V$, i.e., $\deg f = \deg \mathrm{ab}\, f = 1$ if $V \in SO(n)$, and $\deg f = \deg \mathrm{ab}\, f = -1$ if $V \in O(n) \smallsetminus SO(n)$. Thus, the degree of the antipodal map $S^{n-1} \to S^{n-1}$, $x \mapsto -x$, equals 1 if n is even and -1 if n is odd.

The Nonoriented Case

5. The discussion in 1, 2, and 3 can be carried over to non-oriented manifolds if we replace integers by integers modulo 2. This enables us to define $\deg f \in \mathbb{Z}_2$ for any continuous map $f: (X, \partial X) \to (Y, \partial Y)$, where X and Y are smooth, compact manifolds, and Y is connected (no orientability needed). All the properties of the integral degree listed in 4 are preserved. For the case of oriented manifolds, when both degrees (the integral and mod 2) are defined, we continue to use the same notation for both, because misunderstandings are usually eliminated by the context.

Applications

6. <u>Smooth closed manifolds of positive dimension are not contractible.</u>

This is plain if the given manifold is not connected. In the connected case, the identity map of a closed manifold is 1, whereas the degree of any map which takes the whole manifold into one of its points is zero (here we use the $\mathbb{Z}_2$ -degree defined in 5).

7. <u>If</u> $n \neq m$, <u>then</u> S^n <u>and</u> S^m <u>are not homotopy equivalent.</u>

Indeed, if $m < n$, then every continuous map $S^m \to S^n$ is homotopic to a constant map (see 2.3.2.3 and 2.3.1.6), whereas id: $S^n \to S^n$ is not homotopic to a constant map (see 6).

8. <u>The boundary of a nonempty, compact, smooth manifold is not a retract of the manifold.</u>

It suffices to assume that the given manifold X is compact, smooth, connected, and with $\partial X \neq \emptyset$. Let $\rho: X \to \partial X$ be a retraction, and let Z be any component of ∂X. Consider the composite map $X \xrightarrow{\rho} X \xrightarrow{\text{in}} X$. Since its image is not all of X, its degree is 0 (see 4 and 5). On the other hand, this degree equals the degree of ab(in $\circ$ ρ): $Z \to Z$, which is 1, the last map being id Z .

9. <u>Every continuous map $D^n \to D^n$ has a fixed point.</u>

PROOF. Suppose that $f: D^n \to D^n$ is continuous and has no fixed points. Then the map $D^n \to S^{n-1}$ taking each point $x \in D^n$ into its projection on S^{n-1} from the point $f(x)$ is a retraction, and hence S^{n-1} is a retract of D^n: contradiction (see Theorem 8).

10. <u>If an m-dimensional locally Euclidean space is homeomorphic to an n-dimensional locally Euclidean space, then n = m.</u>
(Cf. 3.1.1.4).

PROOF. Every point of $\mathbb{R}^q$ can be covered (in $\mathbb{R}^q$) by a Euclidean q-simplex. Therefore, every point of a q-dimensional locally Euclidean space lies in the interior of a finitely-triangulated subset, and its link in this subset is homeomorphic to S^{q-1}. By Theorem 2.2.6.4, this link is a homotopy invariant, and 7 shows that the spheres S^n and S^m cannot have the same homotopy type unless m = n.

11. Theorem 10 clarifies not only the definition of a locally Euclidean space, but also that of a cellular space. Namely, it shows that the dimension of a cell is uniquely determined by this cell.

Therefore, the dimension function which we introduced into the definition of the cellular decomposition as an additional element of its structure, is actually redundant, being completely determined by the decomposition itself.

12. <u>The boundary of the half space</u> $\mathbb{R}^n_-$ <u>is</u> $\mathbb{R}^{n-1}_1$.

(Cf. 3.1.1.4.)

PROOF. It suffices to show that the point 0 has in $\mathbb{R}^n_-$ no neighborhood homeomorphic to $\mathbb{R}^n$;   see 3.1.1.4.

Assume that such a neighborhood exists. Then 0 is an interior point of a finitely-triangulated subset of this neighborhood, where its link is homeomorphic to S^{n-1} (cf. the proof of Theorem 10). On the other hand, 0 is an interior point of a finitely-triangulated subset, where its link is homeomorphic to D^{n-1}: take any Euclidean n-simplex which lies in $\mathbb{R}^n_-$ and contains 0 in the interior of one of its (n-1)-faces. Since S^{n-1} is not contractible, whereas D^{n-1} is, we contradict Theorem 2.2.6.4.

6. <u>Exercises</u>

1. Let $r \leqslant \infty$, and let $f: X \to Y$ be a C^r-submersion, where X and Y are C^r-manifolds, X compact and Y closed. Show that (X,f,Y) is a C^r-bundle. (Combined with Theorem 1.3, this result shows that for $r \leqslant \infty$, (X,f,Y) is a C^r-bundle whenever X is a compact C^r-manifold and Y is a C^r-manifold, while $f: X \to Y$ is a C^r-submersion.)

2. Let $1 \leqslant r \leqslant \infty$, and let ξ be a C^r-bundle with closed base. Show that there is a collaring $k: \partial tl\, \xi \times I \to tl\, \xi$ such that $k(z \times I) \subset pr\, \xi^{-1}(pr\, \xi(z))$ for every point $z \in \partial tl\, \xi$.

3. Let $1 \leqslant r \leqslant \infty$, and let ξ be a C^r-bundle with $\partial tl\, \xi = pr\, \xi^{-1}(\partial bs\, \xi)$. Show that there are collarings $k: \partial bs\, \xi \times I \to bs\, \xi$ and $l: \partial tl\, \xi \times I \to tl\, \xi$ such that the diagram

$$
\begin{array}{ccc}
\partial tl\, \xi \times I & \xrightarrow{\;\;l\;\;} & tl\, \xi \\
{\scriptstyle ab\, pr\, \xi \times id\, I} \downarrow & & \downarrow {\scriptstyle pr\, \xi} \\
\partial bs\, \xi \times I & \xrightarrow{\;\;k\;\;} & bs\, \xi
\end{array}
$$

commutes.

4. Show that if $r \leqslant \infty$, then for every $C^{\geqslant r}$-bundle ξ with compact bs ξ and tl ξ, $\mathrm{Sec}^r\xi$ is dense in $\mathrm{Sec}^s\xi$ for any $s < r$. (This generalizes Theorem 2.7 for $r \neq a$.)

5. Show that every $C^{\geqslant r}$-bundle ξ with compact bs ξ and tl ξ is C^r-isomorphic to a C^∞-bundle (cf. 2.8).

6. Show that su tang $\mathbb{R}P^n$ is C^a-$\mathrm{GL}\mathbb{R}^{n+1}$-equivalent to the sum of $n+1$ copies of $\mathrm{Gra}(n+1, \mathrm{GL}(1,\mathbb{R}))$, while su tang $\mathbb{C}P^n$ is C^a-$\mathrm{GL}\mathbb{C}^{n+1}$-equivalent to the sum of $n+1$ copies of $\mathrm{Gra}(n+1, \mathrm{GL}(1,\mathbb{C}))$.

7. Show that the normal bundle of the C^a-embedding $G(m,n) \to G(m+1,n)$, described in 3.2.2.3, is C^a-$\mathrm{GL}\mathbb{R}^n$-equivalent to $\mathrm{Gra}(m,\mathrm{GL}(n,\mathbb{R}))$, while the normal bundle of the C^a-embedding $\mathbb{C}G(m,n) \to \mathbb{C}G(m+1,n)$, described in 3.2.2.7, is C^a-$\mathrm{GL}\mathbb{C}^n$-equivalent to $\mathrm{Gra}(m,\mathrm{GL}(n,\mathbb{C}))$.

8. Let $p_1,\ldots,p_{n+1}$ be homogeneous complex polynomials of degree m in $n+1$ variables, whose only common zero is the point 0. Show that the map $\mathbb{C}P^n \to \mathbb{C}P^n$ given by

$$(z_1 : \ldots : z_{n+1}) \mapsto (p_1(z_1,\ldots,z_{n+1}) : \ldots : p_{n+1}(z_1,\ldots,z_{n+1}))$$

has degree m^n.

9. Show that for $n \geqslant 1$ every continuous map $S^n \to S^n$ whose degree is not $(-1)^{n+1}$ has a fixed point.

10. Show that for $n \geqslant 1$ every continuous map $S^n \to S^n$ having odd degree transforms some pair of antipodal points into another such pair.

11. Show that for odd $n > 1$ the degree of any map $S^n \to \mathbb{R}P^n$ is even.

12. Let f be a simplicial map of the standard 2-simplex onto the standard 1-simplex. Show that the symplicial mapping cylinder, $\mathrm{Scyl}\,f$, is not homeomorphic to $\mathrm{Cyl}\,f$.

Chapter 5. Homotopy Groups

§1. THE GENERAL THEORY

1. Absolute Homotopy Groups

1. Let (X,x_0) be a pointed space, and let $r \geq 0$ be an integer. To simplify the notation, let us agree to write $\mathrm{Sph}_r(X,x_0)$ for the set $C(I^r,\mathrm{Fr}\, I^r;X,x_0)$ of all continuous maps $(I^r,\mathrm{Fr}\, I^r) \to (X,x_0)$, and denote the set of homotopy classes of such maps (i.e., $\pi(I^r,\mathrm{Fr}\, I^r;X,x_0)$) by $\pi_r(X,x_0)$. The elements of $\mathrm{Sph}_r(X,x_0)$ will be referred to as <u>r-dimensional spheroids</u> (or simply <u>r-spheroids</u>) <u>of the space</u> X <u>with origin</u> x_0.

For $r > 0$ and two arbitrary spheroids $\phi,\psi \in \mathrm{Sph}_r(X,x_0)$, we define their <u>product</u>, $\phi\psi$, as the spheroid in $\mathrm{Sph}_r(X,x_0)$ given by

$$\phi\psi(t_1,t_2,\ldots,t_r) = \begin{cases} \phi(2t_1,t_2,\ldots,t_r), & \text{if } 0 \leq t_1 \leq 1/2, \\ \\ \psi(2t_1-1,t_2,\ldots,t_r), & \text{if } 1/2 \leq t_1 \leq 1. \end{cases} \qquad (1)$$

For $r > 0$ and $\phi \in \mathrm{Sph}_r(X,x_0)$, the spheroid ϕ^{-1}, called the <u>inverse</u> of ϕ, is defined by $\phi^{-1}(t_1,t_2,\ldots,t_r) = \phi(1-t_1,t_2,\ldots,t_r)$. Obviously, if $\phi,\phi_1,\psi,\psi_1 \in \mathrm{Sph}_r(X,x_0)$ are such that ϕ_1 is homotopic to ϕ and ψ_1 is homotopic to ψ, then the spheroids $\phi_1\psi_1$ and $\phi\psi$ are homotopic. Therefore, (1) defines a multiplication on $\pi_r(X,x_0)$. It turns out that <u>this multiplication is associative,</u> that <u>the homotopy class of the constant spheroid</u> const (<u>which takes</u> I^r <u>into</u> x_0) <u>is a two-sided identity element</u> and that <u>the homotopy classes of the spheroids</u> ϕ <u>and</u> ϕ^{-1} <u>are inverses of one another.</u>

The associativity of the multiplication means that the products $(\phi\psi)\chi$ and $\phi(\psi\chi)$ are homotopic for any spheroids $\phi,\psi,\chi \in \mathrm{Sph}_r(X,x_0)$. Indeed, the formula

$$((t_1,t_2,\ldots,t_r),t) \mapsto$$

$$\mapsto \begin{cases} \phi\left(\dfrac{4t_1}{1+t},t_2,\ldots,t_r\right), & \text{if } 0 \leqslant t_1 \leqslant \dfrac{1+t}{4}, \\[2ex] \psi(4t_1-t-1,t_2,\ldots,t_r), & \text{if } \dfrac{1+t}{4} \leqslant t_1 \leqslant \dfrac{2+t}{4}, \\[2ex] \chi\left(\dfrac{4t_1-t-2}{2-t},t_2,\ldots,t_r\right), & \text{if } \dfrac{2+t}{4} \leqslant t_1 \leqslant 1, \end{cases} \qquad (2)$$

defines a homotopy $I^r \times I \to X$ from $(\phi\psi)\chi$ to $\phi(\psi\chi)$.

To prove the second claim, we have to show that the products $\phi(\mathrm{const})$ and $(\mathrm{const})\phi$ are both homotopic to ϕ, for any $\phi \in \mathrm{Sph}_r(X,x_0)$. The formulas

$$((t_1,t_2,\ldots,t_r),t) \mapsto$$

$$\mapsto \begin{cases} \phi\left(\dfrac{2t_1}{1+t},t_2,\ldots,t_r\right), & \text{if } 0 \leqslant t_1 \leqslant \dfrac{1+t}{2}, \\[2ex] x_0, & \text{if } \dfrac{1+t}{2} \leqslant t_1 \leqslant 1, \end{cases} \qquad (3)$$

and

$$((t_1,t_2,\ldots,t_r),t) \mapsto$$

$$\mapsto \begin{cases} x_0, & \text{if } 0 \leqslant t_1 \leqslant \dfrac{1-t}{2}, \\[2ex] \phi\left(\dfrac{2t_1-1+t}{1+t},t_2,\ldots,t_r\right), & \text{if } \dfrac{1-t}{2} \leqslant t_1 \leqslant 1, \end{cases} \qquad (4)$$

define homotopies from $\phi(\mathrm{const})$ and $(\mathrm{const})\phi$ to ϕ.

Finally, the third claim is that the products $\phi\phi^{-1}$ and $\phi^{-1}\phi$ are both homotopic to const: indeed, a homotopy from $\phi\phi^{-1}$ to const is given by

$$((t_1,t_2,\ldots,t_r),t) \mapsto$$

$$\mapsto \begin{cases} \phi(2t_1,t_2,\ldots,t_r), & \text{if } 0 \leqslant t_1 \leqslant \dfrac{1-t}{2}, \\[2ex] \phi(1-t,t_2,\ldots,t_r), & \text{if } \dfrac{1-t}{2} \leqslant t_1 \leqslant \dfrac{1+t}{2}, \\[2ex] \phi(2-2t_1,t_2,\ldots,t_r), & \text{if } \dfrac{1+t}{2} \leqslant t_1 \leqslant 1. \end{cases} \qquad (5)$$

The set $\pi_r(X,x_0)$, $r > 0$, with this group structure is called the $\underline{r\text{-th homotopy group of the space}}$ X $\underline{\text{at the point}}$ x_0.

If $r > 0$, then each r-spheroid maps I^r into the component X_0 of X containing x_0. Consequently, for $r > 0$ the groups $\pi_r(X,x_0)$ and $\pi_r(X_0,x_0)$ are isomorphic.

By Theorem 2.3.4.3, for a countable cellular space X all the sets $\pi_r(X,x_0)$ are countable.

The Case $r = 0$

2. Since I^0 is a point and $\text{Fr } I^0 = \emptyset$, $\text{Sph}_0(X,x_0)$ and $\pi_0(X,x_0)$ can be identified with X and with the set $\text{comp } X$ of components of X, respectively. $\pi_0(X,x_0)$ has no natural group structure. However, it does have a distinguished element which, in analogy with the higher-dimensional case, will be referred to as an identity: this is the homotopy class of the 0-spheroid const, i.e., the component of X containing x_0.

To be able to use the same language for the cases $r > 0$ and $r = 0$, we shall call $\pi_0(X,x_0)$ the $\underline{\text{0-th homotopy group of}}$ X $\underline{\text{at}}$ x_0, and we shall apply the group-theoretic terminology to sets with distinguished elements and their maps. In particular, by a direct product we understand the usual product, a homomorphism is a map preserving distinguished elements, the kernel of a homomorphism is the preimage of the distinguished element, and an isomorphism is an invertible homomorphism.

The Case $r = 1$

3. One-dimensional spheroids are nothing else but closed paths, and the multiplication, inversion, and homotopy of spheroids, as defined in 1, coincide with the multiplication, inversion, and homotopy of paths, as defined in 1.3.2.1 and 1.3.2.3. The 1-st homotopy group is alternatively known as the $\underline{\text{fundamental group.}}$ It was defined some decades before the higher homotopy groups were introduced, and we shall see below that it holds a special position amongst the homotopy groups.

4. When $r = 1$ the homotopies (2)-(5) are defined not only for loops ϕ,ψ,χ with common origin: the only condition that the paths ϕ,ψ,χ must satisfy is that the products involved be meaningful. As in the case of loops, the homotopy class of a product is uniquely

determined by the homotopy classes of its factors, provided that the origin of the paths in the second class coincides with the end of the paths in the first class. This multiplication is associative; the class of the constant path is a left identity element for the class of paths with the same origin, and a right identity element for the class of paths with the same end; and the classes of the paths s and s^{-1} are inverses of one another, i.e., their product, taken in any order, is homotopic with the corresponding identity element.

The Case $r > 1$

5. <u>For $r > 1$ the group $\pi_r(X,x_0)$ is Abelian.</u>

We have to verify that the products $\phi\psi$ and $\psi\phi$ are homotopic, for any spheroids $\phi,\psi \in \mathrm{Sph}_r(X,x_0)$, $r > 1$. Consider the following three homotopies $I^r \times I \to X$:

$$((t_1,t_2,t_3,\ldots,t_r),t) \mapsto$$

$$\mapsto \begin{cases} \phi(2t_1,(1+t)t_2,t_3,\ldots,t_r), & \text{if } 0 \leqslant t_1 \leqslant \tfrac{1}{2},\ 0 \leqslant t_2 \leqslant \tfrac{1}{1+t}, \\ \psi(2t_1-1,(1+t)t_2-t,t_3,\ldots,t_r), & \text{if } \tfrac{1}{2} \leqslant t_1 \leqslant 1,\ \tfrac{t}{1+t} \leqslant t_2 \leqslant 1, \\ x_0, & \text{otherwise;} \end{cases}$$

$$((t_1,t_2,t_3,\ldots,t_r),t) \mapsto$$

$$\mapsto \begin{cases} \phi(2t_1-t,2t_2,t_3,\ldots,t_r), & \text{if } \tfrac{t}{2} \leqslant t_1 \leqslant \tfrac{1+t}{2},\ 0 \leqslant t_2 \leqslant \tfrac{1}{2}, \\ \psi(2t_1+t-1,2t_2-1,t_3,\ldots,t_r), & \text{if } \tfrac{1-t}{2} \leqslant t_1 \leqslant \tfrac{2-t}{2},\ \tfrac{1}{2} \leqslant t_2 \leqslant 1, \\ x_0, & \text{otherwise} \end{cases}$$

and

$$((t_1,t_2,t_3,\ldots,t_r),t) \mapsto$$

$$\mapsto \begin{cases} \phi(2t_1-1,(2-t)t_2,t_3,\ldots,t_r), & \text{if } \tfrac{1}{2} \leqslant t_1 \leqslant 1,\ 0 \leqslant t_2 \leqslant \tfrac{1}{2-t}, \\ \psi(2t_1,(2-t)t_2+t-1,t_3,\ldots,t_r), & \text{if } 0 \leqslant t_1 \leqslant \tfrac{1}{2},\ \tfrac{1-t}{2-t} \leqslant t_2 \leqslant 1, \\ x_0, & \text{otherwise.} \end{cases}$$

Their successive product is a homotopy from $\phi\psi$ to $\psi\phi$. [These homotopies are pictured in Fig. 13, where the shaded regions are mapped into x_0.]

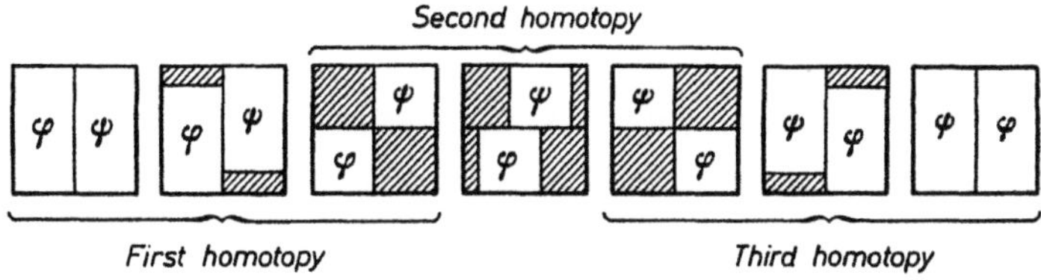

Fig. 13   (r = 2)

Behavior Under Continuous Maps

6.   Let $f: (X,x_0) \to (X',x_0')$ be a continuous map of pointed spaces. Then to each spheroid $\phi: (I^r, \text{Fr } I^r) \to (X,x_0)$ there corresponds the spheroid $f \circ \phi: (I^r, \text{Fr } I^r) \to (X',x_0')$. This defines a map

$$f_\# = f_{\#r}: \text{Sph}_r(X,x_0) \to \text{Sph}_r(X',x_0'),$$

and clearly $f_\#$ takes homotopic spheroids into homotopic ones, and takes the constant spheroid into the constant one. Moreover, $f_{\#r}(\phi\psi) = f_{\#r}(\phi)f_{\#r}(\psi)$ for $r > 0$. Therefore, $f_{\#r}$ defines a homomorphism $\pi_r(X,x_0) \to \pi_r(X',x_0')$ for each $r > 0$, called the <u>homomorphism induced by the map</u> f, and denoted f_* or, more specifically, f_{*r}.

7.   <u>For any two continuous maps,</u> $f: (X,x_0) \to (Y,y_0)$ <u>and</u> $g: (Y,y_0) \to (Z,z_0)$, <u>and any</u> $r > 0$,

$$(g \circ f)_{*r} = g_{*r} \circ f_{*r}.$$

<u>If</u> $f = \text{id}(X,x_0)$, <u>then</u> $f_{*r} = \text{id}\,\pi_r(X,x_0)$.

PROOF. $(g \circ f)_{\#r} = g_{\#r} \circ f_{\#r}$ and $\text{id}(X,x_0)_{\#r} = \text{id}\,\text{Sph}_r(X,x_0)$.

8.   <u>If the continuous maps</u> $f,f': (X,x_0) \to (Y,y_0)$ <u>are homotopic, then</u> $f_{*r} = f'_{*r}$ <u>for all</u> r. <u>If</u> $f: (X,x_0) \to (Y,y_0)$ <u>is a homotopy equivalence, then</u> f_{*r} <u>is an isomorphism for all</u> r.

PROOF.   The spheroids $f \circ \phi$ and $f' \circ \phi$ are homotopic for any $\phi \in \text{Sph}_r(X,x_0)$, which proves the first assertion. The second assertion follows from the equalities $g_{*r} \circ f_{*r} = (g \circ f)_{*r} = \text{id}$ and $f_{*r} \circ g_{*r} = (f \circ g)_{*r} = \text{id}$, where g is any homotopy inverse of f (see 7).

A Multiplication Theorem

9. <u>Let</u> (X,x_0) <u>and</u> (Y,y_0) <u>be arbitrary pointed spaces.</u> <u>Then for any</u> $r \geq 0$ <u>the homotopy group</u> $\pi_r(X \times Y, (x_0,y_0))$ <u>is canonically isomorphic to the direct product</u> $\pi_r(X,x_0) \times \pi_r(Y,y_0)$. <u>The canonical isomorphism</u> $\pi_r(X \times Y, (x_0,y_0)) \to \pi_r(X,x_0) \times \pi_r(Y,y_0)$ <u>is given by</u> $\alpha \mapsto (pr_{1*}(\alpha), pr_{2*}(\alpha))$. <u>If</u> (X',x_0') <u>and</u> (Y',y_0') <u>is another pair of pointed spaces and</u> $f: (X,x_0) \to (X',x_0')$ <u>and</u> $g: (Y,y_0) \to (Y',y_0')$ <u>are continuous, then the diagram</u>

$$
\begin{array}{ccc}
\pi_r(X \times Y, (x_0,y_0)) & \longrightarrow & \pi_r(X,x_0) \times \pi_r(Y,y_0) \\
(f \times g)_* \downarrow & & \downarrow f_* \times g_* \\
\pi_r(X' \times Y', (x_0',y_0')) & \longrightarrow & \pi_r(X',x_0') \times \pi_r(Y',y_0')
\end{array}
$$

<u>commutes</u> (<u>the horizontal maps are the canonical homomorphisms</u>).

The proof is immediate.

2. A Digression: Local Systems

1. We say that on the topological space X there is given a <u>local system of groups</u> if for each point $x \in X$ there is a group G_x, and for each path $s: I \to X$ there is a homomorphism $T_s: G_{s(0)} \to G_{s(1)}$, such that three conditions are satisfied:

(i) if $s_1(0) = s(1)$, then $T_{ss_1} = T_{s_1} \circ T_s$;

(ii) if s is a constant path, then T_s is the identical automorphism of $G_{s(0)}$;

(iii) if s and s_1 are homotopic paths, then $T_s = T_{s_1}$.

Condition (iii) shows that we may write T_σ instead of T_s, where σ is the homotopy class of the path s. Moreover, from (i)-(iii) it follows that all the homomorphisms T_s (T_σ) are actually isomorphisms, and that $T_s^{-1} = T_{s^{-1}}$ (respectively, $T_\sigma^{-1} = T_{\sigma^{-1}}$): indeed, using 1.4, the paths ss^{-1} and $s^{-1}s$ are homotopic to a constant path, and hence $T_{s^{-1}} \circ T_s = T_{ss^{-1}} = id\, G_{s(0)}$ and $T_s \circ T_{s^{-1}} = T_{s^{-1}s} = id\, G_{s(1)}$.

The isomorphism T_s is called the _translation along_ s.

2. In particular, if s is a loop with origin x or, equivalently, if σ is an element of the fundamental group $\pi_1(X,x)$, then $T_s = T_\sigma$ is an automorphism of G_x. Comparing this with conditions (i) and (ii) in 1, we see that the rule $\sigma \mapsto T_\sigma$ defines a right group-action of $\pi_1(X,x)$ on G_x.

To each path $s: I \to X$ there correponds a natural isomorphism $t_s: \pi_1(X,s(0)) \to \pi_1(X,s(1))$, given by $t_s\omega = \sigma^{-1}\omega\sigma$, where σ is the homotopy class of s; sometimes we denote t_s by t_σ. One can check directly that $T_s: G_{s(0)} \to G_{s(1)}$ is a t_σ-map (see 4.2.3.1).

3. Let $(X,\{G_x\},\{T_s\})$ and $(X',\{G'_x\},\{T'_{s'}\})$ be local systems of groups, given on two spaces, X and X', and let $f: X \to X'$ be continuous. Let us assume further that for each point $x \in X$ we are given a homomorphism $h_x: G_x \to G'_{f(x)}$. We say that the homomorphisms h_x and the map f form a _homomorphism of the first local system into the second_ if $h_{s(1)} \circ T_s = T'_{f \circ s} \circ h_{s(0)}$ for any path $s: I \to X$. A homomorphism $(f,\{h_x\})$ is an _isomorphism_ if f is a homeomorphism and all h_x are isomorphisms; $(f,\{h_x\})$ is an _equivalence_ if it is an isomorphism and, in addition, $X' = X$ and $f = \mathrm{id}\, X$.

If $(X',\{G'_x\},\{T'_{s'}\})$ is a local system of groups and $f: X \to X'$ is continuous, then the _induced local system_ $(X,\{G_x\},\{T_s\})$ arises on X: set $G_x = G'_{f(x)}$ and $T_s = T'_{f \circ s}$. Obviously, $(f,\{\mathrm{id}\, G_x\})$ is a homomorphism of the induced local system into the original one.

4. _Let_ X _be a connected space with base point_ x_0. _Two local systems of groups,_ $(X,\{G_x\},\{T_s\})$ _and_ $(X,\{G'_x\},\{T'_{s'}\})$, _are equivalent if and only if the two corresponding actions of_ $\pi_1(X,x_0)$ _on_ G_{x_0} _and_ G'_{x_0} _are isomorphic, i.e., if and only if there is a group isomorphism_ $G_{x_0} \to G'_{x_0}$ _which is also a_ $\pi_1(X,x_0)$-_map._

That the actions of $\pi_1(X,x_0)$ on G_{x_0} and G'_{x_0} arising from equivalent local systems are isomorphic is obvious. To prove the converse, fix a $\pi_1(X,x_0)$-isomorphism $h: G_{x_0} \to G'_{x_0}$ and choose, for each $x \in X$, some path s_x with origin x_0 and end x. It is readily verified that $(\mathrm{id},\{h_x\})$, where $h_x = T'_{s_x} \circ h \circ T^{-1}_{s_x}$, is an equivalence.

5. A local system of groups on a topological space X is said to be _simple_ if it is equivalent to a _canonical simple local system_

$(X,\{G_x\},\{T_s\})$, where all the G_x are equal to some fixed group G, and all the homomorphisms T_s are the identical automorphism of G.

By Theorem 4, a local system of groups on a connected topological space X with base point x_0 is simple if and only if the induced action of $\pi_1(X,x_0)$ on G_{x_0} is the identical action. In particular, a local system is simple whenever $\pi_1(X,x_0)$ is trivial or the groups G_x are all isomorphic to $\mathbb{Z}_2$.

6. It is readily seen that the discussion above may be extended from local systems of groups to local systems of other algebraic objects, such as vector spaces or rings.

In the present section we shall encounter, in addition to local systems of groups, local systems of sets with an identity (a distinguished element).

3. <u>Local Systems of</u>
<u>Homotopy Groups of a Topological Space</u>

1. Two spheroids, $\phi_0 \in \mathrm{Sph}_r(X,x_0)$ and $\phi_1 \in \mathrm{Sph}_r(X,x_1)$, are said to be <u>freely homotopic</u> if the maps $\mathrm{abs}\,\phi_0, \mathrm{abs}\,\phi_1 : I^r \to X$ can be connected by a homotopy consisting only of spheroids. More precisely, ϕ_0 and ϕ_1 are freely homotopic if there is a continuous map $h: I^r \times I \to X$, constant on each set $\mathrm{Fr}\,I^r \times t$ $(t \in I)$, and such that $h(y,0) = \phi_0(y)$, $h(y,1) = \phi_1(y)$ for all $y \in I^r$.

An essential element of such a homotopy is the path described by the origin of the spheroid, i.e., $t \mapsto h(\mathrm{Fr}\,I^r \times t)$. We say that h is a <u>free homotopy connecting the spheroids</u> ϕ_0 <u>and</u> ϕ_1 <u>along this path</u>.

2. <u>Every spheroid with origin</u> x_0 <u>admits a free homotopy along any path with origin</u> x_0. <u>Free homotopies of homotopic spheroids along homotopic paths produce homotopic spheroids.</u>

Let $\phi \in \mathrm{Sph}_r(X,x_0)$, and let s be a path with $s(0) = x_0$. To exhibit a free homotopy of ϕ along s, is is enough to extend somehow the homotopy $(y,t) \mapsto s(t)$ $[y \in \mathrm{Fr}\,I^r, t \in I]$ of the constant map $\phi\big|_{\mathrm{Fr}\,I^r}$ to a homotopy of $\phi: I^r \to X$. (That such an extension exists follows from 2.3.1.3.)

To prove the second claim, let ϕ_0 and ϕ_0' be homotopic spheroids, and let $h,h': I^r \times I \to X$ be free homotopies of ϕ_0 and ϕ_0' along the homotopic paths s and s'. To show that the spheroids

ϕ_1 and ϕ_1' are homotopic, where $\phi_1(t_1,\ldots,t_r) = h((t_1,\ldots,t_r),t)$ and $\phi_1'(t_1,\ldots,t_r) = h'((t_1,\ldots,t_r),t)$, pick some homotopies, $f: I^r \times I \to X$, from ϕ_0 to ϕ_0', and $g: I \times I \to X$, from s to s', and define a subset K of the cube $I^{r+2} = I^r \times I \times I$ and a map $H: K \to X$, as follows:

$$K = \mathrm{Fr}_{\mathbb{R}^{r+2}} I^{r+2} \smallsetminus [(\mathrm{Int}_{\mathbb{R}^{r+1}} I^{r+1}) \times 1],$$

and

$$H(y,u,v) = \begin{cases} f(y,u), & \text{if } v = 0, \\ g(u,v), & \text{if } y \in \mathrm{Fr}_{\mathbb{R}^r} I^r, \\ h(y,v), & \text{if } u = 0, \\ h'(y,v), & \text{if } u = 1 \end{cases}$$

(where $y \in I^r$, $u \in I$, $v \in I$). There exists a homeomorphism $k: I^r \times I \to K$ such that $k(y,u) = (y,u,1)$ for all $(y,u) \in \mathrm{Fr}(I^r \times I)$; for example, take the inverse of the homeomorphism $k_1: K \to I^r \times I$,

$$k_1(y,u,v) = (y_0,\tfrac{1}{2}) + \tfrac{1}{2}(1+v)(y-y_0,u-\tfrac{1}{2}),$$

where $y = (\mathrm{ort}_1 + \ldots + \mathrm{ort}_r)/2$. Now it is clear that $H \circ k: I^r \times I \to X$ is a homotopy from ϕ_1 to ϕ_1'.

    3. According to the previous theorem, the free homotopies along a path $s: I \to X$ define a map $T_s: \pi_r(X,s(0)) \to \pi_r(X,s(1))$ for any $r \geqslant 0$. The same theorem demonstrates that the maps T_s fulfil property 2.1(iii), and it is obvious that they enjoy also the properties 2.1(i), 2.1(ii), and are homomorphisms. The resulting local system, $(X,\{\pi_r(X,x)\},\{T_s\})$, is a local system of groups for any $r \geqslant 1$, and a local system of sets with distinguished elements for $r = 0$; $(X,\{\pi_r(X,x)\},\{T_s\})$ is called the _local system of the r-th homotopy groups of_ X. In particular, $\pi_1(X,x)$ acts naturally from the right on $\pi_r(X,x)$, for any $x \in X$ and $r \geqslant 1$.

    4. _If_ $r = 1$, _then the isomorphism_ T_s _acts by the rule_ $T_s = \sigma^{-1}\omega\sigma$, _where_ σ _is the homotopy class of the path_ s _(i.e.,_ T_s _coincides with the homomorphism_ t_s _from 2.2)._ _In particular, the right action of_ $\pi_1(X,x)$ _on_ $\pi_1(X,x)$ _is the inner right action._

    PROOF. Let w be any loop in the class ω, and define a path $s_t: I \to X$, $t \in I$, by $s_t(y) = s(ty)$. Consider the loop $w_t = (s_t^{-1}w)s_t$. Since s_0 is the constant path, w_0 belongs to ω, and the formula $(y,t) \mapsto w_t(y)$ defines a free homotopy $I \times I \to X$ from

w_0 to the loop $w_1 = (s^{-1}w)s$ along s; w_1 belongs to the class $\sigma^{-1}\omega\sigma$.

5. As 3 shows, for arbitrary fixed r all the homotopy groups $\pi_r(X, x_0)$ of a connected topological space X are isomorphic. For $r = 1$, this was already a corollary of 2.2.

A space X is r-<u>simple</u> if it is connected and the local system of its r-th homotopy groups is simple. In this case, the groups $\pi_r(X, x)$ are not only isomorphic, but are manifestly canonically isomorphic, and hence they may be identified with a unique group, $\pi_r(X)$, referred to as the <u>r-th homotopy group of</u> X <u>without base point.</u> The elements of $\pi_r(X)$ are classes of freely homotopic spheroids. A space is <u>simple</u> if it is r-simple for all r.

If X is not r-simple, then one cannot use the isomorphisms T_s to identify the groups $\pi_r(X, x)$ with different x. In this situation one can speak of the group $\pi_r(X)$ of X only as an abstract group.

Obviously, the local system of the 0-th homotopy groups of a topological space is always simple, and for connected spaces, it becomes a local system of sets, each reduced to one point.

According to 4, a space is 1-simple if and only if it is connected and its fundamental group is Abelian (see 2.5).

6. As 1.6 shows, every continuous map $f: X \to X'$ induces a a homomorphism $f = (f_*)_x: \pi_r(X, x) \to \pi_r(X', f(x))$, for any $x \in X$. If h is a free homotopy from the spheroid ϕ_0 to the spheroid ϕ_1 along the path s, then $f \circ h$ is a free homotopy from $f \circ \phi_0$ to $f \circ \phi_1$ along the path $f \circ s$, and so $(f_*)_{s(1)} \circ T_s = T_{f \circ s} \circ (f_*)_{s(0)}$. Thus, given any $r \geqslant 0$, the induced homomorphisms $(f_{*r})_x$ combine with f to define a homomorphism of the local system of the r-th homotopy groups of X into the local system of the r-th homotopy groups of X'.

Three special cases deserve to be mentioned: (a) X is r-simple; (b) X' is r-simple; and (c) X and X' are both r-simple. In case (a), the homomorphisms $(f_{*r})_x$ take the same group, $\pi_r(X)$, into $\pi_r(X', f(x))$, $x \in X$, and for any path $s: I \to X$ the diagram

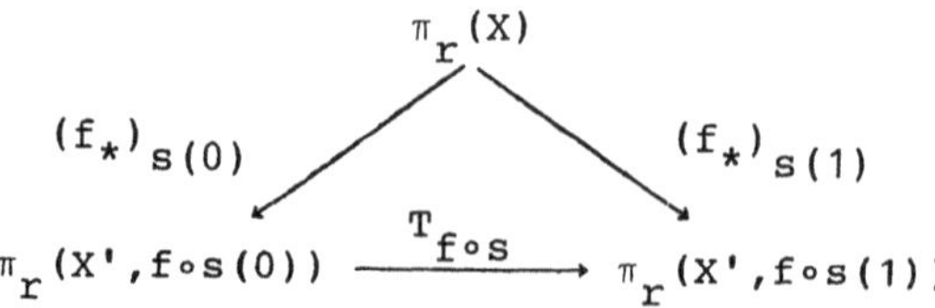

commutes. In case (b), the homomorphisms $(f_{*r})_x$ map the groups $\pi_r(X,x)$, $x \in X$, into the same group, $\pi_r(X')$, and for any path $s: I \to X$ the diagram

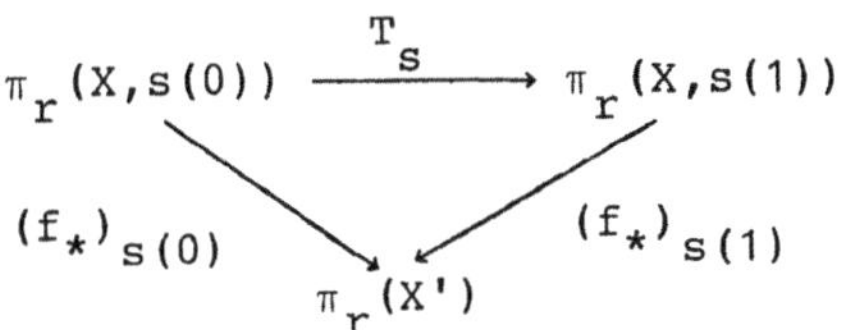

commutes. Finally, in case (c), the local systems of the r-th homotopy groups of X and X' reduce to two groups, $\pi_r(X)$ and $\pi_r(X')$, and the homomorphisms $(f_{*r})_x$ become one homomorphism $f_{*r}: \pi_r(X) \to \pi_r(X')$.

7. __If__ $f: X \to X'$ __is a homotopy equivalence, then all induced homomorphisms__ $(f_{*r})_x: \pi_r(X,x) \to \pi_r(X',f(x))$ __are isomorphisms.__

PROOF. Let $f': X' \to X$ be a homotopy inverse of f. If $H: X \times I \to X$ is a homotopy from $f' \circ f$ to $\mathrm{id}\, X$, then given any spheroid $\phi \in \mathrm{Sph}_r(X,x)$, the map $I^r \times I \to I$, $(y,t) \mapsto H(\phi(y),t)$, is a free homotopy from ϕ to the spheroid $f' \circ f \circ \phi \in \mathrm{Sph}_r(X,f'\circ f(x))$, along the path s, $s(t) = H(x,t)$. Therefore, the homomorphism $(f'_*)_{f(x)} \circ (f_*)_x = ((f' \circ f)_*)_x: \pi_r(X,x) \to \pi_r(X,f'\circ f(x))$ is simply the translation T_s; in particular, it is an isomorphism, implying that $(f'_*)_{f(x)}$ is an epimorphism. On the other hand, $(f_*)_{f'\circ f(x)} \circ (f'_*)_{f(x)}$ is also an isomorphism, and so $(f'_*)_{f(x)}$ is a monomorphism. We conclude that $(f'_*)_{f(x)}$ is an isomorphism, and hence so is $(f_*)_x = [(f'_*)_{f(x)}]^{-1} \circ T_s$.

8. __Let__ (X,x_0) __be a pointed topological space, and let__ $0 \leqslant k < \infty$. __The homotopy groups__ $\pi_r(X,x_0)$ __are trivial for all__ $r \leqslant k$ __if and only if__ X __is k-connected. The homotopy groups__ $\pi_r(X,x_0)$ __are all trivial if and only if__ X __is ∞-connected.__

If X is k-connected, then $\pi_r(X,x_0)$ is trivial for all $r \leqslant k$: to see this, compare the definition of $\pi_r(X,x_0)$ and that of k-connectedness (see 1.3.3.7; in 1.3.3.6 one can replace (D^{r+1},S^{r+1}) by the homeomorphic pair $(I^{r+1},\mathrm{Fr}\, I^{r+1})$). The same two definitions prove the converse statement, since the triviality of $\pi_r(X,x_0)$ for all $r \leqslant k$ implies the triviality of $\pi_r(X,x)$ for all $r \leqslant k$ and any $x \in X$ (see 1.2 and 5).

4. <u>Relative Homotopy Groups</u>

1. Set $J^{r-1} = \mathrm{Fr}_{\mathbb{R}^r} I^r \smallsetminus \mathrm{Int}_{\mathbb{R}^{r-1}} I^{r-1}$. Given any topological pair (X,A) with base point $x_0 \in A$ and any positive integer r, we let $\mathrm{Sph}_r(X,A,x_0)$ denote the set $\mathcal{C}(I^r,\mathrm{Fr}\,I^r,J^{r-1};X,A,x_0)$ of all continuous maps $(I^r,\mathrm{Fr}\,I^r,J^{r-1}) \to (X,A,x_0)$. The elements of $\mathrm{Sph}_r(X,A,x_0)$ are called <u>r-dimensional spheroids</u> (or <u>r-spheroids</u>) <u>with</u> origin x_0 of the pair (X,A). The set $\pi(I^r,\mathrm{Fr}\,I^r,J^{r-1};X,A,x_0)$ of homotopy classes of such spheroids is simply denoted by $\pi_r(X,A,x_0)$.

Notice that every spheroid $\phi \in \mathrm{Sph}_r(X,A,x_0)$, such that $\phi(I^r) \subset A$, is homotopic to the constant spheroid. In fact, there is even a standard homotopy $I^r \times I \to I$ from ϕ to the constant spheroid:

$$((t_1,\ldots,t_{r-1},t_r),t) \mapsto (t_1,\ldots,t_{r-1},(1-t)t_r + t).$$

A 1-spheroid with origin x_0 of the pair (X,A) is simply a path with origin in A and <u>end</u> x_0. Warning: a homotopy of such a spheroid is stationary at the point 1, but if A is not reduced to x_0, it is not necessarily stationary at the point 0.

When $r \geqslant 2$, formula (1) defines a multiplication on $\mathrm{Sph}_r(X,A,x_0)$, and this induces a multiplication on $\pi_r(X,A,x_0)$, which turns $\pi_r(X,A,x_0)$ into a group. The identity element of $\pi_r(X,A,x_0)$ is the homotopy class of the constant spheroid, while the class of the spheroid ϕ^{-1}, with $\phi^{-1}(t_1,t_2,\ldots,t_r) = \phi(1-t_1,t_2,\ldots,t_r)$, is the inverse of the class of ϕ. The proof of these assertions is entirely analogous to that given in the case of absolute homotopy groups.

For $r \geqslant 2$, $\pi_r(X,A,x_0)$ is called the <u>r-th homotopy group of the pair</u> (X,A) <u>at the point</u> x_0. The first homotopy group of (X,A) at x_0 is defined to be the set $\pi_1(X,A,x_0)$ with an identity (a distingushed element), namely the homotopy class of the constant 1-spheroid.

If $A = x_0$, then $\pi_r(X,A,x_0)$ equals $\pi_r(X,x_0)$ (i.e., $\pi_r(X,x_0,x_0)$ and $\pi_r(X,x_0)$ coincide as groups for $r \geqslant 2$, and as sets with distingushed elements for $r = 1$).

For $r > 1$, $\pi_r(X,A,x_0)$ is canonically isomorphic to $\pi_r(X_0,A_0,x_0)$, where X_0 and A_0 are the components of X and A which contain x_0.

For $r > 2$, $\pi_r(X,A,x_0)$ is Abelian; the proof entails obvious modifications of the proof of Theorem 1.5.

2. Every continuous map $f: (X,A,x_0) \to (X',A',x_0')$ yields the <u>induced homomorphism</u> $f_*: \pi_r(X,A,x_0) \to \pi_r(X',A',x_0')$, $r \geqslant 1$, defined as in the absolute case. If $A = x_0$ and $A' = x_0'$, we recover the absolute induced homomorphism, $f_*: \pi_r(X,x_0) \to \pi_r(X',x_0')$. As with the absolute case, $(g \circ f)_* = g_* \circ f_*$ and $id_* = id$. If f and f' are homotopic, then $f_* = f_*'$. If f is a homotopy equivalence, then f_* is an isomorphism.

The Boundary Homomorphism

3. Given a spheroid $\phi \in Sph_r(X,A,x_0)$, its compression, $ab\,\phi : (I^{r-1}, Fr\,I^{r-1}) \to (A,x_0)$, is a spheroid belonging to $Sph_{r-1}(A,x_0)$, called the boundary of ϕ, and denoted $\partial\phi$. The resulting map, $\partial: Sph_r(X,A,x_0) \to Sph_{r-1}(A,x_0)$, takes homotopic spheroids into homotopic ones, takes the sum of two spheroids into the sum of their boundaries, and takes the constant spheroid into the constant one. Therefore, it defines, for every $r \geqslant 1$, a homomorphism $\partial: \pi_r(X,A,x_0) \to \pi_{r-1}(A,x_0)$, called the <u>boundary homomorphism.</u>

Given any continuous map, $f: (X,A,x_0) \to (X',A',x_0')$, the diagram

$$
\begin{array}{ccc}
\pi_r(X,A,x_0) & \xrightarrow{\ \partial\ } & \pi_{r-1}(A,x_0) \\
f_* \downarrow & & \downarrow (ab\,f)_* \\
\pi_r(X',A',x_0') & \xrightarrow{\ \partial\ } & \pi_{r-1}(A',x_0')
\end{array}
$$

commutes for any $r \geqslant 1$. Indeed, we already know that the similar diagram with Sph instead of π, and with $f_\#$ and $(ab\,f)_\#$ instead of f_* and $(ab\,f)_*$, commutes.

Local Systems of
Homotopy Groups of a Topological Pair

4. Two speroids, $\phi_0 \in Sph_r(X,A,x_0)$ and $\phi_1 \in Sph_r(X,A,x_1)$, are said to be <u>freely homotopic</u> if there is a homotopy from ϕ_0 to ϕ_1 consisting only of spheroids, where ϕ_0 and ϕ_1 are viewed as maps $(I^r, Fr\,I^r) \to (X,A)$. In other words, ϕ_0 and ϕ_1 are freely homotopic if there is a map $h: I^r \times I \to I$ such that $h(Fr\,I^r \times I) \subset A$, h is constant on every set $J^{r-1} \times t$ $(t \in I)$, and $h(y,0) = \phi_0(y)$,

386

$h(y,1) = \phi_1(y)$ for all $y \in I^r$. We say that h is a free homotopy from ϕ_0 to ϕ_1, along the path $t \mapsto h(J^{r-1} \times t)$.

Every spheroid with origin x_0 of (X,A) admits a free homotopy along any given path, $s: I \to A$, with $s(0) = x_0$. Moreover, free homotopies of homotopic spheroids along homotopic paths of A produce homotopic spheroids. The proof differs from that of Theorem 3.2 in two details: in the first part of the proof, we have to start with ab $\phi: J^{r-1} \to A$ (instead of $\phi|_{\text{Fr } I^r}$) and extend its homotopy, $(y,t) \mapsto s(t)$, initially to a homotopy of ab $\phi: \text{Fr } I^r \to A$, and then to a homotopy of $\phi: I^r \to X$; in the second part of the proof, we must replace $g: I \times I \to X$ by $g: I \times I \to A$.

This theorem shows that given a path $s: I \to A$, the free homotopies along s define a map $T_s: \pi_r(X,A,s(0)) \to \pi_r(X,A,s(1))$ (for each $r \geqslant 1$). As in the absolute case, T_s are homomorphisms and enjoy properties 2.1(i)-(iii). Therefore, a local system, $(A,\{\pi_r(X,A,x)\},\{T_s\})$, arises on A, which is a local system of groups for $r \geqslant 2$, and a local system of sets with distingushed elements for $r = 1$. This is the <u>local system of the r-th homotopy groups of the pair</u> (X,A). In particular, for any $x \in A$ and $r \geqslant 1$, $\pi_1(A,x_0)$ acts naturally from the right on $\pi_r(X,A,x)$; this is a group-action for $r > 1$, and it fixes the distingushed element for $r = 1$.

From the existence of this local system it follows that, for any $r \geqslant 1$, the r-th homotopy groups $\pi_r(X,A,x)$, $x \in A$, are all isomorphic whenever A is connected.

A pair (X,A) with A connected is said to be <u>r-simple</u> if the local system of its r-th homotopy groups is simple. In this case all the homotopy groups $\pi_r(X,A,x)$, $x \in A$, can be identified with a single group, the <u>r-th homotopy group of the pair</u> (X,A) <u>without base point,</u> $\pi_r(X,A)$; the elements of $\pi_r(X,A)$ are classes of freely homotopic spheroids. A pair is <u>simple</u> if it is r-simple for any $r \geqslant 1$. For example, every pointed space is a simple pair.

5. Given any pair (X,A) and any path $s: I \to A$, the diagram

$$
\begin{array}{ccc}
\pi_r(X,A,s(0)) & \xrightarrow{\ \partial\ } & \pi_{r-1}(A,s(0)) \\
\downarrow{\scriptstyle T_s} & & \downarrow{\scriptstyle T_s} \\
\pi_r(X,A,s(1)) & \xrightarrow{\ \partial\ } & \pi_{r-1}(A,s(1))
\end{array}
$$

obviously commutes. Therefore, the boundary homomorphisms,

$\partial = \partial_x \colon \pi_r(X,A,x) \to \pi_{r-1}(A,x)$, combine with $\mathrm{id}\,A$ to define a homomorphism of the local system of the r-th homotopy groups of the pair (X,A) into the local system of the $(r-1)$-th homotopy groups of the space A.

Further, given any continuous map, $f\colon (X,A) \to (X',A')$, the homomorphisms $f_* = (f_*)_x \colon \pi_r(X,A,x) \to \pi_r(X',A',f(x))$, combine with f to define a homomorphism of the local system of the r-th homotopy groups of the pair (X,A) into the corresponding local system of (X',A'). As in 3.6, we mention three special cases: (a) (X,A) is r-simple; (b) (X',A') is r-simple; and (c) both (X,A) and (X',A') are r-simple. In cases (a) and (b), any path $s\colon I \to A$ yields commutative diagrams similar to those in 3.6; in case (c), the local systems of the r-th homotopy groups of the pairs (X,A) and (X',A') reduce to single groups, $\pi_r(X,A)$ and $\pi_r(X',A')$, while the homomorphisms $(f_*)_x$ reduce to a single homomorphism, $f_* \colon \pi_r(X,A) \to \pi_r(X',A')$.

If f is a homotopy equivalence, then all the homomorphisms $(f_*)_x$ are isomorphisms. The proof is similar to that given in the absolute case (see 3.7).

6. <u>Let</u> (X,A) <u>be a pair with base point</u> $x_0 \in A$. <u>If</u> X, A <u>are connected, then the triviality of all the homotopy groups</u> $\pi_r(X,A,x_0)$ <u>is equivalent to the ∞-connectedness of</u> (X,A); <u>the triviality of the homotopy groups</u> $\pi_r(X,A,x_0)$ <u>for</u> $1 \leqslant r \leqslant k$ <u>is equivalent to the k-connectedness of</u> (X,A).

The proof is a repetition of the proof of Proposition 3.8, with obvious modifications (instead of referring to 1.3.3.7, we refer to 1.3.3.9).

The Group $\pi_2(X,A,x_0)$

7. <u>If</u> $\alpha,\beta \in \pi_2(X,A,x_0)$, <u>then</u> $\alpha^{-1}\beta\alpha = T_{\partial\alpha}\beta$.

PROOF (for an alternative proof, see Subsection 10). We have to check that, given two arbitrary spheroids, $\phi,\psi \in \mathrm{Sph}_2(X,A,x_0)$, there is a free homotopy from ψ to $\phi^{-1}\psi\phi$ along the loop $\partial\phi$. We shall exhibit such a homotopy as a family of maps, $\chi_t \colon I^2 \to X$ $(t \in I)$, constructed as follows.

Set $f(t) = \frac{1}{16} - \frac{1}{8}|t - \frac{1}{2}|$, and divide I^2 into eight parts, as shown in Fig. 14: the points $A_1(t)$, $A_2(t)$, $A_3(t)$, $A_4(t)$ have abscissae $f(t)$, $t/4$, $1 - (t/2)$, $1 - f(t)$, respectively, while the points $B_1(t)$, $B_2(t)$, $B_3(t)$, $B_4(t)$ lie above these points at the height

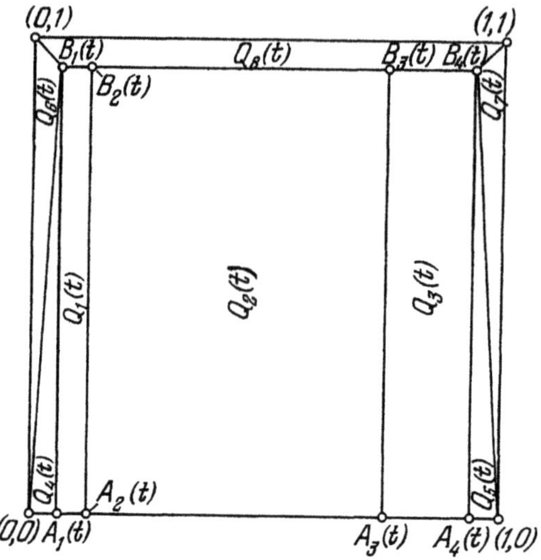

Fig. 14

$1 - f(t)$. Further, let $\alpha_j : Q_j(t) \to I^2$, $j = 1,2,3,$ be the affine maps defined by the conditions

$$\alpha_1(A_1(t)) = (t,0), \quad \alpha_1(A_2(t)) = (0,0), \quad \alpha_1(B_1(t)) = (t,1),$$

$$\alpha_2(A_2(t)) = (0,0), \quad \alpha_2(A_3(t)) = (1,0), \quad \alpha_2(B_2(t)) = (0,1),$$

$$\alpha_3(A_3(t)) = (0,0), \quad \alpha_3(A_4(t)) = (t,0), \quad \alpha_3(B_3(t)) = (0,1).$$

Now set

$$\chi_t\big|_{Q_1(t)} = \phi \circ \alpha_1, \quad \chi_t\big|_{Q_2(t)} = \psi \circ \alpha_2, \quad \chi_t\big|_{Q_3(t)} = \phi \circ \alpha_3,$$

and consider the resulting continuous map $Q_1(t) \cup Q_2(t) \cup Q_3(t) \to X$. Extend it firstly to a map $\cup_1^7 Q_i(t) \to X$ which is constant on the horizontals in $Q_4(t) \cup Q_5(t)$ and on the verticals in $Q_6(t) \cup Q_7(t)$, and then to a map $\cup_1^8 Q_i(t) \to X$ which is constant on the horizontals in $Q_8(t)$. (The latter is possible, since the already extended map $\cup_1^7 Q_i(t) \to X$ is constant on the segment $[B_1(t), B_4(t)]$, and assumes the same values at those points of the segments $[B_1(t), (0,1)]$ and $[B_4(t), (1,1)]$ which lie at the same height.) The continuity of the map $I^2 \times I \to X$ defined by the family $\chi_t : I^2 \to X$ follows from its continuity on each of the eight polyhedrons $\cup_{t \in I}(Q_i(t) \times t)$, $1 \leqslant i \leqslant 8$.

The Action of the Group $\pi_1(X,x_0)$ on $\pi_1(X,A,x_0)$

8. Given a spheroid $w \in Sph_1(X,A,x_0)$ and a loop $s \in Sph_1(X,x_0)$, the product ws is well defined, and obviously $ws \in Sph_1(X,A,x_0)$. Moreover, the homotopy class of ws is uniquely determined by the homotopy classes of w and s, and hence we may define the product $\omega\sigma$ for any $\omega \in \pi_1(X,A,x_0)$ and $\sigma \in \pi_1(X,x_0)$. Using again the homotopies described in 1.1, we see that $\omega(\sigma\sigma') =$ $= (\omega\sigma)\sigma'$ and $\omega e_{\pi_1(X,x_0)} = \omega$ for any $\omega \in \pi_1(X,A,x_0)$ and $\sigma,\sigma' \in \pi_1(X,x_0)$. That is to say, the rule $(\omega,\sigma) \mapsto \omega\sigma$ defines a right action of $\pi_1(X,x_0)$ on $\pi_1(X,A,x_0)$.

If $A = x_0$, then we clearly recover the canonical right action of the group $\pi_1(X,x_0)$ (on itself). It is readily seen that the translation $T_s: \pi_1(X,A,s(0)) \to \pi_1(X,A,s(1))$ is a $[T_{in\circ s}: \pi_1(X,s(0)) \to \pi_1(X,s(1))]$-map for any path $s: I \to A$, i.e., $T_s(\omega\sigma) = T_s(\omega) T_{in\circ s}(\sigma)$ for all $\sigma \in \pi_1(X,s(0))$ and $\omega \in \pi_1(X,A,s(0))$. Furthermore, given any continuous map $f: (X,A,x_0) \to (X',A',x_0')$, the homomorphism $f_*: \pi_1(X,A,x_0) \to \pi_1(X',A',x_0')$ is a $[f_*: \pi_1(X,x_0) \to \pi_1(X',x_0')]$-map, i.e., $f_*(\omega\sigma) = f_*(\omega) f_*(\sigma)$ for all $\omega \in \pi_1(X,A,x_0)$ and $\sigma \in \pi_1(X,x_0)$. In particular, if we consider the map $rel = [in: (X,x_0,x_0) \to (X,A,x_0)]$, then

$$(rel_*\omega)\sigma = rel_*(\omega\sigma)$$

for all $\omega,\sigma \in \pi_1(X,x_0)$.

9. <u>For any</u> $\omega \in \pi_1(X,A,x_0)$ <u>and</u> $\sigma \in \pi_1(A,x_0)$ <u>we have</u>

$$T_\sigma\omega = \omega(in_*\sigma),$$

<u>where</u> $in = [in: (A,x_0) \to (X,x_0)]$.

PROOF. Let w and s be spheroids in the classes ω and σ. Consider, for each fixed $t \in I$, the path $s_t: I \to A$ given by $s_t(y) = s(ty)$, and set $w_t = w(in \circ s_t)$. Since s_0 is a constant path, w and w_0 are homotopic. On the other hand, the formula $(y,t) \mapsto w_t(y)$ defines a free homotopy $I \times I \to I$ from w_0 to $w_1 = w(in \circ s)$, along s.

5. A Digression: Sequences of Groups
and Homomorphisms, and π-Sequences

1. A <u>sequence of groups and homomorphism</u> is a finite or infinite (on one or both sides) sequence of groups such that for each two adjacent groups, G_i and G_{i+1}, there is given a homomorphism $G_i \to G_{i+1}$.

A <u>homomorphism</u> of a sequence of groups and homomorphisms, $\{G_i, h_i : G_i \to G_{i+1}\}$, into another such sequence, $\{G_i', h_i' : G_i' \to G_{i+1}'\}$, is a sequence of homomorphisms $\{H_i : G_i \to G_i'\}$ such that the diagram

$$\begin{array}{ccccccc}
\cdots \; G_{i-1} & \xrightarrow{\ h_{i-1}\ } & G_i & \xrightarrow{\ h_i\ } & G_{i+1} & \xrightarrow{\ h_{i+1}\ } & \cdots \\
\downarrow{\scriptstyle H_{i-1}} & & \downarrow{\scriptstyle H_i} & & \downarrow{\scriptstyle H_{i+1}} & & \\
\cdots \; G_{i-1}' & \xrightarrow{\ h_{i-1}'\ } & G_i' & \xrightarrow{\ h_i'\ } & G_{i+1}' & \xrightarrow{\ h_{i+1}'\ } & \cdots
\end{array}$$

commutes. A homomorphism $\{H_i\}$ such that each $H_i : G_i \to G_i'$ is an isomorphism is called an <u>isomorphism.</u>

A sequence of groups and homomorphisms, $\{G_i, h_i\}$, is <u>exact</u> if for each group G_i, excepting the initial and final ones, the kernel $\mathrm{Ker}\, h_i$ of the homomorphism h_i equals the image $\mathrm{Im}\, h_{i-1}$ of the homomorphism h_{i-1}.

The following three properties are common to all exact sequences $\{G_i, h_i\}$.

(i) If G_i is an inner (i.e., neither initial, nor final) term of the given sequence, then h_{i-1} is trivial if and only if h_i is a monomorphism, while h_i is trivial if and only if h_{i-1} is an epimorphism; h_{i-1} and h_i are both trivial if and only if the group G_i is trivial.

(ii) If G_i and G_{i+1} are inner terms, then h_{i-1} and h_{i+1} are simultaneously trivial if and only if h_i is an isomorphism.

(iii) In particular, the triviality of G_{i-1} and G_{i+1} implies the triviality of G_i, while the triviality of G_{i-1} and G_{i+2} implies that h_i is an isomorphism.

3. An exact sequence of the form $1 \to F \xrightarrow{f} G \xrightarrow{q} H \to 1$ is called <u>short</u> (here 1 denotes the trivial group). An example is

$1 \longrightarrow F \xrightarrow{\text{in}} G \xrightarrow{\text{pr}} G/F \longrightarrow 1$, where F is a normal subgroup of G. This example has a universal character: every short exact sequence, $1 \to F \xrightarrow{f} G \xrightarrow{g} H \to 1$ is canonically isomorphic to a sequence of this type, namely, to $1 \longrightarrow \operatorname{Im} f \xrightarrow{\text{in}} G \xrightarrow{\text{pr}} G/\operatorname{Im} f \longrightarrow 1$; the canonical isomorphism is obviously $\{\operatorname{id} 1, \operatorname{ab} f: F \to \operatorname{Im} f, \operatorname{id} G, h \mapsto \operatorname{pr}(g^{-1}(h)), \operatorname{id} 1\}$.

Splitting

4. Let $\{G_i, h_i\}$ be a sequence of groups and homomorphisms. We say that this sequence is <u>split from the right at the term</u> G_α <u>by the homomorphism</u> $\zeta: G_{\alpha+1} \to G_\alpha$ if $h_\alpha \circ \zeta = \operatorname{id} G_{\alpha+1}$. Such a splitting is said to be <u>normal</u> if $\operatorname{Im} \zeta$ is a normal subgroup of G_α.

Similarly, $\{G_i, h_i\}$ is <u>split from the left at the term</u> G_α <u>by the homomorphism</u> $\zeta: G_\alpha \to G_{\alpha-1}$ if $\zeta \circ h_{\alpha-1} = \operatorname{id} G_{\alpha-1}$. We sometimes say simply that the given sequence is <u>split</u>, or that it <u>splits</u> (at right or at left) at G_α.

5 (LEMMA). <u>Let A and B be groups, and let $u: A \to B$ and $v: B \to A$ be homomorphisms. If $\operatorname{Im} v$ is a normal subgroup of A and $u \circ v = \operatorname{id} B$, then $A = \operatorname{Ker} u \times \operatorname{Im} v$.</u>

PROOF. Every element $a \in A$ can be represented as $[a(v \circ u(a))^{-1}](v \circ u(a))$, and obviously $a(v \circ u(a))^{-1} \in \operatorname{Ker} u$ and $v \circ u(a) \in \operatorname{Im} v$. If $a \in \operatorname{Ker} u \cap \operatorname{Im} v$, then $u(a) = e_B$, and there is $b \in B$ with $v(b) = a$. Thus, $b = u \circ v(b) = u(a) = e_B$ and $a = v(b) = e_A$.

6. <u>If the exact sequence $\{G_i, h_i\}$ splits normally from the right at G_α and splits from the right at $G_{\alpha-3}$, then it also splits from the left at G_α, and $G_\alpha \cong G_{\alpha-1} \times G_{\alpha+1}$. More precisely, under these hypotheses, $h_{\alpha-1}$ is a monomorphism, h_α is an epimorphism, and G_α decomposes into the direct product of $\operatorname{Im} h_{\alpha-1}$ and a subgroup which is mapped isomorphically onto $G_{\alpha+1}$ by h_α. Moreover, every homomorphism $\zeta: G_{\alpha+1} \to G_\alpha$ which splits the sequence from the right is a monomorphism, and if ζ is also normally splitting, then for a direct complement of $\operatorname{Im} h_{\alpha-1}$ one may take $\operatorname{Im} \zeta$.</u>

PROOF. The equality $G_\alpha = \operatorname{Im} h_{\alpha-1} \times \operatorname{Im} \zeta$ is a consequence of Lemma 5 and of the exactness of the given sequence. From $h_\alpha \circ \zeta = \operatorname{id} G_{\alpha+1}$ it follows that h_α is an epimorphism and that ζ is a monomorphism. Since $\{G_i, h_i\}$ splits from the right at $G_{\alpha-3}$, $h_{\alpha-3}$ is an epimorphism, and now the exactness of $\{G_i, h_i\}$ implies that

$h_{\alpha-2}$ is trivial, while $h_{\alpha-1}$ is a monomorphism. Finally, as a homomorphism splitting the given sequence at G_α from the left one can take the homomorphism which is the inverse of $h_{\alpha-1}$ on $\operatorname{Im} h_{\alpha-1}$ and equals the identity on $\operatorname{Im} \zeta$.

7. <u>If the exact sequence</u> $\{G_i, h_i\}$ <u>splits from the left at</u> G_α <u>and</u> $G_{\alpha+3}$, <u>then it splits normally from the right at</u> G_α, <u>and</u> $G_\alpha \cong G_{\alpha-1} \times G_{\alpha+1}$. <u>More precisely, under these hypotheses,</u> $h_{\alpha-1}$ <u>is a monomorphism,</u> h_α <u>is an epimorphism, and</u> G_α <u>decomposes into the direct product of</u> $\operatorname{Im} h_{\alpha-1}$ <u>and a subgroup which is mapped isomorphically onto</u> $G_{\alpha+1}$ <u>by</u> h_α. <u>Moreover, every homomorphism</u> $\zeta: G_\alpha \to G_{\alpha-1}$ <u>which splits the sequence from the left is an epimorphism, and for a direct complement of</u> $\operatorname{Im} h_{\alpha-1}$ <u>one can take</u> $\operatorname{Ker} \zeta$.

PROOF. The equality $G_\alpha = \operatorname{Im} h_{\alpha-1} \times \operatorname{Ker} \zeta$ is a consequence of Lemma 5. From $\zeta \circ h_{\alpha-1} = \operatorname{id} G_{\alpha-1}$ it follows that $h_{\alpha-1}$ is a monomorphism and that ζ is an epimorphism. Since $\{G_i, h_i\}$ splits from the left at $G_{\alpha+3}$ and is exact, h_α is an epimorphism. As a homomorphism splitting the given sequence normally from the right at G_α, one can take the composition $G_{\alpha+1} = G_\alpha/\operatorname{Ker} h_\alpha = G_\alpha/\operatorname{Im} h_{\alpha-1} = (\operatorname{Im} h_{\alpha-1} \times \operatorname{Ker} \zeta)/\operatorname{Im} h_{\alpha-1} = \operatorname{Ker} \zeta \xrightarrow{\text{in}} G_\alpha$.

8. <u>An exact sequence</u> $1 \to F \xrightarrow{f} G \xrightarrow{g} H \to 1$ <u>splits at</u> G <u>from the left if and only if it splits at</u> G <u>normally from the right, and this happens if and only if the subgroup</u> $\operatorname{Im} f = \operatorname{Ker} g$ <u>of</u> G <u>has a direct complement.</u>

This is a corollary of Lemma 5.

Five Lemma

9. <u>If</u>

$$
\begin{array}{ccccccccc}
G_1 & \xrightarrow{\ h_1\ } & G_2 & \xrightarrow{\ h_2\ } & G_3 & \xrightarrow{\ h_3\ } & G_4 & \xrightarrow{\ h_4\ } & G_5 \\
\downarrow{\scriptstyle \phi_1} & & \downarrow{\scriptstyle \phi_2} & & \downarrow{\scriptstyle \phi_3} & & \downarrow{\scriptstyle \phi_4} & & \downarrow{\scriptstyle \phi_5} \\
G_1' & \xrightarrow{\ h_1'\ } & G_2' & \xrightarrow{\ h_2'\ } & G_3' & \xrightarrow{\ h_3'\ } & G_4' & \xrightarrow{\ h_4'\ } & G_5'
\end{array}
$$

<u>is a homomorphism of exact sequences, and if</u> ϕ_1 <u>is an epimorphism,</u> ϕ_2, ϕ_4 <u>are isomorphisms, and</u> ϕ_5 <u>is a monomorphism, then</u> ϕ_3 <u>is an isomorphism.</u>

Let us show first that ϕ_3 is a monomorphism. If $a \in \operatorname{Ker} \phi_3$, then $\phi_4 \circ h_3(a) = h_3' \circ \phi_3(a) = e_{G_4'}$, and so $h_3(a) = e_{G_4}$, i.e., $a \in \operatorname{Ker} h_3 = \operatorname{Im} h_2$. Let $a = h_2(b)$, $b \in G_2$. Since $h_2' \circ \phi_2(b) = \phi_3 \circ h_2(b) = \phi_3(a) = e_{G_3'}$, there is $c \in G_1'$ such that $h_1'(c) = \phi_2(b)$. Therefore, there is $d \in G_1$ such that $h_1' \circ \phi_1(d) = \phi_2(b)$. On the other hand, $h_1' \circ \phi_1(d) = \phi_2 \circ h_1(d)$, and hence $\phi_2(b) = \phi_2 \circ h_1(d)$. Consequently, $b = h_1(d)$ and $a = h_2 \circ h_1(d) = e_{G_3}$.

Now let us verify that ϕ_3 is an epimorphism. Let $a \in G_3'$. Then $\phi_5 \circ h_4 \circ \phi_4^{-1} \circ h_3'(a) = h_4' \circ h_3'(a) = e_{G_5'}$, and so $h_4 \circ \phi_4^{-1} \circ h_3'(a) = e_{G_5}$, i.e., $\phi_4^{-1} \circ h_3'(a) \in \operatorname{Ker} h_4 = \operatorname{Im} h_3$. Let $\phi_4^{-1} \circ h_3'(a) = h_3(b)$, $b \in G_3$. Since $h_3'(a(\phi_3(b))^{-1}) = (\phi_4 \circ \phi_4^{-1} \circ h_3'(a))(h_3' \circ \phi_3(b))^{-1} = \phi_4((\phi_4^{-1} \circ h_3'(a))(h_3(b))^{-1}) = e_{G_4'}$, there is $c \in G_2'$ such that $h_2'(c) = a(\phi_3(b))^{-1}$, and hence there is $d \in G_2$ such that $h_2' \circ \phi_2(d) = a(\phi_3(b))^{-1}$. On the other hand, $h_2' \circ \phi_2(d) = \phi_3 \circ h_2(d)$, and hence $a(\phi_3(b))^{-1} = \phi_3 \circ h_2(d)$. Consequently, $a = \phi_3(h_2(d)b)$.

$$\pi\text{- Sequences}$$

10. In the next subsections we shall handle the so-called <u>homotopy sequences.</u> These are rather cumbersome entities which are similar to sequences of groups and homomorphisms, but possess additional properties and structures. Such sequences are encountered in various geometric situations, but they are all algebraically related. The rest of the present subsection is devoted to a preliminary, purely algebraic description and study of these sequences.

11. Consider a left-infinite sequence

$$\cdots \Pi_7 \xrightarrow{\rho_6} \Pi_6 \xrightarrow{\rho_5} \Pi_5 \xrightarrow{\rho_4} \Pi_4 \xrightarrow{\rho_3} \Pi_3 \xrightarrow{\rho_2} \Pi_2 \xrightarrow{\rho_1} \Pi_1 \xrightarrow{\rho_0} \Pi_0, \tag{6}$$

where:

Π_0, Π_1, Π_2 are sets with an identity (distinguished element),

Π_3, Π_4, Π_5 are groups,

Π_6, Π_7, $\cdots$ are Abelian groups,

ρ_0, ρ_1, ρ_2 are homomorphisms in the sense of 1.2,

ρ_3, ρ_4, ... are group homomorphism.

Then (6) is called a π-<u>sequence</u> if there are given

right group-actions of Π_3 on the groups Π_{3k} with $k \geqslant 2$,

right group-actions of Π_4 on the groups Π_{3k+1} with $k \geqslant 2$,

right group-actions of Π_4 on the groups Π_{3k-1} with $k \geqslant 2$,

and a right action of Π_3 on the set Π_2,

such that

(i) ρ_{3k} is a ρ_3-homomorphism for all $k \geqslant 2$;

(ii) ρ_{3k+1} is a Π_4-homomorphism for all $k \geqslant 2$;

(iii) ρ_{3k-1} is a Π_4-homomorphism for all $k \geqslant 2$, with respect to the right group-action of Π_4 on Π_{3k} induced by the given action of Π_3 on Π_{3k} via ρ_3;

(iv) ρ_4 is a Π_4-homomorphism with respect to the right inner action of Π_4;

(v) the transformation of the group Π_5 induced by the image $\rho_4(\alpha) \in \Pi_4$ of an arbitrary element $\alpha \in \Pi_5$ is the inner automorphism $\beta \mapsto \alpha^{-1}\beta\alpha$;

(vi) the transformation of the set Π_2 induced by an arbitrary element $\sigma \in \Pi_3$ coincides on $\rho_2(\Pi_3)$ with the transformation $\rho_2(\omega)\sigma = \rho_2(\omega\sigma)$.

A <u>homomorphism</u> of the π-sequence $\{\Pi_i, \rho_i\}_{i=0}^{\infty}$ into the π-sequence $\{\Pi_i', \rho_i'\}_{i=0}^{\infty}$ is a sequence of homomorphisms $h_i : \Pi_i \to \Pi_i'$ such that

$\rho_i' \circ h_{i+1} = h_i \circ \rho_i$ for all $i \geqslant 0$;

h_{3k}, h_{3k+1}, and h_{3k-1} ($k \geqslant 2$) are h_3- , h_4- , and h_5-homomorphisms, respectively;

$h_2(\omega)h_3(\sigma) = h_2(\omega\sigma)$ for all $\omega \in \Pi_2$ and $\sigma \in \Pi_3$.

An <u>isomorphism</u> is a homomorphism such that all h_i's are isomorphisms.

12. Among conditions (i)-(vi) above, two refer to ρ_4, namely (iv) and (v). From (iv) it follows that if Π_4 acts identically on Π_5, then $\mathrm{Im}\,\rho_4$ is contained in the center of the group Π_4. From (v) it follows that if Π_4 acts identically on Π_5, then Π_5 is Abelian, and that the converse is true provided ρ_4 is an epimorphism.

In general, (v) implies that $\operatorname{Ker}\rho_4$ is contained in the center of Π_5.

13. The π-sequence (6) is <u>exact</u> if $\operatorname{Ker}\rho_i = \operatorname{Im}\rho_{i+1}$ for all $i \geqslant 0$ and, in addition, the preimages of the elements of Π_1 under ρ_1 are nothing but the orbits of the action of Π_3 on Π_2.

If the π-sequence (6) is exact and $i \geqslant 0$ is arbitrary, then obviously the homomorphism ρ_i is trivial if and only if ρ_{i+1} is an epimorphism, while $\operatorname{Ker}\rho_i$ is trivial if and only if ρ_{i+1} is trivial. In general, when $i \geqslant 3$, $\operatorname{Ker}\rho_i$ is trivial if and only ρ_i is injective, because ρ_i is a group homomorphism. Further, the triviality of $\operatorname{Ker}\rho_2$ means that ρ_2 is injective: if $\rho_2(\alpha) = \rho_2(\beta)$, then
$$\rho_2(\alpha\beta^{-1}) = \rho_2(\alpha)\beta^{-1} = \rho_2(\beta)\beta^{-1} = \rho_2(\beta\beta^{-1}) = \rho_2(e_{\Pi_3}) \quad \text{[see condition (vi)}$$
in 11], and hence $\alpha = \beta$. The triviality of $\operatorname{Ker}\rho_0$ does not imply the injectivity of ρ_0, and this is also valid for $\operatorname{Ker}\rho_1$ and ρ_1. However, in the case of an exact π-sequence (6), the injectivity of ρ_1 is guaranteed if the group Π_3 is trivial, or if it acts identically on Π_2.

The above discussion makes clear that, in the case of an exact π-sequence (6) and for $i \geqslant 1$, the triviality of ρ_i and ρ_{i+2} is equivalent to the invertibility of ρ_{i+1}, that the triviality of Π_i and Π_{i+2} implies the triviality of Π_{i+1}, and, finally, that the triviality of Π_{i-1} and Π_{i+2} implies the invertibility of ρ_i (cf. 2).

14. <u>Let (6) be an exact π-sequence. If the action of Π_4 on Π_2 induced by the action of Π_3 via ρ_3 is identical, then $\operatorname{Im}\rho_3$ is a normal subgroup of Π_3. The converse is true provided ρ_2 is an epimorphism.</u>

PROOF. Assume that Π_4 acts identically on Π_2. If $\alpha \in \Pi_4$, $\beta \in \Pi_3$, then $\rho_2(\beta)\rho_3(\alpha) = \rho_2(\beta)$, and hence $\rho_2(\beta\rho_3(\alpha)\beta^{-1}) =$
$$= \rho_2(\beta\rho_3(\alpha))\beta^{-1} = [\rho_2(\beta)\rho_3(\alpha)]\beta^{-1} = \rho_2(\beta)\beta^{-1} = \rho_2(\beta\beta^{-1}) = e_{\Pi_2} \quad \text{[see}$$
condition (vi) in 11]. Therefore, $\beta\rho_3(\alpha)\beta^{-1} \in \operatorname{Ker}\rho_2 = \operatorname{Im}\rho_3$, and hence $\operatorname{Im}\rho_3$ is a normal subgroup of Π_3.

Now assume that ρ_2 is an epimorphism and that $\operatorname{Im}\rho_3$ is a normal subgroup of Π_3. If $\alpha \in \Pi_4$ and $\gamma \in \Pi_2$, then there is $\beta \in \Pi_3$ such that $\rho_2(\beta) = \gamma$, and since $\beta\rho_3(\alpha)\beta^{-1} \in \operatorname{Im}\rho_3 = \operatorname{Ker}\rho_2$, we have
$$\gamma\rho_3(\alpha) = \rho_2(\beta)\rho_3(\alpha) = \rho_2(\beta\rho_3(\alpha)) = \rho_2(\beta\rho_3(\alpha)\beta^{-1}\beta) = \rho_2(\beta\rho_3(\alpha)\beta^{-1})\beta =$$
$$= e_{\Pi_2}\beta = \rho_2(e_{\Pi_3})\beta = \rho_2(\beta) = \gamma \quad \text{[see again condition (vi) in 11].}$$
Consequently, Π_4 acts identically on Π_2.

15. Every sequence of <u>Abelian</u> groups and <u>group</u> homomorphisms of the form (6) can be viewed as a π-sequence, where the action of Π_3 on Π_{3k}, $k \geqslant 2$, and the actions of Π_4 on Π_{3k+1} and Π_{3k-1}, $k \geqslant 2$, are identical, while the action of Π_3 on Π_2 is given by $\omega\sigma = \omega\rho_2(\sigma)$ $[\omega \in \Pi_2, \; \sigma \in \Pi_3]$. Then it is readily seen that a sequence which is exact in the sense of 2 is also exact as a π-sequence, and that the homomorphisms of sequences in the sense of 1 are also homomorphisms in the sense of 11.

Splitting of π-Sequences

16. We say that the π-sequence (6) is <u>split from the right at the term</u> Π_α <u>by the homomorphism</u> $\zeta: \Pi_{\alpha-1} \to \Pi_\alpha$ if $\rho_{\alpha-1} \circ \zeta = \mathrm{id}\,\Pi_{\alpha-1}$. This splitting is <u>normal</u> if $\alpha = 0,1,2$ or $\alpha \geqslant 3$ and $\mathrm{Im}\,\zeta$ is a normal subgroup of Π_α. We say that the π-sequence (6) is <u>split from the left at the term</u> Π_α <u>by the homomorphism</u> $\zeta: \Pi_\alpha \to \Pi_{\alpha+1}$ if $\zeta \circ \rho_\alpha = \mathrm{id}\,\Pi_{\alpha+1}$. ($\zeta$ is a group homomorphism when this makes sense, and a homomorphism of sets with identity elements otherwise.)

17. <u>If</u> $\alpha \geqslant 5$, <u>then any right splitting of the π-sequence</u> (6) <u>at</u> Π_α <u>is normal. If</u> (6) <u>is exact and</u> Π_4 <u>acts identically on</u> Π_5, <u>then any right splitting at</u> Π_4 <u>is normal.</u>

Since for $\alpha > 5$ the groups Π_α are Abelian, we need consider only Π_4 and Π_5. Suppose that the homomorphism $\zeta: \Pi_4 \to \Pi_5$ splits (6) from the right at Π_5. If $\alpha, \beta \in \Pi_5$, then

$$\beta\alpha\beta^{-1} = [\beta\zeta(\rho_4(\beta^{-1}))][\zeta(\rho_4(\beta))\alpha\beta^{-1}] =$$

$$= [\zeta(\rho_4(\beta))\alpha\beta^{-1}][\beta\zeta(\rho_4(\beta^{-1}))] = \zeta(\rho_4(\beta))\alpha\zeta(\rho_4(\beta^{-1}));$$

permuting the factors is permissible because $\beta\zeta(\rho_4(\beta^{-1})) \in \mathrm{Ker}\,\rho_4$ (as shown by the equality $\rho_4 \circ \zeta = \mathrm{id}\,\Pi_4$), and hence $\beta\zeta(\rho_4(\beta^{-1}))$ belongs to the center of Π_5 (see 12). This representation of $\beta\alpha\beta^{-1}$ shows that if $\alpha \in \mathrm{Im}\,\zeta$, then $\beta\alpha\beta^{-1} \in \mathrm{Im}\,\zeta$ for all $\beta \in \Pi_5$.

Now suppose that $\zeta: \Pi_3 \to \Pi_4$ splits the exact π-sequence (6) from the right at Π_4, and that Π_4 acts identically on Π_5. If $\alpha, \beta \in \Pi_4$, then

$$\beta\alpha\beta^{-1} = [\beta\zeta(\rho_3(\beta^{-1}))][\zeta(\rho_3(\beta))\alpha\beta^{-1}] =$$

$$= [\zeta(\rho_3(\beta))\alpha\beta^{-1}][\beta\zeta(\rho_3(\beta^{-1}))] = \zeta(\rho_3(\beta))\alpha\zeta(\rho_3(\beta^{-1}));$$

permuting the factors is permissible because $\beta\zeta(\rho_3(\beta^{-1})) \in \text{Ker}\,\rho_3 = \text{Im}\,\rho_4$

(as shown by the equality $\rho_3 \circ \zeta = \text{id}\,\Pi_3$), and hence $\beta\zeta(\rho_3(\beta^{-1}))$

belongs to the center of Π_4 (see 12). This representation of $\beta\alpha\beta^{-1}$

shows that if $\alpha \in \text{Im}\,\zeta$, then $\beta\alpha\beta^{-1} \in \text{Im}\,\zeta$ for all $\beta \in \Pi_4$.

18. Let the π-sequence (6) be exact, normally split from the right at Π_α and split from the right at $\Pi_{\alpha+3}$. If $\alpha \geqslant 4$, then, according to 6, sequence (6) also splits from the left at Π_α, and Π_α decomposes into the product of a subgroup canonically isomorphic to $\Pi_{\alpha+1}$ and a subgroup isomorphic to $\Pi_{\alpha-1}$. When $\alpha = 1,2,3$, (6) also splits from the left at Π_α (to see this, repeat the arguments of 6), but obvious examples demonstrate that the above isomorphism $\Pi_\alpha \cong \Pi_{\alpha+1} \times \Pi_{\alpha-1}$ is not necessarily valid.

Now let the π-sequence (6) be exact and split from the left at Π_α and $\Pi_{\alpha-3}$. If $\alpha \geqslant 6$, then, according to 7, (6) is normally split from the right at Π_α, and Π_α decomposes into the product of a subgroup canonically isomorphic to $\Pi_{\alpha+1}$ and a subgroup isomorphic to $\Pi_{\alpha-1}$. A word-for-word repetition of the arguments in 7 shows that this holds also for $\alpha = 4,5$. If $\alpha = 3$, all we can say is that (6) splits from the right at Π_3.

In what follows, we shall often encounter π-sequences which are exact, and split at every third term. The discussion above shows that if the π-sequence (6) is exact and normally split from the right (split from the left) at every term Π_{i_0+3k} with $i_0+3k \geqslant 1$, then it also splits from the left at these terms (respectively, it also splits from the right at Π_{i_0+3k} for $i_0+3k \geqslant 3$, and splits normally from the right at Π_{i_0+3k} for $i_0+3k \geqslant 4$).

The π-Variant of the Five Lemma

19. _Let $\{\Pi_i, \rho_i\}_{i=0}^{\infty}$ and $\{\Pi_i', \rho_i'\}_{i=0}^{\infty}$ be exact π-sequences, and let $\{h_i : \Pi_i \to \Pi_i'\}_{i=0}^{\infty}$ be a homomorphism of the first sequence into the second. If $h_{\alpha-1}$ and $h_{\alpha+1}$ are isomorphisms and $\text{Ker}\,h_{\alpha-2} = e_{\Pi_{\alpha-2}}$, $\text{Im}\,h_{\alpha+2} = \Pi_{\alpha+2}'$, then $\text{Ker}\,h_\alpha = e_{\Pi_\alpha}$ and $\text{Im}\,h_\alpha = \Pi_\alpha'$ (and hence h_α is a group isomorphism for all $\alpha \geqslant 3$)._

If $\alpha \geqslant 5$, then this is contained in 9. The proof for $\alpha = 2,3,4$ is similar.

6. The Homotopy Sequence of a Pair

1. Let (X,A) be a topological pair with base point $x_0 \in A$. According to Subsections 1 and 4, the homotopy groups $\pi_r(X,x_0)$ and $\pi_r(A,x_0)$ are defined for any $r \geqslant 0$, whereas the homotopy groups $\pi_r(X,A,x_0)$ are defined for any $r \geqslant 1$. Moreover, by 4.3, there are the homomorphisms $\partial: \pi_r(X,A,x_0) \to \pi_{r-1}(A,x_0)$. To these we add the homomorphisms $in_*: \pi_r(A,x_0) \to \pi_r(X,x_0)$ and $rel_*: \pi_r(X,x_0) \to \pi_r(X,A,x_0)$, induced by the inclusions $in: A \to X$ and $rel: (X,x_0,x_0) \to (X,A,x_0)$. These three series of homotopy groups and three series of homomorphisms can be assembled into the left-infinite sequence

$$\ldots \xrightarrow{\ \partial\ } \pi_2(A,x_0) \xrightarrow{\ in_*\ } \pi_2(X,x_0) \xrightarrow{\ rel_*\ } \pi_2(X,A,x_0) \xrightarrow{\ \partial\ } \pi_1(A,x_0) \xrightarrow{\ in_*\ }$$

$$\pi_1(X,x_0) \xrightarrow{\ rel_*\ } \pi_1(X,A,x_0) \xrightarrow{\ \partial\ } \pi_0(A,x_0) \xrightarrow{\ in_*\ } \pi_0(X,x_0) \tag{7}$$

Here, all the terms, except for the last six, are Abelian groups, all the terms, except for the last three, are groups, the last three terms are sets with an identity, all the maps, except for the last three, are group homomorphisms, and the last three maps are homomorphisms of sets with identity. By 3.3 and 4.4, $\pi_1(X,x_0)$ acts from the right on $\pi_r(X,x_0)$, while $\pi_1(A,x_0)$ acts from the right on $\pi_r(A,x_0)$ and $\pi_r(X,A,x_0)$, and all these are group-actions. Furthermore, the homomorphisms in_*, rel_*, and ∂ are compatible with these actions, as required in Definition 5.11 (see 3.6, 4.5, and 3.4), and $\alpha^{-1}\beta\alpha = T_{\partial\alpha}\beta$ for all $\alpha,\beta \in \pi_2(X,A,x_0)$ (see 4.7). From 4.8 and 4.9 it follows that $\pi_1(X,x_0)$ acts from the right on $\pi_1(X,A,x_0)$ in such a manner that $rel_*(\omega)\sigma = rel_*(\omega\sigma)$ for all $\omega,\sigma \in \pi_1(X,x_0)$. Therefore, (7) is a π-sequence, called the <u>homotopy sequence of the pair</u> (X,A) <u>with base point</u> x_0.

2. <u>Sequence</u> (7) <u>is exact.</u>

The proof is a routine, direct verification of the six inclusions $\mathrm{Im}\, in_* \subset \mathrm{Ker}\, rel_*$, $\mathrm{Ker}\, rel_* \subset \mathrm{Im}\, in_*$, $\mathrm{Im}\, rel_* \subset \mathrm{Ker}\, \partial$, $\mathrm{Ker}\, \partial \subset \mathrm{Im}\, rel_*$, $\mathrm{Im}\, \partial \subset \mathrm{Ker}\, in_*$, and $\mathrm{Ker}\, in_* \subset \mathrm{Im}\, \partial$, plus a justification of the fact that, given $\alpha,\beta \in \pi_1(X,A,x_0)$, there is a $\sigma \in \pi_1(X,x_0)$ such that $\alpha\sigma = \beta$ if and only if $\partial\alpha = \partial\beta$.

3. <u>Given any path</u> $s: I \to A$, <u>the vertical isomorphisms</u>

$$\cdots \; \pi_r(A,s(0)) \xrightarrow{\;\text{in}_*\;} \pi_r(X,s(0)) \xrightarrow{\;\text{rel}_*\;} \pi_r(X,A,s(0)) \xrightarrow{\;\partial\;} \pi_{r-1}(A,s(0)) \cdots$$

$$\Big\downarrow T_s \qquad\qquad \Big\downarrow T_{\text{in}\circ s} \qquad\qquad \Big\downarrow T_s \qquad\qquad \Big\downarrow T_s$$

$$\cdots \; \pi_r(A,s(1)) \xrightarrow{\;\text{in}_*\;} \pi_r(X,s(1)) \xrightarrow{\;\text{rel}_*\;} \pi_r(X,A,s(1)) \xrightarrow{\;\partial\;} \pi_{r-1}(A,s(1)) \cdots$$

<u>define an isomorphism of the first sequence onto the second.</u>

The commutativity of the first square has been established in 3.6, while the commutativity of the second and the third in 4.5. From the fact that the local systems $(A,\{\pi_r(A,x)\},\{T_s\})$, $(X,\{\pi_r(X,x)\},\{T_s\})$, and $(A,\{\pi_r(X,A,x)\},\{T_s\})$ satisfy property 2.1(i), and from the equality

$T_s(\omega\sigma) = T_s(\omega)T_{\text{in}\circ s}(\sigma)$ $\quad[\sigma \in \pi_1(X,s(0)), \quad \omega \in \pi_1(X,A,s(0))]$, deduced in 4.8, it follows that the vertical homomorphisms are compatible with the actions of the fundamental groups.

4. <u>Given a continuous map</u> $\; f: (X,A,x_0) \to (X',A',x_0')$, <u>the vertical homomorphisms</u>

$$\cdots \; \pi_r(A,x_0) \xrightarrow{\;\text{in}_*\;} \pi_r(X,x_0) \xrightarrow{\;\text{rel}_*\;} \pi_r(X,A,x_0) \xrightarrow{\;\partial\;} \pi_{r-1}(A,x_0) \cdots$$

$$\Big\downarrow (\text{ab } f)_* \qquad\quad \Big\downarrow f_* \qquad\qquad \Big\downarrow f_* \qquad\qquad \Big\downarrow (\text{ab } f)_* \qquad (8)$$

$$\cdots \; \pi_r(A',x_0') \xrightarrow{\;\text{in}_*\;} \pi_r(X',x_0') \xrightarrow{\;\text{rel}_*\;} \pi_r(X',A',x_0') \xrightarrow{\;\partial\;} \pi_{r-1}(A',x_0') \cdots$$

<u>induced by</u> f <u>yield a homomorphism of the first sequence into the second.</u>

The commutativity of the first two squares follows from 1.7 and 4.2. The commutativity of the third square has been established in 4.3, and the compatibility of the vertical homomorphisms with the actions of the fundamental groups - in 3.6, 4.5, and 4.8.

The Most Important Special Cases

5. If X is ∞-connected, then all the homomorphisms $\partial: \pi_r(X,A,x_0) \to \pi_{r-1}(A,x_0)$ are isomorphisms. If X is k-connected and $k < \infty$, then $\partial: \pi_r(X,A,x_0) \to \pi_{r-1}(A,x_0)$ is an isomorphism for all $r \leqslant k$, while $\partial: \pi_{k+1}(X,A,x_0) \to \pi_k(A,x_0)$ is an epimorphism. The converse of both statements is true provided that X is connected.

If A is ∞-connected, then all the homomorphisms $\text{rel}_*: \pi_r(X,x_0) \to \pi_r(X,A,x_0)$ are isomorphisms. If A is k-connected, and $k < \infty$, then $\mathrm{rel}_*: \pi_r(X,x_0) \to \pi_r(X,A,x_0)$ is an isomorphism for all $r \leqslant k$, while $\mathrm{rel}_*: \pi_{k+1}(X,x_0) \to \pi_{k+1}(X,A,x_0)$ is an epimorphism. If one of the spaces X, A is connected, then again, the converse of both statements is true.

If the pair (X,A) is ∞-connected, then all the homomorphisms $\mathrm{in}_*: \pi_r(A,x_0) \to \pi_r(X,x_0)$ are isomorphisms. If (X,A) is k-connected and $k < \infty$, then $\mathrm{in}_*: \pi_r(A,x_0) \to \pi_r(X,x_0)$ is an isomorphism for all $r < k$, while $\mathrm{in}_*: \pi_k(A,x_0) \to \pi_k(X,x_0)$ is an epimorphism. The converse is true in both cases (with no supplementary conditions). In particular, if $\mathrm{in}: A \to X$ is a homotopy equivalence, then the pair (X,A) is ∞-connected (see 3.7 and cf. 1.3.3.9).

6. If A is a retract of X, then the sequence (7) splits from the left at $\pi_r(X,x_0)$, and any retraction $\rho: X \to A$ induces splitting homomorphisms $\rho_*: \pi_r(X,x_0) \to \pi_r(A,x_0)$.

PROOF. Since $\rho \circ \mathrm{in} = \mathrm{id}\, A$, $\rho_* \circ \mathrm{in}_* = \mathrm{id}\, \pi_r(A,x_0)$.

7. Suppose that (X,x_0) can be contracted to (A,x_0), i.e., $\mathrm{id}\, X$ is x_0-homotopic to some map $h: X \to X$ such that $h(X) \subset A$. Then the sequence (7) splits from the right at $\pi_r(A,x_0)$, and as splitting homomorphisms one may take $(\mathrm{ab}\, h)_*: \pi_r(X,x_0) \to \pi_r(A,x_0)$.

PROOF. The composition $\mathrm{in} \circ \mathrm{ab}\, h$ is x_0-homotopic to $\mathrm{id}\, X$, and hence $\mathrm{in}_* \circ (\mathrm{ab}\, h)_* = \mathrm{id}\, \pi_r(X,x_0)$.

8. Suppose that (A,x_0) is contractible in (X,x_0), i.e., the inclusion $A \to X$ is x_0-homotopic to the constant map. Then the sequence (7) splits from the right at $\pi_r(X,A,x_0)$. As splitting homomorphisms $\pi_r(A,x_0) \to \pi_{r+1}(X,A,x_0)$ one may take those induced by the maps $\gamma_r: \mathrm{Sph}_r(A,x_0) \to \mathrm{Sph}_{r+1}(X,A,x_0)$, given by $[\gamma_r(\phi)](t_1,\ldots,t_{r+1}) = h(\phi(t_1,\ldots,t_r),t_{r+1})$ $[\phi \in \mathrm{Sph}_r(A,x_0)]$, where $h: A \times I \to X$ is any homotopy from $\mathrm{in}: A \to X$ to the constant map.

This is a corollary of the obvious equality

$$\partial \circ \gamma_r = \mathrm{id}\, \mathrm{Sph}_r(A,x_0).$$

9. The following remarks are concerned not with the homotopy sequences of pairs themselves, but with the homomorphism (8) induced by a map $f: (X,A,x_0) \to (X',A',x_0')$ between pairs with base point. The π-variant of the Five Lemma (see 5.19) shows that: if $(\mathrm{ab}\, f)_*: \pi_r(A,x_0) \to \pi_r(A',x_0')$, all $r \geqslant 0$, and $f_*: \pi_r(X,A,x_0) \to \pi_r(X',A',x_0')$, all $r \geqslant 1$, are isomorphisms, then so are $f_*: \pi_r(X,x_0) \to \pi_r(X',x_0')$, for all $r \geqslant 1$; if $f_*: \pi_r(X,x_0) \to \pi_r(X',x_0')$,

all $r \geq 1$, and $f_* : \pi_r(X,A,x_0) \to \pi_r(X',A',x_0')$, all $r \geq 1$, are isomorphisms, then so are $(ab\,f)_* : \pi_r(A,x_0) \to \pi_r(A',x_0')$, for all $r \geq 1$; finally, if $f_* : \pi_r(X,x_0) \to \pi_r(X',x_0')$, all $r \geq 0$, and $(ab\,f)_* : \pi_r(A,x_0) \to \pi_r(A',x_0')$, all $r \geq 0$, are isomorphisms, then so are $f_* : \pi_r(X,A,x_0) \to \pi_r(X',A',x_0')$, for all $r \geq 2$, while $f_* : \pi_1(X,A,x_0) \to \pi_1(X',A',x_0')$ is an epimorphism with trivial kernel.

In the last case, $f_* : \pi_1(X,A,x_0) \to \pi_1(X',A',x_0')$ is not necessarily injective; see 3.3.8. However, this map is certainly injective (and hence, an isomorphism) if we assume, in addition, that all the homomorphisms $(ab\,f)_* : \pi_1(A,x) \to \pi_1(A',f(x))$, $x \in A$, are epimorphic. To see this, let $\omega_1, \omega_2 \in \pi_1(X,A,x_0)$ with $f_*(\omega_1) = f_*(\omega_2)$, and let w_1 and w_2 be spheroids in the classes ω_1 and ω_2. Then there is a path $s' : I \to A'$ such that $s'(0) = f \circ w_1(1)$, $s'(1) = f \circ w_2(1)$, and the loop

$$((f \circ w_1)([in: A' \to X'] \circ s'))(f \circ w_2)^{-1} \tag{9}$$

is homotopic to the constant loop. Since $(ab\,f)_* : \pi_0(A,x_0) \to \pi_0(A',x_0')$ is an isomorphism and $(ab\,f)_* : \pi_1(A,w(1)) \to \pi_1(A',f(w_1(1)))$ is an epimorphism, there is a path $s : I \to A$ such that $s(0) = w_1(1)$, $s(1) = w_2(1)$, and the path $ab\,f \circ s : I \to A'$ is homotopic to s'. For such a choice of s, f takes the loop

$$(w_1([in: A \to X] \circ s))w_2^{-1} \tag{10}$$

into a loop homotopic to (9), and therefore homotopic to the constant loop. Finally, from the fact that $f_* : \pi_1(X,x_0) \to \pi_1(X',x_0')$ is an isomorphism, it follows that (10) itself is homotopic to the constant loop, i.e., the spheroids w_1 and w_2 are homotopic, and $\omega_1 = \omega_2$.

The Homotopy Sequence of a Triple

10. Let (X,A,B) be a topological triple with base point $x_0 \in B$. According to Subsection 4, when $r \geq 1$ the homotopy groups $\pi_r(X,A,x_0)$, $\pi_r(X,B,x_0)$, and $\pi_r(A,B,x_0)$, and the homomorphisms $in_* : \pi_r(A,B,x_0) \to \pi_r(X,B,x_0)$ and $rel_* : \pi_r(X,B,x_0) \to \pi_r(X,A,x_0)$, induced by the inclusions $in: (A,B) \to (X,B)$ and $rel: (X,B) \to (X,A)$, are well defined. If $r \geq 2$, we define an additional homomorphism, $\partial : \pi_r(X,A,x_0) \to \pi_{r-1}(A,B,x_0)$, as the composition of the boundary homomorphism $\pi_r(X,A,x_0) \to \pi_{r-1}(A,x_0)$ and the homomorphism $\pi_{r-1}(A,x_0) \to \pi_{r-1}(A,B,x_0)$, induced by the inclusion $(A,x_0,x_0) \to (A,B,x_0)$.

Now we may assemble these three series of groups and three series of homomorphisms into a left-infinite sequence

$$\cdots \xrightarrow{\partial} \pi_2(A,B,x_0) \xrightarrow{\text{in}_*} \pi_2(X,B,x_0) \xrightarrow{\text{rel}_*} \pi_2(X,A,x_0) \xrightarrow{\partial}$$

$$\pi_1(A,B,x_0) \xrightarrow{\text{in}_*} \pi_1(X,B,x_0) \xrightarrow{\text{rel}_*} \pi_1(X,A,x_0) \tag{11}$$

As was (7), (11) is a π-sequence: the right group-actions of $\pi_2(X,A,x_0)$ on the groups $\pi_r(X,A,x_0)$, and the right group-actions of $\pi_2(X,B,x_0)$ on the groups $\pi_r(A,B,x_0)$ and $\pi_r(X,B,x_0)$, are induced by the actions of $\pi_1(A,x_0)$ and $\pi_1(B,x_0)$ via the homomorphisms $\partial: \pi_2(X,A,x_0) \to \pi_1(A,x_0)$ and $\partial: \pi_2(X,B,x_0) \to \pi_1(B,x_0)$; similarly, the right action of $\pi_2(X,A,x_0)$ on $\pi_1(A,B,x_0)$ is induced by the action of $\pi_1(A,x_0)$ via the homomorphism $\partial: \pi_2(X,A,x_0) \to \pi_1(A,x_0)$; finally, 4.4, 4.5, 4.7, and 4.8 show that the conditions imposed by Definition 5.11 are satisfied. π-Sequence (11) is called the <u>homotopy sequence of the triple</u> (X,A,B) <u>with base point</u> x_0.

Sequence (11) is exact; cf. 2.

Given any path $s: I \to B$, the translations $\pi_r(X,A,s(0)) \to \pi_r(X,A,s(1))$, $\pi_r(X,B,s(0)) \to \pi_r(X,B,s(1))$, and $\pi_r(A,B,s(0)) \to \pi_r(A,B,s(1))$ define an isomorphism of the homotopy sequence of the triple (X,A,B) with base point $s(0)$ into the homotopy sequence of the triple (X,A,B) with base point $s(1)$; cf. 3.

Given any continuous map f from a triple (X,A,B) with base point x_0 into a triple (X',A',B') with base point x_0' ($x_0 \in B$, $x_0' \in B'$), the homomorphisms $f_*: \pi_r(X,A,x_0) \to \pi_r(X',A',x_0')$, $f_*: \pi_r(X,B,x_0) \to \pi_r(X',B',x_0')$, and $(\text{ab } f)_*: \pi_r(A,B,x_0) \to \pi_r(A',B',x_0')$, constitute a homomorphism from the homotopy sequence of the first triple into the homotopy sequence of the second triple; cf. 4.

7. The Local System of Homotopy
Groups of the Fibers of a Serre Bundle

1. Suppose that ξ is a Serre bundle, F_0 and F_1 are fibers of ξ, and $x_0 \in F_0$, $x_1 \in F_1$. Two spheroids, $\phi_0 \in \text{Sph}_r(F_0,x_0)$ and $\phi_1 \in \text{Sph}_r(F_1,x_1)$, are said to be <u>fiber homotopic</u> if the spheroids $[\text{in}: F_0 \to \text{tl}\,\xi] \circ \phi_0 \in \text{Sph}_r(\text{tl}\,\xi,x_0)$ and $[\text{in}: F_1 \to \text{tl}\,\xi] \circ \phi_1 \in \text{Sph}_r(\text{tl}\,\xi,x_1)$ can be connected by a free homotopy consisting of spheroids of $\text{tl}\,\xi$ which take I^r into fibers of ξ. In other words, ϕ_0 and ϕ_1 are fiber homotopic if there is a map $h: I^r \times I \to \text{tl}\,\xi$,

such that: h is constant on each set $\text{Fr } I^r \times t$, $t \in I$, $h(y,0) = $
$= \phi_0(y)$, $h(y,1) = \phi_1(y)$, $y \in I^r$, and the map $\text{pr } \xi \circ h$ is constant
on each set $I^r \times t$, $t \in I$. We say that h is <u>a fiber homotopy from</u>
ϕ_0 <u>to</u> ϕ_1 <u>along the path</u> $t \mapsto h(\text{Fr } I^r \times t)$.

2. <u>Given any spheroid</u> ϕ <u>with origin</u> x_0 <u>of the fiber</u> F_0,
<u>there is a fiber homotopy of</u> ϕ <u>along any path with origin</u> x_0 <u>in</u>
$\text{tl } \xi$. <u>Fiber homotopies of homotopic spheroids along homotopic paths of</u>
$\text{tl } \xi$ <u>always lead to homotopic spheroids. Fiber homotopies of freely</u>
<u>homotopic spheroids along paths which cover homotopic paths of</u> $\text{bs } \xi$
<u>lead to freely homotopic spheroids.</u>

PROOF. Let $s: I \to \text{tl } \xi$ be a path with $s(0) = x_0$, and let
$\phi \in \text{Sph}_r(F_0, x_0)$. Define homotopies $H: I^r \times I \to \text{bs } \xi$ and
$G: I^r \times I \to \text{tl } \xi$ by $H(y,t) = \text{pr } \xi \circ s(t)$ and $G(y,t) = s(t)$. Since
$H(y,0) = \text{pr } \xi(\phi(y))$ for all $y \in I^r$, and $G(y,0) = \phi(y)$ for all
$y \in \text{Fr } I^r$, there is a homotopy, $\tilde{H}: I^r \times I \to \text{tl } \xi$, which covers H
and satisfies $\tilde{H}(y,t) = G(y,t)$ for all $y \in \text{Fr } I^r$, $\tilde{H}(y,0) = \phi(y)$
for all $y \in I^r$ (see 4.1.3.6). $\tilde{H}$ is manifestly a fiber homotopy of
ϕ along s.

To prove the second assertion of the theorem, it suffices to
show that two spheroids, $\phi, \psi \in \text{Sph}_r(F_0, x_0)$, which are fiber homotopic
along a loop $s: I \to \text{tl } \xi$ homotopic with the constant loop, are
homotopic in the usual sense. Choose a fiber homotopy $\Phi: I^r \times I \to \text{tl } \xi$
from ϕ to ψ along s, and a homotopy $h: I \times I \to \text{tl } \xi$ from s
to the constant loop. Now define $\tilde{f}: I^{r+1} \to \text{tl } \xi$ and homotopies
$H: I^{r+1} \times I \to \text{bs } \xi$ and $G: \text{Fr } I^{r+1} \times I \to \text{tl } \xi$, by the formulas
$\tilde{f}(t_1, \ldots, t_{r+1}) = \Phi((t_1, \ldots, t_r), t_{r+1})$, $H((t_1, \ldots, t_{r+1}), t) =$
$= \text{pr } \xi \circ h(t_{r+1}, t)$, and

$$G((t_1, \ldots, t_{r+1}), t) = \begin{cases} \phi(t_1, \ldots, t_r), & \text{if } t_{r+1} = 0, \\ \psi(t_1, \ldots, t_r), & \text{if } t_{r+1} = 1, \\ h(t_{r+1}, t), & \text{if } (t_1, \ldots, t_r) \in \text{Fr } I^r. \end{cases}$$

Since $H(y,0) = \text{pr } \xi(\tilde{f}(y))$ for $y \in I^{r+1}$, and $G(y,0) = \tilde{f}(y)$ for
$y \in \text{Fr } I^{r+1}$, there is a homotopy $\tilde{H}: I^{r+1} \times I \to \text{tl } \xi$ which covers
H and satisfies $\tilde{H}(y,t) = G(y,t)$ for $y \in \text{Fr } I^{r+1}$ and $\tilde{H}(y,0) =$
$= \tilde{f}(y)$ for $y \in I^{r+1}$. Now let $\Psi((t_1, \ldots, t_r), t) = \tilde{H}((t_1, \ldots, t_r, t), 1)$,
and note that $\Psi: I^r \times I \to F_0$ is a (usual) homotopy from ϕ to ψ.

Let us prove the last part of the theorem. Suppose that the
spheroids $\phi_0 \in \text{Sph}_r(F_0, x_0)$ and $\phi_1 \in \text{Sph}_r(F_1, x_1)$ are fiber homotopic

along the path $u: I \to tl\,\xi$, and that the same holds for the spheroids $\psi_0 \in Sph_r(F_0,x_0)$, $\psi_1 \in Sph_r(F_1,x_1)$, and the path $v: I \to tl\,\xi$. Further, suppose that the paths $pr\,\xi \circ u, pr\,\xi \circ v: I \to bs\,\xi$ are homotopic, and that ϕ_0 and ψ_0 are freely homotopic along a path $w: I \to F_0$. The last means that there is a fiber homotopy from ϕ_0 to ψ_0 along the path $w_0 = [in: F_0 \to tl\,\xi] \circ w$. It is clear that the loop

$pr\,\xi \circ (u^{-1}(w_0 v)): I \to bs\,\xi$ is homotopic to the constant loop, which in turn implies that the path $u^{-1}(w_0 v)$ is homotopic to some path $w_1: I \to tl\,\xi$ with $w_1(I) \subset F_1$ (see 4.1.3.6). By the first part of the theorem, there exists a fiber homotopy of ϕ_1 along w_1, and now the second part guarantees that this homotopy yields a spheroid which is homotopic to ψ_1 [ϕ_1 is fiber homotopic to ψ_1 along the path $u^{-1}(w_0 v)$]. Consequently, ϕ_1 is fiber homotopic to ψ_1 along a path in F_1, i.e., the spheroids ϕ_1 and ψ_1 are freely homotopic.

3. By Theorem 2, the fiber homotopies along a given path $s: I \to tl\,\xi$ define (for any $r \geqslant 0$) a map $T_s: \pi_r(F_0,s(0)) \to \pi_r(F,s(1))$, where $F_0 = pr\,\xi^{-1}(pr\,\xi(s(0)))$, $F_1 = pr\,\xi^{-1}(pr\,\xi(s(1)))$. The maps T_s are obviously homomorphisms and satisfy conditions (i)-(iii) in 2.1. Therefore, a local system $(tl\,\xi, \{\pi_r(pr\,\xi^{-1}(pr\,\xi(x)),x)\}, \{T_s\})$ arises on $tl\,\xi$. This is a local system of groups for $r \geqslant 1$, and a local system of sets with identity for $r = 0$. It is called the <u>upper local system of the r-th homotopy groups of the fibers of</u> ξ. In particular, given any $x \in tl\,\xi$ and $r \geqslant 1$, there is a natural right group-action of $\pi_1(tl\,\xi,x)$ on $\pi_r(pr\,\xi^{-1}(pr\,\xi(x)),x)$.

Clearly, by restricting this local system to any fiber of ξ we obtain the local system of the r-th homotopy groups of the given fiber. Moreover, the homomorphisms $in_*: \pi_r(pr\,\xi^{-1}(pr\,\xi(x)),x) \to \pi_r(tl\,\xi,x)$ combine with $id\,tl\,\xi$ to define a homomorphism of the upper local system into the local system of the r-th homotopy groups of $tl\,\xi$.

4. Suppose now that the fibers of ξ are r-simple. Then, given any point $b \in bs\,\xi$, all the homotopy groups $\pi_r(pr\,\xi^{-1}(b),x)$, $x \in pr\,\xi^{-1}(b)$, may be identified with a unique group, $\pi_r(pr\,\xi^{-1}(b))$ (see 3.5). In this case, for any path $s: I \to bs\,\xi$, we can define $T_s: \pi_r(pr\,\xi^{-1}(s(0))) \to \pi_r(pr\,\xi^{-1}(s(1)))$ to be the translation $T_{\tilde{s}}: \pi_r(pr\,\xi^{-1}(s(0)),\tilde{s}(0)) \to \pi_r(pr\,\xi^{-1}(s(1)),\tilde{s}(1))$ along some path $\tilde{s}: I \to tl\,\xi$ which covers s; from 2 it follows that T_s does not depend upon the choice of $\tilde{s}$. The maps T_s are obviously homomorphisms and satisfy properties (i)-(iii) in 2.1. Therefore, a local system

$(bs\ \xi,\{\pi_r(pr\ \xi^{-1}(b))\},\{T_s\})$ arises on $bs\ \xi$, which consists of groups (sets with identity) for any $r \geqslant 1$ (respectively, for $r = 0$). This is called the <u>lower local system of the r-th homotopy groups of the</u> <u>fibers of</u> ξ. In particular, for any $r \geqslant 1$ and any $b \in bs\ \xi$, there is a natural right group-action of $\pi_1(bs\ \xi,b)$ on $\pi_r(pr\ \xi^{-1}(b))$.

It is readily seen that the lower local system induces the upper local system, defined in 3 on $tl\ \xi$, via the projection $pr\ \xi$.

5.   Let $\varphi: \xi \to \xi_1$ be a map of Serre bundles. Then $tl\ \varphi$ and the homomorphisms

$$ab\ tl\ \varphi_*: \pi_r(pr\ \xi^{-1}(pr\ \xi(x)),x) \to \pi_r(pr\ \xi_1^{-1}(pr\ \xi_1(tl\ \varphi(x))),tl\ \varphi(x))$$

$[x \in tl\ \xi]$ combine to define a homomorphism of the upper local system of the r-th homotopy groups of the fibers of ξ into the similar system for ξ_1. Furthermore, if the fibers of ξ and ξ_1 are r-simple, then $bs\ \varphi$ and $ab\ tl\ \varphi_*: \pi_r(pr\ \xi^{-1}(b)) \to \pi_r(pr\ \xi_1^{-1}(bs\ \varphi(b)))$ $[b \in bs\ \xi]$ combine to define a homomorphism of the lower local system of the r-th homotopy groups of the fibers of ξ into the similar system for ξ_1.

8.   The Homotopy Sequence of a Serre Bundle

1 (LEMMA).   <u>Let ξ be a Serre bundle with a base point</u> $x_0 \in tl\ \xi$, <u>and let B be a subset of</u> $bs\ \xi$ <u>with</u> $b_0 = pr\ \xi(x_0) \in B$. <u>Then</u> $pr\ \xi_*: \pi_r(tl\ \xi,pr\ \xi^{-1}(B),x_0) \to \pi_r(bs\ \xi,B,x_0)$ <u>and, in particular,</u> $pr\ \xi_*: \pi_r(tl\ \xi,pr\ \xi^{-1}(b_0),b_0) \to \pi_r(bs\ \xi,b_0)$ <u>are isomorphisms for any</u> $r \geqslant 1$.

PROOF. $pr\ \xi_*$ is epimorphic.

Let $\phi \in Sph_r(bs\ \xi,B,b_0)$. Define $\tilde{f}: I^{r-1} \to tl\ \xi$ and two homotopies, $H: I^{r-1} \times I \to bs\ \xi$ and $G: Fr\ I^{r-1} \times I \to tl\ \xi$, by

$\tilde{f}(I^{r-1}) = x_0$, $H((t_1,\ldots,t_{r-1}),t) = \phi(t_1,\ldots,t_{r-1},1-t)$, and

$G(Fr\ I^{r-1} \times I) = x_0$. Since $H(y,0) = pr\ \xi(\tilde{f}(y))$ for $y \in I^{r-1}$, and

$G(y,0) = \tilde{f}(y)$ for $y \in Fr\ I^{r-1}$, there is a homotopy $\tilde{H}: I^{r-1} \times I \to tl\ \xi$

which covers H and equals G on $Fr\ I^{r-1} \times I$ (see 4.1.3.6). Now

the formula $\psi(t_1,\ldots,t_r) = \tilde{H}((t_1,\ldots,t_{r-1}),1-t_r)$ defines a spheroid

$\psi \in Sph_r(tl\ \xi,pr\ \xi^{-1}(B),x_0)$ such that $pr\ \xi_\#(\psi) = \phi$.

2) $pr\ \xi_*$ is monomorphic.

Let $\psi \in Sph_r(tl\ \xi,pr\ \xi^{-1}(B),x_0)$, and suppose that the spheroid $pr\ \xi_\#(\psi) \in Sph_r(bs\ \xi,B,b_0)$ is homotopic to the constant spheroid.

Choose a homotopy $\Phi: I^r \times I \to \text{bs}\,\xi$ from $\text{pr}\,\xi_\#(\psi)$ to the constant spheroid, and define $\tilde{f}: I^r \to \text{tl}\,\xi$ and two homotopies, $H: I^r \times I \to \text{bs}\,\xi$ and $G: \text{Fr}\,I^r \times I \to \text{tl}\,\xi$, by $\tilde{f}(I^r) = x_0$, $H((t_1,\ldots,t_r),t) = \Phi((t_1,\ldots,t_{r-1},1-t),t_r)$, and

$$
G((t_1,\ldots,t_r),t) = \begin{cases} \psi(t_1,\ldots,t_{r-1},1-t), & \text{if } t_r = 0, \\[2mm] x_0, & \text{if } (t_1,\ldots,t_r) \in \text{Fr}\,I^r, \quad t_r \neq 0. \end{cases}
$$

Since $H(y,0) = \text{pr}\,\xi(\tilde{f}(y))$ for $y \in I^r$, and $G(y,0) = \tilde{f}(y)$ for $y \in \text{Fr}\,I^r$, there exists a homotopy $\tilde{H}: I^r \times I \to \text{tl}\,\xi$ which covers H and equals G on $\text{Fr}\,I^r \times I$. Now it is plain that $\Psi((t_1,\ldots,t_r),t) = \tilde{H}((t_1,\ldots,t_{r-1},t),1-t_r)$ defines a homotopy $\Psi: I^r \times I \to \text{tl}\,\xi$ from ψ to the constant spheroid.

The Action of $\pi_1(\text{bs}\,\xi,b_0)$ on $\text{comp}(F_0)$

2. Let ξ be a Serre bundle with base point $b_0 \in \text{bs}\,\xi$. Set $F_0 = \text{pr}\,\xi^{-1}(b_0)$ and define a right action of the group $\pi_1(\text{bs}\,\xi,b_0)$ on $\text{comp}(F_0)$ as follows: for $C \in \text{comp}(F_0)$ and $\sigma \in \pi_1(\text{bs}\,\xi,b_0)$, $C\sigma$ is the component of F_0 which contains the origins of those paths which end in C and cover loops in the class σ. That this action is well defined follows from Lemma 1: a path which ends in C and covers a loop in the class σ can be regarded as a spheroid of the pair $(\text{tl}\,\xi,F_0)$ with origin in C, and which is carried into a loop in class σ by $\text{pr}\,\xi_\#$. If $s_1 \in \text{Sph}_1(\text{tl}\,\xi,F_0,x_1)$ and $s_2 \in \text{Sph}_1(\text{tl}\,\xi,F_0,x_2)$ are two such spheroids, and w is a path in C with $w(0) = x_1$ and $w(1) = x_2$, then the loops $\text{pr}\,\xi_\#(s_1 w)$ and $\text{pr}\,\xi_\#(s_2)$ are homotopic. Now Lemma 1 implies that the spheroids $s_1 w, s_2 \in \text{Sph}_1(\text{tl}\,\xi,F_0,x_2)$ are homotopic, which, in turn, implies that the components of F_0 containing $s_1(0)$ and $s_2(0)$ coincide. It is readily seen that this is indeed a right action.

This action is compatible with the action of the fundamental group of $\text{tl}\,\xi$ on the homotopy groups of the fibers of ξ (see 7.3), namely $C\,\text{pr}\,\xi_*(\sigma) = T_\sigma C$ for all $C \in \text{comp}(F_0) = \pi_0(F_0,x_0)$, $\sigma \in \pi_1(\text{tl}\,\xi,x_0)$, and $x_0 \in F_0$. Moreover, if $f: \xi \to \xi'$ is a map of Serre bundles, then

$$
\text{fact}\,\text{ab}\,\text{tl}\,f : \text{comp}(F_0) \to \text{comp}(\text{pr}\,\xi'^{-1}(\text{bs}\,f(b_0))),
$$

where $\text{ab}\,\text{tl}\,f = [\text{ab}\,\text{tl}\,f: F_0 \to \text{pr}\,\xi'^{-1}(\text{bs}\,f(b_0))]$, is a [bs f_* : $\pi_1(\text{bs}\,\xi, b_0)$ $\to$ $\pi_1(\text{bs}\,\xi', \text{bs}\,f(b_0))$]-map.

      3. <u>If</u> $C \in \text{comp}(F_0)$ <u>and</u> $x_0 \in C$, <u>then the isotropy subgroup</u> <u>of</u> $\pi_1(\text{bs}\,\xi, b_0)$ <u>at</u> x_0 (<u>see</u> 4.2.3.4) <u>equals the image of the homo-</u> <u>morphism</u> $\text{pr}\,\xi_*$: $\pi_1(\text{tl}\,\xi, x_0)$ $\to$ $\pi_1(\text{bs}\,\xi, b_0)$.

      In fact, the equality $C\sigma = C$ means that there is a path $s: I \to \text{tl}\,\xi$ such that $s(0), s(1) \in C$ and s covers a loop in the class σ. This, in turn, guarantees the existence of a loop with origin x_0 which covers a loop in the class σ.

Construction of the Sequence

      4. Let ξ be a Serre bundle with base point $x_0 \in \text{tl}\,\xi$. Let $b_0 = \text{pr}\,\xi(x_0)$, $F_0 = \text{pr}\,\xi^{-1}(b_0)$, and apply Lemma 1 to transform the homotopy sequence of the pair $(\text{tl}\,\xi, F_0)$ with base point x_0 into a new sequence. Namely, for each $r \geqslant 1$, we replace the homotopy group $\pi_r(\text{tl}\,\xi, F_0, x_0)$ by $\pi_r(\text{bs}\,\xi, b_0)$, the homomorphism rel_* : $\pi_r(\text{tl}\,\xi, x_0)$ $\to$ $\pi_r(\text{tl}\,\xi, F_0, x_0)$ - by its composition with the isomorphism $\text{pr}\,\xi_*$: $\pi_r(\text{tl}\,\xi, F_0, x_0)$ $\to$ $\pi_r(\text{bs}\,\xi, b_0)$, and the homomorphism ∂ : $\pi_r(\text{tl}\,\xi, F_0, x_0)$ $\to$ $\pi_{r-1}(F_0, x_0)$ - by the composition $\Delta = \partial \circ (\text{pr}\,\xi_*)^{-1}$: $\pi_r(\text{bs}\,\xi, b_0)$ $\to$ $\pi_{r-1}(F_0, x_0)$. Since the composition of the inclusion $\text{rel}: (\text{tl}\,\xi, x_0, x_0)$ $\to$ $(\text{tl}\,\xi, F_0, x_0)$ with the projection $\text{pr}\,\xi$: $(\text{tl}\,\xi, F_0, x_0)$ $\to$ $(\text{bs}\,\xi, b_0, b_0)$ is simply $\text{pr}\,\xi: (\text{tl}\,\xi, x_0, x_0)$ $\to$ $(\text{bs}\,\xi, b_0, b_0)$, we see that $[\text{pr}\,\xi_*$: $\pi_r(\text{tl}\,\xi, F_0, x_0)$ $\to$ $\pi_r(\text{bs}\,\xi, b_0)] \circ \text{rel}_*$ is nothing else but $\text{pr}\,\xi_*$: $\pi_r(\text{tl}\,\xi, x_0)$ $\to$ $\pi_r(\text{bs}\,\xi, b_0)$. Finally, if we attach the homotopy group $\pi_0(\text{bs}\,\xi, b_0)$ to the right of the resulting sequence by means of the homomorphism $\text{pr}\,\xi_*$: $\pi_0(\text{tl}\,\xi, x_0)$ $\to$ $\pi_0(\text{bs}\,\xi, b_0)$, we obtain the sequence

$$
\begin{array}{l}
\cdots\ \pi_2(F_0, x_0) \xrightarrow{\ \text{in}_*\ } \pi_2(\text{tl}\,\xi, x_0) \xrightarrow{\ \text{pr}\,\xi_*\ } \pi_2(\text{bs}\,\xi, b_0) \xrightarrow{\ \Delta\ } \\[2ex]
\pi_1(F_0, x_0) \xrightarrow{\ \text{in}_*\ } \pi_1(\text{tl}\,\xi, x_0) \xrightarrow{\ \text{pr}\,\xi_*\ } \pi_1(\text{bs}\,\xi, b_0) \xrightarrow{\ \Delta\ } \\[2ex]
\pi_0(F_0, x_0) \xrightarrow{\ \text{in}_*\ } \pi_0(\text{tl}\,\xi, x_0) \xrightarrow{\ \text{pr}\,\xi_*\ } \pi_0(\text{bs}\,\xi, b_0)\ .
\end{array}
\tag{12}
$$

      By 3.3, 7.3, and 2, there are right group-actions of $\pi_1(\text{bs}\,\xi, b_0)$ on $\pi_r(\text{bs}\,\xi, b_0)$, and of $\pi_1(\text{tl}\,\xi, x_0)$ on $\pi_r(\text{tl}\,\xi, x_0)$ and $\pi_r(F_0, x_0)$, and also a right action of $\pi_1(\text{bs}\,\xi, b_0)$ on the set $\pi_0(F_0, x_0)$. The homomorphisms in_*, $\text{pr}\,\xi_*$, and Δ are compatible with these actions, as required by Definition 5.11 (see 3.6, 4.5, 7.3, and 2). Therefore, (12) is a π-sequence, called the <u>homotopy sequence of the</u>

bundle ξ __with base point__ x_0.

     5. __Sequence (12) is exact.__

     This is a corollary of the exactness of the homotopy sequence of the pair $(tl\,\xi, F_0)$ and of two additional and evident facts: the kernel of $pr\,\xi_* : \pi_0(tl\,\xi, x_0) \to \pi_0(bs\,\xi, b_0)$ equals the image of $in_* : \pi_0(F_0, x_0) \to \pi_0(tl\,\xi, x_0)$; and given $\alpha, \beta \in \pi_0(F_0, x_0)$, there is $\sigma \in \pi_1(bs\,\xi, b_0)$ such that $\beta = \alpha\sigma$ if and only if $in_*(\alpha) = in_*(\beta)$.

     6. __Given a map__ $f: \xi \to \xi'$ __of Serre bundles, the vertical__ __homomorphisms__

$$\cdots \pi_r(F_0, x_0) \xrightarrow{\ in_*\ } \pi_r(tl\,\xi, x_0) \xrightarrow{\ pr\,\xi_*\ } \pi_r(bs\,\xi, b_0) \xrightarrow{\ \Delta\ } \pi_{r-1}(F_0, x_0) \cdots$$

$$\downarrow (ab\,tl\,f)_* \qquad \downarrow tl\,f_* \qquad \downarrow bs\,f_* \qquad \downarrow (ab\,tl\,f)_*$$

$$\cdots \pi_r(F_0', x_0') \xrightarrow{\ in_*\ } \pi_r(tl\,\xi', x_0') \xrightarrow{\ pr\,\xi_*'\ } \pi_r(bs\,\xi', b_0') \xrightarrow{\ \Delta\ } \pi_{r-1}(F_0', x_0') \cdots$$

__where__ $x_0' = tl\,f(x_0)$, $b_0' = bs\,f(b_0)$, __and__ $F_0' = pr\,\xi'^{-1}(b_0')$, __constitute a homomorphism of the first homotopy sequence into the second.__

     The commutativity of the first two squares follows from 1.7, while the commutativity of the third follows from 4.2 and 4.3. The compatibility of the vertical homomorphisms with the actions of the fundamental groups was established in 3.6, 4.5, 7.5, and 2.

The Most Important Special Cases

     7. If $tl\,\xi$ is ∞-connected, then all the homomorphisms $\Delta : \pi_r(bs\,\xi, b_0) \to \pi_{r-1}(F_0, x_0)$ are isomorphisms. If $tl\,\xi$ is k-connected and $k < \infty$, then $\Delta : \pi_r(bs\,\xi, b_0) \to \pi_{r-1}(F_0, x_0)$ is an isomorphism for all $r \leqslant k$, while $\Delta : \pi_{k+1}(bs\,\xi, b_0) \to \pi_k(F_0, x_0)$ is an epimorphism. If $tl\,\xi$ is connected, then the converse is true in both cases.

     If $bs\,\xi$ is ∞-connected, then all the homomorphisms $in_* : \pi_r(F_0, x_0) \to \pi_r(tl\,\xi, x_0)$ are isomorphisms. If $bs\,\xi$ is k-connected and $k < \infty$, then $in_* : \pi_r(F_0, x_0) \to \pi_r(tl\,\xi, x_0)$ is an isomorphism for all $r \leqslant k$, while $in_* : \pi_{k+1}(F_0, x_0) \to \pi_{k+1}(tl\,\xi, x_0)$ is an epimorphism. If $bs\,\xi$ is connected, then the converse is true in both cases.

     If F_0 is ∞-connected, then all the homomorphisms $pr\,\xi_* : \pi_r(tl\,\xi, x_0) \to \pi_r(bs\,\xi, b_0)$ are isomorphisms. If F_0 is k-connected and $k < \infty$, then $pr\,\xi_* : \pi_r(tl\,\xi, x_0) \to \pi_r(bs\,\xi, b_0)$ is an isomorphism for all $r \leqslant k$, while $pr\,\xi_* : \pi_{k+1}(tl\,\xi, x_0) \to \pi_{k+1}(bs\,\xi, b_0)$

is an epimorphism. The converse is true in both cases (with no supplementary conditions).

8. <u>If the bundle</u> ξ <u>has a section</u> s <u>such that</u> $s(b_0) = x_0$, <u>then sequence</u> (12) <u>splits from the right at the terms</u> $\pi_r(tl\,\xi, x_0)$, <u>and</u> $s_*: \pi_r(bs\,\xi, b_0) \to \pi_r(tl\,\xi, x_0)$ <u>are splitting homomorphisms for any such</u> <u>section,</u> $s: (bs\,\xi, b_0) \to (tl\,\xi, x_0)$.

PROOF. Since $pr\,\xi \circ s = id\,bs\xi$, $pr\,\xi_* \circ s_* = id\,\pi_r(bs\,\xi, b_0)$.

9. <u>If</u> F_0 <u>is a retract of</u> $tl\,\xi$, <u>then sequence</u> (12) <u>splits</u> <u>from the left at the terms</u> $\pi_r(tl\,\xi, x_0)$, <u>and any retraction</u> $\rho: tl\,\xi \to F_0$ <u>induces splitting homomorphisms</u> $\rho_*: \pi_r(tl\,\xi, x_0) \to \pi_r(F_0, x_0)$.

PROOF. Since $\rho \circ in = id\,F_0$, $\rho_* \circ in_* = id\,\pi_r(F_0, x_0)$.

10. <u>If the inclusion</u> $in: F_0 \to tl\,\xi$ <u>is</u> x_0-<u>homotopic to the</u> <u>constant map, then sequence</u> (12) <u>splits from the right at the terms</u> $\pi_r(bs\,\xi, b_0)$. <u>Moreover, given any</u> x_0-<u>homotopy</u> $h: F_0 \times I \to tl\,\xi$ <u>from</u> in <u>to the constant map, consider the maps</u> $\gamma_r: Sph_r(F_0, x_0) \to$ $\to Sph_{r+1}(bs\,\xi, b_0)$ <u>given by</u> $[\gamma_r(\phi)](t_1, \ldots, t_{r+1}) = pr\,\xi \circ h(\phi(t_1, \ldots, t_r), t_{r+1})$ $(\phi \in Sph_r(F_0, x_0))$. <u>Then the homomorphisms</u> $\pi_r(F_0, x_0) \to \pi_{r+1}(bs\,\xi, b_0)$ <u>induced by</u> γ_r <u>split the sequence.</u>

PROOF. Given an arbitrary spheroid $\phi \in Sph_r(F_0, x_0)$, it suffices to find a spheroid $\psi \in Sph_{r+1}(tl\,\xi, F_0, x_0)$ such that $\partial\psi = \phi$ and $pr\,\xi_\#(\psi) = \gamma_r(\phi)$. We can set, for example, $\psi(t_1, \ldots, t_{r+1}) = $ $= h(\phi(t_1, \ldots, t_r), t_{r+1})$.

11. <u>If</u> $pr\,\xi$ <u>is</u> x_0-<u>homotopic to the constant map, then</u> <u>sequence</u> (12) <u>splits from the left at</u> $\pi_r(F_0, x_0)$. <u>Moreover, given any</u> x_0-<u>homotopy</u> $h: tl\,\xi \times I \to bs\,\xi$ <u>from</u> $pr\,\xi$ <u>to the constant map,</u> <u>consider the maps</u> $\gamma_r: Sph_r(F_0, x_0) \to Sph_{r+1}(bs\,\xi, b_0)$ <u>defined by</u> $[\gamma_r(\phi)](t_1, \ldots, t_{r+1}) = h(\phi(t_1, \ldots, t_r), t_{r+1})$ $(\phi \in Sph_r(F_0, x_0))$. <u>Then</u> <u>the homomorphisms</u> $\pi_r(F_0, x_0) \to \pi_{r+1}(bs\,\xi, b_0)$ <u>induced by</u> γ_r <u>split the</u> <u>sequence.</u>

PROOF. Given an arbitrary spheroid $\psi \in Sph_{r+1}(tl\,\xi, F_0, x_0)$, it suffices to show that the spheroids $\gamma_r \circ \partial(\psi)$ and $pr\,\xi_\#(\psi)$, which belong to $Sph_{r+1}(bs\,\xi, b_0)$, are homotopic. Clearly the formula $((t_1, \ldots, t_{r+1}), t) \mapsto h(\psi(t_1, \ldots, t_r, tt_{r+1}), (1-t)t_{r+1})$ defines such a homotopy.

12. <u>If</u> ξ <u>is a covering in the broad sense, then</u> $pr\,\xi_*: \pi_r(tl\,\xi, x_0) \to \pi_r(bs\,\xi, b_0)$ <u>is an isomorphism for</u> $r \geqslant 2$, <u>and a</u> <u>monomorphism for</u> $r = 1$. <u>If</u> ξ <u>is a covering (in the narrow sense),</u> <u>then, in addition, the map</u> $fact\Delta: \pi_1(bs\,\xi, b_0)/Im\,pr\,\xi_* \to F_0$, <u>induced</u>

<u>by</u> $\Delta: \pi_1(\text{bs}\,\xi,b_0) \to \pi_0(F_0,x_0) = F_0$,   <u>is invertible.</u>

This is a corollary of the exactness of the homotopy sequence of the bundle ξ, and of the fact that $\pi_r(F_0,x_0) = 0$ for all $r > 0$ and $\pi_0(\text{tl}\,\xi,x_0) = 0$ whenever ξ is a covering in the narrow sense.

13. Let ξ and ξ' be Serre bundles with base points $x_0 \in \text{tl}\,\xi$ and $x_0' \in \text{tl}\,\xi'$, and let $f: \xi \to \xi'$, with $\text{tl}\,f(x_0) = x_0'$. Set $b_0 = \text{pr}\,\xi(x_0)$, $b_0' = \text{pr}\,\xi'(x_0')$, $F_0 = \text{pr}\,\xi^{-1}(b_0)$, $F_0' = \text{pr}\,\xi'^{-1}(b_0')$. From Propositions 6 and 5.19, we derive the following conclusions. If $\text{bs}\,f_*: \pi_r(\text{bs}\,\xi,b_0) \to \pi_r(\text{bs}\,\xi',b_0')$, all $r \geqslant 1$, and $(\text{ab tl}\,f)_*: \pi_r(F_0,x_0) \to \pi_r(F_0',x_0')$, all $r \geqslant 0$, are isomorphisms, then so are $\text{tl}\,f_*: \pi_r(\text{tl}\,\xi,x_0) \to \pi_r(\text{tl}\,\xi',x_0')$, for all $r \geqslant 1$. If $\text{tl}\,f_*: \pi_r(\text{tl}\,\xi,x_0) \to \pi_r(\text{tl}\,\xi',x_0')$, all $r \geqslant 0$, and $(\text{ab tl}\,f)_*: \pi_r(F_0,x_0) \to \pi_r(F_0',x_0')$, all $r \geqslant 0$, are isomorphisms, then so are $\text{bs}\,f_*: \pi_r(\text{bs}\,\xi,b_0) \to \pi_r(\text{bs}\,\xi',b_0')$, for all $r \geqslant 1$. Finally, if $\text{bs}\,f_*: \pi_r(\text{bs}\,\xi,b_0) \to \pi_r(\text{bs}\,\xi',b_0')$, all $r \geqslant 0$, and $\text{tl}\,f_*: \pi_r(\text{tl}\,\xi,x_0) \to \pi_r(\text{tl}\,\xi',x_0')$, all $r \geqslant 0$, are isomorphisms, then so are $(\text{ab tl}\,f)_*: \pi_r(F_0,x_0) \to \pi_r(F_0',x_0')$, for all $r \geqslant 1$, while $(\text{ab tl}\,f)_*: \pi_0(F_0,x_0) \to \pi_0(F_0',x_0')$ is an epimorphism with trivial kernel.

We remark that in the last case, $(\text{ab tl}\,f)_*: \pi_0(F_0,x_0) \to \pi_0(F_0',x_0')$ is also an isomorphism if we make the additional assumption that all the homomorphisms $\text{tl}\,f_*: \pi_1(\text{tl}\,\xi,x) \to \pi_1(\text{tl}\,\xi',\text{tl}\,f(x))$ with $x \in F_0$ are epimorphisms. Indeed, let $x_1,x_2 \in F_0$ be such that $\text{tl}\,f(x_1)$ and $\text{tl}\,f(x_2)$ lie in the same component of the fiber F_0', and let $s': I \to F_0'$ be a path with $s'(0) = \text{tl}\,f(x_1)$, $s'(1) = \text{tl}\,f(x_2)$. Since $\text{tl}\,f_*: \pi_0(\text{tl}\,\xi,x_0) \to \pi_0(\text{tl}\,\xi',x_0')$ is an isomorphism, and $\text{tl}\,f_*: \pi_1(\text{tl}\,\xi,x_1) \to \pi_1(\text{tl}\,\xi',\text{tl}\,f(x_1))$ is an epimorphism, there is a path $s: I \to \text{tl}\,\xi$ such that $s(0) = x_1$, $s(1) = x_2$, and $\text{tl}\,f \circ s: I \to \text{tl}\,\xi'$ is homotopic to the path $[\text{in}: F_0' \to \text{tl}\,\xi'] \circ s'$. Then the loop $\text{pr}\,\xi' \circ \text{tl}\,f \circ s$ is homotopic to the constant loop, and from the fact that $\text{bs}\,f_*: \pi_1(\text{bs}\,\xi,b_0) \to \pi_1(\text{bs}\,\xi',b_0')$ is an isomorphism it follows that $\text{pr}\,\xi \circ s$ is also homotopic to the constant loop. Now, applying Theorem 4.1.3.6 to the map s, an arbitrary homotopy from $\text{pr}\,\xi \circ s$ to the constant loop, and the constant homotopy of the map $s|_{\text{Fr}\,I}$, we obtain a homotopy from s to a path $u: I \to \text{tl}\,\xi$ such that $u(I) \subset F_0$, $u(0) = x_1$, and $u(1) = x_2$.

9. The Influence of Other
Structures Upon Homotopy Groups

1. In this subsection we discuss the most elementary
properties of homotopy groups which are due to the presence of an
additional, group-like structure, compatible with the topology of the
space under consideration. The most important such property we shall
consider is simplicity.

The Case of Topological Groups

2. _If_ X _is a topological group and_ $s: I \to X$ _is an
arbitrary path, then the translation_ $\pi_r(X,s(0)) \to \pi_r(X,s(1))$ _coincides,
for any_ $r \geqslant 0$, _with the isomorphism induced by the left group
translation by the element_ $[s(1)][s(0)]^{-1}$.

In fact, there is even a canonical free homotopy from any
spheroid $\phi \in \mathrm{Sph}_r(X,s(0))$ to $[s(1)][s(0)]^{-1}\phi$ along s, given by
$((t_1,\ldots,t_r),t) \mapsto [s(t)][s(0)]^{-1}\phi(t_1,\ldots,t_r)$.

3 (COROLLARY). _The components of a topological group are
simple spaces. In particular, the fundamental groups of these
components are Abelian._

4. If X is a topological group, then, besides the
multiplication on the sets $\mathrm{Sph}_r(X,e=e_X)$ defined in 1.1, there is
another one, resulting from the group operation on X [the product of
two spheroids $\phi,\psi \in \mathrm{Sph}_r(X,e)$ is given by $y \mapsto \phi(y)\psi(y)$]. Moreover,
the second product makes also sense for $r = 0$, when the first product
is not even defined. Obviously, this new multiplication turns $\mathrm{Sph}_r(X,e)$,
$r \geqslant 0$, into a group; the spheroids homotopic to the constant one form
a normal subgroup, and the resulting quotient group equals, as a set,
$\pi_r(X,e)$. When $r = 0$, $\pi_0(X,e)$ equals the quotient group X/X_0, where
X_0 is the component of e. When $r \geqslant 1$, the new group structure on
$\pi_r(X,e)$ coincides with the original one; in fact, given $\phi,\psi \in \mathrm{Sph}_r(X,e)$,
the formula

$$((t_1,\ldots,t_r),t) \mapsto$$

$$\mapsto \phi\left(\min\left[1,\frac{2t_1}{1+t}\right],t_2,\ldots,t_r\right)\psi\left(\max\left[0,\frac{2t_1+t-1}{1+t}\right],t_2,\ldots,t_r\right)$$

defines a homotopy $I^r \times I \to X$ from the original product of ϕ and ψ to the new one.

Let us add that the existing multiplication on X also turns the set $\cup_{x \in X} \mathrm{Sph}_r(X,x)$ of <u>all</u> spheroids of X into a group. The spheroids homotopic to the constant ones form a normal subgroup, and the resulting quotient group is canonically isomorphic to $\pi_r(X,e)$, for any $r \geqslant 1$.

5. Every inner automorphism of the topological group X induces automorphisms of the groups $\pi_r(X,e)$, and thus the inner right action of X defines a right group-action of X on each group $\pi_r(X,e)$. The transformations induced by the elements of the subgroup X_0 (see 4) are all identical: if $w: I \to X$ is a path with $s(0) = e$ and $s(1) = x$, and $\phi \in \mathrm{Sph}_r(X,e)$, then the formula $(y,t) \mapsto [w(t)]^{-1}\phi(y)w(t)$ defines a homotopy $I^r \times I \to X$ from ϕ to the spheroid $y \mapsto x^{-1}\phi(y)x$. Therefore, there are natural right group-actions of $\pi_0(X,e) = X/X_0$ on the groups $\pi_r(X,e)$.

The Case of Homogeneous Spaces

6. <u>Let G be a topological group, and let H be a connected subgroup of G. If $(G,\mathrm{pr},X=G/H)$ is a Serre bundle, then given any path $s: I \to X$, the translation $\pi_r(X,s(0)) \to \pi_r(X,s(1))$ coincides, for any $r \geqslant 0$, with the isomorphism induced by any transformation of X (under the canonical action) which is given by an element of G which takes $s(0)$ into $s(1)$.</u>

PROOF. Let $g \in G$ be any element such that $gs(0) = s(1)$, and let $\tilde{s}: I \to G$ be any path covering s. Since $\tilde{s}(1)$ and $g\tilde{s}(0)$ lie in the same coset of H, and since the cosets of a connected group are connected, there is a path w in the coset containing $\tilde{s}(1)$ and $g\tilde{s}(0)$ such that $w(0) = \tilde{s}(1)$ and $w(1) = g\tilde{s}(0)$. Now given any $\phi \in \mathrm{Sph}_r(X,s(0))$, the formula

$$((t_1,\ldots,t_r),t) \mapsto [s_1(t)][s_1(0)]^{-1}\phi(t_1,\ldots,t_r)$$

where $s_1 = \tilde{s}w$, defines a free homotopy $I^r \times I \to X$ from ϕ to $[s_1(1)][s_1(0)]^{-1}\phi$, along a path homotopic to s (namely, the product of s and the constant path).

7 (COROLLARY). <u>If H is a connected subgroup of the topological group G and $(G,\mathrm{pr},G/H)$ is a Serre bundle, then the components of G/H are simple spaces. In particular, the fundamental groups of</u>

<u>the components of</u> G/H <u>are Abelian.</u>

8. Henceforth, (G,pr,G/H) will automatically be a Serre
bundle (as required in Theorems 6 and 7): G and its quotient space
G/H will always be closed smooth manifolds, while the projection
G → G/H will be a submersion and, by Theorems 4.6.1.3 and 4.1.3.4,
these properties imply that (G,pr,G/H) is a Serre bundle.

The Case of H-Spaces

9. A pointed topological space (X,e) is called an H-<u>space</u>
if there exists a continuous map $\mu: X \times X \to X$ such that $\mu(e,e) = e$,
and the maps $X \to X$, given by $x \mapsto \mu(e,x)$ and $x \mapsto \mu(x,e)$, are
e-homotopic to $\mathrm{id}\, X$. The map μ is called a <u>multiplication</u>, and e
is called an <u>identity</u> (or a <u>homotopy identity</u>). Usually, one writes
xy instead of $\mu(x,y)$.

An H-space X is <u>homotopy associative</u> if the maps
$X \times X \times X \to X$, given by $(x_1,x_2,x_3) \mapsto (x_1 x_2)x_3$ and $(x_1,x_2,x_3) \mapsto$
$\mapsto x_1(x_2 x_3)$, are homotopic, and <u>homotopy commutative</u> (or <u>Abelian</u>) if the
maps $X \times X \to X$, given by $(x_1,x_2) \mapsto x_1 x_2$ and $(x_1,x_2) \mapsto x_2 x_1$, are
homotopic.

Given a H-space X, a continuous map $\nu: X \to X$ is said to
be a <u>homotopy inverse</u> if the maps $X \to X$, given by $x \mapsto x\nu(x)$ and
$x \mapsto \nu(x)x$, are homotopic to the constant map which takes X into e.
Usually, one writes x^{-1} instead of $\nu(x)$.

The primary examples of H-spaces are topological groups.
Every topological group (viewed as a H-space) is homotopy associative
and has a homotopy inverse, and every Abelian topological group is
homotopy Abelian.

10. The spaces of spheroids provide another important class
of H-spaces. For a given pointed topological space (X,x_0), the sets
$\mathrm{Sph}_r(X,x_0)$ with $r \geqslant 1$ become H-spaces if we equip them with the
topology they inherit as subsets of $C(I^r,X)$ and with the usual
multiplication, and take the constant spheroid as the identity. In fact,
the map $\phi \mapsto (\mathrm{const})\phi$ is homotopic to $\mathrm{id}\,\mathrm{Sph}_r(X,x_0)$ via the
const-homotopy $\mathrm{Sph}_r(X,x_0) \times I \to \mathrm{Sph}_r(X,x_0)$, given by $(\phi,t) \mapsto \phi_t$,
where ϕ_t is the spheroid whose value at the point $(t_1,\ldots,t_r) \in I^r$
is given by the right-hand side of formula (3); further, the map
$\phi \mapsto \phi(\mathrm{const})$ is homotopic to $\mathrm{id}\,\mathrm{Sph}_r(X,x_0)$ via the const-homotopy
which is similarly defined by formula (4). The H-spaces $\mathrm{Sph}_r(X,x_0)$
are homotopy associative and have a homotopy inverse for $r \geqslant 1$, and

are homotopy Abelian for $r \geqslant 2$; the formulas in Subsection 1 again provide us with the necessary homotopies.

Similarly, given any topological pair (X,A) with base point x_0 , the sets $\mathrm{Sph}_r(X,A,x_0)$ with $r \geqslant 2$ are homotopy associative H-spaces possessing a homotopy inverse. If $r \geqslant 3$, these H-spaces are homotopy Abelian.

11. <u>Every connected H-space is simple and, in particular, has an Abelian fundamental group.</u>

PROOF. Suppose X is a connected H-space with identity e , $\phi \in \mathrm{Sph}_r(X,e)$ with arbitrary r , and $s \in \mathrm{Sph}_1(X,e)$ is a loop. Then, the formula $(y,t) \mapsto s(t)\phi(y)$ defines a free homotopy $I^r \times I \to X$ from the spheroid $y \mapsto e\phi(y)$ (which is homotopic to ϕ) to the same spheroid $y \mapsto e\phi(y)$, along the loop $t \mapsto s(t)e$ (which is homotopic to s). Thus, $\pi_1(X,e)$ acts identically on the groups $\pi_r(X,e)$.

12. It is clear that the second description of the homotopy groups $\pi_r(X,e)$, given in 4 for the case of topological groups, carries over to H-spaces X with identity e . Generally speaking, the multiplication that each set $\mathrm{Sph}_r(X,e)$ inherits from X does not turn this set into a group. However, this multiplication does induce the usual group structure on $\pi_r(X,e)$, $r \geqslant 1$. Moreover, if X is homotopy associative and has a homotopy inverse, then the set $\pi_0(X,e)$ is a group.

The Local System of Homotopy Groups
of the Total Space of a Principal Bundle

13 (LEMMA). <u>Let</u> ξ <u>be a principal bundle with structure group</u> G , <u>and let</u> $u,v: I \to \mathrm{tl}\,\xi$ <u>be paths satisfying</u> $\mathrm{pr}\,\xi \circ u = \mathrm{pr}\,\xi \circ v$. <u>If</u> $g_0, g_1 \in G$ <u>are such that</u> $u(0)g_0 = v_0$ <u>and</u> $u(1)g_1 = v(1)$ [<u>here</u> G <u>acts canonically from the right on</u> $\mathrm{tl}\,\xi$; <u>see</u> 4.3.2.10], <u>then the diagram</u>

$$
\begin{array}{ccc}
\pi_r(\mathrm{tl}\,\xi, u(0)) & \xrightarrow{\;T_u\;} & \pi_r(\mathrm{tl}\,\xi, u(1)) \\
\downarrow & & \downarrow \\
\pi_r(\mathrm{tl}\,\xi, v(0)) & \xrightarrow{\;T_v\;} & \pi_r(\mathrm{tl}\,\xi, v(1))
\end{array}
$$

<u>where the vertical isomorphisms are induced by the transformations</u> $x \mapsto xg_0$ <u>and</u> $x \mapsto xg_1$, <u>commutes.</u>

PROOF. Recall that the canonical right action $\mathrm{tl}\,\xi \times G \to \mathrm{tl}\,\xi$

is free and its orbits coincide with the fibers of ξ. Since $u(t)$ and $v(t)$ lie in the same fiber, for each $t \in I$ there is a unique $g(t) \in G$ such that $u(t)g(t) = v(t)$. Therefore, if $h: I^r \times I \to tl\,\xi$ is a free homotopy from $\phi_0 \in Sph_r(tl\,\xi, u(0))$ to $\phi_1 \in Sph_r(tl\,\xi, u(1))$ along u, then $(y,t) \mapsto h(y,t)g(t)$ yields a free homotopy from the spheroid $y \mapsto \phi_0(y)g_0$ to the spheroid $y \mapsto \phi_1(y)g_1$ along v.

14. Suppose that ξ is a principal bundle, $b \in bs\,\xi$, and $r \geqslant 0$. The right canonical action of the structure group of ξ on $tl\,\xi$ induces isomorphisms between the homotopy groups $\pi_r(tl\,\xi, x)$ with $x \in pr\,\xi^{-1}(b)$, and so we may identify these groups. We shall call the resulting group (which actually is a group for $r > 0$ and a set with identity for $r = 0$) the <u>r-th homotopy group of the space</u> $tl\,\xi$ <u>over</u> b, and we shall write $\pi_r(tl\,\xi, b)$.

Given a path $s: I \to bs\,\xi$, we define $T_s: \pi_r(tl\,\xi, s(0)) \to \pi_r(tl\,\xi, s(1))$ as the translation $T_{s^\sim}: \pi_r(tl\,\xi, s^\sim(0)) \to \pi_r(tl\,\xi, s^\sim(1))$ along any path $s^\sim$ which covers s. That this is well defined follows from Lemma 13. Obviously, T_s are homomorphisms and satisfy conditions 2.1 (i)-(iii). Therefore, we have produced a local system on $bs\,\xi$, $(bs\,\xi, \{\pi_r(tl\,\xi, b)\}, \{T_s\})$, which we call the <u>lower local system of the</u> <u>r-th homotopy groups of the total space of</u> ξ.

It is clear that the local system on $tl\,\xi$ induced by this system via $pr\,\xi$ is nothing but the usual local system of the r-th homotopy groups of $tl\,\xi$.

Given a monomorphism $\varphi: G \to G'$ and $r \geqslant 0$, every φ-map of the Steenrod G-bundle ξ into the Steenrod G'-bundle ξ' induces a homomorphism of the lower local system of the r-th homotopy groups of $tl\,\xi$ into the corresponding system of $tl\,\xi'$.

The Homotopy Sequence of a Principal Bundle

15. Let ξ be a principal G-bundle with base point $x_0 \in tl\,\xi$. If in sequence (12) we replace (F_0, x_0) by the pair $(G, e=e_G)$ [which is canonically homeomorphic to (F_0, x_0) via $g \mapsto x_0 g$] and attach 1 to the right of the resulting sequence, we obtain

$$\ldots \pi_2(G, e) \xrightarrow{in_*} \pi_2(tl\,\xi, x_0) \xrightarrow{pr\,\xi_*} \pi_2(bs\,\xi, b_0) \xrightarrow{\Delta} \pi_1(G, e) \xrightarrow{in_*}$$

$$\pi_1(tl\,\xi, x_0) \xrightarrow{pr\,\xi_*} \pi_1(bs\,\xi, b_0) \xrightarrow{\Delta} \pi_0(G, e) \xrightarrow{in_*} \qquad (13)$$

$$\pi_0(tl\,\xi, x_0) \xrightarrow{pr\,\xi_*} \pi_0(bs\,\xi, b_0) \longrightarrow 1$$

$(b_0 = pr\,\xi(x_0))$.

Recall that $\pi_0(G,e)$ is a group (see 4) and that $\pi_1(G,e)$ is Abelian (see 3). It is immediate that $\Delta: \pi_1(\mathrm{bs}\,\xi,b_0) \to \pi_0(G,e)$ is a group homomorphism. Moreover, $\pi_0(G,e)$ acts from the right on $\pi_r(G,e)$ (see 5), while $\pi_1(\mathrm{bs}\,\xi,b_0)$ acts similarly on both $\pi_r(\mathrm{bs}\,\xi,b_0)$ (see 3.3) and $\pi_r(\mathrm{tl}\,\xi,x_0) = \pi_r(\mathrm{tl}\,\xi,b_0)$ (see 14). The canonical right action $\mathrm{tl}\,\xi \times G \to \mathrm{tl}\,\xi$ induces a right action of G on $\pi_0(\mathrm{tl}\,\xi,x_0) = \mathrm{comp}(\mathrm{tl}\,\xi)$, and thus a right action of $\pi_0(G,e)$ on $\pi_0(\mathrm{tl}\,\xi,x_0)$. [We regard $\pi_0(G,e)$ as the quotient group of G by the component of e; see 4. The action of this component on $\pi_0(\mathrm{tl}\,\xi,x_0)$ is identical.] Finally, it is clear that the homomorphisms in_*, pr_*, and Δ are compatible with the above actions, as required in Definition 5.11. Consequently, (13) is a π-sequence, called the <u>homotopy sequence of the</u> <u>G-bundle</u> ξ <u>with base point</u> x_0.

Obviously, $\mathrm{pr}\,\xi_*: \pi_0(\mathrm{tl}\,\xi,x_0) \to \pi_0(\mathrm{bs}\,\xi,b_0)$ is an epimorphism, and the partition of $\pi_0(\mathrm{tl}\,\xi,x_0)$ into the orbits of $\pi_0(G,e)$ is exactly $\mathrm{zer}(\mathrm{pr}\,\xi_*)$. Therefore, sequence (13) is exact.

Given a monomorphism $\varphi: G' \to G$, every φ-map f of the principal G'-bundle ξ' with base point $x_0' \in \mathrm{tl}\,\xi'$ into the principal G-bundle ξ with base point $x_0 \in \mathrm{tl}\,\xi$, such that $\mathrm{tl}\,f(x_0') = x_0$, induces a homomorphism of the homotopy sequence of ξ' into the homotopy sequence of ξ.

10. <u>Alternative Descriptions</u> <u>of the Homotopy Groups</u>

1. The spheroid $\mathrm{DS} \circ \mathrm{ID} \in \mathrm{Sph}_r(S^r,\mathrm{ort}_1)$ (see 1.2.8.9) is called the <u>fundamental spheroid of the sphere</u> S^r, denoted IS, and we let sph_r denote the element of $\pi_r(S^r,\mathrm{ort}_1)$ that it defines. The spheroid $\mathrm{ID} \in \mathrm{Sph}_r(D^r,S^{r-1},\mathrm{ort}_1)$ is called the <u>fundamental spheroid</u> <u>of the ball</u> D^r, and we let kug_r denote the element of $\pi_r(D^r,S^{r-1},\mathrm{ort}_1)$ that it defines.

Obviously, $\partial(\mathrm{ID}) = \mathrm{IS}$, whence $\partial(\mathrm{kug}_r) = \mathrm{sph}_{r-1}$.

We let $\mathrm{Sph}_r^O(X,x_0)$ denote the set of all continuous maps from the pointed space (S^r,ort_1) into the pointed space (X,x_0), and define $\mathrm{IS}^{\#}: \mathrm{Sph}_r^O(X,x_0) \to \mathrm{Shp}_r(X,x_0)$ by $\mathrm{IS}^{\#}(\phi) = \phi \circ \mathrm{IS}$. Clearly, this map is invertible, and 1.3.7.6 implies that two maps, $\phi,\psi \in \mathrm{Sph}_r^O(X,x_0)$, are homotopic if and only if the spheroids $\mathrm{IS}^{\#}(\phi)$, $\mathrm{IS}^{\#}(\psi)$ are homotopic. Consequently, replacing our "cubic" spheroids and their homotopies by the "spheric" spheroids from $\mathrm{Sph}_r^O(X,x_0)$ and their homotopies, we are led to an equivalent description of the set $\pi_r(X,x_0)$.

It is readily seen that the identity spheroid $\mathrm{id}\,S^r$ belongs to the class sph_r, and that the element of $\pi_r(X,x_0)$ given by a spheroid $f\colon (S^r,\mathrm{ort}_1) \to (X,x_0)$ equals $f_*(\mathrm{sph}_r)$.

If $r \geqslant 1$, then $\mathrm{IS}^{\#}$ transfers the multiplication in $\mathrm{Sph}_r(X,x_0)$ to $\mathrm{Sph}_r^O(X,x_0)$ The resulting multiplication in $\mathrm{Sph}_r^O(X,x_0)$ may also be described directly: the product

$$\phi\psi\colon (S^r,\mathrm{ort}_1) = (S^1,\mathrm{ort}_1) \otimes \dots \otimes (S^1,\mathrm{ort}_1) \to (X,x_0)$$

of the spheroids $\phi,\psi\colon (S^r,\mathrm{ort}_1) = (S^1,\mathrm{ort}_1) \otimes \dots \otimes (S^1,\mathrm{ort}_1) \to (X,x_0)$ is given by

$$\phi\psi(y_1,y_2,\dots,y_r) = \begin{cases} \phi(y_1^2,y_2,\dots,y_r), & \text{if } \operatorname{im} y_1 \geqslant 0, \\[2em] \psi(y_1^2,y_2,\dots,y_r), & \text{if } \operatorname{im} y_1 \leqslant 0, \end{cases} \tag{14}$$

where $y_1,y_2,\dots,y_r$ are complex numbers of modulus 1, and im denotes the imaginary part. The multiplication that this operation induces on $\pi_r(X,x_0)$ coincides with the existing one. One can use (14) to study directly the homotopy properties of the multiplication in $\mathrm{Sph}_r^O(X,x_0)$ and get an independent description of the homotopy groups $\pi_r(X,x_0)$ in the language of spheric spheroids.

It is particularly simple to describe in this language the spheroid $\phi^{-1} = (\mathrm{IS}^{\#})^{-1}([\mathrm{IS}^{\#}(\phi)]^{-1})$, that is, the inverse of the spheroid $\phi \in \mathrm{Sph}_r^O(X,x_0)$: $\phi^{-1}(x_1,x_2,x_3,\dots,x_{r+1}) = \phi(x_1,-x_2,x_3,\dots,x_{r+1})$.

3. Let $\mathrm{Sph}_r^O(X,A,x_0)$ be the set of all continuous maps $(D^r,S^{r-1},\mathrm{ort}_1) \to (X,A,x_0)$, and define $\mathrm{ID}^{\#}\colon \mathrm{Sph}_r^O(X,A,x_0) \to \mathrm{Sph}_r(X,A,x_0)$ by $\mathrm{ID}^{\#}(\phi) = \phi \circ \mathrm{ID}$. This map is invertible, and 1.3.7.6 implies that two maps, $\phi,\psi \in \mathrm{Sph}_r^O(X,A,x_0)$, are homotopic, if and only if the spheroids $\mathrm{ID}^{\#}(\phi)$ and $\mathrm{ID}^{\#}(\psi)$ are homotopic. If $r \geqslant 2$, then $\mathrm{ID}^{\#}$ transfers the multiplication from $\mathrm{Sph}_r(X,A,x_0)$ to $\mathrm{Sph}_r^O(X,A,x_0)$. The resulting multiplication in $\mathrm{Sph}_r^O(X,A,x_0)$ may also be described directly: given two maps,

$$\phi,\psi\colon (D^r,\mathrm{ort}_1) = (S^1,\mathrm{ort}_1) \otimes \dots \otimes (S^1,\mathrm{ort}_1) \otimes (I,1) \to (X,x_0),$$

formula (14), where $y_1,\dots,y_{r-1}$ are complex numbers of modulus 1 and $y_r \in I$, defines a map $\phi\psi\colon (D^r,\mathrm{ort}_1) \to (X,x_0)$, and this map belongs to $\mathrm{Sph}_r^O(X,A,x_0)$ whenever $\phi,\psi \in \mathrm{Sph}_r^O(X,A,x_0)$. As in the absolute case, $\phi^{-1} = (\mathrm{ID}^\#)^{-1}([\mathrm{ID}^\#(\phi)]^{-1})$, that is, the inverse of the spheroid $\phi \in \mathrm{Sph}_r^O(X,A,x_0)$, is given by $\phi^{-1}(x_1,x_2,x_3,\ldots,x_r) = \phi(x_1,-x_2,x_3,\ldots,x_r)$.

Therefore, by replacing the cubic spheroids by spheric ones, we get an adequate description of the homotopy groups $\pi_r(X,A,x_0)$.

Obviously, the identity spheroid $\mathrm{id}\, D^r$ belongs to the class kug_r, and the element of $\pi_r(X,A,x_0)$ given by a spheroid

$f: (D^r,S^{r-1},\mathrm{ort}_1) \to (X,A,x_0)$ equals $f_*(\mathrm{kug}_r)$.

Unlike the cubic sets $\mathrm{Sph}_r(X,x_0,x_0)$ and $\mathrm{Sph}_r(X,x_0)$, the sets $\mathrm{Sph}_r^O(X,x_0,x_0)$ and $\mathrm{Sph}_r^O(X,x_0)$ are not identical, being merely related through the canonical invertible map $(\mathrm{ID}^\#)^{-1} \circ (\mathrm{IS}^\#)$:

$: \mathrm{Sph}_r^O(X,x_0) \to \mathrm{Sph}_r^O(X,x_0,x_0)$; this map can be described more directly as $\phi \mapsto \phi \circ \mathrm{DS}$.

The <u>boundary</u> $\partial\phi$ of a spheric spheroid $\phi \in \mathrm{Sph}_r^O(X,A,x_0)$ is the element $\partial\phi \in \mathrm{Sph}_{r-1}^O(A,x_0)$ given by $\partial\phi = [\mathrm{ab}\,\phi: (S^{r-1},\mathrm{ort}_1) \to (A,x_0)]$. Clearly, $\mathrm{IS}^\# \circ \partial = \partial \circ \mathrm{ID}^\#$, which demonstrates that this definition of the boundary leads to the same boundary homomorphism $\partial: \pi_r(X,A,x_0) \to \pi_{r-1}(A,x_0)$ as the cubic theory does.

4. For each continuous map, $f: (X,x_0) \to (X',x_0')$ or $f: (X,A,x_0) \to (X',A',x_0')$, we have the map $f_\#: \mathrm{Sph}_r^O(X,x_0) \to \mathrm{Sph}_r^O(X',x_0')$ and respectively $f_\#: \mathrm{Sph}_r^O(X,A,x_0) \to \mathrm{Sph}_r^O(X',A',x_0')$, given by $f_\#(\phi) = f \circ \phi$. Trivially, $\mathrm{IS}^\# \circ f_\# = f_\# \circ \mathrm{IS}^\#$ and $\mathrm{ID}^\# \circ f_\# = f_\# \circ \mathrm{ID}^\#$, so that these maps $f_\#$ lead to the same induced homomorphisms $f_*: \pi_r(X,x_0) \to \pi_r(X',x_0')$ and $f_*: \pi_r(X,A,x_0) \to \pi_r(X',A',x_0')$ as the cubic theory does.

A free homotopy from a spheroid $\phi_0 \in \mathrm{Sph}_r^O(X,x_0)$ to a spheroid $\phi_1 \in \mathrm{Sph}_r^O(X,x_1)$ along a path $s: I \to X$ is defined as a usual homotopy $S^r \times I \to X$ from ϕ_0 to ϕ_1 which takes (ort_1,t) into $s(t)$ for any $t \in I$. If h is such a homotopy, then $h \circ (\mathrm{IS} \times \mathrm{id}\,I)$ is a free homotopy from $\mathrm{IS}^\#(\phi_0) \in \mathrm{Sph}_r(X,x_0)$ to $\mathrm{IS}^\#(\phi_1) \in \mathrm{Sph}_r(X,x_1)$ along s. Therefore, the spheric free homotopies along s yield the same isomorphism $T_s: \pi_r(X,x_0) \to \pi_r(X,x_1)$ as do the cubic ones. It is readily seen that all this carries over to the relative case.

An Alternative Proof of Theorem 4.7

5. Using spheric spheroids and the homotopy sequence of a
pair, one can give a second proof of Theorem 4.7, which is less direct
but, in return, more transparent.

Consider first a model situation: $X = (D^2, \text{ort}_1) \vee (D^2, \text{ort}_1)$,
$A = (S^1, \text{ort}_1) \vee (S^1, \text{ort}_1)$, x_0 is the center of both bouquets,
$\alpha = \text{imm}_{1*}(\text{kug}_2)$, and $\beta = \text{imm}_{2*}(\text{kug}_2)$. Since $\partial(\alpha^{-1}\beta\alpha) = (\partial\alpha)^{-1}(\partial\beta)(\partial\alpha) =$
$= T_{\partial\alpha}\partial\beta$ (see 3.4) and $T_{\partial\alpha}\partial\beta = \partial T_{\partial\alpha}\beta$ (see 4.5), we have $\partial(\alpha^{-1}\beta\alpha) =$
$= \partial T_{\partial\alpha}\beta$. Moreover, since X is contractible, ∂ is an isomorphism
(see 6.5), and hence $\alpha^{-1}\beta\alpha = T_{\partial\alpha}\beta$.

In the general case, pick two arbitray spheroids
$\phi, \psi \in \text{Sph}_r^O(X, A, x_0)$ in the classes α, β, and define

$$f: ((D^2, \text{ort}_1) \vee (D^2, \text{ort}_1), (S^1, \text{ort}_1) \vee (S^1, \text{ort}_1)) \to (X, A)$$

by

$$f(\text{imm}_1(x)) = \phi(x), \quad f(\text{imm}_2(x)) = \psi(x), \quad x \in D^2.$$

Since $\phi = f \circ \text{imm}_1$, $\psi = f \circ \text{imm}_2$, we have

$$\alpha = (f \circ \text{imm}_1)_*(\text{kug}_2), \quad \beta = (f \circ \text{imm}_2)_*(\text{kug}_2),$$

and hence

$$\alpha^{-1}\beta\alpha = f_*([\text{imm}_{1*}(\text{kug}_2)]^{-1}[\text{imm}_{2*}(\text{kug}_2)][\text{imm}_{1*}(\text{kug}_2)]) =$$
$$= f_*(T_{\partial \circ \text{imm}_{1*}(\text{kug}_2)}\text{imm}_{2*}(\text{kug}_2)) = T_{\partial\alpha}\beta.$$

Spheroids in Spheroid Spaces

6. Let (X, x_0) be a pointed space. With every spheroid
$\phi \in \text{Sph}_{r+s}(X, x_0)$ we may associate an r-spheroid of the space
$\text{Sph}_s(X, x_0)$ [see 9.10] with base point the constant spheroid, by means
of each of the formulas

$$[\psi(t_1, \ldots, t_r)](u_1, \ldots, u_s) = \phi(t_1, \ldots, t_r, u_1, \ldots, u_s),$$

and

$$[\psi(t_1, \ldots, t_r)](u_1, \ldots, u_s) = \phi(u_1, \ldots, u_s, t_1, \ldots, t_r).$$

This leads to two maps, Cub, Buc: $\text{Sph}_{r+s}(X, x_0) \to \text{Sph}_r(\text{Sph}_s(X, x_0), \text{const})$.
These maps are invertible and both they and their inverses take
homotopic spheroids into homotopic ones. If $r > 0$, then the

multiplication in $\mathrm{Sph}_{r+s}(X,x_0)$ is transferred by Cub into the usual multiplication in $\mathrm{Sph}_r(\mathrm{Sph}_s(X,x_0),\mathrm{const})$ [that is, the multiplication of spheroids]. If $s > 0$, then the multiplication in $\mathrm{Sph}_{r+s}(X,x_0)$ is transferred by Buc into the multiplication in $\mathrm{Sph}_r(\mathrm{Sph}_s(X,x_0),\mathrm{const})$ which arises from the fact that $\mathrm{Sph}_s(X,x_0)$ is an H-space with identity element const (see 9.11 and 9.10). Therefore, when $r+s > 0$, the maps Cub and Buc define group isomorphisms, cub,buc: $\pi_{r+s}(X,x_0) \to$ $\to \pi_r(\mathrm{Sph}_s(X,x_0),\mathrm{const})$ [the group structure of $\pi_0(\mathrm{Sph}_s(X,x_0),\mathrm{const})$, $s > 0$, was explained in 9.12]. Therefore, one can identify $\pi_r(X,x_0)$ with any of the groups $\pi_{r-q}(\mathrm{Sph}_q(X,x_0),\mathrm{const})$ with $q \leqslant r$.

Finally, note that the isomorphism cub: $\pi_r(X,x_0) \to$ $\to \pi_{r-1}(\mathrm{Sph}_1(X,x_0),\mathrm{const})$ appears also as the isomorphism $\Delta: \pi_r(X,x_0) \to \pi_{r-1}(\mathrm{Sph}_1(X,x_0),\mathrm{const})$ in the homotopy sequence of the Serre bundle $\xi = (C(I,0;X,x_0),\mathrm{ab}\, C([\mathrm{in:Fr}\, I \to I],\mathrm{id}),X=C(\mathrm{Fr}\, I,0;X,x_0))$, whose fiber over the point x_0 is $\mathrm{Sph}_1(X,x_0)$. That ξ is a Serre bundle follows from 4.1.4.2.

11. Additional Theorems

1. **Let** $d_1,\ldots,d_m$ **be pairwise disjoint balls in** $\mathbb{R}^r$ **such that** $d_1,\ldots,d_m \subset D^r$, **and let** $g \in \mathrm{Sph}_r^0(X,A,x_0)$ **be a spheroid with** $g(C) \subset A$, **where** $C = D^r \smallsetminus U_{i=1}^m \mathrm{Int}\, d_i$. **Let** $\tau_i: D^r \to D^r$ **denote the map** $\tau_i(y) = (\text{center of } d_i) + (\text{radius of } d_i)y$. **Suppose further that the segments joining the points** $\tau_1(\mathrm{ort}_1),\ldots,\tau_m(\mathrm{ort}_1)$ **to** ort_1 **are contained in** C. **Then, for** $r > 2$,

$$\gamma = (T_{s_1}\gamma_1)(T_{s_2}\gamma_2)\ldots(T_{s_m}\gamma_m),$$

where $\gamma \in \pi_r(X,A,x_0)$ **and** $\gamma_i \in \pi_r(X,A,g\circ\tau_i(\mathrm{ort}_1))$ **are the elements represented by the spheroids** g **and** $g \circ \tau_i \in \mathrm{Sph}_r^0(X,A,g\circ\tau_i(\mathrm{ort}_1))$, **and** s_i **is the path in** A **given by** $s_i(t) = g((1-t)\tau_i(\mathrm{ort}_1) + t\,\mathrm{ort}_1)$, $i = 1,\ldots,m$. **The same conclusion holds true for** $r = 2$, **provided** $d_1,\ldots,d_m$ **are indexed naturally, i.e., each of the 2-frames** $(\tau_i(\mathrm{ort}_1) - \mathrm{ort}_1,\tau_{i+1}(\mathrm{ort}_1) - \mathrm{ort}_1)$ **defines the natural orientation of** $\mathbb{R}^2$.

PROOF. We proceed by induction on m. We make two preliminary remarks, denoting by ℓ_i the (rectilinear) path in C given by $\ell_i(t) = (1-t)\tau_i(\mathrm{ort}_1) + t\,\mathrm{ort}_1$.

First remark: for given $d_1,\ldots,d_m$, it suffices to prove the theorem when $(X,A,x_0) = (D^r,C,\mathrm{ort}_1)$, $g = \mathrm{rel\ id\ } D^r$, and $s_i = \ell_i$. Indeed, $g_* : \pi_r(D^r,C,\mathrm{ort}_1) \to \pi_r(X,A,x_0)$ takes the class of the spheroid $\mathrm{rel\ id\ } D^r$ into γ, while it takes the class of the spheroid τ_i, translated along ℓ_i, into $T_{s_i}\gamma_i$.

Second remark: for a given m, it suffices to prove the theorem for a standard choice of $d_1,\ldots,d_m$, namely, for the balls of radius $1/2m$ centered at $\frac{m-1}{m}\mathrm{ort}_2$, $\frac{m-3}{m}\mathrm{ort}_2$, $\ldots$, $\frac{3-m}{m}\mathrm{ort}_2$, $\frac{1-m}{m}\mathrm{ort}_2$. To see this, consider, along with these standard balls and the corresponding C, τ_i, ℓ_i, arbitrary balls $d_1',\ldots,d_m'$ satisfying the conditions of the theorem, with the corresponding C', τ_i', ℓ_i'. Clearly, there exists a continuous map $h : D^r \to D^r$, which is S^{r-1}-homotopic to $\mathrm{id}\ D^r$ and satisfies $h(C) \subset C'$, $h \circ \tau_i = \tau_i'$, and $h \circ \ell_i = \ell_i'$. Then $\mathrm{rel\ } h_* : \pi_r(D^r,C,\mathrm{ort}_1) \to \pi_r(D^r,C',\mathrm{ort}_1)$ takes the class of the spheroid $\mathrm{id}\ D^r$ into the class of $\mathrm{id}\ D^r$, while taking the class of the spheroid τ_i, translated along ℓ_i, into the class of the spheroid τ_i', translated along ℓ_i', $i = 1,\ldots,m$.

Now to our induction. The cases $m = 0$ and $m = 1$ are trivial; consider $m = 2$. By our remarks, we may assume that $(X,A,x_0) = (D^r,C,\mathrm{ort}_1)$, $g = \mathrm{id}$, and d_1,d_2 are the standard balls (with radius $1/4$ and centers $\mathrm{ort}_2/2$ and $-\mathrm{ort}_2/2$). Let ρ_ϕ denote the rotation of D^r by an angle ϕ around the subspace given by the equations $x_1 = x_2 = 0$. Further, consider the homotopies $D^r \times I \to D^r$ given by

$$(y,t) \mapsto [(1+t)\rho_{\pi t/2}(y) - 2\,\mathrm{ort}_2]/4$$

and

$$(y,t) \mapsto [(1+t)\rho_{\pi t/2}(y) + 2\,\mathrm{ort}_2]/4$$

$(y \in D^r,\ t \in I)$. These can be viewed as free homotopies of spheroids of the pair (D^r,C); as such, they connect τ_1 and τ_2 with two spheroids with origin 0, σ_1 and σ_2, along two paths, which we call u_1 and u_2. Obviously, $u_1^{-1}\ell_1$ and $u_2^{-1}\ell_2$ are both homotopic to the rectilinear path ℓ which joins 0 to ort_1. Consequently, the product of the classes obtained by translating the classes of τ_1 and τ_2 to ort_1 along ℓ_1 and ℓ_2 is the same as the product of the classes obtained by translating the classes of σ_1 and σ_2 to ort_1 along ℓ, that is, the class of the product of the spheroids σ_1 and σ_2, translated to ort_1 along ℓ. It remains to observe that there is a free homotopy from the product $\sigma_1\sigma_2$ to $\mathrm{id}\, D^r$ along ℓ, for example, a rectilinear homotopy.

Finally, let $m \geqslant 3$. As in the case $m = 2$, we shall assume that $(X,A,x_0) = (D^r,C,\mathrm{ort}_1)$, $g = \mathrm{id}$, and $d_1,\ldots,d_m$ are the standard balls (with radius $1/2m$ and centers $\frac{m-1}{m}\mathrm{ort}_2$, $\frac{m-3}{m}\mathrm{ort}_2,\ldots,$ $\frac{3-m}{m}\mathrm{ort}_2$, $\frac{1-m}{m}\mathrm{ort}_2$). Let d be the ball of radius $3/2m$ centered at $\frac{2-m}{m}\mathrm{ort}_2$ (note that $d_{m-1},d_m \subset d$), and let $\tau : D^r \to D^r$ be defined by $\tau(y) = (\text{center of } d) + (\text{radius of } d)y$. Further, let ℓ, u, and v be the rectilinear paths joining $\tau(\mathrm{ort}_1)$ to ort_1, $\tau_{m-1}(\mathrm{ort}_1)$ to $\tau(\mathrm{ort}_1)$, and $\tau_m(\mathrm{ort}_1)$ to $\tau(\mathrm{ort}_1)$, respectively. We let $\delta \in \pi_r(D^r,C^r,\tau(\mathrm{ort}_1))$ denote the class of τ. Since the products $u\ell$, $v\ell$ are clearly homotopic to the paths ℓ_{m-1}, ℓ_m,

$$T_{\ell_1}\gamma_1\cdots T_{\ell_m}\gamma_m = T_{\ell_1}\gamma_1\cdots T_{\ell_{m-2}}\gamma_{m-2}T_\ell(T_u\gamma_{m-1}T_v\gamma_m). \tag{16}$$

Now apply the theorem, first for the case of two factors, and then for the case of $m-1$ factors, to conclude that

$$T_u\gamma_{m-1}T_v\gamma_m = \delta, \tag{17}$$

and

$$T_{\ell_1}\gamma_1\cdots T_{\ell_{m-2}}\gamma_{m-2}T\,\delta = \gamma. \tag{18}$$

(In the first case, the theorem is applied to $(X,A,x_0) = (D^r,C,\tau(\mathrm{ort}_1))$, $g = \tau$, and the balls $\tau^{-1}(d_{m-1}),\tau^{-1}(d_m)$, while in the second case we take $(X,A,x_0) = (D^r,C,\mathrm{ort}_1)$, $g = \mathrm{id}$, and the balls $d_1,\ldots,d_{m-2},d)$. At last, (16), (17), and (18) yield (15).

2. <u>Let</u> $X = \lim(X_k,\phi_k)$, <u>where</u> X_k <u>are</u> T_1-<u>spaces, and let</u> $x \in X$, $x_0 \in X_0$, $x_1 \in X_1,\ldots$ <u>be points such that</u> $\mathrm{imm}_k(x_k) = x$. <u>If for</u> <u>some</u> r <u>all the homomorphisms</u> $(\phi_k)_{*r} : \pi_r(X_k,x_k) \to \pi_r(X_{k+1},x_{k+1})$ <u>are</u> <u>isomorphism, then so are</u> $(\mathrm{imm}_k)_{*r} : \pi_r(X_k,x_k) \to \pi_r(X,x)$ (<u>with the same</u> r).

PROOF. Notice that, according to Theorem 1.2.4.5, every spheroid $I^r \to X$ may be expressed as the composition of a spheroid $I^r \to X_1$ with the embedding imm_1, for 1 large enough; similarly, every homotopy $I^r \times I \to X$ is the composition of some homotopy $I^r \times I \to X_1$ with imm_1, for 1 large enough. Now the fact that $(\mathrm{imm}_k)_{*r}$ are epimorphisms and monomorphisms is seen to be a consequence of the analogous properties of the compositions

$$(\phi_{1-1})_{*r} \circ (\phi_{1-2})_{*r} \circ \cdots \circ (\phi_k)_{*r} : \pi_r(X_k, x_k) \to \pi_r(X_1, x_1).$$

12. Exercises

1. Let (X, x_0) be a pointed space and suppose that there is given a right group-action of $\pi_1(X, x_0)$ on a group G. Show that there exists a local system of groups, $(X, \{G_x\}, \{T_s\})$, with $G_{x_0} = G$, which determines the given action.

2. Let (X, A) be a cellular pair with base point $x_0 \in A$, and suppose that X is countable. Show that all the groups $\pi_r(X, A, x_0)$ are countable.

3. Let (X, A) be a cellular pair with base point $x_0 \in A$, and suppose that the groups $\pi_r(X, x_0)$ and $\pi_r(A, x_0)$ are finitely generated (for all $r \geqslant 1$). Show that if X is simply connected, then the groups $\pi_r(X, A, x_0)$ with $r \geqslant 2$ are also finitely generated.

4. Let ξ be a Serre bundle, and let E be a subspace of $\mathrm{tl}\,\xi$ such that $(E, \mathrm{pr}\,\xi|_E, \mathrm{bs}\,\xi)$ is also a Serre bundle. Show that for any point $x \in E$ and any $r \geqslant 1$

$$\mathrm{in}_* : \pi_r(\mathrm{pr}\,\xi^{-1}(\mathrm{pr}\,\xi(x)), \mathrm{pr}\,\xi^{-1}(\mathrm{pr}\,\xi(x)) \cap E, x) \to \pi_r(\mathrm{tl}\,\xi, E, x)$$

is an isomorphism.

5. Show that if the base of a covering is k-simple, then its total space is also k-simple.

6. Let $r > 0$ and $s > 0$. Show that the homomorphisms

$$\mathrm{cub}, \mathrm{buc} : \pi_{r+s}(X, x_0) \to \pi_r(\mathrm{Sph}_s(X, x_0), \mathrm{const})$$

differ only by the constant factor $(-1)^{rs}$.

§2. THE HOMOTOPY GROUPS OF
SPHERES AND OF CLASSICAL MANIFOLDS

1. Suspension in the Homotopy Groups of Spheres

1. The <u>suspension of a spheroid</u> $\phi \in \mathrm{Sph}_r(X,x_0)$ is the
spheroid $\mathrm{su}\,\phi \in \mathrm{Sph}_{r+1}(\mathrm{su}(X,x_0),\mathrm{bp})$, given by $\mathrm{su}\,\phi(t_1,\ldots,t_{r+1}) =$
$= \mathrm{pr}(\phi(t_1,\ldots,t_r),t_{r+1})$, where $\mathrm{pr} = [\mathrm{pr}\colon X \times I \to \mathrm{su}(X,x_0)]$. Obviously,
suspensions of homotopic spheroids are homotopic, the suspension of the
product of two spheroids of positive dimensions equals the product of
their suspensions, and the suspension of the constant spheroid is again
the constant spheroid. Consequently, the mapping $\phi \mapsto \mathrm{su}\,\phi$ yields a
homomorphism $\pi_r(X,x_0) \to \pi_{r+1}(\mathrm{su}(X,x_0),\mathrm{bp})$, for any $r \geqslant 0$. This
homomorphism is also called <u>suspension</u> and is denoted by su.

Recall that we have already defined the suspension of a
continuous map on two occasions: in 1.2.6.2, for maps of topological
spaces, and in 1.2.8.5, for maps of pointed topological spaces. The
present, third definition, is more special; it concerns maps from the
pair $(I^r, \mathrm{Fr}\,I^r)$ into pointed spaces, and has no intersection with the
previous ones. At the same time, it is compatible with the second
definition, in the sense that we may obtain the third definition from
the latter by shifting from spheric spheroids to cubic ones. More
precisely, the spheroids in $\mathrm{Sph}_r^O(X,x_0)$, being maps between pointed
spaces, have suspensions in the sense of 1.2.8.5, and the diagram

$$
\begin{array}{ccc}
\mathrm{Sph}_r^O(X,x_0) & \xrightarrow{\ \mathrm{su}\ } & \mathrm{Sph}_{r+1}^O(\mathrm{su}(X,x_0),\mathrm{bp}) \\
\mathrm{IS}^{\#} \downarrow & & \downarrow \mathrm{IS}^{\#} \\
\mathrm{Sph}_r(X,x_0) & \xrightarrow{\ \mathrm{su}\ } & \mathrm{Sph}_{r+1}(\mathrm{su}(X,x_0),\mathrm{bp})
\end{array}
$$

commutes.

Let us add two important, yet obvious remarks. Firstly, if
$f\colon (X,x_0) \to (Y,y_0)$ is continuous, then the diagram

$$
\begin{array}{ccc}
\pi_r(X,x_0) & \xrightarrow{\ \mathrm{su}\ } & \pi_{r+1}(\mathrm{su}(X,x_0),\mathrm{bp}) \\
f_* \downarrow & & \downarrow (\mathrm{su}\,f)_* \\
\pi_r(Y,y_0) & \xrightarrow{\ \mathrm{su}\ } & \pi_{r+1}(\mathrm{su}(Y,y_0),\mathrm{bp})
\end{array}
$$

(where $\mathrm{su}\,f$ is understood as in 1.2.8.5) commutes for any $r \geqslant 0$.

Secondly, since $(su(S^n, ort_1), bp) = (S^{n+1}, ort_1)$, our construction, when applied to spheres, yields a homomorphism $\pi_r(S^n, ort_1) \to \pi_{r+1}(S^{n+1}, ort_1)$. The main theorem of this subsection, Theorem 4, is devoted precisely to this homomorphism.

2. Alternatively, one can describe $su: \pi_r(X, x_0) \to \pi_{r+1}(su(X, x_0), bp)$ by means of the map $lp: (X, x_0) \to Sph_1(su(X, x_0), bp)$ given by $[lp(x)](t) = (x, t)$. Namely, every spheroid $\phi \in Sph_r(X, x_0)$ is taken into $su\,\phi$ by the composition

$$Sph_r(X, x_0) \xrightarrow{\ lp_{\#}\ } Sph_r(Sph_1(su(X, x_0), bp), const) \xrightarrow{\ Cub\ } Sph_{r+1}(su(X, x_0), bp)$$

(see 1.10.6), and hence the homomorphism $su: \pi_r(X, x_0) \to \pi_{r+1}(su(X, x_0), bp)$ may be defined as $su = cub \circ lp_*$ (to check this facts is routine).

A new description of the homomorphism su emerges if we view $su(X, x_0)$ as the quotient space of the cone $con(X, x_0)$ by its base (which is identified with X). Indeed, given any $\phi \in Sph_r(X, x_0)$, consider the spheroid in $Sph_{r+1}(con(X, x_0), X, x_0)$ defined as $(t_1, \ldots, t_{r+1}) \mapsto (\phi(t_1, \ldots, t_r), t_{r+1})$. The latter is taken into $su\,\phi$ by the map $pr_{\#}: Sph_{r+1}(con(X, x_0), X, x_0) \to Sph_{r+1}(su(X, x_0), bp)$, where $pr = [pr: con(X, x_0) \to su(X, x_0)]$, and is transformed back into ϕ by the map $\partial: Sph_{r+1}(con(X, x_0), X, x_0) \to Sph_r(X, x_0)$. Consequently, $su: \pi_r(X, x_0) \to \pi_{r+1}(su(X, x_0), bp)$ is nothing but the composition

$$\pi_r(X, x_0) \xrightarrow{\ \partial^{-1}\ } \pi_{r+1}(con(X, x_0), X, x_0) \xrightarrow{\ pr_*\ } \pi_{r+1}(su(X, x_0), bp)$$

($\partial: \pi_{r+1}(con(X, x_0), X, x_0) \to \pi_r(X, x_0)$ is invertible because the cone is contractible; see 1.6.5).

If, for example, $(X, x_0) = (S^n, ort_1)$, then $con(X, x_0) = D^{n+1}$, $su(X, x_0) = S^{n+1}$, and $pr = DS$.

The Suspension Theorem

3 (LEMMA). <u>Let</u> K <u>and</u> L <u>be closed disjoint subsets of</u> I^m, <u>and assume that</u> K <u>is covered by a finite number of k-dimensional planes, that</u> L <u>is covered by a finite number of l-dimensional planes, and that</u> $K \cap Fr\,I^m \subset I^{m-1} \times [0, 1/2)$ <u>and</u> $L \cap Fr\,I^m \subset I^{m-1} \times (1/2, 1]$. <u>If</u> $k+l \leqslant m-2$, <u>then there is a</u> $Fr\,I^m$<u>-homotopy,</u> $F: I^m \times I \to I^m$, <u>such that:</u>

(i) <u>the maps</u> $I^m \to I^m$ <u>which form</u> F <u>are homeomorphisms, and their inverses also form a homotopy;</u>

(ii) F __connects__ $\mathrm{id}\, I^m$ __with a map__ $I^m \to I^m$ __which takes__ K __into__ $I^{m-1} \times [0,1/2]$ __and__ L __into__ $I^{m-1} \times [1/2,1]$.

PROOF. Suppose first that $L \subset I^{m-1} \times [1/2,1]$. Since the lines which intersect both K and L constitute a set which can be covered by a finite number of planes of dimension $k+l+1 \leqslant m-1$, there is a point $a \in \mathrm{Int}\, I^{m-1} \times (0,1/2)$ which does not lie on any such line. Let us project K and L on $\mathrm{Fr}\, I^m$ from a. Their images are closed disjoint subsets of $\mathrm{Fr}\, I^m$, and so there is a Urysohn function, $\alpha\colon \mathrm{Fr}\, I^m \to I$, for these images. Choose $\varepsilon > 0$ such that the similarity transformation with center a and coefficient ε pulls I^m into $I^{m-1} \times [0,1/2]$, while the similarity transformation with the same center and coefficient $1/(1-\varepsilon)$ does not take $K \cap (I^m \times [1/2,1])$ out of I^m. Now let $\phi_x\colon I \to I^m$ denote the rectilinear path which joins a to a given point $x \in \mathrm{Fr}\, I^m$, and let $\psi_t\colon I \to I$ denote the homeomorphism which takes linearly $[0,(1+t)/2]$ onto $[0,(1-t)/2]$ and $[(1+t)/2,1]$ onto $[(1-t)/2,1]$. The formula

$$F(\phi_x(u),t) = \phi_x \circ \psi_{t(1-\varepsilon)\alpha(x)}(u),$$

where $x \in \mathrm{Fr}\, I^m$ and $t,u \in I$, defines the desired homotopy.

To reduce everything to this special case, we produce, in the general case, a $\mathrm{Fr}\, I^m$-homotopy $F\colon I^m \times I \to I^m$ which satisfies property (i) and connects $\mathrm{id}\, I^m$ with a map which takes L into $I^{m-1} \times [1/2,1]$. We can define such a homotopy F by

$$F((x,u),t) = (x,\psi_{-t\delta}(u)) \qquad [x \in I^{m-1}, \quad t,u \in I],$$

where $\delta \in (0,1)$ is any number such that $L \subset I^{m-1} \times [(1-\delta)/2,1]$.

4. $\mathrm{su}\colon \pi_r(S^n,\mathrm{ort}_1) \to \pi_{r+1}(S^{n+1},\mathrm{ort}_1)$ __is an isomorphism for__ $r \leqslant 2n-2$ __and an epimorphism for__ $r = 2n-1$.

PROOF. a) To see that su is epimorphic for $r \leqslant 2n-1$, we have to verify that, given any spheroid $\phi \in \mathrm{Sph}_{r+1}(S^{n+1},\mathrm{ort}_1)$, there is a spheroid $\psi \in \mathrm{Sph}_r(S^n,\mathrm{ort}_1)$ such that $\mathrm{su}\,\psi$ is homotopic to ϕ.

The proof is quite simple when $\phi(I^r \times [0,1/2])$ is contained in the upper half $\{x_{n+2} \geqslant 0\}$ of S^{n+1}, while $\phi(I^r \times [1/2,1])$ is contained in the lower half $\{x_{n+2} \leqslant 0\}$ of S^{n+1} (here $x_1,\ldots,x_{n+2}$ are the standard coordinates in $\mathbb{R}^{n+2}$). In this case, $\phi(I^r \times (1/2))$ lies in the intersection of these two hemispheres, i.e., $\phi(I^r \times (1/2)) \subset S^n$, and the required ψ is given by

$$\psi(t_1,\ldots,t_r) = \phi(t_1,\ldots,t_r,1/2).$$

The formula

$$((t_1,\ldots,t_{r+1}),t) \mapsto$$

$$\mapsto \begin{cases} \phi(t_1,\ldots,t_r,\dfrac{t_{r+1}}{1-t}), & \text{if } 0 \leqslant t_{r+1} \leqslant \dfrac{1-t}{2}, \\[2ex] \mathrm{pr}(\phi(t_1,\ldots,t_r,\tfrac{1}{2}),t_{r+1}), & \text{if } \dfrac{1-t}{2} \leqslant t_{r+1} \leqslant \dfrac{1+t}{2}, \\[2ex] \phi(t_1,\ldots,t_r,\dfrac{t_{r+1}-t}{1-t}), & \text{if } \dfrac{1+t}{2} \leqslant t_{r+1} \leqslant 1, \end{cases}$$

where $\mathrm{pr} = [\mathrm{pr}\colon S^n \times I \to \mathrm{su}(S^n,\mathrm{ort}_1)]$, defines a homotopy from ϕ to $\mathrm{su}\,\psi$.

Consider now the more general case where $\phi^{-1}(\mathrm{ort}_{n+2}) \subset$ $\subset I^r \times [0,1/2)$ and $\phi^{-1}(-\mathrm{ort}_{n+2}) \subset I^r \times (1/2,1]$. This case is readily reduced to the previous one. Indeed, pick $\varepsilon > 0$ such that the last coordinate of $\phi(y)$ is $\leqslant 1-\varepsilon$ ($\geqslant -(1-\varepsilon)$) for all $y \in I^r \times [1/2,1]$ (respectively, for all $y \in I^r \times [0,1/2]$). Now define $h\colon (S^{n+1},\mathrm{ort}_1) \to (S^{n+1},\mathrm{ort}_1)$ by $h(x_1,\ldots,x_{n+2}) =$

$$= \begin{cases} (x_1,\ldots,x_{n+1},(x_{n+2}^2 - (1-\varepsilon^2))^{1/2})/(1 - (1-\varepsilon^2))^{1/2}, \\[1ex] \hspace{4cm} \text{if } |x_{n+2}| \geqslant 1-\varepsilon, \\[1.5ex] (x_1,\ldots,x_{n+1},0)/(x_1^2 + \ldots + x_{n+1}^2)^{1/2}, \quad \text{if } |x_{n+2}| \leqslant 1-\varepsilon. \end{cases}$$

[This map stretches the polar caps of S^{n+1}, defined by the inequalities $x_{n+2} \geqslant 1-\varepsilon$ and $x_{n+2} \leqslant -(1-\varepsilon)$, over the upper and lower hemispheres, respectively, and contracts the equatorial belt $-(1-\varepsilon) < x_{n+2} < 1-\varepsilon$ to the equator S^n.] It is clear that h is ort_1-homotopic to $\mathrm{id}\,S^{n+1}$ [such a homotopy moves each point $x \in S^{n+1}$ uniformly towards $h(x)$ along the shortest arc of the great circle passing through x and $h(x)$]. Moreover, $h \circ \phi$ takes $I^r \times [0,1/2]$ into the upper hemisphere, and takes $I^r \times [1/2,1]$ into the lower hemisphere.

Finally, in the most general case, we triangulate S^{n+1} in such a way that ort_1 becomes a vertex, while ort_{n+2} and $-\mathrm{ort}_{n+2}$ become interior points of $(n+1)$-simplices, and then consider a rectilinear triangulation of I^n which ensures that ϕ has simplicial approximations. Let ϕ_1 be such an approximation. Then the sets $K = \phi_1^{-1}(\mathrm{ort}_{n+2})$ and $L = \phi_1^{-1}(-\mathrm{ort}_{n+2})$ satisfy the conditions of Lemma 3, with $m = r+1$ and $k = l = r-n$ [here the intersections $K \cap \mathrm{Fr}\,I^{r+1}$ and $L \cap \mathrm{Fr}\,I^{r+1}$ are actually empty]. Let $G\colon I^{r+1} \times I \to I^{r+1}$ be the homotopy made up of the inverses of the homeomorphisms which form the homotopy F provided by Lemma 3. The spheroid ϕ_1 is obviously homotopic to ϕ, and we need only remark that $\phi_1 \circ \mathrm{rel}\, G : (I^{r+1} \times I,$ $\mathrm{Fr}\, I^{r+1} \times I) \to (S^{n+1}, \mathrm{ort}_1)$ is a homotopy from ϕ_1 to a spheroid ϕ_2 such that $\phi_2^{-1}(\mathrm{ort}_{n+2}) \subset I^r \times [0, 1/2)$ and $\phi_2^{-1}(-\mathrm{ort}_{n+2}) \subset I^r \times (1/2, 1]$.

b) To see that su is a monomorphism for $r \leqslant 2n+2$, we must show that every spheroid $\psi : (I^r, \mathrm{Fr}\, I^r) \to (S^n, \mathrm{ort}_1)$ whose suspension su $\psi : (I^{r+1}, \mathrm{Fr}\, I^{r+1}) \to (S^{n+1}, \mathrm{ort}_1)$ is homotopic to the constant spheroid is itself homotopic to the constant spheroid.

Let $\Phi : (I^{r+1} \times I, \mathrm{Fr}\, I^{r+1} \times I) \to (S^{n+1}, \mathrm{ort}_1)$ be a homotopy from su ψ to const. Consider a triangulation of S^{n+1} with the properties: the equator S^n is a simplicial subspace, ort_1 is a vertex, and ort_{n+2}, $-\mathrm{ort}_{n+2}$ are interior points of $(n+1)$-simplices. Further, triangulate rectlinearly $I^{r+1} \times I = I^r \times I \times I$ so that $I^r \times (1/2) \times 0$ is a simplicial subspace and Φ admits simplicial approximations. If Φ_1 is such an approximation, then $\Phi_1 (I^r \times (1/2) \times 0) \subset S^n$, and the formula $\psi_1(y) = \Phi_1(y, 1/2, 0)$ defines a spheroid $\psi_1 : (I^r, \mathrm{Fr}\, I^r) \to (S^n, \mathrm{ort}_1)$ which is homotopic to ψ. Consider the map $\mathrm{perm} : I^{r+2} \to I^{r+2}$, $\mathrm{perm}(t_1, \ldots, t_{r+2}) =$ $= (t_1, \ldots, t_r, t_{r+2}, t_{r+1})$. Clearly, $K = \mathrm{perm}(\psi_1^{-1}(\mathrm{ort}_{n+2}))$ and $L = \mathrm{perm}(\psi_1^{-1}(-\mathrm{ort}_{n+2}))$ satisfy the conditions of Lemma 3, with $m = r+2$ and $k = l = r-n+1$. Thus, let $G : I^{r+2} \times I \to I^{r+2}$ be the homotopy made up of the inverses of the homeomorphisms which form the homotopy F provided by this lemma, and let $\rho : S^{n+1} \smallsetminus (\mathrm{ort}_{n+2} \cup (-\mathrm{ort}_{n+2})) \to S^n$ be a retraction. One may verify directly that

$$(y, t) \mapsto \rho(\Phi_1 \circ \mathrm{perm} \circ G(y, t, 1/2), 1)$$

(where $(y, t, 1/2) \in I^{r+2} = I^r \times I \times I$) is a homotopy $I^r \times I \to S^n$ from ψ_1 to const.

The Series $\{\pi_{n+k}(S^n, \mathrm{ort}_1)\}$

5. The main content of Theorem 4 is that each of the series

$$\ldots \; \pi_r(S^n, \mathrm{ort}_1) \xrightarrow{\;\mathrm{su}\;} \pi_{r+1}(S^{n+1}, \mathrm{ort}_1) \xrightarrow{\;\mathrm{su}\;} \pi_{r+2}(S^{n+2}, \mathrm{ort}_1) \xrightarrow{\;\mathrm{su}\;} \ldots$$

of homotopy groups of spheres, connected by the suspension, stabilizes. That is to say, in the k-th series $\{\pi_{n+k}(S^n, \mathrm{ort}_1)\}$, the groups $\pi_{n+k}(S^n, \mathrm{ort}_1)$ with $n \geqslant k+2$ are isomorphic via suspension. This canonical isomorphism enables us to identify the groups $\pi_{n+k}(S^n, \mathrm{ort}_1)$,

$n \geqslant k+2$, with a single group, called the <u>stable group of the series</u> $\{\pi_{n+k}(S^n, \text{ort}_1)\}$; we denote it by $\text{Stab}(k)$.

2. The Simplest Homotopy Groups of Spheres

1. <u>The groups</u> $\pi_r(S^n)$ <u>with</u> $r < n$ <u>are trivial. In particular,</u> $\text{Stab}(k) = 0$ <u>whenever</u> $k < 0$.

This is a corollary of 2.3.2.2 and 1.3.8.

The Homotopy Groups of the Circle

2. <u>If</u> $r > 1$, $\pi_r(S^1)$ <u>is trivial.</u> $\pi_1(S^1, \text{ort}_1)$ <u>is an infinite cyclic group with generator</u> sph_1.

The proof uses the covering $(\mathbb{R}, \text{hel}, S^1)$ [see 4.1.2.6]. First, notice that $\text{hel}(0) = \text{ort}_1$ and $\text{hel}^{-1}(\text{ort}_1) = \mathbb{Z}$. Since the line $\mathbb{R}$ is contractible, its homotopy groups are trivial, and hence, by 1.8.12 so is $\pi_r(S^1, \text{ort}_1)$ for any $r > 1$, and $\Delta\colon \pi_1(S^1, \text{ort}_1) \to \pi_0(\mathbb{Z}, 0) = \mathbb{Z}$ is invertible. Moreover, $(\mathbb{R}, \text{hel}, S^1)$ is obviously a principal bundle with structure group $\mathbb{Z}$, so that Δ is a group homomorphism (see 1.9.15). Therefore, Δ is a group isomorphism, and it remains to observe that $\Delta(\text{sph}_1) = -1$.

3. The proof above computes the homotopy groups of the circle as quick as ligthning by applying the general theory to the covering $(\mathbb{R}, \text{hel}, S^1)$. Such an approach is natural as the general theory is already available to us. However, it conceals the fact that the computation is in fact quite elementary. Since the fundamental group of the circle has both a historical and intrinsic importance, we shall not spare a few more lines, and we shall redo the proof of the equality $\pi_1(S^1, \text{ort}_1) = \mathbb{Z}$ in an elementary way, which reveals those simple aspects of the general theory that are really necessary. Namely, we need only the following two facts: every path in S^1 with origin ort_1 is covered by a path in $\mathbb{R}$ with origin 0; two paths in $\mathbb{R}$, with origin 0, which cover homotopic paths in S^1 have the same end. (Cf. 6.2.1).

Now to the proof: consider the powers of the fundamental loop $\text{IS}\colon I \to S^1$ (with the natural order of multiplication), and let u_n denote the n-th power $(n \in \mathbb{Z})$. Let $\tilde{u}_n\colon I \to \mathbb{R}$ be the path with origin 0 covering the loop u_n. Obviously, $\tilde{u}_n(1) = n$, and so the loops u_n are pairwise nonhomotopic. Moreover, given any loop

$u\colon I \to S^1$ with origin ort_1, the covering path $\tilde{u}\colon I \to \mathbb{R}$ with origin 0 ends at an integer. Consequently, $\tilde{u}$ is homotopic to one of the paths $\tilde{u}_n$, and thus u is homotopic to one of the loops u_n. In other words, the classes $(\mathrm{sph}_1)^n$ with $n \in \mathbb{Z}$ are pairwise distinct and exhaust $\pi_1(S^1,\mathrm{ort}_1)$.

Corollaries

4. <u>The pair (D^2,S^1) is simple.</u> $\pi_r(D^2,S^1)$ <u>is trivial for any</u> $r \neq 2$. $\pi_2(D^2,S^1,\mathrm{ort}_1)$ <u>is an infinite cyclic group with generator</u> kug_2.

These are all consequences of Theorem 2 and of the fact that D^2 is contractible, which implies that $\partial\colon \pi_r(D^2,S^1,\mathrm{ort}_1) \to \pi_{r-1}(S^1,\mathrm{ort}_1)$ is an isomorphism for any $r \geqslant 1$ (see 1.6.5).

5. It follows from 1 and 2 that for $n \geqslant 1$, S^n is a simple space. Similarly, 1 and 4 show that for $n \geqslant 2$ the pair (D^n,S^{n-1}) is simple.

In particular, given any point $y \in S^n$ with $n \geqslant 1$, the group $\pi_n(S^n,y)$ can be identified with $\pi_n(S^n,\mathrm{ort}_1)$, and given any point $y \in S^{n-1}$ with $n \geqslant 2$, the group $\pi_n(D^n,S^{n-1},y)$ can be identified with $\pi_n(D^n,S^{n-1},\mathrm{ort}_1)$. Therefore, for $r \geqslant 1$, a continuous map $f\colon S^r \to X$, with X a topological space, defines an element of $\pi_r(X,x)$ for any point $x \in f(S^r)$, and not only for $x = f(\mathrm{ort}_1)$. Similarly, for $r \geqslant 2$, a continuous map $f\colon (D^r,S^{r-1}) \to (X,A)$, with (X,A) a topological pair, defines an element of $\pi_r(X,A,x)$ for any $x \in f(S^{r-1})$, and not only for $x = f(\mathrm{ort}_1)$.

The Groups $\pi_n(S^n)$

6. <u>For</u> $n \geqslant 1$, $\mathrm{su}\colon \pi_n(S^n,\mathrm{ort}_1) \to \pi_{n+1}(S^{n+1},\mathrm{ort}_1)$ <u>is an isomorphism and</u> $\mathrm{su}(\mathrm{sph}_n) = \mathrm{sph}_{n+1}$.

PROOF. By Theorem 1.4, $\mathrm{su}\colon \pi_n(S^n,\mathrm{ort}_1) \to \pi_{n+1}(S^{n+1},\mathrm{ort}_1)$ is an isomorphism for $n \geqslant 2$, and an epimorphism for $n = 1$. To prove that this epimorphism is also a monomorphism, we use the homotopy properties of the Hopf bundle (S^3,pr,S^2) [see 4.6.1.4]: the segment

$$\pi_2(S^3) = 0 \to \pi_2(S^2) \to \pi_1(S^1) \to \pi_1(S^3) = 0$$

of its homotopy sequence (see 1) demonstrates that $\pi_2(S^2) \cong \pi_1(S^1)$.

Since $\pi_1(S^1) = \mathbb{Z}$ (see 2), su: $\pi_1(S^1,\text{ort}_1) \to \pi_2(S^2,\text{ort}_1)$ cannot have a nontrivial kernel. The equality $\text{su}(\text{sph}_n) = \text{sph}_{n+1}$ is obvious.

7. **If** $n \geq 1$, $\pi_n(S^n,\text{ort}_1)$ **is an infinite cyclic group with generator** sph_n. **In particular,** $\text{Stab}(0) = \mathbb{Z}$.

For $n = 1$, this statement is a repetition of a part of Theorem 2, while for $n > 1$ it results from 2 and 6.

8 (COROLLARY). **If** $n \geq 2$, $\pi_n(D^n,S^{n-1},\text{ort}_1)$ **is an infinite cyclic group with generator** kug_n.

9. For each $n \geq 1$, Theorem 7 establishes a canonical isomorphism $\pi_n(S^n) \to \mathbb{Z}$. In particular, it associates with each continuous map $f: S^n \to S^n$ an integer, and it is not hard to see that this is nothing but the degree $\deg f$, as defined in Subsection 4.6.5. This is a consequence of three evident facts: $\deg(\text{su}\,f) = \deg f$; the class $k\,\mathbf{sph}_1$ is represented by the spheroid hel_k (see 4.1.2.6); and $\deg(\text{hel}_k) = k$.

Similarly, for each $n \geq 2$, Theorem 8 establishes a canonical isomorphism $\pi_n(D^n,S^{n-1}) \to \mathbb{Z}$. In particular, it associates with each continuous map $f: (D^n,S^{n-1}) \to (D^n,S^{n-1})$ an integer, which coincides with $\deg f$, as defined in Subsection 4.6.9.

Further Information Obtained From the Hopf Bundles

10. **If** $r \geq 3$, **the homomorphism** $\text{pr}_*: \pi_r(S^3) \to \pi_r(S^2)$ **induced by the Hopf map** $\text{pr}: S^3 \to S^2$ **is an isomorphism. In particular,** $\pi_3(S^2)$ **is canonically isomorphic to** $\mathbb{Z}$, **and is generated by** $\text{pr}_*(\text{sph}_3)$, **i.e., by the class of the Hopf map itself.**

This is plain from the segment

$$\pi_r(S^1) = 0 \to \pi_r(S^3) \to \pi_r(S^2) \to \pi_{r-1}(S^1) = 0$$

of the homotopy sequence of the Hopf bundle (S^3,pr,S^2).

11. **If** $r \geq 1$, **the homomorphism** $\text{pr}_*: \pi_r(S^7) \to \pi_r(S^4)$ **induced by the Hopf map** $\text{pr}: S^7 \to S^4$ **maps** $\pi_r(S^7)$ **isomorphically onto a subgroup of** $\pi_r(S^7)$ **which has a direct complement isomorphic to** $\pi_{r-1}(S^3)$. **In particular,** $\pi_7(S^4) \cong \mathbb{Z} \oplus \pi_6(S^3)$.

This is a consequence of Theorem 1.8.10, when applied to the Hopf bundle (S^7,pr,S^4).

12. <u>If $r \geqslant 1$, the homomorphism $pr_* : \pi_r(S^{15}) \to \pi_r(S^8)$ induced by the Hopf map $pr: S^{15} \to S^8$ maps $\pi_r(S^{15})$ isomorphically onto a subgroup of $\pi_r(S^8)$ which has a direct complement isomorphic to $\pi_{r-1}(S^7)$. In particular, $\pi_{15}(S^8) \cong \mathbb{Z} \oplus \pi_{14}(S^7)$.</u>

This is a consequence of Theorem 1.8.10, when applied to the Hopf bundle (S^{15}, pr, S^8).

13. <u>The composite maps $\pi_{r-1}(S^1) \xrightarrow{su} \pi_r(S^2) \xrightarrow{\Delta} \pi_{r-1}(S^1)$, $\pi_{r-1}(S^3) \xrightarrow{su} \pi_r(S^4) \xrightarrow{\Delta} \pi_{r-1}(S^3)$, $\pi_{r-1}(S^7) \xrightarrow{su} \pi_r(S^8) \xrightarrow{\Delta} \pi_{r-1}(S^7)$, where the homomorphisms Δ correspond to the Hopf bundles, coincide for any $r \geqslant 1$ with $id\,\pi_{r-1}(S^1)$, $id\,\pi_{r-1}(S^3)$, and $id\,\pi_{r-1}(S^7)$, respectively.</u>

PROOF. Let q be 2, 4, or 8, and consider the map $\chi: D^q \to S^{2q-1}$ given by

$$\chi(x_1, \ldots, x_q) = (x_1, \ldots, x_q, (1 - x_1^2 - \ldots - x_q^2)^{1/2}, 0, \ldots, 0).$$

Its restriction to S^{q-1} is simply the inclusion $S^{q-1} \to S^{2q-1}$, while $pr \circ \chi$, where pr is the Hopf map $pr: S^{2q-1} \to S^q$, is simply $DS: D^q \to S^q$. Therefore, the diagram

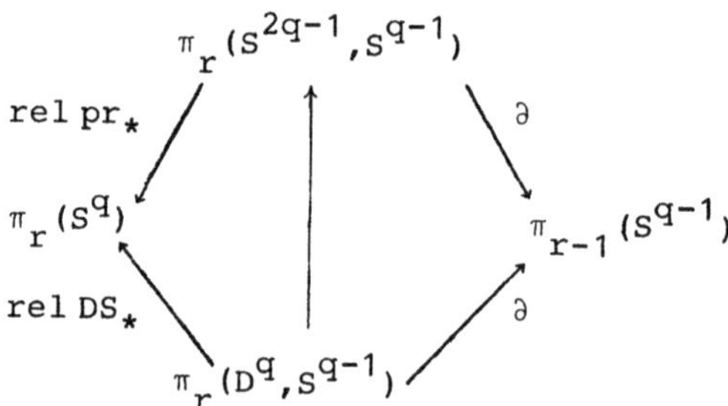

commutes; here the vertical homomorphisms are induced by $rel\,\chi : (D^q, S^{q-1}) \to (S^{2q-1}, S^{q-1})$. Moreover, $rel\,pr_*$ and the lower ∂ are isomorphisms (see 1.8.1 and 1.6.5), and from the above commutativity it follows that the composite homomorphism

$$\pi_{r-1}(S^{q-1}) \xrightarrow{\partial^{-1}} \pi_r(D^q, S^{q-1}) \xrightarrow{rel\,DS_*} \pi_r(S^q) \xrightarrow{(rel\,pr_*)^{-1}}$$

$$\pi_r(S^{2q-1}, S^{q-1}) \xrightarrow{\partial} \pi_{r-1}(S^{q-1})$$

equals $id\,\pi_{r-1}(S^{q-1})$. To complete the proof, notice that $rel\,DS_* \circ \partial^{-1} = su$ (see 1.2) and $\partial \circ (rel\,pr_*)^{-1} = \Delta$ (see 1.8.4).

14. It follows from 13 and 1.8.7 that $su: \pi_5(S^3) \to \pi_6(S^4)$ and $su: \pi_{13}(S^7) \to \pi_{14}(S^8)$ are isomorphisms, that is, in the two series

$\{\pi_{n+2}(S^n)\}$ and $\{\pi_{n+6}(S^n)\}$ [as in the series $\{\pi_n(S^n)\}$] stabilization begins at least one step earlier than guaranteed by Theorem 1.4.

3. The Composition Product

1. Let X be a space with base point x_0. Given two spheroids, $\phi \in Sph_p^O(X,x_0)$ and $\psi \in Sph_q^O(S^p,ort_1)$, the composition $\phi \circ \psi: (S^q,ort_1) \to (X,x_0)$ is a spheroid in $Sph_q^O(X,x_0)$, and the homotopy class of the latter is uniquely determined by the homotopy classes of ϕ and ψ. Therefore, for any two classes, $\alpha \in \pi_p(X,x_0)$ and $\beta \in \pi_q(S^p)$, one may define the <u>composition</u> $\alpha \circ \beta \in \pi_q(X,x_0)$. Equivalently, we can set $\alpha \circ \beta = \phi_*(\beta)$, where ϕ is any representative of α.

The following facts need no proof: if $\alpha \in \pi_p(X,x_0)$, then $\alpha \circ sph_p = \alpha$; if $\alpha \in \pi_p(X,x_0)$, $\beta \in \pi_q(S^p)$, and $\gamma \in \pi_r(S^q)$, then $\alpha \circ (\beta \circ \gamma) = (\alpha \circ \beta) \circ \gamma$; if $\alpha \in \pi_p(X,x_0)$, $\beta \in \pi_q(S^p)$, and $f: X \to Y$ is continuous, then $f_*(\alpha \circ \beta) = (f_*(\alpha)) \circ \beta$; if $\alpha \in \pi_p(X,x_0)$ and $\beta \in \pi_q(S^p)$, then $su(\alpha \circ \beta) = su\,\alpha \circ su\,\beta$; if $\alpha \in \pi_p(X,x_0)$ and $\beta_1,\beta_2 \in \pi_q(S^p)$, then $\alpha \circ (\beta_1 + \beta_2) = \alpha \circ \beta_1 + \alpha \circ \beta_2$ [in particular, $\alpha \circ k\,sph_p = k\,\alpha$ for all $\alpha \in \pi_p(X,x_0)$ and $k \in \mathbb{Z}$]. The last property is called <u>right</u> distributivity, to distinguish it from the left distributivity, which amounts to $(\alpha_1 + \alpha_2) \circ \beta = \alpha_1 \circ \beta + \alpha_2 \circ \beta$ for any $\alpha_1,\alpha_2 \in \pi_p(X,x_0)$ and $\beta \in \pi_q(S^p)$. In general, left distributivity does not hold usually; see 2 and 5, and also 9.1.

2. <u>Given any</u> $\alpha_1,\alpha_2 \in \pi_p(X,x_0)$ <u>and</u> $\beta \in \pi_{q-1}(S^{p-1})$,

$$(\alpha_1 + \alpha_2) \circ su\,\beta = \alpha_1 \circ su\,\beta + \alpha_2 \circ su\,\beta.$$

<u>In particular,</u>

$$(k\,sph_p) \circ su\,\beta = k\,su\,\beta$$

<u>for any</u> $\beta \in \pi_{q-1}(S^{p-1})$ <u>and integer</u> k.

PROOF. Pick representatives $\phi_1,\phi_2 \in Sph_p^O(X,x_0)$ and $\psi \in Sph_{q-1}^O(S^{p-1},ort_1)$ of the classes α_1,α_2 and β, respectively, and let p_1 and p_2 denote the projections $pr: S^{p-1} \times I \to$ $\to su(S^{p-1},ort_1) = S^p$ and $pr: S^{q-1} \times I \to su(S^{q-1},ort_1) = S^q$. By definition (see 1.10.2), the class $\alpha_1 + \alpha_2$ is represented by the spheroid

$$p_1(x,t) \mapsto \begin{cases} \phi_1 \circ p_1(x,2t), & \text{if } 0 \leqslant t \leqslant 1/2, \\[2ex] \phi_2 \circ p_2(x,2t-1), & \text{if } 1/2 \leqslant t \leqslant 1, \end{cases}$$

while $su\,\beta$ is represented by the spheroid $p_2(x,t) \mapsto p_1(\psi(x),t)$ (see 1.1). This shows that both sides of (1) are represented by the spheroid

$$p_2(x,t) \mapsto \begin{cases} \phi_1 \circ p_1(\psi(x),2t), & \text{if } 0 \leqslant t \leqslant 1/2, \\[2ex] \phi_2 \circ p_1(\psi(x),2t-1), & \text{if } 1/2 \leqslant t \leqslant 1. \end{cases}$$

The Ring Stab

3 (LEMMA). <u>Let</u> k, l, <u>and</u> n <u>be nonnegative integers, with</u> $n > 0$. <u>Then for any</u> $\alpha \in \pi_{n+k}(S^n)$ <u>and</u> $\beta \in \pi_{n+1}(S^n)$,

$$su^n\alpha \circ su^{n+k}\alpha = (-1)^{kl}su^n\beta \circ su^{n+1}\alpha.$$

PROOF. Pick representatives $\phi \in \mathrm{Sph}^O_{n+k}(S^n,\mathrm{ort}_1)$ and $\psi \in \mathrm{Sph}^O_{n+1}(S^n,\mathrm{ort}_1)$ of the classes α and β, and for nonnegative integers, p and q, let $\mathrm{perm}(p,q)$ be the (auto)homeomorphism of the sphere $S^{p+q} = (S^1,\mathrm{ort}_1) \otimes \ldots \otimes (S^1,\mathrm{ort}_1)$ (see 1.2.8.9) which permutes the factors according to the rule

$$((1,\ldots,p+q) \mapsto (p+1,\ldots,p+q,1,\ldots,p).$$

One may check directly that the following two compositions are equal:

$$su^n\phi \circ \mathrm{perm}(n,n+k) \circ su^{n+k}\psi \circ \mathrm{perm}(n+k,n+1) : S^{2n+k+1} \to S^n,$$

and

$$\mathrm{perm}(n,n) \circ su^n\phi \circ \mathrm{perm}(n,n+1) \circ su^{n+1}\psi : S^{2n+k+1} \to S^n.$$

Since $\deg \mathrm{perm}(p,q) = (-1)^{pq}$, this yields

$$su^n\alpha \circ [(-1)^{n(n+k)}\mathrm{sph}_{2n+k}] \circ su^{n+k}\beta \circ [(-1)^{(n+k)(n+1)}\mathrm{sph}_{2n+k+1}] =$$

$$= [(-1)^{n^2}\mathrm{sph}_{2n}] \circ su^n\beta \circ [(-1)^{n(n+1)}\mathrm{sph}_{2n+1}] \circ su^{n+1}\alpha.$$

Using the right and left distributivities (see 1 and 2), it is not hard to reduce this equality to the form

$$(-1)^{n(n+k)+(n+k)(n+1)} \, su^n\alpha \, \circ \, su^{n+k}\beta \; = \; (-1)^{n^2+n(n+1)} \, su^n\beta \, \circ \, su^{n+1}\alpha,$$

and now we note that $[n(n+k)+(n+k)(n+1)] - [n^2+n(n+1)] \equiv kl \pmod 2$.

4. We set $\mathrm{Stab} = \bigoplus_{k=0}^{\infty} \mathrm{Stab}(k)$ and identify each group $\mathrm{Stab}(k)$ with its image under the natural embedding $\mathrm{Stab}(k) \to \mathrm{Stab}$. The operation $\circ$ transforms Stab into a ring: if $\alpha \in \pi_{n+k}(S^n)$ and $\beta \in \pi_{n+k+1}(S^n)$, then $su(\alpha \circ \beta) = su\,\alpha \circ su\,\beta$ (see 1). Therefore, $\circ$ is well defined as a distributive multiplication $\mathrm{Stab}(k) \times \mathrm{Stab}(l) \to \mathrm{Stab}(k+l)$ (see 1 and 2), and can be extended bidistributively to a multiplication $\mathrm{Stab} \times \mathrm{Stab} \to \mathrm{Stab}$. It results from 1, 2, and 3 that the ring Stab is associative, has the identity $\mathrm{sph} = \mathrm{sph}_1 = \mathrm{sph}_2 = \ldots$, and is skew-commutative, meaning that $\beta \circ \alpha = (-1)^{kl}\alpha \circ \beta$ for any $\alpha \in \mathrm{Stab}(k)$ and $\beta \in \mathrm{Stab}(l)$.

An Application

5 (LEMMA). <u>The composition</u> $(-\mathrm{sph}_2) \circ pr_*(\mathrm{sph}_3)$, <u>where</u> $pr: S^3 \to S^2$ <u>is the Hopf map, equals</u> $pr_*(\mathrm{sph}_3)$.

This is a consequence of the commutativity of the diagram

$$
\begin{array}{ccc}
S^3 & \longrightarrow & S^3 \\
pr \downarrow & & \downarrow pr \\
S^2 & \longrightarrow & S^2
\end{array}
$$

where the horizontal maps are the spheroids $(x_1,x_2,x_3,x_4) \mapsto (x_1,-x_2,x_3,-x_4)$ and $(x_1,x_2,x_3) \mapsto (x_1,-x_2,x_3)$, which represent the classes sph_3 and $-\mathrm{sph}_2$, respectively.

6. <u>The group</u> $\mathrm{Stab}(1)$ <u>has at most two elements.</u>

Since $\pi_4(S^3)$ is already stable, and $su: \pi_3(S^2) \to \pi_4(S^3)$ is epimorphic (see 1.4), it suffices to show that $\mathrm{Ker}\, su$ contains the element $2[pr: S^3 \to S^2]_*(\mathrm{sph}_3)$, i.e., twice the generator of $\pi_3(S^2)$ (see 2.10). Indeed, since

$$2\,pr_*(\mathrm{sph}_3) = pr_*(\mathrm{sph}_3) + pr_*(\mathrm{sph}_3) = \mathrm{sph}_2 \circ pr_*(\mathrm{sph}_3) +$$
$$+ (-\mathrm{sph}_2) \circ pr_*(\mathrm{sph}_3)$$

(see 5), we get

$$su(2\,pr_*(\mathrm{sph}_3)) = \mathrm{sph}_3 \circ su(pr_*(\mathrm{sph}_3)) + (-\mathrm{sph}_3) \circ (pr_*(\mathrm{sph}_3))$$

(see 1), and we note that the right-hand side of the last equality is 0 (according to 2).

4. Information: Homotopy Groups of Spheres

1. For a long time the study and computation of homotopy groups of spheres was at the center of the attention of topologists. It was hoped that one could succeed in solving this problem and that other, more difficult problems in homotopy theory could be reduced to a considerable extent to it. Deep results have actually been obtained in both these directions; the initial hopes, however, have not been realized. Gradually it became clear that from the homotopy point of view the sphere is not elementary, but rather an intricate, complicated object. On the other hand, the informationa acquired about the homotopy groups of spheres found unexpected applications, first of all in differential topology.

Below we discuss a (rather small) part of these results: general results in 2, and those of tabular character in 3. For more complete information, references, and proofs, see [7].

2. $\pi_{4m-1}(S^{2m})$, $m = 1,2,\ldots,$ are the only infinite groups among $\pi_r(S^n)$ with $r > n$. Each of these infinite groups is isomorphic to a direct sum $\mathbb{Z} \oplus$ (finite group).

For an odd prime p, the order of the group $\mathrm{Stab}(2m(p-1)-1)$ with $1 \leqslant m \leqslant p-1$ is divisible by p, but not by p^2, while the order of $\mathrm{Stab}(k)$ with $k < 2p(p-1)-2$ is not divisible by p if $k \not\equiv -1 \bmod 2(p-1)$.

3. Among the groups $\pi_r(S^n)$ which have been computed are all $\pi_{n+k}(S^n)$ with $k \leqslant 22$, and all $\mathrm{Stab}(k)$ with $k \leqslant 37$. The groups $\pi_{n+k}(S^n)$ with $n \geqslant 2$ and $1 \leqslant k \leqslant 7$ are displayed in Table 1.

In Table 2, where pr always denotes one of the Hopf maps $S^3 \to S^2$, $S^7 \to S^4$, or $S^{15} \to S^8$, we indicate the generators of the groups $\mathrm{Stab}(k)$ with $k = 1,\ldots,7$.

We add the relations

$$su^3(pr_*(sph_3)) \circ su^4(pr_*(sph_3)) \circ su^5(pr_*(sph_3)) = 12\, su(pr_*(sph_7)),$$

$$su^7(pr_*(sph_3)) \circ su^6(pr_*(sph_7)) \circ su^9(pr_*(sph_7)) = 120\, su(pr_*(sph_{15})),$$

which give, together with Table 1, a complete description of $\oplus_{k=1}^{7} \mathrm{Stab}(k)$, a part of Stab.

n \ k	1	2	3	4	5	6	7
2	$\mathbb{Z}$	$\mathbb{Z}_2$	$\mathbb{Z}_2$	$\mathbb{Z}_{12}$	$\mathbb{Z}_2$	$\mathbb{Z}_2$	$\mathbb{Z}_3$
3	$\mathbb{Z}_2$	$\mathbb{Z}_2$	$\mathbb{Z}_{12}$	$\mathbb{Z}_2$	$\mathbb{Z}_2$	$\mathbb{Z}_3$	$\mathbb{Z}_{15}$
4		$\mathbb{Z}_2$	$\mathbb{Z}\oplus\mathbb{Z}_{12}$	$\mathbb{Z}_2\oplus\mathbb{Z}_2$	$\mathbb{Z}_2\oplus\mathbb{Z}_2$	$\mathbb{Z}_2\oplus\mathbb{Z}_{24}$	$\mathbb{Z}_{15}$
5			$\mathbb{Z}_{24}$	$\mathbb{Z}_2$	$\mathbb{Z}_2$	$\mathbb{Z}_2$	$\mathbb{Z}_{30}$
6				0	$\mathbb{Z}$	$\mathbb{Z}_2$	$\mathbb{Z}_{60}$
7					0	$\mathbb{Z}_2$	$\mathbb{Z}_{120}$
8						$\mathbb{Z}_2$	$\mathbb{Z}\oplus\mathbb{Z}_{120}$
9							$\mathbb{Z}_{240}$

Table 1

Groups	Generators
$\mathrm{Stab}(1) = \pi_4(S^3)\ [\cong\mathbb{Z}_2]$	$su(pr_*(sph_3))$
$\mathrm{Stab}(2) = \pi_6(S^4)\ [\cong\mathbb{Z}_2]$	$su^2(pr_*(sph_3)) \circ su^3(pr_*(sph_3))$
$\mathrm{Stab}(3) = \pi_8(S^5)\ [\cong\mathbb{Z}_{24}]$	$su(pr_*(sph_7))$
$\mathrm{Stab}(4) = \pi_{10}(S^6)\ [=0]$	$-$
$\mathrm{Stab}(5) = \pi_{12}(S^7)\ [=0]$	$-$
$\mathrm{Stab}(6) = \pi_{14}(S^8)\ [\cong\mathbb{Z}_2]$	$su^4(pr_*(sph_7)) \circ su^7(pr_*(sph_7))$
$\mathrm{Stab}(7) = \pi_{16}(S^9)\ [\cong\mathbb{Z}_{240}]$	$su(pr_*(sph_{15}))$

Table 2

k	Stab(k)	k	Stab(k)
8	$\mathbb{Z}_2\oplus\mathbb{Z}_2$	12	0
9	$\mathbb{Z}_2\oplus\mathbb{Z}_2\oplus\mathbb{Z}_2$	13	$\mathbb{Z}_3$
10	$\mathbb{Z}_2$	14	$\mathbb{Z}_6\oplus\mathbb{Z}_2$
11	$\mathbb{Z}_{504}$	15	$\mathbb{Z}_{480}\oplus\mathbb{Z}_2$

Table 3

The groups $\text{Stab}(k)$ with $k = 8,\ldots,15$ are listed in Table 3.

5. The Homotopy Groups
of Projective Spaces and Lenses

1. **Let** $2 \leqslant n \leqslant \infty$. $\pi_1(\mathbb{R}P^n,(1:0:0:\ldots))$ **has two elements and is generated by the class of the loop** $rp: I \to \mathbb{R}P^n$ **given by** $rp(t) = (\cos \pi t:\sin \pi t:0:0:\ldots)$. $\pi_r(\mathbb{R}P^n,(1:0:0:\ldots))$ **is isomorphic to** $\pi_r(S^n)$ **for all** $r \neq 1$ [**in particular, this group is trivial for** $n = \infty$], **and the isomorphism is induced by the projection** $S^n \to \mathbb{R}P^n$. $\mathbb{R}P^n$ **is simple for** n **odd, and is not n-simple for** n **even.**

(The case $n = 1$ has been considered in 2.2.)

PROOF. All assertions concerning the groups $\pi_r(\mathbb{R}P^n,(1:0:0:\ldots))$ follow from Theorem 1.8.12, when applied to the covering $(S^n,pr,\mathbb{R}P^n)$. Since the fundamental group of $\mathbb{R}P^n$ is Abelian, $\mathbb{R}P^n$ is 1-simple. Now let $r \geqslant 2$, and consider the automorphism $T_{rp}: \pi_r(\mathbb{R}P^n,(1:0:0:\ldots)) \to \pi_r(\mathbb{R}P^n,(1:0:0:\ldots))$. Let $\widetilde{rp}$ be the path in S^n which covers rp and has origin ort_1. Then from the (obvious) commutativity of the diagram

$$
\begin{array}{ccccc}
\pi_r(S^n,\text{ort}_1) & \xrightarrow{\;T_{\widetilde{rp}}\;} & \pi_r(S^n,-\text{ort}_1) & \xleftarrow{\;(-\text{id}\,S^n)_*\;} & \pi_r(S^n,\text{ort}_1) \\
\Big\downarrow{pr_*} & & \Big\downarrow{pr_*} & & \;\;\;pr_* \\
\pi_r(\mathbb{R}P^n,(1:0:0:\ldots)) & \xrightarrow{\;T_{rp}\;} & \pi_r(\mathbb{R}P^n,(1:0:0:\ldots)) & &
\end{array}
$$

it follows that T_{rp} is the identity if and only if the composition $(-\text{id}\,S^n)_* \circ T_{\widetilde{rp}}$ is the identity. If we make the identification $\pi_r(S^n,-\text{ort}_1) = \pi_r(S^n,\text{ort}_1) = \pi_r(S^n)$ (see 2.5), then we see that T_{rp} is the identity if and only if $(-\text{id}\,S^n)_*: \pi_r(S^n) \to \pi_r(S^n)$ is the identity. Finally, note that if n is odd, then $-\text{id}\,S^n$ and $\text{id}\,S^n$ are homotopic, and hence $(-\text{id}\,S^n)_*: \pi_r(S^n) \to \pi_r(S^n)$ is the identity automorphism for all r, while if n is even,

$$[(-\text{id}\,S^n)_*: \pi_n(S^n) \to \pi_n(S^n)] = -\text{id}\,\pi_n(S^n) .$$

2. **Let** $1 < n \leqslant \infty$. $\pi_2(\mathbb{C}P^n,(1:0:0:\ldots))$ **is isomorphic to** $\mathbb{Z}$ **and is generated by the class of the spheroid** in: $S^2 = \mathbb{C}P^1 \to \mathbb{C}P^n$. $\pi_r(\mathbb{C}P^n,(1:0:0:\ldots))$ **is isomorphic to** $\pi_r(S^{2n+1})$ **for any** $r \neq 2$ [**in**

particular, this group is trivial for $n = \infty$], and the isomorphism is induced by the projection $S^{2n+1} \to \mathbb{C}P^n$.

(For $n = 1$ this theorem repeats Theorem 2.10.)

We make two claims: $\mathrm{in}_* : \pi_2(\mathbb{C}P^1,(1:0)) \to \pi_2(\mathbb{C}P^n,(1:0:0:\ldots))$ is an isomorphism; and $\mathrm{pr}_* : \pi_r(S^{2n+1},\mathrm{ort}_1) \to \pi_r(\mathbb{C}P^n,(1:0:0:\ldots))$ is an isomorphism for all $r \neq 2$. The first follows from the 3-connectedness of the pair $(\mathbb{C}P^n,\mathbb{C}P^1)$ (see 2.3.2.2 and 2.1.3.5), while the second is a consequence of the homotopy sequence of the bundle $(S^{2n+1},\mathrm{pr},\mathbb{C}P^n)$ and Theorem 2.2.

3. Let $1 \leqslant n \leqslant \infty$ and $r \geqslant 1$. The homomorphism induced by the projection $S^{4n+3} \to \mathbb{H}P^n$ maps $\pi_r(S^{4n+3})$ isomorphically onto a subgroup of $\pi_r(\mathbb{H}P^n,(1:0:0:\ldots))$ which has a direct complement isomorphic to $\pi_{r-1}(S^3)$. In particular, $\pi_r(\mathbb{H}P^\infty,(1:0:0:\ldots))$ is isomorphic to $\pi_{r-1}(S^3)$ for all $r \geqslant 1$.

(For $n = 1$ this theorem repeats Theorem 2.11.)

For a proof, we need only apply Theorem 1.8.10 to the bundle $(S^{4n+3},\mathrm{pr},\mathbb{H}P^n)$.

4. The lenses $L(m;\ell_1,\ldots,\ell_n)$ and $L(m;\ell_1,\ell_2,\ldots)$ are simple. $\pi_1(L(m;\ell_1,\ldots,\ell_n))$ and $\pi_1(L(m;\ell_1,\ell_2,\ldots))$ are isomorphic to $\mathbb{Z}_m$. If $r \geqslant 2$, $\pi_r(L(m;\ell_1,\ldots,\ell_n))$ is isomorphic to $\pi_r(S^{2n-1})$, and the isomorphism is induced by the projection $S^{2n-1} \to L(m;\ell_1,\ldots,\ell_n)$. $\pi_r(L(m;\ell_1,\ell_2,\ldots))$ is trivial for all $r \geqslant 2$.

The proof is clearly a generalization of the first part of the proof of Theorem 1.

6. The Homotopy Groups of Classical Groups

1. The inclusion homomorphism $\pi_r(SO(n)) \to \pi_r(SO(n+1))$ is an isomorphism for $r \leqslant n-2$ and an epimorphism for $r = n-1$. $\pi_1(SO(n))$ is isomorphic to $\mathbb{Z}$ for $n = 2$ and to $\mathbb{Z}_2$ for $n \geqslant 3$, and is generated by the class of the inclusion $S^1 = SO(2) \to SO(n)$. $\pi_2(SO(n))$ is trivial for all n. $\pi_3(SO(n))$ is trivial for $n \leqslant 2$, isomorphic to $\mathbb{Z}$ for $n = 3$, to $\mathbb{Z} \oplus \mathbb{Z}$ for $n = 4$, and to the quotient group of $\mathbb{Z} \oplus \mathbb{Z}$ by a cyclic subgroup for $n \geqslant 5$.

PROOF. The triviality of the groups $\pi_2(SO(2))$, $\pi_3(SO(2))$, $\pi_2(SO(3))$, and $\pi_2(SO(4))$, and the isomorphisms $\pi_1(SO(2)) \cong \mathbb{Z}$, $\pi_1(SO(3)) \cong \mathbb{Z}_2$, $\pi_3(SO(3)) \cong \mathbb{Z}$, and $\pi_3(SO(4)) \cong \mathbb{Z} \oplus \mathbb{Z}$ all result from the equalities $SO(2) = S^1$, $SO(3) = \mathbb{RP}^3$, and $SO(4) = \mathbb{RP}^3 \times S^3$ (see 3.2.1.2, 3.2.3.1, and 3.2.3.3), and Theorems 2.2, 2.7, and 5.1. The rest is a consequence of the homotopy sequence of the bundle $(SO(n+1), pr, S^n)$ with base point $id \in SO(n+1)$; see 4.6.1.4.

2. <u>The inclusion homomorphism</u> $\pi_r(U(n)) \to \pi_r(U(n+1))$ <u>is an isomorphism for</u> $r \leqslant 2n-1$ <u>and an epimorphism for</u> $r = 2n$. <u>If</u> $n \geqslant 1$, $\pi_1(U(n))$ <u>is isomorphic to</u> $\mathbb{Z}$ <u>and is generated by the class of the inclusion</u> $S^1 = U(1) \to U(n)$. $\pi_2(U(n))$ <u>is trivial for all</u> n. $\pi_3(U(1))$ <u>is trivial, while</u> $\pi_3(U(n))$ <u>is isomorphic to</u> $\mathbb{Z}$ <u>for all</u> $n \geqslant 2$. <u>The inclusion homomorphism</u> $\pi_1(U(n)) \to \pi_1(SO(2n))$ <u>is epimorphic for all</u> n.

These are corollaries of the equalities [in: $U(1) \to SO(2)$] = = id and $U(2) = S^1 \times S^3$, and of the homotopy sequence of the bundle $(U(n+1), pr, S^{2n+1})$ with base point $id \in U(n+1)$; see 4.6.1.4.

3. <u>The inclusion homomorphism</u> $\pi_r(Sp(n)) \to \pi_r(Sp(n+1))$ <u>is an isomorphism for</u> $r \leqslant 4n+1$ <u>and an epimorphism for</u> $r = 4n+2$. <u>In particular, if</u> $r \leqslant 5$ <u>and</u> $n \geqslant 1$, $\pi_r(Sp(n))$ <u>is isomorphic to</u> $\pi_r(Sp(1)=S^3)$.

This can be seen from the homotopy sequence of the bundle $(Sp(n+1), pr, S^{4n+3})$ with base point $id \in Sp(n+1)$; see 4.6.1.4.

Stabilization

4. Theorems 1–3 show that for $r \geqslant 1$, each series of groups

$$\pi_r(SO(1)) \to \pi_r(SO(2)) \to \pi_r(SO(3)) \to \ldots,$$

$$\pi_r(U(1)) \to \pi_r(U(2)) \to \pi_r(U(3)) \to \ldots,$$

$$\pi_r(Sp(1)) \to \pi_r(Sp(2)) \to \pi_r(Sp(3)) \to \ldots,$$

stabilizes: the first one, starting with $\pi_r(SO(r+2))$, the second one, with $\pi_r(U([r+2]/2]))$, and the third one, with $\pi_r(Sp([(r+2)/4]))$. The groups $\pi_r(SO(n))$ with $n \geqslant r+2$, $\pi_r(U(n))$ with $n \geqslant [(r+2)/2]$, and $\pi_r(Sp(n))$ with $n \geqslant [(r+2)/4]$ are said to be <u>stable</u>, and are denoted by $\pi_r(SO)$, $\pi_r(U)$, and $\pi_r(Sp)$, respectively. By Theorem 1, $\pi_1(SO) \cong \mathbb{Z}$, $\pi_2(SO) = 0$, and $\pi_3(SO) \cong (\mathbb{Z} \oplus \mathbb{Z})/(\text{cyclic subgroup})$. By Theorem 2, $\pi_1(U) \cong \mathbb{Z}$, $\pi_2(U) = 0$, and $\pi_3(U) \cong \mathbb{Z}$. Finally, by Theorem 3, $\pi_1(Sp) = 0$, $\pi_2(Sp) = 0$, and $\pi_3(Sp) \cong \mathbb{Z}$.

The notations $\pi_r(SO)$, $\pi_r(U)$, and $\pi_r(Sp)$ have also a direct meaning: they represent the ordinary r-th homotopy groups of the limit spaces $SO = \lim SO(n)$, $U = \lim U(n)$, and $Sp = \lim Sp(n)$, respectively (see 1.11.2).

Information

5. The homotopy groups $\pi_r(SO)$, $\pi_r(U)$, and $\pi_r(Sp)$ have been explicitly computed. Namely, for any $r \geqslant 1$ there are canonical isomorphisms $\pi_r(SO) \to \pi_{r+8}(SO)$, $\pi_r(Sp) \to \pi_{r+8}(Sp)$, and $\pi_r(U) \to \pi_{r+2}(U)$, and the first seven homotopy groups of SO and Sp, together with the first two homotopy groups of U are displayed in the following tables.

r	1	2	3	4	5	6	7	8
$\pi_r(SO)$	$\mathbb{Z}_2$	0	$\mathbb{Z}$	0	0	0	$\mathbb{Z}$	$\mathbb{Z}_2$
$\pi_r(Sp)$	0	0	$\mathbb{Z}$	$\mathbb{Z}_2$	$\mathbb{Z}_2$	0	$\mathbb{Z}$	0

r	1	2
$\pi_r(U)$	$\mathbb{Z}$	0

For a proof, see [17].

There are also many unstable homotopy groups of the manifolds $SO(n)$, $U(n)$, and $Sp(n)$ which have been computed. For example, $\pi_{2n}(U(n)) \cong \mathbb{Z}_{n!}$, $\pi_{4n+2}(Sp(n)) \cong \mathbb{Z}_{(2n+1)!}$ for n even, and $\pi_{4n+2}(Sp(n)) \cong \mathbb{Z}_{2[(2n+1)!]}$ for n odd. For details and references for the proofs, see [7].

7. The Homotopy Groups
of Stiefel Manifolds and Spaces

1 (LEMMA). Let $k < n$. Then the following are true: the manifold $V(n,k)$ is simple; the inclusion homomorphism $\pi_r(V(n,k)) \to \pi_r(V(n+1,k+1))$ is an isomorphism for $r < n-1$ and an epimorphism for $r = n-1$; and if n is odd and $k = 1$, the last epimorphism is also an isomorphism.

The manifolds $\mathbb{C}V(n,k)$ and $\mathbb{H}V(n,k)$ are all simple. The inclusion homomorphism $\pi_r(\mathbb{C}V(n,k)) \to \pi_r(\mathbb{C}V(n+1,k+1))$ is an isomorphism for $r < 2n$ and an epimorphism for $r = 2n$. The inclusion homomorphism $\pi_r(\mathbb{H}V(n,k)) \to \pi_r(\mathbb{H}V(n+1,k+1))$ is an isomorphism for $r < 4n+2$ and an epimorphism for $r = 4n+2$.

The fact that the Stiefel manifolds are simple may be seen from the equalities $V(n,k) = SO(n)/SO(n-k)$, $\mathbb{C}V(n,k) = U(n)/U(n-k)$, and $\mathbb{H}V(n,k) = Sp(n)/Sp(n-k)$ (see 4.2.3.12 and 1.9.7). To prove the rest, use the homotopy sequences of the bundles $(V(n+1,k+1),pr,S^n)$, $(\mathbb{C}V(n+1,k+1),pr,S^{2n+1})$, and $(\mathbb{H}V(n+1,k+1),pr,S^{4n+3})$ described in 4.6.1.4, taking the inclusions $\mathbb{R}^k \to \mathbb{R}^n$, $\mathbb{C}^k \to \mathbb{C}^n$, and $\mathbb{H}^k \to \mathbb{H}^n$ as base points (in the respective total spaces). In the real case, we take advantage, in addition, of the fact that for n odd, the bundle $(V(n+1,2),pr,S^n)$ admits a section (see 3.1.4.7); for n odd and $k = 1$, this ensures that the first of the aforementioned homotopy sequences splits from the left at the terms $\pi_r(V(n+1,k+1))$ (see 1.8.8).

2. If $k < n$, then the manifold $V(n,k)$ is $(n-k-1)$-connected. $\pi_{n-k}(V(n,k))$ with $0 < k < n$ is cyclic and is generated by the class of the inclusion $S^{n-k} = V(n-k+1,1) \to V(n,k)$; this group is infinite whenever $n-k$ is even or $k = 1$.

This is a corollary of Lemma 1: when $r < n-k$,
$$\pi_r(V(n,k)) \cong \pi_r(V(n-1,k-1)) \cong \ldots \cong \pi_r(V(n-k+1,1)) = \pi_r(S^{n-k}) = 0,$$
while in the sequence $\pi_{n-k}(S^{n-k}) = \pi_{n-k}(V(n-k+1,1)) \to \pi_{n-k}(V(n-k+2,2)) \to$ $\to \ldots \to \pi_{n-k}(V(n,k))$ all the maps are isomorphisms, except for the first, which is an isomorphism for $n-k$ even and an epimorphism for $n-k$ odd.

3. The manifold $\mathbb{C}V(n,k)$ is $2(n-k)$-connected. $\pi_{2n-2k+1}(\mathbb{C}V(n,k))$ is isomorphic to $\mathbb{Z}$ and is generated by the class of the inclusion $S^{2n-2k+1} = \mathbb{C}V(n-k+1,1) \to \mathbb{C}V(n,k)$.

This is a corollary of Lemma 1: when $r \leqslant 2n-2k+1$, in the sequence $\pi_r(S^{2n-2k+1}) = \pi_r(\mathbb{C}V(n-k+1,1)) \to \pi_r(\mathbb{C}V(n-k+2,2)) \to \ldots \to$ $\to \pi_r(\mathbb{C}V(n,k))$ all the arrows are isomorphisms.

4. The manifold $\mathbb{H}V(n,k)$ is $(4n-4k+2)$-connected. $\pi_{4n-4k+3}(\mathbb{H}V(n,k))$ is isomorphic to $\mathbb{Z}$ and is generated by the class of the inclusion $S^{4n-4k+3} = \mathbb{H}V(n-k+1,1) \to \mathbb{H}V(n,k)$.

This is also a corollary of Lemma 1: if $r \leqslant 4n-4k+3$ (actually, if $r \leqslant 4n-4k+5$), then in the sequence $\pi_r(S^{4n-4k+3}) =$ $= \pi_r(\mathbb{H}V(n-k+1,1)) \to \pi_r(\mathbb{H}V(n-k+2,2)) \to \ldots \to \pi_r(\mathbb{H}V(n,k))$ all the arrows are isomorphisms.

5. The spaces $V(\infty,k)$ and $\mathbb{C}V(\infty,k)$ (see 4.5.3.9), as well as $\mathbb{H}V(\infty,k) = \lim(\mathbb{H}V(n,k),in : \mathbb{H}V(n,k) \to \mathbb{H}V(n+1,k))$ are ∞-connected.

This follows from Theorems 2, 3, 4, and 1.11.2.

8. The Homotopy Groups
of Grassman Manifolds and Spaces

1. In this subsection the computation of the most important homotopy groups of the Grassman manifolds $G(n,k)$, $G_+(n,k)$, $\mathbb{C}G(n,k)$, and $\mathbb{H}G(n,k)$, and of the Grassman spaces $G(\infty,k)$, $G_+(\infty,k)$, $\mathbb{C}G(\infty,k)$ (see 4.5.3.2) and $\mathbb{H}G(\infty,k) = \lim(\mathbb{H}G(n,k), \text{in} : \mathbb{H}G(n,k) \to \mathbb{H}G(n+1,k))$ is reduced to the computation of the homotopy groups of the corresponding classical groups.

Grassman manifolds and spaces are taken care of together, and thus n may also take the value ∞.

2. **If** $k > 0$ **and** $0 < r < n-k$, **then** $\pi_r(G_+(n,k))$ **is isomorphic to** $\pi_{r-1}(SO(k))$, **and the inclusion homomorphism** $\pi_r(G_+(n,k)) \to \pi_r(G_+(n',k))$ **is an isomorphism for all** $n' > n$.

The first claim results from Theorems 7.2 and 7.5, and the homotopy sequence of the bundle $(V(n,k),\text{pr},G_+(n,k))$, defined in 4.6.1.4, with the inclusion $\mathbb{R}^k \to \mathbb{R}^n$ as base point. The second claim results from the commutativity of the diagram

$$
\begin{array}{ccc}
\pi_r(G_+(n,k)) & \xrightarrow{\ \Delta\ } & \pi_{r-1}(SO(k)) \\
\downarrow{\scriptstyle \text{in}_*} & & \downarrow{\scriptstyle \text{in}_* = \text{id}} \\
\pi_r(G_+(n',k)) & \xrightarrow{\ \Delta\ } & \pi_{r-1}(SO(k))
\end{array}
$$

(see 1.8.6).

3. $\pi_r(G(n,k))$, **with** $0 < k < n$ **and** $r \geqslant 2$, **is isomorphic to** $\pi_r(G_+(n,k))$. $\pi_1(G(n,k))$ **is isomorphic to** $\mathbb{Z}$ **for** $n = 2$, $k = 1$, **and to** $\mathbb{Z}_2$ **for** $0 < k < n$, $n \geqslant 3$.

Since $G(2,1)$ is homeomorphic to S^1, Theorem 2.2 yields $\pi_1(G(2,1)) \cong \mathbb{Z}$. If we now apply Theorem 1.8.12 to the canonical two-sheeted covering $(G_+(n,k),\text{pr},G(n,k))$, the rest is plain.

4. **If** $0 < r < 2n-2k+1$, **then** $\pi_r(\mathbb{C}G(n,k))$ **is isomorphic to** $\pi_{r-1}(U(k))$, **and the inclusion homomorphism** $\pi_r(\mathbb{C}G(n,k)) \to \pi_r(\mathbb{C}G(n',k))$ **is an isomorphism for all** $n' > n$.

The proof repeats that of Theorem 2, with obvious changes.

5. **If** $0 < r < 4n-4k+3$, **then** $\pi_r(\mathbb{H}G(n,k))$ **is isomorphic to** $\pi_{r-1}(Sp(k))$, **and the inclusion homomorphism** $\pi_r(\mathbb{H}G(n,k)) \to \pi_r(\mathbb{H}G(n',k))$ **is an isomorphism for all** $n' > n$.

Again, the proof repeats that of Theorem 2, with obvious changes.

9. <u>Exercises</u>

1. Let $q = 2,4,8$, and let $pr: S^{2q-1} \to S^q$ be the Hopf map. Show that for any integer k

$$(k \, sph_q) \circ pr_*(sph_{2q-1}) = k^2 pr_*(sph_{2q-1}).$$

2. Show that for any positive integer n, $\mathbb{R}P^n$ is $(n+1)$-simple.

3. Let n be even and k be odd. Show that $G(n,k)$ is simple.

4. Let $3 \leqslant n \leqslant \infty$. Show that $G(n,2)$ is not 2-simple.

5. Show that the inclusion homomorphisms

$$\pi_r(SO(3)) \to \pi_r(SO(4)), \quad \pi_r(SO(7)) \to \pi_r(SO(8)),$$

$$\pi_r(U(1)) \to \pi_r(U(2)), \quad \pi_r(U(3)) \to \pi_r(U(4)),$$

and

$$\pi_r(Sp(1)) \to \pi_r(Sp(2))$$

are monomorphic for any integer r.

6. Consider the map $\mathbb{C}V(n,k) \to V(2n,2k-1)$ which takes each k-frame $(v_1,\ldots,v_k)$ of $\mathbb{C}^n$ into the frame $(v_1,iv_1,\ldots,v_{k-1},iv_{k-1},v_k)$ of $\mathbb{C}^n$, considered as $\mathbb{R}^{2n}$. Show that the homomorphism $\pi_{2n-2k+1}(\mathbb{C}V(n,k)) \to \pi_{2n-2k+1}(V(2n,2k-1))$ induced by this map takes the generator indicated in 7.3 of $\pi_{2n-2k+1}(\mathbb{C}V(n,k))$ into the generator indicated in 7.2 of $\pi_{2n-2k+1}(V(2n,2k-1))$.

§3. HOMOTOPY GROUPS OF CELLULAR SPACES

1. The Homotopy Groups
of One-dimensional Cellular Spaces

1. In this subsection we compute the homotopy groups of a bouquet $B = V_{\mu \in M}(S_\mu = S^1, \text{ort}_1)$ constructed from an arbitrary family, $\{S_\mu = S^1\}_{\mu \in M}$, of circles. As usual, the base point bp will be the center of the bouquet.

To simplify the exposition, we let u_μ and α_μ denote the loop defined by the inclusion $\text{imm}_\mu : S^1 \to B$, i.e., the loop $\text{imm}_\mu \circ \text{IS}: I \to B$, and the homotopy class of u_μ, i.e., $\text{imm}_{\mu *}(\text{sph}_1)$, respectively. A loop will be referred to as <u>standard</u> if it is of the form $(\ldots((v_1 v_2)v_3)\ldots v_{n-1})v_n$, where each of the factors $v_1, \ldots, v_n$ is either one of the loops u_μ or one of their inverses u_μ^{-1} and, in addition, two loops u_μ, u_μ^{-1} with the same μ are not allowed to be adjacent. The case $n = 0$ is not excluded: then, the product is simply the constant loop with origin bp.

2 (LEMMA). <u>There is a covering</u> $(B^\sim, p, B)$ <u>with the following two properties:</u>

(i) $B^\sim$ <u>is contractible;</u>

(ii) <u>the paths which cover standard loops and originate at</u> <u>some point</u> x_0 <u>of the fiber</u> $F_0 = p^{-1}(bp)$ <u>end at distinct points of</u> F_0, <u>and</u> F_0 <u>is exhausted by the ends of these paths.</u>

PROOF. Let us agree to denote by $GF(M)$, as usual, the free group generated by the set M. We equip $GF(M)$ with the discrete topology, form the bouquet $A = V_{\mu \in M}(D_\mu = D^1, 0)$, and then the product $A \times GF(M)$. Further, let p be the partition of $A \times GF(M)$ into the pairs $\{(\text{imm}_\mu(1), g), (\text{imm}_\mu(-1), g\mu)\}$ with $\mu \in M$ and $g \in GF(M)$, and the points which do not appear in any of these pairs, and denote by ρ the composition

$$A \times GF(M) \xrightarrow{\;\text{pr}_1\;} A \xrightarrow{\;V_\mu (DS_\mu = DS)\;} B.$$

Then ρ is obviously constant on the elements of p. Now set

$$B^\sim = [A \times GF(M)]/p, \quad p = [\text{fact}\,\rho : B^\sim \to B], \quad x_0 = \text{pr}(a_0, e),$$

where a_0 is the center of the bouquet A, $e = e_{GF(M)}$, and pr = [pr: $A \times GF(M) \to \tilde{B}$]. Then it is readily seen that $(\tilde{B}, p, B)$ is a covering with $F_0 = \mathrm{pr}(a_0 \times GF(M))$ and $x_0 \in F_0$.

The contractibility of $\tilde{B}$ follows from Lemma 2.3.3.4: in fact, the subspaces $\mathrm{pr}(A \times [GF_n(M) \smallsetminus GF_{n-1}(M)])$ of $\tilde{B}$, where $GF_n(M)$ is the part of $GF(M)$ consisting of words of length $\leqslant n$, satisfy the conditions of this lemma. The path with origin x_0 which covers the standard path $(\ldots(u_{\mu_1}^{\varepsilon_1} u_{\mu_2}^{\varepsilon_2})\ldots)u_{\mu_n}^{\varepsilon_n}$ $[\varepsilon_1, \ldots, \varepsilon_n = \pm 1]$, ends at the point $\mathrm{pr}(a_0, g)$, with $g = \mu_1^{\varepsilon_1} \mu_2^{\varepsilon_2} \ldots \mu_n^{\varepsilon_n}$. Clearly, the ends of of these paths are pairwise distinct and exhaust F_0.

3. <u>The groups</u> $\pi_r(B)$ <u>with</u> $r > 1$ <u>are trivial, whereas</u> $\pi_1(B, bp)$ <u>is a free group with free generators</u> α_μ.

The proof is based on Lemma 2 and uses the same notation. Since $\tilde{B}$ is contractible, all its homotopy groups are trivial, and hence so are the groups $\pi_r(B)$ with $r > 1$; moreover, the map $\Delta: \pi_1(B, bp) \to \pi_0(F_0, x_0)$ is invertible (see 1.8.12). Combining the invertibility of Δ with property (ii) of the covering $(\tilde{B}, p, B)$, we see that the homotopy classes of the standard loops are pairwise distinct and exhaust $\pi_1(B, bp)$. Consequently, $\pi_1(B, bp)$ is a free group with generators α_μ, $\mu \in M$.

4. <u>The fundamental group of a connected one-dimensional cellular space is free, whereas its higher homotopy groups are trivial.</u>

This is a corollary of Theorem 3, because every connected one-dimensional cellular space is homotopy equivalent to a bouquet of circles (see 2.3.3.6).

2. The Effect of Attaching Balls

1. Let $X = A \cup_\varphi [\bigsqcup_{\mu \in M}(D_\mu = D^{k+1})]$, where A is a connected topological space, and φ is a continuous map $\bigsqcup_{\mu \in M}(S_\mu = S^k) \to A$ (see 2.3.2.1), and let $x_0 \in A$. In this subsection we exhibit a system of generators for the group $\pi_{k+1}(X, A, x_0)$ [$k \geqslant 1$].

We remark that the homotopy groups $\pi_r(X, A)$ with $r \leqslant k$ are trivial (see 2.3.2.1), whereas for $r > k+1$, $\pi_r(X, A)$ is already a much more complicated object: in the simplest case, when A is just a point and the family $\{D_\mu\}$ consists of a single ball, $\pi_r(X, A)$ equals $\pi_r(S^{k+1})$.

In Theorem 2 below, f_μ denotes the composite map

$$D^{k+1} \xrightarrow{\ in_\mu\ } \bigsqcup_\nu D_\nu \xrightarrow{\ imm_\mu\ } X,$$

and $\alpha_\mu \in \pi_{k+1}(X,A,f_\mu(ort_1))$ is the class of the spheroid $f_\mu: (D^{k+1},S^k,ort_1) \to (X,A,f_\mu(ort_1))$.

2. *Let* $w_\mu: I \to A$ *be an arbitrary path joining the points* $f_\mu(ort_1)$ *and* x_0. *If* $k \geqslant 1$, *then* $\pi_{k+1}(X,A,x_0)$ *is generated over* $\pi_1(A,x_0)$ *by the classes* $\beta_\mu = T_{w_\mu}\alpha_\mu$ [*i.e., it is generated, in the usual sense, by the classes* $T_\omega\beta_\mu$ *with* $\omega \in \pi_1(A,x_0)$].

We claim that every element $\beta \in \pi_{k+1}(X,A,x_0)$ can be represented as

$$\beta = \prod_{i=1}^{m}[T_{\omega_i}(\beta_{\mu_i})]^{\pm 1} \qquad (\mu_i \in M, \quad \omega_i \in \pi_1(A,x_0)). \tag{1}$$

By Lemma 2.3.2.1, there is a spheroid $g \in Sph_{k+1}^{0}(X,A,x_0)$ in the class μ, and similarity transformations $\sigma_1,\ldots,\sigma_m$ mapping D^{k+1} onto pairwise disjoint balls $d_1,\ldots,d_m \subset Int\, D^{k+1}$, with the following properties: the point of d_i having the largest value of the first coordinate coincides with $\sigma_i(ort_1)$; the segment joining this point with ort_1 lies in $C = D^{k+1} \setminus U_{i=1}^m\, Int\, d_i$; the composition

$$D^{k+1} \xrightarrow{\ \sigma_i\ } d_i \xrightarrow{\ in\ } D^{k+1} \xrightarrow{\ g\ } X$$

is identical with one of the maps f_{μ_i}; and $g(C) \subset A$. Now it is clear that, if we suitably reindex the balls $d_1,\ldots,d_m$, then X, A, x_0, g, and $d_1,\ldots,d_m$ satisfy the conditions of Theorem 1.11.1, and hence we have, in the notation of this theorem, $\gamma = \Pi_{i=1}^m\, T_{s_i}(\gamma_i)$. In our case, $\gamma = \beta$ and $\gamma_i = \alpha_{\mu_i}^{\pm 1}$; the last equality is a consequence of the fact that α_{μ_i} and γ_i are the elements of $\pi_{k+1}(X,A,f_{\mu_i}(ort_1))$ represented by the spheroids f_{μ_i} and $g \circ \tau_i$, which are transformed one into another by the orthogonal transformations of D^{k+1}, $\sigma_i^{-1} \circ ab\, \tau_i$, $(ab\, \tau_i)^{-1} \circ \sigma_i$ (which are inverses of one another). Consequently, $\beta = \Pi_{i=1}^m\, T_{s_i}(\alpha_{\mu_i}^{\pm 1})$, and to obtain (1), we need only write ω_i for the class of the loop $w_{\mu_i}^{-1}s_i$.

3 (COROLLARY). *Under the hypotheses of Theorem 2, the inclusion homomorphism* $\pi_r(A,x_0) \to \pi_r(X,x_0)$ *is an isomorphism for* $r \leqslant k-1$, *and an epimorphism for* $r = k$. *The kernel of this epimorphism*

is generated over $\pi_1(A,x_0)$ by the classes $\partial\beta_\mu = T_{w_\mu}(\partial\alpha_\mu)$ [i.e., by the classes of the attaching spheroids ∂f_μ, translated to x_0].

    4. Let (X,A) be a cellular pair with base point x_0. If A is connected and $A \supset ske_k X$, with $k \geqslant 1$, then $\pi_r(X,A)$ is trivial for all $r \leqslant k$. Moreover, $\pi_{k+1}(X,A,x_0)$ is generated over $\pi_1(A,x_0)$ by the classes of the characteristic maps of the $(k+1)$-cells in $X \smallsetminus A$ (regarded as spheroids), translated to x_0 along arbitrary paths. The inclusion homomorphism $\pi_r(A,x_0) \to \pi_r(X,x_0)$ is an isomorphism for $r \leqslant k-1$ and an epimorphism for $r = k$; the kernel of the latter is generated over $\pi_1(A,x_0)$ by the classes of the attaching spheroids of the $(k+1)$-cells in $X \smallsetminus A$, translated to x_0 along arbitrary paths.

    When $X \smallsetminus A \subset ske_{k+1}$ all these assertions follow from 1, 2, and 3. The general case is reduced to this special situation by Theorem 2.3.2.4.

3. The Fundamental Group of a Cellular Space

    1. In this subsection we pressent an effective method for computing the fundamental group of a cellular space possessing a single 0-cell. This last condition is not a serious limitation, since, firstly, it is fulfilled in the most important cases and, secondly, every connected space can be transformed, by taking a rather simple quotient, into a homotopy equivalent space which meets our requirement (see Subsection 2.3.3). It is by no means difficult to generalize the computation scheme to arbitrary cellular spaces; however, the exposition is cumbersome.

    2. Let X be a cellular space with a single 0-cell x_0. Since x_0 is also the unique 0-cell of $ske_1 X$, this skeleton is homeomorphic to a bouquet of circles. Consequently, $\pi_1(ske_1 X, x_0)$ is the free group generated by the homotopy classes of the characteristic loops, i.e., of the characteristic maps of the 1-cells (see 1.3).

    According to 2.4, $in_*: \pi_1(ske_1 X, x_0) \to \pi_1(X, x_0)$ is an epimorphism whose kernel is generated over $\pi_1(ske_1 X, x_0)$ by the homotopy classes of the attaching maps of the 2-cells of X, translated to x_0 along arbitrary paths. In our case, $\pi_1(ske_1 X, x_0)$ acts as a group of inner automorphisms, and hence $Ker\, in_*$ is the smallest normal subgroup of $\pi_1(X, x_0)$ containing the above elements. Thus, the fundamental group that we want to compute is canonically isomorphic to the quotient group of $\pi_1(ske_1 X, x_0)$ by this normal subgroup.

3. The discussion above shows that in order to compute $\pi_1(X,x_0)$ it suffices to know the 1-skeleton of X and the attaching maps of the 2-cells of X. Given these data, we can exhibit a system of generators and relations for $\pi_1(X,x_0)$: to each 1-cell corresponds a generator, namely the class of the respective characteristic loop; each 2-cells defines a relation, namely that the class of the attaching map of the given 2-cell, when translated to x_0 and expressed in terms of generators, must be equal to the identity element of $\pi_1(X,x_0)$. In a very simplified fashion, we may say that a set of generators of $\pi_1(X,x_0)$ consists of the 1-cells of X, while a system of relations consists of the 2-cells.

We remark that the system of relations is not entirely canonical, because it depends upon the choice of the paths along which we do the translation; consequently, the left-hand sides of the relations are determined only up to conjugation.

4 (COROLLARY). <u>The fundamental group of a finite connected cellular space has a presentation given by a finite number of generators and relations.</u>

An Additional Theorem

5. If A and B are subspaces of the topological space X, with inclusions

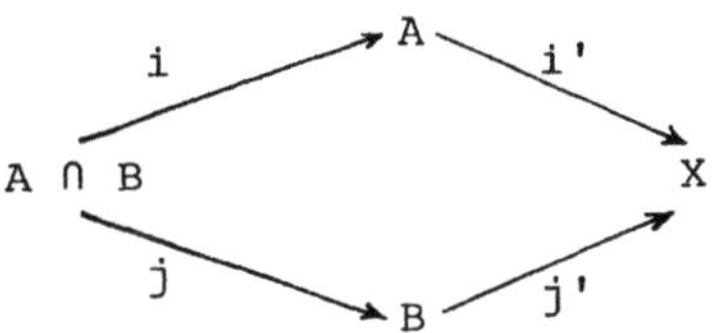

and $x_0 \in A \cap B$, then the rule $\alpha * \beta \to i'_*(\alpha)j'_*(\beta)$ defines a homomorphism $\pi_1(A,x_0) * \pi_1(B,x_0) \to \pi_1(X,x_0)$ [$*$ denotes the free product], whose kernel contains all the elements of the form $i_*(\delta) * j_*(\delta^{-1})$ with $\delta \in \pi_1(A \cap B,x_0)$. Therefore, the same rule defines a homomorphism

$$[\pi_1(A,x_0) * \pi_1(B,x_0)]/vk(X,A,B,x_0) \to \pi_1(X,x_0), \qquad (2)$$

where $vk(X,A,B,x_0)$ designates the smallest normal subgroup of $\pi_1(A,x_0) * \pi_1(B,x_0)$ containing the indicated elements (and is known as the <u>van Kampen subgroup</u>). Furthermore, homomorphism (2) is natural, meaning that the diagram

$$[\pi_1(A,x_0) * \pi_1(B,x_0)]/vk(X,A,B,x_0) \longrightarrow \pi_1(X,x_0)$$

$$\downarrow \qquad\qquad\qquad\qquad\qquad\qquad\qquad\qquad \downarrow$$

$$[\pi_1(A',x_0') * \pi_1(B',x_0')]/vk(X',A',B',x_0) \longrightarrow \pi_1(X',x_0')$$

produced by a continuous map $(X,A,B,x_0) \to (X',A',B',x_0')$ always commutes.

6. <u>Let</u> (X,A,B) <u>be a cellular triad (i.e., a cellular space</u> X <u>with two subspaces,</u> A <u>and</u> B, <u>such that</u> $A \cup B = X$), <u>and let</u> $x_0 \in D = A \cap B$. <u>If</u> A, B, <u>and</u> D <u>are connected, then</u> (2) <u>is an</u> <u>isomorphism.</u>

(This theorem will be generalized in the next section; see 4.3.12.)

PROOF. Let us assume first that X has a single 0-cell x_0. By 3, the fundamental group at x_0 of any of the spaces A, B, C, or D admits a presentation by generators and relations corresponding to its 1-cells and respectively its 2-cells. Therefore, the system of generators and relations of $\pi_1(X,x_0)$ $(\pi_1(D,x_0))$ is the union (respectively, intersection) of the systems of generators and relations of $\pi_1(A,x_0)$ and $\pi_1(B,x_0)$, and the homomorphisms i_*, j_*, i'_*, and j'_* from 5 act as the identity on generators. Using the systems of generators and relations of $\pi_1(A,x_0)$ and $\pi_1(B,x_0)$ we may build a system of generators and relations of the group $\pi_1(A,x_0) * \pi_1(B,x_0)$; however, the generators corresponding to the 1-cells in D must be counted twice. With this choice of generators and relations, the homomorphism $\pi_1(A,x_0) * \pi_1(B,x_0) \to \pi_1(X,x_0)$, $\alpha * \beta \mapsto i'_*(\alpha)j'_*(\beta)$, is the identity on generators, and its kernel is generated by the elements obtained by identifying the generators corresponding to the 1-cells in D. This completes the proof of the case that we considered.

To reduce the general case to this special one, we shall transfer the origin from x_0 to an arbitrary 0-cell e_0 in D (by a translation inside D), and then replace the quadruplet (X,A,B,e_0) by a homotopy equivalent quadruplet (X',A',B',e_0') with a single 0-cell, e_0', and such that $A' \cup B' = X'$. We exhibit such a quadruplet by taking quotients twice: first, the quotient of X by a contractible one-dimensional subspace of D containing ske_0D (see 2.3.3.5 and 1.3.3.7, and cf. 2.3.3.6), and subsequently the quotient of the resulting space by the union (bouquet) of contractible one-dimensional subspaces of the quotients of A and B containing the 0-skeletons of these quotients.

7 (COROLLARY). <u>If</u> A <u>and</u> B <u>are cellular spaces with 0-cells</u> a <u>and</u> b <u>as base points, then</u> $\pi_1((A,a) \vee (B,b), bp) \cong \pi_1(A,a) * \pi_1(B,b)$.

4. <u>Homotopy Groups of Compact Surfaces</u>

1. Recall that a sphere with handles and crosscaps and at least one hole is homotopy equivalent to a bouquet of circles, the number of circles being $2g+1-1$ when g handles and 1 holes are present, and $h+1-1$ when h crosscaps and 1 holes are present (see 3.5.3.9). Thus, the fundamental group of such a surface is free, having $2g+1-1$ or $h+1-1$ generators, respectively, whereas the higher homotopy groups are trivial; see 1.3.

Below we shall discuss the homotopy groups of closed surfaces, i.e., of spheres with handles or crosscaps, but no holes. First (using the cellular decompositions indicated in Subsection 3.5.3, and Theorem 3.3) we compute the fundamental groups, and then (by means of a simple device) we handle the higher homotopy groups. We disregard the sphere and the projective space, whose homotopy groups have been computed in the previous section (see 2.2.7, 2.2.10, 2.4.3, and 2.5.1), and need no further comment.

The Fundamental Groups of Closed Surfaces

2. The cellular decomposition of a sphere with g handles, constructed in 3.5.3.8, contains one 0-cell e_0, $2g$ 1-cells $a_1, b_1, \ldots, a_g, b_g$, and one 2-cell, whose attaching map takes ort_1 into e_0. Let $\alpha_1, \beta_1, \ldots, \alpha_g, \beta_g$ denote the generators of the fundamental group of the 1-skeleton of the given surface which correspond to the 1-cells $a_1, b_1, \ldots, a_g, b_g$. The the homotopy class of the above attaching map (regarded as a loop) is the word

$$\alpha_1 \beta_1 \alpha_1^{-1} \beta_1^{-1} \ldots \alpha_g \beta_g \alpha_g^{-1} \beta_g^{-1}.$$

Therefore, the fundamental group of our surface at the point e_0 may be described as the group with generators $a_1, b_1, \ldots, a_g, b_g$, and the relation

$$a_1 b_1 a_1^{-1} b_1^{-1} \ldots a_g b_g a_g^{-1} b_g^{-1} = 1.$$

The cellular decomposition of a sphere with h crosscaps (see 3.5.3.8) contains one 0-cell e_0, h 1-cells $c_1,\ldots,c_h$, and one 2-cell. Repeating the previous argument with the obvious changes, we see that the fundamental group of this surface at e_0 can be described as the group with generators $c_1,\ldots,c_h$ and the relation

$$c_1 c_1 \ldots c_h c_h = 1.$$

3. It is important to note that the groups computed in 2 are pairwise nonisomorphic. (**To** see this, factor each fundamental group by its commutator subgroup: for a sphere with g handles this yields a free Abelian group of rank $2g$, while in the case of a sphere with h crosscaps the result is the direct sum of a free Abelian group of rank $h-1$ and a group of order 2.) In particular, the closed model surfaces are pairwise nonhomeomorphic.

From this it is readily seen that the compact model surfaces are also pairwise nonhomeomorphic: it suffices to seal up the holes by discs. The number of holes is a topological invariant, because it equals the number of components of the boundary (see 3.1.1.4 and 4.6.5.12).

The Higher Homotopy Groups

4. $\underline{\text{Let } P \underline{ \text{ be a sphere with}} \ g \ \underline{\text{handles. If}} \ g \geqslant 1 \ \underline{\text{and}}}$ $r \geqslant 2$, $\underline{\text{then}} \ \pi_r(P) = 0$.

PROOF. It is clear that P admits as a covering space the infinite garland $\tilde{P}$ constructed from $S^1 \times \mathbb{R}$ by first removing small open discs centered at the points $(\mathrm{ort}_1, 2k)$ $[k = 0, \pm 1, \ldots]$ and then glueing a sphere with $g-1$ handles and one hole in the place of each such disc (see Fig. 15). Denote by $\tilde{P}_n$ the finite garland constructed in the same fashion from the product $S^1 \times [-2n-1, 2n+1]$. Obviously, $\tilde{P}_n$ is a sphere with $(2n+1)(g-1)$ handles and two holes, so that $\pi_r(\tilde{P}_n) = 0$ for all $r \geqslant 2$ (see 1). Since $\tilde{P} = \lim(\tilde{P}_n, \mathrm{in}: \tilde{P}_n \to \tilde{P}_{n+1})$, we also have $\pi_r(\tilde{P})$ for all $r \geqslant 2$ (see 1.11.2). Consequently, $\pi_r(P) = 0$ for all $r \geqslant 2$ (see 1.8.12).

5. $\underline{\text{Let } P \underline{ \text{ be a sphere with}} \ h \ \underline{\text{crosscaps. If}} \ h \geqslant 2 \ \underline{\text{and}}}$ $r \geqslant 2$, $\underline{\text{then}} \ \pi_r(P) = 0$.

Since P admits a sphere with $h-1$ as covering space (see 4.1.2.6), this is a corollary of 4.

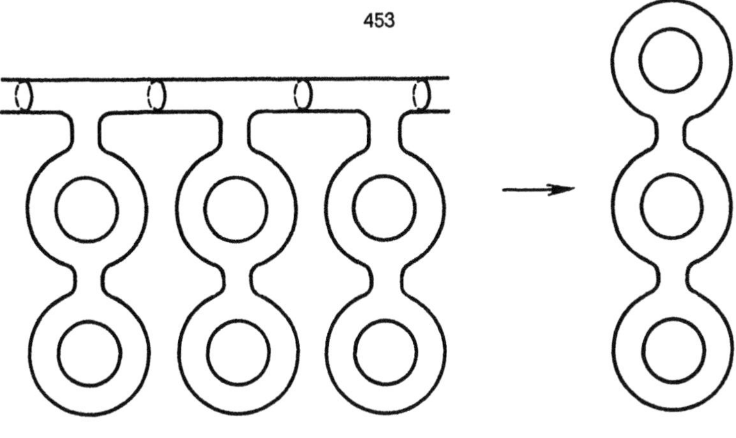

Fig. 15   (g = 3)

5.   <u>The Homotopy Groups of Bouquets</u>

1.   Suppose that we are given a family $\{(X_\mu,x_\mu)\}_{\mu\in M}$ of pointed T_1-spaces, and consider the bouquet $B = V_{\mu\in M}(X_\mu,x_\mu)$. The the formula $\mathrm{Imm}(\{\alpha_\mu\}_{\mu\in M}) = \sum_{\mu\in M} \mathrm{imm}_{\mu*}(\alpha_\mu)$ defines a homomorphism

$$\mathrm{Imm}: \oplus_{\mu\in M} \pi_r(X_\mu,x_\mu) \to \pi_r(B,bp)$$

for any $r \geq 2$. This homomorphism is natural, i.e., if $B' = V_{\mu'\in M'}(X'_{\mu'},x'_{\mu'})$ is another bouquet of T_1-spaces, $\sigma: M' \to M$ is arbitrary, and $f_{\mu'}: (X'_{\mu'},x'_{\mu'}) \to (X_{\sigma(\mu')},x_{\sigma(\mu')})$ are continuous, then the following diagram commutes

$$
\begin{array}{ccc}
\oplus_{\mu'\in M'}\pi_r(X'_{\mu'},x'_{\mu'}) & \xrightarrow{\ \mathrm{Imm}\ } & \pi_r(B',bp) \\
\Big\downarrow & & \Big\downarrow \\
\oplus_{\mu\in M}\pi_r(X_\mu,x_\mu) & \xrightarrow{\ \mathrm{Imm}\ } & \pi_r(B,bp)
\end{array}
\tag{3}
$$

where the left vertical map is the homomorphism

$$\{\alpha'_{\mu'}\}_{\mu'\in M'} \mapsto \{\textstyle\sum_{\mu'\in\sigma^{-1}(\mu)} (f_{\mu'})_*(\alpha'_{\mu'})\}_{\mu\in M'}$$

and the right vertical map is the homomorphism induced by the map $B' \to B$, $\mathrm{imm}_{\mu'}(y'_{\mu'}) \mapsto \mathrm{imm}_{\sigma(\mu')} \circ f_{\mu'}(y'_{\mu'})$ $[y'_{\mu'} \in X'_{\mu'}, \mu' \in M']$.

2 (LEMMA). <u>Given any</u> $\alpha \in \pi_r(B,bp)$ [$r \geqslant 1$], <u>there is a</u> <u>finite set</u> $M' \subset M$ <u>such that:</u> $pr_{\mu*}(\alpha) = 0$ <u>if</u> $\mu \in M \smallsetminus M'$; α <u>lies in</u> <u>the image of the homomorphism</u> $\pi_r(B',bp) \to \pi_r(B,bp)$ <u>induced by the</u> <u>natural embedding of the bouquet</u> $B' = V_{\mu \in M'}(X_\mu, x_\mu)$ <u>in</u> B.

We only have to observe that for any spheroid $\phi \in Sph_r(B,bp)$, $\phi(I^r)$ is covered by a finite number of sets $imm_\mu(X_\mu)$. [Indeed, since $\phi(I^r)$ is compact and every point of X_μ is closed in X_μ, we can choose a point in each nonempty intersection $\phi(I^r) \cap imm_\mu(X_\mu \smallsetminus x_\mu)$ and in this way produce a set which, being both discrete and compact, is finite.]

3. Let $r \geqslant 2$ and define the map $Pr: \pi_r(B,bp) \to \oplus_{\mu \, M}\pi_r(X_\mu, x_\mu)$ by $Pr(\alpha) = \{pr_{\mu*}(\alpha)\}_{\mu \in M}$ (2 shows that this definition is correct). Obviously, Pr is a homomorphism and diagram (3) remains commutative when we replace $\xrightarrow{\;Imm\;}$ by $\xleftarrow{\;Pr\;}$. If M is finite, then Pr equals the composition of the homomorphism $\pi_r(B,bp) \to \pi_r(\times_{\mu \in M}X_\mu, \{x_\mu\})$ induced by the inclusion $B \to \times_{\mu \in M}X_\mu$ (see 1.2.8.3) with the canonical isomorphism $\pi_r(\times_{\mu \in M}X_\mu, \{x_\mu\}) \to \oplus_{\mu \in M}\pi_r(X_\mu, x_\mu)$ (see 1.1.9).

4. $Pr \circ Imm$ <u>equals the identity automorphism of the group</u> $\oplus_{\mu \in M}\pi_r(X_\mu, x_\mu)$. <u>In particular,</u> Pr <u>is epimorphic,</u> Imm <u>is monomorphic,</u> <u>and</u> $\pi_r(B,bp) = Ker\, Pr \oplus Im\, Imm$.

PROOF. Since $pr_\mu \circ imm_\mu = id\, X_\mu$ and $pr_\nu \circ imm_\mu(X_\mu) = x_\nu$ for $\nu \neq \mu$,

$$Pr \circ Imm(\{\alpha_\mu\}_{\mu \in M}) = Pr[\textstyle\sum_{\mu \in M} imm_{\mu*}(\alpha_\mu)] =$$

$$= \{\textstyle\sum_{\mu \in M}(pr_\nu \circ imm_\mu)_*(\alpha_\mu)\}_{\nu \in M} = \{\alpha_\nu\}_{\nu \in M}$$

for any $\alpha_\mu \in \pi_r(X_\mu, x_\mu)$ and $\mu \in M$.

5. <u>Let</u> (X_μ, x_μ) <u>be cellular pairs, and let</u> m, k_μ $(\mu \in M)$ <u>be positive integers such that</u> $k_\mu + k_\nu \geqslant m$ <u>for</u> $\nu \neq \mu$, <u>and for any</u> μ

$$\pi_s(X_\mu, x_\mu) = 0 \quad \underline{for} \quad 1 \leqslant s \leqslant k_\mu.$$

<u>If</u> $2 \leqslant r \leqslant m$, <u>then</u>

$$Imm: \oplus_{\mu \in M}\pi_r(X_\mu, x_\mu) \to \pi_r(B,bp)$$

<u>and</u>

$$Pr: \pi_r(B,bp) \to \oplus_{\mu \in M}\pi_r(X_\mu, x_\mu)$$

<u>are isomorphisms.</u>

(This theorem will be generalized in the next section; see 4.3.1.)

PROOF. Suppose first that M is finite. By Theorem 2.3.3.2 and the fact that Imm is natural (see 1), we may assume that $\mathrm{ske}_k X_\mu$ reduces to the point x_μ for all $\mu \in M$. In this case, $B \supset \mathrm{ske}_{m+1}X$, where X is the cellular product of the spaces X_μ. By Theorem 2.3.2.2, $\mathrm{in}_*\colon \pi_r(B,bp) \to \pi_r(X,bp)$ is an isomorphism for $r \leqslant m$. According to 1.1.9, $\mathrm{in}_* \circ \mathrm{Imm}\colon \oplus_{\mu\in M}\pi_r(X_\mu,x_\mu) \to \pi_r(B,bp)$ is an isomorphism for any r. Therefore, Imm is an isomorphism for $r \leqslant m$.

In the general case, 2 shows that for any $\alpha \in \pi_r(B,bp)$ and $r \leqslant m$, there is a finite subbouquet $B' = V_{\mu\in M'}(X_\mu,x_\mu)$ of B such that α lies in the image of the homomorphism $\pi_r(B',bp) \to \pi_r(B,bp)$ induced by the natural embedding $B' \to B$. Using 1, this homomorphism is part of the commutative diagram

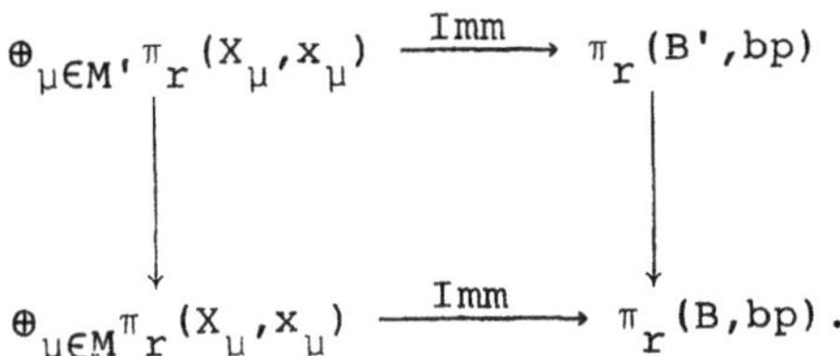

But we have already proved that the upper Imm is an isomorphism, so that α also lies in the image of our (lower) Imm. Therefore, the latter is an epimorphism and this, combined with Theorem 4, implies that Imm and Pr are isomorphisms.

6 (COROLLARY). <u>Let</u> B <u>be a bouquet of n-dimensional spheres, constructed from a family</u> $\{(S_\mu = S^n, \mathrm{ort}_1)\}_{\mu\in M}$. <u>If</u> $n \geqslant 2$, <u>then the groups</u> $\pi_r(B)$ <u>with</u> $r < n$ <u>are trivial, whereas</u> $\pi_n(B,bp)$ <u>is a free Abelian group with free generators</u> $\mathrm{imm}_{\mu*}(\mathrm{sph}_n)$.

6. The Homotopy Groups
of a k-Connected Cellular Pair

1 (ALGEBRAIC LEMMA). <u>Consider the following commutative diagram of groups and homomorphisms</u>

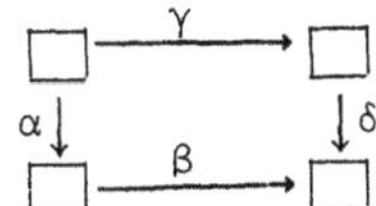

If α, β, <u>and</u> γ <u>are epimorphic and</u> Ker $\beta \subset \alpha(\mathrm{Ker}\,\gamma)$, <u>then</u> δ <u>is</u> <u>epimorphic,</u> Ker $\beta = \alpha(\mathrm{Ker}\,\gamma)$, <u>and</u> Ker $\delta = \gamma(\mathrm{Ker}\,\alpha)$.

PROOF. Since the diagram is commutative and α, β are epimorphic, δ is epimorphic. The commutativity of the diagram implies also that $\alpha(\mathrm{Ker}\,\gamma) \subset$ Ker β and $\gamma(\mathrm{Ker}\,\alpha) \subset$ Ker δ. Let us verify that Ker $\delta \subset \gamma(\mathrm{Ker}\,\alpha)$. Pick $d \in$ Ker δ. If $d = \gamma(a)$, then $\alpha(a) \in$ Ker β (again by commutativity), whence $\alpha(a) \in \alpha(\mathrm{Ker}\,\gamma)$, i.e., there is $c \in \mathrm{Ker}\,\gamma$ such that $\alpha(c) = \alpha(a)$. The last equality yields $ac^{-1} \in$ Ker α, and we have $d = \gamma(a) = \gamma(ac^{-1}) \in \gamma(\mathrm{Ker}\,\alpha)$.

2. <u>Let</u> (X,A) <u>be a cellular pair with base point</u> $x_0 \in A$. <u>If</u> A <u>is connected and, for</u> $r \leqslant k$, $\pi_r(X,A) = 0$, <u>then</u> $\mathrm{pr}_*\colon \pi_{k+1}(X,A,x_0) \to \pi_{k+1}(X/A,\mathrm{pr}(x_0))$ <u>is epimorphic, and for</u> $k \geqslant 1$ $\mathrm{Ker}\,\mathrm{pr}_*$ <u>is the smallest subgroup of</u> $\pi_{k+1}(X,A,x_0)$ <u>containing all the</u> <u>"ratios"</u> $(T_\sigma\alpha)\alpha^{-1}$ <u>with</u> $\alpha \in \pi_{k+1}(X,A,x_0)$ <u>and</u> $\sigma \in \pi_1(A,x_0)$. <u>For</u> $k = 0$, <u>the situation is described by the commutative diagram</u>

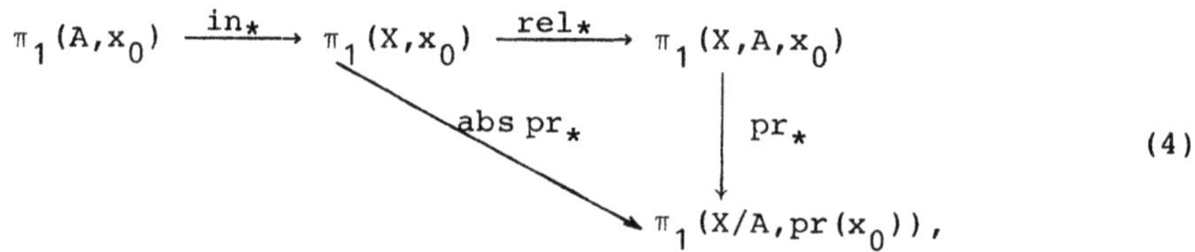

$$(4)$$

where abs pr_* <u>and</u> rel_* <u>are also epimorphic, and</u> $\mathrm{Ker}(\mathrm{abs}\,\mathrm{pr}_*)$ <u>is</u> <u>the smallest normal subgroup of</u> $\pi_1(X,x_0)$ <u>which contains</u> $\mathrm{Ker}\,\mathrm{rel}_* =$ $= \mathrm{Im}\,\mathrm{in}_*$.

(This theorem will be generalized in the next section; see 4.3.14.)

PROOF OF THE CASE $k \geqslant 1$. Suppose first that $X \smallsetminus A$ consists only of $(k+1)$-cells and, as a consequence, X/A is a bouquet of $(k+1)$- -dimensional spheres. For each cell $e \in X \smallsetminus A$, consider the homotopy class of its characteristic map (viewed as a spheroid of the pair (X,A)), and translate it to x_0, denoting the resulting element of $\pi_{k+1}(X,A,x_0)$ by α_e. Set $\beta_e = \mathrm{pr}_*(\alpha_e)$. By Theorem 2.4, the classes $T_\sigma\alpha_e$ with $\sigma \in \pi_1(A,x_0)$ form a system of generators of $\pi_{k+1}(X,A,x_0)$, and by Theorem 5.6, the classes β_e form a system of independent generators of the Abelian group $\pi_{k+1}(X/A,\mathrm{pr}(x_0))$. Moreover, it is obvious that $\mathrm{pr}_*(T_\sigma\alpha) = \mathrm{pr}_*(\alpha)$ for all $\alpha \in \pi_{k+1}(X,A,x_0)$ and $\sigma \in \pi_1(A,x_0)$, and these facts will suffice to complete the proof of the theorem for $k \geqslant 1$. Since $\beta_e = \mathrm{pr}_*(\alpha_e)$ generate $\pi_{k+1}(X/A,\mathrm{pr}(x_0))$, pr_* is epi- morphic. Further, since $\mathrm{pr}_*(T_\sigma\alpha) = \mathrm{pr}_*(\alpha)$, we have $(T_\sigma\alpha)\alpha^{-1} \in \mathrm{Ker}\,\mathrm{pr}_*$. Let us show that the ratios $(T_\sigma\alpha)\alpha^{-1}$ generate $\mathrm{Ker}\,\mathrm{pr}_*$. If $k > 1$

and the class $\xi = \prod_{(e,\sigma)} (T_\sigma \alpha_e)^{\lambda(e,\sigma)}$ [with only a finite number of nonzero integers $\lambda(e,\sigma)$] belongs to $\operatorname{Ker} \operatorname{pr}_*$, then $\sum_\sigma \lambda(e,\sigma) = 0$ for any cell e (because $\sum_e [\sum_\sigma \lambda(e,\sigma)] \beta_e = \operatorname{pr}_*(\alpha) = 0$), and thus $\xi = \prod_{(e,\sigma)} [(T_\sigma \alpha_e) \alpha_e^{-1}]^{\lambda(e,\sigma)}$. When $k = 1$, this argument is valid only after we factor $\pi_{k+1}(X,A,x_0)$ by its commutator subgroup, and it only demonstrates that every element of $\operatorname{Ker} \operatorname{pr}_*$ is a product of the above form multiplied by some commutators. However, since in $\pi_2(X,A,x_0)$ each commutator $\gamma^{-1} \delta \gamma \delta^{-1}$ equals $(T_{\partial\gamma} \delta) \delta^{-1}$ (see 1.4.7), we obtain again the desired decomposition of ξ into ratios $(T_\sigma \alpha) \alpha^{-1}$.

In the general situation, we first transform (X,A) into a k-connected pair, removing those components of X which do not contain x_0, and then replace it by a homotopy equivalent pair (X',A') such that $\operatorname{ske}_k X' \subset A'$ (see 1.4.6 and 2.3.3.1). Thus, we may assume that $\operatorname{ske}_k X \subset A$. Now set $Y = A \cup \operatorname{ske}_{k+1} X$ and consider the commutative diagram

$$
\begin{array}{ccc}
\pi_{k+1}(Y,A,x_0) & \xrightarrow{\;i = \operatorname{in}_*\;} & \pi_{k+1}(X,A,x_0) \\
\Big\downarrow{\scriptstyle p' = \operatorname{pr}_*} & & \Big\downarrow{\scriptstyle p = \operatorname{pr}_*} \\
\pi_{k+1}(Y/A, \operatorname{pr}(x_0)) & \xrightarrow{\;i' = \operatorname{in}_*\;} & \pi_{k+1}(X/A, \operatorname{pr}(x_0)).
\end{array}
$$

Here i, i', and p' are epimorphic: i because $\operatorname{ske}_{k+1} X \subset Y$, i' because $\operatorname{ske}_{k+1}(X/A) \subset Y/A$, and p' because of the proof above. We claim that our diagram also satisfies the last condition of the algebraic lemma: $\operatorname{Ker} i' \subset p'(\operatorname{Ker} i)$.

To see this, note that every $(k+2)$-cell from $(X/A) \smallsetminus (Y/A)$ is the image under p of some cell e from $X \smallsetminus Y$, and its corresponding attaching map can be expressed as $\operatorname{pr} \circ \operatorname{att}_e$. By Theorem 2.4, this implies that $\operatorname{Ker} i' \subset \operatorname{pr}_*(\operatorname{Ker} \operatorname{in}_*)$, where $\operatorname{in}_* = [\operatorname{in}_* : \pi_{k+1}(Y,x_0) \to \pi_{k+1}(X,x_0)]$ and $\operatorname{pr}_* = [\operatorname{pr}_* : \pi_{k+1}(Y,x_0) \to \pi_{k+1}(Y/A, \operatorname{pr}(x_0))]$. Since the diagram

$$
\begin{array}{ccc}
\pi_{k+1}(Y,y_0) & \xrightarrow{\;\operatorname{in}_*\;} & \pi_{k+1}(X,x_0) \\
\Big\downarrow{\scriptstyle \operatorname{rel}_*} & & \Big\downarrow{\scriptstyle \operatorname{rel}_*} \\
\pi_{k+1}(Y,A,x_0) & \xrightarrow{\;i\;} & \pi_{k+1}(X,A,x_0)
\end{array}
$$

commutes and $\operatorname{pr}_* = p' \circ [\operatorname{rel}_* : \pi_{k+1}(Y,x_0) \to \pi_{k+1}(Y,A,x_0)]$, we see that $\operatorname{pr}_*(\operatorname{Ker} \operatorname{in}_*) \subset p'(\operatorname{Ker} i)$, whence $\operatorname{Ker} i' \subset p'(\operatorname{Ker} i)$.

Applying Lemma 1, we conclude that p is epimorphic and $\mathrm{Ker}\,p = i(\mathrm{Ker}\,p')$. We have proved already that $\mathrm{Ker}\,p'$ is generated by the ratios $(T_\sigma \alpha)\alpha^{-1}$ with $\alpha \in \pi_{k+1}(Y,A,x_0)$ and $\sigma \in \pi_1(A,x_0)$. Since $i((T_\sigma \alpha)\alpha^{-1}) = [T_\sigma(i(\alpha))](i(\alpha))^{-1}$ and i is epimorphic, $\mathrm{Ker}\,p$ is generated by the ratios $(T_\sigma \alpha)\alpha^{-1}$ with $\alpha \in \pi_{k+1}(X,A,x_0)$ and $\sigma \in \pi_1(A,x_0)$.

PROOF OF THE CASE $k = 0$. The commutativity of (4) is obvious, while the fact that rel_* is epimorphic results from the connectedness of A. It remains to verify that $\mathrm{abs}\,\mathrm{pr}_*$ is an epimorphism with the indicated kernel. If x_0 is the unique 0-cell of X, this follows from 3.3: indeed, the system of generators and relations for $\pi_1(X/A, \mathrm{pr}(x_0))$ given in 3.3 can be obtained from the system of generators and relations for $\pi_1(X,x_0)$, also appearing in 3.3, by deleting the 1-cells and 2-cells of A. When x_0 is not the unique 0-cell of X, we may still reduce to this special case by translating inside A the origin at some 0-cell e_0, and subsequently replacing the triple (X,A,e_0) by a homotopy equivalent triple (X',A',e_0') having a single 0-cell e_0'. To produce such a triple, we take quotients twice: first the quotient of X by a one-dimensional contractible subspace of A containing $\mathrm{ske}_0 A$, and then the quotient of the resulting space by a one-dimensional contractible subspace which contains all its 0-cells (cf. the proof of Theorem 3.6).

3. <u>Let</u> X <u>be a cellular space, and let</u> A <u>be a simply connected subspace of</u> X. <u>Then</u> X/A <u>is k-connected</u> $(0 \leqslant k \leqslant \infty)$ <u>if and only if the pair</u> (X,A) <u>is k-connected. If this condition is fulfilled for some</u> $k < \infty$, <u>then</u> $\mathrm{pr}_*: \pi_{k+1}(X,A,x_0) \to \pi_{k+1}(X/A, \mathrm{pr}(x_0))$ <u>is an isomorphism.</u>

The second assertion is an obvious corollary of Theorems 2 and 1.4.6. The first assertion follows from the second by induction on k. However, note that to deduce the k-connectedness of X/A from the k-connectedness of (X,A), Theorems 2.3.3.1 and 2.3.2.2 suffice.

4. <u>If the cellular space</u> X <u>with the 0-cell</u> x_0 <u>as base point is k-connected, then</u> $\mathrm{su}: \pi_{k+1}(X,x_0) \to \pi_{k+2}(\mathrm{su}(X,x_0),\mathrm{bp})$ <u>is an isomorphism.</u>

This is a corollary of 3 (see 2.1.2).

5 (INFORMATION). Under the assumptions of Theorem 3, if A is l-connected $(1 \leqslant \mathrm{l} \leqslant \infty)$, then $\mathrm{pr}_*: \pi_r(X,A,x_0) \to \pi_r(X/A, \mathrm{pr}(x_0))$ is an isomorphism for $r \leqslant k+1$ and an epimorphism for $r = k+l+1$.

Under the assumptions of Theorem 4, $\mathrm{su}: \pi_r(X,x_0) \to$

$\to \pi_{r+1}(su(X,x_0),bp)$ is an isomorphism for $r \leqslant 2k$ and an epimorphism
for $r = 2k+1$ (cf. 2.1.4).

7. Spaces With Prescribed Homotopy Groups

1 (LEMMA). _Let_ π _be a group and_ n _a positive integer._
_If_ π _is Abelian or if_ n = 1, _then there is a connected cellular space_
X _such that all the groups_ $\pi_r(X)$ _with_ $r \neq n$ _are trivial, whereas_
$\pi_n(X) \cong \pi$.*

PROOF. We proceed by induction and construct connected
cellular spaces $X_0, X_1, \ldots,$ with base points $x_0, x_1, \ldots,$ and base-point
preserving cellular embeddings $\phi_0: X_0 \to X_1,$ $\phi_1: X_1 \to X_2,$ such that:
(i) the groups $\pi_r(X_k, x_k)$ with $r < n$ and $n < r \leqslant n+k$ are trivial;
(ii) $\pi_n(X_k, x_k)$ is isomorphic to π; $\phi_{k*}: \pi_n(X_k, x_k) \to \pi_n(X_{k+1}, x_{k+1})$
is an isomorphism. Then the space $X = \lim(X_k, \phi_k)$ will have the
desired properties (see 2.1.5.6 and 1.11.2).

To produce $(X_0, x_0),$ write π as a quotient group $F/F',$
where F is a free group if n = 1, and a free Abelian group if n > 1.
Let B and B' be bouquets of n-dimensional spheres such that
$\pi_n(B, bp) = F$ and $\pi_n(B', bp) = F'$ (see 1.3 and 5.6). Further, let
f: $(B', bp) \to (B, bp)$ be a continuous map such that $f_*: \pi_n(B', bp) \to$
$\to \pi_n(B, bp)$ equals the inclusion $F' \to F$ (one can construct such a map
out of a family of spheroids whose classes in $\pi_n(B, bp) = F$ constitute
a free system of generators for F'). Now replace each sphere in B'
by the ball that it bounds and take X_0 to be the result of attaching
this new bouquet (of balls) to B by f. Theorems 1.3, 5.6, and 2.3
show that X_0 and $x_0 = \text{imm}_1(bp) [= \text{imm}_2(bp)]$ satisfy conditions (i)
and (ii) for k = 0.

Assume that for some $i \geqslant 1,$ pointed spaces $(X_k, x_k),$ k < i,
and maps $\phi_k,$ k < i-1, are already constructed and satisfy conditions
(i), (ii), and (iii). Represent $\pi_{n+i}(X_{i-1}, x_{i-1})$ as the quotient group
of a free Abelian group, say G, and then construct a bouquet C of
(n+i)-dimensional spheres, together with a map g: $(C, bp) \to (X_{i-1}, x_{i-1})$
such that $g_*: \pi_{n+i}(C, bp) \to \pi_{n+i}(X_{i-1}, x_{i-1})$ equals the projection
$G \to \pi_{n+i}(X_{i-1}, x_{i-1})$. [To establish the existence of such a C and a
g, one may procced as in the proof of the existence of B' and f
above; however, here Theorem 1.3 is not necessary.] Now replace each

---

* Translator's note. A space with such homotopy groups is known as a
cellular $K(\pi,n)$-space or as a cellular space of type (π,n).

sphere of C by the ball that it bounds and then attach the resulting bouquet to X_{i-1} by g to obtain X_i. Finally, set $x_i = \text{imm}_2(X_{i-1})$ and $\phi_{i-1} = \text{imm}_2$. The fact that (X_i, x_i) satisfies (i), (ii) for $k = i$, and ϕ_{i-1} satisfies (iii) for $k = i-1$, is a consequence of 2.3.

2. <u>Given an arbitrary group π_1 and arbitrary Abelian groups $\pi_2, \pi_3, \ldots,$ there exists a connected cellular space X such that $\pi_r(X) \cong \pi_r$ $(r = 1, 2, \ldots)$.</u>

PROOF. Let $X_1, X_2, \ldots$ be connected cellular spaces with 0-cells $x_1, x_2, \ldots$ as base points, such that the groups $\pi_r(X_k)$ are trivial for $r \neq k$, whereas $\pi_k(X_k) \cong \pi_k$ (see 1). Define inductively cellular spaces $Y_0, Y_1, \ldots$ and cellular embeddings $\psi_k : Y_k \to Y_{k+1}$ by $Y_0 = D^0$, $Y_{k+1} = Y_k \times_c X_{k+1}$, and $\psi_k(y) = (y, x_{k+1})$. Applying Theorems 1.1.9 and 1.11.2, the space $X = \lim(X_k, \psi_k)$ has the desired properties.

8. Eight Instructive Examples

1. If $r > 1$, then the r-th homotopy group of a finite, connected cellular space is not necessarily finitely generated (cf. 3.4). The bouquet $(S^r, \text{ort}_1) \vee (S^1, \text{ort}_1)$ is a simple illustration of this phenomenon: its r-th homotopy group $(r > 1)$ is a free Abelian group of infinite rank. Indeed, $(S^r, \text{ort}_1) \vee (S^1, \text{ort}_1)$ has a covering space which is homotopy equivalent to an infinite bouquet of r-dimensional spheres: to produce such a space, attach one copy of S^r in one point at each integer point of the real line $\mathbb{R}$.

INFORMATION. The homotopy groups of a finite cellular space with finite fundamental group are finitely generated. For a proof, see [19].

2. Under the conditions of Theorem 1.6.6, the subgroup $\text{Ker rel}_* = \text{Im in}_*$ of $\pi_1(X, x_0)$ is not necessarily normal (cf. 1.5.14). Example: X is the bouquet of two circles, A the first circle, x_0 the center of the bouquet, and ρ takes the second circle into x_0.

3. Under the conditions of Theorem 1.6.7, the right splitting of the homotopy sequence of the pair (X, A) at $\pi_1(X, x_0)$ is not necessarily normal (cf. 1.5.17). Example: $X = (D^2, \text{ort}_1) \vee (S^1, \text{ort}_1)$, $A = (S^1, \text{ort}_1) \vee (S^1, \text{ort}_1)$, x_0 is the center of both bouquets, and $h = [\text{imm}_2 : S^1 \to X] \circ [\text{pr}_2 : X \to S^1]$.

4. For any $k \geqslant 0$ there exist k-connected pairs (X,A) with A connected, which are not $(k+1)$-simple; moreover, under the conditions of Theorem 6.2, and for any $k \geqslant 0$, the epimorphism $pr_*: \pi_{k+1}(X,A,x_0) \to \pi_{k+1}(X/A,pr(x_0))$ is not necessarily an isomorphism. Example: $X = (S^{k+1},ort_1) \vee (S^1,ort_1)$, $A = imm_2(S^1)$ (cf. 1 and 3).

5. The second homotopy group of a pair (X,A) with connected A is not necessarily Abelian, even when $\pi_1(A)$ is Abelian and X is simply connected. The simplest example: $A = S^1 \times S^1$, $x_0 = (ort_1,ort_1)$, and X is the result of attaching two copies of D^2 to A by the maps $S^1 \to A$ given by $y \mapsto (y,ort_1)$ and $y \mapsto (ort_1,y)$. Then $\pi_1(A) = \mathbb{Z} \oplus \mathbb{Z}$, $\pi_2(A) = 0$, $\pi_1(X) = 0$, $\pi_2(X) \cong \mathbb{Z}$ (X is homotopy equivalent to S^2), and we have the exact sequence

$$0 \xrightarrow{\ in_* \ } \mathbb{Z} \xrightarrow{\ rel_* \ } \pi_2(X,A,x_0) \xrightarrow{\ \partial \ } \mathbb{Z} \oplus \mathbb{Z} \xrightarrow{\ in_* \ } 0, \tag{5}$$

which shows, in particular, that ∂ is epimorphic. Assuming that $\pi_2(X,A,x_0)$ is Abelian, it follows from 1.4.7 that $\pi_1(A,x_0)$ acts identically on $\pi_2(X,A,x_0)$, whence, by Theorem 2.2, $rank\, \pi_2(X,A,x_0) \leqslant 2$. The latter contradicts the exactness of (5).

6. There exist 1-connected pairs (X,A) such that $pr_*: \pi_3(X,A,x_0) \to \pi_3(X/A,pr(x_0))$ is not even epimorphic. Example: $X = D^2$, $A = S^1$, $x_0 = ort_1$. Here $\pi_r(X,A) = 0$ for $r \neq 2$, whereas $\pi_3(X/A) \cong \mathbb{Z}$ (X/A is homeomorphic to S^2).

7. For any $k \geqslant 2$, there are $(k-1)$-connected but not k-connected cellular pairs (X,A) with X and A connected and X/A contractible. To construct an example, let $\sigma \in \pi_1((S^1,ort_1) \vee (S^k,ort_1),bp)$ and $\alpha \in \pi_k((S^1,ort_1) \vee (S^k,ort_1),bp)$ designate the classes of the spheroids $imm_1: S^1 \to (S^1,ort_1) \vee (S^k,ort_1)$ and $imm_2: S^k \to (S^1,ort_1) \vee (S^k,ort_1)$, respectively. Next, attach D^{k+1} to $(S^1,ort_1) \vee (S^k,ort_1)$ by an arbitrary spheroid $S^k \to (S^1,ort_1) \vee (S^k,ort_1)$ in the homotopy class $2\alpha - T_\sigma\alpha$. Take the resulting cellular space as X, and the circle $ske_1 X$ as A. The quotient space X/A may be described as the result of attaching D^{k+1} to S^k by a map $S^k \to S^k$ homotopic to $id\, S^k$, which implies that X/A is contractible (see 1.3.7.8). It is evident that X and A are connected and that (X,A) is $(k-1)$-connected; therefore, it remains to check that $\pi_k(X)$ is not trivial. By 1, $\pi_k((S^1,ort_1) \vee (S^k,ort_1),bp)$ is a free Abelian group with free generators $\alpha_n = T_\sigma^n\alpha$ $(n = 0,\pm1,\ldots)$, while $\pi_k(X)$ is the quotient

group of $\pi_k((S^1,\mathrm{ort}_1) \vee (S^k,\mathrm{ort}_1),\mathrm{bp})$ by its subgroup generated by the elements $2\alpha_n - \alpha_{n+1}$ (see 6.2). Consequently, $\pi_k(X)$ is isomorphic to the additive group of binary rational numbers.

8. The homomorphism $f_*: \pi_1(X,A,x_0) \to \pi_1(X',A',x_0')$ induced by a continuous map $f: (X,A,x_0) \to (X',A',x_0')$ is not necessarily an isomorphism even if all the homomorphisms $f_*: \pi_r(X,x_0) \to \pi_r(X',x_0')$ and $(\mathrm{ab}\,f)_*: \pi_r(A,x_0) \to \pi_r(A',x_0')$ are isomorphisms. Example: $X = X' = (S^1,\mathrm{ort}_1) \vee (I,0)$, $A = \mathrm{imm}_1(\mathrm{ort}_2) \cup \mathrm{imm}_2(1)$, $A' = \mathrm{imm}_1(S^1) \cup \mathrm{imm}_2(1)$, $x_0 = x_0' = \mathrm{imm}_2(1)$, and $f = \mathrm{rel\,id}\,X$.

9. Exercises

1. Consider the subset of $\mathbb{C}^2$ defined by the equation $x_1^p = x_2^q$, where p and q are comprime integers, and intersect it with the sphere S^3. Show that the fundamental group of the complement of this intersection in S^3 is isomorphic to the group with generators α_1, α_2, which are connected by the relation $\alpha_1^p = \alpha_2^q$.

(The above intersections (with various p,q) are all homeomorphic to a circle and lie on the torus $|x_1| = \sqrt{2}/2$, $|x_2| = \sqrt{2}/2$ which is contained in S^3; they are called <u>torus knots</u>. A torus knot and the torus which carries it are depicted in Fig. 16 as they lie in $\mathbb{R}^3 = S^3 \setminus \mathrm{bp}$.)

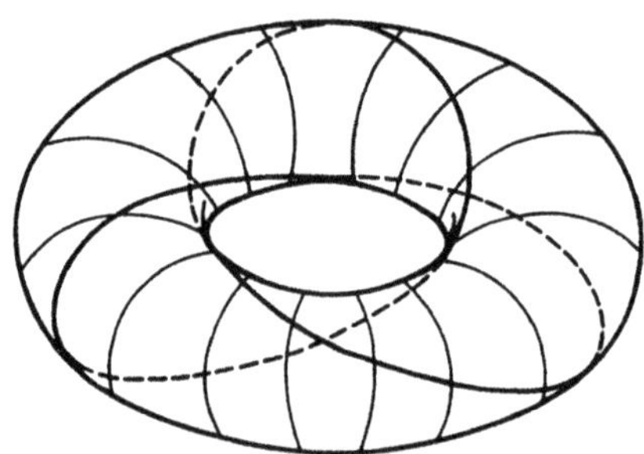

Fig. 16   ($p = 2$, $q = 3$)

2. Consider the submanifold of $\mathbb{C}P^2$ defined by the equation $x_1^m + x_2^m + x_3^m = 0$, with m a positive integer (cf. 3.5.4.2). Show that the fundamental group of the complement of this submanifold is isomorphic to $\mathbb{Z}_m$.

3. Consider the submanifold of nonzero vectors in the total

space of the tangent bundle of a sphere with g handles, and show that its fundamental group is isomorphic to the group with generators $a_1,\ldots,a_g,b_1,\ldots,b_g,d,$ which are connected by the relations

$$a_1 b_1 a_1^{-1} b_1^{-1} \ldots a_g b_g a_g^{-1} b_g^{-1} = d^{2-2g}$$

and

$$a_1 d = da_1,\ldots,a_g d = da_g, \quad b_1 d = db_1,\ldots,b_g d = db_g.$$

4. Consider the submanifold of nonzero vectors in the total space of the tangent bundle of a sphere with h crosscaps, and show that its fundamental group is isomorphic to the group with generators $c_1,\ldots,c_h,d,$ which are connected by the relations

$$c_1 c_1 \ldots c_h c_h = d^{2-h}$$

and

$$c_1 d = d^{-1} c_1,\ldots,c_h d = d^{-1} c_h.$$

5. Let π be any group which can be presented by a finite number of generators and relations. Show that there exists a smooth, closed 4-dimensional manifold whose fundamental group is isomorphic to π.

6. Show that every smooth, closed, oriented manifold of dimension $\neq 0,3,$ is oriented cobordant to a simply connected manifold.

INFORMATION. This is also valid for dimension 3.

7. Show that in Example 8.5, $\pi_2(X,A,x_0)$ is isomorphic to the group with generators $a,b,c,$ which are connected by the relations $ac = ca$, $bc = cb$, and $aba^{-1}b^{-1} = c$.

§4. WEAK HOMOTOPY EQUIVALENCE

1. Fundamental Concepts

1. If X and Y are topological spaces, a continuous map $f: X \to Y$ is called a <u>weak homotopy equivalence</u> if $f_*: \pi_r(X,x) \to \pi_r(Y,f(x))$ is an isomorphism for all $r \geq 0$ and all $x \in X$. To justify this term, we remark that every homotopy equivalence

is a weak homotopy equivalence and that the converse is not true. The first fact was established in 1.3.7; for the second, see 3.5.

The composition of two weak homotopy equivalences is obviously a weak homotopy equivalence.

2. <u>Let</u> $f: X \to Y$ <u>be a weak homotopy equivalence. Then for any cellular pair</u> (K,L) <u>and continuous maps</u> $\phi: K \to Y$ <u>and</u> $\psi: L \to X$ <u>with</u> $f \circ \psi = \phi\big|_L$ <u>there is a continuous map</u> $\chi: K \to X$ <u>such that</u> $\chi\big|_L = \psi$ <u>and</u> $f \circ \chi$ <u>is L-homotopic to</u> ϕ. <u>The converse is also true; moreover, if</u> $f: X \to Y$ <u>is continuous and has the property that for any continuous maps,</u> $\phi: D^r \to Y$ <u>and</u> $\psi: S^{r-1} \to X$ $(r \geqslant 0)$ <u>with</u> $f \circ \psi = \phi\big|_{S^{r-1}}$, <u>there is a continuous map</u> $\chi: D^r \to X$ <u>such that</u> $\chi\big|_{S^{r-1}} = \psi$ <u>and</u> $f \circ \chi$ S^{r-1}<u>-homotopic to</u> ϕ, <u>then</u> f <u>is a weak homotopy equivalence.</u>

To prove the first part, consider the mapping cylinder Cyl f (see 1.2.6.10). By Theorem 2.3.1.3, there exists a homotopy $\Phi: K \times I \to \text{Cyl } f$ of the composition $K \xrightarrow{\phi} Y \xrightarrow{\text{in}} \text{Cyl } f$, such that $\Phi(z,t) = \text{imm}_1(\psi(z),1-t)$ for all $z \in L$, $t \in I$. Since $\text{rt } f \circ [\text{in}: X \to \text{Cyl } f] = f$ and f is a weak homotopy equivalence, while $\text{rt } f$ is a homotopy equivalence, we conclude that in is a weak homotopy equivalence. Hence, $\text{in}_*: \pi_r(X,x) \to \pi_r(\text{Cyl } f,x)$ is an isomorphism for all $r \geqslant 0$ and all $x \in X$. Thus, $(\text{Cyl } f,X)$ is an ∞-connected pair (see 1.6.5), and since $\Phi(L \times 1) \subset X$, there is an L-homotopy $\Psi: K \times I \to \text{Cyl } f$ such that $\Psi(x,0) = \Phi(x,1)$ for all $x \in K$, and $\Psi(K \times 1) \subset X$ (see 2.3.1.6). Now define $\chi: K \to X$ by $\chi(x) = \Phi(x,1)$. Obviously, $\chi\big|_L = \psi$, and the product of the homotopies $\text{rt } f \circ \Phi$ and $\text{rt } f \circ \Psi$ is an L-homotopy from $f \circ \chi$ to ϕ.

To prove the second part, we must check that $f_*: \pi_r(X,x) \to \pi_r(Y,f(x))$ is both epimorphic and monomorphic for all $r \geqslant 0$ and $x \in X$. To see that f_* is epimorphic, set $\psi(S^{r-1}) = x$ and take ϕ to be some spheroid $(D^r,S^{r-1}) \to (Y,f(x))$ of an arbitrarily given class $\beta \in \pi_r(Y,f(x))$; the resulting spheroid $\chi: (D^r,S^{r-1}) \to (X,x)$ defines a class $\alpha \in \pi_r(X,x)$, and it is clear that $f_*(\alpha) = \beta$. Finally, to show that f_* is monomorphic, pick $\alpha \in \pi_r(X,x)$ with $f_*(\alpha) = 0$, take ψ to be some spheroid of class α, and take ϕ to be any continuous map $D^{r+1} \to Y$ such that $\phi\big|_{S^r} = f \circ \psi$; the resulting map $\chi: D^{r+1} \to X$ extends ψ, and hence $\alpha = 0$.

3. <u>If</u> $f: X \to Y$ <u>is a weak homotopy equivalence, then the mapping</u> $\pi(\text{id},f): \pi(M,X) \to \pi(M,Y)$ <u>is invertible for any cellular space</u> M.

PROOF. The first part of Theorem 2 shows that, given a class $\beta \in \pi(M,Y)$, there is an $\alpha \in \pi(M,X)$ such that $[\pi(\mathrm{id},f)](\alpha) = \beta$: we need only set $K = M$, $L = \emptyset$, and take ϕ to be any map in the class β. Further, given arbitrary continuous maps $\phi_0, \phi_1 \colon M \to X$, it follows from the same first part of Theorem 2 that if the compositions $f \circ \phi_0$ and $f \circ \phi_1$ are homotopic, then so are ϕ_0 and ϕ_1: indeed, take $K = M \times I$, $L = (M \times 0) \cup (M \times 1)$, and $\phi \colon (M \times 0) \cup (M \times 1) \to X$, $\phi(x,0) = \phi_0(x)$, $\phi(x,1) = \phi_1(x)$, $x \in M$, and take for ψ any homotopy $M \times I \to X$ from $f \circ \phi_0$ to $f \circ \phi_1$.

The Case of Cellular Spaces

4. <u>If X and Y are cellular spaces, then every weak homotopy equivalence $X \to Y$ is a homotopy equivalence.</u>

PROOF. Suppose $f \colon X \to Y$ is a weak homotopy equivalence. By 3, the mapping $\pi(\mathrm{id},f) \colon \pi(Y,X) \to \pi(Y,Y)$ is invertible, and hence there is a map $g \colon Y \to X$ whose homotopy class is taken by $\pi(\mathrm{id},f)$ into the class of $\mathrm{id}\, Y$. That is to say, $f \circ g$ is homotopic to $\mathrm{id}\, Y$, and it remains to verify that $g \circ f$ is homotopic to $\mathrm{id}\, X$. The latter is a consequence of the invertibility of $\pi(\mathrm{id},f) \colon \pi(X,X) \to \pi(X,Y)$, because this mapping takes the homotopy classes of $g \circ f$ and $\mathrm{id}\, X$ into the same element (indeed, $f \circ g \circ f$ and f are homotopic).

5. Theorem 4 states that two connected cellular spaces, X and Y, are homotopy equivalent whenever there is a continuous map $X \to Y$ which induces isomorphisms of the homotopy groups, but it certainly does not guarantee that X and Y are homotopy equivalent if their homotopy groups are just isomorphic. In fact, we have simple examples to show that the latter is not true. Take $X = S^p \times \mathbb{R}P^q$, $Y = S^q \times \mathbb{R}P^p$, and suppose that $1 < p < q$. By 2.5.1 and 1.1.9, $\pi_r(X) \cong \pi_r(Y)$ for all r. However, X and Y are not homotopy equivalent. Indeed, the map $\mathrm{pr}_1 \colon S^p \times \mathbb{R}P^q \to S^p$ induces a group isomorphism $\pi_p(S^p \times \mathbb{R}P^q) \to \pi_p(S^p)$. We next show that there is no continuous map $f \colon S^q \times \mathbb{R}P^p \to S^p$ which induces a group isomorphism $\pi_p(S^q \times \mathbb{R}P^p) \to \pi_p(S^p)$. Assuming that such an f exists, the composition

$$S^p \xrightarrow{\ \mathrm{pr}\ } \mathbb{R}P^p \xrightarrow{\ x \mapsto (\mathrm{ort}_1,x)\ } S^q \times \mathbb{R}P^p \xrightarrow{\ f\ } S^p \tag{1}$$

induces an automorphism of $\pi_p(S^p)$. On the other hand (by 2.3.2.4), every continuous map $\mathbb{R}P^p \to S^p$ is homotopic to a map which takes

$\mathbb{R}P^{p-1}$ into ort_1, and thus (1) is homotopic to the composition of the composite projection $S^p \to \mathbb{R}P^p \to \mathbb{R}P^p / \mathbb{R}P^{p-1} = S^p$ with some continuous map $S^p \to S^p$. Consequently, (1) cannot have degree ± 1, since the above composite projection has degree 0 when p is even and degree 2 when p is odd. Contradiction.

The following example illustrates the same phenomenon in the simply connected case. Set $X = S^3 \times \mathbb{C}P^\infty$ and $Y = S^2$. By 2.2.10, 2.5.2, and 1.1.9, $\pi_r(X) \cong \pi_r(Y)$ for all r. However, X and Y have not the same homotopy type. Indeed, $\mathrm{pr}_1 : S^3 \times \mathbb{C}P^\infty \to S^3$ is not null homotopic (because it induces a group isomorphism $\pi_3(S^3 \times \mathbb{C}P^\infty) = \mathbb{Z} \to \pi_3(S^3) = \mathbb{Z}$). On the other hand, every continuous map $S^2 \to S^3$ is null homotopic.

6. We say that a topological space is <u>homotopy fit</u> if it is homotopy equivalent to a cellular space. From Theorem 4 it follows that if X and Y are homotopy fit, then every weak homotopy equivalence $X \to Y$ is a homotopy equivalence.

By Theorem 3.5.2.13, all smooth compact manifolds are homotopy fit.

An example of a space which is not homotopy fit was given in 2.3.5.4. This space is not connected. For an 'example of a connected (and even ∞-connected) space which is not homotopy fit, see 4.1 below.

INFORMATION. Every CNRS is homotopy fit, and the same holds true for every topological manifold (compact or not). A product of homotopy fit spaces is homotopy fit. If Y is homotopy fit, then $C(X,Y)$ is homotopy fit for any compact space X. If Y has the homotopy homotopy type of a countable cellular space, then $C(X,Y)$ has the homotopy type of a countable cellular space, for any compact space X with countable base. For proofs, see [16].

k-Equivalence

7. Let X and Y be topological spaces. A continuous map $f : X \to Y$ is a <u>k-equivalence</u> if, for all $x \in X$, $f_* : \pi_r(X,x) \to \pi_r(Y,f(x))$ is an isomorphism for $r < k$ and an epimorphism for $r = k$. Here k is a nonnegative integer; sometimes, weak homotopy equivalences are referred to as ∞-equivalences.

A composition of two k-equivalences is obviously a k-equivalence.

8. <u>Let</u> $f : X \to Y$ <u>be a k-equivalence. Then for any cellular</u>

pair (K,L) <u>with</u> $K \smallsetminus L \subset \mathrm{ske}_k K$ <u>and continuous maps</u> $\phi: K \to Y$ <u>and</u>
$\psi: L \to X$ <u>with</u> $f \circ \psi = \phi\big|_L$ <u>there is a continuous map</u> $\chi: K \to X$ <u>such</u>
<u>that</u> $\chi\big|_L = \psi$ <u>and</u> $f \circ \chi$ <u>is L-homotopic to</u> ϕ. <u>The converse is also</u>
<u>true; moreover, if</u> $f: X \to Y$ <u>is continuous and has the property that</u>
<u>for any continuous maps</u> $\phi: D^r \to Y$ <u>and</u> $\psi: S^{r-1} \to X$ $(0 \leqslant r \leqslant k)$ <u>with</u>
$f \circ \psi = \phi\big|_{S^{r-1}}$ <u>there is a continuous map</u> $\chi: D^r \to X$ <u>such that</u>
$\chi\big|_{S^{r-1}} = \psi$ <u>and</u> $f \circ \chi$ <u>is S^{r-1}-homotopic to</u> ϕ, <u>then</u> f <u>is a</u>
<u>k-equivalence.</u>

The proof repeats that of Theorem 2, with obvious changes:
in the first pair, the pair (Cyl f,X) is now k-connected; in the
second part, to see that f_* is epimorphic (monomorphic), we take $r \leqslant k$
(respectively, $r < k$).

9. <u>If</u> $f: X \to Y$ <u>is a k-equivalence, then the mapping</u>
$\pi(\mathrm{id},f): \pi(M,X) \to \pi(M,Y)$ <u>is invertible (surjective)</u> <u>for any cellular</u>
<u>space</u> M <u>with</u> $\dim M < k$ <u>(respectively,</u> $\dim M = k$).

The proof repeats that of Theorem 3, except that we need
Theorem 8 instead of Theorem 2.

10. <u>If</u> X <u>and</u> Y <u>are cellular spaces with</u> $\dim X < k$ <u>and</u>
$\dim Y \leqslant k$, <u>then every k-equivalence</u> $X \to Y$ <u>is a homotopy equivalence.</u>

The proof repeats that of Theorem 4, except that we need
Theorem 9 instead of Theorem 3.

The Relative Case

11. If (X,A) and (Y,B) are topological pairs, a continuous
map $f: (X,A) \to (Y,B)$ is said to be a <u>weak homotopy equivalence</u> if
$\mathrm{abs}\, f : X \to Y$ and $\mathrm{ab}\, f\ (=\mathrm{ab\,abs}\, f): A \to B$ are weak homotopy equivalences.

We remark that if $f: (X,A) \to (Y,B)$ is a weak homotopy
equivalence, then $f_*: \pi_r(X,A,x) \to \pi_r(Y,B,f(x))$ is an isomorphism for
all $r \geqslant 1$ and all $x \in A$. To see this, apply the 5-Lemma (see 1.5.19)
to the homomorphism induced by f from the homotopy sequence of the pair
(X,A) into the homotopy sequence of the pair (Y,B). As another
corollary of the 5-Lemma, we have the following result: suppose that
$f: (X,A\neq\emptyset) \to (Y,B)$ is continuous, that one of the maps $\mathrm{abs}\, f : X \to Y$,
$\mathrm{ab}\, f : A \to B$ is a weak homotopy equivalence, and that for any $x \in A$ all
the homomorphisms $f_*: \pi_r(X,A,x) \to \pi_r(Y,B,f(x))$, $r \geqslant 1$, $(\mathrm{abs}\, f)_*:$
$\pi_0(X,x) \to \pi_0(Y,f(x))$, and $(\mathrm{ab}\, f)_*: \pi_0(A,x) \to \pi_0(B,f(x))$ are
isomorphisms. Then f is a weak homotopy equivalence.

12. **If** $f: (X,A) \to (Y,B)$ **is a weak homotopy equivalence, then the mapping** $\pi(\mathrm{id},f): \pi(M,N;X,A) \to \pi(M,N;Y,B)$ **is invertible for any cellular pair** (M,N).

PROOF. Let us show first that every continuous map $\phi:(M,N) \to (Y,B)$ is homotopic to the composition of some continuous map $(M,N) \to (Y,B)$ with f. By 2, there is a continuous map $\psi: N \to A$ whose composition with $\mathrm{ab}\, f : A \to B$ is homotopic to $\mathrm{ab}\, \phi : N \to B$. Using 2.3.1.3, any homotopy from $\mathrm{ab}\, \phi$ to $\mathrm{ab}\, f \circ \psi$ may be extended to a homotopy of ϕ, and hence there is a $\phi': (M,N) \to (Y,B)$ homotopic to ϕ and satisfying $\mathrm{ab}\, \phi' = \mathrm{ab}\, f \circ \psi$. Finally, again using 2, we see that there is a continuous map $\chi: M \to X$ extending ψ and such that $f \circ \chi$ is N-homotopic to ϕ'. Obviously, $\chi(N) \subset B$ and the maps $f \circ \chi, \phi: (M,N) \to (Y,B)$ are homotopic.

To complete the proof, we have to show, given two continuous maps $\phi_0, \phi_1: (M,N) \to (X,A)$ such that $f \circ \phi_0$ and $f \circ \phi_1$ are homotopic, that ϕ_0 and ϕ_1 are also homotopic. Let $\Phi: (M \times I, N \times I) \to (Y,B)$ be a homotopy from $f \circ \phi_0$ to $f \circ \phi_1$. By 2, there is a homotopy $\Psi: N \times I \to A$ from $\mathrm{ab}\, \phi_0 : N \to A$ to $\mathrm{ab}\, \phi_1 : N \to A$, such that $\mathrm{ab}\, f \circ \Psi$ is $[(N \times 0) \cup (N \times 1)]$-homotopic to $\mathrm{ab}\, \Phi: N \times I \to B$. Further, by 2.3.1.3, every $[(N \times 0) \cup (N \times 1)]$-homotopy from $\mathrm{ab}\, \Phi$ to $\mathrm{ab}\, f \circ \Psi$ extends to a $[(M \times 0) \cup (M \times 1)]$-homotopy of Φ. Consequently, there is a homotopy $\Phi': (M \times I, N \times I) \to (Y,B)$ from $f \circ \phi_0$ to $f \circ \phi_1$ such that $[\mathrm{ab}\, \Phi': N \times I \to B] = \mathrm{ab}\, f \circ \Psi$. Finally, we apply 2 again to deduce that there exists a continuous map $\Xi: M \times I \to X$ extending Ψ, such that $\Xi(x,0) = \phi_0(x)$, $\Xi(x,1) = \phi_1(x)$ for all $x \in M$. In other words, Ξ is a homotopy from ϕ_0 to ϕ_1.

13. **If** (X,A) **and** (Y,B) **are cellular pairs, then every weak homotopy equivalence** $(X,A) \to (Y,B)$ **is a homotopy equivalence.**

The proof repeats that of Theorem 4, but one must refer to Theorem 12 instead of Theorem 3.

2. Weak Homotopy Equivalence and Constructions

1. Many of the operations on maps which were described in § 1.2 carry weak homotopy equivalences into weak homotopy equivalences. For example, it is clear that $\bigsqcup_\mu f_\mu : \bigsqcup_\mu X_\mu \to \bigsqcup_\mu X'_\mu$ is a weak homotopy equivalence for any family $\{f_\mu : X_\mu \to X'_\mu\}_{\mu \in M}$ of weak homotopy equivalences; similarly, $f_1 \times \ldots \times f_n: X_1 \times \ldots \times X_n \to X'_1 \times \ldots \times X'_n$ is a weak homotopy equivalence for any weak homotopy equivalences

$f_1: X_1 \to X_1', \ldots, f_n: X_n \to X_n'$. Also, the limit of a sequence of of weak homotopy equivalences is a weak homotopy equivalence.

Similar results for other constructions need supplementary arguments, or hold only under additional assumptions. The present subsection is devoted to such results.

2. <u>Let</u> $f: (X,A,B) \to (X',A',B')$ <u>be a map of triads such that</u> $\mathrm{ab}\, f: A \to A'$, $\mathrm{ab}\, f: B \to B'$, <u>and</u> $\mathrm{ab}\, f: A \cap B \to A' \cap B'$ <u>are weak homotopy equivalences. If</u> $\mathrm{Int}\, A \cup \mathrm{Int}\, B = X$ <u>and</u> $\mathrm{Int}\, A' \cup \mathrm{Int}\, B' = X'$, <u>then</u> $\mathrm{abs}\, f: X \to X'$ <u>is a weak homotopy equivalence.</u>

PROOF. Using Theorem 1.2, and given continuous maps $\phi: D^r \to X'$ and $\psi: S^{r-1} \to X$ with $f \circ \psi = \phi\big|_{S^{r-1}}$, it suffices to produce a continuous map $\chi: D^r \to X$ such that $\chi\big|_{S^{r-1}} = \psi$ and $f \circ \chi$ is S^{r-1}-homotopic to ϕ. Obviously,

$$U = [\phi^{-1}(\mathrm{Int}\, A') \cap \mathrm{Int}\, D^r] \cup \psi^{-1}(\mathrm{Int}\, A)$$

and

$$V = [\phi^{-1}(\mathrm{Int}\, B') \cap \mathrm{Int}\, D^r] \cup \psi^{-1}(\mathrm{Int}\, B)$$

are open and cover D^r. Therefore, there is an $\varepsilon > 0$ such that any subset of D^r with diameter less than ε is contained in U or V. Now triangulate S^{r-1} so that the diameter of each simplex is less than ε, and then extend the triangulation to D^r, preserving this property. Let K (L) be the union of all simplices contained in U (respectively, V). It is clear that K and L are simplicial subspaces of D^r such that $\psi(K \cap S^{r-1}) \subset \mathrm{Int}\, A$, $\psi(L \cap S^{r-1}) \subset \mathrm{Int}\, B$, $\phi(K) \subset \mathrm{Int}\, A'$, and $\phi(L) \subset \mathrm{Int}\, B'$. By Theorem 1.2, there is a continuous map $\chi_0: K \cap L \to A \cap B$ such that $\chi_0\big|_{K \cap L \cap S^{r-1}} = \mathrm{ab}\, \psi$ and the composition of χ_0 with $\mathrm{ab}\, f: A \cap B \to A' \cap B'$ is $(K \cap L \cap S^{r-1})$-homotopic to $\mathrm{ab}\, \phi: K \cap L \to A' \cap B'$. By 2.3.1.3, every $(K \cap L \cap S^{r-1})$-homotopy from $\mathrm{ab}\, \phi: K \cap L \to A' \cap B'$ to $\mathrm{ab}\, f \circ \chi_0$ extends to a $(K \cap S^{r-1})$-homotopy of $\mathrm{ab}\, \phi: K \to A'$ and to a $(L \cap S^{r-1})$-homotopy of $\mathrm{ab}\, \phi: L \to B'$. The two resulting homotopies combine to define an S^{r-1}-homotopy from ϕ to a map $\phi': D^r \to X'$ which satisfies $\phi'\big|_{K \cap L} = (f\big|_{A \cap B}) \circ \chi_0$, $\phi'(K) \subset A'$, and $\phi'(L) \subset B'$. Finally, apply Theorem 1.2 again to deduce that there are continuous maps $\chi_1: K \to A$ and $\chi_2: L \to B$ with the following properties: $\chi_0 = \mathrm{ab}\, \chi_1$, $\chi_0 = \mathrm{ab}\, \chi_2$; the composition of χ_1 with $\mathrm{ab}\, f: A \to A'$ is $(K \cap L)$-homotopic to $\mathrm{ab}\, \phi': K \to A'$; and the composition of χ_2 with $\mathrm{ab}\, f: B \to B'$ is $(K \cap L)$-homotopic to $\mathrm{ab}\, \phi': L \to B'$. Now the desired map $\chi: D^r \to X$ is obtained by combining χ_1 and χ_2.

3. <u>Suppose that</u> (X,C) <u>and</u> (X',C') <u>are Borsuk pairs,</u> Y and Y' <u>are topological spaces, and</u> $\varphi: C \to Y$, $\varphi': C' \to Y'$, $f: (X,C) \to (X',C')$, <u>and</u> $g: Y \to Y'$ <u>are continuous, with</u> $g \circ \varphi = \varphi' \circ ab\,f$. <u>If</u> f <u>and</u> g <u>are weak homotopy equivalences, then the</u> <u>formulas</u>

$$F \circ [imm_1: X \to Y \cup_\varphi X] = [imm_1: X' \to Y' \cup_{\varphi'} X'] \circ f$$

<u>and</u>

$$F \circ [imm_2: Y \to Y \cup_\varphi X] = [imm_2: Y' \to Y' \cup_{\varphi'} X'] \circ g$$

<u>define a weak homotopy equivalence</u> $F: Y \cup_\varphi X \to Y' \cup_{\varphi'} X'$.

PROOF. Let us glue X, Y, and $C \times I$, identifying each point $(c,0) \in C \times I$ with $c \in X$, and each point $(c,1) \in C \times I$ with $\varphi(c) \in Y$. Let Z denote the resulting space and, to avoid confusion, denote the maps $imm_1: X \to Z$, $imm_2: Y \to Z$, and $imm_3: C \times I \to Z$ by α, β, and γ. First, we want to show that the formulas

$$h \circ \alpha = [imm_1: X \to Y \cup_\varphi X], \quad h \circ \beta = [imm_2: Y \to Y \cup_\varphi X],$$

and

$$h \circ \gamma = [imm_2: Y \to Y \cup_\varphi X] \circ \varphi \circ [pr_1: C \times I \to C]$$

define a homotopy equivalence $h: Z \to Y \cup_\varphi X$. To this end, extend the the homotopy $\gamma: C \times I \to Z$ to a homotopy $\Gamma: X \times I \to Z$ of the map α and define $k: Y \cup_\varphi X \to Z$ by $k(imm_1(x)) = \Gamma(x,1)$ $[x \in X]$, $k \circ imm_2 = \beta$. Then the formulas

$$\begin{cases} H(imm_1(x),t) = h(\Gamma(x,t)) & \text{for } x \in X, \ t \in I, \\[2mm] H(imm_2(y),t) = imm_2(y) & \text{for } y \in Y, \ t \in I, \end{cases}$$

and

$$\begin{cases} K(\alpha(x),t) = \Gamma(x,t) & \text{for } x \in X, \ t \in I, \\ K(\beta(y),t) = \beta(y) & \text{for } y \in Y, \ t \in I, \\ K(\gamma(c,t),u) = \gamma(c,tu-u+1), & \text{for } c \in C, \ t \in I, \ u \in I, \end{cases}$$

define a homotopy $H: (Y \cup_\varphi X) \times I \to Y \cup_\varphi X$ from $h \circ k$ to $\mathrm{id}(Y \cup_\varphi X)$ and a homotopy $K: Z \times I \to Z$ from $k \circ h$ to $\mathrm{id}\,Z$. Consequently, k is a homotopy inverse to h.

Now repeat all this for X', C', Y', φ',.... We obtain a space Z', continuous maps $\alpha': X' \to Z'$, $\beta': Y' \to Z'$, $\gamma': C' \times I \to Z'$, and a homotopy equivalence $h': Y' \cup_{\varphi'} X' \to Z'$. Let $G: Z \to Z'$ denote the map defined by $G \circ \alpha = \alpha' \circ f$, $G \circ \beta = \beta' \circ g$, $G \circ \gamma = \gamma' \circ (ab\,f \times \mathrm{id}\,I)$. It is clear that G maps the triad

$(Z, \alpha(X) \cup \gamma(C \times [0,1/2]), \beta(Y) \cup \gamma(C \times I))$ into the triad $(Z', \alpha'(X') \cup \gamma'(C' \times [0,1/2]), \beta'(Y') \cup \gamma'(C' \times I))$. At the same time, all the conditions of Theorem 2 are fulfilled. Thus, applying this theorem, G is a weak homotopy equivalence. Finally, the commutativity of the diagram

$$
\begin{array}{ccc}
Y \cup_\varphi X & \xrightarrow{\ F\ } & Y' \cup_{\varphi'} X' \\
\downarrow h & & \downarrow h' \\
Z & \xrightarrow{\ \ G\ \ } & Z'
\end{array}
$$

implies that F is a weak homotopy equivalence.

4. <u>Let</u> $f: (X,A,B) \to (X',A',B')$ <u>be a map of triads such that</u> $ab\,f: A \to A'$, $ab\,f: B \to B'$, <u>and</u> $ab\,f: A \cap B \to A' \cap B'$ <u>are weak homotopy equivalences. If</u> $(A, A \cap B)$ <u>and</u> $(A', A' \cap B')$ <u>are Borsuk pairs, then</u> $abs\,f: X \to X'$ <u>is a weak homotopy equivalence.</u>

This is a corollary of Theorem 3, because $X = B \cup_{in} A$, in = [in: $A \cap B \to B$], while $X' = B' \cup_{in} A'$, in = [in: $A' \cap B' \to B'$].

5. <u>Let</u> (X,A) <u>and</u> (X',A') <u>be Borsuk pairs. If</u> $f: (X,A) \to (X',A')$ <u>is a weak homotopy equivalence, then so is</u> fact $f: X/A \to X'/A'$.

It suffices to apply Theorem 3 for $Y = D^0$, $Y' = D^0$.

6. <u>If</u> $f: X \to X'$ <u>is a weak homotopy equivalence, then so is</u> su: su $X \to$ su X'.

It suffices to apply Theorem 5 to the map rel con $f: (\text{con } X, X) \to (\text{con } X', X')$.

7. <u>If</u> $f: X \to X'$ <u>and</u> $g: Y \to Y'$ <u>are weak homotopy equivalences, then so is</u> $f * g: X * Y \to X' * Y'$.

PROOF. Let A and B be the images of $X \times Y \times [0,2/3]$ and $X \times Y \times [1/3,1]$ under the projection $X \times Y \times I \to X * Y$. Similarly, let A' and B' be the images of $X' \times Y' \times [0,2/3]$ and $X' \times Y' \times [1/3,1]$ under the projection $X' \times Y' \times I \to X' * Y'$. Obviously, $(f * g)(A) \subset A'$, $(f * g)(B) \subset B'$, and rel$(f * g)$: $: (X * Y, A, B) \to (X' * Y', A', B')$ satisfies the conditions of Theorem 2. Therefore, $f * g$ is a weak homotopy equivalence.

8. <u>If</u> $f: X \to Y$ <u>is a weak homotopy equivalence and</u> K <u>is a locally finite cellular space, then</u> $C(\text{id}, f): C(K,X) \to C(K,Y)$ <u>is a weak homotopy equivalence.</u>

Given arbitrary continuous maps $\phi: D^r \to C(K,Y)$ and

$\psi\colon S^{r-1} \to C(K,X)$ with $C(\mathrm{id},f) \circ \psi = \phi\big|_{S^{r-1}}$, we need only exhibit a continuous map $\chi\colon D^r \to C(K,X)$ such that $\chi\big|_{S^{r-1}} = \psi$ and $C(\mathrm{id},f) \circ \chi$ is S^{r-1}-homotopic to ϕ (see 1.2). Consider the maps $\phi^\wedge\colon D^r \times K \to Y$, $\psi^\wedge\colon S^{r-1} \times K \to X$ (see 1.2.7.6 and 2.1.4.3), which are continuous and satisfy $f \circ \psi^\wedge = \phi^\wedge\big|_{S^{r-1}\times K}$. The first part of Theorem 1.2 applies to $\phi^\wedge$ and $\psi^\wedge$ and yields a continuous map $\alpha\colon D^r \times K \to X$ such that $\alpha\big|_{S^{r-1}\times K} = \psi^\wedge$, as well as an $(S^{r-1} \times K)$-homotopy $h\colon (D^r \times K) \times I \to Y$ from $\phi^\wedge$ to $f \circ \alpha$. Denote by H the composition of the canonical homeomorphism $(D^r \times I) \times K \to (D^r \times K) \times I$ with h, and set $\chi = \alpha^\vee$ (see again 1.2.7.6). It is clear that $\chi\big|_{S^{r-1}} = \psi$ and that $H^\vee$ is an S^{r-1}-homotopy from ϕ to $C(\mathrm{id},f) \circ \chi$.

9. <u>Let X_μ and X'_μ be topological spaces with base points</u> x_μ <u>and</u> x'_μ <u>such that</u> (X_μ,x_μ) <u>and</u> (X'_μ,x'_μ) <u>are Borsuk pairs. If</u> $f_\mu\colon (X_\mu,x_\mu) \to (X'_\mu,x'_\mu)$ <u>are weak homotopy equivalences, then so is</u> $\bigvee_\mu f_\mu\colon \bigvee_\mu (X_\mu,x_\mu) \to \bigvee_\mu (X'_\mu,x'_\mu)$.

This is a corollary of 5.

3. Cellular Approximations of Topological Spaces

1. A <u>cellular approximation of the topological space</u> X is any pair (K,ϕ) consisting of a cellular space K and a weak homotopy equivalence $\phi\colon K \to X$.

Example: if X is a Hausdorff space with a cellular decomposition enjoying the property that each compact subset of X intersects only a finite number of cells, then $(X^\sim,\mathrm{id}\,X)$, where $X^\sim$ is the cellular space obtained from X through the cellular weakening of its topology, is a cellular approximation of X. In particular, if $X_1,\dots,X_n$ are cellular spaces, then $(X_1 \times_c \dots \times_c X_n,\mathrm{id})$ is a cellular approximation of $X_1 \times \dots \times X_n$, while $(X_1 *_c \dots *_c X_n,\mathrm{id})$ is a cellular approximation of $X_1 * \dots * X_n$.

2. <u>Every topological space admits cellular approximations.</u>

PROOF. We observe first that cellular approximations of the components of a topological space yield a cellular approximation of the entire space, while the case of an empty space is trivial. Thus, we may assume that the space X we want to approximate is connected and nonempty. We shall construct a sequence $K_0,K_1,\dots$ of cellular spaces with 0-cells $y_0, y_1, \ldots$ as base points, a sequence of continuous maps $\phi_i: (K_i, y_i) \to (X, x_0)$, $i = 0, 1, \ldots$ (here $x_0 \in X$ is an arbitrarily fixed point) and, finally, a sequence of cellular embeddings $\eta_i: (K_{i-1}, y_{i-1}) \to (K_i, y_i)$, $i = 1, 2, \ldots$, such that ϕ_i is an i-equivalence and $\phi_i \circ \eta_i = \phi_{i-1}$. Then the pair $(\lim K_i, \lim \phi_i)$ will be a cellular approximation of X (see 1.11.2).

Proceed by induction. Set $K_0 = D^0$, $y_0 = D^0$, $\phi_0(K_0) = x_0$, and assume that K_i, y_i, ϕ_i, η_i, $i < r$, have been defined and enjoy the required properties. Pick a spheroid $f_\alpha: S^r \to X$ in each homotopy class $\alpha \in \pi_r(X, x_0)$. Further, for any class $\beta \in N$, where $N = \mathrm{Ker}[(\phi_{r-1})_*: \pi_{r-1}(K_{r-1}, y_{r-1}) \to \pi_{r-1}(X, x_0)]$, pick a spheroid $g_\beta: S^{r-1} \to K_{r-1}$ of class β, together with a continuous map $h_\beta: D^r \to X$ satisfying $h_\beta|_{S^{r-1}} = \phi_{r-1} \circ g_\beta$. The maps f_α, g_β, and h_β (with $\alpha \in \pi_r(X, x_0)$ and $\beta \in N$) combine to define three other continuous maps:

$$f: V_{\alpha \in \pi_r(X, x_0)} (S_\alpha = S^r, \mathrm{ort}_1) \to X,$$

$$g: \bigsqcup_{\beta \in N}(S_\beta = S^{r-1}) \to K_{r-1}, \quad \text{and} \quad h: \bigsqcup_{\beta \in N}(D_\beta = D^r) \to X.$$

Now set

$$K_r = [(K_{r-1} \cup_g (\bigsqcup_{\beta \in N} D_\beta)), \mathrm{imm}_2(y_{r-1})] \; V \; [(V_{\alpha \in \pi_r(X, x_0)} (S_\alpha, bp)), bp]$$

and define η_r to be the composition

$$K_{r-1} \xrightarrow{\mathrm{imm}_2} K_{r-1} \cup_g (\bigsqcup_{\beta \in N} D_\beta) \xrightarrow{\mathrm{imm}_1} K_r,$$

$\phi_r: K_r \to X$ to be the map assembled from ϕ_{r-1}, h, and f, and y_r to be $\eta_r(y_{r-1})$. Applying Theorems 3.2.3 and 3.5.5, we see that ϕ_r is an r-equivalence, and it is clear that $\phi_r \circ \eta_r = \phi_{r-1}$ and that η_r is a cellular embedding.

3. <u>Let</u> (K, ϕ) <u>and</u> (K', ϕ') <u>be cellular approximations of the topological spaces</u> X <u>and</u> X', <u>and let</u> $f: X \to X'$ <u>be an arbitrary continuous map. Then there is a continuous map</u> $g: K \to K'$ <u>such that the diagram</u>

$$\begin{array}{ccc} K & \xrightarrow{\;g\;} & K' \\ \downarrow{\scriptstyle \phi} & & \downarrow{\scriptstyle \phi'} \\ X & \xrightarrow{\;f\;} & X' \end{array}$$

<u>is homotopy commutative (i.e., the maps</u> $f \circ \phi$ <u>and</u> $\phi' \circ g$ <u>are</u>

homotopic). _This property uniquely defines the homotopy class of_ g.

PROOF. The mapping $\pi(\mathrm{id},\phi')\colon \pi(K,K') \to \pi(K,X')$ is invertible (see 1.3), and thus in $\pi(K,K')$ there is a unique element which is taken by $\pi(\mathrm{id},\phi')$ into the class of $f \circ \phi$.

4. _Let_ (K,ϕ) _and_ (K',ϕ') _be two cellular approximations of the same topological space_ X. _Then there is a homotopy equivalence_ $g\colon K \to K'$ _such that_ $\phi' \circ g$ _is homotopic to_ ϕ.

To obtain g, apply Theorem 3 to $X' = X$ and $f = \mathrm{id}\,X$. To get a homotopy inverse g' to g, interchange the roles of (K',ϕ') and (K,ϕ). Finally, apply Theorem 3 again to show that $g \circ g'$ and $g' \circ g$ are homotopic to $\mathrm{id}\,K'$ and $\mathrm{id}\,K$, respectively.

Weak Homotopy Equivalence
as an Equivalence Relation

5. Two topological spaces are said to be _weakly homotopy equivalent_ if they admit cellular approximations (K,ϕ) and (L,ψ) with $K = L$. It is clear that this defines an equivalence relation (in the usual, set-theoretic sense).

Let X and Y be topological spaces such that there is a weak homotopy equivalence $f\colon X \to Y$. Then X and Y are weakly homotopy equivalent. Indeed, if (K,ϕ) is a cellular approximation of X, then $(K,f \circ \phi)$ is a cellular approximation of Y. The converse is false: there are examples of weakly homotopy equivalent spaces X, Y such that there is no weak homotopy equivalence $X \to Y$ and no weak homotopy equivalence $Y \to X$. See 4.2 below for such an example.

Two homotopy equivalent spaces are certainly weakly homotopy equivalent. The converse is false: for example, every topological space is weakly homotopy equivalent to a cellular one (see 2), but not every topological space has the homotopy type of a cellular space (see 1.6). On the other hand, Theorem 1.4 shows that two weakly homotopy equivalent cellular spaces are actually homotopy equivalent.

Weak Homotopy Equivalence
of the Fibers of a Serre Bundle

6. _Any two fibers of a Serre bundle with connected base are weakly homotopy equivalent._

PROOF. Let ξ be the given Serre bundle with $bs\,\xi$ connected, and let $b_0, b_1 \in bs\,\xi$. Pick a path $s: I \to bs\,\xi$ with $s(0) = b_0$, $s(1) = b_1$, and a cellular approximation (K,ϕ) of the fiber $pr\,\xi^{-1}(b_0)$. The map $f^{\sim} = [in: pr\,\xi^{-1}(b_0) \to tl\,\xi] \circ \phi$ and the homotopy $F: K \times I \to bs\,\xi$, $F(x,t) = s(t)$, satisfy $F(x,0) = pr\,\xi \circ f^{\sim}(x)$ $[x \in K]$, and hence there is a homotopy $F^{\sim}: K \times I \to tl\,\xi$ of $f^{\sim}$ which covers F (see 4.1.3.6). Now define $\psi: K \to pr\,\xi^{-1}(b_1)$ by $x \mapsto F^{\sim}(x,1)$ and check that ψ is a weak homotopy equivalence.

To do this, given any $x \in K$ and any spheroid $g \in Sph_r(K,x)$, note that the formula $(y,t) \mapsto F^{\sim}(g(y),t)$ defines a fiber homotopy (see 1.7.1) from the spheroid $\phi_{\#}(g)$ to the spheroid $\psi_{\#}(g)$ along the path $s : I \to tl\,\xi$ given by $t \mapsto F^{\sim}(g(ort_1),t)$. Consequently, the diagram

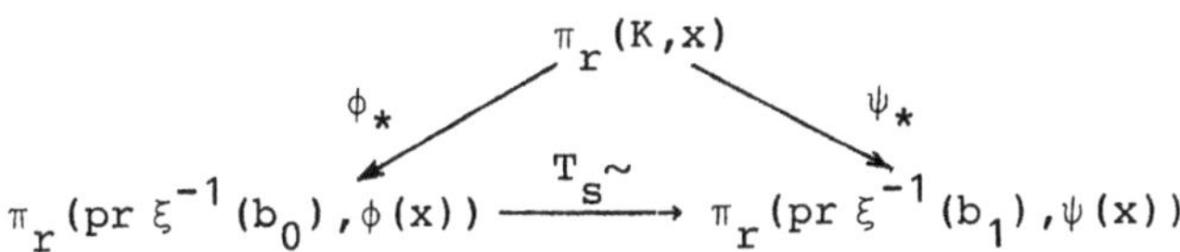

commutes (the translation $T_s{\sim}$ is defined in 1.7.3). Since $T_s{\sim}$ is an isomorphism, the invertibility of ϕ_* implies the invertibility of ψ_*.

Cellular Approximations of Topological Pairs

7. A <u>cellular approximation of the topological pair</u> (X,A) is any pair $[(K,L),\phi]$ consisting of a cellular pair (K,L) and a weak homotopy equivalence $\phi: (K,L) \to (X,A)$.

When A and L are points, a cellular approximation $[(K,L),\phi]$ of (X,A) is termed a <u>cellular approximation of the pointed space</u> (X,A).

8. <u>Every topological pair</u> (X,A) <u>admits cellular approximations. Moreover, given any cellular approximation</u> (L,ψ) <u>of the subspace</u> A, <u>there is a cellular approximation</u> $[(K,L),\phi]$ <u>of</u> (X,A) <u>with</u> $\psi = ab\,\phi$.

PROOF. Let (M,χ) be a cellular approximation of X (see 2), and let $g: L \to M$ be a cellular map such that $\chi \circ g$ is homotopic to $[in: A \to X] \circ \psi$ (see 3 and 2.3.2.4). Set $K = Cyl\,g$ and define ϕ to be the relativization of the map $K \to X$ given by χ and some homotopy $L \times I \to X$ from $[in: A \to X] \circ \psi$ to $\chi \circ g$. Obviously, ϕ is a weak homotopy equivalence and $ab\,\phi = \psi$.

9. <u>Let</u> $[(K,L),\phi]$ <u>and</u> $[(K',L'),\phi']$ <u>be cellular approximations of the topological pairs</u> (X,A) <u>and</u> (X',A'), <u>and let</u> $f: (X,A) \to (X',A')$ <u>be an arbitrary continuous map. Then there is a continuous map</u> $g: (K,L) \to (K',L')$ <u>(unique up to homotopies)</u> <u>such that the diagram</u>

$$\begin{array}{ccc} (K,L) & \xrightarrow{\ \ g\ \ } & (K',L') \\ \Big\downarrow{\phi} & & \Big\downarrow{\phi'} \\ (X,A) & \xrightarrow{\ \ f\ \ } & (X',A') \end{array}$$

<u>is homotopy commutative. If</u> $[(K,L),\phi]$ <u>and</u> $[(K',L'),\phi']$ <u>are cellular approximations of the same topological pair, then there is a homotopy equivalence</u> $g: (K,L) \to (K',L')$ <u>such that</u> $\phi' \circ g$ <u>is homotopic to</u> ϕ.

The proof repeats the proofs of Theorems 3 and 4, except that one has to refer to 1.12 instead of 1.3.

Cellular Approximations and Constructions

10. It is clear that if (K_μ, ϕ_μ) are cellular approximations of the spaces X_μ $(\mu \in M)$, then $(\bigsqcup_{\mu \in M} K_\mu, \bigsqcup_{\mu \in M} \phi_\mu)$ is a cellular approximation of $\bigsqcup_{\mu \in M} X_\mu$. Also, applying Theorem 2.9, we see that if (X_μ, x_μ) are Borsuk pairs $(x_\mu \in X_\mu$ are base points) and $[(K_\mu, y_\mu), \phi_\mu]$ are cellular approximations of these pointed spaces, then $[V_\mu(K_\mu, y_\mu), V_\mu \phi_\mu]$ is a cellular approximation of the bouquet $V_\mu(X_\mu, x_\mu)$.

Further, if $(K_1, \phi_1), \ldots, (K_n, \phi_n)$ are cellular approximations of $X_1, \ldots, X_n$, then $(K_1 \times_c \ldots \times_c K_n, \phi_1 \times \ldots \times \phi_n)$ is a cellular approximation of $X_1 \times \ldots \times X_n$ (see 2.1 and 1). In the same circumstances, $(K_1 *_c \ldots *_c K_n, \phi_1 * \ldots * \phi_n)$ is a cellular approximation of $X_1 * \ldots * X_n$ (see 2.7). In particular, $(\mathrm{su}\,K, \mathrm{su}\,\phi)$ is a cellular approximation of $\mathrm{su}\,X$ whenever (K, ϕ) is a cellular approximation of X.

If $[(K,L),\phi]$ is a cellular approximation of the Borsuk pair (X,A), then $(K/L, \mathrm{fact}\,\phi : K/L \to X/A)$ is a cellular approximation of X/A. This is a corollary of 2.5.

An Application:
Generalization of Theorems 3.3.6, 3.5.5, and 3.6.2

11 (LEMMA). <u>Let</u> (X,A,B) <u>be a triad with the property that</u> <u>either</u> $\text{Int}\,A \cup \text{Int}\,B = X$ <u>or</u> $(A, A \cap B)$ <u>is a Borsuk pair. Then there exist a cellular triad</u> (K,L,M) <u>and a continuous map</u> $f: (K,L,M) \to (X,A,B)$ <u>such that</u> $\text{abs}\,f: K \to X$, $\text{ab}\,f: L \to A$, <u>and</u> $\text{ab}\,f: L \cap M \to A \cap B$ <u>are weak homotopy equivalences.</u>

PROOF. By 2, $A \cap B$ has a cellular approximation, say (N,χ), and the latter can be extended to cellular approximations $[(L',N),\phi]$ and $[(M',N),\psi]$ of the pairs $(A, A \cap B)$ and $(B, B \cap A)$, as shown by 8. Next, attach the cylinder $N \times I$ to $L' \sqcup M'$ by the map $(N \times 0) \cup (N \times 1) \to L \sqcup M$, $(x,0) \mapsto \text{in}_1(x)$, $(x,1) \mapsto \text{in}_2(x)$, and call the resulting cellular space K. Now identify $N \times I$, L', and M' with their images in K and set $L = (N \times I) \cup L'$, $M = (N \times I) \cup M'$. The composite maps

$$N \times I \xrightarrow{\;\text{pr}_1\;} N \xrightarrow{\;\chi\;} A \cap B \xrightarrow{\;\text{in}\;} X,$$

$$L' \xrightarrow{\;\phi\;} A \xrightarrow{\;\text{in}\;} X, \quad \text{and} \quad M' \xrightarrow{\;\psi\;} B \xrightarrow{\;\text{in}\;} X$$

jointly define the map $f: (K,L,M) \to (X,A,B)$. Obviously, $\text{Int}\,L \cup \text{Int}\,M = K$ and $\text{ab}\,f: L \to A$, $\text{ab}\,f: M \to B$, and $\text{ab}\,f: L \cap M \to A \cap B$ are weak homotopy equivalences. Therefore, so is $\text{abs}\,f: K \to X$ (see 2.2 and 2.4).

12. The homomorphism

$$[\pi_1(A,x_0) \star \pi_1(B,x_0)]/vk(X,A,B,x_0) \to \pi_1(X,x_0), \tag{2}$$

defined in 3.3.5, is an isomorphism not only for a cellular triad (X,A,B) with A, B, $A \cap B$ connected (as asserted by Theorem 3.3.6), but also for any triad (X,A,B) such that A, B, $A \cap B$ are connected and either $\text{Int}\,A \cup \text{Int}\,B = X$ or $(A, A \cap B)$ is a Borsuk pair. In fact, this follows from Theorem 3.3.6 and Lemma 11, since the homomorphism (2) is natural.

In particular, we see that the fundamental group of the bouquet of two spaces is canonically isomorphic to the free product of the fundamental groups of these spaces under the only assumption that each space forms, together with its base point, a Borsuk pair (cf. 3.3.7).

13. Concerning Theorem 3.5.5, we can weaken the demand that

the pairs (X_μ, x_μ) be cellular and instead ask only that they be Borsuk pairs. That this is possible is guaranteed by Theorem 8, the discussion of bouquets in 10, an the commutativity of diagram (3) in 3.5.1.

14. Theorem 3.6.2 and its corollary 3.6.3 are valid not only for cellular pairs, but also for arbitrary Borsuk pairs. This generalization follows from Theorem 8 and the last statement on quotients in 10.

4. <u>Exercises</u>

1. Consider the union X of the graph of the function $x \mapsto \sin(1/x)$ on the interval $0 < x \leqslant 1/\pi$ and the broken line made of the four segments with the successive vertices $(1/\pi, 0)$, $(1/\pi, 2)$, $(-1, 2)$, $(-1, 0)$, and $(0, 0)$. Show that X is ∞-connected but not homotopy fit. (Cf. 1.6)

2. Let $A = \{0, 2^n \mid n \in \mathbb{Z}\} \subset \mathbb{R}$ (cf. 4.2.4.2). Show that the spaces $X = \mathbb{Z} \bigsqcup (A \times S^1)$ and $Y = A \bigsqcup (\mathbb{Z} \times S^1)$ are weakly homotopy equivalent, but there is no weak homotopy equivalence $X \to Y$, and no weak homotopy equivalence $Y \to X$. (Cf. 3.5.)

3. Let X denote the subset of $\mathbb{R}^3$ consisting of the segment I and the sequence of segments with endpoints $n\, \mathrm{ort}_2$, $\mathrm{ort}_1 + \mathrm{ort}_3/n$ $(n = 1, 2, \ldots)$. Show that $(X, (x_1, x_2, x_3) \to x_1, I)$ is a Serre bundle, but there exists fibers which are not homotopy equivalent.

4. Suppose (X, x_0), (Y, y_0) are pointed topological spaces, Z is a cellular space with a 0-cell z_0 for base point, and $f \colon X \to Y$ is a weak homotopy equivalence with $f(x_0) = y_0$. Show that $ab\, C(\mathrm{id}, f) \colon C(Z, x_0; X, x_0) \to C(Z, z_0; Y, y_0)$ is a weak homotopy equivalence. (Cf. 2.8.)

§5. THE WHITEHEAD PRODUCT

1. <u>The Class</u> wd(m,n)

1. In this section we define and study some of the properties of an operation on the elements of homotopy groups. In a certain sense, this operation generalizes the action of the fundamental group on the

homotopy groups. The definition assumes that a pair m,n of positive integers is given.

The present subsection is devoted to a very specific preliminary construction. Recall (see 2.1.3.2 and 2.1.5.2) that the cellular decomposition of $S^m \times S^n$, determined by the standard decompositions of S^m and S^n (each having two cells) consists of four cells: an $(m+n)$-cell and three other cells which form the bouquet $(S^m, \mathrm{ort}_1) \vee (S^n, \mathrm{ort}_1)$. We denote this bouquet by $B(m,n)$ or, simply, by B. The standard characteristic map of the $(m+n)$-cell is the composition of the canonical homeomorphism $D^{m+n} \to D^m \times D^n$ (see 1.2.6.9) with the map $DS \times DS$; it takes S^{m+n-1} into B, and takes the point $(\mathrm{ort}_1 + \mathrm{ort}_{m+1})/\sqrt{2}$ into $bp = (\mathrm{ort}_1, \mathrm{ort}_1)$. Therefore, this characteristic map defines an element of the group $\pi_{m+n}(S^m \times S^n, B, bp)$ (see 2.2.5), which we call $Wd(m,n)$ or, simply, Wd. Also, we write $wd(m,n)$ or, simply, wd, for the element $\partial(Wd) \in \pi_{m+n-1}(B, bp)$, i.e., the class of the attaching spheroid $S^{m+n-1} \to B$.

We need two additional notations: θ for the homeomorphism $B(m,n) \to B(n,m)$ which permutes S^m and S^n, and μ for the product of the spheroids $\mathrm{imm}_1, \mathrm{imm}_2 \colon (S^n, \mathrm{ort}_1) \to (B(m,n), bp)$ when $m = n$.

2. <u>The class wd has infinite order.</u>

It is enough to establish that Wd is of infinite order and that

$$\partial \colon \pi_{m+n}(S^m \times S^n, B, bp) \to \pi_{m+n-1}(B, bp)$$

is monomorphic. The first is a consequence of the fact that the homomorphism

$$pr_* \colon \pi_{m+n}(S^m \times S^n, B, bp) \to \pi_{m+n}((S^m \times S^n)/B = S^{m+n}, pr(bp)) = \mathbb{Z}$$

takes Wd into a generator of the right-hand group. The second claim follows from the exactness of the homotopy sequence of the pair $(S^m \times S^n, B)$ with base point bp, because $in_* \colon \pi_{m+n}(B, bp) \to \pi_{m+n}(S^m \times S^n, bp)$ is epimorphic (see 3.5.4).

3. <u>The isomorphism</u> $\theta_* \colon \pi_{m+n-1}(B(m,n), bp) \to \pi_{m+n-1}(B(n,m), bp)$ <u>takes</u> $wd(m,n)$ <u>into</u> $(-1)^{mn} wd(n,m)$.

This results from the commutativity of the diagram

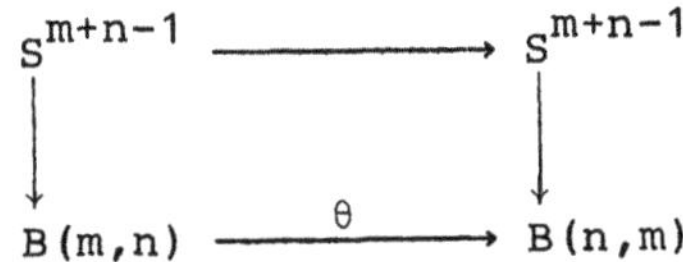

where the vertical maps are the attaching spheroids which represent the classes $wd(m,n)$ and $wd(n,m)$ (see 1), while the upper horizontal map is given by $(x_1,\ldots,x_{m+n}) \mapsto (x_{m+1},\ldots,x_{m+n},x_1,\ldots,x_m)$ (and its degree is $(-1)^{mn}$).

4. **If** $m = 1$, **then**

$$wd(m,n) = imm_{2*}(sph_n)\,[T_{imm_1(sph_1)}\,imm_{2*}(sph_n)]^{-1}.$$

In particular, $wd(1,1) = \alpha_2\alpha_1\alpha_2^{-1}\alpha_1^{-1}$, **where (as in Subsection 3.1),** α_1,α_2 **denote the elements** $imm_{1*}(sph_1), imm_{2*}(sph_1) \in \pi_1(B(1,1),bp)$.

PROOF. According to 1, $wd(1,n)$ is represented by the spheroid $S^n \to B(1,n)$,

$$(x_1,\ldots,x_{n+1}) \mapsto \begin{cases} imm_1 \circ DS(\sqrt{2}\,x_1), & \text{if } |x_1| \leqslant 1/\sqrt{2}, \\[2mm] imm_2 \circ DS(\sqrt{2}\,(x_2,\ldots,x_{n+1})), & \text{if } |x_1| \geqslant 1/\sqrt{2}. \end{cases}$$

This is obviously homotopic to the product of the spheroid

$$(x_1,\ldots,x_{n+1}) \mapsto \begin{cases} bp, & \text{if } x_1 \leqslant 1/\sqrt{2}, \\[2mm] imm_2 \circ DS(\sqrt{2}\,(x_2,\ldots,x_{n+1})), & \text{if } x_1 \geqslant 1/\sqrt{2}. \end{cases} \tag{1}$$

with the spheroid obtained by translating the spheroid

$$(x_1,\ldots,x_{n+1}) \mapsto \begin{cases} imm_2 \circ DS(\sqrt{2}\,(x_2,\ldots,x_{n+1})), & \text{if } x_1 \leqslant -1/\sqrt{2}, \\[2mm] bp, & \text{if } x_1 \geqslant -1/\sqrt{2}, \end{cases} \tag{2}$$

along the path $t \mapsto imm_1 \circ DS(1-2t)$. Now it remains to observe that the class of (1) is $imm_{2*}(sph_n)$, the class of (2) is $[imm_{2*}(sph_n)]^{-1}$, and the class of the above path is $imm_{1*}(sph_1)$.

5. **The class** $wd(m,n)$ **belongs to the kernel of each of the following three homomorphisms:**

$$pr_{1*}: \pi_{m+n-1}(B,bp) \to \pi_{m+n-1}(S^m,ort_1),$$

$$pr_{2*}: \pi_{m+n-1}(B,bp) \to \pi_{m+n-1}(S^n,ort_1),$$

and

$$in_*: \pi_{m+n-1}(B,bp) \to \pi_{m+n-1}(S^m \times S^n,bp).$$

For in_*, this results from the exactness of the homotopy sequence of the pair $(S^m \times S^n,B)$ with base point bp. For the first

and the second homomorphisms, use the equalities $[pr_1 : B \to S^m] =$
$= [pr_1 : S^m \times S^n \to S^m] \circ in$ and $[pr_2 : B \to S^n] = [pr_2 : S^m \times S^n \to S^n] \circ in$.

    6. <u>The homomorphism</u> $(is\ S^m\ V\ \mu)_* : \pi_{m+n-1}(B(m,n),bp) \to$
$\to \pi_{m+n-1}((S^m,ort_1)\ V\ (B(n,n),bp),bp)$ <u>takes</u> $wd(m,n)$ <u>into</u>

$$[(id\ S^m\ V\ imm_1)_*(wd(m,n))][(id\ S^m\ V\ imm_2)_*(wd(m,n))].$$

    PROOF. When $m = 1$ or $n = 1$ this follows from 4 and 3;
now let $m > 1$ and $n > 1$. The bouquet $(S^m,ort_1)\ V\ (B(n,n),bp)$ is
simply connected, and it yields the product $S^m \times B(n,n)$ when we add
two $(m+n)$-cells with attaching spheroids $S^{m+n-1} \to (S^m,ort_1)\ V\ (B(n,n),bp)$
belonging to the classes $(id\ S^m\ V\ imm_1)_*(wd(m,n))$ and
$(id\ S^m\ V\ imm_2)_*(wd(m,n))$. Consequently, the kernel of the homomorphism

$$in_* : \pi_{m+n-1}((S^m,ort_1)\ V\ (B(n,n),bp),bp) \to \pi_{m+n-1}(S^m \times B(n,n),bp)$$

is generated by the indicated classes. This kernel contains also the
class $(id\ S^m\ V\ \mu)_*(wd(m,n))$. To see this, note that $wd(m,n)$ sits in
the kernel of the homomorphism induced by the inclusion $B(m,n) \to S^m \times S^n$
(see 5), while $id\ S^m\ V\ \mu$ is the compression of the map
$id\ S^m \times \mu : S^m \times S^n \to S^m \times B(n,n)$. Therefore,

$$(id\ S^m\ V\ \mu)_*(wd(m,n)) =$$
$$= [(id\ S^m\ V\ imm_1)_*(wd(m,n))]^{k_1}[(id\ S^m\ V\ imm_2)_*(wd(m,n))]^{k_2}, \quad (3)$$

with $k_1, k_2 \in \mathbb{Z}$, and we shall presently show that $k_1 = k_2 = 1$.

    The compositions $(id\ S^m\ V\ pr_1) \circ (id\ S^m\ V\ \mu)$ and
$(id\ S^m\ V\ pr_2) \circ (id\ S^m\ V\ \mu)$, where pr_1, pr_2 are the projections of
$B(n,n)$ onto S^n, are both homotopic to $id\ B(m,n)$. At the same time,
$(id\ S^m\ V\ pr_1) \circ (id\ S^m\ V\ imm_1) = (id\ S^m\ V\ pr_2) \circ (id\ S^m\ V\ imm_2) =$
$= id\ B(m,n)$, while both $(id\ S^m\ V\ pr_1) \circ (id\ S^m\ V\ imm_2)$ and
$(id\ S^m\ V\ pr_2) \circ (id\ S^m\ V\ imm_1)$ equal the composition
$$B(m,n) \xrightarrow{\ pr_1\ } S^m \xrightarrow{\ imm_1\ } B(m,n).$$

Now applying the homomorphisms $(id\ S^m\ V\ pr_1)_*$ and $(id\ S^m\ V\ pr_2)_*$
to both members of (3) and using Proposition 5, we get $wd(m,n) =$
$= wd(m,n)^{k_1}$, $wd(m,n) = wd(m,n)^{k_2}$. Finally, these equalities yield,
by virtue of 2, $k_1 = 1$, $k_2 = 1$.

    7. <u>The class</u> $wd(m,n)$ <u>belongs to the kernel of the</u>
<u>homomorphism</u>

$$su: \pi_{m+n-1}(B(m,n),bp) \to \pi_{m+n}(su(B(m,n),bp)=B(m+1,n+1),bp).$$

PROOF. By 2.1.1, the diagram

$$
\begin{array}{ccccc}
\pi_{m+n-1}(S^m,ort_1) & \xleftarrow{pr_{1*}} & \pi_{m+n-1}(B(m,n),bp) & \xrightarrow{pr_{2*}} & \pi_{m+n-1}(S^n,ort_1) \\
\downarrow su & & \downarrow su & & \downarrow su \\
\pi_{m+n}(S^{m+1},ort_1) & \xleftarrow{pr_{1*}} & \pi_{m+n}(B(m+1,n+1),bp) & \xrightarrow{pr_{2*}} & \pi_{m+n}(S^{n+1},ort_1)
\end{array}
$$

commutes. This, combined with Theorem 5, shows that $su(wd(m,n))$ belongs to the kernels of pr_{1*} and pr_{2*}, and thus to the kernel of the homomorphism

$$\pi_{m+n}(B(m+1,n+1),bp) \to \pi_{m+n}(S^{m+1},ort_1) \oplus \pi_{m+n}(S^{n+1},ort_1)$$

given by pr_{1*} and pr_{2*}. Finally, recall that the last homomorphism is an isomorphism (see 3.5.5).

2. Definition and the Simplest Properties of the Whitehead Product

1. Let (X,x_0) be a pointed topological space, and let $\alpha \in \pi_m(X,x_0)$, $\beta \in \pi_n(X,x_0)$. Clearly, the homotopy class of the map $h: (B,bp) \to (X,x_0)$ defined by arbitrary spheroids $(S^m,ort_1) \to (X,x_0)$ and $(S^n,ort_1) \to (X,x_0)$ representing α and β is independent of the choice of these spheroids. Therefore, the element $h_*(wd(m,n)) \in \pi_{m+n-1}(X,x_0)$ is determined solely by the classes α and β. This element is called the __Whitehead product__ of α and β, denoted $[\alpha,\beta]$.

Notice that in terms of this definition, $wd(m,n)$ itself is the Whitehead product of the classes of the spheroids $imm_1: (S^m,ort_1) \to (B,bp)$ and $imm_2: (S^n,ort_1) \to (B,bp)$, i.e.,

$$wd(m,n) = [imm_{1*}(sph_m),imm_{2*}(sph_n)].$$

It is readily cheked that $f_*([\alpha,\beta]) = [f_*(\alpha),f_*(\beta)]$ for any $\alpha \in \pi_m(X,x_0)$, $\beta \in \pi_n(X,x_0)$, and continuous $f: (X,x_0) \to (Y,y_0)$. Furthermore, $T_s([\alpha,\beta]) = [T_s\alpha,T_s\beta]$ for any $\alpha \in \pi_m(X,x_0)$, $\beta \in \pi_n(X,x_0)$, and any path $s: I \to X$ with $s(0) = x_0$.

2. __If__ $\alpha \in \pi_m(X,x_0)$ __and__ $\beta \in \pi_n(X,x_0)$, __then__

$$[\beta,\alpha] = (-1)^{mn}[\alpha,\beta].$$

Indeed, if $h: (B,bp) \to (X,x_0)$ is the map defined by two spheroids, $(S^m, ort_1) \to (X,x_0)$ and $(S^n, ort_1) \to (X,x_0)$ which represent the classes α and β, then

$$[\beta,\alpha] = (h \circ \theta)_*(wd(n,m)) = h_*(\theta_*(wd(n,m))) =$$

$$= h_*((-1)^{mn}wd(m,n)) = (-1)^{mn}[\alpha,\beta]$$

(see 1.3).

3. If $\alpha \in \pi_m(X,x_0)$ and $\beta_1,\beta_2 \in \pi_n(X,x_0)$ with $n > 1$, then $[\alpha,\beta_1 + \beta_2] = [\alpha,\beta_1] + [\alpha,\beta_2]$. If $\alpha_1,\alpha_2 \in \pi_m(X,x_0)$ with $m > 1$ and $\beta \in \pi_n(X,x_0)$, then $[\alpha_1 + \alpha_2,\beta] = [\alpha_1,\beta] + [\alpha_2,\beta]$.

Because of 2, one has to prove only the first equality. Consider the map $h: ((S^m,ort_1) \vee (B(n,n),bp),bp) \to (X,x_0)$ defined by arbitrary spheroids, $f: (S^m,ort_1) \to (X,x_0)$ and $g_1,g_2: (S^n,ort_1) \to (X,x_0)$, representing the classes α and β_1,β_2, respectively. Then the map $(B(m,n),bp) \to (X,x_0)$ defined by f and g_1 equals $h \circ (id\, S^m \vee imm_1)$, the map defined by f and g_2 equals $h \circ (id\, S^m \vee imm_2)$, and finally the map defined by f and the product of the spheroids g_1 and g_2 equals $h \circ (id\, S^m \vee \mu)$. Hence,

$$[\alpha,\beta_1 + \beta_2] = h_* \circ (id\, S^m \vee \mu)_*(wd(m,n)) =$$

$$= h_*((id\, S^m \vee imm_1)_*(wd(m,n)) + (id\, S^m \vee imm_2)_*(wd(m,n))) =$$

$$= (h \circ (id\, S^m \vee imm_1))_*(wd(m,n)) +$$

$$+ (h \circ (id\, S^m \vee imm_2))_*(wd(m,n)) = [\alpha,\beta_1] + [\alpha,\beta_2].$$

4. If $\alpha \in \pi_1(X,x_0)$ and $\beta \in \pi_n(X,x_0)$ with $n \geqslant 1$, then $[\alpha,\beta] = \beta(T_\alpha\beta)^{-1}$. In particular, $[\alpha,\beta] = \beta\alpha\beta^{-1}\alpha^{-1}$ for any $\alpha,\beta \in \pi_1(X,x_0)$.

This is a corollary of 1.4.

5. For any $\alpha \in \pi_m(X,x_0)$ and $\beta \in \pi_n(X,x_0)$, the product $[\alpha,\beta]$ belongs to the kernel of the homomorphism $su: \pi_{m+n-1}(X,x_0) \to \pi_{m+n}(su(X,x_0),bp)$.

This is a corollary of 1.7.

6. For any $\alpha \in \pi_m(X,x_0)$ and $\beta \in \pi_n(Y,y_0)$, the product $[imm_{1*}(\alpha),imm_2(\beta)]$ belongs to the kernel of each of the homomorphisms:

$$pr_{1*}\colon \pi_{m+n-1}((X,x_0) \vee (Y,y_0),bp) \to \pi_{m+n-1}(X,x_0),$$

$$pr_{2*}\colon \pi_{m+n-1}((X,x_0) \vee (Y,y_0),bp) \to \pi_{m+n-1}(Y,y_0),$$

and

$$in_*\colon \pi_{m+n-1}((X,x_0) \vee (Y,y_0),bp) \to \pi_{m+n-1}(X \times Y,bp).$$

This is a corollary of 1.5.

7. Generally speaking, the Whitehead product is not associative. This was already implicit in 4: if we let (as in Subsection 3.1)
$\alpha_1,\alpha_2,\alpha_3$ denote the elements $imm_{1*}(sph_1)$, $imm_{2*}(sph_1)$, $imm_{3*}(sph_1)$ of
$\pi_1((S^1,ort_1) \vee (S^1,ort_1) \vee (S^1,ort_1),bp)$, then $[[\alpha_1,\alpha_2],\alpha_3] =$
$= \alpha_3\alpha_2\alpha_1\alpha_2^{-1}\alpha_1^{-1}\alpha_3^{-1}\alpha_1\alpha_2\alpha_1^{-1}\alpha_2^{-1}$, whereas $[\alpha_1,[\alpha_2,\alpha_3]] =$
$= \alpha_3\alpha_2\alpha_3^{-1}\alpha_2^{-1}\alpha_1\alpha_2\alpha_3\alpha_2^{-1}\alpha_3^{-1}\alpha_1^{-1}$. A second example can be found in 4.2 below.

The Case of H-Spaces

8. **If** X **is a H-space, then** $[\alpha,\beta] = 0$ **for any** $x_0 \in X$,
$\alpha \in \pi_m(X,x_0)$, $\beta \in \pi_n(X,x_0)$.

(This generalizes Theorem 1.9.11.)

It is enough to consider the case where x_0 is the identity.
Let $f\colon (S^m,ort_1) \to (X,x_0)$ and $g\colon (S^n,ort_1) \to (X,x_0)$ be spheroids in
the classes α and β. Define $h\colon (S^m \times S^n,bp) \to (X,x_0)$ and spheroids
$f_1\colon (S^m,ort_1) \to (X,x_0)$, $g_1\colon (S^n,ort_1) \to (X,x_0)$ by $h(x,y) = f(x)g(y)$,
$f_1(y) = f(y)x_0$, $g_1(y) = g(y)x_0$. Obviously, f_1 and g_1 are homotopic
to f and g, while the map $(B,bp) \to (X,x_0)$ defined by f_1 and g_1
equals $h|_B = h \circ [in\colon B \to S^m \times S^n]$. Consequently, $[\alpha,\beta] =$
$= (h \circ in)_*(wd(m,n)) = h_* \circ in_*(wd(m,n))$. Since $in_*(wd(m,n)) = 0$,
we get $[\alpha,\beta] = 0$.

3. Applications

1. **Let** (X,x_0) **and** (Y,y_0) **be pointed spaces, and let**
k **and** 1 **be nonnegative integers. If** X **is k-connected,** Y **is**
1-connected, and (X,x_0), (Y,y_0) **are Borsuk pairs, then the kernel of**
the homomorphism

$$in_*: \pi_{k+l+1}((X,x_0) \vee (Y,y_0),bp) \to \pi_{k+l+1}(X \times Y,bp)$$

is generated, as a subgroup of $\pi_{k+l+1}((X,x_0) \vee (Y,y_0),bp)$, by the products $[imm_{1*}(\alpha),imm_{2*}(\beta)]$ with $\alpha \in \pi_{k+1}(X,x_0)$ and $\beta \in \pi_{l+1}(Y,y_0)$. (Cf. 3.5.5 and 4.3.13.)

PROOF. By 2.6, $[imm_{1*}(\alpha),imm_{2*}(\beta)] \in Ker\, in_*$ for all $\alpha \in \pi_{k+1}(X,x_0)$ and $\beta \in \pi_{l+1}(Y,y_0)$. To see that these products actually generate $Ker\, in_*$, note that 4.2.1, 4.2.9, and 2.3.3.2 together guarantee that it suffices to examine the case where (X,x_0) and (Y,y_0) are cellular spaces with $ske_k X = x_0$ and $ske_l Y = y_0$. Under these circumstances, $ske_{k+l+1}(X \times Y) \subset (X,x_0) \vee (Y,y_0)$, and the classes of the attaching maps of the $(k+l+2)$-cells in $(X \times Y) \smallsetminus [(X,x_0) \vee (Y,y_0)]$ are Whitehead products of the classes of the characteristic maps of the $(k+1)$-cells in $imm_1(X)$ and the classes of the charactetistic maps of the $(l+1)$-cells in $imm_2(Y)$ (this is an immediate consequence of Definition 2.1). Therefore, when $k > 0$ and $l > 0$, $Ker\, in_*$ is generated by the classes $[imm_{1*}(\alpha),imm_{2*}(\beta)]$ with $\alpha \in \pi_{k+1}(X,x_0)$ and $\beta \in \pi_{l+1}(Y,y_0)$ (see 3.2.4); when $k > 0$ and $l = 0$, $Ker\, in_*$ is generated by the classes

$$T_{imm_{2*}(\sigma)}[imm_{1*}(\alpha),imm_{2*}(\beta)]$$

with $\alpha \in \pi_{k+1}(X,x_0)$ and $\beta,\sigma \in \pi_1(Y,y_0)$ (see 3.2.4); when $k = 0$ and $l > 0$, $Ker\, in_*$ is generated by the classes

$$T_{imm_{1*}(\sigma)}[imm_{1*}(\alpha),imm_{2*}(\beta)]$$

with $\alpha,\sigma \in \pi_1(X,x_0)$ and $\beta \in \pi_{l+1}(Y,y_0)$ (see 3.2.4); finally, when $k = l = 0$, $Ker\, in_*$ is generated by the classes

$$T_{imm_{1*}(\sigma_1)imm_{2*}(\omega_1)\ldots imm_{1*}(\sigma_q)imm_{2*}(\omega_q)}[imm_{1*}(\alpha),imm_{2*}(\beta)]$$

with $\alpha,\sigma_1,\ldots,\sigma_q \in \pi_1(X,x_0)$ and $\beta,\omega_1,\ldots,\omega_q \in \pi_1(Y,y_0)$ (see Subsection 3.3). Now all it remains is to observe that

(i) $T_{imm_{1*}(\sigma)}[imm_{1*}(\alpha),imm_{2*}(\beta)] =$
$$= -[imm_{1*}(\sigma),imm_{2*}(\beta)] + [imm_{1*}(\alpha\sigma),imm_{2*}(\beta)]$$

for any $\alpha,\sigma \in \pi_1(X,x_0)$, $\beta \in \pi_{l+1}(Y,y_0)$ with $l > 0$;

(ii) $T_{imm_{2*}(\sigma)}[imm_{1*}(\alpha),imm_{2*}(\beta)] =$

486

$$= -[\mathrm{imm}_{1*}(\alpha),\mathrm{imm}_{2*}(\sigma)] + [\mathrm{imm}_{1*}(\alpha),\mathrm{imm}_{2*}(\beta\sigma)]$$

for any $\alpha \in \pi_{k+1}(X,x_0)$, $\beta,\sigma \in \pi_1(Y,y_0)$ with $k > 0$; and

$$\text{(iii)} \quad T_{\mathrm{imm}_{1*}(\sigma)\,\mathrm{imm}_{2*}(\omega)}\,[\mathrm{imm}_{1*}(\alpha),\mathrm{imm}_{2*}(\beta)] =$$

$$= [\mathrm{imm}_{1*}(\sigma),\mathrm{imm}_{2*}(\omega\beta)]^{-1}[\mathrm{imm}_{1*}(\sigma\alpha),\mathrm{imm}_{2*}(\omega\beta)][\mathrm{imm}_{1*}(\sigma\alpha),\mathrm{imm}_{2*}(\omega)]^{-1}[\mathrm{imm}_{1*}(\sigma),\mathrm{imm}_{2*}(\omega)]$$

for any $\alpha,\sigma \in \pi_1(X,x_0)$, $\beta,\omega \in \pi_1(Y,y_0)$ (see 2.4).

2. <u>The class</u> $[\mathrm{sph}_n,\mathrm{sph}_n]$ <u>has infinite order in</u> $\pi_{2n-1}(S^n)$, <u>for every even positive integer</u> n. <u>In particular, the groups</u> $\pi_{4k-1}(S^{2k})$ <u>with</u> $k \geqslant 1$ <u>are infinite.</u>

(Cf. Subsections 2.2 and 2.4.)

PROOF. Since

$$\mu_*[\mathrm{sph}_n,\mathrm{sph}_n] = [\mu_*(\mathrm{sph}_n),\mu_*(\mathrm{sph}_n)] =$$

$$= [\mathrm{imm}_{1*}(\mathrm{sph}_n) + \mathrm{imm}_{2*}(\mathrm{sph}_n),\mathrm{imm}_{1*}(\mathrm{sph}_n) + \mathrm{imm}_{2*}(\mathrm{sph}_n)] =$$

$$= [\mathrm{imm}_{1*}(\mathrm{sph}_n),\mathrm{imm}_{1*}(\mathrm{sph}_n)] + [\mathrm{imm}_{2*}(\mathrm{sph}_n),\mathrm{imm}_{2*}(\mathrm{sph}_n)] +$$

$$+ 2[\mathrm{imm}_{1*}(\mathrm{sph}_n),\mathrm{imm}_{2*}(\mathrm{sph}_n)] =$$

$$= \mathrm{imm}_{1*}[\mathrm{sph}_n,\mathrm{sph}_n] + \mathrm{imm}_{2*}[\mathrm{sph}_n,\mathrm{sph}_n] + 2\,\mathrm{wd}(n,n)\,,$$

we obtain

$$2\,\mathrm{wd}(n,n) = \mu_*[\mathrm{sph}_n,\mathrm{sph}_n] - \mathrm{imm}_{1*}[\mathrm{sph}_n,\mathrm{sph}_n] - \mathrm{imm}_{2*}[\mathrm{sph}_n,\mathrm{sph}_n].$$

Now assuming that $[\mathrm{sph}_n,\mathrm{sph}_n]$ has finite order, the class $\mathrm{wd}(n,n)$ would have finite order too, contradicting 1.2.

3. <u>The kernel of</u> $\mathrm{su}\colon \pi_{2n-1}(S^n) \to \pi_{2n}(S^{2n+1})$ <u>is infinite for every positive integer</u> n.

This is a corollary of 2 and 2.5.

4.   Exercises

1. Compute the third homotopy group of a bouquet of two--dimensional spheres.

2. Show that if

$$\alpha = \mathrm{imm}_{1*}(\mathrm{sph}_1) \in \pi_1(B(1,2),\mathrm{bp})$$

and

$$\beta = \gamma = \text{imm}_{2*}(\text{sph}_2) \in \pi_2(B(1,2),bp),$$

then

$$[[\alpha,\beta],\gamma] \neq [\alpha,[\beta,\gamma]].$$

3. Show that

$$(-1)^{pm}[[\alpha,\beta],\gamma] + (-1)^{mn}[[\beta,\gamma],\alpha] + (-1)^{np}[[\gamma,\alpha],\beta] = 0,$$

for any $\alpha \in \pi_m(X,x_0)$, $\beta \in \pi_n(X,x_0)$, and $\gamma \in \pi_p(X,x_0)$ with $m > 1$, $n > 1$, and $p > 1$.

§6. CONTINUATION OF THE THEORY OF BUNDLES

1. Weak Homotopy Equivalence and Steenrod Bundles

1. Two Steenrod bundles, ξ_1 and ξ_2, with the same standard fiber F, are said to be k-<u>equivalent</u> if there exist a cellular space B and two k-equivalences, $\phi_1: B \to bs\,\xi_1$ and $\phi_2: B \to bs\,\xi_2$, such that the bundles $\phi_1^!\xi_1$ and $\phi_2^!\xi_2$ are F-equivalent. Here $0 \leqslant k \leqslant \infty$; the most important case is $k = \infty$, and two ∞-equivalent Steenrod bundles are also referred to as <u>weakly homotopy equivalent.</u>

Clearly, every bundle induced from a Steenrod bundle by a k-equivalence is k-equivalent with the original bundle. Moreover, bases of weakly homotopy equivalent Steenrod bundles are weakly homotopy equivalent, and by Lemma 1.5.9 this holds true for their total spaces too. [To see this, consider the maps $\text{ad}\,\phi_1: \phi_1^!\xi_1 \to \xi_1$ and $\text{ad}\,\phi_2: \phi_2^!\xi_2 \to \xi_2$ and the homomorphisms that they induce from the homotopy sequences of the bundles $\phi_1^!\xi_1$ and $\phi_2^!\xi_2$ into the homotopy sequences of the bundles ξ_1 and ξ_2, respectively, and apply to these homomorphisms Lemma 1.5.19.]

2. <u>Let ζ_1 and ζ_2 be Steenrod bundles with the same standard fiber F. Suppose that ζ_1 is universal. Then ζ_2 is universal if and only if it is weakly homotopy equivalent to ζ_1.</u>

We first show that the condition is sufficient, i.e., that given any Steenrod bundle ξ with standard fiber F and cellular base, any (cellular) subspace B of $bs\,\xi$, and any continuous map

$\phi: B \to bs\, \zeta_2$ such that $\phi^!\zeta_2$ is F-equivalent to $\xi|_B$, there exists a continuous map $\psi: bs\, \xi \to bs\, \zeta_2$ such that $\psi|_B = \phi$ and the bundle $\psi^!\zeta_2$ is F-equivalent to ξ (see 4.4.2.2). By 1, we can produce a cellular space K together with weak homotopy equivalences, $f_1: K \to bs\, \zeta_1$ and $f_2: K \to bs\, \zeta_2$, such that $f_1^!\zeta_1$ and $f_2^!\zeta_2$ are F-equivalent. Using 4.1.2, there is a continuous $g: B \to K$ such that $f_2 \circ g$ is homotopic to ϕ. Therefore, in the chain of F-bundles:

$$(f_1 \circ g)^!\zeta_1 = g^!(f_1^!\zeta_1), \quad g^!(f_2^!\zeta_2) = (f_2 \circ g)^!\zeta_2, \quad \phi^!\zeta_2, \quad \xi|_B, \quad \text{the}$$

adjacent bundles are F-equivalent, and hence so are $(f_1 \circ g)^!\zeta_1$ and $\xi|_B$. Since ζ_1 is universal, there is a continuous $h: bs\, \xi \to bs\, \zeta_1$ such that $h|_B = f_1 \circ g$ and the bundles $h^!\zeta_1$ and ξ are F-equivalent (see again 4.4.2.2). Further, since f is a weak homotopy equivalence, there is a continuous $k: bs\, \xi \to K$ such that $k|_B = g$ and $f_1 \circ k$, h are homotopic (see 4.1.2). The restriction of $f_1 \circ k: bs\, \xi \to bs\, \zeta_2$ to B equals $f_2 \circ g$, and hence is homotopic to ϕ. Consequently, there is a continuous map $\psi: bs\, \xi \to bs\, \zeta_2$ homotopic to $f_2 \circ k$ such that $\psi|_B = \phi$. Now the fact that the adjacent bundles in the chain $\psi^!\zeta_2$,

$$(f_2 \circ k)^!\zeta_2 = k^!(f_2^!\zeta_2), \quad k^!(f_1^!\zeta_1) = (f_1 \circ k)^!\zeta_1, \quad h^!\zeta_1, \quad \xi, \quad \text{are}$$

F-equivalent implies the F-equivalence of the bundles $\psi^!\zeta_2$ and ξ.

We may use what we just proved to show that the condition of the theorem is also necessary. In fact, assume that ζ_1 and ζ_2 are universal, and let (K, ϕ_1) and (K, ϕ_2) be cellular approximations of $bs\, \zeta_1$ and $bs\, \zeta_2$. Since $\phi_1^!\zeta_1$ and $\phi_2^!\zeta_2$ are weakly homotopy equivalent, they are both universal. Hence, K_1 and K_2 are homotopy equivalent (see 4.4.2.4).

3 (COROLLARY). <u>If X and Y are classifying spaces of the same topological group, then X and Y are weakly homotopy equivalent.</u>

4. <u>Given any topological group G, there is a universal G-bundle with cellular base.</u>

This follows from 4.4.3.4, 4.3.2, and 2.

5. <u>A principal bundle is universal if and only if its total space is ∞-connected.</u>

PROOF. First, let ξ be a universal G-bundle. Applying 2, ξ is weakly homotopy equivalent to $Mi\, G$. Since $tl\, Mi\, G$ is ∞-connected (see 2.3.3.10, 4.2.7, and 1.11.2), $tl\, \xi$ is also ∞-connected (see 1).

Now let ξ be a Steenrod G-bundle with ∞-connected total space $tl\, \xi$. Pick a cellular approximation (K, ϕ) of $bs\, \xi$ and consider the G-bundle $\phi^!\xi$. Since the bundle $Mi\, G$ is universal, there is a continuous $\psi: K \to bs\, Mi\, G$ such that $\psi^!\xi$ is G-equivalent to

$\psi^! \mathrm{Mi}\, G$. Furthermore, $\mathrm{tl}\, \mathrm{Mi}\, G$ and $\mathrm{tl}(\phi^! \xi)$ are ∞-connected, and the latter implies the ∞-connectedness of $\mathrm{tl}(\psi^! \mathrm{Mi}\, G)$. Applying Lemma 1.5.19 (to the homomorphism from the homotopy sequence of the bundle $\psi^! \mathrm{Mi}\, G$ into the homotopy sequence of the bundle $\mathrm{Mi}\, G$ induced by the map $\mathrm{ad}\,\psi : \psi^! \mathrm{Mi}\, G \to \mathrm{Mi}\, G$), we see that ψ is a weak homotopy equivalence. Thus, ξ is weakly homotopy equivalent to $\mathrm{Mi}\, G$, and so ξ is universal (see 2).

6. <u>If</u> X <u>is a classifying space of the topological group</u> G, <u>then for any</u> $r \geqslant 1$ <u>the groups</u> $\pi_r(X)$ <u>and</u> $\pi_{r-1}(G)$ <u>are isomorphic.</u>

Indeed, by Theorem 5, the homomorphisms Δ figuring in the homotopy sequence of the universal G-bundle with base X are isomorphisms (see 1.8.7).

7. <u>If</u> X_1 <u>and</u> X_2 <u>are classifying spaces of the topological groups</u> G_1 <u>and</u> G_2, <u>then</u> $X_1 \times X_2$ <u>is a classifying space of</u> $G_1 \times G_2$.

PROOF. Given a universal G_1-bundle ξ_1 and a universal G_2-bundle ξ_2, it is enough to verify that the $(G_1 \times G_2)$-bundle $\xi_1 \times \xi_2$ is universal. But this follows from 5 and 1.1.9.

8. Theorems 2 and 5 carry over to k-universal bundles. Thus, if ζ_1 and ζ_2 are Steenrod bundles with the same standard fiber and ζ_1 is k-universal, then ζ_2 is k-universal if and only if it is k-equivalent to ζ_1, and a principal bundle is k-universal if and only if its total space is k-connected. The proofs repeat those of Theorems 2 and 5, with obvious modifications.

An Application: Universal Principal
Bundles for Finitely Generated Abelian Groups

9. <u>For arbitrary</u> n <u>and</u> $m_1, \ldots, m_1$,

$$(\underbrace{\mathbb{R} \times \ldots \times \mathbb{R}}_{n} \times \underbrace{S^\infty \times \ldots \times S^\infty}_{1}, \underbrace{\mathrm{hel} \times \ldots \times \mathrm{hel}}_{n} \times \underbrace{\mathrm{pr} \times \ldots \times \mathrm{pr}}_{1},$$

$$S^1 \times \ldots \times S^1 \times L(m_1) \times \ldots \times L(m_1))$$

<u>is a universal</u> $(\mathbb{Z}^n \oplus \mathbb{Z}_{m_1} \ldots \oplus \mathbb{Z}_{m_1})$-<u>bundle . In particular,</u> S^1 <u>is a classifying space for</u> $\mathbb{Z}$, <u>while</u> L(m) <u>is a classifying space for</u> $\mathbb{Z}_m$.

Since $\mathbb{R}$ and S^∞ are ∞-connected, this follows immediately from 5 and 1.1.9.

2. Theory of Coverings

1. The main achievement of the present section is a clear enumeration of the covering spaces of nonpathological connected spaces and a criterion for the equivalence of two given coverings. Our instrument will be the fundamental group. The analysis is elementary enough: in fact, all we use from the whole theory of bundles may be concentrated in the following two theorems, where it is assumed that a covering ξ together with a base point $x \in tl\,\xi$ are given. Then:

(i) every path in $bs\,\xi$ with origin $pr\,\xi(x)$ is covered by one and only one path in $tl\,\xi$ with origin x; and

(ii) if two paths in $bs\,\xi$ with common origin $pr\,\xi(x)$ are homotopic, then the paths in $tl\,\xi$ which cover them are also homotopic and, in particular, have the same end.

(These assertions are straightforward corollaries of Theorems 4.1.3.6 and 4.1.3.8.)

It is true that here and there we do refer to other facts from the theory of bundles (for example, to Theorem 1.8.12), but all these can be readily deduced from (i) and (ii).

At the heart of the theory of coverings lies the definition of the <u>group of a covering.</u> Recall that, according to 1.8.12, $pr_*\colon \pi_1(tl\,\xi,x) \to \pi_1(bs\,\xi,pr\,\xi(x))$ is a monomorphism for any covering ξ with base point $x \in tl\,\xi$. We call the image of pr_* the <u>group of the covering</u> ξ <u>with base point</u> x, and denote it by $gr\,\xi(x)$.

2. <u>Let $pr\,\xi(x_1) = pr\,\xi(x_0)$ and let s be a path in $tl\,\xi$ beginning at x_0 and ending at x_1. Then $gr\,\xi(x_1) = \sigma[gr\,\xi(x_0)]\sigma^{-1}$, where σ is the homotopy class of the loop $pr\,\xi \circ s$. In particular, if $pr\,\xi(x_1) = pr\,\xi(x_0)$, then $gr\,\xi(x_0)$ and $gr\,\xi(x_1)$ are conjugate subgroups of $\pi_1(bs\,\xi,x_0)$. The converse is also true: the groups $gr\,\xi(x)$ with $x \in pr\,\xi^{-1}(pr\,\xi(x_0))$ exhaust the subgroups of $\pi_1(bs\,\xi,pr\,\xi(x_0))$ which are conjugate to $gr\,\xi(x_0)$.</u>

In fact, $gr\,\xi(x_1) = pr\,\xi_*(\pi_1(tl\xi,x_1)) = pr\,\xi_* \circ T_s(\pi_1(tl\,\xi,x_0))$
$= T_{pr\,\xi\circ s} \circ pr\,\xi_*(\pi_1(tl\,\xi,x_0)) = T_{pr\,\xi\circ s}(gr\,\xi(x_0)) = \sigma[gr\,\xi(x_0)]\sigma^{-1}$.
These equalities demonstrate that for every $\sigma \in \pi_1(bs\,\xi,pr\,\xi(x_0))$ the group $\sigma[gr\,\xi(x_0)]\sigma^{-1}$ equals $gr\,\xi(s(1))$, where s is a path in $tl\,\xi$ with origin x_0 and which covers some path in the class σ.

The Hierarchy of Coverings

3. We say that the covering ξ with base point $x_0 \in \mathrm{tl}\,\xi$ is <u>subordinate</u> to the covering ξ' with base point $x_0' \in \mathrm{tl}\,\xi'$, if $\mathrm{bs}\,\xi' = \mathrm{bs}\,\xi$ and there exists a continuous map $\phi: \xi' \to \xi$ such that $\mathrm{bs}\,\phi = \mathrm{id}\,\mathrm{bs}\,\xi$ and $\mathrm{tl}\,\phi(x_0') = x_0$. In this case, the map ϕ is called a <u>subordination.</u>

Obviously, if $\phi: \xi' \to \xi$ is a subordination, then $(\mathrm{tl}\,\xi', \mathrm{tl}\,\phi, \mathrm{tl}\,\xi)$ is a covering.

4. <u>If a subordination exists, then it is unique.</u>

PROOF. Suppose that the covering ξ with base point x_0 is subordinate to the covering ξ' with base point x_0', and let ϕ and ψ be two such subordinations. Then, if $x' \in \mathrm{tl}\,\xi'$ is arbitrary and $s: I \to \mathrm{tl}\,\xi'$ is such that $s(0) = x_0'$, $s(1) = x'$, then the paths $\mathrm{tl}\,\phi \circ s$ and $\mathrm{tl}\,\psi \circ s$ cover the same path, $\mathrm{pr}\,\xi' \circ s$, in $\mathrm{bs}\,\xi$ and have the same origin. Therefore, $\mathrm{tl}\,\phi \circ s = \mathrm{tl}\,\psi \circ s$ and $\mathrm{tl}\,\phi(x') = \mathrm{tl}\,\phi \circ s(1) = \mathrm{tl}\,\psi \circ s(1) = \mathrm{tl}\,\psi(x')$.

5. <u>If two coverings with base points are mutually subordinate, then the corresponding subordinations are equivalences which are inverses of one another.</u>

PROOF. Let $\phi: \xi' \to \xi$ and $\phi': \xi \to \xi'$ be subordinations. Then $\phi' \circ \phi: \xi' \to \xi'$ and $\phi \circ \phi': \xi \to \xi$ are also subordinations. By 4, $\phi' \circ \phi = \mathrm{id}\,\xi'$ and $\phi \circ \phi' = \mathrm{id}\,\xi$.

6 (LEMMA). <u>Let ξ and ξ' be coverings with base points $x_0 \in \mathrm{tl}\,\xi$ and $x_0' \in \mathrm{tl}\,\xi'$, and such that $\mathrm{bs}\,\xi' = \mathrm{bs}\,\xi$, $\mathrm{pr}\,\xi'(x_0') = \mathrm{pr}\,\xi(x_0)$, and $\mathrm{gr}\,\xi'(x_0') \subset \mathrm{gr}\,\xi(x_0)$. If the paths $s_1', s_2': I \to \mathrm{tl}\,\xi'$ have the common origin x_0' and a common end, then the paths in $\mathrm{tl}\,\xi$ which cover $\mathrm{pr}\,\xi' \circ s_1'$ and $\mathrm{pr}\,\xi' \circ s_2'$ and have origin x_0 also have a common end.</u>

PROOF. Let s_1 and s_2 be the path in $\mathrm{tl}\,\xi$ with origin x_0 which cover $\mathrm{pr}\,\xi' \circ s_1'$ and $\mathrm{pr}\,\xi' \circ s_2'$. The class of the loop $(\mathrm{pr}\,\xi' \circ s_1')(\mathrm{pr}\,\xi' \circ s_2')^{-1}$ belongs to $\mathrm{gr}\,\xi'(x_0')$, and hence to $\mathrm{gr}\,\xi(x_0)$. Consequently, the path in $\mathrm{tl}\,\xi$ with origin x_0 which covers this loop is closed; moreover, it can be expressed as the product of a path which covers $\mathrm{pr}\,\xi' \circ s_1'$ (and which, having origin x_0, coincides with s_1) with a path which covers $(\mathrm{pr}\,\xi' \circ s_2')^{-1}$ (and which, having origin x_0, coincides with s_2^{-1}). Therefore, s_1 and s_2 have the same end.

7. <u>Let</u> ξ <u>and</u> ξ' <u>be coverings with base points</u> $x_0 \in tl\,\xi$ <u>and</u> $x_0' \in tl\,\xi'$, <u>such that</u> $bs\,\xi' = bs\,\xi$ <u>and</u> $pr\,\xi'(x_0') = pr\,\xi(x_0)$.

(i) <u>If</u> ξ <u>is subordinate to</u> ξ', <u>then</u> $gr\,\xi'(x_0') \subset gr\,\xi(x_0)$;

(ii) <u>If</u> $gr\,\xi'(x_0') \subset gr\,\xi(x_0)$ <u>and</u> $bs\,\xi$ <u>is locally connected,</u> <u>then</u> ξ <u>is subordinate to</u> ξ'.

Assertion (i) is trivial, so let us prove (ii). For each point $x' \in tl\,\xi'$, consider the common end of all paths in $tl\,\xi$ which start at x_0 and cover paths of the form $pr\,\xi' \circ u'$, where u' is a path in $tl\,\xi'$ such that $u'(0) = x_0'$, $u'(1) = x'$ (see 6). This defines a map $F: tl\,\xi' \to tl\,\xi$ satisfying $pr\,\xi \circ F = pr\,\xi'$, and all we have to check is the continuity of F. So let $x' \in tl\,\xi'$ and U be a neighborhood of $F(x')$. We shall produce a neighborhood U' of x' such that $F(U') \subset U$.

To do this, let $W \subset bs\,\xi$ be a neighborhood of $pr\,\xi'(x')$ such that there are neighborhoods, V' of x' and V of $F(x')$, which are homeomorphically mapped by $pr\,\xi'$ and respectively $pr\,\xi$ onto W. Since $bs\,\xi$ is locally connected, we can find a neighborhood $W_1 \subset W$ of $pr\,\xi'(x')$ such that every point of W_1 can be joined to $pr\,\xi'(x')$ by a path in W. Now set $V_1' = V' \cap (pr\,\xi')^{-1}(W_1)$, $V_1 = V \cap pr\,\xi^{-1}(W_1)$, and $U' = V_1' \cap (pr\,\xi')^{-1}(pr\,\xi(U))$. Obviously, $x' \in U'$. Let us check that $F(U') \subset U$.

Let $y' \in U'$. Since $ab\,pr\,\xi': V' \to W$ is a homeomorphism, there is a path $v': I \to tl\,\xi'$ such that $v'(I) \subset V'$ and $v'(0) = x'$, $v'(1) = y'$. Define $v: I \to tl\,\xi$ by

$$v(t) = (pr\,\xi\big|_V)^{-1}(pr\,\xi' \circ v'(t)),$$

pick a path $u': I \to tl\,\xi'$ with the property that $u'(0) = x_0'$, $u'(1) = x'$, and consider the path $u: I \to tl\,\xi$ with $u(0) = x_0$, $u(1) = F(x')$, and u covers the path $pr\,\xi' \circ u'$. Clearly, $(u'v')(0) = x_0$, $(u'v')(1) = y'$, while the path uv covers $pr\,\xi' \circ (u'v')$ and $(uv)(0) = x_0$. Therefore, $(uv)(1) = F(y')$, and hence $F(y') = v(1) = (pr\,\xi\big|_V)^{-1}(pr\,\xi'(y')) \in V \cap pr\,\xi^{-1}(pr\,\xi'(U')) \subset U$.

8 (COROLLARY). <u>Two coverings,</u> ξ <u>and</u> ξ', <u>with</u> $bs\,\xi' = bs\,\xi$ <u>a locally connected space, are equivalent if and only if for some points</u> $x_0 \in tl\,\xi$, $x_0' \in tl\,\xi'$, <u>such that</u> $pr\,\xi'(x_0') = pr\,\xi(x_0)$, <u>the groups</u> $gr\,\xi(x_0)$ <u>and</u> $gr\,\xi'(x_0')$ <u>are conjugate in</u> $\pi_1(bs\,\xi, pr\,\xi(x_0))$.

The Group of Automorphisms of a Covering

9. As with the automorphisms (i.e., self-equivalences) of an arbitrary bundle ξ, the automorphisms of a covering ξ form a group $\mathrm{Aut}\,\xi$.* Its structure is described in Theorem 10 below, where x_0 designates, as usual, a base point in $\mathrm{tl}\,\xi$, and two new notations are used. Namely, let ev be the map from $\mathrm{Aut}\,\xi$ into the fiber $\mathrm{pr}\,\xi^{-1}(\mathrm{pr}\,\xi(x_0))$ given by $\mathrm{ev}(\phi) = \mathrm{tl}\,\phi(x_0)$, and let Reg be the set of all points $x \in \mathrm{pr}\,\xi^{-1}(\mathrm{pr}\,\xi(x_0))$ such that $\mathrm{gr}\,\xi(x) = \mathrm{gr}\,\xi(x_0)$. From 2 it follows that the preimage of Reg under the map
$$\Delta: \pi_1(\mathrm{bs}\,\xi, \mathrm{pr}\,\xi(x_0)) \to \pi_0(\mathrm{pr}\,\xi^{-1}(\mathrm{pr}\,\xi(x_0)), x_0) = \mathrm{pr}\,\xi^{-1}(\mathrm{pr}\,\xi(x_0)) \quad \text{is}$$
nothing else but the normalizer $\mathrm{Nr}(\mathrm{gr}\,\xi(x_0))$ of the group $\mathrm{gr}\,\xi(x_0)$ in $\pi_1(\mathrm{bs}\,\xi, \mathrm{pr}\,\xi(x_0))$. [Recall that, given a subgroup H of a group G, its normalizer $\mathrm{Nr}(H)$ is the set of all $g \in G$ such that $gHg^{-1} = H$; $\mathrm{Nr}(H)$ is a subgroup of G and contains H as a normal subgroup.] Therefore, Δ induces an invertible map

$$\mathrm{fact\ ab}\ \Delta: \mathrm{Nr}(\mathrm{gr}\,\xi(x_0))/\mathrm{gr}\,\xi(x_0) \to \mathrm{Reg}.$$

10. ev <u>is injective,</u> $\mathrm{ev}(\mathrm{Aut}\,\xi) \subset \mathrm{Reg}$, <u>and the composition</u>

$$\mathrm{Aut} \xrightarrow{\ \mathrm{ab}\ \mathrm{ev}\ } \mathrm{Reg} \xrightarrow{\ (\mathrm{fact\ ab}\ \Delta)^{-1}\ } \mathrm{Nr}(\mathrm{gr}\,\xi(x_0))/\mathrm{gr}\,\xi(x_0)$$

<u>is an antihomomorphism. If the base</u> $\mathrm{bs}\,\xi$ <u>is locally connected, then</u> $\mathrm{ev}(\mathrm{Aut}\,\xi) = \mathrm{Reg}$, <u>and hence the group</u> $\mathrm{Aut}\,\xi$ <u>is antiisomorphic to</u> $\mathrm{Nr}(\mathrm{gr}\,\xi(x_0))/\mathrm{gr}\,\xi(x_0)$. <u>If, in addition,</u> ξ <u>is regular (see 11 below),</u> <u>then</u> $\mathrm{Nr}(\mathrm{gr}\,\xi(x_0)) = \pi_1(\mathrm{bs}\,\xi, \mathrm{pr}\,\xi(x_0))$ <u>and</u> $\mathrm{Aut}\,\xi$ <u>is antiisomorphic to</u> $\pi_1(\mathrm{bs}\,\xi, \mathrm{pr}\,\xi(x_0))/\mathrm{gr}\,\xi(x_0)$.

PROOF. Since every automorphism $\phi \in \mathrm{Aut}\,\xi$ may be thought of as a subordination of the covering ξ with base point $\mathrm{tl}\,\phi(x_0)$ to the covering ξ with base point x_0, the injectivity of ev is immediate from 5. Similarly, the inclusion $\mathrm{ev}(\mathrm{Aut}\,\xi) \subset \mathrm{Reg}$ follows from part (i) of Theorem 7, while the equality $\mathrm{ev}(\mathrm{Aut}\,\xi) = \mathrm{Reg}$ results, when $\mathrm{bs}\,\xi$ is locally connected, from part (ii) of the same theorem. Finally, that $(\mathrm{fact\ ab}\ \Delta)^{-1} \circ \mathrm{ev}$ is an antihomomorphism is plain.

Regular Coverings

11.   A covering ξ is <u>regular</u> if for some point $x_0 \in tl\,\xi$, $gr\,\xi(x_0)$ is a normal subgroup of $\pi_1(bs\,\xi, pr\,\xi(x_0))$.   In this case $gr\,\xi(x)$ is a normal subgroup of $\pi_1(bs\,\xi, pr\,\xi(x))$ for all $x \in tl\,\xi$, as seen from 2.

If ξ is regular, then again using 2, the groups $gr\,\xi(x)$ with $x \in pr\,\xi^{-1}(b)$ are all equal for each fixed $b \in bs\,\xi$. ,

We remark that every two-sheeted covering is regular (because every subgroup of index 2 is normal). More examples of regular coverings are $(\mathbb{R}, hel, S^1)$, (S^1, hel_m, S^1) [see 4.1.2.6], $(S^{2n-1}, pr, L(m; \ell_1, \ldots, \ell_n))$ [see 4.2.3.15], and $(S^3, pr, S^3/G\tilde{\,}P)$, where P is a tetrahedron, a cube, or a dodecahedron [see 4.2.3.17]. An example of a nonregular covering is given in Fig. 17 (where the two points marked A are identified, as are the two points marked B); its nonregularity is a result of the following theorem.

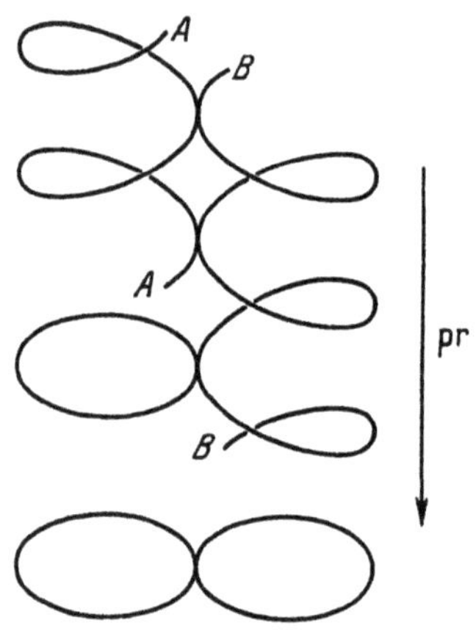

Fig. 17

12.   <u>A covering ξ is regular if and only if there is a point $x_0 \in tl\,\xi$ with the property that given any loop $s: I \to tl\,\xi$ with $s(0) = x_0$ and any point $x \in pr\,\xi^{-1}(pr\,\xi(x_0))$, the path with origin x and covering $pr\,\xi \circ s$ is closed.</u>

Indeed, the last condition means that each element of $gr\,\xi(x_0)$ also belongs to every group $gr\,\xi(x)$ with $x \in pr\,\xi^{-1}(pr\,\xi(x_0))$, and this implies that every subgroup of $\pi_1(bs\,\xi, pr\,\xi(x_0))$ which is conjugate with $gr\,\xi(x_0)$ actually equals $gr\,\xi(x_0)$ [see 2].

13. _A covering_ ξ _is regular if and only if it is equivalent to a principal bundle. The structure group of such a bundle is discrete and isomorphic to_ $G = \pi_1(\text{bs}\,\xi, \text{pr}\,\xi(x_0))/\text{gr}\,\xi(x_0)$. _Two Steenrod G-bundles which are equivalent to_ ξ _are themselves G-equivalent._

PROOF. By 4.3.2.10 and 4.3.2.11, ξ is equivalent to a Steenrod G-bundle if and only if there exists a free, continuous right action of G on $\text{tl}\,\xi$ whose orbits are precisely the fibers of ξ; in particular, G is necessarily discrete. Moreover, the transformation of $\text{tl}\,\xi$ determined by such an action yield the automorphisms of ξ. Finally, by 9, such an action exists if and only if ξ is regular and G is antiisomorphic to $\text{Aut}\,\xi$, i.e., is isomorphic to the group $\pi_1(\text{bs}\,\xi, \text{pr}\,\xi(x_0))/\text{gr}\,\xi(x_0)$.

Existence of Coverings

14. A topological space X is said to be _semilocally simply connected_ if every point $x \in X$ has a neighborhood U such that $\text{in}_*: \pi_1(U,x) \to \pi_1(X,x)$ is trivial.

Obviously, the class of semilocally simply connected spaces contains all simply connected spaces and all locally contractible spaces [and, among the latter, all locally Euclidean spaces, all CNRS's (see 1.3.6.8), in particular, all cellular spaces (see 2.1.4.5)].

15. _Every space which has a simply connected covering space is semilocally simply connected._

In fact, let ξ be a covering projection with $\text{tl}\,\xi$ simply connected and let $b \in \text{bs}\,\xi$. Then the homomorphism $\text{in}_*: \pi_1(U,b) \to \to \pi_1(\text{bs}\,\xi, b)$ is trivial for every neighborhood U of b such that $\xi\big|_U$ is trivial.

16. _Let_ B _be a connected, locally connected, and semilocally simply connected space, let_ $b_0 \in B$, _and let_ π _be any subgroup of_ $\pi_1(B, b_0)$. _Then there exists a covering_ ξ _with base point_ $x_0 \in \text{tl}\,\xi$ _such that_ $\text{bs}\,\xi = B$, $\text{pr}\,\xi(x_0) = b_0$, _and_ $\text{gr}\,\xi(x_0) = \pi$.

PROOF. Consider $C(I, 0; B, b_0)$, identify any two paths s_1, s_2 in this set whenever $s_1(1) = s_2(1)$ and the homotopy class of $s_1 s_2^{-1}$ belongs to ξ, and denote the resulting quotient space by E. Further, given open subsets U of B and V of U, and any path $s: I \to B$, with $s(0) = b_0$, $s(1) \in V$, let $\text{Nb}(U,V;s)$ be the subset of E consisting of the equivalence classes of the paths sw with $w(I) \subset U$ and $w(1) \in V$. The sets $\text{Nb}(U,V;s)$ satisfy the conditions of

Proposition 1.1.1.9, and thus yield a base for a topology on E. [Information: this topology coincides with the quotient topology that E inherits as a quotient of the topological space $C(I,0;B,b_0)$.] The map $p: E \to B$ which takes each point from E into the common end of the paths which represent it is clearly continuous and open. We set $\xi = (E,p,B)$ and take for x_0 the point of E represented by the constant path. Then $p(x_0) = b_0$. Let us show that ξ is a covering and that $\operatorname{gr}\xi(x_0) = \pi$.

We first verify that ξ is a covering in the broad sense. Pick an arbitrary point $b \in B$ and find a neighborhood U of b such that the inclusion homomorphism $\pi_1(U,b) \to \pi_1(B,b)$ is trivial, and then a neighborhood V of b in U such that b can be joined to every point of V by a path in U. We claim that $\xi|_V$ is a trivial bundle with discrete fiber.

To see this, consider an arbitrary path $s: I \to B$ such that $s(0) = b_0$, $s(1) = b$, and for each coset $\alpha \in \pi_1(B,b_0)/\pi$ choose a loop $u_\alpha: I \to B$ representing some element of α. The sets $Nb(U,V;u_\alpha s)$ are open and pairwise disjoint [if the paths $(u_\alpha s)w$ and $(u_{\alpha_1} s)w_1$ with $w(I) \subset U$, $w_1(I) \subset U$, define the same point of E, then $\alpha = \alpha_1$: indeed, the loop $((u_\alpha s)w)((u_{\alpha_1} s)w_1)^{-1}$, and hence the loop $u_\alpha u_{\alpha_1}^{-1}$ homotopic to it, are elements of a class that belongs to π]. Also, the sets $Nb(U,V;u_\alpha s)$ cover $p^{-1}(V)$ [every path s' with $s'(0) = b_0$, $s'(1) \in V$, is homotopic to a path of the form $(us)w$, where u is a loop and $w(I) \subset U$: and example of such a path is $(((s'w^{-1})s^{-1})s)w$, where $w: I \to U$ is any path with $w(0) = b$, $w(1) = s'(1)$]. Finally, the maps $\operatorname{ab}p: Nb(U,V;u\,s) \to V$ are open (because p is) and invertible [if the paths $(u_\alpha s)w$, $(u_\alpha s)w_1$ with $w(I) \subset U$, $w_1(I) \subset U$, have the same end, then they are homotopic]; that is, $\operatorname{ab}p$ are homeomorphisms. Consequently, $\xi|_V$ is a trivial bundle with discrete fiber, as asserted.

The last thing to prove is that E is connected and that a path in E which has origin x_0 and covers a loop from a homotopy class $\sigma \in \pi_1(B,b_0)$ is closed if and only if $\sigma \in \pi$. Given $s: I \to B$ with $s(0) = b_0$, let $s^\sim: I \to E$ be the path which takes each $t \in I$ into the point of E represented by the path $\tau \mapsto s(t\tau)$. Clearly, $s^\sim(0) = x_0$, $s^\sim(1)$ is the point of E represented by s, and $s^\sim$ covers s. This has two consequences: the first is that every point of E can be joined to x_0 by a path, and the second that given a loop with origin b_0, the end of the path which starts at x_0 and covers it is precisely the point of E represented by the given loop. The first shows that E is connected, while the second implies that, given a loop with origin b_0, the path which covers it and has origin b_0 ends at x_0 if and only if the homotopy class of the given loop belongs to π.

17. The previous theorem completes the theory of coverings with a fixed base. Combined with Theorem 7, it establishes a one-to-one correspondence between the equivalence classes of coverings over a connected, locally connected, and semilocally simply connected space B with base point b_0 and the classes of conjugate subgroups of $\pi_1(B,b_0)$. It transforms the hierarchy of coverings into the usual, set-theoretic hierarchy of subgroups, and the normal subgroups correspond to the regular coverings. The covering corresponding to the trivial subgroup has a simply connected total space. Since every covering is subordinate to this one, it is called <u>universal.</u> (Warning: do not confuse this universality with the notion of universality defined in the theory of Steenrod bundles. Note, however, that among the universal Steenrod bundles one finds also universal coverings, namely the universal principal bundles with discrete structure groups; see 1.5 and cf. 18).

An Application:
Classifying Spaces of Discrete Groups

18. <u>In order that a connected topological space X be a classifying space of a given discrete group π, it is necessary that the groups $\pi_r(X)$ be trivial for all $r \geqslant 2$ and that $\pi_1(X)$ be isomorphic to π. If X is locally connected and semilocally simply connected, then this condition is also sufficient.</u>

Necessity results from 1.5, 13, and 1.9.15. Sufficiency results from 16, 1.8.12, and 1.5.

Covering Maps

19. <u>Let ξ and ξ' be coverings with base points $x_0 \in tl\,\xi$ and $x_0' \in tl\,\xi'$, and let $f: (bs\,\xi',pr\,\xi'(x_0')) \to (bs\,\xi,pr\,\xi(x_0))$ be a continuous map. Then the inclusion $f_*(gr\,\xi'(x_0')) \subset gr\,\xi(x_0)$ is a necessary condition for the existence of a continuous map $\phi: \xi' \to \xi$ such that $bs\,\phi = f$ and $tl\,\phi(x_0') = x_0$. If $bs\,\xi'$ is locally connected, then this condition is also sufficient. If such a ϕ exists, then it is unique.</u>

PROOF. Assume that such a ϕ exists. Then from the

498

commutative diagram (see 1.8.6)

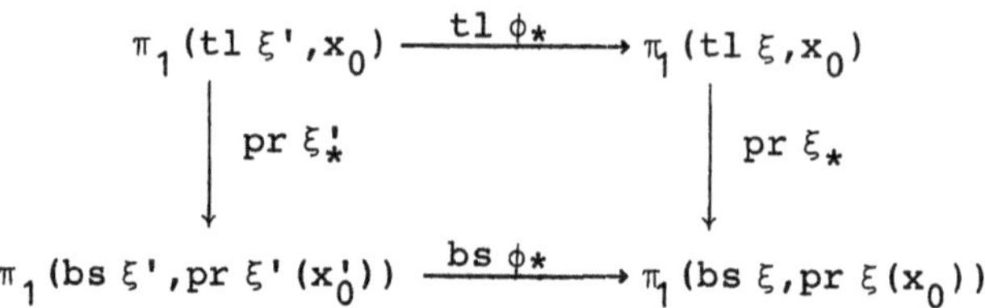

it follows that $f_*(gr\,\xi'(x_0')) \subset gr\,\xi(x_0)$. Now let us prove the converse.
Consider the bundle $f^!\xi$ together with the map $ad\,f: f^!\xi \to \xi$. Let
$y_0' \in (pr\,f^!\xi)^{-1}(pr\,\xi'(x_0'))$ be such that $tl\,ad\,f(y_0') = x_0$. Further, let
Y' be the component of $tl(f^!\xi)$ containing y_0', and let p' be the
restriction of $pr\,f^!\xi$ to Y'. Obviously, $(Y',p',bs\,\xi')$ is a covering,
$tl\,ad\,f$ and f combine to define a map $(Y',p',bs\,\xi') \to \xi$, and
$tl\,ad\,f|_{Y'}$ is injective on every fiber of $(Y',p',bs\,\xi')$. In the diagram

$$\begin{array}{ccccc}
\pi_1(Y',y_0') & \xrightarrow{p_*} & \pi_1(bs\,\xi',pr\,\xi'(x_0')) & \xrightarrow{\Delta} & \pi_0((p')^{-1}(pr\,\xi'(x_0')),y_0') \\
\downarrow{\scriptstyle ab\,tl\,ad\,f_*} & & \downarrow{\scriptstyle f_*} & & \downarrow{\scriptstyle ab\,tl\,ad\,f_*} \\
\pi_1(tl\,\xi,x_0) & \xrightarrow{pr\,\xi_*} & \pi_1(bs\,\xi,pr\,(x_0)) & \xrightarrow{\Delta} & \pi_0(pr\,\xi^{-1}(pr\,\xi(x_0)),x_0)
\end{array}$$

the rows are exact, while the left vertical homomorphism is monomorphic.
Consequently, $f_*^{-1}(Im(pr\,\xi_*)) = Im\,p_*$, whence $Im\,p_* \supset gr\,\xi'(x_0')$. Since
$bs\,\xi'$ is locally connected, the last inclusion shows that the covering
$(Y',p',bs\,\xi')$ with base point y_0' is subordinate to the covering ξ'
with base point x_0'. If ψ' is such a subordination, then
$(tl\,ad\,f|_{Y'}) \circ tl\,\psi'$ and f combine to define the desired map ϕ.

Finally, we claim that ϕ is unique. Indeed, suppose
$\phi_1: \xi' \to \xi$ is another map with $bs\,\phi_1 = f$ and $tl\,\phi_1(x_0') = x_0$. Then
for every point $x' \in tl\,\xi'$ and every path $s': I \to tl\,\xi'$ with $s'(0) =$
$= x_0'$, $s'(1) = x'$, the paths $tl\,\phi \circ s$ and $tl\,\phi_1 \circ s'$ cover
$f \circ pr\,\xi' \circ s'$ and have the same origin x_0. Therefore, in these
circumstances $tl\,\phi \circ s' = tl\,\phi_1 \circ s'$, and hence $tl\,\phi(x') = tl\,\phi \circ s'(1) =$
$= tl\,\phi_1 \circ s'(1) = tl\,\phi_1(x')$.

20. <u>Suppose</u> ξ <u>is a covering with base point</u> $x_0 \in tl\,\xi$,
Y <u>is a locally connected space with base point</u> y_0, <u>and</u> $f: (Y,y_0) \to$
$\to (bs\,\xi,pr\,\xi(x_0))$ <u>is continuous. If</u> $f_*(\pi_1(Y,y_0)) \subset gr\,\xi(x_0)$ [<u>in</u>
<u>particular, if</u> Y <u>is simply connected</u>], <u>then there is a map</u>
$F: (Y,y_0) \to (tl\,\xi,x_0)$ <u>which covers</u> f.

(Cf. 4.1.3.8.)

To see this, apply Theorem 19 to ξ and $\xi' = (Y, \mathrm{id}\, Y, Y)$.

Coverings and Additional Structures

21. There is an important general principle asserting that under favorable conditions an additional structure defined on the base of a given covering can be lifted to the covering space. To conclude our study of coverings, we apply this principle to three structures: differentiable, cellular, and simplicial. Further applications appear in Exercises 5.10 and 5.11.

Concerning differentiable structures, we restrict ourselves from the onset to manifolds, i.e., we assume that $\mathrm{bs}\,\xi$ is a C^r-manifold $(1 \leqslant r \leqslant a)$ and that the number of sheets of ξ is countable. We know that every chart $\phi \in \mathrm{Atl}\,\mathrm{bs}\,\xi$ such that the bundle $\xi\big|_{\mathrm{supp}\,\phi}$ is trivial has a family of copies in $\mathrm{tl}\,\xi$ which cover $\mathrm{pr}\,\xi^{-1}(\mathrm{supp}\,\phi)$. Obviously, these copies (taken for all $\phi \in \mathrm{Atl}\,\mathrm{bs}\,\xi$) constitute a C^r-atlas of the space $\mathrm{tl}\,\xi$. This atlas provides $\mathrm{tl}\,\xi$ with a C^r-structure and turns it into a C^r-manifold. The fundamental property of this lifted structure, and the one which defines it uniquely, is that relative to this structure $\mathrm{pr}\,\xi$ is a C^r-submersion, i.e., that ξ is a C^r-bundle. Let us add that the orientability of $\mathrm{bs}\,\xi$ implies the orientability of $\mathrm{tl}\,\xi$.

We can similarly lift a cellular structure. However, to lift the cells and their characteristic maps we need Theorem 20. As a result, the total space of the given covering with cellular base becomes a cellular space (rigged whenever the base is rigged), and the projection becomes a cellular map.

Lifting a simplicial structure may be viewed as a special case of lifting a cellular one. The total space of a covering with simplicial base thus becomes a simplicial space (ordered whenever the base is so), and the projection becomes a simplicial map.

22. In all the three cases considered above, the lifted structure is invariant under the automorphism of the given covering. It is clear, at least if the covering ξ is regular, that every differentiable or cellular structure defined on $\mathrm{tl}\,\xi$ and invariant under the automorphisms of ξ can be lowered to $\mathrm{bs}\,\xi$, i.e., the given structure on $\mathrm{tl}\,\xi$ is the result of lifting a similar structure from $\mathrm{bs}\,\xi$ (this becomes true also for simplicial structures after we effect the barycentric subdivision twice). For example, the lenses and the quotient spaces of S^3 by the binary tetrahedral, cube, or icosahedral groups (see 4.2.3.17) are regularly covered by the corresponding spheres

(see 4.3.2.11 and 13), and hence are C^a-manifolds.

If the given covering ξ is a C^r-bundle, then the C^r-structure on $tl\,\xi$ is automatically invariant under the automorphisms of ξ. In this case, the lowered C^r-structure surely coincides with the original C^r-structure on $bs\,\xi$. This is exactly what happens to all the coverings described in 4.1.2.6 and, in particular, to $(\mathbb{R},hel,S^1)$, (S^1,hel_m,S^1), and $(S^n,pr,\mathbb{R}P^n)$.

3. Orientations

1. In the present subsection the technique developed in the previous one is applied to the orientability problems considered in §§ 3.1 and 4.5 (see Subsections 3.1.3, 4.5.1, 4.5.4). We follow the recipe of Subsection 4.5.1, which applies to bundles over arbitrary spaces, in constrast to the recipe of Subsection 4.5.4, which yields the same results, but only for bundles over a cellular base.

2. Let ξ be an n-dimensional real vector bundle. Consider the associated principal bundle, $asso(\xi,GL(n,\mathbb{R}))$, and construct the orbit space of the right action of $GL_+(n,\mathbb{R})$ on $tl\,asso(\xi,GL(n,\mathbb{R}))$ obtained by the restriction to $GL_+(n,\mathbb{R})$ of the canonical right action of $GL(n,\mathbb{R})$ on $tl\,asso(\xi,GL(n,\mathbb{R}))$ [see 4.3.2.10]. Denote this orbit space by $bs_+\xi$, and observe that it is the total space of a bundle with base $bs\,\xi$ and projection $fact\,pr\,asso(\xi,GL(n,\mathbb{R}))$. It is clear that $(bs_+\xi,fact\,pr\,asso(\xi,GL(n,\mathbb{R})),bs\xi)$ is a two-sheeted covering in the broad sense; we denote it by $or\,\xi$.

One may view the points of $bs_+\xi$ as pairs (x,ε), where $x \in bs\,\xi$ and ε is an orientation of the fiber $pr\,\xi^{-1}(x)$. This permits us to look upon a simultaneous orientation of all the fibers of ξ as a map $s: bs\,\xi \to bs_+\xi$ such that $pr\,or\xi \circ s = id\,bs\,\xi$. Clearly, the compatibility (in the sense of 4.5.1.8) of the orientation given by s on the fibers of ξ is equivalent to the continuity of s. Thus, the orientations of the bundle ξ turn out to be the sections of the bundle $or\,\xi$.

We list the most important corollaries of the above discussion:

(i) a bundle ξ is orientable if and only if the bundle $or\,\xi$ is trivial;

(ii) if the base $bs\,\xi$ is locally connected and the fundamental groups of its components have no subgroups of index 2 (the last happens, in particular, whenever the components of $bs\,\xi$ are simply connected), then ξ is orientable.

(iii) if ξ is orientable, then the number of its orientations equals the number of maps $comp\,bs\,\xi \to S^0$ (and, in particular, equals 2^m if $bs\,\xi$ has $m < \infty$ components).

3. We would like to make another remark concerning the previous construction. Given a real vector bundle ξ, consider the bundle ξ_+ with base $bs_+\xi$, induced from ξ by the projection $pr\,or\,\xi$. Here we note that ξ_+ possesses a canonical orientation. Namely, the orientation of its fiber $pr\,\xi_+^{-1}(x,\varepsilon)$ (where $x \in bs\,\xi$ and ε is an orientation of the fiber $pr\,\xi^{-1}(x)$) is simply the orientation ε, transferred to $pr\,\xi_+^{-1}(x,\varepsilon)$ by the isomorphism

$$ab\,tl\,ad(pr\,or\,\xi): pr\,\xi_+^{-1}(x,\varepsilon) \to pr\,\xi^{-1}(x);$$

that these orientations are compatible in the sense of 4.5.1.8 is plain.

The Case of Smooth Manifolds

4. Orienting a smooth manifold is the same as orienting its tangent bundle (see 4.6.4.4), and hence the discussion in 2 carries over to the orientability and orientations of smooth manifolds. In particular, a manifold X is orientable if and only if the bundle $or\,tang\,X$ (which is a two-sheeted covering in the broad sense with base X) is trivial. Also, a smooth manifold is automatically orientable if the fundamental groups of its components have no subgroups of index 2 (as happens when all its components are simply connected).

Let us add that, according to 2.2.1, the space $tl\,or\,tang\,X$ is a smooth manifold. By 3, the latter carries a canonical orientation. If X is connected and nonorientable, then $or\,tang\,X$ is a (two-sheeted) covering; we call it the <u>orientation-covering</u> of the manifold X.

4. Some Bundles Over Spheres

1. In this subsection we present the most elementary results concerning orientable real vector bundles and complex vector bundles over lower-dimensional spheres. This topic has independent interest and, at the same time, illustrates the general theory.

Recall that the classes of $GL_+\mathbb{R}^n$-equivalent $GL_+\mathbb{R}^n$-bundles over a given cellular base are in a one-to-one correspondence with the homotopy classes of continuous maps of this base into $G_+(\infty,n)$. Similarly, the classes of $GL\mathbb{C}^n$-equivalent $GL\mathbb{C}^n$-bundles over a cellular base are in a one-to-one correspondence with the homotopy classes of continuous maps of this base into $\mathbb{C}G(\infty,n)$. [See 4.5.3.7.] If the base is S^r, $r \geqslant 1$, then the fact that $G_+(\infty,n)$ and $\mathbb{C}G(\infty,n)$ are simple spaces implies that the aforementioned homotopy classes may be thought of as elements of the groups $\pi_r(G_+(\infty,n))$ and $\pi_r(\mathbb{C}G(\infty,n))$, respectively. Since these groups are isomorphic to $\pi_{r-1}(SO(n))$ and $\pi_{r-1}(U(n))$ (see 2.8.2 and 2.8.4), it follows that the classes of $GL_+\mathbb{R}^n$-equivalent $GL_+\mathbb{R}^n$-bundles over S^r $(r \geqslant 1)$ are in a natural one-to-one correspondence with the elements of $\pi_{r-1}(SO(n))$, while the classes of $GL\mathbb{C}^n$-equivalent $GL\mathbb{C}^n$-bundles over S^r $(r \geqslant 1)$ are in a natural one-to-one correspondence with the elements of $\pi_{r-1}(U(n))$. Since we already know $\pi_{r-1}(SO(n))$ and $\pi_{r-1}(U(n))$ for some small values of r (see Subsection 2.6), we get the classification of the corresponding bundles, which we shall outline with minimum of details below. Some supplements appear in Exercises 5.14–5.19.

We note that this method works in a considerably more general situation. Namely, let G be a topological group, and let F be an effective G-space. According to the general theory of bundles (see Subsection 4.4.2), the elements of $\mathrm{Stee}(S^q,F)$ are in a natural one-to--one correspondence with the elements of $\pi(S^q,X)$, where X is any classifying space of G. If X is q-simple (for example, if G is connected), then $\pi(S^q,X)$ coincides with $\pi_q(X)$ and, for $q \geqslant 1$, with $\pi_{q-1}(G)$ (see 1.6).

The Real Oriented Case

2. Since $\pi_0(SO(n))$ is trivial, every $GL_+\mathbb{R}^n$-bundle over S^1 is $GL_+\mathbb{R}^n$-trivial, for any positive integer n.

Since $\pi_1(SO(1))$ is trivial, $\pi_1(SO(2)) \cong \mathbb{Z}$, and $\pi_1(SO(n)) \cong \mathbb{Z}_2$ for all $n \geqslant 3$, it follows that: every $GL_+\mathbb{R}^1$-bundle over S^2 is $GL_+\mathbb{R}^1$-trivial; the pairwise $GL_+\mathbb{R}^2$-nonequivalent $GL_+\mathbb{R}^2$-bundles over S^2 form an infinite and countable collection; for $n \geqslant 3$, the number of pairwise $GL_+\mathbb{R}^n$-nonequivalent $GL_+\mathbb{R}^n$-bundles over S^2 is 2.

Since $\pi_2(SO(n))$ is trivial, every $GL_+\mathbb{R}^n$-bundle over S^3 is $GL_+\mathbb{R}^n$-trivial, for any positive integer n.

Since $\pi_3(SO(1))$ and $\pi_3(SO(2))$ are trivial, whereas $\pi_3(SO(n))$, $n \geqslant 3$, is infinite and countable, it follows that: every

$GL_+\mathbb{R}^1$-bundle over S^4 is $GL_+\mathbb{R}^1$-trivial; every $GL_+\mathbb{R}^2$-bundle over S^4 is $GL_+\mathbb{R}^2$-trivial; for $n \geqslant 3$, the pairwise $GL_+\mathbb{R}^n$-nonequivalent $GL_+\mathbb{R}^n$-bundles over S^4 form an infinite and countable collection.

The Complex Case

3. Since $\pi_0(U(n))$ is trivial, every $GL\mathbb{C}^n$-bundle over S^1 is $GL\mathbb{C}^n$-trivial, for any positive integer n.

Since $\pi_1(U(n)) \cong \mathbb{Z}$ for all $n \geqslant 1$, the classes of pairwise $GL\mathbb{C}^n$-nonequivalent $GL\mathbb{C}^n$-bundles over S^2 form an infinite and countable collection, for any positive integer n.

Since $\pi_2(U(n))$ is trivial, every $GL\mathbb{C}^n$-bundle over S^3 is $GL\mathbb{C}^n$-trivial, for any positive integer n.

Since $\pi_2(U(1))$ is trivial and $\pi_3(U(n)) \cong \mathbb{Z}$ for all $n \geqslant 2$, it follows that: every $GL\mathbb{C}^1$-bundle over S^4 is $GL\mathbb{C}^1$-trivial; if $n \geqslant 2$, the pairwise $GL\mathbb{C}^n$-nonequivalent $GL\mathbb{C}^n$-bundles over S^4 form an infinite and countable collection.

5. <u>Exercises</u>

1. Show that a smooth two-dimensional manifold does not admit the disc D^2 as a covering space (unless the given manifold is homeomorphic to D^2, and then the covering projection is the corresponding homeomorphism).

2. Show that a smooth, compact two-dimensional manifold does not admit the half plane $\mathbb{R}^2_-$ as a covering space.

3. Show that a sphere with g handles admits a sphere with $\tilde{g}$ handles as an m-sheeted covering space if and only if $\tilde{g} - 1 = m(g - 1)$. [Cf. 4.1.5.1.]

4. Show that a sphere with h crosscaps admits a sphere with $\tilde{h}$ crosscaps as an m-sheeted covering space if and only if $\tilde{h} - 2 = m(h-2)$. [Cf. 4.1.5.2.]

5. Show that a sphere with h crosscaps admits a sphere with g handles as a covering space if and only if m is even and $g - 1 = m(h - 2)/2$.

6. Show that the Klein bottle admits a topological space as

a covering space with nonidentical covering projection if and only if the given space is homeomorphic to the Klein bottle, to the interior of a Möbius strip, or to one of the products $\mathbb{R} \times \mathbb{R}$, $\mathbb{R} \times S^1$, $S^1 \times S^1$.

7. Let $\pi_1, \pi_2, \pi_3, \ldots$ be arbitrary groups, with π_k Abelian for $k \geqslant 2$, and suppose that there is given a right group action of π_1 on π_k, $k \geqslant 2$. Show that there is a cellular space X with base point x_0 together with an isomorphism $f_1 : \pi_1 \to \pi_1(X, x_0)$, such that the group $\pi_k(X, x_0)$ is f-isomorphic to π_k for all $k \geqslant 2$.

8. Let ξ be a covering with bs ξ an n-dimensional locally Euclidean space. Prove that tl ξ is an n-dimensional locally Euclidean space with boundary $\partial\, \mathrm{tl}\, \xi = \mathrm{pr}\, \xi^{-1}(\partial\, \mathrm{bs}\, \xi)$.

9. Show that every covering space of a locally finite cellular space is locally finite (see 2.21).

10. Let ξ be a covering with bs ξ a topological group. Prove that, given any point $x \in \mathrm{pr}\, \xi^{-1}(e_{\mathrm{bs}\, \xi})$, on tl ξ there is one and only one group structure, which turns tl ξ into a topological group with identity x, and turns pr ξ into a homomorphism.

11. Let ξ be a covering with a finite number of sheets, such that a connected, compact topological group acts transitively on bs ξ. Prove that one can define a transitive action of a connected, compact topological group on tl ξ.

12. Let ξ be a two-sheeted covering with bs ξ a nonorientable smooth manifold and tl ξ orientable. Show that ξ is equivalent to the orientation-covering of bs ξ.

13. Let X be a nonorientable smooth manifold and assume that X admits an orientable manifold Y as a covering space. Show that Y is also a covering space of tl or tang X.

14. Show that:

(i) for each positive integer n there is a unique (up to $\mathrm{GL}\mathbb{R}^n$-equivalences) $\mathrm{GL}\mathbb{R}^n$-nontrivial $\mathrm{GL}\mathbb{R}^n$-bundle over S^1 and that for $n = 1$ its total space is homeomorphic to the interior of a Möbius strip;

(ii) the pairwise $\mathrm{GL}\mathbb{R}^2$-nonequivalent $\mathrm{GL}\mathbb{R}^2$-bundles over S^2 form an infinite and countable collection;

(iii) for $n \geqslant 3$, the number of pairwise $\mathrm{GL}\mathbb{R}^n$-nonequivalent $\mathrm{GL}\mathbb{R}^n$-bundles over S^2 is equal to 2;

(iv) for any positive integer n, every $\mathrm{GL}\mathbb{R}^n$-bundle over S^3 is $\mathrm{GL}\mathbb{R}^n$-trivial;

(v) every $\mathrm{GL}\mathbb{R}^2$-bundle over S^4 is $\mathrm{GL}\mathbb{R}^2$-trivial;

(vi) for $n \geqslant 3$, the pairwise $\mathrm{GL}\mathbb{R}^n$-nonequivalent $\mathrm{GL}\mathbb{R}^n$-bundles over S^4 form an infinite and countable collection.

15. Show that for $r \geqslant 2$ every $\mathrm{GL}\mathbb{R}^1$-bundle over S^r is $\mathrm{GL}\mathbb{R}^1$-trivial.

16. Let $r \geqslant 3$. Show that:

(i) every $\mathrm{GL}\mathbb{R}^2$-bundle over S^r is $\mathrm{GL}\mathbb{R}^2$-trivial;

(ii) every $\mathrm{GL}_+\mathbb{R}^2$-bundle over S^r is $\mathrm{GL}_+\mathbb{R}^2$-trivial;

(iii) every $\mathrm{GL}\mathbb{C}^1$-bundle over S^r is $\mathrm{GL}\mathbb{C}^1$-trivial.

17. Show that two $\mathrm{GL}_+\mathbb{R}^2$-bundles over S^2 become equivalent by extending the structure group $\mathrm{GL}_+(2,\mathbb{R})$ to $\mathrm{GL}(2,\mathbb{R})$ if and only if the corresponding elements of $\pi_1(\mathrm{SO}(2))$ [see 4.1] are either equal or inverses of one another.

18. Show that for every nontrivial $\mathrm{O}\mathbb{R}^2$-bundle ξ with $\mathrm{bs}\,\xi = S^2$ the space $\mathrm{tl\,asso}(\xi,S^1)$ is homeomorphic to one of the lenses $L(m;1,1)$ [here $\mathrm{O}(2)$ acts canonically on S^1].

19. Show that for $n \geqslant 2$ the number of pairwise $\mathrm{GL}\mathbb{C}^n$-nonequivalent $\mathrm{GL}\mathbb{C}^n$-bundles over S^5 does not exceed 2.

INFORMATION. Actually, the last number is 2 if $n = 2$, and 1 if $n > 2$.

Bibliography

[1]   Adams, J. F.: On the non-existence of elements of Hopf invariant
        one, Ann. Math., II. Ser. 72, No. 1, 20-104 (1960).

[2]   Adams, J. F., Atiyah, M. F.: K-theory and the Hopf invariant,
        Q. J. Math., Oxf. II. Ser. 17, 31-38 (1966).

[3]   Bourbaki, N.: Topologie Générale, Chap. I/II, Hermann, Paris,
        1947; English Transl.: General Topology, Addison-Wesley, Reading,
        Mass., 1966.

[4]   Brown, M.: Locally flat imbeddings of topological manifolds,
        Ann. Math., II. Ser. 75, No. 2, 331-341 (1962).

[5]   Calabi, E., Rosenlicht, M.: Complex-analytic manifolds without
        countable base, Proc. Am. Math. Soc. 4, 335-340 (1954).

[6]   Dowker, C. H.: Topology of metric complexes, Am. J. Math. 74,
        555-577 (1952).

[7]   Fuks, D. B.: Homotopic topology, Itogi Nauki Tekh., Ser. Algebra
        Topologiya Geom., 71-122 (1969); English. Transl.: J. Sov. Math.
        1, No. 3, 333-362 (1973).

[8]   Grauert, H.: On Levi's problem and the imbedding of real analytic
        manifolds, Ann. Math., II. Ser. 68, 460-472 (1958).

[9]   Hodge, W. V. D., Pedoe, D.: Methods of Algebraic Geometry, Vol. I,
        Cambridge Univ. Press, 1947.

[10]   Kampen, E. R., van: Komplexe in Euklidischen Räumen, Abh. Math.
        Semin. Univ. Hamb. 9, 72-78 and 152-153 (1932).

[11]   Kelley, J. L.: General Topology, Van Nostrand, New York, 1955;
        Reprint Graduate Texts in Math., Springer-Verlag, New York,
        Heidelberg, Berlin, 1975.

[12]   Kervaire, M.: A manifold which does not admit any differentiable
        structure, Comment. Math. Helv. 34, 257-270 (1960).

[13]   Kuratowski, C.: Topologie, Vol. I, Monografie Matematyczne,
        Warszawa, 1948; English Transl.: Topology, Vol. I, Academic Press,
        New York, 1966.

[14]   Kuratowski, C.: Topologie, Vol. II, Monografie Matematyczne,
        Warszawa, 1950; English Transl.: Topology, Vol. II, Academic
        Press, New York, 1968.

[15]   Milnor, J.: On manifolds homeomorphic to the 7-sphere, Ann. Math.,
        II. Ser. 64, No. 2, 399-405 (1956).

[16]   Milnor, J.: On spaces having the homotopy type of a CW-complex,
        Trans. Am. Math. Soc. 90, No. 2, 272-280 (1959).

[17]   Milnor, J.: Morse Theory, Princeton Univ. Press, Princeton, N.J.,
        1963.

[18]   Rokhlin, V. A.: The imbedding of non-orientable three-dimensional
        manifolds into a five-dimensional Euclidean space, Dokl. Akad.
        Nauk SSSR 160, No. 3, 549-551; English Transl.: Sov. Math. Dokl.
        6, No. 1, 153-156 (1965).

[19]   Serre, J-P.: Groupes d'homotopie et classes de groupes abéliens,
        Ann. Math., II. Ser. 58, No. 2, 258-294 (1953).

[20]   Smale, S.: A survey of some recent developments in differential
        topology, Bull. Am. Math. Soc. 69, No. 2, 131-145 (1963).

[21]   Sternberg, S.: Lectures on Differential Geometry, Prentice Hall,
        Englewood Cliffs, N. J., 1964.

[22]   Whitney, H.: Differentiable mnaifolds, Ann. Math., II. Ser.
        37, 645-680 (1936).

[23]   Whitney, H.: A function non constant on a connected set of
        critical points, Duke Math. J. 1, 514-517 (1935).

Index

Glossary of Symbols

ab – 2

abs – 2

ad – 266, 275

asso – 305, 306, 315

Atl – 143

Atl_x – 156

att – 85

Aut – 493

Aux – 233, 234

aux – 233, 234

ba – 108, 109

bar – 100, 102

blk – 118

bp – 50

bs – 265

bst – 117

Buc – 419

buc – 420

$\mathbb{C}$ – 38

$\mathcal{C}$ – 46

$\mathbb{C}^n$ – 39

$\mathcal{C}^r$ – 141, 143, 197, 346

$\mathcal{C}^{\geq r}$ – 143

$\mathcal{C}^r_\partial$ – 197

$\mathbb{C}^\infty$ – 41

$\mathbb{C}^*$ – 281

$\mathbb{C}a$ – 38

$\mathbb{C}a^n$ – 39

$\mathbb{C}aP^1$ – 40

$\mathbb{C}aP^2$ – 40

Catl – 150

cell – 83

$\mathbb{C}G(n,k)$ – 182

$\mathbb{C}G(\infty,k)$ – 337

cha – 82, 83

Cl – 6

comp – 62

Con – 46

con – 42, 51

conj – 328

const – 50

cop – 137

corr – 267

$\mathbb{C}P^n$ – 39

$\mathbb{C}P^\infty$ – 41

csc – 114

Cub – 419

cub – 420

$\mathbb{C}V(n,k)$ – 173

$\mathbb{C}V(\infty,k)$ – 342

$\mathbb{C}V'(n,k)$ – 176

$\mathbb{C}V'(\infty,k)$ – 342

Cyl – 45

$\mathbb{C}\xi$ – 335

d – 159

∂ – 133, 143, 385

D_i – 157

D^n – 9

d_x – 156, 159

D^∞ – 41

$d_\tau(x,\rho)$ – 202

deg – 364

diag – 4

Diff – 197

Dist – 9

dist – 8

dopp – 137, 210

DS – 54

DS_- – 63

DS_+ – 63

DT – 100

e_G – 277

El – 244

el – 244

Emb^r – 197

Esi – 99

ev – 493

ext – 325

fact – 3

Fibr – 317

Fr – 6

$G(n,k)$ – 177

$G(\infty,k)$ – 337

$G'(n,k)$ – 185

$G_+(n,k)$ – 178

$G_+(\infty,n)$ – 337

GF – 445

$GL(n,\mathbb{C})$ – 176

$GL(n,\mathbb{H})$ – 176

$GL(n,\mathbb{R})$ – 175

$GL_+(n,\mathbb{R})$ – 176

$GL_+\mathbb{R}^n$ – 324

$GL\mathbb{C}^n$ – 324

$GL\mathbb{R}^n$ – 324

gr – 490

Gra – 340

grad – 241

$\mathbb{H}$ – 38

$\mathbb{H}^n$ – 39

$\mathbb{H}^\infty$ – 41

$\mathbb{H}^*$ – 281

hel – 268

hel_m – 268

$\mathbb{H}G(n,k)$ – 183

$\mathbb{H}P^n$ – 40

$\mathbb{H}P^\infty$ – 41

$\mathbb{H}V(n,k)$ – 175

$\mathbb{H}V'(n,k)$ – 176

$\#$ (e.g. $f_{\#}$) - 156, 377, 418

$\times$ (e.g. $h^{\times}$, $G^{\times}$) - 307, 308

$*$ (e.g. f_{*}) - 377

$\emptyset$ - 1

$\sqcup\!\sqcup$ - 3

$\times$ - 1, 265

V - 50

$*$ - 42, 52

$\otimes$ - 51

$\oplus$ - 334

U_{φ} - 38

$\times_{c}$ - 94

$*_{c}$ - 97, 98

$\otimes_{c}$ - 98

$\times_{s}$ - 110

$*_{s}$ - 113